69.60 POK

DYNAMICS OF FLAMES
AND REACTIVE SYSTEMS

Edited by
J. R. Bowen
University of Washington
Seattle, Washington

N. Manson
Universite de Poitiers
Poitiers, France

A. K. Oppenheim
University of California
Berkeley, California

R. I. Soloukhin
Institute of Heat and Mass Transfer
BSSR Academy of Sciences
Minski, USSR

Volume 95
PROGRESS IN
ASTRONAUTICS AND AERONAUTICS

Martin Summerfield, Series Editor-in-Chief
Princeton Combustion Research Laboratories, Inc.
Princeton, New Jersey

Technical papers presented from the Ninth International Colloquium on Gasdynamics of Explosions and Reactive Systems, Poitiers, France, July 1983, and subsequently revised for this volume.

Published by the American Institute of Aeronautics and Astronautics, Inc.
1633 Broadway, New York, NY 10019

American Institute of Aeronautics and Astronautics, Inc.
New York, New York

Library of Congress Cataloging in Publication Data
Main entry under title:

International Colloquium on Gasdynamics of Explosions and Reactive Systems (9th:1983:Poitiers, France) Dynamics of flames and reactive systems.

(Progress in astronautics and aeronautics; v. 95)
Revised technical papers presented from the Ninth International Colloquium on Gasdynamics of Explosions and Reactive Systems, Poitiers, France, July 1983.
Companion volume to the 9th colloquium's "Dynamics of shock waves, explosions, and detonations."
Includes bibliographies and index.
1. Flame—Congresses.   2. Combustion—Congresses.
I. Bowen, J. R. (J. Ray) II. Title. III. Series.
TL507.P75   vol. 95   629.1 s [621.402'3]   84-21560
[QD516]
ISBN 0-915928-92-2

Copyright © 1984 by the American Institute of Aeronautics and Astronautics, Inc. All rights reserved. Printed in the United States of America. No part of this publication may be reproduced, distributed, or transmitted, in any form or by any means, or stored in any data base or retrieval system, without the prior written permission of the publisher.

Progress in Astronautics and Aeronautics

Series Editor-in-Chief
Martin Summerfield
Princeton Combustion Research Laboratories, Inc.

Series Associate Editors

Burton I. Edelson
*National Aeronautics
and Space Administration*

Allen E. Fuhs
Naval Postgraduate School

J. Leith Potter
Vanderbilt University

Assistant Series Editor
Ruth F. Bryans
Ocala, Florida

Norma J. Brennan
Director, Editorial Department
AIAA

Jeanne Godette
Series Managing Editor
AIAA

Table of Contents

Preface xiii

Introduction to Modern Laminar Flame Theory 1
    P. Clavin, *Université de Provence, Marseille, France*

Chapter I. Premixed Flames 35

    Effects of Chemical Equilibrium on the Structure and Extinction of Laminar Diffusion Flames 37
        N. Peters, *Rheinisch-Westfälische Technische Hochschule, Aachen, Federal Republic of Germany,* and F.A. Williams, *Princeton University, Princeton, New Jersey*

    Stretch Effects in Planar Premixed Hydrogen-Air Flames 61
        J. Warnatz, *Universität Heidelberg, Federal Republic of Germany,* and N. Peters, *Rheinisch-Westfälische Technische Hochschule, Aachen, Federal Republic of Germany*

    Slowly Varying Flames with Chain-Branching/Chain-Breaking Kinetics 75
        G.S.S. Ludford, *Cornell University, Ithaca, New York,* and N. Peters, *Rheinisch-Westfälische Technische Hochschule, Aachen, Federal Republic of Germany*

    Effect of Dissociation on the Near-Stoichiometric Burning of Non-Dilute Mixtures 92
        G.S.S. Ludford, *Cornell University, Ithaca, New York,* and A.K. Sen, *Purdue School of Sciences, Indianapolis, Indiana*

    The Feedback of a Flame Front on Turbulent Flows 103
        G. Searby, F. Sabathier, J. Monreal, P. Clavin, and L. Boyer, *Université de Provence, Marseille, France*

    Flame Front Stability with General Intermolecular Interaction Potential 115
        P.G. Ybarra, *Université de Rouen, Mont-Saint-Aignan, France*

    Stability Limits and Critical Size of Structures in Premixed Flames 129
        J. Quinard, G. Searby, and L. Boyer, *Université de Provence, Marseille, France*

Nonsteady Gasdynamic Effects in the Induction Behind a Strong Shock Wave 142
J.F. Clarke and R.S. Cant, *Cranfield Institute of Technology, Bedford, England*

Structure of Premixed Laminar Methanol-Air Flames: Experimental and Computational Results 164
L.L. Andersson, *Chalmers University of Technology, Göteborg, Sweden,* and B. Christenson, A. Höglund, J.O. Olsson, and L.G. Rosengren, *Volvo, Göteborg, Sweden*

Burning Velocities of Ethanol-Air and Ethanol-Water-Air Mixtures 181
O.L. Gülder, *National Research Council of Canada, Ottawa, Ontario, Canada*

Computer Modeling Study of Acetylene-Oxygen Ignition and Flames Using a Truncated Reaction Mechanism 198
S.-M. Hwang and W.C. Gardiner Jr., *University of Texas, Austin, Texas,* and J. Warnatz, *University of Heidelberg, Federal Republic of Germany*

Prediction of Laminar Flame Properties of Propane-Air Mixtures 211
C.K. Westbrook and W.J. Pitz, *Lawrence Livermore National Laboratory, University of California, Livermore, California*

Homogeneity and Propagation of Autoignited Cool and Blue Flames 236
Y. Ohta, *Nagoya Institute of Technology, Nagoya, Japan,* and H. Takahashi, *Meijo University, Nagoya, Japan*

Stability of Solid Propellant Combustion Subject to Nonplanar Perturbations 248
F.J. Higuera and A. Liñán, *Escuela Técnica Superior de Ingenieros Aeronauticos, Madrid, Spain*

Chapter II.   Diffusion Flames 259

Laminar Diffusion Flames with Cylindrical Symmetry, Arbitrary Values of Diffusion Coefficients and Inlet Velocities, and Chemical Reactions in the Approach Streams 261
S.S. Penner, M.Y. Bahadori, and E.M. Kennedy, *University of California, San Diego, La Jolla, California*

Transition and Transport in the Initial Region of a Turbulent Diffusion Flame 293
A.R. Masri, S.H. Stårner, and R.W. Bilger, *The University of Sydney, Sydney, Australia*

Predicted Structure of Stretched and Unstretched Methane-Air
Diffusion Flames 305
    S.K. Liew and J.B. Moss, *Cranfield Institute of Technology,*
    Bedford, England, and K.N.C. Bray, *University of*
    Southampton, Southampton, England

An Experimental Study of Turbulent Jet Diffusion Flames 320
    O.K. Sønju and J. Hustad, *Norwegian Institute of Technology,*
    Trondheim, Norway

Chapter III.   Turbulent Combustion 341

On Sound Sources in Turbulent Combustion 343
    N. Kidin and V. Librovich, *Institute for Problems in Mechanics,*
    Moscow, USSR, and J. Roberts and M. Vuillermoz, *Polytechnic*
    of the South Bank, London, United Kingdom

Comparisons of Experimental and Computed Length Scales
and Velocities in Turbulent Combustion 356
    A.Y. Abdalla, D. Bradley, S.B. Chin, and C. Lam, *University*
    of Leeds, Leeds, United Kingdom

Flow Rate and Equivalence Ratio Influences on the Thermal
Field of a Turbulent Cool Flame 367
    I. Gökalp, *C.N.R.S., Orléans, France,* and N. Zarrad,
    G.M.L. Dumas, and R.I. Ben Aïm, *Université*
    Pierre et Marie Curie, Paris, France

Turbulent Reacting Concentric Jets: Comparison Between
pdf and Moment Calculations 384
    P. Givi, J.I. Ramos, and W.A. Sirignano, *Carnegie-Mellon*
    University, Pittsburgh, Pennsylvania

Chapter IV.   Constant Volume Combustion 419

Influence of Turbulent Motion on Spark Ignition 421
    P.S. Tromans and S.J. O'Connor, *Shell Research Ltd.,*
    Chester, Great Britain

Vibratory Combustion Triggered by a Small Cavity in the Wall
of a Constant Volume Combustion Chamber 433
    A. Girard, F. Fisson, and J.C. Leyer, *Université de Poitiers,*
    Poitiers, France

Direct Measurement of the Head-on Flame Quenching
Distance in Closed Chambers 443
    A. Girard and J.C. Leyer, *Université de Poitiers,*
    Poitiers, France

Chapter V. Spray Combustion 453

**Timed Ignition of Explosives and Flammables from
Desensitized Solutions** 455
    M. Gerstein and P.R. Choudhury, *University of Southern California,
Los Angeles, California*

**Comparative Study of Droplet Heating and Vaporization at
High Reynolds and Peclet Numbers** 464
    H.A. Dwyer and B.R. Sanders, *Sandia National Laboratories,
Livermore, California*

Comparisons of Computed and Measured Dense Spray Jets 484
    L. Martinelli and F.V. Bracco, *Princeton University, Princeton,
New Jersey,* and R.D. Reitz, *General Motors Technical Center,
Warren, Michigan*

**A Study of the Motion of Vaporizing Droplets in a
Turbulent Flow** 513
    A.A. Mostafa and S.E. Elghobashi, *University of California,
Irvine, California*

Simulations of Two-Dimensional Fuel Droplet Flows 540
    M.J. Fritts, D.E. Fyfe, and E.S. Oran, *Naval Research
Laboratory, Washington, D.C.*

**Induction Time Measurements for Ignition of Liquid Fuel Jets
in Air at High Temperatures and Pressures** 554
    V.K. Baev, A.N. Bazhaikin, A.A. Buzukov, *University of
Novosibirsk, Academy of Sciences, Novosibirsk, USSR*

Spray Characteristics of Simplex Swirl Atomizers 563
    N.K. Rizk and A.H. Lefebvre, *Purdue University,
West Lafayette, Indiana*

Chapter VI. Nonequilibrium Flows 581

Flows in Laval Nozzles with a High-Temperature Diatomic Gas ... 583
    N.K. Mitra and M. Fiebig, *Ruhr-Universität, Bochum,
Federal Republic of Germany*

**Unsteady Aerodynamics of Chemically Reacting Flows Past
Oscillating Thin Bodies** 593
    L. Librescu, *Tel-Aviv University, Tel-Aviv, Israel*

**Uniform Solutions for Characteristics and Weak Shock Waves
in a Reactive Medium** 610
    B.D. Pandey and D.C. Chou, *University of New Mexico,
Albuquerque, New Mexico*

Chapter VII. Combustion Diagnostics 629

CARS Instrument for Practical Combustion Measurements 631
G.M. Dobbs, J.H. Stufflebeam, and A.C. Eckbreth,
United Technologies Research Center, East Hartford, Connecticut,
and P.A. Tellex, *Pratt & Whitney Aircraft Group,*
West Palm Beach, Florida

Study of OH-Saturated Laser-Induced Fluorescence in Low-Pressure Flame 642
D. Stepowski and M.J. Cottereau, *Faculté des Sciences de Rouen, Mont Saint Aignan, France*

Flame Concentrations and Temperatures by Spontaneous Raman Spectroscopy 658
R. Michael-Saade, J.P. Sawerysyn, L.-R. Sochet, G. Buntinx, M. Crunelle-Cras, F. Grase, and M. Bridoux, *Université des Sciences et Techniques de Lille, France*

The Application of Rotational Raman Spectroscopy to Dynamic Measurements in Gas Flowfields 672
P.P. Yaney, R.J. Becker, P.T. Danset, M.R. Gallis, and J.I. Perez, *University of Dayton, Dayton, Ohio*

Flash X-Ray Tomographic System for Diagnostics of Microsecond Phenomena................................. 700
C.K. Zoltani and K.J. White, *Ballistic Research Laboratory, Aberdeen Proving Ground, Maryland*

Two-Dimensional Imaging of Flame Temperature Using Laser-Induced Fluorescence............................... 714
R.J. Cattolica and D.A. Stephenson, *Sandia National Laboratories, Livermore, California*

LDV Measurements of Gas Flow Behind Reflected Shocks 722
M. Frenklach and C.K. Li Kwok Cheong, *Louisiana State University, Baton Rouge, Louisiana,* and E.S. Oran, *Naval Research Laboratory, Washington, D.C.*

Droplet Size Distributions from Diffracted Light Intensities........ 736
A. Tardieu, S.M. Candel, and E. Esposito, *l' Ecole Centrale des Arts et Manufactures, Châtenay-Malabry, France*

Measurement of NO in Methane-Air Flames by Tunable Atomic Line Molecular Spectroscopy............................ 750
E. Cuellar and N.J. Brown, *University of California Berkeley, California*

Author Index for Volume 95 767

List of Series Volumes 769

Table of Contents for Companion Volume 94

Chapter I.   Detonations in Gaseous Mixtures 1

Direct Initiation of Planar Detonation Waves in Methane/
Oxygen/Nitrogen Mixtures 3
    S. Ohyagi, T. Yoshihashi, and Y. Harigaya, *Saitama University,*
    Saitama-Ken, Japan

Measurements of Cell Size in Hydrocarbon-Air Mixtures and
Predictions of Critical Tube Diameter, Critical Initiation
Energy, and Detonability Limits 23
    R. Knystautas, C. Guirao, J.H. Lee, and A. Sulmistras, *McGill*
    University, Montreal, Canada

Power-Energy Relations for the Direct Initiation of
Gaseous Detonations 38
    K. Kailasanath and E.S. Oran, *Naval Research Laboratory,*
    Washington, D.C.

Detonation Length Scales for Fuel-Air Explosives 55
    I.O. Moen, J.W. Funk, S.A. Ward, and G.M. Rude, *Defence*
    Research Establishment Suffield, Ralston, Canada and
    P.A. Thibault, *University of Toronto Institute for Aerospace*
    Studies, Ontario, Canada

The Influence of Yielding Confinement on Large-Scale
Ethylene-Air Detonations.. 80
    S.B. Murray, *Defence Research Establishment Suffield, Ralston,*
    Canada, and J.H. Lee, *McGill University, Montreal, Canada*

Cellular Structure in Detonation of Acetylene-Oxygen Mixture 104
    M. Vandermeiren and P.J. Van Tiggelen, *Université*
    Catholique de Louvain, Louvain-la-Neuve, Belgique

The Influence of Initial Pressure on Critical Diameters of
Gaseous Explosive Mixtures... 118
    P. Bauer, C. Brochet, and H.N. Presles, *Université de Poitiers,*
    Poitiers, France

"Galloping" Gas Detonations in the Spherical Mode 130
    J.E. Elsworth, P.J. Shuff, and A. Ungut, *Shell Research Ltd.,*
    Chester, Great Britian

Chemical Kinetics of Propane Oxidation in Gaseous
Detonations 151
    C.K. Westbrook, W.J. Pitz, and P.A. Urtiew, *Lawrence Livermore*
    National Laboratory, University of California,
    Livermore, California

High-Speed Deflagration with Compressibility Effects 175
    J.F. Clarke, *Cranfield Institute of Technology, Bedford, England,*
    and D.R. Kassoy, *University of East Anglia, Norwich, England*

Numerical Simulations on the Establishment of Gaseous
Detonation 186
    S. Taki, *Fukui University, Fukui, Japan,* and T. Fujiwara,
    Nagoya University, Nagoya, Japan

A Shock Tube Study of the Chlorine Azide Decomposition 201
    C. Paillard and G. Dupré, *C.N.R.S. et Université d'Orléans, France,*
    and N.A. Fomin, *Academy of Sciences, Minsk, USSR*

Chapter II.   Detonations in Two-Phase Systems 219

Dust, Hybrid, and Dusty Detonations 221
    C.W. Kauffman, P. Wolański, A. Arisoy, P.R. Adams,
    B.N. Maker, and J.A. Nicholls, *University of Michigan,*
    Ann Arbor, Michigan

The Structure of Dust Detonations 241
    P. Wolański, D. Lee, M. Sichel, C.W. Kauffman, and
    J.A. Nicholls, *University of Michigan, Ann Arbor, Michigan*

"Double-Front" Detonations in Gas-Solid Particles Mixtures 264
    B. Veyssiére, *Centre National de la Recherche Scientifique,*
    Poitiers, France

Unconfined Aluminum Particle Two-Phase Detonation in Air 277
    A.J. Tulis and J.R. Selman, *Illinois Institute of Technology,*
    Chicago, Illinois

Dynamics of Dispersion and Ignition of Dust Layers by a
Shock Wave 293
    V.M. Boiko and A.N. Papyrin, *Institute of Theoretical and Applied*
    Mechanics, Novosibirsk, USSR, and M. Woliński and P. Wolański,
    Technical University of Warsaw, Poland

Detonations in Explosive Foams 302
    J.P. Saint-Cloud and O. Peraldi, *Université de Poitiers,*
    Poitiers, France

Propagation Velocity and Mechanism of Bubble Detonation 309
    T. Hasegawa, *Nagoya Institute of Technology, Japan,* and
    T. Fujiwara, *Nagoya University, Japan*

Nonsteady Shock Wave Propagating in a Bubble-Liquid System 320
    T. Sugimura, *Meijo University, Nagoya, Japan,* K. Tokita, *Nippon*
    Oil & Fats Co. Ltd., Taketoyo, Chita, Aichi, Japan, and
    T. Fujiwara, *Nagoya University, Nagoya, Japan*

Ignition of Dust Suspensions Behind Shock Waves 332
    A.A. Borisov, B.E. Gel'fand, E.I. Timofeev, S.A. Tsyganov, and
    S.V. Khomic, *Academy of Sciences, Moscow, USSR*

Chapter III.   Condensed Explosives 341

Characterization of an Overdriven Detonation State in
Nitromethane 343
    L. Hamada, H.N. Presles, C. Brochet, and R. Bouriannes,
    Université de Poitiers, Poitiers, France, and R. Cheret, *C.E.A.,*
    Vaujours, Sevran, France

The Effects of Grain Size on Shock Initiation Mechanisms in
Hexanitrostilbene (HNS) Explosive 350
    R.E. Setchell and P.A. Taylor, *Sandia National Laboratories,*
    Albuquerque, New Mexico

Theoretical Modeling of Converging and Diverging Detonation
Waves in Solid and Gaseous Explosives 369
    C.M. Tarver and P.A. Urtiew, *Lawrence Livermore National*
    Laboratory, University of California, Livermore, California

Model Similarity Solutions for Shock Initiation Containing a
Realistic Constitutive Relationship for Condensed Explosive 387
    M. Cowperthwaite, *SRI International, Menlo Park, California*

The Simulation of Shock-Induced Energy Flux in Molecular
Solids 405
    A.M. Karo, F.E. Walker, and T.M. DeBoni, *Lawrence Livermore
National Laboratory, University of California, Livermore,
California,* and J.R. Hardy, *Behlen Laboratory of Physics,
University of Nebraska, Lincoln, Nebraska*

Detonation Temperatures of Nitromethane Aluminum Gels 416
    Y. Kato, *Fukui Institute of Technology, Fukui, Japan,*
and C. Brochet, *Université de Poitiers, Poitiers, France*

Chapter IV. Explosions 427

Theory of Vorticity Generation by Shock Wave and Flame
Interactions 429
    J.M. Picone, E.S. Oran, J.P. Boris, and T.R. Young Jr., *Naval
Research Laboratory, Washington, D.C.*

The Interaction of Explosively Produced Shock Waves with
Internal Discontinuities and External Objects 449
    M.A. Fry, *Science Applications, Inc., McLean, Virginia,* and
D.L. Book, *Naval Research Laboratory, Washington, D.C.*

Flame Propagation and Pressure Buildup in a Free Gas-Air
Mixture Due to Jet Ignition 474
    M. Schildknecht, W. Geiger, and M. Stock, *Battelle-Institut e.V.,
Frankfurt, Federal Republic of Germany*

Flame Acceleration by a Postflame Local Explosion 491
    M. Stock, M. Schildknecht, and W. Geiger, *Battelle-Institut e.V.,
Frankfurt, Federal Republic of Germany*

Flame Acceleration of Propane-Air in a Large-Scale Obstructed
Tube 504
    B.H. Hjertager, K. Fuhre, S.J. Parker, and J.R. Bakke,
Chr. Michelsen Institute, Bergen, Norway

Initiation of Unconfined Gaseous Detonation by Diffraction
of a Detonation Front Emerging from a Pipe 523
    A. Ungut, P.J. Shuff, and J.A. Eyre, *Shell Research Ltd., Chester,
Great Britain*

Large-Scale Experiments on the Transmission of Fuel-Air
Detonations from Two-Dimensional Channels 546
    W.B. Benedick, *Sandia National Laboratories, Albuquerque,
New Mexico,* and R. Knystautas and J.H. Lee, *McGill University,
Montreal, Canada*

Air Blast from Unconfined Gaseous Detonations 556
    J. Brossard, *University of Orleans, Bourges, France,* J.C. Leyer,
D. Desbordes, and J.P. Saint-Cloud, *University of Poitiers, Poitiers,
France,* S. Hendrickx, *E.D.F., Paris, France,* J.L. Garnier,
C.E.A., Fontenay aux Roses, France, A. Lannoy, *E.D.F.,
Saint-Denis, France,* and J. Perrot, *C.E.A., Le Barp, France*

Chapter V. Interactions 567

Collapse of Gas-Filled Cavities in Water 569
    M. Holt and M.J. Djomehri, *University of California,
Berkeley, California*

Crack Propagation in Burning Solid Propellants 575
    J.G. Siefert and K.K. Kuo, *The Pennsylvania State University,
University Park, Pennsylvania*

Preface

This and a companion volume include revised and edited versions of papers presented at the Ninth International Colloquium on the Dynamics of Explosions and Reactive Systems held in Poitiers, France, in July 1983.

These Colloquia originated in 1966 as a result of the widely-held belief among leading researchers that revolutionary advances in the understanding of detonation wave structure warranted a forum for the discussion of important findings in the gasdynamics of flows associated with the exothermic process—the essential feature of detonation waves—as well as for the discussion of other, associated phenomena.

The contributions to this, the Ninth Colloquium, have been assembled into two volumes: *Dynamics of Shock Waves, Explosions, and Detonations* and *Dynamics of Flames and Reactive Systems*. The dynamics of explosions, which the former addresses, is concerned principally with the interrelationship between the rate processes of energy deposition in a compressible medium and the concurrent nonsteady flow as it occurs typically in explosion phenomena. The dynamics of reactive systems, which is the focus of the latter volume, is a broader area encompassing the processes of coupling between the dynamics of fluid flow and molecular transformations in reactive media occurring in any combustion system. The Colloquium, then, in addition to embracing the usual topics of explosions, detonations, shock phenomena, and reactive flow, included the presentation of papers that dealt primarily with the gasdynamic aspect of nonsteady flow in combustion systems, the fluid mechanics aspects of combustion, with particular emphasis on the effects of turbulence, and with diagnostic techniques employed in the study of combustion phenomena.

In this volume, *Dynamics of Flames and Reactive Systems,* the papers have been arranged into chapters on premixed flames, diffusion flames, turbulent combustion, constant volume combustion, spray combustion, nonequilibrium flows, and combustion diagnostics. Many of the 82 papers comprising these two volumes provoked interesting discussions during the Colloquium. While the brevity of this Preface does not permit the editors to do justice to all of the stimulating papers, highlights of some of the more noteworthy contributions among them follow.

This volume includes the Colloquium's plenary lecture given by *P. Clavin* which focuses on the development of modern laminar flame theory. *Clavin* discusses applications of asymptotic analysis of laminar flame to the coupling that occurs in cellular flames between fluid mechanics, transport phenomena, and chemical kinetics. The lecture sets the theme for sessions which are covered in Chapter I, Premixed Flames. *Peters* and *Williams* use activation energy asymptotics to analyze the influence of finite rate chemistry with equilibrium dissociation on flame structure and extinction. Equilibrium effects are shown to promote extinction, the degree of which decreases as the heat of dissociation increases for a given equilibrium constant. *Warnatz* and *Peters* discuss the analysis of a plane premixed flame that is stretched by a diverging flowfield. Numerical analysis is performed for detailed kinetics and the solutions are compared to the analytical results for large activation engines with single-step kinetics. *Ludford* and *Sen* investigate the effect of dissociation on near-stoichiometric combustion, while *Ludford* and *Peters* examine the effects of chain branching and chain termination kinetics on slowly varying flames. *Searby* and coworkers report on an application of laser tomography and laser doppler anemometry to study the stability of a premixed flame in a turbulent flowfield. They have rationalized the observed interaction of a stable flame with the upstream turbulent flow through fluid mechanical consideration of the flow in the flame. *Ybarra* analyzes downward flame propagation in a gravity field for arbitrary general intermolecular interaction potential and reports analytical expressions for Markstein scale and stability limits. *Quinard* and coworkers report an experimental investigation of stability limits and critical cellular dimensions in premixed flames. The observed critical dimensions agree approximately with the zeroth-order predictions of the model. *Clarke* and *Cant* report on an application of small perturbation analysis to reactive flows behind a shock wave. Numerical analyses with detailed kinetics are reported by *Andersson* and coworkers (for application to the structure of methanol flames), by *Hwang* and coworkers (for determination of the effect of truncated kinetics for lean acetylene-oxygen mixtures on shock initiated combustion), and by *Westbrook* and *Pitz* (for characterization of properties of flames in propane-air mixtures). *Ohta* and *Takahashi* report on observations of homogeneity and propagation of autoignited cool and blue flames in an engine.

In Chapter II, Diffusion Flames, *Penner* and coworkers extend well-established flame theories to develop "a simplified procedure for the analytical solution of coupled diffusion and laminar decomposition flames with cylindrical symmetry." *Masri* and coworkers report the results of an experimental investigation of the initial region of a diffusion flame. Their results indicate that molecular transport effects are important in the initial region even at large Reynolds number. *Liew* and coworkers report their numerical solutions of stretched and unstretched laminar diffusion flames and indicate the application of these results to turbulent flames when the phenomenon can be considered as a laminar flamelet in a nonreactive turbulent flowfield. *Sønju* and *Hustad* report on their investigations of turbulent jet diffusion flames and compare results for small-scale laboratory experiments and field tests.

In Chapter III, Turbulent Combustion, *Sirignano* and coworkers report on extensive numerical calculations to test turbulence models for reactive flows. For asymmetric concentric jet configurations with dilute reactants and incompressible flow, chemistry is decoupled from fluid mechanics. As a consequence, the system presents an ideal situation for the testing of pdf and moment calculations of the extent of reaction. *Gökap et al.* report experimental observations on turbulent cool flames as an example of turbulent flames in the low Damköhler number limit. Their results indicate that a TCF is an extended stabilized flame front with slow chemistry and low exothermicity. *Bradley* and coworkers compare experiment and simulation for a conical single jet stirred reactor. *Librovich, Roberts* and coworkers report on their work on the mechanisms of sound emission from turbulent combustion.

In Chapter IV, Constant Volume Combustion, *Tromans* and *O'Connor* report on ignition experiments that show that turbulence increases the minimum energy required for combustion and that turbulence intensity broadens the band of spark energies over which ignition is uncertain. *Leyer* and coworkers at E.N.S.M.A. report observations that small cavities in the side walls can be sources of knock. In a second paper, *Girard* and *Leyer* report the use of CH radical emission from the flame front to monitor flame trajectory and head-on quenching distance in closed combustion chambers.

In Chapter V, Spray Combustion, *Gerstein* and *Choudhury* report on a study of the ignition of explosives or flammable solutes upon evaporation of a desensitizing solvent. *Dwyer* and *Sanders* show that droplet heating and vaporization at high Reynolds and Peclet numbers manifest highly unsteady behavior whose analysis requires full solution of time-dependent conservation equations. Their numerical solutions provide interesting insight into the phenomena. *Bracco* and coworkers report on a two-dimensional, time-dependent model of spray jets issuing from a single cylindrical orifice. The model uses a k-ϵ description of turbulence and a stochastic algorithm to account for droplet motion including collisions and coalescence. The mean quantities predicted are shown to have achieved good agreement with those observed. *Baev* and coworkers discuss their observations of induction times of gasoline and diesel oil sprays injected into a reservoir filled with high temperature air under high pressure. Activation energies determined from these observations appear to be strongly influenced by fluid dynamical phenomena and consequently are not activation energies in the chemical kinetic sense. The study of *Rizk* and *Lefebvre* focuses on the effects of air and liquid properties and atomizer dimensions on spray characteristics of simplex pressure swirl atomizers.

In Chapter VI, Nonequilibrium Flows, *Mitra* and *Fiebig* discuss the calculation of supersonic nozzle flows with coupled vibration dissociation. *Librescu* contributes a study of time-dependent pressure fields on oscillating elastic thin bodies in a reactive flow; closed-form solutions for flat thin bodies have application in flutter analysis. *Pandey* and *Chou* develop uniform solutions for weak axisymmetric shock waves in nonequilibrium flows.

In Chapter VII, Combustion Diagnostics, *Eckbreth* and coworkers report on the application of coherent anti-Stokes Raman spectroscopy (CARS) to measure temperature and species concentrations in combustion fields—in particular to gas turbine exhausts. In their paper, *Stepowski* and *Cottereau* deal with observations of fluorescence from OH radicals in low pressure flames at near saturation conditions. *Michael-Saade et al.* apply multichannel pulsed Raman spectrometry to determine spatial dependence of concentration and temperature in reactive systems. *Zoltani* and *White* describe the design of a flash tomographic x-ray facility to observe transient ballistic events. *Cattolica* and *Stephenson* report the application of two-dimensional imaging of laser-induced fluorescence from OH molecules to determine the postflame temperature field of a flat flame. *Candel* and coworkers report on their efforts to develop an optical system that will give accurate and fast measurements of droplet size distribution. *Cuellar* and *Brown* demonstrate that tunable atomic line molecular spectroscopy (TALMS) can be used to observe NO in the postflame regions at atmospheric conditions. With TALMS, measurement of NO over concentration ranges that differ by several orders of magnitude and at concentrations as low as a few ppm is possible.

The companion volume includes papers on gaseous detonations, heterogeneous detonations, condensed explosives, explosions, and interactions (Volume 94 in the *AIAA Progress in Astronautics and Aeronautics* series). Both volumes, we trust, will help to satisfy that need first articulated in 1966 and will continue the tradition of contribution to our understanding of the dynamics of explosions and reactive systems begun the following year in Brussels with the first Colloquium. Subsequent Colloquia have been held on a biennial basis since then (1969 in Novosibirsk, 1971 in Marseilles, 1973 in La Jolla, 1975 in Bourges, 1977 in Stockholm, 1979 in Göttingen, 1981 in Minsk, and 1983 in Poitiers). They have now achieved the status of a prime international meeting on these topics and attract contributions from scientists and engineers throughout the world. The *Proceedings* of the First through the Sixth Colloquia have appeared as part of the journal, *Acta Astronautica,* or its predecessor, *Astronautica Acta*. With the publication of the Seventh Colloquium, the *Proceedings* now appear as part of the *AIAA Progress in Astronautics and Aeronautics* series.

Acknowledgments

The Ninth Colloquium was held under the auspices of the Ecole Nationale Supérieure de Mécanique et d'Aérotechnique, Université de Poitiers, France, July 3-8, 1983. Arrangements in Poitiers were made by Dr. J. C. Bellet. The publication of the *Proceedings* has been made possible by grants from the National Science Foundation (USA) and the Army Research Office (USA).

Preparations for the Tenth Colloquium are under way. The meeting is scheduled to take place in August 1985 at the University of California, Berkeley, California.

J. Ray Bowen
Numa Manson
Antoni K. Oppenheim
R. I. Soloukhin
June 1984

Introduction to Modern Laminar Flame Theory

Paul Clavin*
Université de Provence, Marseille, France

Abstract

A survey of the milestones in the evolution of the theory of premixed laminar flames is presented. The current status of the subject is summarized, including the most recent advances in the study of the coupling occurring in wrinkled flame fronts between the fluid mechanical processes, transport phenomena, and chemical kinetic effects of combustion.

Historical Background

The first attempt at interpreting physical properties of flames is generally attributed to Greek philosophers (c. 500 B.C.) who considered that fire was one of the four elements that, along with air, water, and earth, made up the universe. The first study of flames was carried out during the Renaissance, motivated by the interest in the structure of a candle flame. The question was obscured for a long period of time by the phlogiston theory based on ancient alchemical ideas. It was only after the publication by Lavoisier of his <u>Reflexions sur le Phlogistique</u> (1777) that the combustion phenomena were definitively attributed to a chemical reaction between two components [for more details, see Fristrom and Westenberg (1965)].

At the same time, another aspect of combustion was discovered. As reported in a book by John Tyndall entitled

---

Presented at the 9th ICODERS, Poitiers, France, July 3-8, 1983. Copyright © 1984 by Paul Clavin. Published by the American Institute of Aeronautics and Astronautics, Inc. with permission.
*Professor, Departement de Combustion.

Sound (1867), in 1777, soon after the discovery of the physical nature of the gaseous phase, Higgins studied the sound produced by a hydrogen flame when the burner was surrounded by a vertical tube. Much later, in 1818 Faraday showed that this phenomenon was not restricted to hydrogen flames [see Rayleigh (1945) Vol. II, p.227].

Later, in 1857, the reverse phenomenon was observed by Tyndall: a voice pitched to the note of the tube caused the flame to quiver, an effect that could lead to extinction. This action of sound was also observed in unconfined flames close to the condition of flaring. This was first observed by John Leconte (1858) at a musical party in the United States: "The flame exhibited pulsations in height which were exactly synchronous with audible beats. The phenomena was very striking to everyone in the room and especially so when the strong notes of the violoncello came in.... A deaf man might have seen the harmony." Studying the interaction of acoustic modes and combustion, Lord Rayleigh (1877) defined a criterion that is still used by engineers to rationalize instabilities occurring in combustion chambers. These early investigations point out that, in addition to chemical processes, combustion science must deal with fluid mechanics.

The first quantitative studies of flames in premixed gases were carried out by two members "du Corps des Mines," Mallard and Le Chatelier (1883), who were charged by "la Commission du grisou" to develop proper means to improve safety in mine galleries. They also were the first to explore flame propagation in tubes. At the same time, Berthelot and Vielle discovered detonation waves. Some time before, in 1816, Sir Humphrey Davy investigated flame quenching in narrow tubes [see Mallard and Le Chatelier (1883)]. In their 1883 treatise, Mallard and Le Chatelier considered flames in premixed gases as reaction-diffusion waves and developed the basic thermal propagation theory of flame propagation. Introducing the concept of a temperature of inflammation below which chemical reaction does not take place, they interpreted the propagation as the result of sequential ignition of fresh gas layers, the thermal energy being transported primarily by heat conduction from a reacting layer to the next unreacted layer, which became ignited when it reached the inflammation temperature. However, in the absence of precise mathematical formulation, they were not able to derive an appropriate expression for the flame speed.

Thermal Wave Propagation

The differential equation for the local energy balance that controls the thermal propagation of a planar combustion wave has been derived independently by Mikhelson (1889), Taffanel (1913), and Jouguet (1913). Expressed with respect to the front, this equation takes the form of the Fourier equation for steady heat conduction associated with convection and production:

$$mC\, dT/dx - \lambda\, d^2T/dx^2 = qw \qquad (1)$$

$$\text{Convection} \quad \text{Heat conduction} \quad \text{Production}$$

The heat conductivity λ and the specific heat of the reactive mixture C are usually taken to be constant; q is the heat released per unit mass of reactants, w is the mass rate at which reactants are consumed per unit volume, while m is the mass flow per unit area associated with the wave propagation velocity u_o relative to the unburnt mixture (usually called the laminar flame speed), whence

$$m = \rho u = \rho_o u_o = \rho_b u_b \qquad (2)$$

In the above, ρ is the gas density, while u is the gas velocity; and subscripts o and b refer to the fresh mixture at $x = -\infty$ and the burnt mixture at $x = +\infty$, respectively. Because of gas expansion, $\gamma = (\rho_o - \rho_b)/\rho_o$, $0 < \gamma < 1$, the flow velocity u undergoes a change across the front. The temperature profile $T(x)$ as well as m are the unknowns of the problem. The mass rate of reactant consumption per unit volume of the reactive mixture w is a function of T, can be expressed in a nondimensional form as

$$w = \rho_b \omega / \tau_r \qquad (3)$$

where τ_r is the characteristic time of the reaction. The nondimensional production term ω is a function of the reduced temperature

$$\theta = (T-T_o)/(T_b-T_o)$$

where the adiabatic flame temperature T_b is obtained from the overall energy balance $C(T_b-T_o) = q$ with $\int_{-\infty}^{+\infty} w\, dx = m$.

The diffusivity is defined as $D_T = \lambda/\rho C$, while ρD_T is assumed to be constant. Thus, the thermal energy balance can be written as:

$$m \, d\theta/dx - \rho D_T \, d^2\theta/dx^2 = \rho_b \omega/\tau_r \qquad (4a)$$

The boundary conditions for Eq. (4) are

$$x = -\infty : \theta = 0 \quad \text{at} \quad x = +\infty : \theta = 1 \qquad (4b)$$

Consideration of the dimensional form of Eq. (4) leads to the conclusion that the laminar flame speed must be proportional to the square root of the ratio of diffusivity and reaction time:

$$u_o \propto \sqrt{D_T/\tau_r} \qquad (5a)$$

whence the flame thickness

$$d \propto D_T/u_o = \sqrt{D_T \tau_r} \qquad (5b)$$

One should observe that Eqs. (5a) and (5b) express simply the balance betweeen the diffusion time $d^2/D_T (=D_T/u_o^2)$ and the reaction time, τ_r. This result was first obtained by Taffanel (1913). The difficult problem is to find the coefficient of proportionality, a function of ω.

Thus, the thermal propagation of a planar flame is seen to be simply controlled by a single reaction-diffusion equation. These types of equations have been also studied extensively in other sciences, such as biology [see, for example, Murray (1977) and Fife (1979)], where θ represents population instead of temperature. As a consequence of the strong sensitivity of the consumption term to temperature, the combustion wave represents a special case.

Steady-state progressive waves can exist only if the upstream mixture is at steady state, whereas the burnt mixture is at equilibrium. Thus,

$$x = -\infty : \theta = 0, \, \omega(\theta) = 0 \qquad (6a)$$
(fresh mixture: frozen flow)

$$x = +\infty : \theta = 1, \, \omega(\theta) = 0 \qquad (6b)$$
(burnt mixture: equilibrium flow)

However, as shown by Fisher (1937), as well as Kolmogorov et al. (1937), who studied, in a biological context, progressive waves for the particular case of $\omega(\theta) = \theta(1-\theta)$, the boundary conditions (6) do not assure the uniqueness of solution for Eq. (4a). In fact, there exists a continuous set of possible values of m with a lower bound m_{KPP} completely determined by the behavior of the consumption term at the upstream condition $\theta = 0$:

$$m_{KPP} = 2\sqrt{\rho_b(\rho D_T)\omega_o'/\tau_r} \qquad (7)$$

where

$$\omega_o' \equiv d\omega/d\theta \bigg|_{\theta=0} > 0$$

The particular solution (7) is, moreover, proved to be physically relevant. In fact, this solution is attained asymptotically in time for a class of initial conditions, including a heaviside step function [see Fife (1979) and Zel'dovich (1948)], by the time-dependent version of Eq. (4a). Such a result has been originally derived for a concave and strictly positive function $\omega(\theta)$ (within the interval $0<\theta<1$) and has been also generalized to nonconcave situations, although the lower bound of the spectrum may have, in this case, a higher value than that specified by Eq. (7) [Fife (1979)]. This situation occurs in combustion where the reaction rate w contains an Arrhenius factor with an activation energy, E, which is significantly larger than the thermal energy, i.e., $E \gg kT$. In a recent paper, Aldushin et al. (1981) showed that the nature of the solution associated with the lower bound of the spectrum changes when the activation energy becomes sufficiently large. The corresponding modification can be easily observed in the phase space ($p = d\theta/dx, \theta$) of Eq. (4a), where the behavior of the orbit at the node ($p=0, \theta=0$) changes above a critical value of the activation energy, E^* having the same order of magnitude as kT_b. For $E<E^*$, the KPP solution expressed by Eq. (7) holds true. But when $E>E^*$, the lower bound of the spectrum becomes larger than m_{KPP}, and the corresponding solution is not controlled any more by the shape of the consumption term at conditions of unburnt medium ($\theta=0$). In fact, at the limit of high values of the reduced activation energy $E/kT_b \gg 1$, this solution depends only on the behavior of ω at conditions of the burnt medium ($\theta=1$). Properties of this transition have been considered in detail by Clavin and Liñán (1984), where an exact solution is presented for a simplified model.

In the case of the ignition-temperature model defined by

$$0 < \theta_i < 1 \quad : \quad \omega(\theta) = 0 \tag{8}$$

the uniqueness of the solution has been proved by Zel'dovich (1948) [see also Gelfand (1959) and Johnson and Nachbar (1963)]. This is also true when $\omega(\theta) < 0$ for $0 < \theta < \theta_i$ and when $\omega(\theta)$ has only one zero in the interval $[0,1]$ [Frank-Kamenetskii (1969); Fife (1979)]. It turns out that in the limit of $E/kT_b \to \infty$ the value of $\dot\omega(\theta)$ tends to zero everywhere except close to $\theta=1$, and the result corresponding to the lower bound tends to coincide with that of an ignition temperature θ_i model when $\theta_i \to 1$.

As noted by Frank-Kamenetskii (1969), the condition of $\omega'_o > 0$ corresponds to unburnt mixtures that are unstable with respect to even the smallest fluctuations of θ around 0. Thus, when θ represents the temperature, the KPP solution of Eq. (7) cannot be meaningful, because it corresponds to waves propagating with a characteristic time scale of an order of magnitude similar to that associated with the growth rate of the fluctuations in the unstable medium. In such a case ($\omega_o > 0$), the steady propagation can be meaningful only if the characteristic time at which the fluctuations grow in the fresh mixture is infinitely large in comparison to the characteristic time of the propagation that, as will be shown later, is expressed in terms of $\int_0^1 \omega(\theta)d\theta$. This conclusion is less drastic in biology, where θ represents the population of species that can be so rare that they are completely absent in the upstream medium (no possible fluctuations). The above considerations imply that different time scales must be involved in the domain $\theta \in [0,1]$ for the reaction rate $\omega(\theta)$. This is the case when, in the Arrhenius law, $E/kT_b \gg 1$.

From a purely mathematical point of view, however, the Arrhenius law is associated with an even worse defect. The fresh mixture is not at a steady state, $\omega(\theta=0) \neq 0$—a feature referred to as the cold boundary problem. Strictly speaking, Eq. (4a) does not admit any steady solution. Burnt gases are at equilibrium because the consumption of the reactants makes the production rate vanishing in the burnt gas mixture, i.e., $\omega(\theta=1)=0$. However, the reaction is not completely quenched at $\theta=0$, where the reaction rate is proportional to e^{-E/kT_o}. A good deal of effort was made during the 1950's to obviate the cold boundary problem by either modifying the Arrhenius law (Friedman and Burke 1953) or by assuming the existence of a flame-holder

(Hirschfelder, Curtiss, and Campell, 1953). None of these techniques is satisfactory, because the final result depends too much on these artificial assumptions. A comprehensive review of these "laminar flame" theories (or, more correctly, constant pressure deflagration theories) has been presented by Oppenheim (1963).

As noted by Frank-Kamenetskii (1969), the cold boundary problem is really of an academic nature, and it cannot play any role in the physical world, as can be demonstrated by the following simple order of magnitude estimate. For ordinary values of activation energy ($E \cong 30$ kcal/mole), flame temperature ($T_b \cong 2000$ K), and initial temperature ($T_o \cong 300$ K), the ratio of the reaction rates $\tau_r(T_o)/\tau_r(T_b)$ is of an order of $\approx 2.8 \times 10^{18}$. Thus, for $\tau_r(T_b) = 10^{-6}$ s, about 900 centuries will be required for completion of the reaction. For $E \sim 20$ kcal/mole, this time is reduced to a month. A situation of this sort is typical for a two-time-scale problem. The time-dependent version of Eq. (4a) must be considered, and one must look for a quasisteady flame propagation in a fresh medium in which the reaction takes place homogeneously at a rate of $\tau_r^{-1}(T_o)$ [see Aldushin et al. (1981)]. On the laboratory time scale $t \ll \tau_r(T_o)$, the problem is reduced to the solution of the steady-state Eq. (4a), where $\omega(\theta)$ is, in effect, proportional to the difference of the reaction rate at T and the initial reaction rate at T_o. In combustion theory, such refinements are not necessary because, to quote Frank-Kamenetskii (1969), "the reaction rates, as a consequence of their remarkable exponential temperature dependence, fall so sharply with a decrease in the temperature that we can consider them as negligibly small even at temperatures considerably above T_o." This demonstrates the possibility of keeping a reactive mixture in a frozen composition far from equilibrium and ensures the uniqueness of the combustion wave. A pertinent solution for this unique combustion wave was published in 1938 by Zel'dovich and Frank-Kamenetskii.

The Zel'dovich and Frank-Kamenetskii Solution

In their outstanding paper of 1938, so ahead of its time that a more advanced development did not occur over the following 30 years, Zel'dovich and Frank-Kamenetskii (ZFK) developed an approximate theory for flame propagation that was proved in 1970 to correspond to the dominant order of the asymptotic expansion in large values of the reduced activation energy ($\beta \to \infty$). Following Lewis and von

Elbe (1934) and Jost and Muffling (1938), who demonstrated the essential role of molecular diffusion, ZFK developed a flame model based on a one-step overall reaction of the decomposition type:

$$R \rightarrow P + Q \qquad (9)$$

where R stands for reactants, P for products, and Q for heat release. Thus, the energy conservation equation (4a) is coupled with the continuity equation for the reactants via the production term $\omega(\theta,\Psi)$ that also depends on the reduced mass fraction Ψ of reactive species R, $0<\Psi<1$. The mass rate of the reactant consumption w is expressed by the Arrhenius law:

$$w = \rho B \Psi^n \exp(-E/kT) \qquad (10)$$

where n is the order of reaction and B the frequency factor. The factor Ψ^n corresponds to the effect of the consumption of reactants, while the exponential term reflects the high sensitivity to temperatures. The nondimensional form of Eq. (10) is given by Eq. (3), where

$$\tau_r^{-1} = \beta^{-n} B_b \exp(-E/kT_b) \qquad (11)$$

while

$$\omega(\theta,\Psi) = (\rho B/\rho_b B_b) \beta^n \Psi^n \exp\{-\beta(1-\theta)/1+\gamma(\theta-1)\} \qquad (12a)$$

In the above, β is the reduced activation energy to be referred to as the Zel'dovich number †, i.e.,

$$\beta = \gamma E/kT_b \text{ and } \gamma = T_b-T_o/T_b \qquad (12b)$$

One should note that $e^{+\beta}$ measures the sensitivity of the reaction rate to a temperature change on the order of T_b-T_o. The characteristic reaction time τ_r is defined by Eq. (11) at the burnt gas temperature; the use of the factor β^n is justified later. When β is large ($\beta \gg 1$), ω is, according to its definition, Eq. (12a), transcenden-

---

† This number was introduced first by Zel'dovich and Frank-Kamenetskii in their 1938 paper. To associate it with the name of Zel'dovich was decided at the 1983 meeting in Poitiers.

tally small unless θ is sufficiently close to unity to make β(1-θ)=O(1). The equation for the temperature has the same form as Eq. (4a), and the continuity equation for the reactants controlling Ψ is

$$m \, d\Psi/dx - \rho D_\Psi \, d^2\Psi/dx^2 = -(\rho_b/\tau_r)\omega(\theta,\Psi) \quad (13a)$$

$$\text{Convection} \qquad \text{Diffusion} \qquad \text{Consumption}$$

whereas the boundary conditions are

$$x = -\infty : \Psi = 1 \quad \text{(fresh mixture)}$$
$$x = +\infty : \Psi = 0 \quad \text{(burnt gas)} \quad (13b)$$

Here, D_Ψ is the molecular diffusion coefficient of the reactant, while, as before, ρD_Ψ is assumed to be constant. In this model, the planar flame propagation is reduced to two-coupled equations of diffusion-reaction Eqs. (4a) and (13a). However, as noted in the ZFK paper, when the thermal and the molar diffusion coefficients are equal, $D_\Psi = D_T$ (Lewis Number $L = D_T/D_\Psi = 1$), the two equations (4a) and (13a) are identical. Thus, according to boundary conditions (4b) and (13b), one must have complete similarity between the temperature and the concentration profiles, as displayed in Fig. 1, whereas

$$\Psi + \theta = 1 \quad (14)$$

When L=1, the problem is reduced, therefore, to the simpler model of thermal propagation, as prescribed by Eq. (4a), with boundary conditions (4b), where $\omega(\theta)$ is obtained from Eqs. (14) and (12):

$$\omega(\theta) = (\rho B/\rho_b B_b) \, \beta^n (1-\theta)^n \, \exp\{-\beta(1-\theta)/1+\gamma(\theta-1)\} \quad (15)$$

For large Zel'dovich numbers (β>>1), the reaction is quenched, except close to the flame temperature [1-θ = O(1/β)]. Because of Eq. (14), however, there is a small amount of reactants left [Ψ=1-θ=O(1/β); see Fig. 1], and the reaction must stop soon after its initiation. Thus, the reaction zone is limited to a thin layer of thickness d/β, separating a larger preheated zone of thickness d (frozen flow) from the burnt gases (equilibrium flow). The time τ_r, as defined by Eq. (11), is then, in effect, the characteristic chemical time in the reaction zone where Ψ=1-θ=O(1/β).

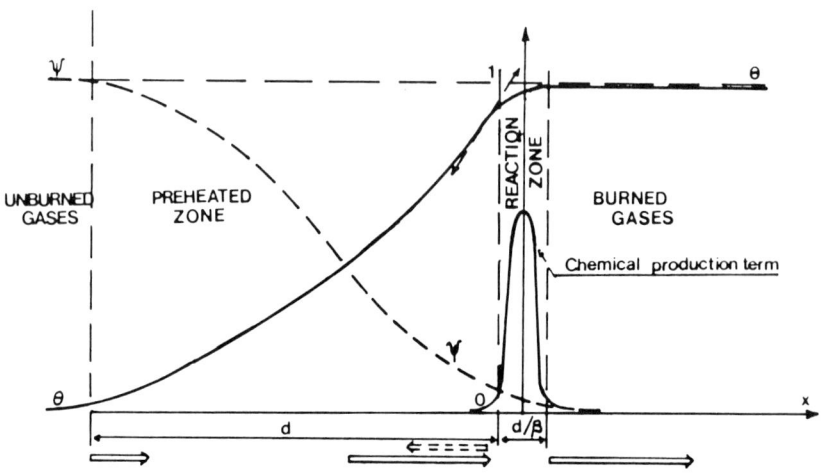

Fig.1 Structure of a laminar flame in a premixed gas of nonstoichiometric composition, sustained by a single reaction expressed in terms of an Arrhenius relation, with a large value of the Zel'dovich number.

It appears clearly from Eq. (1) that the heat released by the reaction, qw, is partially consumed by heating the reacting gas, mC dT/dx, and partially carried away by heat conduction to the neighboring elements of the gas, $\lambda\ d^2T/dx^2$. If the temperature of the zone in which the reaction actually proceeds is already near T_b, the amount of heat consumed by heating the reacting gas up to its final temperature T_b (behind the front of the flame) is small. With good approximation it can be assumed that all the heat from the reaction zone is carried away by conduction (Zel'dovich and Frank-Kamenetskii 1938). Thus, in the reaction zone, the thermal balance expressed by Eq. (1) is approximately reduced to

$$- \lambda\ d^2T/dx^2 = qw \qquad (16a)$$

that is,

$$- (\rho D_T)\ d^2\theta/dx^2 = (\rho_b/\tau_r)\omega(\theta) \qquad (16b)$$

where $\omega(\theta)$ is given by Eq. (15). This means that, in the reaction zone, the convective terms, m (d/dx), are small in comparison to those expressing diffusive effects, $\rho D_T\ (d^2/dx^2)$, which have to be balanced by the production term $\rho_b\omega/\tau_r$. After multiplication by $d\theta/dx$, Eq. (16b) can be integrated, subject to the burnt boundary condition $\theta=1$, $d\theta/dx = 0$ (see Fig. 1), yielding

$$(\rho D_T)\ 1/2\ (d\theta/dx)^2 = - (\rho_b/\tau_r) \int_1^\theta \omega(\theta')d\theta' \qquad (17)$$

By using a dummy variable $\beta(1-\theta) = \Xi$, the integral of Eq. (17) can be expressed as

$$(\rho D_T)_b \, d\theta/dx = \sqrt{2I\rho_b \, (\rho D_T)_b/\beta\tau_r} \qquad (18a)$$

where

$$I = \int_0^{\beta(1-\theta)} d\Xi \, \Xi^n \exp\{-\Xi/(1-\gamma\Xi/\beta)\} \qquad (18b)$$

while the frequency factor B has been assumed to obey the condition: $\rho B = \rho_b B_b$. Equation (18a) can be used to compute the energy flux leaving the reaction zone by heat conduction toward the preheated zone, as shown in Fig. 1.

The one and only difficulty left is the accurate determination of the upper bound of Eq. (18b) corresponding to the boundary between the reaction zone and the preheated region. This is a matching problem that can be solved systematically by the appropriate techniques of asymptotic expansion (Van Dyke 1975; Cole 1968) that was not known in 1938. In spite of that, with an excellent sense of physics, in 1938 ZFK gave the correct answer for the limiting case of $\beta = \infty$. Noting that

$$\lim_{\beta \to \infty} I = \int_0^\infty d\Xi \, \Xi^n e^{-\Xi} = \Gamma(n+1) \quad (=n! \text{ when } n \in Z^+) \qquad (19)$$

ZFK used the following approximate expression for the heat conduction flux:

$$(\rho D_T) \, d\theta/dx = \rho_b \sqrt{2n! \, D_{Tb}/\beta\tau_r} \qquad (20)$$

where D_{Tb} is the heat diffusivity of the reactive mixture at T_b.

In the preheated region located just upstream of the thin reaction zone, as depicted in Fig. 1, the chemical rate is negligible (ω is transcendentally small), and the temperature, as well as species concentrations, evolve under frozen flow conditions, as prescribed by the balance of the diffusive and convective fluxes. Thus, the heat conduction flux, compressed by Eq. (20), that enters the preheated zone must be balanced by the energy flux leaving this zone by the convective flow associated with flame motion relative to the fresh mixture. This convective flux is simply $m\theta_b - m\theta_o = m$. In fact, when $\beta \to \infty$, one obtains by integrating Eq. (4) (with $\omega = 0$) from the unburnt gases

($x = -\infty$) to the reaction front located at $x=0$:

$$m = (\rho D_T) \, d\theta/dx \Big|_{x=0-} \qquad (21)$$

The two equations (20) and (21) provide the expression for the laminar flame speed specified by ZKF:

$$u_{oZFK} = (\rho_b/\rho_o) \sqrt{2(n!)D_{Tb}/\beta\tau_r} \qquad (22)$$

As pointed out by Zel'dovich (1948), Landau included the effect of the difference between the diffusivities $D_T \neq D_\psi$, so that $L \neq 1$. This can be done by noting that, when the convective terms $m(d/dx)$ are neglected in the reaction zone, it follows that

$$\Psi/L = 1-\theta \quad \text{(reaction zone)} \qquad (23)$$

When $L \neq 1$, Eq. (23) is valid only in the reaction zone where it replaces Eq. (14). When Eq. (23) is combined with Eq. (12a), one obtains for ω the expression given by Eq. (15) times L^n. The same calculation as above leads then to

$$u_o = (\rho_b/\rho_o) \sqrt{(2n!)L^n D_{Tb}/\beta\tau_r} \qquad (24)$$

Zel'dovich (1941) also extended this analysis to take into account the effects of heat losses and was able to describe, on this basis, the thermal extinction phenomena that were observed by Davy at the beginning of the 19th century [see Zel'dovich and Barenblatt (1959)].

It should be noted that, due to the localization of the integral on the right hand side of Eq. (17), $\int d\theta\omega(\theta)$ inside the thin reaction zone, the characteristic time of flame propagation is not governed by just the chemical time in the reaction zone, τ_r, but by the product, $\beta\tau_r$.

One should note also that, according to Eqs. (22) and (24), no modification of the Arrhenius law is required for the derivation of this result. The laminar flame speed u_o is thus completely determined by the physical and chemical properties defined at the state of burnt gases. This is quite different from the Kolmogorov-Petrovsky-Piskunov (KPP) solution, Eq. (7). The ZFK solution is, in fact, an approximate expression for the solution that applies when

KPP is no longer valid, i.e., when $E>E^*$ (see the preceding section). The corresponding critical value, β^*, has been computed by Higuera (1983) for the model $\omega(\theta)=(1-\theta)^n \left(e^{\beta(\theta-1)} - e^{-\beta}\right)$, where the term $-e^{-\beta}$ is added to ensure the existence of a steady solution of flame propagation (Aldushin et al. 1981). The results are $\beta^*=3.04$ for $n=1$, and $\beta^*=5.11$ for $n=2$. This result is in agreement with the time dependent study of Aldushin et al. (1981).

Indeed, because of the approximations involved in by Eqs. (16a) and (19), the results of ZFK and the results of Landau, expressed by Eqs. (22) and (24), are strictly valid only in the limit where $\beta \to \infty$. Concomitantly, one is led to the conclusion that the quantity $m^2 \beta \tau_r$ [where τ_r is given by Eq. (11)] remains at the level of unity when the activation energy tends to infinity. This behavior and the concept of a thin reaction zone form the basic background for the asymptotic expansion analysis developed after 1970 to describe the flame structure when $\beta > \beta^*$.

Further Developments - The Asymptotic Analysis

With the development of numerical techniques, considerable effort was spent during the 1950's by Hirschfelder and his co-workers (1953) and by von Kármán (1957) to derive complete sets of equations for the conservation of species, momentum, and energy in a multicomponent mixture of reactive gases. This was associated with investigations to identify the multistep kinetic scheme involved in the chemical reaction controlling the heat release. The results of these studies point out clearly the limitations of the simple model of Eqs. (4a) and (13a) to describe the actual behavior of laminar flames. Nevertheless, as one cannot expect to solve more complex models without a complete understanding of the simpler case, many investigations were still conducted after 1950 on the basis of Eqs. (4a) and (13a). Thus, different iterative procedures and variational methods have been used (Rosen 1959; Johnson and Nachbar 1963), but none of them succeeded in circumventing the cold boundary problem. Von Kármán (1957) developed an approximate method corresponding, in the first order, to the result of ZFK where the exact value of the integral I in Eq. (18a) was used instead of its limiting value expressed by Eq. 19). Such a procedure is not self-consistent because the corrective terms due to the finite value of β are retained in some places and omitted in other places. For example, Eqs. (16a) and (16b), on which the result is based, are strictly valid in the reaction zone. Thus, the unsystematic procedure of this perturbative approach does not lead to appropriate expressions for the corrective terms of the ZFK theory (Eq. 22) (strictly valid when $\beta \to \infty$). In an earlier attempt of Boys and Corner (1949), a similar approximate method was used, but the iterative procedure was less convergent for large values of β; in particular, even their first-order solution did not correspond to the ZFK result. Among all the attempts made before 1970, the results obtained by von Kármán have been proved to be most accurate (see, for example, Bush and Fendell 1970).

A systematic treatment of flame structure based on asymptotic expansion in large values of β was carried out at first independently by Bush and Fendell (1970, 1971) for a one-step model, and by Liñan (1971) for a chain branching reactions scheme. The technique of asymptotic expansions around a singular limit was developed in the 1950's for the Navier-Stokes equations (in particular, by Kaplan and Lagerstrom at the California Institute of Technology) to improve the Prandtl boundary layer theory. Its first application to combustion was accomplished by Bowen et al. (1963), as a refinement to the quasisteady-state theory of chemical kinetics. In the application to flames, the original purpose of analyses based on the technique of asymptotic expansions was to obtain an accurate analytical solution for the structure and speed of a plane commbustion front. Another important application of this method was made to wrinkled flames (Joulin and Clavin, 1979), leading to a fairly complete description of the dynamic properties of such flames, including the effects of coupling between fluid mechanics, diffusion, and reaction (Clavin and Williams 1982; Pelcé and Clavin 1982; Frankel and Sivashinsky 1982; Matalon and Matkowsky 1982; Clavin and Joulin 1983; Liñan and Clavin 1984). These results are briefly presented in the last section of this paper, a more detailed account having been given in a recent review article (Clavin 1984).

To give a flavor of the method, first and second orders of the asymptotic expansion obtained for the solution of the thermal model (4) with L=1 is outlined below (see Joulin and Clavin 1976, 1979). Here, ρD_T is not assumed to be necessarily constant. By the introduction of the transformed independent variable, $\xi = \int_0^x (m/\rho D_T)dx$, the problem is reduced to the following equation:

$$d\theta/d\xi - d^2\theta/d\xi^2 = \Lambda \tilde{w}(\theta) \qquad (25a)$$

where

$$\tilde{w}(\theta) = A(\theta)\, \beta^{n+1}(1-\theta)^n \exp\{-\beta(1-\theta)/1+\gamma(\theta-1)\} \quad (25b)$$

while

$$A(\theta) = \rho D_T \rho B/(\rho D_T \rho B)_b \quad (25c)$$

The coefficient

$$\Lambda = (\rho D_T)_b\, \rho_b/m^2 \beta \tau_r \quad (26)$$

is the eigenvalue of the problem, whereas the frequency factor B and the chemical time τ_r are defined respectively by Eqs. (10) and (11). According to the ZFK Eq. (22), Λ must remain finite in the limit $\beta \to \infty$. One seeks, therefore, a solution expanded as follows:

$$\Lambda = \Lambda_0 + \Lambda_1/\beta + O(1/\beta^2) \quad (27)$$

In the preheated zone, $\tilde{w}(\theta)$ is transcendentally small and Eq. (25a) can be solved exactly, yielding

$$\theta(\xi) = e^\xi\, \theta(0) + \text{tst} \quad \text{valid for } \xi < 0$$

where tst stands for transcendentally small terms.

The origin of the axis ($\xi=0$) can be chosen so that $\theta(0)=1$:

Outer Solution: $\quad \theta(\xi) = e^\xi + \text{tst} \quad \text{valid for } \xi < 0 \quad (28)$

To solve the problem in the reaction zone of thickness d/β located around $\xi=0$, the stretched variable η is introduced and considered as the independent variable. One has then for $\eta = \beta\xi$:

$$d/d\xi = \beta\, d/d\eta$$

and the solution is expanded as follows:

$$(1-\theta) = \Xi_0(\eta)/\beta + \Xi_1(\eta)/\beta^2 + O(1/\beta^3) \quad (29)$$

Using a corresponding expansion of $\Lambda\tilde{w}(\theta)$, Eq. (25), provides then the equations for Ξ_0, Ξ_1.

Inner Equations:
$$d^2\Xi_0/d\eta^2 = \Lambda_0 \quad (30a)$$

$$-d\Xi_0/d\eta + d^2\Xi_0/d\eta^2 = \begin{Bmatrix} \Lambda_1 \Xi_0 e^{-\Xi_0} - \Lambda_0 \gamma b_1 \Xi_0^{n+1} e^{-\Xi_0} - \Lambda_0^\gamma \Xi_0^{n+2} e^{-\Xi_0} \\ +\Lambda_0 \Xi_1 (n\Xi_0^{n-1} e^{-\Xi_0} - \Xi_0^n e^{-\Xi_0}) \end{Bmatrix}$$

$$b_1 = \partial \ln(\rho D_T \rho B)/\partial \ln T \Big|_{T=T_b} \qquad A = 1 - \gamma b_1 \Xi_0/\beta + 0(1/\beta^2) \quad (30b)$$

Boundary conditions on the side of burnt gases [see Eq. (4b)] are

$$\eta \to +\infty : \quad \Xi_0 \to 0, \; \Xi_1 \to 0 \quad (30c)$$

According to Eq. (30a), solutions Ξ_0 and Ξ_1 present the following limiting behavior in the direction of the preheated zone:

$$\eta \to -\infty : \; d\Xi_0/d\eta \to C^t \neq 0, \; d^2\Xi_1/d\eta^2 \to C^t \neq 0 \quad (30d)$$

Matching conditions (Van Dyke 1975; Cole 1968) of the inner solution (29) with the outer solution $\theta(\xi) = \theta_0(\xi) + 1/\beta \; \theta_1(\xi) + 0(1/\beta^2)$ of the preheated zone, yields for $\eta \to -\infty$:

$$\Xi_0(\eta) = \left(\partial\theta_0/\partial\xi\right)\Big|_{\xi=0^-} \eta + \theta_1(\xi=0) + tst$$

$$\Xi_1(\eta) = \left(\partial^2\theta_0/\partial\xi^2\right)\Big|_{\xi=0^-} \eta^2/2 + \left(\partial\theta_1/\partial\xi\right)\Big|_{\xi=0^-} \eta + \theta_2(\xi=0) + tst \quad (31)$$

whence, according to Eq. (28), for $\eta \to -\infty$:

$$\Xi_0(\eta) = \eta + tst$$
$$\Xi_1(\eta) = \eta^2/2 + tst \quad (32)$$

The eigenvalues Λ_0 and Λ_1 appearing in inner equations are determined by prescribing boundary conditions (30c) and (32) (see Joulin and Clavin 1976):

$$\Lambda_o = 1/2\ \Gamma_{n+1} \qquad (33a)$$

$$\Lambda_1 = \Lambda_o \{b_1\ \gamma\ \Gamma_{n+2}/\Gamma_{n+1} + \gamma\ \Gamma_{n+3}/\Gamma_{n+1} - J_n/\Lambda_o\Gamma_{n+1}\}$$

where

$$J_n = \int_0^\infty d\Xi \left(1 - \sqrt{\int_0^\Xi dx\ x^n e^{-x}/\Gamma_{n+1}}\right) \qquad (33b)$$

As already mentioned, the dominant order (33a) is identical to the ZFK solution given by Eq. (22). Numerical evaluation of J_n yields $J_{n=1} = 1.344$ (see Fendell 1972).

For n=1 and $\gamma=0.85$, the flame velocity is, according to Eqs. (26), (33a), and (33b):

$$u_o = u_{oZFK}\{1 - 0.57/\beta + O(1/\beta^2)\} \quad \text{for}\ b_1 = -3/2$$
$$u_o = u_{oZFK}\{1 - 1.2/\beta + O(1/\beta^2)\} \quad \text{for}\ b_1 = 0 \qquad (34)$$

According to values of β given in Table 1, Eqs. (34) indicate that the first correction is not always negligible. By comparison with the numerical solutions of Eq. (25), expressions (34) are found to be accurate within less than 15% error for $\beta>3$ (Bush and Fendell 1970). For $\beta>5$, the error is less than 5%. In fact, the very limitation of the asymptotic method toward the small value of the Zel'dovich number β is the critical value $\beta*$ corresponding to the KPP solution.

The coefficient γ appearing in the first term in the bracket of (33) is not included in the result of Fendell (1972), corresponding to n=1 while $L\neq 1$. In order to deal with the cold boundary problem, $A(\theta)$ was approximated by $\theta^b 1$ in the model used by Fendell. As already mentioned, such a modification of the Arrhenius law is not necessary in the framework of the asymptotic expansion when $\beta\to\infty$. It may lead to irrelevant corrective terms when the hot boundary is modified.

The asymptotic method has been successfully applied to other one-dimensional and steady cases, such as the evaluation of the effects of the departure of the Lewis number from unity (Bush and Fendell 1970), of volumetric heat losses leading to thermal extinction (Buckmaster 1976; Joulin and Clavin 1976), of monopropellant droplet

Table 1 Values of β computed from Eq. (12b) for $T_o = 300$ K

Ekcal/mole T_b,K	30	40	50
1500 K $\gamma=0.8$	8	10.7	13.3
2000 K $\gamma=0.85$	6.4	8.5	10.7
2500 K $\gamma=0.88$	5.3	7	8.8

burning (Fendell 1972; Liñan 1976), of stretch effect of a planar front stabilized in a stagnation point flow (Buckmaster and Mikolaitis 1982; Libby and Williams 1982), of the effects of two limiting components for reactive mixtures close to the stoichiometric composition (Joulin and Mitani 1981), and of the influence of flameholders and the consequences of spherical geometry (Deshaies et al. 1981, Clarke and McIntosh 1980). In the last case, the temperature in the burnt gas can decrease to room temperature, a very small value in comparison to the flame temperature, and the corresponding structure of the inner zone becomes more complex than that presented above. These aspects were first studied by Liñan (1974) in the context of diffusion flames.

Multiple-Step Chemistry

From the early studies of Semenov (1935), it was known that the reactants are not converted directly into products. Many elementary steps involving different intermediate species are required. In the simplest case of the H_2-O_2 system, 60 elementary reactions and five intermediate species have been identified. Most of the intermediates are free atoms and radicals that are short-lived, active species, rapidly removed by recombination. Active species are produced initially by the decomposition of the reactants (chain initiation) and are also involved in chain reactions that can be either of the branching or nonbranching type. Chain reactions are autocatalytic (with a rate proportional to the concentration of radical species), and, in the case of chain branching, they enhance the total amount of radicals. Thus, in addition to thermal feedback, most of the combustion phenomena are also characterized by another feedback mechanism of a purely chemical-kinetic nature, owing to the action of radicals and chain branching reactions.

This mechanism was identified very early (in the 1940's) to be responsible for the propagation of cool flames and for the phenomena of two-stage ignition and the oscillatory phenomena that are sometimes observed at low temperatures (T~600 K) in the course of hydrocarbon oxidation [see Lewis and Von Elbe (1961) for a review of early studies on this subject]. Thus, in 1940, Frank-Kamenetskii [see Frank-Kamenetskii (1969)] proposed a simplified kinetic model, of the same type as the famous Lokte-Volterra scheme, involving two intermediate species that are capable of sustaining isothermal kinetic oscillations. In 1939, Voronkov and Semenov used a simple kinetic model involving only one radical (produced by a chain branching reaction of second order for the radical and removed by a first-order chain termination) to compute the isothermal propagation of cool flames in extremely lean mixtures of hydrogen sulfide in air [see Frank-Kamenetskii, (1969)]. An even simpler model, similar to the Fisher (1937) scheme, was studied in 1937 by Skalov and Todes [cited by Zel'dovich (1948)], where one radical is involved in a first-order (for the radical) chain branching reaction. In these models, the reactants are also consumed by the chain reaction mechanism, while the molecular diffusion of the radical is responsible for the isothermal propagation of chemical reaction. Because of the lower order of the chain termination reaction in the model used by Voronkov and Semenov, the propagation is only possible when the ratio of the rate constant of the branching reaction to that of the chain termination step is larger than 2. When, as is usually the case, the order of the chain terminating reaction is higher than that of chain branching, the effect of the former is less important, and, for slow recombination, it can be completely neglected, as in the model of Skalov and Todes. One should note that in both of these models no chain initiation is involoved. In fact, because of the high activation energy of the initiation reactions, their effect upon the production of radicals can be neglected in comparison to the effect of chain branching, the initiation of reaction in the fresh mixture being due entirely to the diffusion of radicals. These early models are quite useful for understanding possible mechanisms of isothermal propagation. Typical models for the propagation of reaction-diffusion waves are controlled by the KPP solution when the spectrum of solutions is continuous (see the section on thermal wave propagation).

Because of the sensitivity to temperature of chain reactions, thermal conduction cannot be neglected in ordinary hot flames. Nevertheless, it appears from the literature (see, for example, the review article of M. Evans 1952) that a good deal of attention has been paid to the so-called "diffusion theory" which assumes that the propagation velocity of such flames is determined by the diffusion of active intermediate species. Among all these approaches, the most popular were those of Tanford and Pease, Van Tiggelen, Gaydon and Wolfhard, [see Evans (1952)]. These empirical studies underestimated the importance of thermal conductivity and exothermicity, having been based on phenomenological analysis from which it is not easy to extract the relevant effects. More instructive are studies based on the solutions of the conservation equations, even when they are associated with an oversimplified kinetic scheme.

A different approach was developed by von Kármán and Penner (1954), who solved conservation equations for a full kinetic scheme using a quasisteady-state approximation for intermediate species. The essence of this approximation is the assumption that the production and removal of every radical species compensate each other. This approximation holds true when characteristic chemical times for the production and consumption of radicals are shorter than those required for diffusion and convection controlling the flame velocity. When this condition is applied to all the radicals, definite expressions for their concentrations are obtained as functions of the temperature and mole fractions of reactants. The system of equations is thus reduced to a set similar to Eqs. (13a) and (14), where only the temperature and the reactant species are taken into account. In such an approximation, the diffusion of radicals cannot affect the flame propagation. In general, this is not usually true for all the radicals; the steady-state assumption can only be used for radicals whose recombination rate is sufficiently fast.

The steady-state approximation is very useful in carrying out relevant reductions of the full kinetic scheme by eliminating from the conservation equations the intermediates whose diffusion does not essentially affect the flame propagation. When the reduced kinetic scheme is simple enough, the asymptotic method can be used to solve the conservation equations as soon as the chemical production of the diffusing radicals is sufficiently sensitive to the temperature. In this case radical production is limited to a zone that is thin (in first approximation) in comparison with the total flame thickness, but not necessarily located (as in Fig. 1) at the maximum temperature. The recombination reactions, being exothermic and nonsensitive to the temperature, may cause heat to be released outside the thin production zone. Such an analysis was carried out by Liñan (1971) using the chain reaction model proposed by Zel'dovich (1948) to describe flames in the H_2-O_2-N_2 mixture. But, as shown by Liñan, this model is only valid at very restricted conditions (see Clavin 1984). This is a general property. The simplified schemes used for describing the flame structure are expected to be valid only in a restricted domain of the temperature, pressure, and composition of the initial mixture. More recently, the asymptotic method has been applied by Liñan and Clavin (1984) to a simplified kinetic scheme introduced by Adams and Stocks (1953) to model the hydrazine decomposition flame and used afterwards by Spalding (1956) to model the reaction of hydrogen with halogens. These approaches are very promising, but, so far, they have been limited to cases where the complete chemical scheme can be reduced to a few steps. For example, such a drastic reduction has not yet been performed successfully for the simplest hydrocarbons. At present, there are numerical methods available to determine the structure of flames sustained by most complex kinetic scheme. But because of the huge amount of physicochemical parameters necessary to completely describe the kinetic scheme, it is not easy to analyze the numerical results and to check their validity. A promising avenue is to combine numerical and analytical treatments. This is, in fact, necessary to check the limits of the validity of both techniques (see, e.g., Peters and Warnatz 1982) and must constitute the most effective way to advance the knowledge on the structure of laminar flames (especially in the case of hydrocarbon-air mixtures). Such an approach is even more necessary in the case of multidimensional problems, as in the case of wrinkled front propagation in a turbulent flowfield.

Front Dynamics

Let us consider the motion of flame fronts in unsteady and nonuniform flowfields. When the length and time scales of the initial flowfield are larger than those associated with planar flames (i.e., the flame thickness d and the transit time d/u_o, where u_o is the laminar flame speed), the flame front can be considered, in first approximation, as a surface of discontinuity whose motion is controlled by two distinct factors: the normal burning velocity u_n, associated with the mass flux of fresh mixture crossing the front, and the magnitude velocity vector at the front. Each point of the front moves at a velocity equal to the difference between the upstream flow velocity and the normal burning speed.

As soon as one is confronted with wrinkled fronts or inhomogeneous flows, the flame cannot be described by a pure reaction-diffusion model. The streamlines are deflected across the tilted front as a consequence of gas expansion produced by the temperature increase in the preheated zone, and a strong coupling with hydrodynamic phenomena ensues. When the size λ of the front wrinkles is large in comparison to flame thickness d the fluid mechanical effecs can be split into two distinct components: 1) the influence of the convective transfer produced in the preheated zone by the deflection of streamlines upon the flame structure; and 2) the modification of the flowfield upstream and downsteam of the flame at a distance λ from the front, where the gas density ρ and the temperature T are uniform. This flow modification is due to the difference of density between unburnt and burnt gases.

The first effect results in a change of the normal burning speed so that $u_n \neq u_o$. It must be noted that, in addition to the convective flux produced by gas expansion, the diffusion fluxes of heat and mass exert a profound effect upon the structure of wrinkled flames. The second effect results in a change of the flow velocity at the front; it becomines dominant in the range of long wavelengths, whereas at short wavelengths it is subdued by the first. Both exert profound influence upon the flame motion.

When the modification of the flame structure is neglected, $u_n = u_o$, the flame may be considered as a passive interface in the sense that the motion is completely prescribed by the flowfield at its front. Even in this case, the second effect provides a strong hydrodynamic feedback for the motion of the front. This effect was first described in the pioneering studies of Darrieus (1938) and Landau (1944), who computed the flowfield induced by front wrinkling when $u_n = u_o$ and d=0. The analysis was carried out for a linear approximation in the amplitude of the front corrugation. The induced flow velocity was found, then, to be in phase with the front wrinkles. The resulting motion of the flame revealed the existence of a strong instability mechanism for planar uniform flows. The smallest deviation from the planar steady-state solution, described here before, is amplified by the second hydrodynamic effect specified above. For shorter wavelengths, the front is more unstable. Darrieus and Landau concluded, therefore, that planar flames propagating freely in a homogenous mixture cannot exist. The instability mechanism appeared to be so strong that they surmised, moreover, that combustion must be a self-turbilizing phenomenon. In fact, with only two parameters involved in their theory, the gas expansion parameter γ and the laminar flame speed u_o, the growth rate σ of the instability was found to be proportional to the modulus k of the wave vector of the front wrinkles, i.e.,

$$\sigma = \sigma_1(\gamma)u_o k \qquad (35)$$

where $\sigma_1(\gamma)$ is a positive nondimensional quality vanishing only in the unrealistic limit of $\gamma \to 0$ (zero gas expansion, i.e., $\rho_o = \rho_b$). The first effect specified above, which they overlooked, is expected to modify this result, especially in the domain of the short wavelengths.

The first attempt to take into account the effect of the modification of flame structure was made during the 1950's by Markstein (1964), who assumed that the normal burning speed u_n is related to the mean radius of curvature of the front, R, a positive quantity when the front is concave toward the unburnt gases, as follows:

$$(u_n - u_o)/u_o = L/R \qquad (36)$$

where L is a phenomenological length (referred to as the Markstein length)** that was assumed to be proportional to the flame thickness d, a characteristic property of the reactive mixture. According to the semianalytical theory of Eckhaus (1961), Markstein modified Eq. (36) in order to take into account the fact that the flame structure must be modified also by the inhomogeneity of the local flow conditions. Such an effect is well established in the case of planar flames stabilized at stagnation point. It can be easily anticipated from Eq. (36) that the effects associated with the modification of the flame structure can only change the dispersion relation (35) through a k^2 term. Thus, at large wavelengths, the flame cannot be stabilized by such a mechanism. This is easily understood for the diffusion process which, as is well known, is associated with a relaxation time of the order of $(Dk^2)^{-1}$,

---

** Professor M. Barrère has brought to my attention a private note of November 1951 signed by Darrieus. In this note, Darrieus explains that in November 1950 he suggested to Markstein the idea of taking into account the curvature effect.

where D is the diffusivity. In regard to the first effect specified above, it is sufficient to note that, according to Eq. (35), the amplitude of the induced flowfield, as computed by Darrieus and Landau, is proportional to k. However, it is the gradient of the flow velocity that is the relevant quantity involved in the modification of the flame structure by convective fluxes. The first effect introduced thus is also a k^2 term.

The first attempt to evaluate analytically the wrinkled flame structure was made by Barenblatt et al. (1962). It was restricted, however, to the thermo-diffusional model where gas expansion effects i and ii are neglected. This model was extensively employed over the last decade, culminating with the derivation by Sivashinsky (1977) of a nonlinear differential equation†† for the flame motion, describing self-turbulization of a cellular flame structure (Mickelson and Sivashinksy 1977). The main virtue of this model is that it provides us with a simple framework for a systematic study of all the dynamic effects that can be produced by the diffusion of heat and mass. To solve this model in the limit of large values of the Zel'dovich number, $\beta \to \infty$, Joulin and Clavin (1979) applied the asymptotic technique. They succeeded thereby in revealing the dynamic effects of flames in the presence of heat losses that can produce thermal extinction. The model predicted the existence of traveling and spinning waves, as well as oscillatory fronts [(for a review see Buckmaster and Ludfors (1982) and Sivashinsky (1983)]. However, even as modified by Sivashinsky to take into account weak gas expansion, this model underestimates the hydrodynamic effects that then play a particularly dominant role [see Eq. (35)].

Recently, the coupling between diffusion and fluid mechanic effects was taken into account by Clavin and Williams (1982) in the analysis of wrinkled flame structure. The asymptotic expansion $\beta \to \infty$ was used together with a multiscale method based on the assumption that $\varepsilon = d/\lambda$ is smaller than unity. The result was used by Pelcé and Clavin (1982) to study stability limits of planar flames propagating downward. Their conclusions can be summarized as follows:

1) In the approximation of a one-step overall chemical reaction, the modification to flame structure by wrinkling has a stabilizing effect for most ordinary

---

†† The same equation was derived independently by Kuramoto (1978) in another context. The same author showed later that this equation is generic for unstable reaction-diffusion waves.

hydrocarbon mixtures, irrespective of the equivalence ratios; the only exception may be encountered in the case of mixtures containing a very light reactive component, as, for example, hydrogen diluted in nitrogen. This conclusion contradicts the result of the thermodiffusion model and appears as a typical outcome of the hydrodynamic effect i.

2) Gas viscosity has no effect on the stability of planar flames.

3) Acceleration of gravity, g, associated with the modification of flame structure by wrinkling, can counterbalance the hydrodynamic instability for all wavenumbers when the flame velocity is low enough. The cellular threshold is predicted to occur at flame velocities, u_o, within the interval between 5 and 17 cm/s, for rich mixtures of ordinary fuels at the lower side of $u_o < 12$ cm/s, and for lean mixtures at the higher side.

4) Cell size at the threshold can be expressed in terms of variables g, u_o, and Y only; detailed properties (chemical kinetics, transport processes, etc.) do not have to be taken into account.

These predictions are in good agreement with recent experimental results of Quinard et al. (1984). The induced velocity field has been also measured recently in the unburnt mixture by Searby et al. (1983, 1984) in the case of stable fronts stabilized in a weakly turbulent flowfield. As predicted by the theory of planar stable flames, the induced velocity field is found to be out of phase with respect to front corrugations, leading to the blocking of low frequencies of turbulence ahead of the flame.

The analysis of the flame structure of wrinkled fronts in a nonhomogeneous flow presented by Clavin and Williams (1982) has been extended independently by Matalon and Matkowsky (1982) and by Clavin and Joulin (1983) to the nonlinear case of finite amplitude of flame corrugations. As anticipated by the early phenomenological analysis of Karlowitz et al. (1953), the modification of the normal burning velocity u_n, produced by the curvature of the front and by flow inhomogeneities, can be expressed in terms of only one geometrical scalar, the total flame stretch

$$u_n - u_o = - L(1/\sigma)(d\sigma/dt) \qquad (37)$$

where σ is a local infinitesimal area of the front flame, while d/dt expresses its time derivative when each point of the flame surface moves as prescribed here previously;

L is the Markstein length, which, together with the laminar flame velocity u_o, depends on chemical kinetics of the reaction. The corresponding expression for L has been obtained for different cases. The effects of the dependence of the transport coefficients on the temperature have been explored by Clavin and Garcia (1983); the nonadiabatic case and the behavior near the flammability limits were studied by Clavin and Nicoli (1984). Some consequences of multiple step chemistry were investigated by Liñan and Clavin (1984).

Except for its limitation to weak stretch, this surprisingly simple result is quite general, and it can be used in any flow configuration, stagnation point flows, spherical flames, turbulent flames, etc. The effects of strong stretch have been recently studied by Libby et al. (1983) for the particular case of a planar flame at the stagnation point. Once again, the effect of gas expansion has been proved to be of crucial importance. For example, flame extinction under strong stretch predicted by the thermodiffusion model ($\gamma=0$) in the case of one overall chemical reaction rate expression is unpredictable when effect i is properly taken into account.

It is of interest to express eq.(37) in terms of the mean radius of curvature of the front, R. At the same time, the modification to normal burning velocity can be expressed as (see Clavin and Joulin 1983):

$$u_n - u_o = Lu_o \{1/R + (1/u_o) \mathbf{n} \cdot \nabla \mathbf{u} \cdot \mathbf{n}\} \quad (38)$$

where $\mathbf{n}$ is the unit vector normal to the front and $\nabla \mathbf{u}$ is the "rate of the strain tensor" of the upstream flow evaluated at the flame front. Each of the two terms on the right hand side of Eq. (38) represents a contribution to the total flame stretch. The term $-u_o/R$ expresses the stretch of the front moving in a uniform flow at a constant normal velocity u_o, due to the effect of the nonplanar geometry of the front. The term $\mathbf{n} \cdot \nabla \mathbf{u} \cdot \mathbf{n}$ is the stretch of the front convected by the flowfield, $\mathbf{u}$, whereas outside the flame $\nabla \cdot \mathbf{u} = 0$. This term expresses the effect of the nonhomogeneity of the flow. Equations (37) and (38) reconcile the theories of Karlowitz et al. (1953) and Markstein (1964). As noted by Frankel and Sivashinsky (1983), according to Eqs. (37) or (38), L has a different value in burnt gases than in the unburnt mixture. This is associated with the fact that, because of unsteady or multidimensional effects, the normal mass flux is not constant across the flame when its thickness is not negligible. This explains, in particular, the difference in the behavior of converging and diverging spherical fronts.

Finally, it should be pointed out that in the case of unsteady flows, Eqs. (37) and (38) are nonlinear evolution equations for the flame front. However, as already mentioned, the velocity field appearing in these equations is a function of the flame front because of the hydrodynamic effect ii. A purely hydrodynamic analysis has to be carried out to obtain this function necessary to close the equation of evolution. This aspect of flame dynamics has been treated only in the linear approximation [see Pelcé and Clavin (1982) and Searby and Clavin (1984)]. Nonetheless, Eqs. (37) and (38) can yield the turbulent flame speed in the case of turbulent wrinkled flame if the corresponding random process is stationary and homogeneous. The time average of the modification to normal burning velocity is practically nil, and the turbulent flame speed turns out to be given simply by the laminar flame speed multiplied by the mean area increase of the front. This result has been known for a long time [see Lewis and Von Elbe (1961)], but it was considered to be associated with the assumption of constant normal burning velocity on the flame front. Our analysis demonstrated that this simple result of turbulent flame theory holds true also for smooth wrinkled flame fronts and is not restricted to the case of constant normal velocity.

More details concerning flame dynamics can be found in a recent review article of Clavin (1984), while a more general exposition of the modern theory of combustion will be included in the new edition of a book by F.A. Williams (1984).

Conclusions

Asymptotic analysis in the case of large Zel'dovich numbers has been very significant over the last decade in advancing our knowledge of the structure and dynamics of wrinkled flame fronts. The flame models tractable by this method are now realistic enough to yield theoretical results that are in satisfactory agreement with experimental observations. Insofar as the detailed structure associated with the actual multiple-step chemistry is concerned, a concentrated effort will be needed to establish chemical kinetic schemes and concomitant rate constants that are indispensible for a meaningful solution. This must be associated with the development of numerical techniques capable of taking into account full kinetic schemes. As a consequence of the significant amount of parameters that must be dealt with in this connection, development of simplified models to reveal the essential aspects of the chemical kinetic mechanism is highly desirable.

The modern theory of flames is associated with equations describing the evolution of the front, providing information on the deformation of the flame front and the concomitant development of flow inhomogeneities. However, in general, the solution of the nonlinear hydrodynamic problem of flow distortion produced by the flame cannot be found analytically.

Thus, the understanding of nonplanar and nonsteady flames in nonhomogeneous and nonsteady flows can be improved only by analytical studies and numerical analyses carried out in harmony with each other.

Acknowledgments

The author is greatly indebted to A. Liñan, M. Barrère, H. Guénoche, N. Manson, and A.K. Oppenheim for the original manuscripts of the pioneering publications, as well as for most helpful comments and suggestions. The author is also grateful to Mrs. Karen Wolf for technical assistance in the preparation of the manuscript.

References

Adams, G. K. and Stocks, G. V. (1953) The combustion of hydrazine. Fourth (International) Symposium on Combustion, pp. 239-248. The William and Wilkins Company, Baltimore, Md.

Aldushin, A. P., Khudyaev, S. I., and Zel'dovich, Y. B. (1981) Flame propagation in the reacting gaseous mixture. Arch. Combust. 1(1/2), 9-21.

Barenblatt, G. I., Zel'dovich, Y. B., and Istratov, A. G. (1962) On diffusional thermal instability of laminar flame. Zh. Prikl. Mekh. Tekh. Fiz. 2(4), 21-26.

Bowen, J. R., Acrivos, A., and Oppenheim, A. K. (1963) Singular perturbation refinement to quasi-steady-state approximation in chemical kinetics. Chemical Engineering Science 18, 177-188, Pergamon Press, Oxford, England.

Boys, S. F. and Corner, J. (1949) The structure of the reaction zone in a flame. Proc. R. Soc. Ser.A 197(1048), 90-106.

Buckmaster, J. D. (1976) The quenching of deflagration waves. Combust. Flame 26(2), 151-162.

Buckmaster, J. D. and Ludfors, G. S. S. (1982) Theory of Laminar Flames. Cambridge University Press, Cambridge, England.

Buckmaster, J. D. and Mikolaitis, D. (1982) The premixed flame in counterflow. Combus. Flame 47(2), 191-204.

Bush, W. B. and Fendell, F. E. (1970) Asymptotic analysis of laminar flame propagation. Combust. Sci. Tech. 1(6), 421-428.

Bush, W. B. and Fendell, F. E. (1971) Asymptotic analysis of the structure of steady planar detonation. Combust. Sci. Tech 2(5-6), 271-285.

Clarke, J. F. and McIntosh, A. C. (1980) The influence of a flame holder on a plane flame, including its static stability. Proc. R. Soc. London Ser.A 372(1750), 367-392.

Clavin, P. and Williams, F. A. (1982) Effects of molecular diffusion and of thermal expansion on the structure and dynamics of premixed flames. J. Fluid. Mech. 116(1-3), 251-282.

Clavin, P. and Garcia, P. (1983) The influence of the temperature dependence of diffusivities on the dynamics of flame fronts. J. Mech. Theor. Appl. 2(2), 245-263.

Clavin, P. and Joulin, G. (1983) Premixed flames in large scale and high intensity turbulent flow. J. Phys. Lett. 44(1), 11-12.

Clavin, P. and Liñan, A. (1984) Theory of gaseous combustion. NASA ASI Series: Nonequilibrium Cooperative Phenomena in Physics and Related Fields (edited by M. G. Velarde). Plenum Press, New York. (to be published).

Clavin, P. and Nicoli, C. (1984) Effects of the heat losses on the limits of stability of premixed flames propagating downward. (submitted to Combust. Flame)

Clavin, P. (1984) Dynamical behavior of premixed fronts in laminar and turbulent flows. Prog. Energy Combust. Sci. (to be published).

Cole, J. D. (1968) Perturbation Methods in Applied Mathematics. Blaisdell Press, Waltham, Massachusetts.

Darrieus, G. (1938) Propagation d'un front de flamme. Presented at La Technique Moderne (1938) and Congrs de Mecanique Appliquee (1945), Paris (unpublished).

Darrieus, G. (1951) Complement a la communication de Mr. Darrieus au colloque du laboratoire de recherche techniques de St. Louis sur la stabilite de l'onde de deflagration du 29 octobre 1951. (unpublished).

Deshaies, B., Joulin, G. and Clavin, P. (1981) Etude asymptotique des flammes spheriques non adiabatiques. J. Mec. 20(4), 691-735.

Eckhaus , W. (1961) Theory of flame-front stability. J. Fluid. Mech. 10(1-3), 80-100.

Evans, M. W. (1952) Current theoretical concepts of steady-state flame propagation. Chem. Rev. 51(3), 363-429.

Fendell, F. E. (1972) Asymptotic analysis of premixed burning with large activation energy. J. Fluid Mech. 56(1-3), 81-95.

Fife, P. C. (1979) Mathematical aspects of reacting and diffusing systems. Lecture Notes in Biomathematics No. 28. Springer-Verlag, Berlin.

Fisher, R. A. (1937) The wave of advance of advantageous genes. Ann. of Eugen. 7, 355-369.

Frankel, M. L. and Sivashinsky, G. I. (1982) The effect of viscosity on hydrodynamic stability of a plane flame front. Combust. Sci. Tech. 29(3-6), 207-224.

Frankel, M. L. and Sivashinsky, G. I. (1983) On effects due to thermal expansion and Lewis number in spherical flame propagation. Combust. Sci Tech. 31(3-4), 131-138.

Frank-Kamenetskii, D. A. (1969) Diffusion and Heat Transfer in Chemical Kinetics; Second Edition. Plenum Press, New York.

Friedman, R. and Burke, E. (1953) A theoretical model of gaseous combustion wave governed by a first-order reaction. J. Chem. Phys. 21(4), 710-714.

Fristrom, R. M. and Westenberg, A. A. (1956) Flame Structure. McGraw-Hill, New York.

Gelfand, I. M. (1959) Some Problems in the Theory of Quasilinear Equations. Usp. Mat. Nauk 14(2), 87-158. Translated in: American Mathematical Translation (1963) Ser 2., 29, pp. 295-381.

Hirschfelder, J. O. Curtiss, C. F. and Campbell, D. E. (1953) The theory of flames and detonations. Fourth (International) Symposium on Combustion, pp. 190-211. The Williams and Wilkins Company, Baltimore, Maryland.

Higuera, P. (1983) Private communication, E.T.S.I. Aeronautics Univ. Politecnica, Madrid.

Johnson, W. E. and Nachbar, W. (1963) Laminar flame theory and the steady linear burning of a monopropellant. Arch. Rational Mech. Anal. 12(1), 58-92.

Jost, W. and Muffling, L. (1938) Untersuchungen uber Flammengeschwin-digkeiten. Z. Phys. Chem. 181-3-A, 208-214.

Jouquet, E. (1913) Sur la propagation des deflagrations dans les melanges gazeux. C. R. Acad. Sci. Paris 156(11), 872-876; (see also La mecanique des explosifs. Douin, Paris, 1917).

Joulin, G. and Clavin, P. (1976) Analyse asymptotique des conditions d'extinction des flammes laminaires. Acta Astronaut. 3(3/4), 223-240.

Joulin, G. and Clavin, P. (1979) Linear stability of non-adiabatic flames. Combust. Flame 35(2), 139-153.

Joulin, G. and Mitani, T. (1981) Linear stability of two reactant flames. Combust. Flame 40(3), 235-246.

Kolmogorov, A. N., Petrovskii, I. G., and Piskunov, N. S. (1937) A study of the equation of diffusion with increase in the quantity of matter and its application to a biological problem. Byul. Mosk. Gos. Univ. 1(7), 1-72.

Karlowitz, B., Denniston, J. R., Knapschaefer, D. H., and Wells, F. E. (1953) Studies on turbulent flames. Fourth (International) Symposium on Combustion, pp. 613-620. The William and Wilkins Company, Baltimore, Maryland.

Kuramoto, Y. (1978) Diffusion-induced chaos in reaction systems. Supplement of the Progress of Theoretical Physics, Kyoto, 64, pp. 346-347; (see also Synergetics. (1979) Bielefeld, Springer-Verlag, Berlin).

Landau, L. (1944) On the theory of slow combustion. Acta. Phys. URSS 19(1), 77-85.

Leconte, J. M. D. (1858) On the influence of musical sounds on the flame of a jet of coal gas. Philos. Mag., March 15, 235-239.

Lewis, B. and Von Elbe, G. (1934) On the theory of flame propagation. J. Chem. Phys 2(8), 537-568.

Lewis, B. and Von Elbe, G. (1961) Combustion Flames and Explosions of Gases, Second Edition. Academic Press, New York.

Libby, P A. and Williams, F. A. (1982) Structure of laminar flamelets in premixed turbulent flames. Combust. Flame 44(1-3), 287-303.

Libby, P. A., Liñan, A., and Williams, F. A. (1983) Strained premixed flames with nonunity Lewis numbers. Combust. Sci. Tech. 34(1-6), 257-293.

Liñan, A. (1971) A theoretical analysis of premixed flame propagation with an isothermal chain reaction. INTA EOOAR 68-0031, Technical Report 1.

Liñan, A. (1974) The asymptotic structure of counter flow diffusion flames for large activation energies. Acta Astronaut. 1(7-8), 1007-1039.

Liñan, A. (1976) Monopropellant droplet decomposition for large activation energies. Acta Astronaut. 2(11-12), 1009-1029.

Liñan, A., and Clavin, P. (1984) Premixed flames with non-branching chain reaction. J. Chem. Phys. (to be published).

Mallard, L., and Le Chatelier, H. L. (1883) Recherches experimentales et theoriques sur la combustion des melanges gazeux. Ann. Mines 4, 274-553.

Markstein, G. H. (1964) Nonsteady Flame Propagation. AGARD Monograph 75, Pergamon Press, New York.

Matalon, M. and Matkowsky, B. J. (1982) Flames as gasdynamic discontinuities. J. Fluid Mech. 124(1-3), 239-259.

Mickelson, D. M. and Sivashinsky, G. I. (1977) Non-linear analysis of hydrodynamic instabilty in laminar flames II. Acta Astronaut. 4(11-12), 1207-1221.

Mikhelson, V. A. (1889) On the normal ignition rate of fulminating gas mixtures. Thesis, Moscow, (see Combustion Theory F.A. Williams 1984).

Murray, J. D. (1977) Lectures on Nonlinear Differential Equations Model in Biology. Clarendon Press, Oxford, England.

Oppenheim, A. K. (1963) Conferences sur l'aerothermique des ondes dans les milieux reactifs. Publications Scientifiques et Techniques du Ministee de l'Air, Le Service de Documentation et d'Information Technique de l'Aeronautique, Paris, France.

Pelcé, P. and Clavin, P. (1982) Influence of hydrodynamics and diffusion upon the stability of limits of laminar premixed flames. J. Fluid Mech. 124(1-3), 219-237.

Peters, N. and Warnatz, J. (1982) Numerical methods in laminar flame propagation. Notes on Numerical Fluid Mechanics, Vol. 6. Vieweg, Braunschweige.

Quinard, J., Searby, G. and Boyer, L. (1983) The stability limits and critical size of structures in premixed flames. Proceedings of the IX International Colloquium on Dynamics of Explosions and Reactive Systems, July 3-8, 1983, Poitiers, France.

Rayleigh, J. W. S. (1877) The Theory of Sound. Second Edition. 2 Volumes, 1945, Dover Publications, New York.

Rosen, G. (1959) An action principle for the laminar flame. Seventh International) Symposium on Combustion, Butterworths, London, 339-341.

Searby, G., Sabathier, F., Clavin, P. and Boyer, L. (1983) The hydrodynamical coupling between the motion of a flame front and the upstream gas flow. Phys. Rev. Lett. 51(16), 1450-1453.

Searby, G., Sabathier, F., Monreal, J., Clavin, P. and Boyer, L. (1983) The feed-back of a flame front on turbulent flows. Proceedings of the IX International Colloquium on Dynamics of Explosions and Reactive Systems, July 3-8, 1983, Poitiers, France.

Searby, G. and Clavin, P. (1984) Hydrodynamics of wrinkled fronts in premixed turbulent combustion. Combust. Sci. Tech. (to be published).

Semenov, N. N. (1935) Chemical Kinetics and Chain Reaction. Clarendon Press, Oxford, England.

Sivashinsky, G. I. (1977) Non-linear analysis of hydrodynamic instability in laminar flames. Acta Astronaut. 4(11-12), 1177-1206.

Sivashinsky, G. I. (1983) Instabilities, pattern formation, and turbulence in flames. Ann. Rev. Fluid Mech 15, 179-199.

Spalding, D. B. (1956) The theory of flame phenomena with a chain reaction. Phil. Trans. R. Soc. London Ser. A 249(957), 1-25.

Taffanel, M. (1913) Sur la combustion des melanges gazeux et les vitesses de reaction. C. R. Acad. Sci. Paris 27(X), 714-717; 29(XII), 42-45.

Tyndall, J. (1867) Sound, p. 226. Appleton & Company, New York.

Van Dyke, M. (1975) The Perturbation Methods in Fluid Mechanics. The Parabolic Press, Stanford, California.

Von Karman, T. (1957) The present status of the theory of laminar flame propagation. Sixth (International) Symposium on Combustion, pp. 1-11. Reinhold Publishing Corporation, New York.

Von Karman, T. and Penner, S. S. (1954) Selected Combustion Problems, pp. 1-41. AGARD, Butterworths, London, England.

Williams, F. A. (1984) Combustion Theory (new edition). Addison Wesley, New York.

Zel'dovich, Y. B. (1941) Theory of slow flame propagation. Zh. Eksp. Theor. Fiz. 11(1), 159-168.

Zel'dovich, Y. B. (1948) Theory of flame propagation. Zh. Fiz. Khim. SSSR 22(1), 27-49.

Zel'dovich, Y. B. and Barenblatt, G. I. (1959) Theory of flame propagation. Combust. Flame 3(1), 61-74.

Zel'dovich, Y. B. and Frank-Kamenetskii, D. A. (1938) A theory of thermal propagation of flame. Acta Phys. URSS IX(2), 341-350.

Chapter I. Premixed Flames

Effects of Chemical Equilibrium on the Structure and Extinction of Laminar Diffusion Flames

N. Peters[*]
*Rheinisch-Westfälische Technische Hochschule
Aachen, Federal Republic of Germany*
and
F. A. Williams[†]
Princeton University, Princeton, New Jersey

Abstract

Methods of activation-energy asymptotics are applied to diffusion flames with a one-step forward and backward reaction. The reversible kinetics lead to two types of flame broadening: equilibrium and reaction-rate broadening. The interaction between these two effects takes place within the inner reactive-diffusive zone of the flame. Limits of small and large equilibrium constants are considered to derive approximate solutions. The objective is to develop means for calculating flame structure and extinction in the presence of dissociative equilibria. The results identify new types of equations for reactive-diffusive zones and indicate how these equations may be employed in flame calculations.

Introduction

Diffusion flames near the Burke-Schumann limit experience both equilibrium broadening and reaction-rate broadening of the thin reaction zone. Most previous investigations have addressed one or the other of these phenomena and have not considered their simultaneous presence. The analysis of equilibrium broadening for the reaction $\nu_F F + \nu_O O \rightleftarrows$ products is a relatively straight-

---

Presented at the 9th ICODERS, Poitiers, France, July 3-8, 1983. Copyright © 1984 by the American Institute of Aeronautics and Astronautics, Inc. All rights reserved.
[*]Professor, Institut für Allgemeine Mechanik.
[†]Professor, Dept. of Mechanical and Aerospace Engineering.

forward problem in regular perturbations that yields a rounding of the temperature profile at the flame sheet and a small region of overlap of the fuel and oxidizer concentration profiles. (Chung and Blankenship 1966; Fendell 1967). It is an approximation to more general results obtainable numerically from full equilibrium chemistry (Bilger 1977; Peters 1975) and says nothing about the important phenomena of ignition and extinction, which require the consideration of reaction-rate broadening.

The asymptotic analysis of reaction-rate broadening, giving ignition and extinction in the absence of equilibrium dissociation, has been developed by Liñán (1974). This analysis has been applied for extracting overall reaction-rate parameters from measurements of diffusion-flame extinction (Krishnamurthy et al. 1976; Williams 1981). Considerable uncertainty remains in interpreting results of the measurements for situations in which the extinction occurs at flame temperatures high enough for equilibrium dissociation to be significant (Sohrab and Williams 1981). To be able to analyze these important high-temperature extinctions with reasonable accuracy, thereby obtaining overall rate parameters relevant at the high flame temperatures of greatest practical interest, we need theoretical predictions for extinction in the presence of equilibrium dissociation. An objective of the present study is to develop these predictions.

One study has been published in which both equilibrium and finite-rate chemistry have been considered by asymptotic methods (Peters 1979). This work, directed toward the hydrogen-oxygen flame, addressed a very high-temperature situation in which there is a broad equilibrium region bounded on each side by a narrow region of transition from equilibrium to frozen flow. Inner equations for the structure of the transition region were derived and analyzed, and extinction was equated to the condition under which the location of one of the transition regions reaches the point of maximum temperature in the equilibrium region. This analysis is reconsidered herein as a limiting behavior of the regimes analyzed. However, attention is focused more directly on situations in which the finite-rate chemistry and the equilibrium dissociation both occur in the same thin zone and there is a maximum of the temperature profile in this zone.

In developing the analysis, it is convenient for greatest generality to work with the mixture fraction Z, introduced earlier (Bilger 1980). By treating various general functional forms of the equilibrium dependence of the temperature T and of the fuel and oxidizer mass fractions, Y_F and Y_O on Z, one can develop asymptotic analyses for effects of finite-rate chemistry without specifically stating an equilibrium reaction scheme. The equilibrium chemistry consistent with the equilibrium functions adopted may then be considered later. Instead of following this general approach, since we have specific classes of equilibrium reaction schemes in mind, we introduce these at the outset. This helps to facilitate useful selections of ordering in the analysis. Our notation and formulation parallel that which we used earlier (Peters and Williams 1983) and we retain the assumption of Lewis numbers of unity, an assumption that is appearing increasingly reasonable (Libby et al. 1983).

Formulation

We address here problems in which there exists one thin zone in which reaction occurs. In analyzing such problems, instead of defining a specific configuration, such as the counterflow diffusion flame, it is both convenient and of greater generality to introduce Z as an independent variable and to perform the analysis in terms of Z rather than in terms of spatial coordinates for specific flows. In this respect, we follow our earlier formulation (Peters and Williams 1983) introducing the generalizations needed to account for chemical equilibria.

We assume that the supply flow consists of two feeds, the fuel and the oxidizer, and that no-flux conditions apply on all other bounding surfaces. Furthermore, we assume that the feeds are not partially premixed, and that no products are present in either the fuel or oxidizer supply streams. Therefore the mass fractions Y_{FF} and Y_{OO} may differ from unity only because of the presence of inerts that are not products. Then the mixture fraction Z may be defined as

$$Z = \frac{Y_{Ffr}}{Y_{FF}} = 1 - \frac{Y_{Ofr}}{Y_{OO}} \qquad (1)$$

where Y_{Ffr} and Y_{Ofr} are the mass fractions of fuel and oxidizer in the corresponding nonreacting flow problem.

The general conservation equation for Z is the source free equation applicable to a "conserved scalar," viz.,

$$\rho \partial Z/\partial t + \rho \vec{v} \cdot \nabla Z = \nabla \cdot (\rho D \nabla Z) \qquad (2)$$

In chemically frozen flow, the temperature is $T_{fr}(Z)$, where $T_{fr} = T_O$ at $Z = 0$ (the oxidizer temperature) and

$T_{fr} = T_F$ at $Z = 1$ (the fuel temperature). With variable specific heat the function $T_{fr}(Z)$ need not be linear, but (with a Lewis number of unity) the corresponding thermal enthalpy $H_{T_{fr}}$ is linear in Z, as are the frozen fuel and oxidizer mass fractions.

The chemical mechanism considered here is written as

$$\nu_F F + \nu_O O \underset{k_b}{\overset{k_f}{\rightleftarrows}} \nu_{P1} P_1 + \nu_{P2} P_2 \qquad (3)$$

The balance equations for fuel and oxygen may be combined to show that the quantity $Y_F - \nu Y_O$ is also a conserved scalar which obeys a conservation equation of the form of Eq. (2) and is therefore a linear function of Z. By evaluating the constants from the boundary values we find that

$$Y_O/Y_{OO} = 1 - Z/Z_{st} + Y_F/(\nu Y_{OO}) \qquad (4)$$

where Z_{st} is the stoichiometric mixture fraction to be calculated from

$$Z_{st} = [1 + Y_{FF}/(\nu Y_{OO})]^{-1} \qquad (5)$$

In Eqs. (4) and (5) $\nu = (\nu_F W_F)/(\nu_O W_O)$ where the W_i's denote molecular weights. The product mole fractions are related according to $X_{P1} = \nu_{P1} X_{P2}/\nu_{P2}$. The corresponding mass fractions then obey the relationships

$$\left. \begin{array}{l} Y_{P1} \equiv Y_P = \mu_{P1}(Y_{FF}Z - Y_F) \\[6pt] Y_{P2} = \dfrac{\mu_{P2}}{\mu_{P1}} Y_P = \mu_{P2}(Y_{FF}Z - Y_F) \end{array} \right\} \qquad (6)$$

where $\mu_{Pj} = \nu_{Pj} W_{Pj}/(\nu_F W_F)$, $j = 1, 2$.

At chemical equilibrium the temperature is $T_{eq}(Z)$ and typically has a peak in the vicinity of the stoichiometric point, $Z = Z_{st}$, as illustrated in Fig. 1. For ideal gas mixtures in general, the total enthalpy per unit mass is

$$H = \sum_i Y_i H_i = \sum_i Y_i H_i^0 + \sum_i Y_i \int_{T^0}^{T} c_{pi} dT \qquad (7)$$

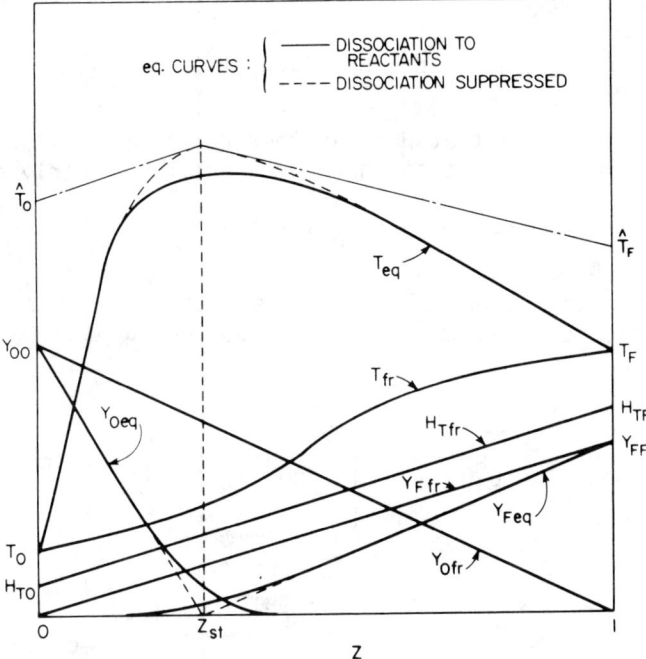

Fig. 1 Schematic illustration of frozen and equilibrium profiles in the mixture-fraction variable.

where H_i^o is the heat of formation per unit mass for species i at temperature T^o and c_{pi} is the specific heat at constant pressure for species i. For purposes of the present analysis it will be convenient to select T^o to be a value near $T_{eq}(Z_{st})$, the equilibrium temperature at stoichiometric conditions. Then for changes occurring near the stoichiometric equilibrium point to first order we may replace H_i by H_i^o in

$$dH = \sum_i H_i dY_i + c_p dT \qquad (8)$$

with

$$c_p = \sum_i Y_i c_{pi} \qquad (9)$$

to obtain with Eqs. (4) and (6) the relation

$$dH = Q_F dY_F + c_p dT \\ + [Y_{FF}(\mu_{P1}H_{P1}^o + \mu_{P2}H_{P2}^o) - Y_{OO}H_O^o/Z_{st}]dZ \qquad (10)$$

where

$$Q_F = H_F^0 + H_O^0/\nu - \mu_{P1}H_{P1}^0 - \mu_{P2}H_{P2}^0 \qquad (11)$$

It may be noted that H is another conserved scalar that obeys Eq. (2) and that therefore depends linearly on Z,

$$H = H_0 + (H_F - H_0)Z \qquad (12)$$

In this expression, if c_{pN} is the specific heat of the neutral species, and if we set $H_N^0 = 0$, then at the boundaries we have

$$\left.\begin{array}{l} H_0 = Y_{OO}H_O^0 + Y_{OO}\displaystyle\int_{T^0}^{T_0} c_{pO}dT + (1-Y_{OO})\displaystyle\int_{T^0}^{T_0} c_{pN}dT \\[2ex] H_F = Y_{FF}H_F^0 + Y_{FF}\displaystyle\int_{T^0}^{T_F} c_{pF}dT + (1-Y_{FF})\displaystyle\int_{T^0}^{T_F} c_{pN}dT \end{array}\right\} \qquad (13)$$

From Eq. (7) and the frozen profiles of Eq. (1) one obtains an equation for the frozen temperature,

$$\begin{aligned} H_{fr} &= Y_{OO}H_O^0 + (Y_{FF}H_F^0 - Y_{OO}H_O^0)Z \\ &+ Z\left[Y_{FF}\int_{T^0}^{T_{fr}} c_{pF}dT + (1-Y_{FF})\int_{T^0}^{T_{fr}} c_{pN}dT\right] \\ &+ (1-Z)\left[Y_{OO}\int_{T^0}^{T_{fr}} c_{pO}dT + (1-Y_{OO})\int_{T^0}^{T_{fr}} c_{pN}dT\right] \end{aligned} \qquad (14)$$

To account for variable specific heats in our analysis, the terms involving integrals in Eq. (14) may be replaced by $c_{pst}(T_{fr}^*-T^0)$ where c_{pst} is to be evaluated from Eq. (9) at the equilibrium stoichiometric conditions that are used to define T^0. By T_{fr}^* an effective frozen temperature suitable for the use in the vicinity of the stoichiometric point is defined. It may be calculated with Eqs. (12-14) as

$$T^*_{fr} = T^O - \left\{ Z\left[Y_{FF}\int_{T_F}^{T^O} c_{pF}dT + (1-Y_{FF})\int_{T_F}^{T^O} c_{pN}dT\right] \right.$$
$$\left. + (1-Z)\left[Y_{OO}\int_{T_0}^{T^O} c_{pO}dT + (1-Y_{OO})\int_{T_0}^{T^O} c_{pN}dT\right]\right\} / c_{pst} \quad (15)$$

Also, using Eqs. (12-14), the equation for the frozen temperature may be cast in a form resembling that for constant specific heats by defining effective boundary temperatures as

$$\hat{T}_0 = T^O - \left[Y_{OO}\int_{T_0}^{T^O} c_{pO}dT + (1-Y_{OO})\int_{T_0}^{T^O} c_{pN}dT\right]/c_{pst}$$
$$\hat{T}_F = T^O - \left[Y_{FF}\int_{T_F}^{T^O} c_{pF}dT + (1-Y_{FF})\int_{T_F}^{T^O} c_{pN}dT\right]/c_{pst} \quad (16)$$

thereby obtaining

$$T^*_{fr} = \hat{T}_0 + (\hat{T}_F - \hat{T}_0)Z \quad (17)$$

Using Eq. (10) and the conservation equations for H and Y_F one obtains the temperature equation,

$$c_p\rho\frac{\partial T}{\partial t} + c_p\rho\vec{v}\cdot\nabla T - \nabla\cdot(\rho D c_p \nabla T) = Q_F w_F \quad (18)$$

where Eq. (2) for Z has been employed and where w_F denotes the net mass rate of consumption of fuel per unit volume.

To write an expression for w_F we further assume that the reaction is of first order with respect to each F and O in the forward direction and of first order with respect to each P_1 and P_2 in the backward direction. If $k_b \neq 0$, then the reaction may be viewed as an elementary step, and strictly speaking these reaction-order assumptions then require $\nu_F = \nu_O = \nu_{P1} = \nu_{P2}$. If $k_b = 0$, then there is no difficulty in viewing the reaction as a one-step approximation to a complex kinetic scheme with ν_F and ν_O arbitrary and not related to reaction orders. Although we are not concerned here with $k_b = 0$, we shall retain the general stoichiometry for potential applications in which the forward and backward steps both are approximations to complex kinetic schemes, modifying the equilibrium-constant definition if necessary by expansion about stoichiometric equilibrium. The specific reaction-rate constant k_f for the second-order forward reaction may be written as $k_f = (B_F/\nu_F)e^{-E/T}$, where B_F is the pre-exponential factor in the expression for the molar rate of consumption of fuel and where E is the effective overall activation temperature for the first step. Then

$$w_F = (B_F \rho^2/W_0)e^{-E/T}\left[Y_F Y_0 - (Y_p/\mu_{p1})^2 K e^{-B/T}\right] \quad (19)$$

where the equilibrium constant for the first step has been written as $k_f/k_b = \nu_{p1}\nu_{p2}W_0/(\nu_F^2 W_F)K^{-1}e^{B/T}$ with K and B taken to be positive constants. Equations (4) and (6) may be used to eliminate Y_0 and Y_p in terms of Z and Y_F in Eq. (19), and the H expression is then used to express Y_F in terms of Z and T.

In considering the flame structure we adopt equilibrium profiles as outer solutions and stretch coordinates about the point where $Z = Z_{st}$ to describe the inner (reaction) zone. The inner equations are simplified if the coordinate normal to the reaction sheet is transformed to Z. The general transformation has been given earlier (Peters and Williams 1983); in the first approximation in the stretched variable Eq. (18) becomes (with ρD evaluated at equilibrium stoichiometric conditions)

$$d(c_p dT/dZ)/dZ = -Q_F w_F/(\rho D |\nabla Z|^2) \quad (20)$$

Outer Structure

A number of different problems could be considered on the basis of the equations formulated here. For example, equilibrium structure in the outer zones could be addressed. However, it appears that especially when conditions approach extinction, the finite-rate effects should extend over the region of equilibrium dissociation. The situation can be addressed by treating E/T and B/T as being large. In this case, the variations associated with all aspects of the chemistry are localized in a narrow zone, and outside

DIFFUSION-FLAME STRUCTURE AND EXTINCTION   45

of this zone the chemistry is unimportant. For Y_F and Y_O the outer solutions become the same as those derived earlier (Peters and Williams 1983). They correspond to those of the Burke-Schumann flame and are given by

$$\left. \begin{array}{ll} Y_F = \nu Y_{OO}(Z/Z_{st}-1), & Y_O = 0, \quad Z > Z_{st}, \\ Y_O = Y_{OO}(1-Z/Z_{st}), & Y_F = 0, \quad Z < Z_{st} \end{array} \right\} \quad (21)$$

Since c_p was not assumed constant, the outer solution for the temperature is not piecewise linear in Z but is to be obtained from Eqs. (7) and (12).

Use of Eq. (21) as outer solutions involves neglecting both finite-rate chemistry and equilibrium dissociation in the first approximation. This selection provides a good starting point for including both of these effects simultaneously. The corresponding appropriate value of T^o, to be denoted herein by T_{st}, is the undissociated equilibrium temperature at stoichiometric conditions and may be shown to be defined by

$$Z_{st}\left[Y_{FF}\int_{T_F}^{T^o}c_{pF}dT + (1-Y_{FF})\int_{T_F}^{T^o}c_{pN}dT\right]$$

$$+ (1-Z_{st})\left[Y_{OO}\int_{T_O}^{T^o}c_{pO}dT + (1-Y_{OO})\int_{T_O}^{T^o}c_{pN}dT\right]$$

$$= Y_{FF}Q_F Z_{st}$$

With this selection for T^o, the proper way to use Eq. (9) for calculating c_{pst} is to evaluate c_{pi} at T_{st} and to evaluate Y_i at undissociated stoichiometric equilibrium, so that only species N, P_1 and P_2 contribute to c_{pst}. With the definitions that have been introduced, in the vicinity of the stoichiometric point the outer solutions for temperature become

$$\left. \begin{array}{ll} T = T_{fr}^* + [Q_F \nu Y_{OO}/c_{pst}](1-Z), & Z > Z_{st} \\ T = T_{fr}^* + (Q_F Y_{FF}/c_{pst})Z, & Z < Z_{st} \end{array} \right\} \quad (22)$$

in the first approximation, as illustrated in Fig. 1. Equations (21) and (22) will provide matching conditions for the inner solutions. The introduction of T_{fr}^* thus has enabled us to cast the problem in the same form that would have been obtained more simply if all specific heats were constant and equal.

Equations for the Structure of the Reaction Zone

In analyzing the reaction zone we restrict our attention to what has been termed the diffusion-flame regime (Liñán 1974) and do not address questions associated with establishment of conditions in the premixed-flame regime. We seek a distinguished limit in which the reaction-zone equations exhibit effects of dissociation and of finite-rate chemistry. In terms of the stoichiometric temperature T_{st}, the small parameter of the problem may be written as $\varepsilon = 2T_{st}^2 c_{pst}(1-Z_{st})/(EQ_F Y_{FF})$. The stretched mass-fraction variable is $\zeta = (Z-Z_{st})/\varepsilon$, and the dependent variable y for the inner region, related to the stretched temperature, is $y = (T_{st}-T)E/T_{st}^2 - \gamma\zeta$, where

$$\gamma = 2Z_{st} - 1 - 2(1-Z_{st})c_{pst}(\hat{T}_F - \hat{T}_0)/(Q_F Y_{FF}) \qquad (23)$$

The matching conditions become $dy/d\zeta \to \pm 1$ as $\zeta \to \pm \infty$. The inner equation is found from Eq. (20) in lowest order to be

$$d^2 y/d\zeta^2 = \delta(y_F y_0 - be^{-\beta y_T})e^{-y_T} \qquad (24)$$

where

$$y_T = y + \gamma\zeta, \quad y_F = y + \zeta = 2Y_F(1-Z_{st})/(\varepsilon Y_{FF}),$$
$$y_0 = y - \zeta = 2Y_0 Z_{st}/(\varepsilon Y_{00}) \qquad (25)$$

and the constants

$$\beta = \frac{B}{E}, \quad b = \left(\frac{EQ_F Y_{FF} Z_{st}}{c_{pst} T_{st}^2}\right)^2 \nu K e^{-B/T_{st}},$$

$$\delta = \frac{4(1-Z_{st})^2 Y_{FF} \rho_{st} B_{Fst} e^{-E/T_{st}}}{\nu W_0 D_{st} |\nabla Z|^2} \left(\frac{c_{pst} T_{st}^2}{Y_{FF} Q_F E}\right)^3 \qquad (26)$$

all have been assumed to be of order unity. The first approximation to the inner solution is obtained by solving Eq. (24) subject to the stated boundary conditions. The solution depends on the four parameters b, β, γ, and δ and, of course, must be found numerically. We shall restrict our attention to situations in which $|\gamma| < 1$. Then, by analogy with the results of Liñán (1974) it may be expected that there is a critical extinction value δ_E of the parameter δ, dependent on the other parameters, such that no solutions exist for $\delta < \delta_E$ and two solutions exist for $\delta > \delta_E$. When two solutions exist, the one with the lower value of the maximum temperature is assumed to be unstable and not of physical interest. It may also be expected from the results of Liñán (1974) that there will be a leakage of both reactants through the flame, in that Y_F will not approach zero as $\zeta \to -\infty$ and y_O will not approach zero as $\zeta \to +\infty$. The asymptotic solutions for $\zeta \to \pm\infty$, and therefore the leakage, can be shown to be influenced substantially by taking $b \neq 0$. These leakage effects will not be studied thoroughly here. Instead, attention will be focused mainly on the extinction criterion δ_E. For $b = 0$ Liñán (1974) has shown that in an excellent approximation

$$\delta_E = e[(1 - |\gamma|) - (1 - |\gamma|)^2 + 0.26(1 - |\gamma|)^3 + 0.055(1 - |\gamma|)^4] \qquad (27)$$

We are interested in finding how Eq. (27) is modified to depend on b and β when $b \neq 0$.

Extinction for a Moderate Equilibrium Constant

In general, the critical value $\delta_E(b, \beta, \gamma)$ must be obtained from numerical integrations. However, if $1 - |\gamma|$ is small, then an approach analogous to that of Liñán (1974) may be followed to obtain formulas for δ_E in various limiting cases. To illustrate the procedure, let us put $\gamma = -1 + \phi$ and treat ϕ as a small positive parameter.

When ϕ is small and δ is sufficiently small, a large amount of fuel leaks through the flame while nearly equilibrium conditions prevail as $\zeta \to +\infty$. Finite-rate chemistry occurs now mainly over a range of ζ of order unity centered at relatively large positive values of ζ, and since an appreciable amount of fuel has leaked to smaller values of ζ than this, the reaction zone resembles that of a premixed flame. The analysis of this reaction zone is facilitated by introducing the dependent variable $z = y - \zeta$, related to the oxidizer concentration, and by translating the origin of the independent variable through the transformation $\eta = 2\zeta - \kappa$, where κ is $(y+\zeta)_{-\infty}$ as ζ goes to minus infinity. A plot of κ vs. δ will eventually provide a C-shaped response curve from with the critical value δ_E can be calculated. We anticipate κ to be large but to be determined in the course of the analysis. It is intended that the reaction primarily will occur in the vicinity of $\eta = 0$. The boundary conditions for Eq. (24) imply that $z \to -\eta$ as $\eta \to -\infty$ and $dz/d\eta \to 0$ as $\eta \to \infty$, while the differential equation itself becomes

$$2\frac{d^2z}{d\eta^2} = \frac{\delta}{2}[z(\kappa+\eta+z) - be^{-\beta(z+\phi\kappa/2+\phi\eta/2)}]e^{-(z+\phi\kappa/2+\phi\eta/2)} \quad (28)$$

Since the changes in z for Eq. (28) are expected to occur over a range of η of order unity and with values of z of order unity, the right-hand side of the equation may be approximated as

$$\frac{\delta}{2}\kappa e^{-\phi\kappa/2}[z - \frac{b}{\kappa}e^{-\beta\phi\kappa/2}e^{-\beta z}]e^{-z}$$

which may be justified a posteriori. To see best how to find δ_E, consider first $\beta = 0$. The dependent variable $\hat{z} = z - b/\kappa$ then obeys the equation

$$2d^2\hat{z}/d\eta^2 = \Lambda\hat{z}e^{-\hat{z}} \quad (29)$$

with $d\hat{z}/d\eta \to -1$ as $\eta \to -\infty$ and $d\hat{z}/d\eta \to 0$ as $\eta \to \infty$, where

$$\Lambda = \frac{\delta}{2}\kappa e^{-\phi\kappa/2-b/\kappa}$$

Equation (29) is the well-known equation describing the reaction-zone structure for premixed flame propagation and is known to have a solution satisfying the specified boundary conditions only for $\Lambda = 1$, a burning-rate eigenvalue (Liñán 1974). Thus, with $x = \kappa\phi/2$, we have

$$x \exp[-(x + \frac{\phi b}{2x})] = \frac{\phi}{\delta} \quad (30)$$

For a fixed value of $\phi b/2$ the left-hand side of Eq. (30) is a function of x that vanishes at $x = 0$ and at $x = \infty$ and that is positive in between, attaining its maximum value,

$$\tfrac{1}{2}(1 + \sqrt{1+2\phi b})e^{-\sqrt{1+2\phi b}} \quad \text{at} \quad x = \tfrac{1}{2}(1 + \sqrt{1+2\phi b})$$

Equation (30) clearly possesses no solution for x if ϕ/δ exceeds this maximum value; if ϕ/δ is below the maximum, then there are two solutions for x, and the smaller of these provides the physically relevant value of κ associated with the given δ. The minimum value of δ below which Eq. (30) ceases to possess solutions (for given, fixed values of ϕ and b) is the extinction value δ_E. Thus, with ϕ re-expressed in terms of γ,

$$\delta_E = \frac{2(1 - |\gamma|)\exp[\sqrt{1 + 2b(1 - |\gamma|)}]}{1 + \sqrt{1 + 2b(1 - |\gamma|)}} \tag{31}$$

It may be noted that with ϕ small b must be large for the reverse reaction to have a substantial effect on the extinction conditions. By introducing $\hat{z}$ with b large we account for the oxidizer concentration approaching its equilibrium value, not zero, on the fuel side of the reaction zone; the fuel concentration changes negligibly across the reaction zone. The introduction of $-|\gamma|$ for the negative quantity $+\gamma$ in writing Eq. (31) has enabled the formula to be used for positive values of γ as well, pertaining to a reaction zone at large negative values of ζ with fuel depletion and nearly constant oxidizer concentration. The expansion of Eq. (31) for small values of ϕb is

$$\delta_E = e(1 - |\gamma|)[1 + \tfrac{b}{2}(1 - |\gamma|) + \ldots] \tag{32}$$

The agreement of this result for b = 0 with the expansion of Eq. (27) for small ϕ is evident. It is seen from Eq. (32) that increasing b increases δ_E; according to Eq. (31) this increase becomes exponential when ϕb becomes large. Thus, the reverse reaction promotes extinction.

These results have been restricted to $\beta = 0$, a condition under which the influence of the reverse reaction might be expected to be greatest. If β is of order unity then the full expression that follows Eq. (28) must be retained on the right-hand side. A first integral again is readily obtained to show, upon application of the matching conditions, that

$$1 = \frac{\delta}{2} \kappa e^{-\phi\kappa/2 - z_\infty} \left[1 + z_\infty - \frac{b}{\kappa(1+\beta)} e^{-\beta(\phi\kappa/2 + z_\infty)} \right]$$

where z_∞ is the equilibrium value of z, now given by $z_\infty = (b/\kappa) e^{-\beta(\phi\kappa/2 + z_\infty)}$, as may be seen from the expression following Eq. (28). From the differentials of these two expressions it may be shown that the extremum (or extinction) condition corresponds to $\phi\kappa/2 = 1 + z_\infty$, whence the Damköhler number at extinction may be expressed as

$$\delta_E = \phi(1+z_\infty)^{-1}(1 + \frac{\beta z_\infty}{1+\beta})^{-1} e^{1+2z_\infty} \qquad (33)$$

in which z_∞ is to be obtained from

$$z_\infty(1+z_\infty) e^{\beta(1+2z_\infty)} = \phi b/2 \qquad (34)$$

For $\beta = 0$ Eq. (34) gives $z_\infty = (\sqrt{1+2\phi b} - 1)/2$, which when substituted into Eq. (33) produces Eq. (31). As β approaches infinity with ϕb not too large, Eq. (34) shows that z_∞ approaches $(\phi b/2)e^{-\beta}$, so that δ_E approaches $e\phi$, and the effect of equilibrium on extinction disappears. For small values of β, if ϕb is not too large, it may be shown from Eqs. (33) and (34) that approximately

$$\delta_E \approx \frac{2(1 - |\gamma|)\exp[\sqrt{1 + 2b(1-\beta)(1 - |\gamma|)}]}{[1 + \sqrt{1 + 2b(1-\beta)(1 - |\gamma|)}][1 + b\beta(1 - |\gamma|)/2]}$$

which becomes

$$\delta_E \approx e(1 - |\gamma|)[1 + b(\tfrac{1}{2} - \beta)(1 - |\gamma|) + \ldots]$$

when ϕb is small. These formulas indicate that increasing β does indeed decrease δ at any fixed value of b. However, because of the exponential dependence of the equilibrium constant on β, the numerical results obtained from these last expansions will be inaccurate unless β is the rather small, e.g., $\beta < 0.1$.

The results for δ_E that have been given here are accurate only if $(1 - |\gamma|)$ is small. The analysis can be carried to second order in ϕ to produce better agreement at the smaller values of b in Fig. 6 (shown later). The restriction to small ϕ in fact applies in many cases of practical interest. However, it is also of interest to have parametric results for other values of γ. Parametric results in the whole range of γ will be obtained in the limit of large b in the following section.

Extinction for a Large Equilibrium Constant

With increasing b and sufficiently small β the equilibrium region in the inner flame zone will broaden. Eventually, as b becomes very large, there will be an equilibrium zone in the center, surrounded by two premixed-flame zones that accomplish the transition to frozen-flow layers outside. This structure has been analyzed in the context of the Libby and Economos (1963) flame-zone model (Peters 1979) where the equilibrium and frozen solutions appeared as outer solutions. The same procedure may be applied to Eq. (28) to obtain the structure of the transition zone. When this is done, it is found that in suitably transformed variables the structure is described by the equation for Liñán's (1974) premixed-flame regime. It is also found, as anticipated from the work of Liñán (1974), that extinction corresponds to attainment of a zero temperature gradient on the equilibrium side of the transition zone. This corresponds to the previous statement (Peters 1979) that extinction occurs when one of the transition layers reaches the point of maximum equilibrium temperature. To address the extinction problem here, instead of applying the procedure of Peters (1979), we present an alternative development that is simpler algebraically and that leads to the same result.

Let us consider a transition zone with equilibrium to the right and frozen conditions on the left. Let the departure from equilibrium occur in the vicinity of $y = y_c$, $\zeta = \zeta_c$, which are taken to satisfy the equilibrium condition. Thus, from Eqs. (24) and (25), we have

$$y_c^2 - \zeta_c^2 = b e^{-\beta(y_c + \gamma \zeta_c)} \tag{35}$$

When b is large, clearly y_c is large, and in the transition zone appropriate scalings of the inner variables can be shown to be the same as those of y and ζ. Thus, as inner variables introduce $\hat{y} = y - y_c$ and $\hat{\zeta} = \zeta - \zeta_c$. By use of Eq. (35), the first approximation to Eq. (24) in the inner zone is then found to be

$$d^2\hat{y}/d\hat{\zeta}^2 = \delta e^{-(y_c + \gamma \zeta_c)}[2y_c\hat{y} - 2\zeta_c\hat{\zeta} + \beta(y_c^2 - \zeta_c^2)(\hat{y} + \gamma\hat{\zeta})]e^{-(\hat{y} + \gamma\hat{\zeta})} \tag{36}$$

The matching condition for Eq. (36) as $\hat{\zeta} \to -\infty$ is $d\hat{y}/d\hat{\zeta} \to -1$, as obtained from the relationship above Eq. (24). For $\hat{\zeta} \to +\infty$, the slope must now match to the equilibrium slope, which is found from the derivative of Eq. (35) to be $y'_c \equiv -A/B$, where

$$\left. \begin{array}{l} A = -2\zeta_c + \beta\gamma(y_c^2-\zeta_c^2) \\ B = 2y_c + \beta(y_c^2-\zeta_c^2) \end{array} \right\} \quad (37)$$

thus, with this definition, we need $d\hat{y}/d\hat{\zeta} \to y'_c$ as $\hat{\zeta} \to \infty$. In this formulation βy_c evidently must be of order unity, so that β is small if a broad equilibrium zone exists.

Introduction of the variables $\tilde{\zeta} = \hat{\zeta}(1 - A/B)$ and $\tilde{y} = \hat{y} + \hat{\zeta}(A/B)$ serves to transform Eq. (36) into

$$2d^2\tilde{y}/d\tilde{\zeta}^2 = \Delta\tilde{y}e^{-(\tilde{y}+m\tilde{\zeta})} \quad (38)$$

with the boundary conditions that $d\tilde{y}/d\tilde{\zeta} \to -1$ as $\tilde{\zeta} \to -\infty$ and $d\tilde{y}/d\tilde{\zeta} \to 0$ as $\tilde{\zeta} \to \infty$, where $m = (\gamma B-A)/(B-A)$ and

$$\Delta = 2\delta B^3(B-A)^{-2}e^{-(y_c+\gamma\zeta_c)} \quad (39)$$

From Liñán (1974), we know that here extinction requires $m = 0$, and then $\Delta = 1$. From Eq. (37) it is seen that $m = 0$ gives $\zeta_c = -\gamma y_c$. This may be used in Eq. (35) to provide an expression for y_c, after which Eq. (37) and $\Delta = 1$ may be employed in Eq. (39) to obtain a formula for the extinction Damköhler number δ_E. To express the result in a convenient form it is helpful to define $X = \beta y_c(1-\gamma^2)/2$, so that Eq. (35) becomes

$$Xe^X = \beta\sqrt{b}\sqrt{1-\gamma^2}/2 \quad (40)$$

After an appreciable amount of algebra it is then found that Eq. (39) gives

$$\delta_E = \sqrt{\frac{2(1-|\gamma|)}{b}}\left[1 - \tfrac{1}{2}(1 - |\gamma|)\right]^{5/2}(1+X)^{-1}e^{X(1+2/\beta)} \quad (41)$$

where the prevailing restriction to negative values of γ has been removed through suitable introduction of absolute values. Equation (40) provides X in terms of γ, β, and b, after which Eq. (41) gives δ_E in terms of these same parameters.

The derivation given here requires b to be large and $\beta\sqrt{b}$ not to be large, but $\phi \equiv 1 - |\gamma|$ need not be small; if ϕ is small, then the factor raised to the 5/2 power in Eq. (41) may be replaced by unity. As $\beta\sqrt{b}$ approaches zero, Eq. (40) shows that X approaches $\beta\sqrt{b}\sqrt{1-\gamma^2}/2$, a small quantity, and Eq. (41) then becomes

$$\delta_E = \sqrt{\frac{2(1-|\gamma|)}{b}} \, e^{\sqrt{b}\sqrt{1-\gamma^2}} \left[1 - \tfrac{1}{2}(1 - |\gamma|)\right]^{5/2} \quad (42)$$

which reduces to $\sqrt{2\phi/b} \, e^{\sqrt{2\phi b}}$ at small values of ϕ, in agreement with the expansion of Eq. (31) for larger values of ϕb. The first correction to this result for $\beta \neq 0$, obtained from Eqs. (40) and (41), is the factor $(1 - \beta\phi b)$, which differs from the corresponding factor $(1 - \beta\phi b/2)$, obtained from an expansion of the approximation appearing after Eq. (34) for small β and large b; the latter is incorrect because of the restriction on the size of b underlying the derivation of the approximation, a restriction that can be removed (to produce agreement) at the expense of complicating the expression appreciably. If β is large, then there is a narrow inner zone of equilibrium, and analyses may be presented to show, for example, that

$$\delta_E = 4\phi \left(\frac{\phi b}{2}\right)^{1/\beta} \left[1 + \frac{1}{\beta} \ln\left(\frac{\phi b}{2}\right)\right]^{-2}$$

for ϕ small and ϕb large and that $\delta_E = (\beta/4)b^{1/(2\beta)}[\ln(\beta^2 b)]^{-1}$ for $\gamma = 0$ and b large; these derivations are not shown because there is little or no practical interest in large values of β. For $\gamma = 0$ and $\beta = 0$, Eq. (42) gives $\delta_E = e^{\sqrt{b}}/(4\sqrt{b})$, which agrees with the first term of the expansion to be developed independently below.

For $\gamma < 0$ the relevant transition layer is that on the left, the oxidizer side, as considered here, although near extinction it moves to the fuel side of the stoichiometric point (see Fig. 3 later). The case $\gamma > 0$ is obtained from considerations of symmetry. As $\phi \to 1$, $\gamma \to 0$ and the symmetric case is approached. With decreasing δ the transition layers move to the center where extinction should occur if the transition layers reach the maximum temperature (minimum y). For $\gamma = 0$ this means, however, that the equilibrium zone has decreased to zero and that the two transition zones merge. It is evident that the analysis of the transition layer, which requires matching to the equilibrium zone, is no longer valid as $\gamma \to 0$.

To investigate this situation further, let us consider the special case $\gamma = 0$, $\beta = 0$, $b \to \infty$, so that

$$d^2y/d\zeta^2 = \delta(y^2 - \zeta^2 - b)e^{-y} \qquad (43)$$

with the boundary conditions $y = y_0$, $dy/d\zeta = 0$ at $\zeta = 0$ and $dy/d\zeta \to 1$ as $\zeta \to \infty$. Here we expect a C-shaped behavior for y_0 as a function of δ, from which the extinction condition can be obtained.

We expand y and δ to second order in $1/\sqrt{b}$,

$$\left. \begin{array}{l} y = \sqrt{b} + z_1(\zeta) + z_2(\zeta)/\sqrt{b} \ldots \\ \delta = \delta_1(1 + \delta_2/\sqrt{b}) \ldots \end{array} \right\} \qquad (44)$$

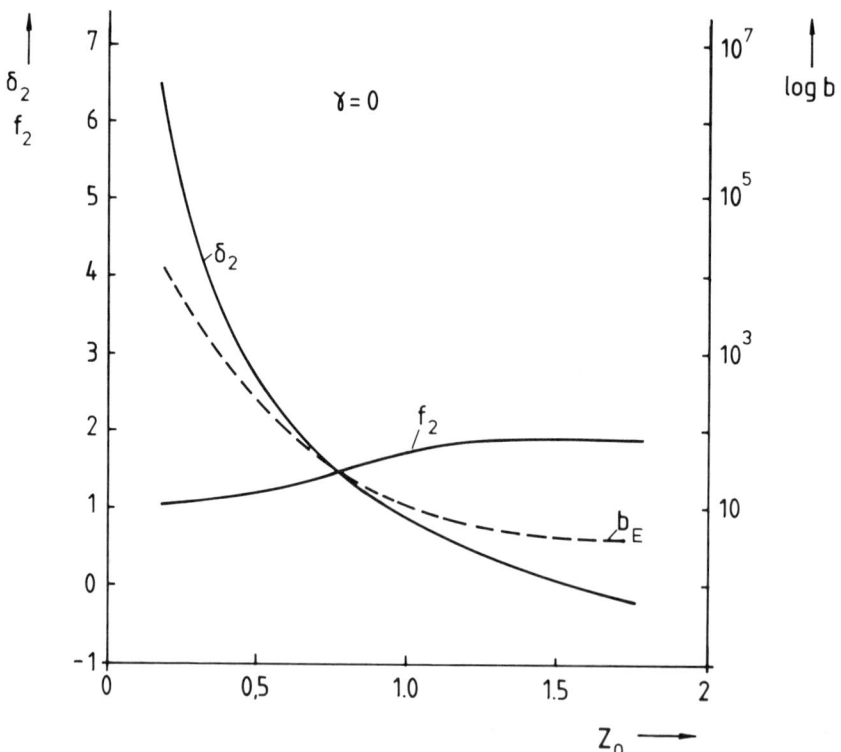

Fig. 2 Second-order Damköhler number and related parameters at extinction as functions of z_0 for $\gamma = 0$.

which gives, for the equilibrium function,

$$y^2 - \zeta^2 - b = 2\sqrt{b}\left[z_1 + \frac{1}{\sqrt{b}}(\frac{z_1^2-\zeta^2}{2} + z_2)\right] \cdots \quad (45)$$

To leading order Eq. (43) becomes

$$d^2z_1/d\zeta^2 = \Lambda_1 z_1 e^{-z_1} \quad (46)$$

$$\Lambda_1 = 2\delta_1\sqrt{b}\, e^{-\sqrt{b}} \quad (47)$$

with $z_1 = z_0$, $dz_1/d\zeta = 0$ at $\zeta = 0$, and $dz_1/d\zeta \to 1$ as $\zeta \to \infty$, where $z_0 = y_0 - \sqrt{b}$. Equation (46) may be integrated once between $\zeta = 0$ and $\zeta \to \infty$ to obtain

$$\Lambda_1 = \tfrac{1}{2}\frac{e^{z_0}}{1 + z_0} \quad (48)$$

The quenching condition $d\delta_1/dy_0 = d\Lambda_1/dz_0 = 0$ provides $z_{0,E} = 0$ and δ_E to leading order as

$$\delta_{1,E} = \tfrac{1}{4}\frac{e^{\sqrt{b}}}{\sqrt{b}} \quad (49)$$

For two reasons the leading-order result is not entirely satisfactory.

1) It differs by a factor of about 2 from precise numerical calculations for b below 10.

2) The stable branch resulting from Eq. (48) lies at $0 > z_0 > -1$ and therefore according to Eq. (45) below the equilibrium solutions. This is physically unrealistic.

Therefore the second-order problem

$$d^2z_2/d\zeta^2 = \Lambda_1(z_1 - z_2 z_1 + (z_1^2-\zeta^2)/2 + \delta_2 z_1)e^{-z_1} \quad (50)$$

with $z_2 = 0$, $dz_2/d\zeta = 0$ at $\zeta = 0$, and $dz_2/d\zeta \to 0$ as $\zeta \to \infty$ was considered. Here the eigenvalue δ_2 was determined as a function of z_0 numerically (Fig. 2). The extinction condition leads then to

$$\frac{d\delta}{dy_0} = \frac{e^{\sqrt{b}}}{2\sqrt{b}} \left[\frac{d\Lambda_1}{dz_0} (1 + \frac{\delta_2}{\sqrt{b}}) + \frac{\Lambda_1}{\sqrt{b}} \frac{d\delta_2}{dz_0} \right]_E = 0 \qquad (51)$$

and therefore to

$$\frac{z_{0,E}}{1 + z_{0,E}} + \frac{1}{\sqrt{b} + \delta_{2,E}} \frac{d\delta_{2,E}}{dz_{0,E}} = 0 \qquad (52)$$

This allows us to relate the value of b at extinction to z_0, with $\delta_2(z_0)$ given (Fig. 2). Finally, the extinction Damköhler number is obtained as

$$\delta_E = \delta_{1,E} \, f_2(z_{0,E}) \qquad (53)$$

where the term

$$f_2 = \frac{\exp z_{0,E}}{1 + z_{0,E}} (1 + \delta_{2,E}/\sqrt{b}) \qquad (54)$$

describes the correction to the leading-order term.

The results of the numerical calculations are shown in Fig. 2. They show that $1 \leq f_2 \leq 2$ and that $\delta_{2,E}$ and b grow very rapidly as z_0 approaches zero. Since $\delta_{2,E}$ is not independent of b but increases rapidly as b increases, the contribution of the second-order term is significant even for large b. Also, since $z_{0,E}$ is always positive, physical intuition about the behavior of the solution is no longer offended.

Numerical Results

Equation (24) was solved numerically for $\gamma = 0, -0.5$, and -0.9 for various values of b and β. For the cases $\gamma = 0$ and $\gamma = -0.5$ a shooting method was used. The value of κ was prescribed and δ was obtained as an eigenvalue. Then κ was increased until the turning point at a minimum $\delta = \delta_E$ was reached. The shooting method failed to converge for $\gamma = -0.9$ and $b > 2.0$. For these cases a time-dependent finite-difference method was used. With a prescribed δ the steady-state solution was obtained for large times. Then δ was decreased until, at a value identified as δ_E, a steady-state solution could no longer be found.

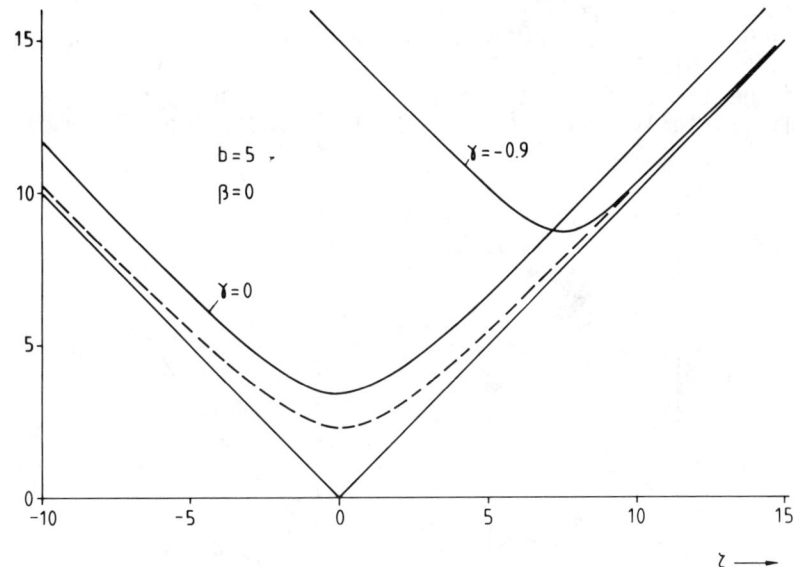

Fig. 3 Typical profiles in the reactive-diffusive zone: ——— nonequilibrium, --- equilibrium.

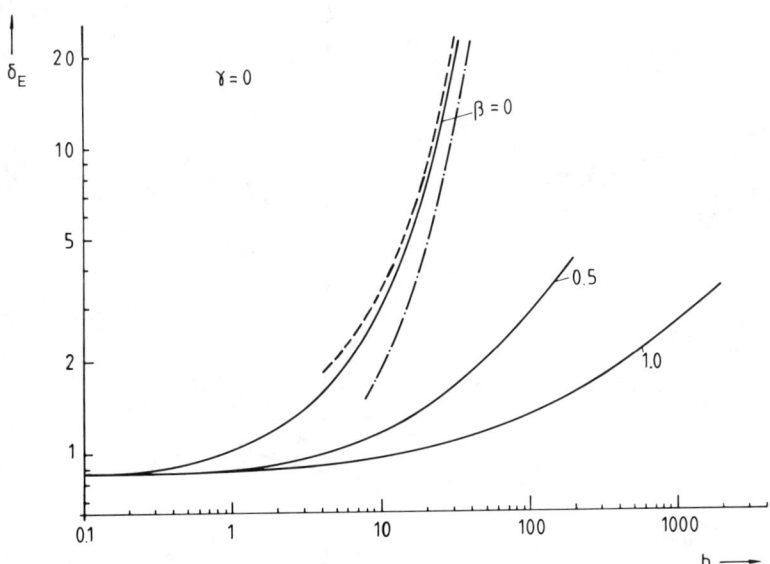

Fig. 4 Extinction Damköhler number for $\gamma = 0$. Limiting curves for large b: - • - first order approximation, Eq. (49); --- second-order approximation, Eq. (53); ——— numerical results.

Typical profiles of y vs ζ are shown for $b = 5$, $\beta = 0$, and the two extreme cases $\gamma = 0$ and $\gamma = -0.9$ at extinction in Fig. 3. Also shown is the equilibrium solution. In the symmetric case the nonequilibrium solution differs substantially from the equilibrium solution for the whole

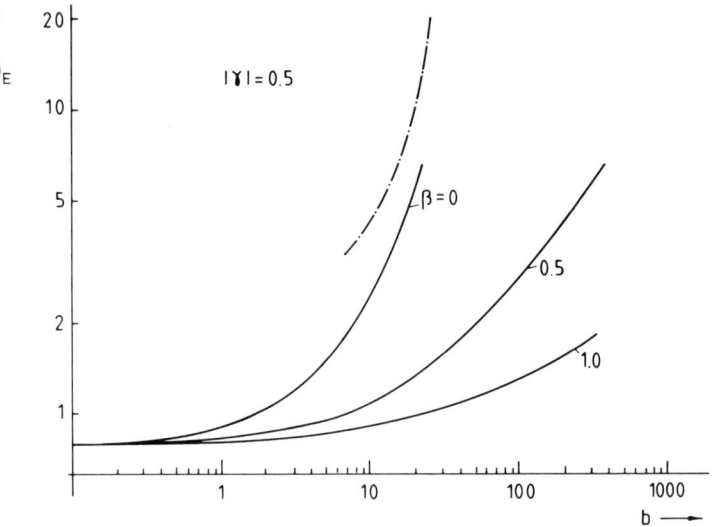

Fig. 5 Extinction Damköhler number for $|\gamma| = 0.5$.
- · - · - limiting curve for large b, Eq. (41); ——— numerical results.

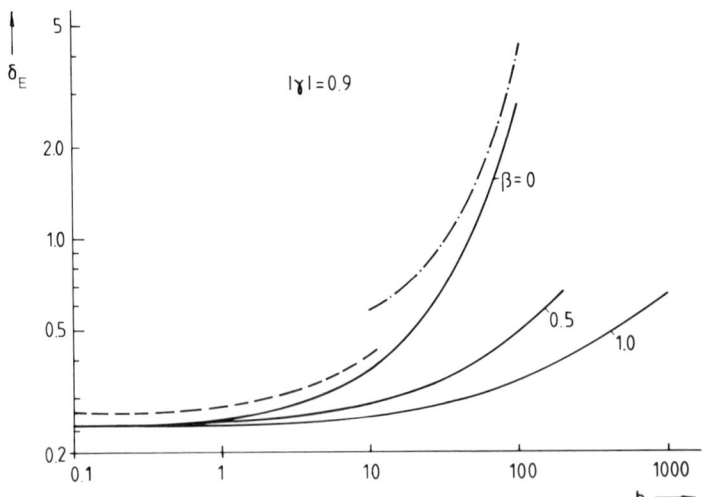

Fig. 6 Extinction Damköhler number for $|\gamma| = 0.9$.
- · - · - limiting curve for large b, Eq. (42); - - - limiting curve for moderate b, Eq. (31); ——— numerical results.

range of ζ. Contrary to this, the solution for $\gamma = -0.9$ is frozen for $\zeta < 5$, exhibits a nonequilibrium transition zone around $\zeta = 8$ and follows the equilibrium solution for large ζ. Extinction Damköhler numbers δ_E are shown for $|\gamma| = 0$, 0.5, and 0.9 in Figs. 4-6. For $\beta = 0$ the limiting solutions for large b and in the case $|\gamma| = 0.9$ for moderate b are also shown in these figures. For $\gamma = 0$ it is shown that the two-term expansion follows very closely the numerical results, while the leading order result is rather inaccurate. This is even worse for $|\gamma| = 0.5$, where the equilibrium region decreases, while the comparison is quite good for moderate and large b in the case $|\gamma| = 0.9$. In all cases it is seen that the Damköhler number at extinction increases easily by an order of magnitude, if b is sufficiently large.

Conclusions

Influences of equilibrium dissociation on flame structure with finite-rate chemistry and on extinction have been studied. The results expose flame-structure modifications caused by the equilibrium and provide new extinction criteria that can be used in analyzing extinction experiments. Application of these results to experiments entails first identifying the most appropriate equilibrium chemistry and next evaluating the relevant constants describing the equilibrium prior to addressing the experimental data. A general conclusion that can be drawn from the study is that equilibrium effects tend to promote extinction. The extent to which extinction is enhanced decreases as the heat of dissociation is increased at a fixed value of the equilibrium constant. Further studies of equilibrium effects would be desirable, e.g., for mechanisms different from that of Eq. (3), to aid in interpreting experimental results on high-temperature extinctions.

Acknowledgment

One of us (FAW) would like to thank the Alexander von Humboldt Foundation for providing him with a U.S. Senior Scientist Award that made possible his participation in this research.

References

Bilger, R. W. (1977) Reaction rates in diffusion flames. Combust. Flame 30, 277-284.

Bilger, R. W. (1980) Turbulent flows with nonpremixed reactants. Turbulent Reacting Flows (edited by P. A. Libby and F. A. Williams) 65-113. Springer-Verlag, Berlin.

Chung, P. M. and Blankenship, V. D. (1966) Equilibrium structure of a thin diffusion flame zone. Phys. Fluids 9, 1569-1577.

Fendell, F. E. (1967) Combustion in initially unmixed reactants for one-step reversible chemical kinetics. Acta Astronautica 13, 183-191.

Krishnamurthy, L., Williams, F. A., and Seshadri, K. (1976) Asymptotic theory of diffusion flame extinction in the stagnation-point boundary layer. Combust. Flame 26, 363-377.

Libby, P. A. and Economos, C. (1963) A flame zone model for chemical reaction in a laminar boundary layer with application to the injection of hydrogen-oxygen mixtures. Int. J. Heat Mass Transf. 6, 113-128.

Libby, P. A., Liñán, A., and Williams, F. A. (1983) Strained laminar flames with nonunity Lewis numbers. Combust. Science and Technology 34, 257-293.

Liñán, A. (1974) The asymptotic structures of counter flow diffusion flames for large activation energies. Acta Astronautica 1, 1007-1039.

Peters, N. (1975) Berechnung einer Methan-Luft-Diffusionsflamme im örtlichen Gleichgewicht und im Nichtgleichgewicht. VDI-Berichte 246, 5-12.

Peters, N. (1979) Premixed burning in diffusion flames - The flame zone model of Libby and Economos. Int. J. Heat Mass Transf. 22, 691-703.

Peters, N. and Williams, F. A. (1983) Liftoff characteristics of turbulent jet diffusion flames. AIAA J. 21, 423-429.

Sohrab, S. H. and Williams, F. A. (1981) Extinction of diffusion flames adjacent to flat surfaces of burning polymers. J. Polymer Science: Polymer Chemistry Edition 19, 2955-2976.

Williams, F. A. (1981) Review of flame extinction. Fire Safety Journal 3, 163-175.

Stretch Effects in Planar Premixed Hydrogen-Air Flames

J. Warnatz*
*Physikalisch-Chemisches Institut
der Universität Heidelberg, Federal Republic of Germany*
and
N. Peters†
*Institut fuer Allgemeine Mechanik
der Rheinisch-Westfaelische Technische Hochschule
Aachen, Federal Republic of Germany*

Abstract

This work reports a study of the effects of flame stretching due to a steady diverging flow of a planar premixed flame with detailed kinetics. Numerical results for detailed kinetics are compared to the analytic result obtained from large activation energy asymptotics for one-step kinetics. A rich flame with a hydrogen mole fraction in the unburnt gas of $X_{H_2} = 0.6$ and a lean flame with $X_{H_2} = 0.15$ is considered. As predicted by the asymptotic theory, the mass flow rate through the flame is decreased owing to stretch for the rich flame, while it is increased for the lean flame. Extinction due to flame stretch occurs for the rich flame only.

Nomenclature

B_{fk}, B_{bk} = frequency factors of the k-th forward and backward reaction
c_p = mean specific heat capacity
c_{pi} = heat capacity of species i
$D_{i,M}$ = diffusion coefficient with respect to the mixture
E = global activation energy
E_{fk}, E_{bk} = activation energies of the k-th forward and backward reaction

---

Paper presented at the 9th ICODERS, Poitiers, France, July 3-8, 1983. Copyright © American Institute of Aeronautics and Astronautics, Inc., 1984. All rights reserved.
*Professor, Department of Physical Chemistry.
+Professor, Department of Mechanics.

h_i = specific enthalpy of species i
j_i = diffusion mass flux of species i
K = stretch parameter defined in Eq.(2)
Ka = Karlovitz number defined in Eq.(12)
Le = Lewis number = $\lambda/(\rho c_p \mathcal{D}_{i,N_2})_{x=0}$
M = inert body in reaction scheme
$\dot{M}$ = mass flow rate at x = 0
n = total number of species
n_f, n_b = exponents in Eq.(9)
r = total number of reactions
T = temperature
T_b = flame temperature at x = 0, K = 0
u,v,w = velocities in the Cartesian coordinate system
v_F = flame velocity in the unburnt combustible
$\dot{w}_i$ = reaction rate
x,y,z = Cartesian coordinates
β = nondimensional activation energy defined in Eq.(11)
λ = heat conductivity
ν'_{ik}, ν''_{ik} = stoichiometric coefficients of the i-th species in the k-th forward and backward reaction,

$$\nu_s = \sum_{i=1}^{r} (\nu''_{ik} - \nu'_{ik})$$

ψ = Lagrangian coordinate, defined in Eq.(4)
ρ = density

Subscripts

ex = extinction
i,j = species index
k = reaction index
ref = reference state, K = 0

Introduction

Karlovitz et. al(1953) proposed that a differential increase of the flame surface area, called flame stretch, would lead to a lowering of the temperature in the reaction zone and eventually to extinction. This general picture was placed in doubt by the results of activation energy asymptotics when Sivashinsky (1976) showed that the effect of stretch is related to differential diffusion of heat and reactant. If the Lewis number, Le, defined as the ratio of thermal diffusivity to the molecular diffusivity of the deficient reactant, is greater than 1, the rate of heat loss from the reaction zone exceeds the rate at which the deficient reactant is transported into the reaction zone. Hence, the enthalpy balance over the preheat zone is negative, and the temperature in the reaction zone decreases. If the activation energy is large, the flame velocity is very sensitive to changes in the flame temperature. Therefore, a small decrease in flame temperature leads to a large decrease in convective flow of enthalpy through the preheat zone. This enhances the differential diffusion effect relative to convection leading to even lower flame temperatures. Thus, if flame stretch exceeds a critical value, the two mechanisms cannot be balanced and the flame is extinguished. The opposite argument holds if the Lewis number is less than 1; then, differential diffusion increases the flame temperature, and the two mechanisms are balanced at an increased flame velocity.

Sivashinsky's (1976) analysis was based on the assumption that $(Le-1)/Le$ was of the order $O(1)$, while the nondimensional stretch parameter K was of order $O(\beta^{-1})$, where β is the nondimensional activation energy. This case is referred to as weak stretch. Buckmaster (1979a) analyzed stretch due to flow divergence in front of a bluff body for the case $(Le-1)/Le = O(\beta^{-1})$ and $K = O(1)$ (strong stretch) under the constant density assumption. From kinematic arguments, he also derived an expression between stretch and the rate of fractional flame area increase. This result demonstrates that both flow divergence and curvature effects contribute to flame stretch (Buckmaster 1979 b).

Both Sivashinsky and Buckmaster used the asymptotic limit of large activation energy to obtain closed formed results. Numerical calculations were performed for stagnation point flows in front of a solid body by Smith et al. (1974), Saitoh (1974), and Sato and Tsuji (1983) and for counterflow by Libby and Williams (1982, 1983).

The aforementioned analyses are based on the assumption that the complex flame chemistry can be described by a one-step, irreversible, global reaction. The validity of the one-step kinetics has often been challenged by chemical kineticists. Those devoted to activation energy asymptotics generally believe that a qualitatively correct picture can be drawn from the one-step results and that more complex kinetics would only modify this picture quantitatively. This depends, however, on the validation of the order of magnitude assumptions that enter into the analysis. If, for instance, the concentration of intermediate species is of the same order of magnitude as the deficient reactant, differential diffusion effects between reactant and intermediate as well as the effect of the intermediate on the flame temperature should be considered (Seshadri and Peters (1978).

The intent of this work is to compare the asymptotic results with numerical calculations of stretched plane premixed hydrogen-air flames with complex kinetics. Since the most simple asymptotic results apply to situations where one of the reactant is deficient, mixtures far from stoichiometry are considered. Hydrogen flames were chosen for the study, because the kinetic mechanism is sufficiently well known. Since the Lewis numbers are not close to 1, the weak stretch results should hold. Consequently, attention will be restricted to the case of a one-dimensional flame in a slowly diverging flow. In this case, the flowfield is characterized by a single parameter, and it is not necessary to specify the entire flowfield.

Numerical Results

A steady, planar, premixed flame in a diverging gas flow as shown in Fig. 1 is considered. The flow is presumed to be expanding or contracting. A Cartesian coordinate system x^*, y^*, z^* is used. The unburnt premixed combustible is assumed to flow from $x^* = -\infty$. Since the propagation velocity of the flame is much less than the velocity of sound, the momentum equation implies that the pressure p is approximately constant. Owing to the plane flame assumption, the den-

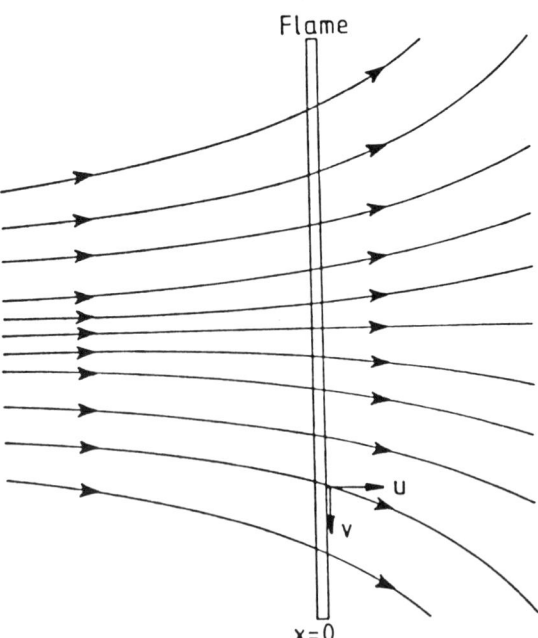

Fig. 1   Schematic representation of the flowfield.

sity is a function of x only insofar as the continuity equation yields

$$\frac{d(\rho u)}{dx} + \rho\left(\frac{dv}{dy} + \frac{\partial w}{\partial z}\right) = 0 \qquad (1)$$

If the stretch parameter

$$K = \frac{dv}{dy} + \frac{\partial w}{\partial z} \qquad (2)$$

and the mass flow rate

$$M = \rho u \quad \text{at} \quad x = 0 \qquad (3)$$

are used with the mass-weighted coordinate

$$\psi = \int_0^x \rho \, dx \qquad (4)$$

the continuity may be integrated:

$$\rho u = M - K\psi \qquad (5)$$

The flow variation is assumed to be slow enough that K can be assumed constant. This can be justified for the case of weak stretch. The steady-state energy and species conservation equations are

$$(M - K\psi)\frac{dT}{d\psi} = \frac{1}{c_p}\frac{\partial}{\partial \psi}\left(\rho\lambda\frac{\partial T}{\partial \psi}\right) - j_H\frac{\partial T}{\partial \psi} - \frac{\Sigma h_i \dot{w}_i}{\rho c_p} \qquad (6)$$

$$(M - K\psi)\frac{\partial Y_i}{\partial \psi} = -\frac{\partial j_i}{\partial \psi} + \frac{\dot{w}_i}{\rho} \qquad (7)$$

where

$$j_i = -\rho^2 D_{i,M}\frac{\partial Y_i}{\partial \psi}, \quad j_H = \sum_{i=1}^n \frac{c_{p_i} j_i}{c_p} \qquad (8)$$

$$\dot{w}_i = M_i \sum_{i=1}^r (\nu''_{ik} - \nu'_{ik})\left\{k_{fk}\prod_{j=1}^n \left(\frac{\rho Y_j}{M_j}\right)^{\nu'_{jk}} - k_{bk}\prod_{j=1}^n \left(\frac{\rho Y_j}{M_j}\right)^{\nu''_{jk}}\right\}$$

Table 1 Reaction mechanism

	Reaction		$B_{f,}$ cm,mole,s	n_f	$E_{f,}$ kJ/mole	$B_{b,}$ cm,mole,s	n_b	$E_{b,}$ kJ/mole
1.	$H + O_2$	$\rightleftarrows$ $OH + O$	1.20×10^{17}	-0.91	69.0	8.10×10^{15}	-0.91	0.0
2.	$O + H_2$	$\rightleftarrows$ $OH + H$	1.50×10^{7}	2.00	31.6	6.70×10^{6}	2.00	23.3
3.	$OH + H_2$	$\rightleftarrows$ $H_2O + H$	1.00×10^{8}	1.60	13.8	4.60×10^{8}	1.60	77.6
4.	$OH + OH$	$\rightleftarrows$ $H_2O + O$	1.48×10^{9}	1.14	0.0	1.50×10^{10}	1.14	72.1
5.	$H + H + M$	$\rightarrow$ $H_2 + M$	1.83×10^{18}	-1.00	0.0	...	...	...
6.	$H + OH + M$	$\rightarrow$ $H_2O + M$	2.15×10^{22}	-2.00	0.0	...	...	...
7.	$H + O_2 + M$	$\rightarrow$ $HO_2 + M$	2.00×10^{18}	-0.80	0.0	...	...	...
8.	$H + HO_2$	$\rightarrow$ $OH + OH$	1.50×10^{14}	0.00	4.2	...	...	...
9.	$H + HO_2$	$\rightarrow$ $H_2 + O_2$	2.50×10^{13}	0.00	2.9	...	...	...
10.	$O + HO_2$	$\rightarrow$ $OH + O_2$	2.00×10^{13}	0.00	0.0	...	...	...
11.	$OH + HO_2$	$\rightarrow$ $H_2O + O_2$	2.00×10^{13}	0.00	0.0	...	...	...

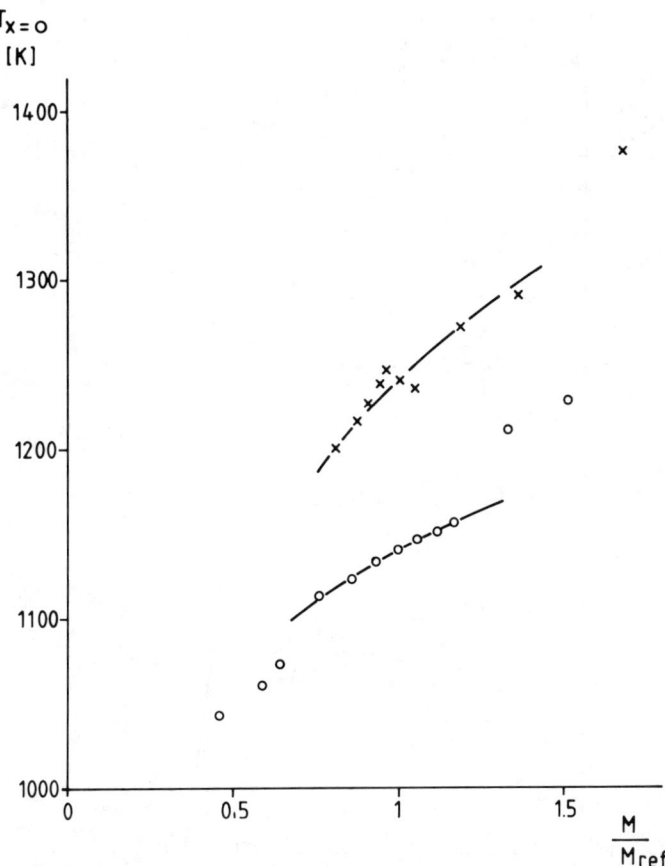

Fig. 2 Plot of the stretch parameter as a function of the temperature at the point of maximum O_2 consumption rate. o: rich flame ($x_{H_2} = 0.60$); x: lean flame ($x_{H_2} = 0.15$).

The upstream boundary conditions were specified by T = 298 K and the equivalence ratio of the hydrogen-air flames. Open boundary conditions were used downstream. This choice, where the value at the last grid point is determined by the outflow from the previous grid point, is widely used in one-dimensional flame calculations (Warnatz 1982). It is justified for slowly diverging flowfields by the fact that behind the reaction zone, convection and diffusion have the same direction. Since the stagnation point is very far downstream, boundary conditions, if applied there, would not affect the reaction zone.

The calculation of the properties λ and D_{i,N_2}, c_{pi}, h_i has been discussed elsewhere (Warnatz 1978). The system of

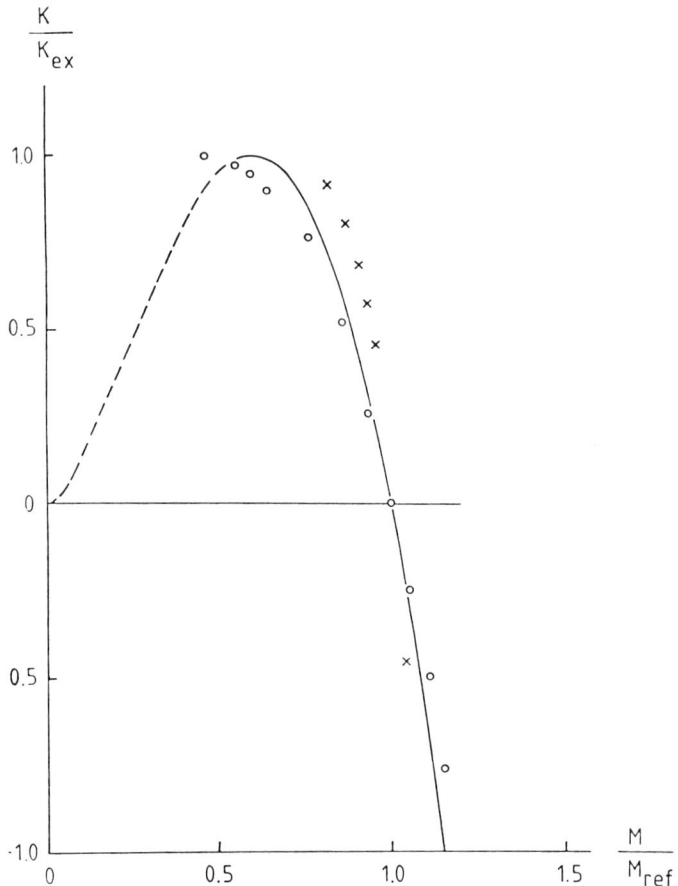

Fig. 3 Normalized plot of the stretch parameter as a function of the mass flow rate. (--- ——): asymptotic result; o: numerical result for rich flame; x: numerical result for lean flame.

equations is solved by prescribing K and adjusting the eigenvalue M such that a steady solution is obtained. The origin $\psi = 0$ was chosen at the point within the flame where the consumption rate of oxygen $(-\dot{w}_{O2})$ attains a maximum. Since K is small, this choice, although arbitrary, has little effect on the resulting values of M.

The kinetic scheme in terms of

$$k_{fk} = B_{fk} \, T^{n_{fk}} \, \exp(-E_{fk}/RT)$$
$$k_{bk} = B_{bk} \, T^{n_{bk}} \, \exp(-E_{bk}/RT) \tag{9}$$

where T is in degrees Kelvin, the activation energies E_f and E_b are in kJ/mole and B_f and B_b have the units

$$(cm^3/mole)^{\nu} s^{-1} / s$$

as shown in Table 1.

Two cases were analyzed: a fuel-rich flame with the mole fraction of hydrogen in the oncoming stream of

$$X_{H_2} = 0.60 \ (Y_{H_2} = 0.095)$$

and a lean flame with

$$X_{H_2} = 0.15 \ (Y_{H_2} = 0.0122)$$

The pressure was 1 bar. The properties at $x = 0$, $\psi = 0$, as well as the mass flow rate and the flame velocity for the flames without stretch ($K = 0$), are given in Table 2. For the stretched flames, the temperature T at $\psi = 0$ and the reduced mass burning rate M/M_{ref} are shown in Figs. 2 and 3. Owing to interpolation, the numerical error in the temperature data is significantly larger than the error in the mass flow rates. This explains a certain inconsistency in the temperature values for the lean flame between $K = 100$ and $K = -125$ s^{-1} in Fig. 2. For comparison with the asymptotic theory, calculations with negative values of K (flame compression) were included in this steady-state calculation. This is an unstable case, since an unsteady flame propagation would lead to the lower velocity region being upstream in a converging flowfield.

Comparison with Asymptotic Results

A special form of the weak stretch result (Sivashinsky 1976) is

$$Ka\beta \frac{Le-1}{Le} = - \frac{M^2}{M_{ref}^2} \ln \frac{M^2}{M_{ref}^2} \qquad (10)$$

where the nondimensional activation energy and the nondimensional stretch parameter, the Karlovitz number, are defined by

$$\beta = \frac{E(T_b - T_u)}{R T_b^2} \qquad (11)$$

Table 2   Properties of the unstretched flames at x = 0

Quantity	Units	Rich flame	Lean flame
ρ	kg/m^3	0.149	0.269
λ	kJ/m·s·K	2.19×10^{-4}	9.2×10^{-5}
c_p	kJ/kg·K	2.402	1.297
D_{i,N_2}	m^2/s	2.03×10^{-4}	8.39×10^{-4}
Le	–	3.018	0.3133
E/R	K	25764	16368
T_b	K	1140	1240
β	–	16.7	10.0
M_{ref}	kg/m^{2}s	0.958	0.3947
v_F	m/s	1.90	0.402
K_{ex}	1/s	1954.2	-218.19
K_{ex}, Eq.(16)	1/s	2221.6	-136.5

$$Ka = \frac{K}{M_{ref}^2}\left(\frac{\lambda \rho}{c_p}\right)_{x=0} \tag{12}$$

Equation (10) implies the following relation between the flame temperature perturbations and the mass burning rate:

$$\frac{M^2}{M_{ref}^2} = \exp\left(\frac{T_{x=0} - T_b}{T_b} \frac{E}{RT_b}\right) \tag{13}$$

In Eqs. (10) and (13), T_b denotes the temperature at x = 0 for the unstretched case. To obtain the special form Eq. (10), the thermal diffusivity $\lambda\rho/c_p$ was assumed to be constant. Since the asymptotic expansion is performed around x = 0, all properties will be evaluated at this point.

From Eq. (10) it can be shown that the Karlovitz number has an extremum at

$$(M/M_{ref})_{ex} = e^{-1/2} \tag{14}$$

where

$$Ka_{ex} \beta (Le-1)/Le = e^{-1} \tag{15}$$

This implies that for Le > 1, a further increase of the Karlovitz number beyond the value given by Eq. (15) will extinguish the flame. For Le < 1, a decrease, corresponding to compression, beyond the (negative) value obtained from Eq. (15) will also extinguish the flame. The Karlovitz number in Eq. (10) has been normalized by Ka_{ex} and is plotted over M/M_{ref} in Fig. 3, where the dotted part of the curve denotes the unstable branch. Since the extinction point could only be reached approximately in the numerical calculations, a parabola was fitted to the last three points in the tables in order to extrapolate to the extinction point, thereby obtaining the values

$$K_{ex} = 1954.2 \text{ s}^{-1}, (M/M_{ref})_{ex} = 0.4338 \text{ for the rich flame;}$$

and

$$K_{ex} = -218.19 \text{ s}^{-1}, (M/M_{ref})_{ex} = 0.7228 \text{ for the lean flame.}$$

Figure 3 shows the predicted general behavior of both the rich and the lean flame as far as acceleration, deceleration, and extinction are concerned. However, the numerical values of M, particular at extinction, deviate from the asymptotic prediction by up to 30 % for the rich flame and 20 % for the lean flame.

A similar descrepancy in the stretch parameters at extinction is obtained if K_{ex} is evaluated from Eqs. (11), (12), and (15) as

$$K_{ex} = \left(\frac{Le}{Le-1} \frac{c_p}{\lambda \rho} \right)_{x=0} \frac{M_{ref}^2}{e\beta} \quad (16)$$

Here the data in Table 2 were used, where the binary diffusion coefficient of the deficient species was that of oxygen for the rich flame and hydrogen for the lean flame, both with respect to the nitrogen. The average activation energy was calculated from Eq. (13) with the data from Figs. 2 and 3. The comparison of the last two lines in Table 2 shows that the calculated stretch parameter at extinction differs from the predicted value by 14 % for the rich flame and 38% for the lean flame.

Discussion and Conclusions

Two theoretical methods are compared to analyze flow divergence effects in premixed hydrogen flames. In the numerical calculation nonconstant transport properties and elementary kinetics were used, while the asymptotic treatment uses essentially constant transport properties and a one-step overall reaction. The intention of the comparison was to check the claim of the asymptotic theory to be able to predict the response of real flames to stretch. This claim implies that nonconstant properties and multi-step kinetics are second-order effects. In view of the differences in the formulation the results show a surprisingly good agreement: rich flames decelerate and are extinguished with increasing stretch while lean flames accelerate. The opposite is true for flame compression by flow contraction. Although this is not a stable physical situation, it was included to substantiate the comparison.

The quantitative data seem to differ in a systematic way from the asymptotic results: close to extinction, the flow rate through the flame is smaller for the rich flame and larger for the lean flame. The absolute values of the stretch parameter are larger in the rich case and smaller in the lean case. This can partly be explained by the influence of intermediates such as the radicals H, OH, and O. A recent analysis with a two-step mechanism (Seshadri and Peters, 1983) shows that additional positive terms must be added to $(Le-1)/Le$ to account for the differential diffusion of intermediates. Thus, the effective Lewis number would increase in the fuel-rich case and decrease in the fuel-lean case. In view of Eq. (16), this would shift the values of the stretch parameter predicted from asymptotic theory toward those obtained from the kinetic calculation.

A referee of the paper suggested further comments about the open downstream boundary conditions used in the numerical calculation. The choice of open boundary conditions is justified for one-dimensional flame calculations as long as downstream convection largely dominates over upstream diffusion. Then the downstream boundary condition will not influence the flame structure. In our case the velocity decreases with x for $K > 0$, such that there exists a stagnation point beyond which the flow is directed upstream. However, the flow field is only slowly diverging, since we are in the regime of weak stretch. This implies that the stagnation point appears many flame thicknesses downstream of the flame, virtually at $+\infty$.

It is concluded that asymptotic theory in the limit of weak stretch (implying Lewis numbers that differ from 1 by an $O(1)$ amount) is able to predict stretch effects in hydrogen flames. In this case differential diffusion is the dominating effect, while nonconstant properties and multi-step kinetics appear to be of smaller influence. In this respect hydrogen flames may even be considered as an exception and care must be taken not to generalize this conclusion to hydrocarbon flames.

Acknowledgments

One of the authors (J.W.) is grateful to Sandia National Laboratories for their hospitality and to Karl Winnacker-Stiftung for making possible this work. The financial support of Deutsche Forschungsgemeinschaft and Fonds der Chemischen Industrie is also acknowledged.

References

Buckmaster, J. (1979a) The quenching of a deflagration wave held in front of a bluff body. 17th Symposium (International) on Combustion, pp. 835-842. The Combustion Institute, Pittsburgh, Pa.

Buckmaster, J. (1979b) The quenching of two-dimensional premixed flames. Acta Astronaut. (6), 741-769.

Karlovitz, B.,Denniston, D.W., Jr., Knapschaefer, D.H., and Wells, F.E. (1953) Studies on turbulent flames. Fourth Symposium (International)on Combustion, pp. 613-620. The Combustion Institute, Pittsburgh, Pa.

Libby, P.A. and Williams, F.A. (1982) Structure of laminar flamelets in premixed turbulent flames. Combust. Flame (44), 287-303.

Libby, P.A. and Williams, F.A. (1983) Strained premixed laminar flames under nonadiabatic conditions. Combust. Sci. Technol. (31), 1-42.

Saitoh, T. (1974) Extinction analysis of premixed flames for counter flow and blunt body forward stagnation region flow. Int.J. Heat Mass Transfer (17), 1063-1077.

Sato, J. and Tsuji, H. (1983) Extinction of premixed flames in a stagnation flow considering general Lewis number. Combust.Sci. Technol. (33), 193-205.

Seshadri, K., and Peters, N. (1983) The influence of stretch on a premixed flame with two-step kinetic. Combust. Sci. Technol. (33), 35-63.

Sivashinsky, G.I. (1976) On a distorted flame front as a hydrodynamic discontinuity. Acta Astronaut.(3), 889-918.

Smith, H.W., Schmitz, R.A., and Ladd, R.G. (1971) Combustion of a premixed system in stagnation flow: I. Theoretical. Combust. Sci.Technol. (4), 131-142.

Warnatz, J. (1978) Calculation of the structure of laminar flat flames: 11. Flame velocity and structure of freely propagating hydrogen-oxygen and hydrogen-air flames. Ber. Bunsenges. Phys. Chem. (82), 643-649.

Warnatz, J. (1982) Discussion of test problem B, Numerical Methods in Laminar Flame Propagation, Notes on Numerical Fluid Mechanics, Vol. 6, Vieweg, Braunschwig/Wiesbaden, 49-64.

Slowly Varying Flames with Chain-Branching/Chain-Breaking Kinetics

G.S.S. Ludford*
Cornell University, Ithaca, New York
and
N. Peters†
*Rheinisch-Westfälische Technische Hochschule
Aachen, Federal Republic of Germany*

Abstract

The plane premixed flame with two-step kinetics is analyzed within the framework of slowly varying flames. The kinetics consist of chain-branching step, $A + X \to 2X$, with large activation energy and a chain-breaking step, $2X + M \to 2P + M$, with zero activation energy. An asymptotic analysis is performed for large activation energies and large and small values of the Damköhler number of the chain-breaking reaction. The evolution equation for the burning rate of the flame is given explicitly for small Damköhler numbers and in terms of a certain well-defined function for large Damköhler numbers. In both cases, the plane flame is unstable if the Lewis number of the reactant A exceeds a critical value, which is found to be equal to 1 as in the case of a one-step reaction.

Nomenclature

c_p = specific heat of mixture
$\mathcal{D}_1$ = Damköhler number of chain-branching reaction
$\mathcal{D}_2$ = Damköhler number of chain-breaking reaction
K = Lewis number of radical
L = Lewis number of reactant
M = dimensionless burning rate
M_r = reference mass flux

Presented at the 9th ICODERS, Poitiers, France, July 3-8, 1983. Copyright © by the American Institute of Aeronautics and Astronautics, Inc.,1984. All rights reserved.
*Professor of Applied Mathematics, Department of Theoretical and Applied Mechanics.
†Professor, Institut für Allgemeine Mechanik.

q_1 = proportion of heat released by chain-branching reaction
q_2 = proportion of heat released by chain-breaking reaction
T = temperature of mixture
v = speed of mixture
X = mass fraction of radical
Y = mass fraction of reactant
$\delta( ) = ( )_{+0} - ( )_{-0}$
θ = dimensionless activation energy
λ = thermal conductivity coefficient of mixture
ρ = density of mixture

Suffixes

f = value in fresh mixture
* = value at flame sheet

Introduction

The slow evolution of a plane premixed flame was first discussed by Buckmaster (1977), who, following Sivashinsky (1974), found that for a Lewis number $L < 1$ its speed is driven toward the steady value, while for $L > 1$ it is driven away. Thus, even if a flame survives its initial development, it is eventually unstable for $L > 1$. No conclusion of stability can be drawn for $L < 1$ since only plane disturbances are considered and the initial development could be unstable. In fact, only for a narrow band of Lewis numbers close to 1 is there stability. The results for slow-flame evolution can be interpreted as predicting the existence of the upper limit of this band.

The analyses cited were based on the assumption that the complex chemistry of a flame can be described by a one-step irreversible reaction. The effect of radicals is thereby neglected, an important omission that needs to be rectified. Accordingly, the object of this paper is to examine a more complex model, suggested by Zel'dovich (1948), containing a chain-branching step and a chain-breaking step, namely

$$A + X \to 2X, \quad 2X + M \to 2P + M \tag{1}$$

where A is the reactant, X the radical, P the product, and M a third body. The activation energy of the chain-branching step is very large, while that of the chain-breaking step is small and taken to be zero. Zel'dovich considered only the recombination reaction to be exothermic, but we shall generalize by allowing heat to be released in both reactions. The model was discussed by Linan (1971) and has been used by Seshadri and Peters (1983), which should be consulted for more details.

Seshadri and Peters were concerned with the effect of weak stretch. For the one-step reaction, the flame speed is increased by stretch for $L < 1$ and decreased for $L > 1$, as result that can be used to predict the existence of the lower limit of the stability band (Buckmaster and Ludford 1982, p. 205). The critical value of the appropriate Lewis number for the two-step reaction, as determined by Seshadri and Peters (1983), turns out to be identical to ours, so that once more a narrow band of Lewis numbers may be expected for stability.

The analysis follows that for the one-step reaction, but is found to have its limitation: it works here only when the recombination step is either sufficiently slow or sufficiently rapid. Intermediate recombination rates do not lead to slow variations; their treatment will be given in a separate paper.

Governing Equations

The unsteady one-dimensional flow of a mixture reacting according to the generalized Zel'dovich (1948) model satisfies

$$\frac{\partial \rho}{\partial t} + \frac{\partial (\rho v)}{\partial z} = 0 \tag{2}$$

$$\rho \frac{\partial X}{\partial t} + \rho v \frac{\partial X}{\partial z} - K^{-1} \frac{\partial (\lambda \partial X/c_p \partial z)}{\partial z} = \rho D_1 XY e^{-\theta/T} - \rho D_2 X^2 \tag{3}$$

$$\rho \frac{\partial Y}{\partial t} + \rho v \frac{\partial Y}{\partial z} - L^{-1} \frac{\partial (\lambda \partial Y/c_p \partial z)}{\partial z} = -\rho D_1 XY e^{-\theta/T} \tag{4}$$

$$\rho \frac{\partial T}{\partial t} + \rho v \frac{\partial T}{\partial z} - \frac{\partial (\lambda \partial T/c_p \partial z)}{\partial z} = \rho D_1 q_1 XY e^{-\theta/T} + \rho D_2 q_2 X^2 \tag{5}$$

These equations were developed by Seshadri and Peters (1983) in a different notation. The notation adopted here follows Buckmaster and Ludford (1982), but with the Damköhler numbers D replaced by ρD. The unit of temperature is formed from the total heat release in the two reactions, so that the proportionate heat releases satisfy

$$q_1 + q_2 = 1 \tag{6}$$

The continuity equation (2) enables us to introduce the particle function $\psi(z,t)$ defined by

$$\frac{\partial \psi}{\partial z} = \rho, \quad \frac{\partial \psi}{\partial t} = -\rho v \tag{7}$$

Replacing z as an independent variable by ψ gives

$$\frac{\partial X}{\partial t} - K^{-1} \frac{\partial^2 X}{\partial \psi^2} = D_1 XY e^{-\theta/T} - D_2 X^2 \tag{8}$$

$$\frac{\partial Y}{\partial t} - L^{-1} \frac{\partial^2 Y}{\partial \psi^2} = - D_1 XY e^{-\theta/T} \tag{9}$$

$$\frac{\partial T}{\partial t} - \frac{\partial^2 T}{\partial \psi^2} = D_1 q_1 XY e^{-\theta/T} + D_2 q_2 X^2 \tag{10}$$

when the realistic assumption

$$\rho \lambda / c_p = 1 \tag{11}$$

is made. Introducing the spatial coordinate

$$x = \psi - \psi(z_*, t) \tag{12}$$

measured from the flame sheet $z = z_*(t)$ shows that these equations may be written

$$\frac{\partial X}{\partial t} + M \frac{\partial X}{\partial x} - K^{-1} \frac{\partial^2 X}{\partial x^2} = D_1 XY e^{-\theta/T} - D_2 X^2 \tag{13}$$

$$\frac{\partial Y}{\partial t} + M \frac{\partial Y}{\partial x} - L^{-1} \frac{\partial^2 Y}{\partial x^2} = - D_1 XY e^{-\theta/T} \tag{14}$$

$$\frac{\partial T}{\partial t} + M \frac{\partial T}{\partial x} - \frac{\partial^2 T}{\partial x^2} = D_1 q_1 XY e^{-\theta/T} + D_2 q_2 X^2 \tag{15}$$

where

$$M = \rho_*(V + v_*) \quad \text{with} \quad V = - \dot{z}_*(t) \tag{16}$$

is the mass flux through the flame sheet, i.e., the burning rate. The equation of continuity in the variables x,t will not be needed. Note that the conditions in the fresh mixture require the solution to satisfy

$$X, Y, T \to 0, Y_f, T_f \quad \text{as} \quad x \to \infty \tag{17}$$

Perturbation of the Flame Temperature

Slow variations change the flame temperature by $O(\theta^{-1})$ and thereby the burning rate $O(1)$. To determine this effect for a one-step reaction, it is sufficient to calculate the difference in enthalpy between the fresh mixture and the burned mixture, since that difference manifests itself as the perturbation of the temperature at the flame. For the two-step reaction considered here, some of the difference remains unreleased (as heat), being locked up as chemical enthalpy in the radical [of which there is actually an $O(1)$ amount]. Therefore, we shall determine the flame temperature correct to $O(\theta^{-1})$ by integrating upstream from the known temperature of the burned mixture. To that end, we must consider the first two terms in the expansion

$$u = u_0(x,\tau) + \theta^{-1} u_1(x,\tau) + \ldots \tag{18}$$

of the generic variable u ahead of and behind the flame sheet, i.e., for $x \lessgtr 0$; here

$$\tau = t/\theta \tag{19}$$

is the slow time.

Linear combination of Eqs. (13-15), chosen so as to eliminate the reaction terms, shows that the enthalpy

$$H = T + q_2 X + Y \tag{20}$$

satisfies the equation

$$\theta^{-1} \frac{\partial H}{\partial \tau} + M \frac{\partial H}{\partial x} = \frac{\partial Q}{\partial x} \quad \text{with} \quad Q = \frac{\partial (T + q_2 K^{-1} X + L^{-1} Y)}{\partial x} \tag{21}$$

Integration from $-\infty$ to any positive x then gives

$$\theta^{-1} \frac{\partial}{\partial \tau} \int_{-\infty}^{x} (H - H_f) dx + M(T + q_2 X - H_f) = \frac{\partial}{\partial x} (T + q_2 K^{-1} X) \tag{22}$$

since

$$Y = 0 \quad \text{to all orders behind the flame sheet} \tag{23}$$

[In Eq. (22) $H_f = T_f + Y_f$ has been introduced to insure a convergent integral.]

$$M(T_0 + q_2 X_0 - T_f - Y_f) = \frac{\partial}{\partial x}(T_0 + q_2 K^{-1} X_0) \tag{24}$$

since $X_f = 0$. Once $X_0(x,\tau)$ has been determined (see following sections), this is an equation for T_0; its solution is

$$T_0 = q_2[(K^{-1}-1)M \int_x^\infty e^{M(x-\bar{x})} X_0(\bar{x},\tau)d\bar{x} - K^{-1} X_0(x,\tau)] + T_f + Y_f$$

$$\text{for } x > 0 \tag{25}$$

under the requirement of boundedness as $x \to \infty$. In particular, the final temperature of the mixture is $T_f + Y_f$, a result that follows from purely thermodynamic considerations.

To order θ^{-1}, Eq. (22) gives

$$\frac{\partial}{\partial \tau} \int_{-\infty}^x (H_0 - H_f)dx + M(T_1 + q_2 X_1) = \frac{\partial}{\partial x}(T_1 + q_2 K^{-1} X_1) \tag{26}$$

where

$$H_0 - H_f = M^{-1} Q_0 \quad \text{for all } x \tag{27}$$

according to the original Eq. (21). The integral in Eq. (26) is therefore

$$\int_{-\infty}^x (H_0 - H_f)dx = M^{-1}(T_0 + q_2 K^{-1} X_0 - T_f - L^{-1} Y_f) \quad \text{for } x > 0 \tag{28}$$

in view of the result (23); so the Eq. (26) yields

$$T_1 = \int_x^\infty e^{M(x-\bar{x})} f(\bar{x},\tau) d\bar{x} \tag{29}$$

under the requirement of boundedness, where

$$f(x,\tau) = q_2 K^{-1} \frac{\partial X_1}{\partial x} - M q_2 X_1 + (T_f + L^{-1} Y_f) \frac{dM^{-1}}{d\tau}$$
$$- \frac{\partial M^{-1}(T_0 + q_2 K^{-1} X_0)}{\partial \tau} \quad (30)$$

The formula (29) shows, incidentally, that there is a perturbation

$$M^{-1} f(\infty, \tau) = (1 - L^{-1}) Y_f M^{-3} \frac{dM}{dt} \quad (31)$$

in the temperature of the burned mixture due to the unsteadiness. Thus, once $X_1(x,\tau)$ has been determined in addition to $X_0(x,\tau)$, T_1 can be calculated and, in particular, its value is

$$T_{1*} = \int_0^\infty e^{-Mx} f(x,\tau) dx \quad (32)$$

at the flame sheet.

The procedure is to determine X_0, T_0, and X_1 behind the flame sheet successively and then to evaluate the perturbation of the flame temperature from the formula (32), which represents an integration upstream from the burned mixture. The determination of X_0 and X_1 requires their consideration for all x and hence a discussion of their jumps at the flame sheet (where the radical is produced).

The leading terms in the expansion of the reactant fraction ahead of the flame sheet are

$$Y_0 = Y_f (1 - e^{LMx}), \quad Y_1 = \left(\frac{Y_f \dot{M}}{M^2}\right)\left(\frac{x - LMx^2}{2}\right) e^{LMx} \quad (33)$$

These results follow from the equation

$$\theta^{-1} \frac{\partial Y}{\partial \tau} + M \frac{\partial Y}{\partial x} - L^{-1} \frac{\partial^2 Y}{\partial x^2} = 0 \quad \text{for } x < 0 \quad (34)$$

i.e.,

$$L^{-1} \frac{\partial^2 Y_0}{\partial x^2} - M \frac{\partial Y_0}{\partial x} = L^{-1} \frac{\partial^2 Y_1}{\partial x^2} - M \frac{\partial Y_1}{\partial x} - \frac{\partial Y_0}{\partial x} = 0 \quad (35)$$

and the boundary conditions

$$Y_0(-\infty,\tau) = Y_f, \quad Y_0(0,\tau) = Y_1(-\infty,\tau) = Y_1(0,\tau) = 0 \quad (36)$$

(The last of these conditions is a consequence of selecting the origin where Y vanishes to order θ^{-1}, a step that is allowed by the invariance of the problem under translations in x.)

In general, the above formulation leads to an inconsistency: the leading-order temperature (25) depends on M and therefore varies by $O(1)$ amounts, whereas the theory is based on the premise that changes in the flame temperature are $O(\theta^{-1})$. The origin of the difficulty lies in the fact that $O(1)$ changes in M will, in general, produce changes in the radical distribution (and hence the flame temperature) of the same order.

Exceptions will occur when the leading-order flame temperature, given by Eq. (25) with $x = 0$, is independent of M; it is not difficult to show that X_0 must then be constant behind the flame sheet, which can be ensured in two ways. When $\mathcal{D}_2$ is small (of order θ^{-1}), convection and diffusion lead to an unbounded X_0 unless it is constant. When $\mathcal{D}_2$ is large (of order θ^3), no radical at all escapes from the reaction zone because the recombination step is too rapid.

These two possibilities will now be treated in detail.

The Evolution Equation for M when $\mathcal{D}_2 = \theta^{-1}D$

The radical fractions X_0 and X_1 are determined in Appendix B; here just their values downstream from the flame sheet are needed, i.e.,

$$X_0 = Y_f, \quad X_1 = Y_f \dot{M}(L-K)/KLM^3 - Y_f^2 D(KMx + 3/2)KM^2$$

$$\text{for } x > 0 \quad (37)$$

From the formulas (25) and (32) we can now calculate

$$T_0 = T_f + q_1 Y_f \quad \text{for } x > 0 \quad (38)$$

$$T_{1*} = \frac{q_1 Y_f (1 - L^{-1})\dot{M}}{M^3} + \frac{q_2 Y_f^2 (2 + K^{-1})D}{2M^2} \quad (39)$$

The functions X_0, X_1, and $(T_0 - H_f)$ do not vanish as $x \to \infty$; expansions on a scale $O(\theta)$ in x are required to accomplish such decays (cf. Appendix B), which has no effect on the value of T_0 at $x = 0$ or of T_{1*}.

Equation (39) gives one relation between the flame temperature perturbation and the burning rate. Another, namely

$$T_{1*} = 2T_{0*}^2 \ln M \quad \text{with} \quad T_{0*} = T_f = q_1 Y_f \tag{40}$$

comes from an examination of the structure of the reaction zone presented in Appendix A. In deriving this formula, the respresentative mass flux (which has not yet been chosen) is taken to be

$$M_r = \sqrt{2LD_1}\, T_{0*}^2 \exp(-\theta/2T_{0*})/q_1 Y_f^{\frac{1}{2}} \theta \tag{41}$$

where $D_1 = M_r^2 \mathcal{D}_1$ is the pre-exponential factor in the reaction rate of the chain-branching step. The choice (41) is the steady burning rate when there is no recombination of the radicals: $\partial/\partial\tau = D = 0$ make $T_{1*} = 0$ and hence $M = 1$, see Eqs. (39) and (40).

The evolution equation

$$b\dot{M} = M^3 \ln M^2 - cM \tag{42}$$

where

$$b = \frac{q_1 Y_f (1 - L^{-1})}{T_{0*}^2}, \quad c = \frac{q_2 Y_f^2 (2 + K^{-1}) D}{2 T_{0*}^2} > 0 \tag{43}$$

is now obtained by eliminating T_{1*} between these two relations. It reduces to that for a single reactant when $q_1 = 1$ and $q_2 = 0$ [Buckmaster and Ludford (1982, p. 48) —the definition of the Damköhler numbers adopted here leads to a simpler coefficient b than theirs]. The reason is that the recombination step is then thermally neutral, the presence of the radical having no effect on the flame temperature. The steady-state burning rate $M_s(c)$ is given by

$$M_s^2 \ln M_s^2 = c \tag{44}$$

Since c is positive, there is a unique solution with

$$M_s > 1 \tag{45}$$

As expected, the heat release upon the recombination of the radical enhances the burning rate.

The following conclusion may be drawn from the evolution equation: when the burning rate is disturbed from its steady-state value, it is driven back toward M_s if b is negative and away from M_s if b is positive. Thus, for slowly varying disturbances, the steady state is

$$\text{stable or unstable accordingly as} \quad L \lessgtr 1 \qquad (46)$$

The Evolution Equation for M when $\mathcal{D}_2 = \theta^3 \mathcal{D}$

Equation (13) shows that the radical fraction vanishes to all orders outside the flame sheet; from

$$X_0 = X_1 = 0 \quad \text{for} \quad x \lessgtr 0 \qquad (47)$$

we find that

$$T_0 = T_f + Y_f \quad \text{for} \quad x > 0, \quad T_{1*} = Y_f(1-L^{-1})\dot{M}/M^3 \qquad (48)$$

See Eqs. (25), (30), and (32). Only the coefficient of $\dot{M}$ in T_{1*} is needed to conclude that the steady state is

$$\text{stable or unstable accordingly as} \quad L \lessgtr 1 \qquad (49)$$

but we shall take the analysis far enough to show how the second relation between the flame temperature perturbation and the burning rate--the analog of Eq. (40)--is determined.

To investigate the structure of the flame sheet, expansions of the form (A1) are introduced for T, X, and Y with

$$\tilde{T}_0 = T_{0*} \equiv T_f + Y_f, \quad \tilde{X}_0 = \tilde{Y}_0 = 0 \qquad (50)$$

The perturbations are found to satisfy the equations

$$K^{-1} \frac{\partial^2 \tilde{T}_1}{\partial \xi^2} - D\tilde{X}_1^2 = -L^{-1} \frac{\partial^2 \tilde{Y}_1}{\partial \xi^2}$$

$$= q_1^{-1} \frac{\partial^2 \tilde{T}_1}{\partial \xi^2} + Dq_2 q_1^{-1} \tilde{X}_1^2 = -\mathcal{D}\tilde{X}_1 \tilde{Y}_1 e^{\tilde{T}_1/T_{0*}^2} \qquad (51)$$

where

$$\tilde{D} = D_1 \theta^{-3} e^{-\theta/T_{0*}} \qquad (52)$$

and the boundary conditions

$$\tilde{X}_1 = \begin{cases} o(1) \\ o(1) \end{cases} \quad \tilde{Y}_1 = \begin{cases} -Y_f L M \xi + o(1) \\ o(1) \end{cases}$$

$$\tilde{T}_1 = \begin{cases} Y_f M \xi + T_{1*} + o(1) \\ T_{1*} + o(1) \end{cases} \quad \text{as } \xi \to \mp \infty \qquad (53)$$

which come from matching with the combustion field on either side of the flame sheet.

The integral

$$\tilde{T}_1 + q_2 K^{-1} \tilde{X}_1 + L^{-1} \tilde{Y}_1 = T_{1*} \qquad (54)$$

enables us to eliminate $\tilde{X}_1$ from the problem, which under the transformation

$$\tilde{T}_1 = T_{1*} - T_{0*}^2 u, \quad \tilde{Y}_1 = L T_{0*}^2 v, \quad \xi = T_{0*}^2 \eta / Y_f M \qquad (55)$$

then becomes (Joulin and Liñán 1980)

$$\frac{\partial^2 u}{\partial \eta^2} = q_1 r \Delta (u-v) v e^{-u} + \Delta (u-v)^2, \quad \frac{\partial^2 v}{\partial \eta^2} = r \Delta (u-v) v e^{-u} \qquad (56)$$

$$u, v = \begin{cases} -\eta + o(1) \\ o(1) \end{cases} \quad \text{as } \eta \to \mp \infty \qquad (57)$$

where

$$\Delta = \frac{K^2 T_{0*}^6 D}{q_2 Y_f^2 M^2}, \quad r = \frac{L D e^{T_{1*}/T_{0*}^2}}{KD} \qquad (58)$$

For given q_1 and Δ, this problem determines r; we may therefore write

$$r = F(\Delta; q_1) \tag{59}$$

and suppose the function F to be known. Thus,

$$T_{1*} = T_{0*}^2 f(M) \quad \text{with} \quad f = \ln[(KD/L\tilde{D})F(K^2 T_{0*}^6 D/q_2 Y_f^2 M^2; q_1)] \tag{60}$$

and

$$f(1) = 0 \tag{61}$$

if the representative mass flux in $\mathcal{D}$ is taken to be that in the steady state ($M = 1$, $T_{1*} = 0$), i.e.,

$$\tilde{\mathcal{D}} = (KD/L) F(K^2 T_{0*}^6 D/q_2 Y_f^2; q_1) \tag{62}$$

The evolution equation is now seem to be

$$b\dot{M} = M^3 f(M) \quad \text{with} \quad b = Y_f(1 - L^{-1})/T_{0*}^2 \tag{63}$$

The only property of f required to draw the conclusion (49) is

$$f(M) \lessgtr 0 \quad \text{accordingly as} \quad M \lessgtr 1 \tag{64}$$

which is ensured by F being a decreasing function of Δ. This property of F, among others, is the subject of a paper by Joulin et al. (1984).

Conclusion

The critical value is

$$L = 1 \quad \text{for both} \quad \mathcal{D}_2 = \theta^{-1} D \quad \text{and} \quad \theta^3 D \tag{65}$$

and stability changes to instability as L increases through 1. The same critical value arises when considering the effect of a weak stretch on the burning rate (Seshadri and Peters 1983); as L increases through 1, the stretch changes from a decelerating influence to an accelerating one. A simple argument shows that a plane flame is therefore unstable to multidimensional disturbances for $L < 1$ (Buckmaster and Ludford 1982, p. 205).

The conclusion to be drawn from these stability results is that only for $L - 1 = O(\theta^{-1})$, where the present analysis breaks down, can a plane flame be stable. This is the analog of what is found for a single-step reaction; examination of the θ^{-1} neighborhood of 1 then uncovers a stability band of Lewis numbers (Buckmaster and Ludford 1982, p. 202). Determination of the corresponding stability band for the two-step reaction considered here will be the subject of a later paper.

Appendix A
Structure of the Reaction Zone for $\mathcal{D}_2 = \theta^{-1}\mathcal{D}$

The jump conditions across the flame sheet are obtained by considering its structure, i.e., introducing the expansion

$$u = \tilde{u}_0(\tau) + \theta^{-1}\tilde{u}_1(\xi,\tau) + \theta^{-2}\tilde{u}_2(\xi,\tau) \quad \text{with} \quad \xi = \theta x \tag{A1}$$

of the generic variable u. Note that $\tilde{u}_0$ is independent of ξ and equals both $u_0(-0,\tau)$ and $u_0(+0,\tau)$, which are in fact independent of τ^0, implying that the jump in u across the flame sheet is zero to leading order, in particular the condition (B3b); otherwise a balance in the structure cannot be attained.

Equation (23) therefore gives

$$\tilde{Y}_0 = 0 \tag{A2}$$

in view of which the first-order structure equations become

$$K^{-1}\frac{\partial^2 \tilde{X}_1}{\partial \xi^2} = -L^{-1}\frac{\partial^2 \tilde{Y}_1}{\partial \xi^2} = q_1^{-1}\frac{\partial^2 \tilde{T}_1}{\partial \xi^2} = -\tilde{\mathcal{D}}\tilde{Y}_1 \exp\left(\frac{\tilde{T}_1}{T_{0*}^2}\right) \tag{A3}$$

where

$$\tilde{\mathcal{D}} = \mathcal{D}_1 \tilde{X}_0 \theta^{-2} e^{-\theta/T_{0*}}, \quad T_{0*} = \tilde{T}_0 \tag{A4}$$

These equations show that certain linear combinations of $\tilde{X}_1$, $\tilde{Y}_1$, and $\tilde{Y}_1$, $\tilde{T}_1$ are linear functions of ξ; the coefficients can be found in terms of functions outside the flame sheet by matching. Thus

$$K^{-1}\tilde{X}_1 + L^{-1}\tilde{Y}_1 = [K^{-1}\frac{\partial X_0}{\partial x}(\mp 0,\tau) + L^{-1}\frac{\partial Y_0}{\partial x}(\mp 0,\tau)]\xi$$

$$+ K^{-1}X_1(\mp 0,\tau) + L^{-1}Y_1(\mp 0,\tau) \quad (A5)$$

$$L^{-1}\tilde{Y}_1 + q_1^{-1}\tilde{T}_1 = [L^{-1}\frac{\partial Y_0}{\partial x}(\mp 0,\tau) + q_1^{-1}\frac{\partial T_0}{\partial x}(\mp 0,\tau)]\xi$$

$$+ L^{-1}Y_1(\mp 0,\tau) + q_1^{-1}T_1(\mp 0,\tau) \quad (A6)$$

Since

$$\frac{\partial Y_0}{\partial x}(-0,\tau) = -Y_f LM, \quad \frac{\partial Y_0}{\partial x}(+0,\tau) = 0, \quad Y_1(+\bar{0},\tau) = 0 \quad (A7)$$

according to the results (23) and (33), the first of these so-called Shvab-Zel'dovich integrals yields the jump conditions (B3c) and (B4b). (The second gives similar jumps on T_0 and T_1, but these are not needed.)

The remaining problem at this order is to solve

$$\frac{\partial^2 \tilde{Y}_1}{\partial \xi^2} = L\mathcal{D}\tilde{Y}_1 \exp(-\frac{\tilde{T}_1}{T_{0*}^2}) \quad (A8)$$

with

$$\tilde{T}_1 = -q_1 L^{-1}\tilde{Y}_1 + \frac{\partial T_0}{\partial x}(\mp 0,\tau)\xi + T_1(\mp 0,\tau) \quad (A9)$$

under the boundary conditions

$$\tilde{Y}_1 = -Y_f LM\xi + o(1) \quad \text{as} \quad \xi \to -\infty, \quad \tilde{Y}_1 = o(1) \quad \text{as} \quad \xi \to +\infty \quad (A10)$$

obtained by matching with the expansions (23) and (33). For $\mathcal{D}_2 = \theta^{-1}\mathcal{D}$ we find

$$\frac{\partial T_0}{\partial x}(+0,\tau) = 0 \quad (A11)$$

according to the result (38), so that the problem reduces to that for a plane flame with a single reactant (see

Buckmaster and Ludford 1982, p. 40). As there, a single integration using the boundary conditions show that

$$M^2 = \frac{2L\tilde{D}T_{0*}^4}{q_1^2 Y_f^2} \exp\left(\frac{T_{1*}}{T_{0*}^2}\right) \quad \text{with} \quad T_{1*} = T_1(+0,\tau) \quad \text{(A12)}$$

The choice (41) for the respesentative mass flux M_r is now seen to correspond to $T_{1*} = 0$, and our result becomes

$$M = \exp(T_{1*}/2T_{0*}^2) \quad \text{(A13)}$$

Although the second-order structure must be considered in order to find the jump condition (B4c) on $\partial X_1/\partial x$, only the equation

$$K^{-1}\frac{\partial^2 \tilde{X}_2}{\partial \xi^2} - M\frac{\partial \tilde{X}_1}{\partial \xi} = -L^{-1}\frac{\partial^2 \tilde{Y}_2}{\partial \xi^2} + M\frac{\partial \tilde{Y}_1}{\partial \xi} \quad \text{(A14)}$$

need be examined. The integral

$$K^{-1}\frac{\partial \tilde{X}_2}{\partial \xi} + L^{-1}\frac{\partial \tilde{Y}_2}{\partial \xi} - M(\tilde{X}_1 + \tilde{Y}_1) = \text{const} \quad \text{(A15)}$$

shows, by matching on the two sides of the reaction zone, that (in particular)

$$K^{-1}\delta\left(\frac{\partial X_1}{\partial x}\right) + L^{-1}\delta\left(\frac{\partial Y_1}{\partial x}\right) = M[\delta(X_1) + \delta(Y_1)] \quad \text{(A16)}$$

The conclusion (B4c) now follows from the results (33b) and (B4b).

Appendix B
Determination of X_0 and X_1 for $\mathcal{D}_2 = \theta^{-1}\mathcal{D}$

Seshadri and Peters (1983) have given formulas for X_0 but we shall, for completeness and because of the difference in notation, repeat their analysis to that extent. Our determination of X_1 is not the same as theirs, even when their analysis is corrected (Tam and Ludford, 1984).

The radical fraction satisfies the equation

$$\theta^{-1}\frac{\partial X}{\partial \tau} + M\frac{\partial X}{\partial x} - K^{-1}\frac{\partial^2 X}{\partial x^2} = -\theta^{-1}DX^2 \quad \text{for } x \lessgtr 0 \quad (B1)$$

so that

$$K^{-1}\frac{\partial^2 X_0}{\partial x^2} - M\frac{\partial X_0}{\partial x} = K^{-1}\frac{\partial^2 X_1}{\partial x^2} - M\frac{\partial X_1}{\partial x} - \frac{\partial X_0}{\partial \tau} - DX_0^2 = 0 \quad (B2)$$

The boundary conditions are

$$X_0(\pm\infty,\tau) = 0, \quad \delta(X_0) = 0, \quad \delta(\frac{\partial X_0}{\partial x}) = -Y_f KM \quad (B3)$$

and

$$X_1(\pm\infty,\tau) = 0, \quad \delta(X_1) = 0, \quad \delta(\frac{\partial X_1}{\partial x}) = Y_f \frac{K\dot{M}}{LM^2} \quad (B4)$$

If we set

$$X_0 = Y_f F(y) \quad \text{with} \quad y = KMx \quad (B5)$$

then

$$F = \begin{cases} e^y \\ 1 \end{cases} \quad \text{for } y \lessgtr 0 \quad (B6)$$

Similarly,

$$X_1 = (Y_f \dot{M}/KM^3)G(y) + (Y_f^2/KM^2)H(y) \quad (B7)$$

leads to

$$G = \begin{cases} (y^2/2 - y + 1 - K/L)e^y \\ 1 - K/L \end{cases} \quad H = \begin{cases} De^{2y}/2 - 2De^y \\ -Dy - 3D/2 \end{cases} \quad \text{for } y \lessgtr 0 \quad (B8)$$

None of the functions F, G, or H vanishes as $y \to +\infty$, but this just means that expansions with the variable y/θ are required to describe their eventual decay. These expansions play no role in our analysis, so we omit them.

Acknowledgment

This work was supported by the U.S. Army Research Office and by the Alexander von Humboldt Foundation (through a Senior U.S. Scientist Award).

We are indebted to Richard Tam for uncovering an error in an earlier version of this paper.

References

Buckmaster, J. (1977) Slowly varying laminar flames. Combust. Flame 28, 225-239.

Buckmaster, J. D. and Ludford, G. S. S. (1982) Theory of Laminar Flames, University Press, Cambridge.

Joulin, G. and Liñán, A. (1980) Private communication.

Joulin, G., Liñán, A., Lainé, C., Ludford, G. S. S., and Peters, N. (1984) Plane flames with chain-branching/chain-breaking kinetics. Submitted for publication.

Liñán, A. (1971) A theoretical analysis of premixed flame propagation with an isothermal chain reaction. Instituto Nacional de Tecnica Aerospatial "Esteban Terradas" (Madrid), USAFOSR Contract EOOAR68-0031, Tech. Rept. 1.

Seshadri, K. and Peters, N. (1983) The influence of stretch on a premixed flame with two step kinetics. Combust. Sci. Technol. 33, 35-64.

Sivashinsky, G. I. (1974) On a converging spherical flame front. Int. J. Heat Mass Transfer 17, 1499-1506.

Tam, R. and Ludford, G. S. S. (1984) Comment on the stretch-resistant flames of Seshadri and Peters. Combust. Sci. Technol. (in press).

Zel'dovich, Y. B. (1948) Teorii rasprostranenia plameni. Zh. Fiz. Khim. 22, 27-49. English translation (1951): Theory of flame propagation, NACA TM 1282.

Effect of Dissociation on the Near-Stoichiometric Burning of Non-Dilute Mixtures

G.S.S. Ludford*
Cornell University, Ithaca, New York
and
A. K. Sen†
Purdue School of Sciences, Indianapolis, Indiana

Abstract

The effect of dissociation on the near-stoichiometric behavior of a combustible mixture has been examined analytically on the basis of the multicomponent diffusion laws and a one-step reversible reaction. The results can be expressed in terms of "effective" Lewis numbers of a mixture whose reactants are dilute in an inert; the effect of dissociation on such dilute mixtures has already been established by the authors. The main conclusion is that, for all practical purposes, dissociation plays exactly the same (minor) role in nondilute mixtures as it does in dilute mixtures, a role that may be summarized as follows. For all parameter values, dissociation can produce a fuel-rich maximum in the flame temperature, but has a subsidiary effect on the fuel-rich maximum of the burning rate established by mass diffusion.

Nomenclature

a, b	= parameters in eq. (9)
a_1, a_2, a_3, a_4	= constants (36, 37)
c	= 1 - a
c_p	= specific heat at constant pressure
d	= parameter (8b)
D	= Damköhler number

Presented at the 9th ICODERS, Poitiers, France, July 3-8, 1983. Copyright © American Institute of Aeronautics and Astronautics, Inc., 1984. All rights reserved.
*Professor of Applied Mathematics, Department of Theoretical and Applied Mechanics.
†Assistant Professor of Mathematical Sciences, Department of Mathematical Sciences.

DISSOCIATION IN NEAR-STOICHIOMETRIC BURNING

D_{ij}	= binary diffusion coefficient of species i and j
k	= equilibrium constant in the reaction term (2)
$\hat{k}$	= $\theta^2 k$
K, L	= effective Lewis numbers (28) and (29)
L_{ij}	= Lewis number $\lambda/\rho D_{ij} c_p$
M	= burning rate ρv
N	= mass fraction of inert
T	= dimensionless temperature
T_*	= flame-temperature perturbation (6)
v	= mass-average velocity of mixture
x	= distance variable in units of $\lambda/c_p M$
x_1, y_1	= coefficient functions in the expansions (23)
y_*	= measure (5) of closeness to stoichiometry
X, Y, Z	= mass fractions of oxidant, fuel, and product
$\hat{X}$	= θX
ξ, η, ζ	= mass-flux fractions of oxidant, fuel, and product
θ	= activation temperature in Ω
λ	= thermal conductivity coefficient
Λ	= DM^{-2}
$\hat{\Lambda}$	= parameter (30)
ρ	= density of mixture
τ	= $\theta(T - T_\infty)$
Ω	= reaction term (2)

Subscripts

i, j	= O, F, P, N
O, F, P, N	= oxidant, fuel, product, and inert
o	= leading term in θ^{-1}-expansion
$\pm\infty$	= value at $x = \pm\infty$

Introduction

In a series of papers culminating in one at the Eighteenth Combustion Symposium (Sen and Ludford 1981a, hereafter called S&L 1), we have shown that diffusion of the reactants is capable by itself of producing the observed fuel-rich maximum in the burning rate of combustible mixtures. S&L 1 removed the restriction to dilute reactants in the earlier work, the first time that the full multicomponent diffusion laws, previously thought to be intractable, had been subjected to mathematical analysis. This series justifies the assertion made in its very first paper (Sen and Ludford 1979), that dissociation is not the primary cause of the fuel-rich maximum. A similar conclusion can be drawn from the later paper of Clarke and McIntosh (1980).

Our original neglect of dissociation drew criticism, misguidely (in our opinion) since the object was to isolate the effect of mass diffusion. At the very least the results suggest that an explanation based solely on kinetics, of whatever kind and however complex, is incomplete: it is difficult to see how the introduction of more complex kinetics would eleminate the existing diffusion effect. To strengthen this suggestion we undertook an investigation (Sen and Ludford 1981b, hereafter called S&L 2) of the effect of dissociation, the mechanism invoked by the simpler kinetic explanations (even though it is absent in cool flames). However, S&L 2 was itself criticized for being based on a dilute-mixture theory, the implication being that the sub-sidiary nature of dissociation was then a creature of the (generalized) Fick's law used. Accordingly, the object of the present paper is to remove the restriction to dilute mixtures and, in doing so, we shall be involved once more with the supposedly intractable multicomponent diffusion laws.

It is important to distinguish between the flame temperature, which follows from purely thermostatic considerations, and the burning rate (or flame speed), which is the result of dynamic processes. Most explanations are concerned with the flame temperature, on the supposition that its maximization ensures that the highest burning rate is achieved. S&L 2 shows that, for all parameter values, dissociation does indeed account for the fuel-rich maximum of the flame temperature, when such occurs, but that it has a subsidiary effect on the burning rate. The results were generalized and completed in a paper at the Nineteenth Combustion Symposium (Sen and Ludford 1983). Our main conclusion here is that, for all practical purposes, dissociation plays the same subsidiary role in nondilute mixtures.

In view of these results, any explanation of the fuel-rich maximum of the burning rate that does not involve mass diffusion should be treated with skepticism. The precise location of the maximum can, however, be expected to depend on the chemical kinetics.

Another distinction that is sometimes blurred concerns the conditions under which experiments are generally performed. There are two cases, and the one that is being considered should always be specified. As the fuel and oxidant concentrations in the fresh mixture are varied, either the inert concentrations is held fixed (case I) or the inert-to-oxidant ratio is held fixed (case II, as for air).

DISSOCIATION IN NEAR-STOICHIOMETRIC BURNING

The reaction model used is the same as in S&L 2, namely

$$F + O \rightleftarrows 2P$$

but now the mixture is not necessarily dilute in the reactants and their product. (The dilute-mixture assumption reduces the multicomponent diffusion laws to a separate Fick's law for each species except the inert.) As in S&L 1, the same molecular mass and the same specific heat are taken for all species. Only the near-stoichiometric behavior of the mixture is investigated (as in S&L 2), since that is where the phenomena of interest occur. (Dissociation, being weak, clearly has just a perturbation effect away from stoichiometry.)

The results are obtained by asumptotic analysis in the limit of large activation energy, the method used in all our previous investigations.

Governing Equations

The derivation of the equations governing the combustion of a nondilute mixture given in S&L 1 will not be repeated here. We merely note that the reaction term

$$\Omega = \Lambda XY \exp(-\theta/T) \tag{1}$$

is replaced by

$$\Omega = \Lambda (XY - kZ) \exp(-\theta/T) \tag{2}$$

when dissociation is taken into account. (Justification for treating dissociation so simply is given in S&L 2.) Note that the equilibrium constant k is a (given) function of T.

The state behind the flame, i.e., at $x = +\infty$, is related to that in the fresh mixture at $x = -\infty$ by the equations

$$T_\infty = T_{-\infty} + Z_\infty, \quad Y_\infty = Y_{-\infty} - X_{-\infty} + X_\infty \tag{3}$$

$$Z_\infty = 2(X_{-\infty} - X_\infty), \quad X_\infty Y_\infty = k(T_\infty) Z_\infty \tag{4}$$

which come from purely thermostatic considerations (see S&L 2). In general, the final temperature T_∞ can only be found by iteration from the four equations (3) and (4) for the unknowns T_∞, X_∞, Y_∞, and Z_∞. However, when the dissociation is weak, as is always the case in practice, an asymptotic approximation to T_∞ can be given. In particular, for a near-stoichiometric mixture that is fuel rich, i.e., for

$$Y_{-\infty} - X_{-\infty} = \theta^{-1} y_* \quad \text{with} \quad y_* > 0 \qquad (5)$$

S&L 2 found

$$T_\infty = T_{\infty_0} - \theta^{-1} T_* \qquad (6)$$

with

$$T_{\infty_0} = T_{-\infty} + 2c/(b+2), \quad T_* = 2\left(\hat{X}_\infty + y_*/(b+2)\right) \qquad (7)$$

where $\hat{X}_\infty$ is to be determined from the quadratic equation

$$\hat{X}_\infty (y_* + \hat{X}_\infty) = d \quad \text{with} \quad d = 2ck(\hat{T}_{\infty_0})/(b+2) \qquad (8)$$

Here $\theta^{-1}\hat{X}_\infty$ is the small amount of oxidant produced by the weak dissociation, represented by the small equilibrium constant $\theta^{-2} k(T\infty)$, and b, c are constants describing the composition of the fresh mixture:

$$N_{-\infty} = a + bX_{-\infty} \quad \text{with} \quad a = 1 - c. \qquad (9)$$

If $N_{-\infty}$ is kept fixed as the ratio of fuel to oxidant is varied (case I), then $b = 0$. If the ration $X_{-\infty}/N_{-\infty}$ is kept fixed (case II, as for air), then $a = 0$.

These results are based on purely thermodynamic principles and, therefore, hold whether the mixture is dilute or not (as here). The nature of the diffusion process enters only when details of the transition from the fresh state to the burnt state are required, a question to which we now turn our attention.

When all the species in the mixture may be considered to have equal molecular masses and equal specific heats, the dimensionless form of the relevant governing equations (in one space dimension) is

$$2d\xi/dx = 2d\eta/dx = (d^2T/dx^2 - dT/dx) = -\Omega \qquad (10)$$

$$dX/dx = L_{ON}(X - \xi) + (L_{OF} - L_{ON})(X\eta - Y\xi)$$

$$+ (L_{OP} - L_{ON})(X\zeta - Z\xi) \qquad (11)$$

$$dY/dx = L_{FN}(Y - \eta) + (L_{OF} - L_{FN})(Y\xi - X\eta)$$

$$+ (L_{FP} - F_{FN})(Y\zeta - Z\eta) \qquad (12)$$

Here ξ, η, and ζ are the mass-flux fractions of the oxidant, fuel and product, respectively. Equations (10) are statements of mass (oxidant and fuel) and enthalpy balances; Eqs. (11) and (12) are multicomponent diffusion laws (Williams 1965, p. 2) for oxidant and fuel, in a form derived in the Appendix of S&L 1. Once again the diffusion of product and inert need not be considered since they play no role in the sequel (even though ζ and Z appear here). The system (10-12) is to be solved under the boundary conditions

$$X = \xi = X_{-\infty}, \quad Y = \eta = Y_{-\infty}, \quad T = T_{-\infty} \quad \text{at} \quad x = -\infty \qquad (13)$$

Near-Stoichiometric Flame Temperature and Burning Rate

From the determination (6) of the flame temperature T_∞, S&L 2 concludes that there is a maximum at

$$Y_* = b\left(2\hat{c}k(T_{\infty_0})/(b + 1)(b + 2)\right)^{1/2} \qquad (14)$$

For $b = 0$ (case I), the maximum flame temperature is exactly at stoichiometry; but, for $b \neq 0$ (case II), it lies on the fuel-rich side. Analogous formulas show that there is no maximum on the fuel-lean side, in accordance with experimental observations of Khitrin (1962, p. 166), for example.

We turn now to the determination of the burning rate, for which it is convenient to reformulate the system (10-12). In view of the boundary conditions (13), Eqs. (10a and b) immediately give

$$\eta = \xi + Y_{-\infty} - X_{-\infty}, \quad dT/dx = T - T_{-\infty} + 2(\xi - X_{-\infty}) \qquad (15)$$

where, according to the thermostatic results (3a) and (4a),

$$T_{-\infty} + 2X_{-\infty} = T_\infty + 2X_\infty \qquad (16)$$

If x is now eliminated by taking T as the new independent variable, the remaining equations of the system may be written

$$\frac{dX}{dT} = \frac{L_{ON}(X-\xi) + (L_{OF}-L_{ON})(X\eta-Y\xi) + (L_{OP}-L_{ON})(X\zeta-Z\xi)}{T - T_\infty + 2(\xi - X_\infty)} \quad (17)$$

$$\frac{dY}{dT} = \frac{L_{FN}(Y-\eta) + (L_{OF}-L_{FN})(Y\xi-X\eta) + (L_{FP}-L_{FN})(Y\zeta-Z\eta)}{T - T_\infty + 2(\xi - X_\infty)} \quad (18)$$

$$\frac{d\xi}{dT} = -\frac{\Lambda}{2} \frac{(XY - kZ)\exp(-\theta/T)}{T - T_\infty + 2(\xi - X_\infty)} \quad (19)$$

In the absence of dissociation, i.e., for $k = X_\infty = 0$, these reduce to Eqs. (2.10) in S&L 1, as they should. For a dilute mixture, i.e., X, Y, Z, ξ, η, ζ small, the quadratic terms in Eqs. (17) and (18) may be neglected, to yield all but one of the equations (5.6) in S&L 2 with

$$K = L_{ON}, \quad L = L_{FN} \quad (20)$$

(The remaining equation, for Z, is unnecessary and $\zeta = 2(X_{-\infty} - \xi)$ is used in place of ξ.)

The solution of the system (17-19) is expected to satisfy

$$X, Y, \xi = \begin{cases} X_{-\infty}, Y_{-\infty}, X_{-\infty} & \text{at } T = T_{-\infty} \\ X_\infty, Y_\infty, X_\infty & \text{at } T = T_\infty \end{cases} \quad (21)$$

but, for any finite θ, is unable to do so (cold-boundary difficulty). The reaction term Ω must be modified so as to vanish near $T = T_{-\infty}$ (Williams 1965, p. 109). In the limit $\theta \to \infty$ the difficulty disappears (Buckmaster and Ludford 1982, p. 28); no modification of the reaction term is necessary and we may deal with the asymptotic problem directly. To that end, two parts of the T interval must be considered: an outer region where $T - T_\infty = 0(1)$ and an inner region where $T - T_\infty = 0(\theta^{-1})$. The reason is that, for large activation energy, virtually all the reaction takes place at the temperature of the burnt mixture.

In the outer region the right side of Eq. (19) is exponentially small in the limit, so that

$$\xi = X_{-\infty} \quad (22)$$

to all orders in θ^{-1}. (Since η and ζ are then also constant, Eqs. (11) and (12) and the corresponding equation for dZ/dx show that X, Y, and Z are linear combinations of exponentials whose arguments are proportional to x. We shall not write down these complicated formulas.)

So far there has been no assumption of near stoichiometry. That comes with writing

$$X = \theta^{-1} x_1(\tau) + \ldots, \quad Y = \theta^{-1} y_1(\tau) + \ldots \quad (23)$$

in the inner region, where

$$\tau = \theta(T - T_\infty) \quad (24)$$

is the scaled variable there. When these, together with

$$Z = 2X_{-\infty} + \ldots, \quad \xi = \xi_0(\tau) + \ldots \quad (25)$$

are substituted into the system (17-19), we find

$$dx_1/d\tau = -K/2, \quad dy_1/d\tau = -L/2 \quad (26)$$

$$d\xi_0/d\tau = -(\hat{\Lambda}_o/4\xi_0)\left(x_1 y_1 - 2\hat{k}(T_\infty) X_{-\infty}\right) \exp(\tau/T_\infty^2) \quad (27)$$

where

$$K = L_{ON} + 2X_{-\infty}(L_{OP} - L_{ON}) > 0 \quad (28)$$

$$L = L_{FN} + 2X_{-\infty}(L_{FP} - L_{FN}) > 0 \quad (29)$$

and $\hat{\Lambda}_o$ is the leading term in the expansion of

$$\hat{\Lambda} = \Lambda \theta^{-3} \exp(-\theta/T_\infty) \quad (30)$$

Here

$$X_{-\infty} = c/(b + 2) \quad (31)$$

its stoichiometric value, to sufficient accuracy; and the initial conditions

$$x_1(0) = \hat{X}_\infty, \quad y_1(0) = y_* + \hat{X}_\infty, \quad \xi_0(0) = 0 \quad (32)$$

must be satisfied. In addition, matching with the outer solutions requires

$$\xi_0 \to X_{-\infty} \quad \text{as} \quad \tau \to -\infty \quad (33)$$

as can be seen from the result (22). For every value of Λ_o, the third-order system of differential equations (26) and (27) has an unique solution satisfying the three initial conditions (32); but the matching condition (33) is only satisfied for one particular value of Λ_o. The burning rate $M = (D/\Lambda)^{1/2}$ is thereby determined to leading order.

The problem (26, 27, 32, and 33) is seen to be identical to the problem (5.9, 10, 12, 14, and 16) considered in S&L 2 when the equation and initial condition for z_1 are discarded as unnecessary and ζ_0 is replaced by $2(X_{-\infty} - \xi_0)$. We conclude that, near stoichiometry, a nondilute mixture behaves like a dilute mixture with Lewis numbers (28, 29). As $X_{-\infty} \to 0$, these effective Lewis numbers tend to L_{ON}, L_{FN} as is appropriate to a dilute mixture. This correspondence was one of the results found in S&L 1; the present analysis show that it also holds when weak dissociation is taken into account.

It follows that the burning rate is given, to leading order, by

$$M^2 = \frac{DT_{\infty_0}^4 e^{-\theta/T_{\infty_0}}}{4\theta^3 c^2/(b+2)^2} (KLT_{\infty_0}^2 + L\hat{X}_\infty + Kd/\hat{X}_\infty) \exp(-T_*/T_{\infty_0}^2) \quad (34)$$

where T_* and d have the values (7b) and (8b). Since $\hat{X}_\infty$ can be expressed in terms of y_* from the quadratic equation (8a), this formula determines the burning rate for any small departure from stoichiometry measured by y_* on the θ^{-1}-scale. It is, however, more convenient to keep X_∞, representing the small amount of oxidant produced by the dissociation, as a measure of nearness to stoichiometry. Then, as shown in S&L 2, the stationary values of the burning rate are determined by the roots of the quartic equation

$$L\hat{X}_\infty^4 + a_1 \hat{X}_\infty^3 + a_2 k \hat{X}_\infty^2 + a_3 k \hat{X}_\infty - a_4 k^2 = 0 \quad (35)$$

where

$$a_1 = L\left(K - \frac{b+2}{2(b+1)}\right)T_{\infty_0}^2, \quad a_2 = \frac{2c}{b+2}\left(K - \frac{L}{b+1}\right) \quad (36)$$

$$a_3 = \frac{2c}{(b+1)(b+2)} K\left(\frac{b+2}{2} - L\right)T_{\infty_0}^2, \quad a_4 = \frac{4c^2}{(b+1)(b+2)^2} K \quad (37)$$

Similar formulas hold when the mixture is fuel lean (see S&L 2).

Conclusion

We have seen that, in case I, dissociation has no effect on the location of the maximum temperature but that, in case II, it shifts it onto the fuel-rich side of stoichiometry. Details of the effect of dissociation on the location of the burning rate maximum are given in S&L 2; here we shall repeat the main result for

$$K > (b + 2)/2(b + 1) \, , \quad L < 1 + (b/2) \quad (38)$$

conditions that are invariably satisfied in practice. In that connection we note that, when

$$L_{OP} = L_{ON} \quad \text{and} \quad L_{FP} = L_{FN} \quad (39)$$

(conditions that are closely approximated in most mixtures of practical interest), the effective Lewis numbers are just

$$K = L_{ON} \quad \text{and} \quad L = L_{FN} \quad (40)$$

the values for a dilute mixture.

The inequalities (38) ensure that, in the absence of dissociation, there is just one maximum of the burning rate, located at

$$y_* = (1 + b/2 - L) \, T_\infty^2 \quad (41)$$

on the fuel-rich side of stoichiometry. As $\hat{k}$ increases from zero in case I, this maximum moves towards or further away from stoichiometry accordingly as $K + L$ is less or greater than 2, approaching the position

$$y_* = (K - L) \, T_\infty^2 / (K + L) \quad (42)$$

asymptotically as $\hat{k} \to \infty$. For $K + L = 2$, a condition very nearly satisfied by a large number of fuel-oxidant mixtures, the dissociation has absolutely no effect on the burning rate maximum. In case II, the maximum continually moves further away from stoichiometry as k increases. For either case, no other maximum appears.

In short, dissociation merely shifts the single fuel-rich maximum that exists in its absence, without creating another maximum.

Acknowledgment

This work was supported by the U.S. Army Research Office and by the Alexander von Humboldt Foundation (through a Senior US-Scientist Award).

References

Buckmaster, J. D. and Ludford, G. S. S. (1982) Theory of laminar flames. 266 pp. University Press, Cambridge, England.

Clarke, J. F. and McIntosh, A. C. (1980) The influence of a flameholder on a plane flame, including its static stability. Proc. R. Soc. A372, pp. 367-392.

Khitrin, L. N. (1962) The physics of combustion and explosion. 488 pp. U. S. Department of Commerce, Washington, D.C.

Sen, A. K. and Ludford, G. S. S. (1979) The near-stoichiometric behavior of combustible mixtures, Part I: Diffusion of the reactants. Combust. Sci. Technol. 21, pp. 15-23.

Sen, A. K. and Ludford, G. S. S. (1981a) Effects of mass diffusion on the burning rate of non-dilute mixtures. 18th Symposium (International) on Combustion, pp. 417-424. The Combustion Institute, Pittsburgh, Pa.

Sen, A. K. and Ludford, G. S. S. (1981b) The near-stoichiometric behavior of combustible mixtures. Part II: Dissociation of the products. Combust. Sci. Technol. 26, pp. 183-191.

Sen, A. K. and Ludford, G. S. S. (1983) Maximum flame temperature and burning rate of combustible mixtures. 19th Symposium (International) on Combustion, pp. 267-274. The Combustion Institute, Pittsburgh, Pa.

Williams, F. A. (1965) Combustion theory. 477 pp. Addison-Wesley Publishing Company, Reading, Mass.

The Feedback of a Flame Front on Turbulent Flows

G. Searby,* F. Sabathier,† J. Monreal,† P. Clavin,‡ and L. Boyer‡
Université de Provence, Centre St. Jérôme
Marseille, France

Abstract

The principal mechanisms governing the stability (or instability) of a premixed flame in a turbulent gas flow are reviewed. The hydrodynamic discontinuity associated with the flame front is shown to give rise to a feedback effect on the fluctuations of the upstream gas flow. The response of a stable flame to a turbulent flow is calculated as a function of spatial and temporal frequencies. Reported here is an experimental study of the response of a stable planar flame, 7 cm in diameter, to a weakly turbulent gas flow. The gas velocity field is determined by Laser Doppler Anemometry and the dynamics of the flame front by Laser Tomography. It is shown that the upstream gas flow is substantially modified by the presence of the flame. Some Fourier components of the turbulence are almost completely blocked by a negative hydrodynamic feedback while others are amplified. Agreement with the theoretical model is within experimental uncertainties.

Introduction

This paper reports a theoretical and experimental study of the interaction of a flame front and a weakly turbulent premixed gas flow. The domain of interest is

---

Presented at the 9th ICODERS, Poitiers, France, July 3-8, 1983. Copyright © American Institute of Aeronautics and Astronautics, Inc., 1984. All rights reserved.
  *Chargé de Recherche au CNRS Laboratoire de Dynamique et Thermophysique des fluides.
  †Graduate student, Laboratoire de Dynamique et Thermophysique des Fluides.
  ‡Professor, Laboratoire de Dynamique et Thermophysique des Fluides.

limited to the case where the free flame front is intrinsically stable. As will be shown below, stable flames exist only for downwards propagating fronts at low propagation velocities. The intensity of turbulence was voluntarily limited so as to permit valid comparison with the linearized analysis presented below. For that reason, we are concerned with premixed flames of propane-air diluted with nitrogen to produce flame speeds in the range 7 cm/s (extinction) to around 12 cm/s (the instability threshold). The study of the instability threshold itself is presented in a second paper (Quinard et al. 1983).

Theoretical Results

The main theoretical results that are pertinent to this work will be recalled. The flame front can be considered, as a first approximation, to be a simple surface of discontinuity separating the fresh and burnt gases. This is justified because the spatial scale characterizing the turbulence and the wrinkling of the flame ($\lambda \sim 1$ cm) is very large compared to the thickness of the flame (d ~ 0.04 cm). The exact chemical kinetics within the reaction zone will have, at worst, only a slight quantitative effect on the actual flame speed for a given mixture.

The average flame front is assumed to lie in the horizontal y-z plane (see Fig. 1) and the average gas flow is adjusted so that the average flame front is stationary in the laboratory frame. If $x = \alpha(y,z,t)$ is the surface which describes the wrinkled flame front at time t ($\alpha = 0$ for an unperturbed front), then Clavin and Williams (1982) have given an expression for the instantaneous local velocity of the flame front in the laboratory frame, taking into account the transverse thermal and molecular diffusion effects produced by curvature of the flame front

$$\dot{\alpha}(y,z,t) = u_T(x=\alpha) + K u_L d \left\{ \nabla^2 \alpha - \frac{1}{u_L} \frac{\partial u_T}{\partial x} \bigg|_{x=\alpha} \right\} \quad (1)$$

where $u_T(x=\alpha)$ is the x component of the fluctuation of the total upstream gas flow measured at the flame front, K is a dimensionless constant (a function of the equivalence ratio and the thermal and molecular diffusion coefficients) given explicitly by Clavin and Garcia (1983), u_L is the propagation speed of an unperturbed laminar flame, and d is the flame thickness. $K u_L d$ measures the sensitivity

FEEDBACK OF A FLAME FRONT ON TURBULENT FLOWS   105

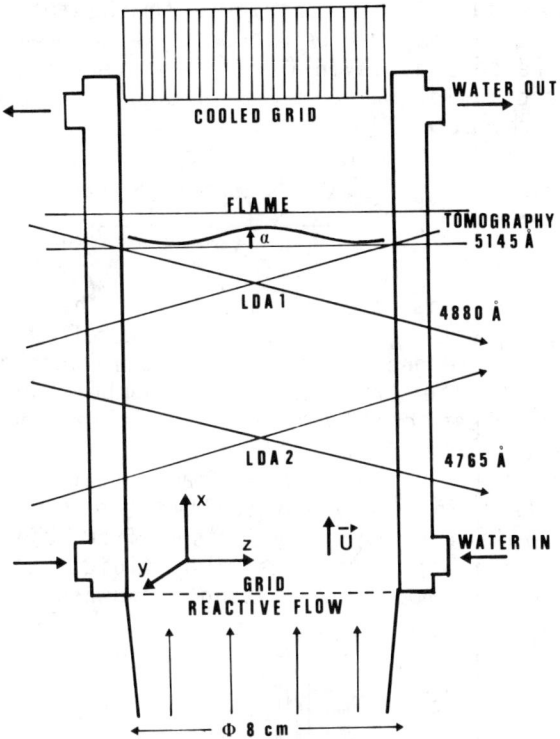

Fig. 1 Details of the combustion chamber and the system of axes used in this work.

of the flame velocity to local flame curvature. The term $(1/u_L)(\partial u_T/\partial x)$ is a linearized measure of the curvature of the flowfield (Karlowitz effect). Eq. (1) describes a diffusional relaxation process in which the relaxation time is given by $(Ku_L dk^2)^{-1}$ where k is the wave number of the wrinkle. For many flames, K is a positive quantity (Clavin 1984) and the diffusional mechanisms can be stabilizing even for rich flames (Quinard et al. 1983).

It must be realized that in Eq. (1) the term $u_T(x=\alpha)$ is the gas flow at the flame including the perturbations to the flow induced by the presence of the front. We will thus write

$$u_T(x,y,z,t) = u_e(x,y,z,t) + u_i(x,y,z,t) \qquad (2)$$

where u_e is the turbulent flow that would have been measured in the absence of the flame (termed the "exciting" velocity field), and u_i is the "induced" velocity field.

This induced field will contribute to the total evolution of the system and may be considered as a hydrodynamic feedback. Pelcé and Clavin (1983) have recently calculated this induced field, which may be found in a more explicit form in Searby et al. (1983)

$$u_i(x,k_y,k_z,\omega) = e^{+k(x-\alpha)} \frac{\left(-\gamma gk + \frac{\gamma}{1-\gamma}k^2 u_L^2 - adk^3 u_L^2 - bi\omega dk^2 u_L\right)\alpha}{(2-\gamma)i\omega + 2ku_L} \quad (3)$$

where $x-\alpha$ ($x \leq \alpha$) is the distance from the flame front, $\gamma = 1 - \rho_b/\rho_0$, ρ_b and ρ_0 are the densities of the burnt and fresh gases, respectively, g is the acceleration of gravity, k is the transverse wave number, and a and b are dimensionless coefficients. The corresponding terms, which contain the flame thickness d, arise because of the dependence of the local flame velocity on flame curvature (see Eq. (1)). Physically, the term containing the acceleration of gravity arises from the fact that the flame surface separates two fluids of different density. For flames propagating downwards, this term

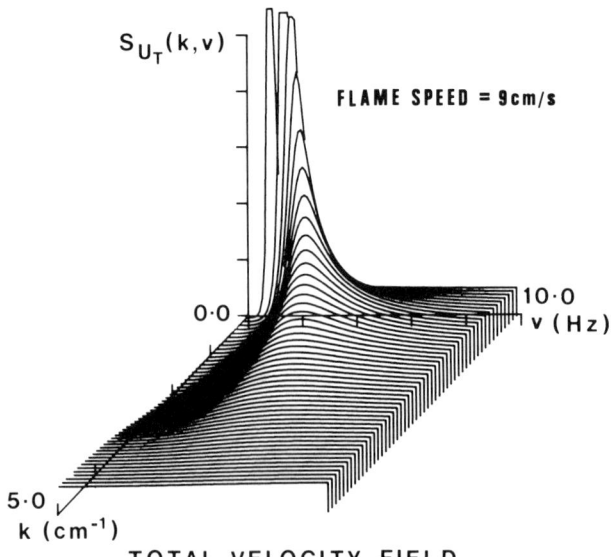

Fig. 2 The calculated total turbulent velocity field spectrum as a function of frequency and wave number as should be measured 0.15 cm from the flame for a constant exciting field. The numerical values used for the flame speed, the diffusion coefficients, and the the gas expansion ratio are those of the flame presented in the experimental part.

is negative, indicating that at low frequencies the velocity field induced by gravity is in opposition to the exciting velocity field. This stabilizing effect of gravity is essential for the overall stability of the flame. The term $k^2 u_L^2 \gamma/(1-\gamma)$ arises from the acceleration of the gas flow by the heat released in the reaction zone. The acceleration is normal to the local flame front and produces a deviation of the flow lines through the flame, which in turn deviates the flow lines upstream of the flame, the flow being incompressible to a good approximation. This effect, in the absence of coupling to gravity and diffusive effects, was described many years ago by Darrieus (1938, 1945) and Landau (1945) who showed that gas expansion tends to destabilize the flame at all wavelengths. In the present analysis this term will destabilize the flame for sufficiently high flame velocities. The physical interpretation of Eq. (3) has been discussed at greater length by Searby et al.(1983).

Equations (1), (2), and (3) form a closed system with four variables, α, u_T, u_e, u_i. If the velocity field

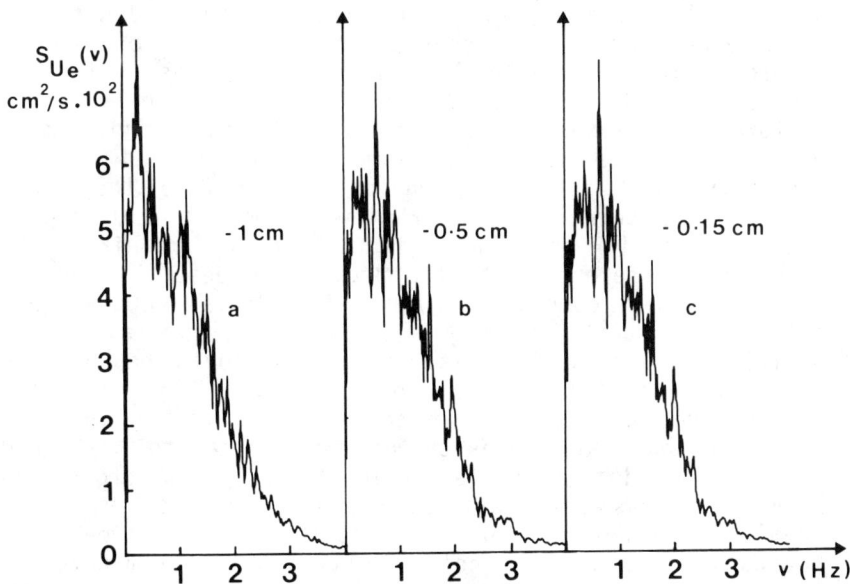

Fig. 3 The frequency spectra of the turbulent flows in the absence of flame. The distance between the observation point and the position of the "flame" are : a) 1 cm, b) 0.5 cm, and c) 0.15 cm. The attenuation of the turbulence arises from a differential buoyancy effect (see text).

in the absence of flame, u_e, is known, it is then possible to express the other quantities as functions of u_e (Clavin and Searby 1984). Fig. 2 shows the total velocity spectrum, $Su_T(k,\nu) \cdot Su_T^*(k,\nu)$, which is calculated at a position 0.15 cm from the flame in a uniform exciting field, $u_e(k,\nu)=1$, and with other parameters corresponding to the flame presented in the experimental part. It can be seen that the hydrodynamic feedback of the flame on the flow produces striking results. For small wave numbers and low frequencies the total velocity field tends to **ZERO,** showing that the induced field is of equal magnitude and opposite phase to the exciting field. This is essentially an effect of gravity. For small wave numbers and slightly higher frequencies there is a sharp resonance which can be identified with propagating gravity waves. For higher wave numbers the positive feedback from the Darrieus-Landau instability is rapidly damped by the diffusive mechanisms and is barely visible for this stable flame.

Experiment and Results

The apparatus is similar to the one already described by Sabathier et al. (1981). The air, propane, and nitrogen flows were regulated by constant mass flow sonic orifices before mixing. The mean flow in the burner was controlled independently by bleeding off a proportion of the reactive mixture. The mixed gases were seeded with a fine mist of oil droplets ~ 1 μm in diameter. The seeded mixture was introduced at high Reynolds number into the lower part of the burner and then slowed down in a divergent cone to obtain the lower flow velocity required. An important difference with the previously described burner is that the combustion zone is in a nondivergent cylindrical duct (see Fig. 1). The flame has a neutral position stability in this burner but could be held to ±1 mm by careful control of the gas flow and the temperature. A wire grid at the entry to the cylindrical zone reduced the intensity of turbulence to ~ 6 % and restored a flat (average) velocity profile across the burner. It was necessary to place a grid at the exit from the burner to prevent instabilities between the hot burnt gases and the cold surrounding air from entering the burner. This grid was cooled to prevent excessive radiation towards the upstream part of the burner.

The flow velocity of the fresh gases was measured by Laser Doppler Anemometry (LDA). The mean sampling frequency was greater than 2000 measurements/s. This was more than two orders of magnitude greater than the highest frequencies present in the turbulent flow (5 Hz), and the analog output of the LDA counter after appropriate low pass filtering could be considered a faithful representation of the instantaneous gas velocity.

The time evolution of the local position and velocity of the flame front were measured by Laser Tomography. This technique, described in detail by Boyer (1980) and by Sabathier et al. (1981), makes use of the fact that the seeding particles are vaporized in the reaction zone and thus the burnt gases do not diffuse laser light.

The analog signals, gas velocities, flame position, and flame velocity were recorded using a multitrack FM magnetic tape recorder. The corresponding power spectra were obtained in real-time or deferred time using a Rockland 512 Channel FFT spectrum analyzer.

Tight thermal control of all the elements of the burner was found to be essential in obtaining coherent reproducible results. The temperature of the incoming gases, the divergent duct, the combustion chamber, and the downstream grid was regulated using circulating water baths. In these experiments the absence of temperature regulation for the upstream wire grid led to an interesting side effect. The downstream walls of the burner, which were in contact with the burnt gases, were stabilized at about 120°C (lower temperatures caused condensation). These walls radiated a constant heat flux towards the upstream part of the burner, particularly the upstream grid which rose in temperature by about 3°C and heated the fresh gases. If the heat flux to the grid is constant, the turbulent gas flow through the grid is not, with the result that large-scale fluctuations in the gas velocity produce fluctuations of opposite sign in the temperature of the gas leaving the grid. Thus, the zones of instantaneous higher velocity in the turbulent flow are slightly denser than the zones of lower velocity. This creates a differential buoyancy effect which slowly damps out large-scale turbulence as the gas is convected away from the grid.

To measure the turbulence field in the absence of flame but under the same conditions, the same thermal effects must be re-created. This was done by replacing the flame with an electrically-heated wire spiral. Fig. 3 shows the spectral density of the gas flow, $Su_e(\nu)$, in the burner without flame (but with the heated spiral) at various distances from the position of the flame front. Fig. 4 shows typical spectra in the presence of a flame, $Su_T(\nu)$, along with the flame velocity spectrum $S_{\dot{\alpha}}(\nu)$. In Fig. 4c we have reproduced the intensity spectrum

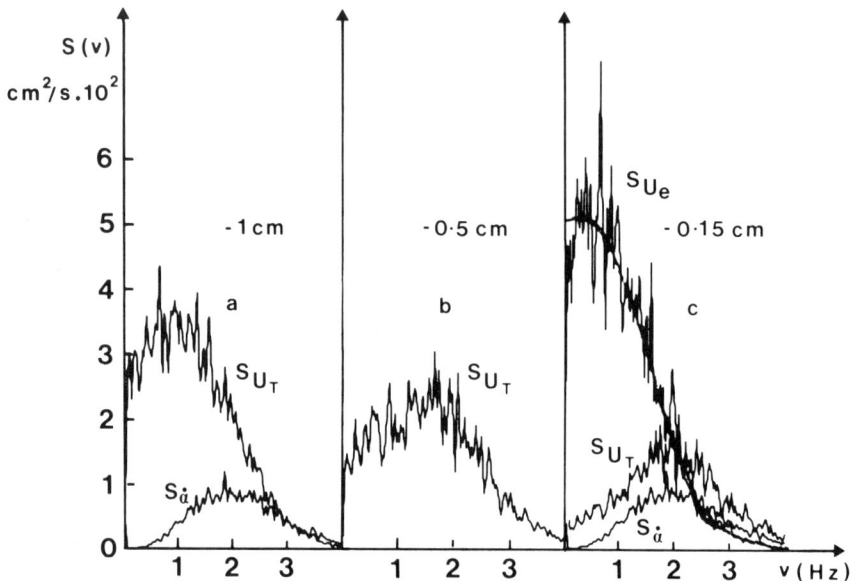

Fig. 4 The frequency spectrum of the upstream turbulent flow in the presence of flame. The distances from the front are: a) 1 cm, b) 0.5 cm, and c) 0.15 cm. The spectrum of the flame velocity (lower curve) is also shown. In c, the exciting field of Fig. 3c is reproduced for comparison.

of Fig. 3c for comparison. From these spectra it can be seen that the velocity field induced by the flame is a function of frequency. At low frequencies the total velocity field is very small; the flame is providing a strong negative hydrodynamic feedback. At higher frequencies, however, the total velocity field is slightly greater than the exciting field; the hydrodynamic feedback is positive. It can also be seen that the flame velocity spectrum differs very little from the total gas velocity spectrum measured just upstream of the flame. This may be anticipated from Eq. (1) where it can be seen that $\alpha = u_T$ plus a term which takes account of curvature effects, which are small compared to the hydrodynamic feedback effects.

To compare these experimental spectra with the theoretical analysis, the spectrum of the turbulent gas flow must be measured as a function of both frequency and transverse wave number. This will be done in the near future. However, preliminary measurements indicate that the frequency and wave number dependencies of the spectra are not strongly inter-correlated, i.e.,

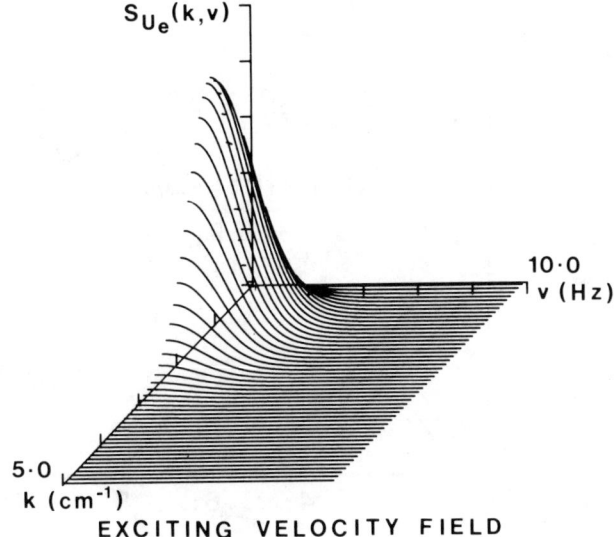

Fig. 5 The modeled structure of the velocity spectrum in the absence of flame as a function of frequency and wave number (see text).

$Su_e(k,\nu) \simeq Su_e(k) \cdot Su_e(\nu)$ and as a first step a mathematical model has been invoked to approximate the turbulent velocity field. The frequency dependence of the field was fitted to the observed frequency spectrum with an empirical power law of the form $Su_e(\nu) = A_0 \exp(\nu/\nu_0)^\beta$, and the wave number dependence was modeled on a transverse spatial correlation function for Gaussian eddies, $C(R)=(1(R/R_0)^2)\exp(-R/R_0)^2$ (Townsend 1976). The characteristic distance R_0 was chosen such that the wave number spectrum had dominant values around 1 cm^{-1}, as suggested by these experimental results. The modeled spectrum, $Su_e(k,\nu)$, of the exciting velocity field is shown in Fig. 5. This function was multiplied by the transfer function relating $Su_T(k,\nu)$ to $Su_e(k,\nu)$. (See the theoretical analysis, this function is illustrated in Fig. 2). It was then integrated numerically over all wave numbers to yield $Su_T(\nu)$. The resulting calculated spectrum is shown in Fig. 6 along with the calculated spectrum of the flame velocity and the modeled representation of the frequency spectrum of the exciting velocity field $Su_e(\nu)$. These theoretical spectra are to be compared with the experimental spectra of Fig. 4c. Although a rather crude model has been used for the exciting velocity field, the calculated spectra are in

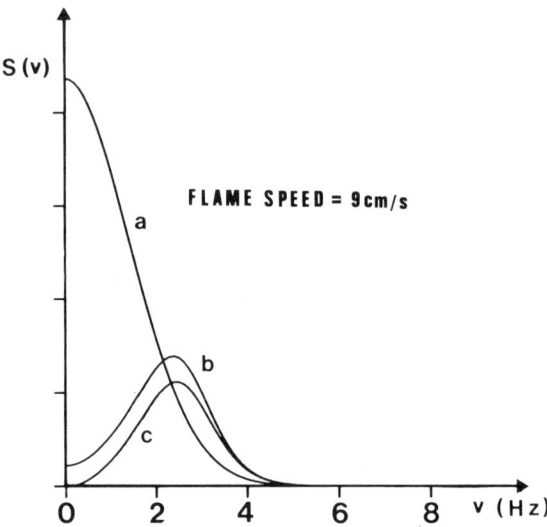

Fig. 6 Spectra calculated using the information presented in Figs. 2 and 5. These spectra are to be compared with Fig. 4c. a) Modeled exciting field, integrated over all wave numbers. b) Total velocity field 0.15 cm ahead of the flame front, integrated over all wave numbers. c) Flame velocity spectrum, integrated over all wave numbers.

good qualitative agreement with the measured spectra. In particular, the total velocity field in front of the flame is strongly attenuated by the presence of the stable front in the low frequency domain. At higher frequencies the gas velocity and flame velocity spectra tend towards the same values which are slightly greater than the spectrum of the exciting field.

Conclusion

The intensity and spectral density of the turbulent flow in the burner has been measured with and without flame. It is seen that the presence of a stable flame produces a strong feedback on the upstream flow. This feedback is negative for low frequencies and becomes slightly positive at "high" frequencies. The penetration distance is of the order of k^{-1}, where k is the wave number of the turbulent fluctuation. This effect is explicable from hydrodynamic considerations of the effect of the change in gas density and velocity as the mixture crosses the reaction zone. Theoretical spectra have been calculated for numerical values corresponding to our flame and reproduce all the major features of the observed spectra.

Acknowledgments

This work was partially supported by the Direction Générale de la Recherche Scientifique et Technologique and by an Action Thématique Programmée No. 3966 from the CNRS. We are indebted to F. Abétino for his invaluable technical assistance and J. Marchisio for help from the electronics workshop.

References

Boyer, L. (1980) Laser tomographic method for flame front movement studies. Comb. Flame 39, 321-323.

Clavin, P. and Williams F.A. (1982) Effects of molecular diffusion and of thermal expansion on the structure and dynamics of premixed flames in turbulent flows of large scale and low intensity. J. Fluid Mech. 116, 251-282.

Clavin, P. (1984) The dynamical behaviour of premixed flame fronts in laminar and turbulent flows. Progress in Energy and Combustion Science, Pergamon Press, Oxford, Engand (to appear).

Clavin, P. and Garcia, P. (1983) The influence of the temperature dependence of diffusivities on the dynamics of flame fronts. J. Mécanique Theor. Appl. 2, 245-263.

Clavin, P. and Searby, G. (1984) The evaluation of hydrodynamical effects in wrinkled flames. Comb. Sci. Technol. (Submitted for publication).

Darrieus, G. Propagation d'un front de flamme. (unpublished work presented at "La Technique Moderne", 1938, and at "Le Congrès de Mécanique Appliquée", 1945).

Landau, L. (1944) On the theory of slow combustion. Acta Physico-Chimica, U.R.S.S. 19, 77-85.

Pelcé, P. and Clavin, P. (1982) The influence of hydrodynamics and diffusion upon the stability limits of laminar premixed flames. J. Fluid Mech. 124, 218-237.

Quinard, J., Searby, G. and Boyer, L. (1983) The stability limits and critical sizes of structures in premixed flames. This Colloquium.

Sabathier, F., Boyer, L., and Clavin, P. (1981) Experimental study of a weak turbulent premixed flame. Combustion in Reactive Systems:Progress in Astronautics and Aeronautics (edited by J.R. Bowen, N. Manson, A.K. Oppenheim, and R.I. Soloukhin), Vol. 76, pp. 246-248. AIAA, New York.

Searby, G., Sabathier, F., Clavin, P. and Boyer, L. (1983) The hydrodynamic coupling between the motion of a flame front and the upstream gas flow. *Phys. Rev. Lett.* 51, 1450-1453.

Townsend, A. (1976) *The structure of turbulent shear flow.* Cambridge University Press, Cambridge, England, p. 9.

Flame Front Stability with General Intermolecular Interaction Potential

P. García Ybarra*
Université de Rouen, Faculté des Sciences et Techniques
Mont-Saint-Aignan, France

Abstract

The stability of a premixed flame propagating downwards in a gravity field is studied without restriction as to the form of the interaction potential between the reactive molecules. For temperature dependent thermal, molecular, and viscous diffusivities, analytical expressions for the Markstein scale and the stability limits are derived. Numerical results are reported for a model flame which burns in a nitrogen-abundant atmosphere and whose constituent molecules are characterized by the Lennard-Jones interaction potential.

Introduction

Deflagration wave propagation is directly dependent, because of diffusive transport, on the interaction potential between the reactive molecules. As is well known from the kinetic theory of gases, the temperature dependence of the molecular, thermal, and viscous diffusivities differs from one interaction potential model to another. Because of the strong temperature increase through the flame, the various potential models may give quite different theoretical predictions.

Pelcé and Clavin (1982) [herein identified as (I)] have presented a complete stability study of a premixed

---

Presented at the 9th ICODERS, Poitiers, France, July 3-8, 1983. Copyright © 1984 by the American Institute of Aeronautics and Astronautics. All rights reserved.

*Associate Researcher, LA No. 230 CNRS. Presently, Assistant Professor, Dept. Física Fundamental UNED.

flame propagating downwards in a gravity field with constant diffusivities. The study shows that cellular flames are a typical interfacial instability for which the "Markstein scale" (Markstein 1964) plays the role of an effective surface tension which dampens the small wavelength perturbation region. The Markstein scale emerges from the diffusive properties of the flame and consits of two parts. The first comes from diffusive relaxation phenomena and accounts for the pure thermal and concentration relaxation of the transverse gradients generated by a perturbation, the flame front behaving as an inert nonreactive interface. The second comes from chemical relaxation phenomena and accounts for the competitive influence of the above-mentioned transverse heat and mass fluxes on the local chemical-reaction velocity. The Markstein scale and the stability properties of the flame must be altered when the temperature-dependence of the diffusivities is taken into account; as has been shown by Clavin and García (1983) [herein identified as (II)] for the hard sphere model. For this simple interaction potential the product of the density and the diffusivity has a square root temperature-dependence for each diffusivity, and the model predicts constant Lewis and Prandtl numbers (see, for example, Chap. 8 of Hirschfelder et al. 1967). When a more complicated potential is used the thermal conductivity λ and the dynamic viscosity η of a monatomic pure substance are expressed as $\lambda \sim \eta \sim \sqrt{T} / \Omega^{(2,2)}$ up to first order in the Enskog-Chapman approximation. The collision integrals $\Omega^{(\ell,s)}$ are also temperature dependent. The binary diffusion coefficient $\mathcal{D}$, to this first approximation is $\rho \mathcal{D} \sim \sqrt{T} / \Omega^{(1,1)}$, where ρ is the gas density. The Lewis number $L = \lambda/\rho \mathcal{D} c_p$, c_p being the specific heat at constant pressure, is $L \sim \Omega^{(1,1)}/\Omega^{(2,2)}$ and will vary in the flame as the collision integrals vary. From the viewpoint of the premixed flame stability theory, this is a basic difficulty because the activation energy asymtotic method, used commonly to solve the reactive region of the flame, requires a Lewis number equal to unity at the lowest order as a necessary condition to have a transition from plane flame to cellular flame (Sivashinsky 1977; Joulin and Clavin 1979). As the ratio $A \equiv \Omega^{(2,2)}/\Omega^{(1,1)}$ for the usual intermolecular potentials is a smooth function of the temperature, the difficulty can be circumvented through the retention of the temperature variation as a first-order correction of the Lewis number. As a consequence, the leading order of the Lewis number is equal to unity.

When more complicated types of molecules with internal degrees of freedom are considered, the Prandtl number will also become temperature dependent because diffusive energy transfer is increased by internal motion, which changes the thermal conductivity (Eucken correction) but no the shear viscosity. The effect of a bulk viscosity also arises in this case. With the simple temperature dependencies assumed in (I) and (II) there is no effect due to bulk viscosity in the resulting dispersion relations. As this dependence is a function of the relaxation times of the internal modes of motion and there are no calculations for realistic molecular models, in the present work no specific functional form is assumed for it. High dilution in an inert component will be supposed here so that the thermal conductivity and the viscosities of the mixture are equal to those of the diluant at the lowest order in the concentration of the diluted species (García, Nicoli, and Clavin 1984).

In the following section the mathematical analysis showing the form of the Markstein scale that leads to the stability criterion is outlined. The results in the particular case of the Lennard-Jones (6-12) potential are also discussed.

Analysis

Following (I) and (II), the flame is supposed to be a discontinuous surface between two isothermal incompressible gaseous media (Darrieus 1938; Landau 1944). The stability limits study of the steady plane propagation of such a surface requires the determination of the hydrodynamic jump conditions. The subscripts ($-\infty$) and ($+\infty$) will characterize the values in the fresh and burned sides, respectively.

The width of the transition zone is of the order

$$\delta \equiv \lambda_{-\infty}/c_{p-\infty}\, \rho_{-\infty}\, u_L$$

where u_L is the laminar propagation velocity of the plane flame relative to the fresh gas. The governing equations must be solved in this zone up to second order in the perturbation parameter $\varepsilon = \delta/\Lambda \ll 1$, where Λ is the wavelength of the flame wrinkle. These ε-expansions permit the solution of the hydrodynamic aspects and must be performed concurrently with the classical activation energy asymptotic expansions. For simplicity, the composition of fuel and oxidizer is assumed to be far from the stoichiometric composition so that the reaction is controlled only by the limiting reactant concentration. The overall reaction rate is assumed to follow an Arrhenius law characterized by the activation energy E_a (see, for example, Appendix B of Williams 1965). The nonisothermal region is then divided into an outer nonreactive preheat zone and an inner reactive zone of δ and δ/β orders of thickness, respectively, where $\beta = \gamma E_a/RT_{+\infty}$ is the nondimensional activation energy, R is the universal gas constant, and $\gamma = (T_{+\infty}-T_{-\infty})/T_{+\infty}$ is a parameter of order of unity introduced for convenience. Because of its isothermal character, the solution of the linearized perturbed flow equations in the fresh and burned incompressible zones parallels exactly that of (II); only the results that are needed will be quoted here.

To study the flame zone, δ and δ/u_L are chosen as units of length and time, respectively. With the dimensionless temperature $\theta = (T-T_{-\infty})/(T_{+\infty}-T_{-\infty})$ and the function defined as $\star(\theta) = (\lambda/c_p)/(\lambda/c_p)_{-\infty}$, the energy equation in the preheating region takes the form

$$r\frac{\partial\theta}{\partial t} + s\frac{\partial\theta}{\partial x} = \frac{\partial}{\partial x}\left(\star\frac{\partial\theta}{\partial x}\right) + \frac{\partial}{\partial y}\left(\star\frac{\partial\theta}{\partial y}\right) - \frac{\partial}{\partial y}\left(\star\frac{\partial\alpha}{\partial y}\frac{\partial\theta}{\partial x}\right) \quad (1)$$

The energy equation is written in a moving frame which advances with the flame in the negative direction of the x axis. Here t and y are the dimensionless time and transverse coordinate[†], respectively. The following nondimensional magnitudes have been introduced: $\alpha(y,t)$ is the flame wrinkle; $r = \rho/\rho_{-\infty}$ (the density ratio); and $s = r(U-\partial\alpha/\partial t)$ is the longitudinal mass flux, where U is the dimensionless longitudinal velocity. Finally, it has been supposed that $\theta(x,y,t) = \theta_L(x) + \hat{\theta}(x,y,t)$ where the subscript L refers to the plane propagation solution and the hat ($\hat{\ }$) refers to the perturbation. Some terms relating to nonlinearities in respect to the perturbation have been neglected in writing Eq.(1).

Except for the specific dependence of the function $\star(\theta)$ on the temperature, Eq.(1) is equivalent at the corresponding case of (II) and, consequently, a similar treatment leads to

$$\theta_L = e^\xi \quad \text{where} \quad d\xi = dx/\star_L \quad (2)$$

for the plane propagation and

$$\frac{\partial\hat{\theta}}{\partial x}(0^-) = \hat{\theta}(0^-) + \int_{-\infty}^{0}\left(\hat{s}+\star_L\frac{\partial^2\alpha}{\partial y^2}\right)\frac{d\theta_L}{dx}dx - \hat{\theta}(0^-)\frac{d\star_L}{d\theta}(0^-) + o(\varepsilon^2) \quad (3)$$

[†] For symmetry considerations only one transverse coordinate is used.

for the perturbation, where 0^- represents the fresh lefthand side of the reactive sheet. To arrive at Eq.(3) an integration of Eq.(1) through the flame thickness must be performed, with the condition of vanishing perturbation in the boundary $x \to -\infty$. Also the expansion in powers of ε has to be considered; the time and the transverse coordinate have been scaled with ε, $\partial/\partial t \sim \partial/\partial y \sim O(\varepsilon)$, and the perturbation with ε^2, $\hat{\theta} \sim O(\varepsilon^2)$ [see (I)].

For the reduced mass fraction Ψ of the limiting reactant, defined such as $\Psi \to 1$ when $x \to -\infty$, the conservation equation in this preheating region can be written as

$$r \frac{\partial \Psi}{\partial t} + s \frac{\partial \Psi}{\partial x} = \frac{\partial}{\partial x}\left(\frac{\hat{x}}{L} \frac{\partial \Psi}{\partial x}\right) + \frac{\partial}{\partial y}\left(\frac{\hat{x}}{L} \frac{\partial \Psi}{\partial y}\right) - \frac{\partial}{\partial y}\left(\frac{\hat{x}}{L} \frac{\partial \alpha}{\partial y} \frac{\partial \Psi}{\partial x}\right) \qquad (4)$$

The temperature dependence of the Lewis number L introduces some differences in respect to the results of (II). Upon introduction of the variable ξ of Eq.(2), Eq.(4) can be integrated for the case of stationary plane propagation with the boundary conditions $x \to -\infty$, $\Psi_L \to 1$ and $x = 0$, $\Psi_L = 0$ to obtain ‡

$$\Psi_L = 1 - \exp\left[\int_0^\xi L \, d\xi'\right] \qquad (5)$$

For the perturbation, a first integration of Eq.(4) with the condition $x \to -\infty$, $\hat{\Psi} \to 0$ gives

$$\frac{1}{L_{+\infty}} \frac{\partial \hat{\Psi}}{\partial x}(0^-) = \int_{-\infty}^0 \left[\hat{s} + \frac{\hat{x}_L}{L} \frac{\partial^2 \alpha}{\partial y^2}\right] \frac{d\Psi_L}{dx} \, dx$$

$$+ \hat{\theta}(0^-) \frac{d\hat{x}_L}{d\theta}(0^-) + \frac{\hat{\theta}(0^-)}{L_{+\infty}} \frac{dL}{d\theta}(0^-) + o(\varepsilon^2) \qquad (6)$$

where $\hat{\Psi} \sim O(\varepsilon^2)$ and the origin is chosen such that $\Psi(0) = 0$ for all orders.

Equations (3) and (6) provide expressions for the outer gradients of θ and Ψ, respectively, in the unburnt side of the reaction zone. With the hypothesis of no heat loss and complete consumption of the reactant, the corresponding gra-

---

‡ In order to satisfy the boundary condition at $x \to -\infty$, a supplementary requirement should be imposed on the Lewis number to assure the divergence of the integral $\int_0^{-\infty} L \, d\xi'$. A sufficient condition is $\lim_{\xi \to -\infty} L \neq 0$, commonly verified in the nature.

dients on the burnt side 0^+ vanish, so

$$\frac{\partial\theta}{\partial x}(0^+) = \frac{\partial\psi}{\partial x}(0^+) = 0 \tag{7}$$

To analyze the inner reactive zone, the Lewis number is only taken to differ from unity by terms of the order β^{-1}. Since in this zone the dimensionless temperature is

$$\theta - 1 = O(\beta^{-1})$$

it follows that here the Lewis number is constant up to the order β^{-1}

$$L = L_{+\infty} + o(\beta^{-1})$$

As a consecuence, the analysis of (II) need not be altered and the corresponding results holds

$$\left|\frac{\partial\hat{\theta}}{\partial x}\right|_{0-}^{0^+} + \frac{1}{L_{+\infty}}\left|\frac{\partial\hat{\psi}}{\partial x}\right|_{0-}^{0^+} = o(\beta^{-1}) \tag{8}$$

$$\beta\,\hat{\theta}(0) = 2\,\frac{\partial\hat{\theta}}{\partial x}(0^-) + o(\beta^0) \tag{9}$$

where $|\ |_{0-}^{0+}$ is the difference of the enclosed quantity between both sides, burnt and unburnt, of the reactive surface.

Now, prior to the insertion of results from equations (3),(6) and (7) into the jump conditions of Eq.(8), the development of Eq.(6) around $L = 1$ up to $O(\beta^{-1})$ is needed. This can be worked out by means of the following relation between the differentials of the plane solutions, easily inferred from Eqs.(2) and (5)

$$d\psi_L = -\left\{L + \int_1^{\theta_L}\frac{L-1}{\theta}\,d\theta\right\}d\theta_L + o(\beta^{-1}) \tag{10}$$

If Eq.(9) is used to eliminate $\hat{\theta}(0)$, the jump condition of Eq.(8) leads to the integral relation

$$\int_0^1\left\{\hat{s}+\lambda_L\frac{\partial^2\alpha}{\partial y^2}\right\}d\theta_L =$$

$$\frac{\beta}{2}\int_0^1\left\{\hat{s}(L-1) + \left[\hat{s}+\lambda_L\frac{\partial^2\alpha}{\partial y^2}\right]\left(\int_1^{\theta_L}\frac{L-1}{\theta}\,d\theta\right)\right\}d\theta_L \tag{11}$$

FLAME STABILITY AND INTERMOLECULAR POTENTIAL

which reduces to that of (II) when L-1 is constant, as required.

This integral relation expresses a necessary condition to make the matching possible. Inserting the results of the associated hydrodynamic flow study concerning the perturbation of the longitudinal mass flux $\hat{s}$ leads to the equation for the dynamic evolution of the flame front. The ε-expansion of $\hat{s}$ in the preheat region [see (I) and (II)] is of the form

$$\hat{s} = \varepsilon\,\hat{S}_{-\infty}(\varepsilon x, \varepsilon y, \varepsilon t) + \varepsilon\,\hat{s}_1(x, \varepsilon y, \varepsilon t) + \varepsilon^2\,\hat{s}_2(x, \varepsilon y, \varepsilon t) + o(\varepsilon^2) \qquad (12)$$

where $\varepsilon\,\hat{S}_{-\infty}(\varepsilon x, \varepsilon y, \varepsilon t)$ represents the solution in the incompressible unburnt zone whose longitudinal variation is on the long scale εx and must be developed in a Taylor series around the origin. The remainder of Eq.(12) is the alteration of this external solution due to heating, which varies longitudinally on the small scale x. With analogous developments for the other hydrodynamical variables, the continuity and momentum equations can be solved up to $O(\varepsilon^2)$ by iteration for a perfect gas state equation and a quasi-isobaric process. Since the analysis of (II) is valid for any form of the function $\tilde{\tau}(\theta)$ and the Lewis number is absent in this part of the calculation, its results apply directly to the present work

$$\hat{s}_1 = 0$$

$$\hat{s}_2\Big|_{x=\alpha} = -\frac{\gamma}{1-\gamma}\left[\frac{\partial \hat{S}_{-\infty}(0,\varepsilon y,\varepsilon t)}{\partial(\varepsilon x)} - \frac{\partial^2 \alpha(\varepsilon y,\varepsilon t)}{\partial(\varepsilon y)^2}\right]\int_0^{\theta_L} \tilde{\tau}_L r_L\, d\theta_L \qquad (13)$$

Introduction of the expansion of Eq.(12) into Eq.(11) gives an equation for the flame front evolution (after evaluation of some integrals by parts)

$$\frac{\partial \alpha}{\partial t} = \hat{U}_{-\infty}\Big|_{x=\alpha} - (\mathcal{L}/\delta)\left\{\frac{\partial \hat{U}_{-\infty}}{\partial x}\Big|_{x=\alpha} - \frac{\partial^2 \alpha}{\partial y^2}\right\} \qquad (14)$$

where $\hat{U}_{-\infty}\big|_{x=\alpha}$ is the value of the upstream longitudinal velocity perturbation on the flame front.

Equation (14) shows that the difference between local flame movement and longitudinal velocity fluctuations is proportional to the local curvature of the flame front relative to that of the flow. The proportionality factor is the nondimensional Markstein scale $\mathcal{L}/\delta$ for which the expression is:

$$\mathcal{L}/\delta = \frac{J}{\gamma} - \frac{1}{2} \int_0^1 \frac{\overset{*}{x}_L}{1+\frac{\gamma}{1-\gamma}\theta_L} \left\{ \int_1^{\theta_L} \frac{\beta(L-1)}{\theta} d\theta \right\} d\theta_L \quad (15)$$

where

$$J \equiv \frac{\gamma}{1-\gamma} \int_0^1 \frac{\overset{*}{x}_L}{1+\frac{\gamma}{1-\gamma}\theta_L} d\theta_L \quad (16)$$

Comparison with the corresponding expression of (II) clearly shows the effect induced by the temperature variation of the Lewis number, manifested by the double integral which now appears in the second term, weighing in a nontrivial manner the value of the Lewis number through the flame thickness.

To determine the stability domain of the plane propagation, this expression for the Markstein scale in Eq.(15) must be introduced into the dispersion relation obtained as the necessary condition for the existence of a nontrivial solution of the preturbed hydrodynamic equations in the incompressible zones. Minor changes must be introduced in the analysis of (II) to apply to the present case, in particular the velocity deflection expression does not change and only the terms containing viscosity effects are altered. Again bulk viscosity does not play any role in the stability properties. For the shear viscosity the final term is of the form $\int_0^1 (n_{+\infty} - n_L) d\theta_L$ where the function $n_+ \equiv n/n_{-\infty}$ substitutes for $\overset{*}{x}_L$ in the corresponding result of (II).

For the stationary marginal states the dispersion relation gives

$$A(\mathcal{L}) k^2 - B(\mathcal{L}) k + (1-\gamma)(g\delta/u_L^2) = 0 \quad (17)$$

where

$$A(\mathcal{L}) \equiv \int_0^1 \overset{*}{x}_L d\theta_L + \frac{2+\gamma}{\gamma} \frac{\mathcal{L}}{\delta} - \frac{2}{\gamma} J + 2 \text{ Pr} \int_0^1 (n_{+\infty} - n_L) d\theta_L$$

$$B(\mathcal{L}) \equiv 1 + (1-\gamma)(g\delta/u_L^2)[(\mathcal{L}/\delta) - (J/\gamma)]$$

Here $k = 2\pi\delta/\Lambda$ is the dimensionless wave number of the perturbation; $Pr = (\eta c_p/\lambda)_{-\infty}$ is the Prandtl number; and g is the gravitational acceleration.

If the Markstein scale is taken as a critical parameter differentiation of Eq.(16) with respect to k with the condition of critical point, $\partial(\mathcal{L}/\delta)/\partial k|_{k=k_c} = 0$ provides the expression of the critical wave number

$$k_c = B(\mathcal{L}_c)/2A(\mathcal{L}_c) \qquad (18)$$

where the subscript c refers to the critical point. Finally, if Eq.(18) is introduced into Eq.(17), a second degree equation is obtained for $(\mathcal{L}/\delta)_c$

$$\frac{g\delta}{u_L^2}(1-\gamma)\left[\frac{\mathcal{L}}{\delta}\right]_c^2 - 2\left(1 + \frac{4}{\gamma} + \frac{1-\gamma}{\gamma}\frac{g\delta}{u_L^2}J\right)\left[\frac{\mathcal{L}}{\delta}\right]_c$$

$$- 4\int_0^1 \overset{\star}{\chi}_L \, d\theta_L + \frac{6}{\gamma}J - 8\,Pr\int_0^1 (n_{+\infty} - n_L)\,d\theta_L$$

$$+ \frac{1-\gamma}{\gamma}\frac{g\delta}{u_L^2}J + \frac{u_L^2}{(1-\gamma)g\delta} = 0 \qquad (19)$$

From this equation it is possible to determine the critical values of the Markstein scale as a function of the plane flame propagation velocity u_L.

The effect of the specific characteristics of each particular intermolecular potential appears in Eq.(19) through the functions $\overset{\star}{\chi}(\theta)$, $n(\theta)$, and $L(\theta)$. As an example, the case of Lennard-Jones (6-12) potential is discussed in the next section.

Discussion of Results and Conclusions

To evaluate quantitatively the modifications induced by the effects studied here, numerical solutions of Eqs.(15) and (19) defining the Markstein scale and the marginal states, respectively, are generated for the Lennard-Jones (6-12) potential. With increasing temperature internal degrees of freedom increase the heat capacity and the thermal conductivity. The last variation is described by

$$\lambda = \lambda_o \, E \qquad (20)$$

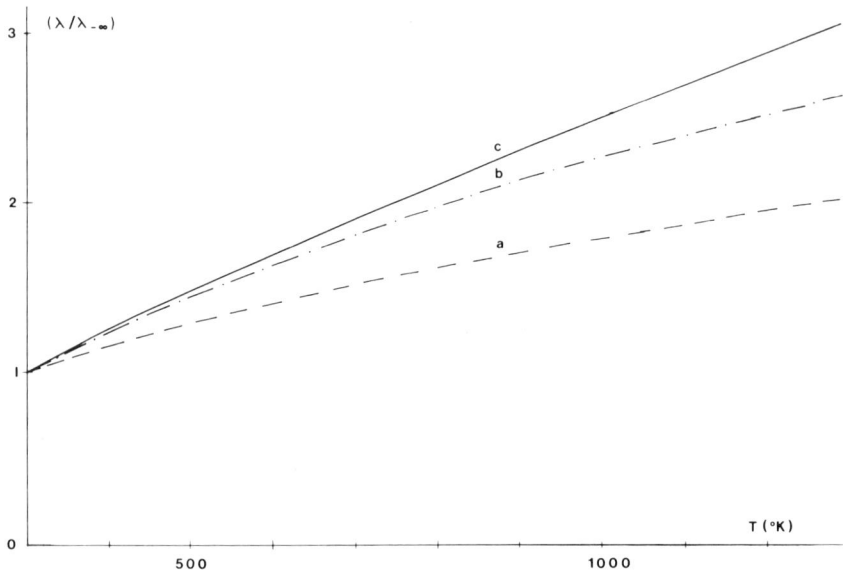

Fig. 1 Reduced thermal conductivity, $\lambda/\lambda_{-\infty}$, vs temperature for the nitrogen N_2. a) Hard sphere potential; b) Lennard-Jones potential, monatomic (c_p=constant); c) Lennard-Jones potential, diatomic [$c_p=c_p(T)$].

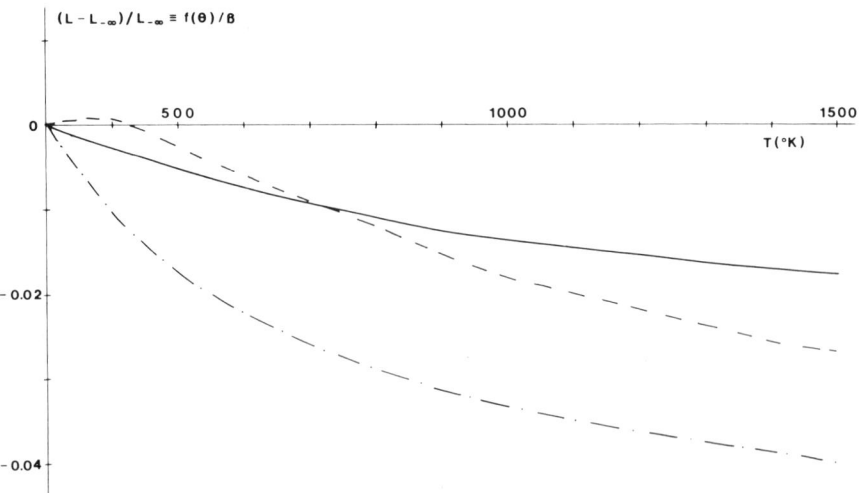

Fig. 2 Relative variation of the Lewis number with temperature, computed with the Lennard-Jones (6-12) potential. a) Monatomic gas, b) N_2, c) C_3H_8.

where the subscript o refers to the monatomic conductivity, E is the Eucken's correction factor

$$E = 0.115 + 0.178 \, M \, c_p \tag{21}$$

(Hirschfelder 1957), and M is the molar mass of the gas. In Fig.1 a comparison is presented between the different thermal variations of λ given by the hard sphere and the Lennard-Jones models in the monatomic and polyatomic cases of a typical molecule (e.g., N_2). Although a successive increase in the predicted values is shown, the relevant physical property is the ratio $\lambda/c_p = \lambda_o E/c_p$. Since E and c_p both increase with temperature in such a way that the relative variations of its ratio never exceed the unity by more than a few hundredths in usual gases, the result is $\lambda/c_p \simeq (\lambda/c_p)_o$. As a consequence, the effect of the internal degrees of freedom is negligible with regard to the heat diffusion.

The temperature dependence of the Lewis number can be written as

$$L = L_{-\infty} F(\theta) = 1 + \frac{\ell_{-\infty}}{\beta} + \frac{f(\theta)}{\beta} + O(\beta^{-2}) \tag{22}$$

where

$$L_{-\infty} \equiv \left(\frac{\lambda_o E/c_p}{\rho D}\right)_{-\infty} = 1 + \frac{\ell_{-\infty}}{\beta} \qquad \ell_{-\infty} = O(\beta^o)$$

$$F(\theta) \equiv \frac{E_{-\infty} A_{-\infty} c_{p-\infty}}{E_{-\infty} A \, c_p} = 1 + \frac{f(\theta)}{\beta} \qquad f(\theta) = O(\beta^o) \tag{23}$$

Table 1 Computed values of the integrals J, D, and N, and the Markstein scale value $\mathcal{L}/\delta$, for a flame burning in an abundant nitrogen atmosphere

	J	D	N	$\mathcal{L}/\delta$
a	1.61	2.40	0	$2.01+0.30\ell_{-\infty}$
b	2.47	3.04	0	$3.09+0.38\ell_{-\infty}$
c	2.96	3.38	-0.34	$3.66+0.42\ell_{-\infty}$

Note: a = results of (I) corresponding to constant diffusivities. b = results of (II) corresponding to a hard sphere-like molecular interaction. c = present results for monatomic molecules with a Lennard-Jones like molecular interaction. When polyatomic effects are incorporated, N takes the value -0.44 which does not alter measurably $\mathcal{L}/\delta$.

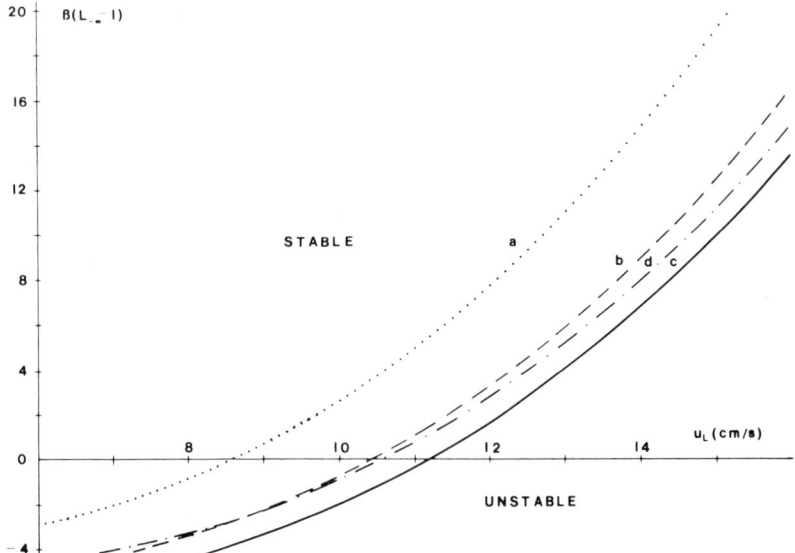

Fig. 3 Neutral stability curves in the $\beta(L_{-\infty}-1)$ vs u_L plane. Curves a, b, and c correspond to the cases explained in Table 1. Curve d corresponds to the nonviscous version of case c.

The function $f(\theta)/\beta$ is depicted in Fig.2 for some monatomic and polyatomic cases. In general this function has negative values, showing that the Lewis number diminishes with increase in temperature. The temperature dependency of the reduced shear viscosity η for negligible effects due to nonspherical symmetry of the molecular interaction is that of the monatomic thermal conductivity $(\lambda/\lambda_{-\infty})_o$ shown in Fig.1. The Markstein scale is:

$$\frac{\mathcal{L}}{\delta} = \frac{J}{\gamma} + \frac{1-\gamma}{\gamma} (\ell_{-\infty}D + N) \qquad (24)$$

where J is defined in Eq.(16) and corresponds to the thermal relaxation and

$$D \equiv -\frac{\gamma}{1-\gamma} \int_0^1 \frac{\lambda_L \ln\theta_L}{1+\theta_L\gamma/(1-\gamma)} d\theta_L$$

$$N \equiv -\frac{\gamma}{1-\gamma} \int_0^1 \frac{\lambda_L}{1+\theta_L\gamma/(1-\gamma)} \left\{ \int_1^{\theta_L} \frac{f(\theta)}{\theta} d\theta \right\} d\theta_L \qquad (25)$$

correspond to the chemical relaxation, which differs from the corresponding result of (II) in the integral N which takes into account the thermal variation of the Lewis number. From Fig.2 it can be seen that absolute value increases with the temperature and with the number of the internal degrees of freedom, but never exceeds a few tenths when it is multiplied by β (usually $\beta \simeq 10$) to evaluate N.

A summary of the typical values of J, D, N, and $\mathcal{L}/\delta$ is given in Table 1 where, by way of comparison, the corresponding results or (I) and (II) are also presented. As can be seen, the net result is an appreciable increase in the Markstein scale, which, in turn, increases the stability of the flame. This is a consequence of the stronger temperature dependence of the thermal conductivity, in spite of the weak decrease of the Lewis number which is shown to be a negligible destabilizing effect.

Solutions of Eq.(19) provide the critical values of the Markstein scale or, alternatively, the Lewis number obtained from $\ell_{-\infty_c} = \beta(L_{-\infty_c}-1)$, which in general is the experimentally-controlled parameter. A plot of the results as a function of the plane propagation velocity is given in Fig.3. The total increase of the stability with respect to the results of (I) and (II), also plotted, is limited to between 10 and 20% approximatively. The viscosity effect turns up only in this part of the computation, through the term $Pr \int_0^1 (n_{+\infty} - n_1) d\theta_1$ which appears in the independent term of the second degree equation, Eq.(19). To show its stabilizing effect the resulting neutral stability curve, computed with Pr = 0, is also depicted in Fig.3 as case d. As was already pointed out in (II), the shift is not dramatic and vanishes for constant viscosity, $n_{+\infty} = n_1 = 1$, in agreement with (I) (see also Frenkel and Sivashinsky 1983).

Acknowledgments

The author thanks Prof. P. Clavin for fruitful discussion and Prof. J. Ray Bowen for his help in the English redaction. This work has been supported by the Stiftung Volkswagenwerk.

References

Clavin, P. and García, P. (1983) The influence of the temperature dependence of the diffusivities on the dynamics of flame fronts. Journal de Mécanique Théorique et Appliquée 2, 245-263.

Clavin, P. and Williams, F. A. (1982) Effects of molecular diffusion and of thermal expansion on the structure and dynamics of premixed flames in turbulents flows of large scale and low intensity. Journal of Fluid Mechanics 116, 251-282.

Darrieus, G. (1938) Propagation d'un front de flamme. Essai de théorie des vitesses anormales de déflagration par développement spontané de la turbulence. Unpublished works presented at La Technique Moderne 1938 and at Le Congrcs de Mécanique Appliquée, 1945.

Frankel, M. L. and Sivashinsky, G. I. (1982) The effect of viscosity on hydrodynamic stability of a plane flame front. Combustion Science and Technology 29, 207-224.

García, P., Nicoli, C. and Clavin, P. (1984) Soret and dilution effects on premixed flames. Combustion Science and Technology (in press).

Hirschfelder, J. O., Curtiss, C. F., and Bird, R. D. (1964) The Molecular Theory of Gases and Liquids. John Wiley and Sons, Inc., New York.

Hirschfelder, J. O. (1957) Heat conductivity in polyatomic or electronically excited gases II. Journal of Chemical Physic 26, 282.

Joulin, G. and Clavin, P. (1979) Linear stability analysis of non-adiabatic flames: Diffusional thermal model. Combustion and Flame 35, 139-153.

Landau, L. (1944) On the theory of slow combustion. Acta Physicochimica 19(1), 77-85.

Markstein, G. H. (1964) Nonsteady Flame Propagation. Pergamon Press, Oxford, England.

Pelcé, P. and Clavin, P. (1982) Influence of hydrodynamics and diffusion upon the stability limits of laminar premixed flame. Journal of Fluid Mechanics 124, 219-237.

Sivashinsky, G. I. (1977) Diffusional thermal theory of cellular flame. Combustion Science and Technology 15, 137-145.

Williams, F. A. (1965) Combustion Theory. Addison-Wesley Publishing Co., Mass.

Stability Limits and Critical Size of Structures in Premixed Flames

J. Quinard,* G. Searby,* and L. Boyer†
Université de Provence—Centre St. Jérôme
Marseille, France

Abstract

An apparatus for the study of the instabilities of a premixed flame in a laminar flow is described. In this burner, the gas flow is laminar, uniform, and homogeneous. The flame front is 9 cm in diameter and, below the instability threshold, is flat to within 1%. This work concerns a propane/air mixture diluted with nitrogen. The flame speeds are measured by laser anemometry close to the front and range from the extinction limit (6 cm/s) to the autoturbulent regime (∼11 cm/s). The dilution ranges from 12% to 16% and the equivalence ratio from 0.8 to 1.2. The isodilution curves and the stability limits are presented and compared with theoretical predictions. The critical size of the cellular structures is obtained from an appropriate analysis of photographic images of the flame front. The experimental results agree fairly well with the relation:

$$\Lambda_c = \pi U_{fc}^2 / (1-\gamma)g$$

where Λ_c is the critical wavelength, U_{fc} is the flame velocity, γ is the thermal expansion ratio, and g is the acceleration of gravity.

Presented at the 9th ICODERS, Poitiers, France, July 3-8, 1983. Copyright © American Institute of Aeronautics and Astronautics, Inc., 1984. All rights reserved.
*Chargé de Recherche au C.N.R.S. Laboratoire de Dynamique et Thermophysique des Fluides.
†Professor, Laboratoire de Dynamique et Thermophysique des Fluides.

Introduction

The study of the stability threshold of a premixed flame front should provide fundamental information about the combustion process, including the coupling of reactive, diffusive, and convective mechanisms. The results of such a study may be used to check theoretical models before they are applied to more realistic situations in combustion. As in other fields, the onset of instability is revealed by the appearance of structures that are stationary at the threshold and that become turbulent as the system becomes more unstable.

Since the first observations of Smithells and Ingle (1892) and the quantitative study of Markstein (1964), which led to a phenomenological insight into the instability process, the understanding of this phenomenon has remained incomplete. The three fundamental mechanisms controlling stability --diffusion, hydrodynamics, and buoyancy-- had been recognized and described independently (Darrieus 1938, 1945; Landau 1944; Markstein 1964; Barrenblatt et al. 1962). Recently Pelcé and Clavin (1982) have developed a model which accounts for these three effects simultaneously and have shown that they are intimately and nontrivially coupled.

This analytical theory is concerned with the stability of an initially flat flame front propagating downwards and in which the reactive properties are approximated by an overall onestep reaction. In this case and for wavelengths large compared to the flame thickness d_f (this assumption is realistic as the cell size is about 1 cm, while the flame thickness is 0.01 cm), instability results from the competition between three mechanisms, two of which are stabilizing:

1) The flame front separates the "heavy" unburnt mixture (at the bottom of the burner) from the "light" hot gas; thus, the force of gravity tends to flatten any deviation from horizontality.

2) Ahead of a wrinkled flame front, the transverse diffusion of heat and reactants modify the local flame speed (Barenblatt et al. 1962) in such a way that an initial perturbation of the flame front is flattened by heat diffusion, whereas it is amplified by the mass diffusion of the reactants. However, advective fluxes created ahead of a wrinkled front interfere with the diffusive fluxes, so that the net effect is globally stabilizing for usual hydrocarbon flames (Clavin 1983).

The third mechanism is destabilizing: The normal acceleration of the gas across the flame front induces a deflection of the streamlines and gives rise, in turn, to a modification of the gas velocity ahead of the flame front. As a consequence, the incoming velocity of the unburnt gas is increased where the flame lags locally behind the average front position. This is the well-known hydrodynamic instability (Darrieus 1938, 1945; Landau 1944).

The strength of these effects is essentially controlled by three parameters:
1) The expansion coefficient,

$$\gamma = 1 - \rho_b/\rho_0 \simeq 1 - T_0/T_b \simeq 1 - U_0/U_b$$

where ρ, T, U are, respectively, the density, temperature, and gas velocity in the fresh mixture (indexed with o) and in the burnt gas (indexed with b). This parameter controls simultaneously the strength of the buoyancy forces and of the hydrodynamic instability.

2) The normal flame speed, taken equal to the normal gas velocity U_0 just ahead of the flame front. Increasing the normal flame velocity increases the strength of the hydrodynamic instability.

3) The Lewis number, $L_e = D_{th}/D_m$, which measures the relative efficiency of the diffusion of heat (D_{th} is the thermal diffusivity) to the diffusion of mass (D_m is the molecular diffusivity). In the present work, the mass fractions of the reactants are small, and the thermal diffusivity is essentially that of the dilutant (N_2). The effective Lewis number, which controls the actual way in which the flame speed depends on the different species present is not necessarily given simply by the diffusivity of the limiting component but may also depend on the concentration of intermediate species produced in the reaction layer. However, we may still expect the flame to be more stable when the concentration of the heavier reactant (propane in this work) is decreased.

The main conclusions of this linear analysis are shown on Fig. 1, where the amplification rate σ of an initial perturbation is plotted as a function of its wavenumber k (the deviation of the flame front position from the planar case is taken as $a(y,t) = a\ e^{iky+\sigma t}$). The stabilization by buoyancy effects is dominant at low wavenumbers, while transverse diffusion gives a contribution proportional to k^2 that stabilizes the perturbations at large wavenumbers. Independently of the equivalence ratio, an increase in the flame speed increases the positive contribution of hydrodynamical effects, and the amplification rate becomes positive for a critical flame speed U_{fc} related to the diffusive properties of the mixture and the expansion coefficient, γ. This instability threshold may be represented in a universal form in the plane of reduced Lewis number[‡] ℓ vs flame speed (for a given value of γ). The original "universal" result of Pelcé and Clavin (1982) for adiabatic flames with constant transport properties has been extended by Clavin and Nicoli (1984) to include the effects of heat losses; and by Clavin and Garcia (1983) to include the effects of temperature-dependent transport coefficients. These latter considerations modify the form of the theoretical instability limit; however, one major conclusion of these authors is that to first-order analysis, the size of the structures appearing at the instability threshold can be expressed in a form that does not depend on the transport coefficients or the chemical kinetics, but is simply given by

$$\Lambda_c = \pi U_{fc}^2/(1-\gamma)g \qquad (1)$$

where g is the acceleration of gravity. This provides an opportunity to check this model without further assumptions concerning the actual chemical kinetics or the transport properties of the mixture.

Apparatus

The stability properties of a premixed flame must be studied in a flowfield that is uniform, homogeneous, and laminar on both sides of the flame front. This requirement eliminates devices commonly used for combustion studies, such as flame holders or porous plugs and also flames stabilized by stagnation flows, which strongly modify the flow conditions and the expected stability limits (Clavin 1983). For this study a flat flame burner (11 cm in diameter) has been constructed in which the flame is far from any obstacle but is dynamically stabilized by adjustment of the mean flow velocity (Fig. 2). The mass flow of each gas (fuel, air, nitrogen) is regulated by a sonic throat following two pressure regulators that reduce the fluctuations to about 0.1%.

[‡] $\ell = \beta(1-1/L_e)$, where β is the reduced activation energy related to the actual activation energy E_a through $\beta=(E_a/kT_b)(T_b T_0)/T_b$. The Boltzmann constant is k.

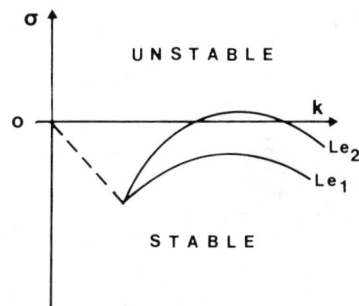

Fig. 1 Amplification rate σ vs wavenumber k for identical hydrodynamic conditions but with different Lewis numbers ($L_{e1} > L_{e2}$).

The secondary flow is regulated with an electrovalve controlled by the tomographic signal (Boyer et al. 1983). The flow velocity can be measured at any point in the burner using a laser Doppler anemometer. Oil droplets are used for tomography measurements and for velocity measurements in the unburnt gas, whereas alumina particles are used for velocity measurements in the burnt gas.

The main experimental difficulty is to maintain a uniform velocity field ahead of the flame front while the upstream part of the burner is nonuniformly heated by heat radiation from the downstream walls. This heating can induce convective flows with velocities of the same order of magnitude as the mean flow velocity. In these experiments, a nearly uniform temperature field is ensured up to 5 cm ahead of the flame front via differential wall heat exchangers (see Fig. 2). The temperatures of three independent cooling circuits are adjusted so that the wall temperature, seen by the reactive flow, is maintained as close as possible to the temperature of the flow itself. Because of thermal conduction along the burner walls, a thin convective boundary layer is established 4 to 5 cm ahead of the flame front, but the velocity field in the center of the burner does not vary more than 1% over a diameter of 9 cm and from 3 cm upstream of the flame front to the preheat zone (Fig. 3).

When the flame front is flat and stable, the normal flame speed is taken as the gas velocity just before the increase of velocity in the preheat zone. When the flame is unstable, the normal flame speed is taken as the mean velocity measured 0.3 to 0.5 cm upstream of the luminous zone of the flame. This measurement procedure is the only one that yields reproducible results for

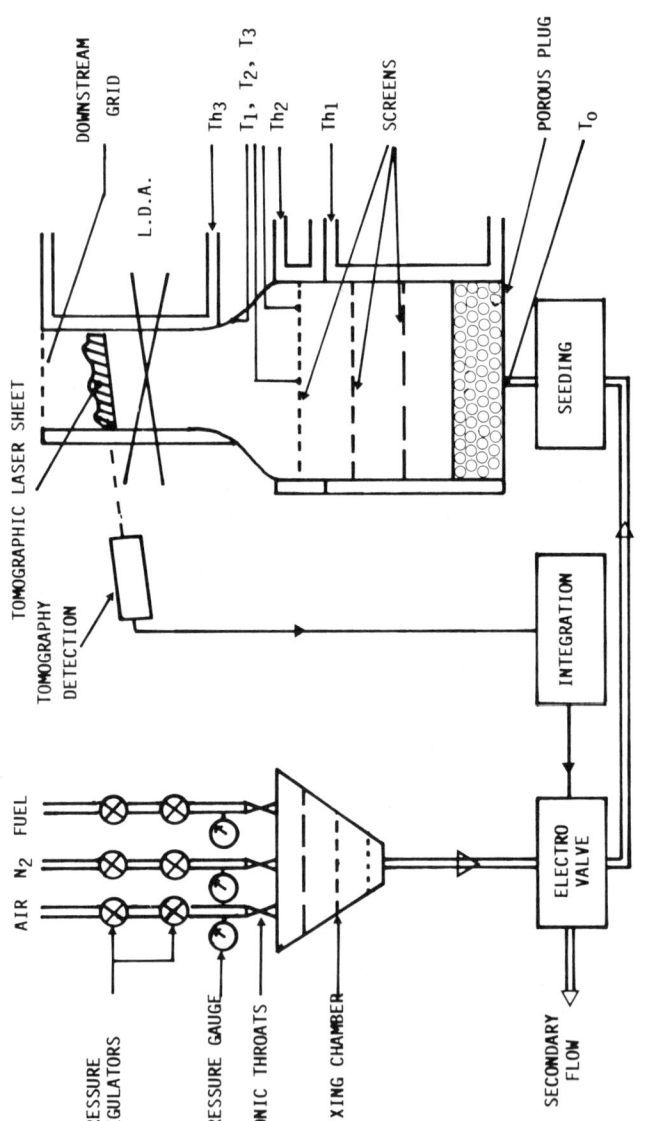

Fig. 2 Apparatus. Th1, Th2, Th3: Water flows. T_0, T_1, T_2, T_3: temperature probes. LDA: laser Doppler anemometry setup.

a given composition in nonuniform flowfields. In this latter case, the velocity field induced by the presence of the quasiplanar front is nonzero (see Searby et al. 1983). In this work, situations have been observed for which the induced field causes the total velocity field to vary by more than 50% in the distance 1 cm upstream of the preheat zone. These experiments suggest the basis of a standard protocol for the measurement of the "fundamental normal flame velocity", which can be defined uniquely as the gas velocity observed at the intersection between the fast varying region (preheat zone) and slow varying region (approach zone) for a free flame stationary in the laboratory frame. The results presented here were obtained at the center of the burner, but it has been checked that the normal flame speed is constant to within 1% over a diameter of 9 cm.

Close to the instability threshold, the wrinkles of the flame front are of small amplitude (less than 0.1 cm in height). The most sensitive technique for measuring the cell size is to produce photographs with a high contrast obtained by full development of a Kodak Tri-X Pan film (400 ASA) up to 1600 ASA. The views are taken from below, at an angle of about 45 deg. The photographs are then reduced to digital form for analysis with an HP85 computer to compensate the perspective effects and the optical deformations due to the cylindrical shape of the burner (Fig. 4). With this technique, it was possible to obtain reproducible results (flame speed, stability limit, critical cell size) to within 1%.

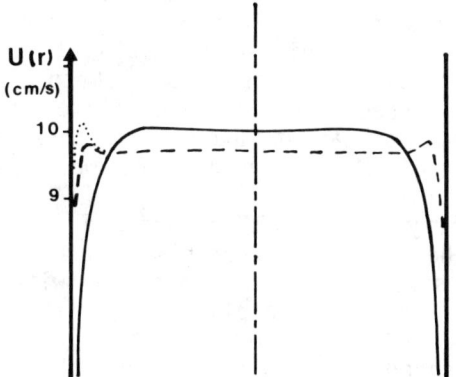

Fig. 3 Radial velocity profile in the presence of a flame (burner diameter = 11 cm). (———): 6 cm upstream of the flame front; (----): 2 cm upstream of the flame front; (····): 0.5 cm upstream of the flame front.

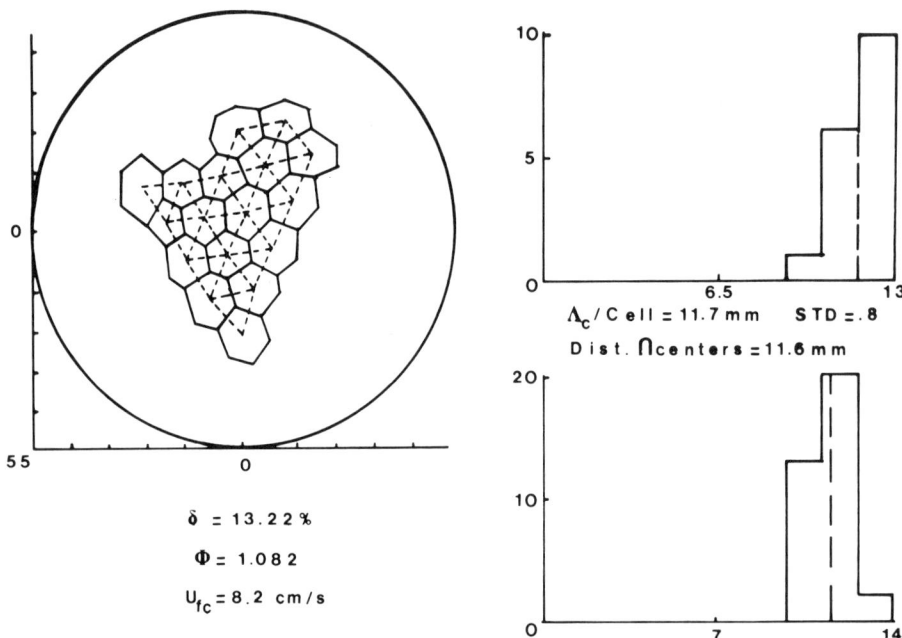

Fig. 4 Left side: reconstitution of the flame front structure. Upper right: histogram of the mean wavelength per cell (Λ_c in mm). Lower right: histogram of the mean distance between centers.

Results

The present work concerns propane/air mixtures diluted with nitrogen. The normal flame speed has been measured as a function of the equivalence ratio (between 0.8 and 1.2) and for various dilutions (from 12% to 16% of oxygen in the mixture) (see Fig. 5). The flame speed at the extinction limit was about 6 cm/s in this burner, and the temperature of the fresh mixture was maintained at 300 ±1 °K.

The stability limit was determined when the flame front showed the first perceptible signs of structure, but the cell size could be measured with sufficient precision only when the amplitude of the corrugation reached about 5% of its characteristic length, that is, when the flame speed was 1% or 2% larger than the critical flame speed at the threshold. However, experimental observations suggest that the averaged cell size remains relatively constant near the threshold. Figure 6 shows the measured cell size for cell amplitudes ranging from

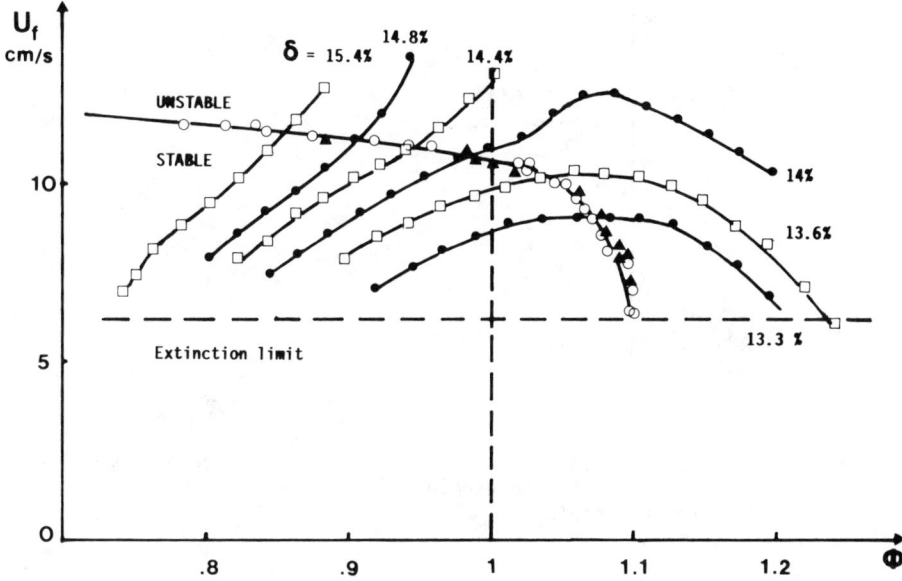

Fig. 5 □, ●: Flame velocity of $C_3H_8/O_2/N_2$ gas mixtures. Dilution $\delta = O_2/(O_2+N_2)$ by volume; ϕ = equivalence ratio; ○, ▲ = stability limit.

5% to 50% of the cell size. In the light of these remarks, these results should be comparable to the results obtained from the linear stability analysis established for the size of the structures at the threshold.

The representation of the stability limit (Fig. 5) is not universal, as it depends upon the diffusive properties of the mixture and the kinetic characteristics of the reactions. However, the comparison with the theoretical predictions (Fig. 7) evaluated for a simplified chemical scheme and with constant diffusion coefficients shows good qualitative agreement, particularly when heat losses are taken into account: The negative slope of the stability limit on the lean side of the mixture can be explained by considering the evolution of the Lewis number from a high value $(L_e(C_3H_8 \to N_2) \simeq 1.7)$ towards smaller values $(L_e(O_2 \to N_2) \simeq 0.9)$. Close to the stochiometric composition, the critical velocity decreases faster due to the joint effect of the decrease in Lewis number and the relative increase of heat losses.

The experimental values of the critical cell size are plotted as a function of the critical velocity in Fig. 8 along with the theoretical values calculated from

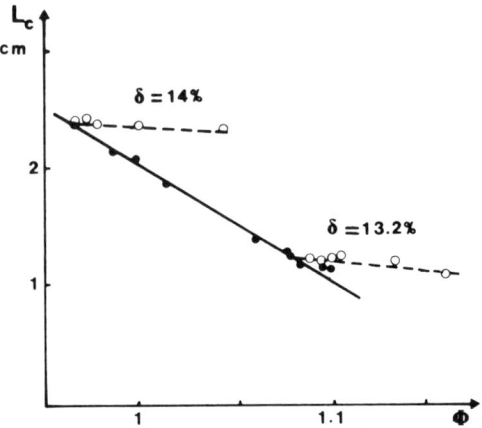

Fig. 6 Cell size as a function of the equivalence ratio. ●: averaged values at the threshold; ○: measured values when moving away from the threshold.

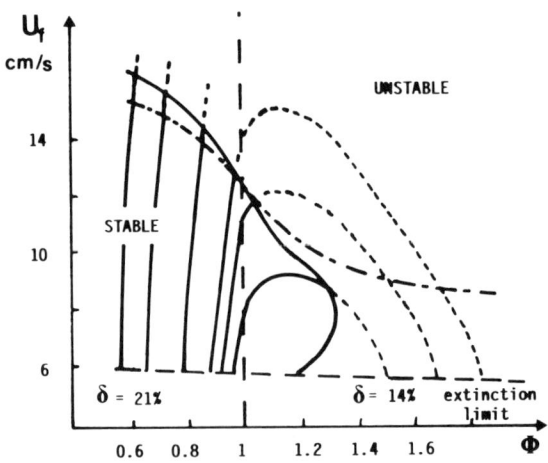

Fig. 7 Calculated values of the flame velocity and the stability limits (Clavin and Nicoli 1983). (—·—): stability limit in the adiabatic case; (———): stability limit with heat losses leading to an extinction velocity of 6 cm/s.

Eq. (1) and the experimental value of the expansion coefficient (Fig. 9). The mean cell size is in quantitative agreement with Eq. (1), and, moreover, it increases effectively with the critical flame speed as predicted. However, the expected parabolic behavior is not well verified. Below 10 cm/s, the critical wavelength tends to a limiting value of about 1 cm. This occurs on the

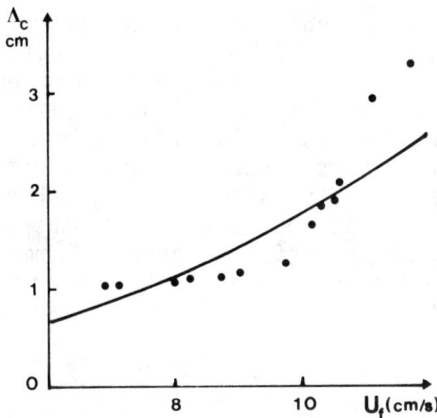

Fig. 8 Experimental values of the critical wavelength vs critical flame velocity. Full line: theoretical prediction with the experimental value of γ.

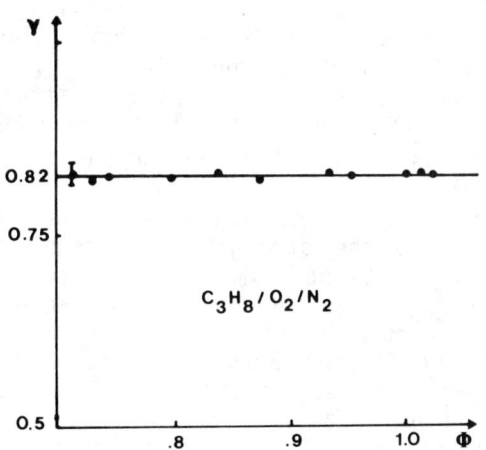

Fig. 9 Measured expansion coefficient as a function of the equivalence ratio along the stability limit.

rich side of the marginal stability curve close to the extinction limit, where heat losses become important and may affect the validity of the linear analysis. The stabilizing higher-order terms in the development of the linear analysis would also lead to a relative increase of the critical wavelength particularly at low flame speeds.

When the critical flame speed is greater than 10 cm/s, the slope is steeper than predicted and varies roughly as U_{fc}^3. Small terms neglected in the derivation of relation (1), would produce a similar deviation but their magnitude is too small to explain the observed discrepancy. However, it is not certain that the finite size of the burner (and hence of the flowfield) has no effect on the size of larger structures. Further experiments are planned in different sized burners to elucidate this point.

Finally, it is possible that these discrepancies arise from nonlinear effects, absent from the theoretical models, but which are necessarily present and limit the amplitude of the cell height in the unstable domain.

Conclusion

These experiments, performed in conditions as close as possible to those considered in the theoretical models, confirm the existence of a stability threshold with the appearance of structures of finite size at the threshold. This finite size results from the competition between three fundamental mechanisms controlling the stability of a planar premixed flame propagating downwards: buoyancy, hydrodynamics, and diffusion. For a given equivalence ratio, that is, for given diffusive properties of the reactants, the instability threshold is crossed when the strength of the hydrodynamic instability is increased following an increase in the normal flame speed. The critical velocity decreases with the effective Lewis number, owing to the strengthening of destabilizing mass fluxes. As a first approximation, the cell size is given by $\Lambda_c = \pi U_{fc}^2/(1-\gamma)g$, in qualitative agreement with the theory; but further experiments are needed with other fuels and in different burners to establish whether the observed deviations of critical cell sizes are caused by finite burner size and heat losses or by transport effects arising from chemical kinetic schemes that cannot be considered to be confined to a thin reaction zone.

Acknowledgments

Thanks are due to J. P. Pahin and F. Abetino for their technical assistance. This work was supported in part by the CNRS and by the Direction des Recherches et Etudes Techniques under Contract No. 80/434.

References

Barenblatt, G. I., Zel'dovich, Y. B., and Istratov, A. G. (1962) On diffusional thermal stability of laminar flame. Prikl. Mekh. Fiz. 2, (n°4) 21-26.

Boyer, L., Searby, G., and Quinard, J. (1983) Some recent applications of laser tomography in combustion. <u>Proceedings of XV ICHMT Symposium on Heat and Mass Transfer Measurement Techniques</u>, Dubrovnik, Hemisphere publishing corporation (to appear).

Clavin, P. (1984) Dynamical behavior of premixed flame fronts in laminar and turbulent flows. <u>Progress in Energy and Combustion Science</u>, Pergamon Press, New York (to appear).

Clavin, P. and Garcia, P. (1983) The influence of the temperature dependence of diffusivities on the dynamics of flame fronts. J. Mec. Théo. Appl. 2, (n°2) 245-263.

Clavin, P. and Nicoli, C. (1984) Effect of the heat-losses on the limits of stability of premixed flames propagating downwards. <u>Combust. Flame</u> (to appear).

Darrieus, G. (1938, 1945) Propagation d'un front de flamme. Essai de théorie des vitesses anormales de déflagration par dévelopement spontané de la turbulence. Presented at <u>La Technique Moderne</u> (1938) and at <u>Le Congrès de Mécanique Appliquée</u> (unpublished).

Landau, L. (1944) On the theory of slow combustion. <u>Acta Physicochim. URSS</u> 19 (n°1), 77-85.

Markstein, G. H. (1964) <u>Nonsteady Flame Propagation</u> (p.15-103). Pergamon Press, New York.

Pelcé, P., and Clavin, P. (1982) Influence of hydrodynamics and diffusion upon the stability limits of laminar premixed flames. J. Fluid Mech. 124, 218-237.

Searby, G., Sabathier, F., Monreal, J., Clavin, P. and Boyer, L. (1983) The feed-back of a flame front on turbulent flows. IXth<u>International Colloquium on Dynamics of Explosions and Reactive Systems</u>, Poitiers, Ed.AIAA Progress in Astronautics and Aeronautics.

Smithell, A. and Ingle, H. (1982) <u>J. Chem. Soc.</u> 61, 204.

Nonsteady Gasdynamic Effects in the Induction Domain Behind a Strong Shock Wave

J. F. Clarke* and R. S. Cant†
Cranfield Institute of Technology, Bedford, England

Abstract

Processes in the region between a driving piston and a strong shock wave (e.g., Mach number > 1.5) can be modeled theoretically as relatively small-perturbation events. The combustion reaction is assumed to be of simple irreversible Arrhenius type; perturbations are then of the order of the dimensionless inverse activation energy (typically 1/25). Events behind the shock are then described by either an integral equation, which is solved by straightforward numerical iterative methods, or by a differential equation. The latter reveals the basic physics as a coupling between four limiting processes, namely, explosion at constant pressure and at constant volume coupled with wave propagation at the isentropic and isothermal sound speeds. Illustrative results demonstrate that induction activity within a fluid element takes place midway between constant-volume and constant-pressure limits, with appropriate influence on the time to ignition, when piston speed is constant. Thermal expansion of fluid elements produces significant gasdynamical activity, whose influence is also evident in the strong dependence of induction time on weak modifications of piston path behavior.

Introduction

The unperturbed atmosphere is assumed to consist of a "cool" spatially uniform combustible gas mixture; its induction time is therefore very long in the context of the

---

Presented at the 9th ICODERS, Poitiers, France, July 3-8, 1983. Copyright © by the American Institute of Aeronautics and Astronautics, Inc., 1983. All rights reserved.
*Professor of Gasdynamics, Department of Aerodynamics.
†Research Student, Department of Aerodynamics.

time intervals that will be of interest in the present analysis. It is assumed that this atmosphere is disturbed by a shock wave of sufficient strength to make the induction time immediately downstream of the shock in its initial condition very much shorter than the "long" time that exists in the ambient atmosphere. Induction times in simple Arrhenius reactions are proportional to $(T/\theta)\exp(\theta/T)$, where T is absolute temperature and θ is an activation-energy "temperature" (Strethlow 1969). Thus a shock of Mach number M_s equal to 1.46 lowers the ambient induction time to less than 1/100th of the ambient value for a θ value equal to twenty times the shocked-gas temperature (specific heats ratio $\gamma = 1.3$). One can think of the shock as having been created initially by the movement of a piston. In the highly practical cases of a shock tube experiment or the bursting of a high-pressure vessel into a combustible atmosphere the role of the piston is played by the contact surface.

The present task is to evaluate events as they evolve for times $t>0$ between the contact surface and the primary shock, whose main function has been to switch on chemical activity between the two surfaces. With a planar configuration it is convenient to use a Lagrangian formulation of the conservation equations, and also to make these equations dimensionless with the following set of typical quantities: temperature T_c, density ρ_c, pressure p_c, sound speed $a_c = (\gamma p_c/\rho_c)^{\frac{1}{2}}$, fuel mass fraction c_c, and time t_c. Distances are measured in units of $a_c t_c$, so that the dimensionless Lagrangian coordinate ψ is given by

$$\rho_c a_c t_c \psi = \int_{x'_p(t')}^{x'} \rho'(\hat{x}, t') d\hat{x} \qquad (1)$$

where $x_p'(t')$ is the piston path in dimensional x',t' space. All subscript c quantities are given by the conditions that exist immediately downstream of the (strong) initial shock, whose Mach number is M_{si}.

If the chemical reaction is a simple decomposition of the form "fuel species F → product species P", the equations can be written as follows:

$$\rho_t + \rho^2 u_\psi = 0 \qquad (2)$$

$$\gamma u_t + p_\psi = O(Re^{-1}) \qquad (3)$$

$$T_t + Qc_t - (\gamma - 1)p_t/\gamma\rho = O(Re^{-1}) \qquad (4)$$

$$c_t + \Omega\exp(-\theta/T)c = O(Re^{-1}) \qquad (5)$$

$$p = \rho T \qquad (6)$$

where Q is the chemical energy available for release by the reaction $F \to P$, measured in units of $C'_p T_c$ (C'_p = specific heat at constant pressure), and

$$\Omega = WB't_c \qquad (7)$$

Here W is the molecular weight of F and B' is the pre-exponential factor or reaction frequency. The quantity θ is the dimensionless activation temperature (molar activation energy divided by RT_c) and Re is a Reynolds number, namely

$$Re = a_c^2 t_c \rho_c/\mu = t_c/t_{coll} \qquad (8)$$

where μ_c is a typical dynamic viscosity and t_{coll} is a typical molecular collision interval.

The terms $O(Re^{-1})$ on the righthand sides of Eqs. (3,4 and 5) consist of molecular transport terms that will be presumed negligible in the present first phase of the analysis here, since t_c will be identified with a chemical time that is essentially much greater than t_{coll}. These $O(Re^{-1})$ terms will eventually have a part to play in some region of the evolving flowfield.

Conditions at the Shock and at the Piston

The righthand side of Eq. (1) is the mass of gas, per unit area of streamtube, that exists between the piston and position x'. When the latter is a position x' on the shock wave, namely $x'_s(t')$, a simple conservation of mass argument shows that $\rho_c a_c t_c \psi_s(t)$ is equal to $\rho_a x'_s(t')$, where $\psi_s(t)$ is the value of ψ on the shock and ρ_a is the ambient atmosphere density ahead of the shock. Since shock Mach number M_s is equal to $(dx'_s/dt')/a_a$ it follows that

$$d\psi_s/dt = (\rho_a a_a/\rho_c a_c)M_s \equiv K_a M_s/M_{si} \qquad (9)$$

Equation (9) also defines K_a, which will be useful in what follows; K is a function of γ and M_{si}, such that $K_a \leq 1$

for $M_{si} \geq 1$;

$$K_a = \{[2 + (\gamma - 1)M_{si}^2]/(2\gamma M_{si}^2 - \gamma + 1)\}^{\frac{1}{2}} \qquad (10)$$

The values of u, p, and ρ immediately downstream of the shock, namely u_s, p_s and ρ_s are functions of M_s, M_{si}, and γ. When $M_s = M_{si}$ both p_s and ρ_s are unity in the present system, while $u_s = u_{si} = u_{pi}$ (u_{pi} is the piston or contact surface velocity); in general

$$u_s = 2M_{si}(M_s - M_s^{-1})\{[2 + (\gamma - 1)M_{si}^2](2\gamma M_{si}^2 - \gamma + 1)\}^{-\frac{1}{2}} \qquad (11)$$

If the shock Mach number M_s differs from M_{si} by only a small $O(\sigma)$ amount, where $\sigma \ll 1$, one can write

$$M_s = M_{si}[1 + \sigma m(t)] \qquad (12)$$

It can then be shown that

$$u_s - u_{pi} \simeq \sigma K_u m(t) \qquad (13a)$$

$$p_s - 1 \simeq \sigma K_p m(t) \qquad (13b)$$

$$\rho_s - 1 \simeq \sigma K_\rho m(t) \qquad (13c)$$

where

$$K_u \equiv 2(M_{si}^2 + 1)\{2 + (\gamma - 1)M_{si}^2 (2\gamma M_{si}^2 - \gamma + 1)\}^{-\frac{1}{2}} \qquad (13d)$$

$$K_p \equiv 4\gamma M_{si}^2 (2\gamma M_{si}^2 - \gamma + 1)^{-1} \qquad (13e)$$

$$K_\rho \equiv 4((\gamma - 1)M_{si}^2 + 2)^{-1} \qquad (13f)$$

These results are important in the subsequent analysis.

At the piston it is only necessary to specify that the gas velocity u is equal to the prescribed piston velocity, recalling that the latter must commence with the value $u_s = u_{si} = u_{pi}$; u_{pi} is found from Eq. (11) with M_s equal to M_{si}.

Perturbation Analysis

It is now proposed to consider small $O(\sigma)$ perturbations of the field from its initial postshock condition. In the early stages of the shock heating processes departures from uniform conditions will always be small, but this is not at all to say that such disturbances are insignificant, as the analysis here will show. To that end one must define the set of coefficient functions $\zeta^{(1)}$ as follows:

$$\zeta \sim \zeta_0 + \zeta^{(1)}(t,\psi), \quad \zeta = p,\rho,T,c,u$$

$$\zeta_0 = 1, \quad \zeta = p,\rho,T,c \quad (14)$$

$$u_0 = u_{pi} = u_{si}$$

together with the limit $\sigma \to 0$ with t,ψ fixed. The quantity u_{pi} is the initial piston speed, given by Eq. (11) with M_{si} written for M_s. The earlier statements at the end of the Introduction about the Reynolds number can now be formalised to read $Re^{-1} = o(1)$.

Behavior of the Arrhenius factor in Eq. (5) will depend on the relative orders of magnitude of θ and σ; if $\theta\sigma = o(1)$,

$$\exp(-\theta/T) \sim \exp(-\theta)[1 + \sigma\theta T^{(1)} + \ldots] \quad (15a)$$

w ile if $\theta\sigma = 1$, and θ is therefore necessarily large,

$$\exp(-\theta/T) \sim \exp(-\theta)\exp(T^{(1)})(1 + \ldots) \quad (15b)$$

Evidently Eq. (5) and the asymptotic forms Eqs (14) demand that σ shall always be proportional (or even equal) to $\Omega \exp(-\theta)$; since Q is $O(1)$ in the σ limit, by hypothesis, it is convenient to choose to make

$$\sigma = Q\Omega \exp(-\theta) \quad (16)$$

Therefore

$$\theta\sigma = o(1) \Rightarrow Q\Omega\theta\exp(-\theta) = o(1) \quad (17a)$$

$$\theta\sigma = 1 \Rightarrow Q\Omega\theta\exp(-\theta) = 1 \quad (17b)$$

and the choice of behavior according to either one of these statements becomes a matter of the choice of value for the "typical" time t_c, given that all of the other factors in the problem, namely Q, Ω [hence W,B'; see Eq. (7)] and θ are all parameters whose magnitudes are fixed by the character of any particular gas mixture. Clearly eq. (17a) refers to a smaller t_c value than Eq. (17b). The latter implies that

$$t_c = \exp(\theta)/Q\theta WB' \equiv t_I \qquad (18)$$

which is the induction time for the reaction under initial postshock conditions.

The value t_I for t_c will be significant in the major part of the analysis here, but it is both interesting and useful to exploit the "short" time forms that derive from the adoption of Eq. (17a). It is important to observe that on the piston

$$u(t,0) = u_{pi} + \sigma u_p^{(1)}(t) \qquad (19)$$

where $\sigma u_p^{(1)}(t)$ is specified. The perturbation scheme limits the choice of $u(t,0) - u_{pi}$ to values that are $O(\sigma)$.

Results for Times $t \ll t_I$

When the orderings are in accord with Eq. (17a) the equations for the coefficient functions become

$$\rho_t^{(1)} + u_\psi^{(1)} = 0 \qquad (20a)$$

$$\gamma u_t^{(1)} + p_\psi^{(1)} = 0 \qquad (20b)$$

$$p^{(1)} = \rho^{(1)} + T^{(1)} \qquad (20c)$$

$$T_t^{(1)} - (\gamma - 1)p_t^{(1)}/\gamma = 1 = -Qc_t^{(1)} \qquad (20d)$$

From Eq. (19), $u^{(1)}(t,0)$ must be equal to $u_p^{(1)}(t)$, while Eqs. (13a and b) combine to give the condition at the shock.

$$\gamma u^{(1)}[t,\psi_s(t)] = \gamma K_u K_p^{-1} p^{(1)}[t,\psi_s(t)] \equiv r p^{(1)}[t,\psi_s(t)] \qquad (21)$$

Equation (21) also defines the factor r which, it should be noted, is close to unity for the whole range of Mach number M_{si}. In fact for M_{si} in $1 \leq M_{si} \leq \infty$, $1 \geq r \geq \gamma/2(\gamma-1)^{\frac{1}{2}}$, with monotonic behavior as M_{si} increases.

The set of equations (20) is easily solved up to the point that one finds expressions for velocity and pressure perturbations,

$$\gamma u^{(1)}(t,\psi) = -f'(\xi) + f'(\eta) + \tfrac{1}{2}\gamma\psi + \gamma u_p^{(1)}(\eta) \quad (22)$$

$$p^{(1)}(t,\psi) = -f'(\xi) - f'(\eta) + \tfrac{1}{2}\gamma t - \gamma u_p^{(1)}(\eta) \quad (23)$$

where

$$\xi = t - \psi \quad (24a)$$

$$\eta = t + \psi \quad (24b)$$

and f' is an unknown function that is to be found by ensuring satisfaction of Eq. (21). This leads to the following functional equation for f', where ξ_s, η_s is written for ξ,η on the shock at a time t, for which ψ is equal to ψ_s [see especially Eq. (9)];

$$f'(\eta_s) = \left(\frac{1-r}{1+r}\right)f'(\xi_s) + \tfrac{1}{4}\gamma\left[\xi_s - \left(\frac{1-r}{1+r}\right)\eta_s\right] \quad (25)$$
$$-\gamma u_p^{(1)}(\eta_s)$$

If r is very close to unity $(1-r)/(r+1)$ is a small number (e.g., 0.1 for $M_{si} \leq 3$, $\gamma = 4/3$) and an accurate estimate of f' is found from Eq. (25) by ignoring any quantities that contain this factor as an element, and by writing 1 for r elsewhere. It is also true that $\psi_s \simeq K_a t$ to order σ accuracy in the early stages of the process; in such circumstances

$$\xi_s \simeq (1 - K_a)\eta_s/(1 + K_a)$$

and, ignoring $u_p^{(1)}$ for the present, it is quite accurate to take

$$f'(\eta) \simeq \tfrac{1}{4}\gamma\left(\frac{1-K_a}{1+K_a}\right)\eta \quad (26)$$

INDUCTION DOMAIN BEHIND A STRONG SHOCK WAVE

Estimates of the early time behavior can then be acquired in the following analytically simple forms (illustrated here on the assumption that the piston speed is constant, for brevity)

$$u^{(1)} = \psi/(1 + K_a) \tag{27a}$$

$$p^{(1)} = \gamma K_a t/(1 + K_a) \tag{27b}$$

$$\rho^{(1)} = \left[(\psi/K_a)(1 + K_a K_\rho K_u^{-1}) - t\right](1 + K_a)^{-1} \tag{27c}$$

$$T^{(1)} = (\gamma K_a + 1)t - (\psi/K_a)(1 + K_a K_\rho K_u^{-1})(1 + K_a)^{-1} \tag{27d}$$

$$m(t) = K_a t K_u^{-1}(1 + K_a)^{-1} \tag{27e}$$

As can be seen from Eqs. (27e) and (12) the shock accelerates linearly with time to this approximation; temperature rises most quickly where ψ is zero, namely at the piston face; pressure increases uniformly with time across the space between piston and shock. In view of the absence of additional piston motions ($u_p^{(1)} = 0$) all of this behavior is brought about, at this stage, by the small and spatially uniform [see Eqs. (20d and e)] amount of chemical energy release. Even in the present short time intervals since the motion was initiated, the true situation is not quite this simple, in view of the adoption of Eq. (26) in place of Eq. (25). Proper account of disturbance behavior at the shock can be restored by reinstating r as a quantity near to but not equal to unity; in this case Eq. (25) is reasonably easily solved by successive direct estimations of f' in the righthand side of Eq. (25) by using the expression itself in an obvious way; if iteration is used to correct ψ_s at each time step a solution of Eq. (25) to any desired accuracy can be achieved and can be readily handled by a computer. It is important to do this as M_{si} rises since $(1 - r)/(1 + r)$ also increases in these circumstances.

Apart from their intrinsic merit as respectable estimates of early time behavior, especially when augmented by the $u_p^{(1)}$ effects where necessary. Eqs. (27) provide useful starting points for the iterative solutions, both for short times and for times approaching t_I, as is to be described in the next section. The present early time results have assumed that $\theta\sigma$ is $o(1)$ [recall Eq. (17a)]. It can be seen from Eq. (27d) that $_\sigma T^{(1)}$ behaves like σt on the piston path; if t increases to values that are of order $1/\theta\sigma$ it is clear that $\exp(-\theta/T)$ can no longer be written as in Eq. (15a). The time scale t_c implicit in Eq. (17a) is then no longer appropriate and one must adopt the value appropriate to Eqs. (15b) and (17b), namely t_I, as defined in Eq. (18).

Developments During Times of Order t_I

When $\theta\sigma$ is equal to one and t_c is equal to t_I, Eq. (5) becomes

$$Qc_t + \theta^{-1}\exp[\theta(1 - 1/T)]c = O(Re^{-1}) \qquad (28)$$

and any perturbation scheme such as Eq. (14) will require $\zeta - \zeta_0 = O(\theta^{-1})$ for $\theta \gg 1$. This small change in the orders of magnitude of disturbances relative to the dimensionless activation energy factor θ has a profound effect upon the physics of the problem. If the $O(Re^{-1})$ transport terms can be neglected, Eq. (4) can be rewritten with the aid of Eqs. (2), (6), and the result that

$$a^2 = T = p/\rho \qquad (29)$$

where a is the local sound speed measured in units of a_c, as follows

$$p_t + \gamma\rho^2 a^2 u_\psi = \gamma\rho\theta^{-1}\exp[\theta(1 - 1/T)]c \equiv S \qquad (30)$$

Adding and subtracting $\gamma\rho a$ times Eqs. (3) to (30) give

$$p_\eta + \gamma\rho a u_\eta = St_\eta \qquad (31a)$$

$$p_\xi - \gamma\rho a u_\xi = St_\xi \qquad (31b)$$

where

$$\psi_\eta = \rho a t_\eta \qquad (32a)$$

$$\psi_\xi = -\rho a t_\xi \qquad (32b)$$

INDUCTION DOMAIN BEHIND A STRONG SHOCK WAVE   151

and these characteristic-parameter versions of the equations will be useful in the subsequent analysis. The use of ξ and η as symbols for the pair of characteristic parameters will be shown not to conflict seriously with Eqs. (24a and b), although some careful interpretation will be called for.

Analysis of the field between $\psi = 0$ and $\psi_s(t)$ can now proceed via the proposition that

$$\zeta \sim \zeta_0 + \theta^{-1}\zeta^{(1)}(\xi,\eta), \quad (33)$$

$$\zeta = p,\rho,a,T,c,\psi,t$$

where the unknown quantities at this stage now include ψ and t, since ξ and η are now the independent variables. Values of ζ_0 are still given in Eq. (14) but ψ_0 and t_0 must be found from Eqs. (31c and d) and (33), which give

$$\psi_{0\eta} = t_{0\eta} \quad (34a)$$

$$\psi_{0\xi} = -t_{0\xi} \quad (34b)$$

Thus $\psi_0 - t_0$ is a function of ξ and $\psi_0 + t_0$ is a function of η. Making

$$\psi_0 - t_0 = -\xi \quad (35a)$$

$$\psi_0 + t_0 = \eta \quad (35b)$$

implies that $\xi = \eta$ when $\psi_0 = 0$, which is clearly the lowest order estimate of the position of the piston. The link between ξ,η here and in the previous section, especially Eqs. (24a and b), is obvious and requires one to view the solutions there as valid only to lowest order in their distributions through ψ,t space.

With Eq. (33) the chemical "source" term S, defined in Eq. (30), reduces to $S \sim \gamma\theta^{-1}\exp[T^{(1)}]$, where Eqs. (31a and b) integrate once to give

$$p^{(1)} + \gamma u^{(1)} = \tfrac{1}{2}\gamma \int_{\xi}^{\eta} \exp[T^{(1)}]\, \partial\hat{\eta} + 2f(\xi) \quad (36a)$$

$$p^{(1)} - \gamma u^{(1)} = \tfrac{1}{2}\gamma \int_{\eta}^{\xi} \exp[T^{(1)}]\, \partial\hat{\xi} + 2g(\eta) \quad (36b)$$

Lower limits on the integrals are chosen so that application of the condition on the piston makes

$$\gamma u_p^{(1)} = f(\eta) - g(\eta) \tag{37}$$

The condition on the shock is the same as Eq. (21), except that it is expedient in this section to choose η and calculate $\xi = \xi_s(\eta)$ on the shock.

The function f is therefore found from

$$f(\eta) - \left(\frac{1-r}{1+r}\right) f(\xi_s) = \frac{1}{4}\gamma \int_{\xi_s}^{\eta} \exp\left[T^{(1)}(\hat{\xi},\eta)\right] \partial\hat{\xi}$$

$$+ \frac{1}{4}\gamma\left(\frac{1-r}{1+r}\right) \int_{\xi_s}^{\eta} \exp\left[T^{(1)}(\xi_s,\eta)\right] \partial\hat{\eta} + \gamma u_p^{(1)}(\eta) \tag{38}$$

Equation (4) can be written in terms of the characteristic parameter variables as follows;

$$\tfrac{1}{2}\rho\gamma\left[T_\xi(t_\xi)^{-1} + T_\eta(t_\eta)^{-1}\right] - \tfrac{1}{2}(\gamma-1)\left[p_\xi(t_\xi)^{-1} + p_\eta(t_\eta)^{-1}\right] = S \tag{39}$$

where S is defined in Eq. (30), and it can then be shown by using Eqs. (33) and (35a and b) that Eq. (39) gives

$$\gamma T_{t_0}^{(1)} - (\gamma-1) p_{t_0}^{(1)} = \exp(T^{(1)}) \tag{40}$$

to lowest order. Integrating Eq. (40) gives

$$T^{(1)} - (\gamma-1)p^{(1)}/\gamma = \int_{G(\psi_0)}^{t_0} \exp\left[T^{(1)}(\psi_0,\hat{t})\right] \partial\hat{t} + F(\psi_0) \tag{41}$$

where $F(\psi_0)$, $G(\psi_0)$ are two arbitrary functions of ψ_0 at this stage the connection between $\psi_0, t_0, \xi,$ and η is given in Eq. (35). It will be shown in a little more detail in

INDUCTION DOMAIN BEHIND A STRONG SHOCK WAVE 153

the next section that relation Eq. (9) for shock position is equivalent to

$$\psi_0 = \psi_{os} \equiv K_a t_0 \Leftrightarrow (1 - K_a)\eta = (1 + K_a)\xi \equiv (1 + K_a)\xi_s$$

(42)

to lowest order. Making $G(\psi_0)$ equal to ψ_0/K_a means that

$$T^{(1)}(\psi_0, \psi_0/K_a) - [(\gamma - 1)/\gamma] p^{(1)}(\psi_0, \psi_0/K_a) = F(\psi_0)$$

so that Eq. (41) can be written as

$$T^{(1)} = \int_{\psi_0/K_a}^{t_0} \exp[T^{(1)}(\psi_0, \hat{t})] \partial \hat{t} + T^{(1)}(\psi_0, \psi_0/K_a)$$

$$+ [(\gamma - 1)/\gamma][p^{(1)} - p^{(1)}(\psi_0, \psi_0/K_a)]$$

(43)

Any perturbation quantity, such as $T^{(1)}$, is a function of either ξ, η or, via Eq. (35), of ψ_0/t_0; thus $T^{(1)}$ will be written as $T^{(1)}(\xi, \eta)$ or $T^{(1)}(\psi_0, t_0)$, for example.
Equations (36a and b) give a value for $p^{(1)}$ in terms of integrals of $\exp[T^{(1)}]$ and the functions $f(\xi)$ and $g(\eta)$; g can be eliminated in terms of f and $u_p^{(1)}$, via Eq. (37), and Eq. (38) can be solved for f, once again in terms of integrals $\exp[T^{(1)}]$. All of this information can be substituted into Eq. (43), which therefore becomes an integral equation for $T^{(1)}$, and the task of solving this equation iteratively can be delegated to the computer (in the present case a VAX-11/782).
Some results from these calculations will be presented below but first it is important to note that $T^{(1)}$ satisfies a very interesting and informative partial differential equation, as well as the integral equation just described which more readily provides the solution for $T^{(1)}(\psi_0, t_0)$. If Eqs. (2) and (3) are transformed into ξ, η coordinates first, and then Eqs. (33, 34, and 35) are used, it can be shown with the aid of Eq. (40) that

$$\{T_{t_0}^{(1)} - \gamma\exp[T^{(1)}]\}_{t_0 t_0} - \{T_{t_0}^{(1)} - \exp[T^{(1)}]\}_{\psi_0 \psi_0} = 0$$

(44)

Equation (44) has been derived before in other, slightly different, circumstances and by other routes (Clarke 1978a, 1981a; it also appears in embryo in other studies of wave propagation in combustible gases, Clarke 1981b, 1983). It evidently has an important part to play in any description of the way in which small gasdynamic disturbances behave in situations for which they provoke, and are provoked by, significant perturbation to the ambient chemical activity, prior to ignition of the ambient gas.

Eq. (44) expresses a balance between four physical events, epitomised by the four operators, namely $(\ )_{t_0}$
- exp $(\ )$ which implies explosion at constant pressure;

$$(\ )_{t_0} - \gamma \exp(\ )$$

which implies explosion at constant volume; $(\ )_{t_0 t_0} - (\ )_{\psi_0 \psi_0}$
which implies wave propagation at the adiabatic isentropic sound speed; and

$$\gamma (\ )_{t_0 t_0} - (\ )_{\psi_0 \psi_0}$$

which implies wave propagation at the isothermal or Newtonian sound speed.

From the character of the wave operators here it is clear that wave propagation is described as a linear, although dispersive, process. At least this is true in the ψ_0, t_0 space. However some pains have been taken to set up the system of equations in a characteristic parameter space which is capable in some circumstances (see the recent review by Kluwick 1981) of accounting fully, that is to say to a uniform order of approximation throughout a field of interest, for nonlinear cumulative gasdynamic effects. The present system is both dispersive and, through the appearance of Arrhenius-inspired terms $\exp[T^{(1)}]$, nonlinear in its modeling of chemistry. Characteristic parameters will not therefore guarantee a uniformly valid approximation everywhere in the present case, but it is still very useful to monitor behavior of the nonlinear straining factors $\psi^{(1)}$ and $t^{(1)}$ for reasons enlarged upon in the next section.

Nonlinear Gasdynamic Effects

It is important to observe a restriction that is implicit in the link between (for example) the exact statement in Eq. (39) and its approximate interpretation in Eq. (40). The proposition that both t_ξ and t_η are equal to $\frac{1}{2}$ has been relied on in arriving at Eq. (40), which can only be satisfactory so long as $\sigma t_\xi^{(1)}$ and $\sigma t_\eta^{(1)}$ remain

$O(\sigma)$. For this reason it is important to monitor the behavior of these factors. They are calculated from Eqs. (32a and b) and (33); noting from Eq. (29) that $\rho^2 a^2 = p\rho$, it can then be shown that

$$\psi_\xi^{(1)} + t_\xi^{(1)} = -\frac{1}{4} A^{(1)}$$

$$\psi_\eta^{(1)} - t_\eta^{(1)} = \frac{1}{4} A^{(1)} \qquad (45)$$

$$A^{(1)} = A^{(1)}(\xi,\eta) \equiv p^{(1)} + \rho^{(1)}$$

Integration of Eq. (45) is straightforward but clearly requires two conditions from which to evaluate the two arbitrary functions of integration. One can continue to assert that $\xi = \eta$ represents the piston path if one demands

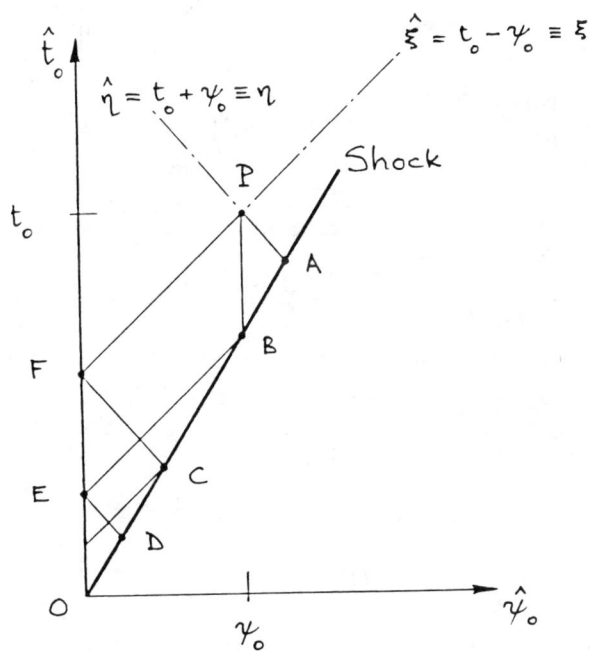

Fig. 1 The coordinates $\hat{\psi}_o$, $\hat{t}_o$ are "source" coordinates that must be used in the integral equation whose solution gives temperature, pressure, etc., at the field point P (ψ_0, t_0). These coordinates provide lowest order estimates of the location of characteristic lines, such as ξ and η. The letters A, B, etc., help to identify the various contributions to the total behavior at P; in particular BP is a particle path; all other paths make up the gasdynamical effect at P.

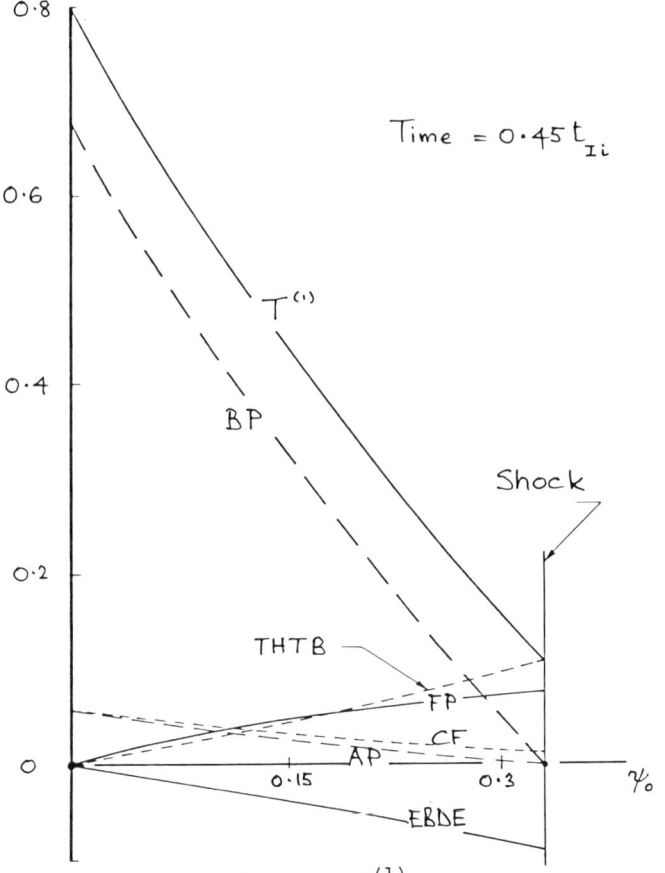

Fig. 2 The temperature increment $T^{(1)}$ at P vs ψ_0 at time $t_0 = 0.45$; piston speed $u_{pi} = $ constant $ = 0.604$, $M_{si} = 1.5$. The letters all relate to Fig. 1, except for THTB which shows the influence of the temperature increment at the (accelerating) shock wave; EBDE is the sum of contributions EB and DE.

$$\psi^{(1)}(\xi = \eta, \eta) = 0 \qquad (46)$$

for all η. Conditions at the shock wave impose their own constraints upon $\psi^{(1)}$ and $t^{(1)}$, as can be seen by first transforming Eqs. (9) and (10) into ξ, η coordinates to give the exact relationship

$$\psi_{\eta s} + \dot{\xi}_s \psi_{\xi s} = K_a (1 + \sigma m \{t [\xi_s(\eta), \overline{\eta}]\})(t_{\eta s} + \dot{\xi}_s t_{\xi s}) \qquad (47)$$

Since $\psi = \psi(\xi,\eta)$, $t = t(\xi,\eta)$; then $\xi = \xi_s(\eta)$ on the shock wave makes both ψ on the shock, ψ_s and t equal to functions of η alone; ψ_η, t_ξ, etc, are the usual partial derivatives of ψ and t and the subscript s indicates that the quantities are evaluated at the shock; with $\dot{\xi}_s \equiv d\xi_s/d\eta$, Eq. (41) follows from Eq. (9). If Eq. (33) is used for ψ and t it can then be seen that

$$\dot{\xi}_s = \dot{\xi}_s(\eta) = \frac{1 - K_a}{1 + K_a} \eta \tag{48a}$$

$$\frac{d}{dt_o}[\psi_s^{(1)} - K_a t^{(1)}] = K_a m(t_o) \tag{48b}$$

where $\eta = (1 + K_a)t_o$ on the shock.

Equation (48b) in the integrated form

$$\psi_s^{(1)} - K_a t_s^{(1)} = K_a \int_0^{t_o} m(\hat{t})d\hat{t} \tag{49}$$

is the constraint referred to after Eq. (46), where $\psi_s^{(1)}$, $t_s^{(1)}$ are $\psi^{(1)}$ and $t^{(1)}$ evaluated on the shock, where ξ is equal to $\xi_s(\eta)$. The full results for $\psi^{(1)}$ and $t^{(1)}$ are straightforward, but lengthy, and will not be given here for brevity. One of the arbitrary functions of integration must be found by solving a functional equation, but once the solution has been initiated from the small time analytical solutions, described above, its continuation by the computer is straightforward.

Results

Figure 1 is a sketch of the configuration in the space of a pair of running coordinates $\hat{\psi}_o, \hat{t}_o$ for the zero-th order estimates described in an earlier section, namely Developments During Times of Order t_I. Point P, with coordinates ψ_o, t_o denotes a typical point to which it is desired to advance the solution in the manner outlined in the paragraph after Eq. (28). A physical interpretation of the integral equation is provided, in part, by the wavelet paths AP, FP, EB, FC, and ED, and the particle path BP; integrals are required along these paths and all conspire to produce the final effect at the point P. When account is taken of wavelet reflections from the shock by retaining the reflection factor $(1 - r)/(1 + r)$ in Eq. (38), more wavelet

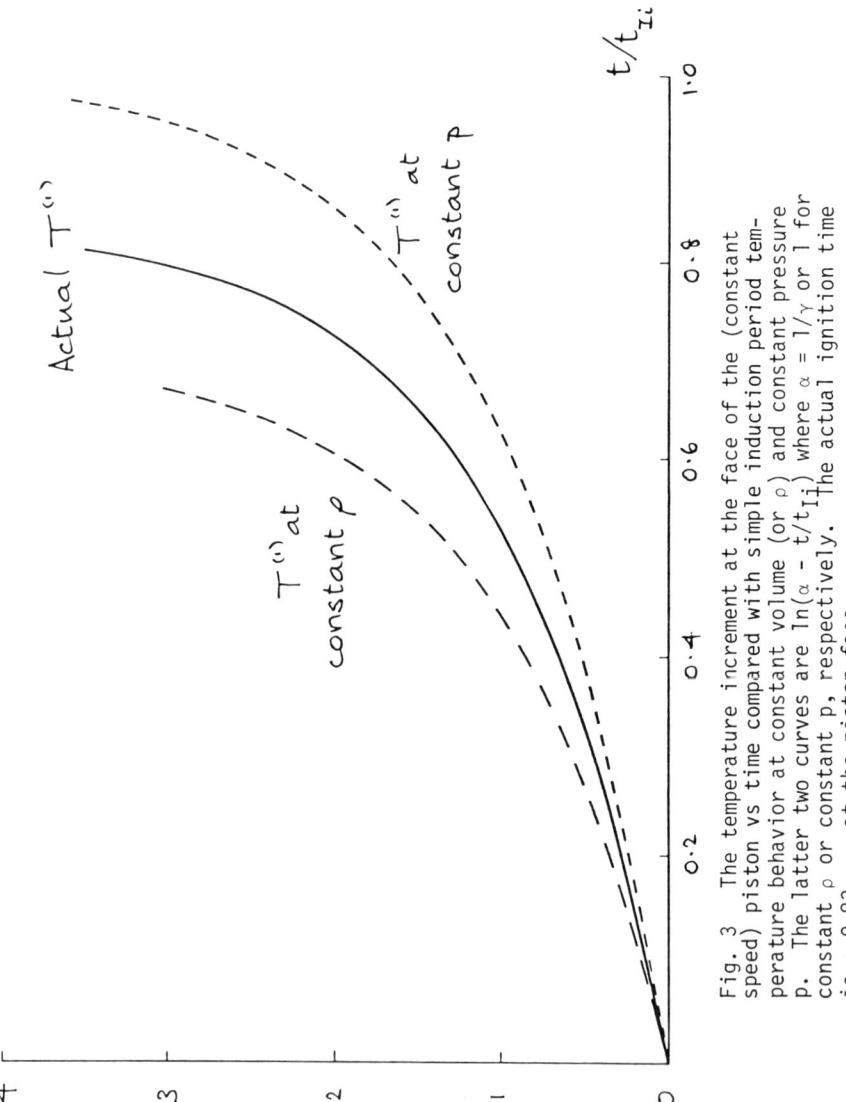

Fig. 3 The temperature increment at the face of the (constant speed) piston vs time compared with simple induction period temperature behavior at constant volume (or ρ) and constant pressure p. The latter two curves are $\ln(\alpha - t/t_{Ii})$ where $\alpha = 1/\gamma$ or 1 for constant ρ or constant p, respectively. The actual ignition time is ≈ 0.83 ... at the piston face.

INDUCTION DOMAIN BEHIND A STRONG SHOCK WAVE 159

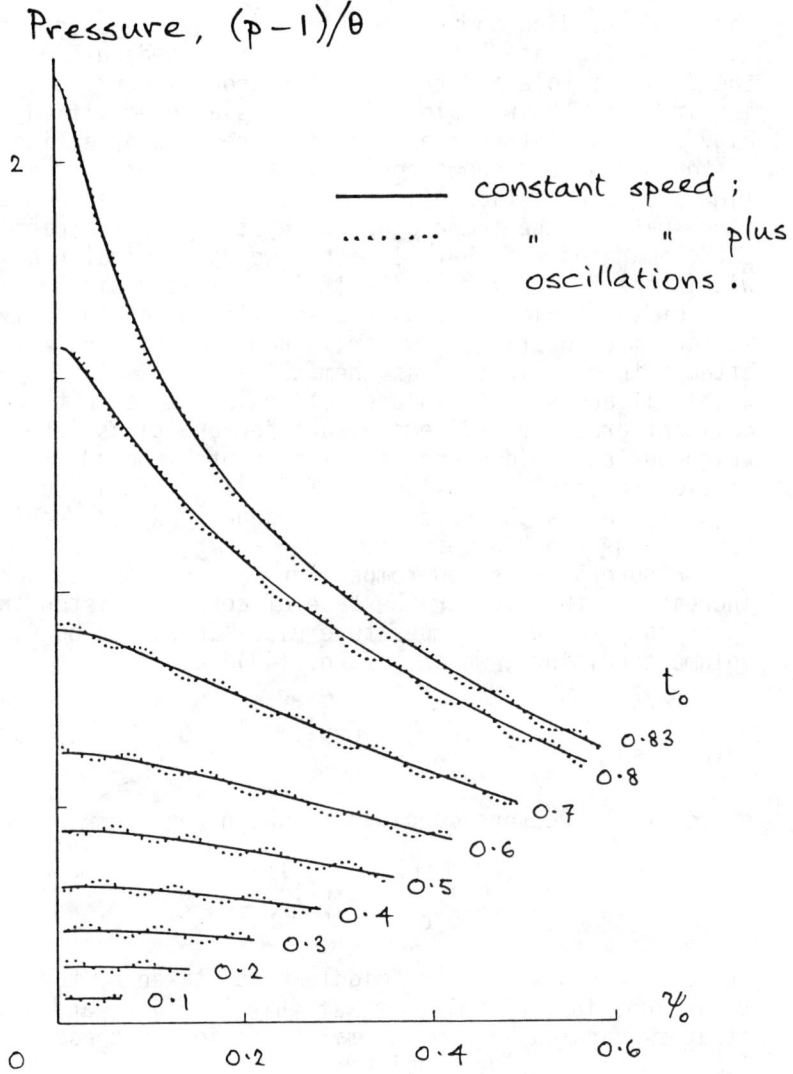

Fig. 4 Pressure increment vs ψ_0 for various t_0 values for two piston speeds, namely u_p = 0.604 (constant) and u_p = 0.604 + a sin wt (a = 1/100θ, w = 0.083). This figure has been prepared by superimposing results for the latter case on the former. The small amplitude evidently admits, effectively, linear superposition. The waves are of "high" frequency type and propagate at the frozen isentropic adiabatic sound speed relative to the fluid in $0 \leq \psi \leq \psi_s(t)$ to the present order of accuracy.

paths, descending towards point P in Fig. 1, go into the summary effect at P, but the paths indicated on Fig. 1 have the dominant role and the computer program permits the isolation of their various influences as exemplified in Fig. 2. This latter figure is for the case of a fixed piston speed, and shows how $T^{(1)}$ varies with ψ_o at a fixed time after initiation of the motion. It can be seen that a large part of the temperature rise at any ψ_o (or particle path) emanates from BP. In other words chemical heating within the fluid element itself is, not suprisingly, a major part of the fluid heating process. It is not the only mechanism of heating, however, since each element will attempt to grow in size as chemical energy is liberated within it and so will induce motion in the gas with concomitant pressure and temperature perturbations into other neighbouring fluid elements. The points labeled THTB on Fig. 2 indicate the influence of $T^{(1)}(\psi_o, \psi_o/K_a)$ [see Eq.(43)] which expresses the role of the gradually strengthening "switch-on" shock wave.

Figure 3 shows how temperature at the contact surface increases with time for the case of constant piston speed. It is compared with temperature rises in either constant volume conditions, which [cf Eq. (43)] requires

$$T^{(1)}_{t_o} - \gamma \exp[T^{(1)}] = 0 \qquad (50)$$

or constant pressure conditions, which require

$$T^{(1)}_{t_o} - \exp[T^{(1)}] = 0 \qquad (51)$$

The comparison shows the "middle road" taken by temperature variations in fluid elements at this location, and illustrates graphically how thermal expansion of these elements leads to behavior near <u>neither</u> extreme of constant pressure and constant volume.

In assessing the length of an induction zone behind a strong shock it is clearly most important to take account of the fact that it cannot be done by "simple" (constant pressure or constant volume) chemistry alone, and that one must take proper account of the chemico-gas dynamical couplings.

Figure 4 illustrates pressure increment $p^{(1)}$ vs ψ_o for a variety of times up to the induction time (just over 0.83 in the present case). Two cases are superimposed in this figure; in the first, piston speed is constant, while in the

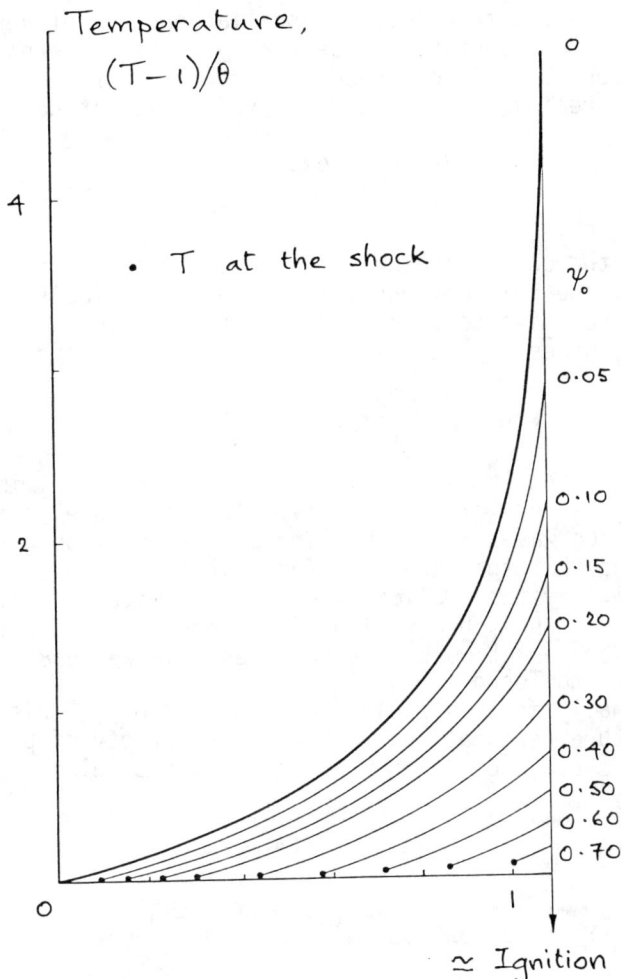

Fig. 5 Temperature increment vs t_o at various ψ_o values for the case of a decelerated piston $u_p = 0.606 - t/\theta$, $M_{si} = 1.5$, $\theta = 25$. The "ignition" time at the piston face has increased to nearly 1.1 from the value of 0.83... given in Fig. 3. It can be seen, from the fact that temperature at the shock barely increases at all, that the shock acceleration is quite small in the present case.

second the piston face is assumed to oscillate with a fixed very small amplitude and a given frequency. One notes the uniform rise of pressure between contact face and shock at early times, with an increasingly rapid local rise of pressure in the neighbourhood of the ignition point at

$\psi_0 = 0$. As the superimposed oscillations are of very small amplitude they appear as essentially linear superpositions on the constant-piston-speed background. For any given locus in real x,t space, say $x = x_f(t)$, it can be shown that

$$\frac{d\psi_f}{dt} = \rho_f(\frac{dx_f}{dt} - u) \equiv \rho_f V_f \qquad (52)$$

so that the corresponding path in ψ,t space has a slope equal to the related density ρ_f times the velocity of the x_f path relative to local fluid speed. To lowest order Eq. (52) gives

$$d\psi_{of}/dt_o = V_{of}$$

(V_f is written as $V_{of} + \sigma V_f^{(1)} + \ldots$) and Fig. 4 indicates that V_{of} is very close to unity. The inference is that the illustrated waves are travelling at the isentropic sound speed and that they are therefore, in the context of the dispersive system implicit in Eq. (44), waves of "high" frequency. This observation is of interest in the light of some other recent work on such waves (Clarke 1978b, 1979; Abousief and Toong 1981).

Finally in this preliminary account Fig. 5 displays temperature pertubations vs time at a variety of positions between shock and piston when the latter's velocity is given by

$$u_o(t) = 0.604 - \frac{1}{\theta}t$$

$$u_{pi} = 0.604 \Rightarrow M_{si} = 1.5 \qquad (53)$$

$$\theta = 25$$

The expansion, and consequent cooling, that is provoked by contact surface deceleration [implied by Eq. (53)] is evident from the fact that temperature at the shock (the left-most point on any given curve of constant ψ_0) increases hardly at all. Much more important is the significant increase in induction time that can be seen by consulting the abscissa scale on Fig. 5; induction is now occupying about 1.1 units of time, as opposed to the 0.83 units required in the no-cooling case illustrated in Fig. 3.

References

Abousieff, G. E. And Toong, T. Y. (1981) Non-linear wave-kinetic interactions in irreversibly reacting media, J. Fluid Mech. 103, 1-19.

Clarke, J. F. (1978a) A progress report on the theoretical analysis of the interaction between a shock wave and an explosive gas mixture, CoA Memo 7801, Cranfield Institute of Technology, Bedford, England.

Clarke, J. F. (1978b) Small amplitude gas dynamic disturbances in an exploding atmosphere, J. Fluid Mech. 89, 343-356.

Clarke, J. F. (1979) On the evolution of compression pulses in an exploding atmosphere; initial behavior, J. Fluid Mech. 94, 195-208.

Clarke, J. F. (1981a) Propagation of gasdynamic disturbances in an explosive atmosphere. Progress in Astronautics and Aeronautics and Combustion in Reactive Systems: (edited by J. R. Bowen, N. Manson, A. K. Oppenheim, and R. I. Soloukhin) Vol. 76, pp. 383-402. AIAA, New York.

Clarke, J. F. (1981b) Plane waves in reacting gases. Part II: waves in explosive atmospheres, Sonderforschungsbereich 27, Wellenfokussierung, RWTH Aachen, FRG, May 1981: also CoA Memo 8106, Cranfield Institute of Technology, Cranfield, Bedford, England.

Clarke, J. F. (1983) On the gas dynamics of a combustible atmosphere, In Festschrift for Prof. L. Crocco, Plenum, New York (in press).

Kluwick, A. (1981) Characteristic parameter perturbations, Prog. Aerospace Sci. 19, 197-313.

Strehlow, R. A. (1969) Fundamentals of Combustion, International Text Book Co., Swanton, Pa.

Structure of Premixed Laminar Methanol-Air Flames: Experimental and Computational Results

L. L. Andersson*
Chalmers University of Technology, Göteborg, Sweden
and
B. Christenson,† A. Höglund,‡ J. O. Olsson,§ and L. G. Rosengren‡
Volvo, Göteborg, Sweden

Abstract

Concentrations of stable and unstable species in a premixed flat methanol/air flame at 13.3 kPa are presented. They were measured using a modulated molecular beam mass spectrometer and calibrated without assuming the mole fraction of any species along the flame. The stable specie profiles are consistent with earlier measurements. It was found from temperature measurements that the sampling probe cools and delays the flame in the precombustion zone. The temperature in the undisturbed postcombustion zone at half the adiabatic flame velocity is about 1600 K. The Westbrook-Dryer model for methanol/air combustion, including chemical kinetics and molecular diffusion, was used to compute flame structure. The calculations in this study reproduced Westbrook-Dryer results at 10 kPa within a few percent when their empirical diffusion coefficients, which were about three times higher at 1000 K than fundamental diffusion coefficients, were used. Only CH_4 and CH_3 of the hydrocarbons were important at these conditions. Fundamental diffusion coefficients and a kinetic mechanism with only these hydrocarbons gave a good agreement with the present experiments. An exception

---

Presented at the 9th ICODERS, Poitiers, France, July 3-8, 1983. Copyright © 1984 by L. L. Andersson. Published by the American Institute of Aeronautics and Astronautics, Inc., with permission.
*Chemist, Physical Chemistry Department.
†Chemist, Now with Hässle, Göteborg, Sweden.
‡Physicist, Applied Physics Department.
§Chemist, Applied Physics Department.

was CH_2O, for which the experimental maximum was five times higher than the computational maximum. The experimental specie profiles showed steeper gradients compared to the computed profiles.

Introduction

This work has been initiated as a part of an alcohol fuel research program related to internal combustion engines. The decreasing access to petroleum has stimulated interest in methanol as an alternative fuel. One advantage of methanol is that the combustion process can be analyzed efficiently compared to motor gasoline and diesel fuel, which are unspecified mixtures. Since the detailed chemical processes important for efficiency and emissions cannot be easily studied in engines, one approach is to study only a part of an engine process and try to extrapolate these results to engines. Laminar premixed flames is a subsystem, which can be important in engines.

Detailed experimental data about combustion of methanol and air are relative scarce. Measurements of stable species in laminar flames of methanol/air at low pressure have previously been made by Akrich et al. (1978). Pauwels et al. (1982) have measured the radicals H, O, and OH. Vandooren et al. (1981) have studied both stable and unstable species at low pressure in a methanol-oxygen flame diluted with about 50% argon. The present experimental study of a methanol/air flame is, to the knowledge of the authors, the first to include both stable and unstable species. Probe perturbation of the flame is also investigated (Biordi et al. 1974; Hayhurst and Kittelson 1977; Stepowski et al. 1981). These results have been compared to the predictions of a detailed model of combustion chemistry of methanol/air combustion.

The generally accepted one-dimensional model for methanol/air combustion in premixed laminar flames, including both chemical kinetics and transport processes, is the model proposed by Westbrook and Dryer (1980). In the present study their model is used 1) to reproduce their study at low-pressure conditions 10 kPa; 2) to calibrate the computer program prepared for study of methanol combustion; 3) to study the effect of modification of the values of diffusion coefficients and alternative reaction mechanisms; and 4) to develop concentration profiles for comparison with the molar fraction profiles from the experimental study.

Experimental

The methanol/air flame studied consisted, at stoichiometric conditions, of 12.3% CH_3OH, 18.4% O_2, 67.3% N_2, and 2% Ar. The flame velocity was 0.25 m/s, which is half the adiabatic flame velocity. The operating pressure was 13.3 kPa.

Apparatus

A recently developed sensitive mass spectrometer with modulated beam (Höglund and Rosengren 1983) was used to monitor the flame species. The sampling cone is made of quartz. It has an apex angle of 60 deg and a tip radius of 0.5 mm. The orifice diameter is 60 μm, and the channel length ~100 μm. The cone was chemically inactivated with a 5% HF solution. The flame was stabilized on a water-cooled porous flat flame burner, 60 mm in diameter. The diffusion of argon to the center of the flame at different distances above the flameholder was measured to check the one dimensionality assumption made for flat flames. In this case the mole fraction of argon was < 0.001 at the beginning of the postcombustion zone, which means that disturbances due to diffusion into and out of the flame can be neglected. Methanol was vaporized before mixing with oxygen and nitrogen. Critical flow through well-defined orifices was used to maintain constant gas flows (Höglund et al. 1981).

The temperature was measured with a (Pt-Pt10%Rh, 50-μm-diam) thermocouple coated with a thin layer of silica to reduce catalytic effects. Kaskan's relation was used to correct the thermocouple readings for radiation losses:

$$T = (1.25\epsilon k T^4 D^{0.75}/\lambda)(\eta/\rho v)^{0.25} \quad (1)$$

where ϵ is the emissivity, k the Stefan Bolzmann's constant, T the wire temperature, D the wire diameter, λ the thermal conductivity of the gas at the wire temperature, η the gas viscosity, ρ the density, and v the gas velocity (Kaskan 1957).

Calibration of Mass Spectrometric Data

Often, mass spectrometer data from flames are calibrated relative to an assumed mole fraction curve, through the flame, for the diluent, i. e., nitrogen or argon (Akrich et al. 1978). A calibration procedure is introduced in this work that is not based on such an approximation. It has some common features with the method used by Vandooren et al. (1981). An expression for the responsivity of the mass spectrometer is introduced by Höglund and Rosengren (1983). The measured intensity $I_i(z)$ is related to the concentration $c_i(z)$ of the species measured, by the responsivity $R_i(z)$:

$$I_i(z) = R_i(z)c_i(z) \qquad (2)$$

where i identifies the species, and z is the distance above burner. The responsivity depends on many variables (Biordi et al. 1974), some of which are specific for the species measure. The responsivity is:

$$R_i(z) = a(z)b_i\sigma_i g_i \qquad (3)$$

where $a(z)$ is a function of the attenuation of the molecular beam, which depends on the total inflow. This flow varies along the flame due to the temperature variation. The variable b_i is determined by the specific beam propagation and ion source parameters for species i. The ionization cross section for species i is denoted σ_i, and the quadropole, multiplier, and amplifier gain is denoted by g_i.

The responsivity is calibrated in four different ways. The concentration in the flame of the reactants O_2 and CH_3OH and the diluents N_2 and Ar are known at $z = 0$. If $a(0) = 1$ and $r_i = b_i\sigma_i g_i$, the responsivity becomes

$$R_i(0) = r_i = I_i(0)/c_i(0) \qquad (4)$$

Known concentrations of the stable products CO and CO_2 are diluted in N_2 at the flame pressure in the absence of a flame but with the same ionization parameters. The mass spectrometer signals then becomes $I_i' = a'b_i\sigma_i g_i'c_i'$ for i = CO, CO_2, and N_2, where the prime denotes calibration measurement. With these equations and r_{N2}, the r_i values in the flame for CO and CO_2 are given by

$$r_i = (I_i'g_i c_{N2}'g_{N2}'/c_i'g_i'I_{N2}g_{N2}) r_{N2} \qquad (5)$$

Some intermediate species and products O, OH, H_2O, etc. are assumed to be in partial equilibrium in the postcombustion zone (Vandooren et al. 1981). They are calibrated at just one position z_1 from the expression

$$r_i = (I_i(z_1)X_{N2}(z_1)/X_i(z_1)I_{N2}(z_1))\, r_{N2} \qquad (6)$$

where $X_{i,N2}$ are calculated mole fractions. Intermediate species CH_3, etc. which do not exist in the postcombustion zone, are calibrated from the ionization cross-section expression used by Vandooren et al. (1981). The ionization probability for different atoms of a specie are added to give the ionization potential for the species measured. The r_i value is related to a similar species at a certain flame position z_2 and becomes

$$r_i = (\sigma_i g_i I_i(z_2)|_{\Delta e}/\sigma_j g_j I_i(z_2)|_{2.6})\, r_j|_{2.6} \qquad (7)$$

where Δe is the electron energy above the ionization potential used in the recording of the profile of species i. It is also assumed that $b_i = b_j$. The mole fraction of the different species along the flame can now be calculated from

$$X_i(z) = (I_i(z)/r_i)/(\Sigma I_k(z)/r_k) \qquad (8)$$

Theory and Physical Background

The model is based on the following assumptions: A premixed gas mixture with a laminar and constant mass flow is ignited at a constant pressure. The burning gas mixture behaves as a continous fluid and the components follows the ideal gas law. No temperature or concentration gradients exist parallel to the direction of the flame front; i e. the flame is one dimensional. For certain conditions, the flame relaxes into a stationary state with the flame velocity equal to the cold gas velocity. This ideal free adiabatic flame is difficult to create in a laboratory. The closest approximation is the cooled wide porous burner on which a nearly one-dimensional flat flame can be maintained for certain combinations of mass flow, pressure, and cooling. In the limit of no cooling, the flame behaves as a free adiabatic flame.

The conservation equation for each specie i is

$$\frac{dX_i}{dt} = (R_i - (\rho v)\frac{dX}{dz} - \frac{d}{dz}(\rho X_i V_i))/\rho \qquad (9)$$

where X_i is the number of mole per massunit, R_i the rate of production by chemical reactions, ρ the density, v the mass-averaged velocity, z the distance from the burner, and V_i the diffusion velocity for specie i. The conservation equation for the energy has an analogous form.

Diffusion

The molecular diffusion of a specie i is modeled as a trace diffusion in nitrogen. This approximation is excellent when air is the oxidant (Fristrom and Westenberg 1965). The binary diffusion coefficients, in cm^2/s,

$$D_{ij} = 1.66 \times 10^{-3} (M_i^{-1} + M_j^{-1})^{1/2} T^{1.67} / (p\sigma_{ij}^2 (\varepsilon_{ij}/k)^{0.17}) \quad (10)$$

were calculated from Lennard Jones 12-6 potential parameters, σ_{ij} and ε_{ij}/k in Å and K, respectively, from viscosity measurements given by Fristrom and Westenberg (1965) for all molecules. For the values of H and O, data were taken from Warnatz (1979). The combination rules $\sigma_{ij} = (\sigma_{ii} + \sigma_{jj})/2$ and $\varepsilon_{ij}/k = (\varepsilon_{ii}/k * \varepsilon_{jj}/k)^{1/2}$ were used. Fristrom and Westenberg (1965) have thoroughly discussed Eq. (10) and many other aspects of diffusion related to combustion. For radicals like OH, HO_2, CH_2OH, and CH_3O, the diffusion coefficients were calculated by interpolation of values for similar molecules. The subscript j in our model stands for the nitrogen gas.

The Westbrook-Dryer empirical approximation D_{iN2}, in cm^2/s, was also used in this investigation.

$$D_{ij} = 9.2 \times 10^{-6} RT^{3/2} / (M_i^{1/2} p) \quad (11)$$

where R is the gas constant (82.05 cm^3, atm/K, mol), and the constant value of 9.2×10^{-6} is determined from a fit of calculated flame velocities to experimental ones (Westbrook and Dryer 1981). The only molecular property is M_i, the molecular weight. The diffusion coefficients from Eq. (11) are about three times higher than corresponding ones from Eq. (10) or from experiments.

Chemical Kinetics

The methanol/air reaction mechanism due to Westbrook and Dryer (1980) contains 26 species and 84 reactions where each reaction includes a forward and a backward direction. The basic species comes from the H_2/air system: H_2, O_2, H, O, OH, HO_2, H_2O_2, H_2O and N_2. Carbon/oxygen compounds CH_3OH, CH_2OH, CH_3O, CH_2O, CHO, CO, and CO_2 is a second type of species. The third type of species is the hydrocarbon molecules and radicals C_2H_6, C_2H_5, C_2H_4, C_2H_3, C_2H_2, C_2H, and CH. Forward reaction rate constants were taken directly from Westbrook and Dryer (1980). The reverse rate constants were calculated, from the forward rate constants and corresponding equilibrium constants, in the same way as Westbrook and Dryer (1980) did.

Method for Flame Calculations

The computer program developed for this study incorporates one-dimensional specie conservation equations, chemical kinetics, mass transport, molecular diffusion. A time-dependent method is used with a stepwise iteration toward the stationary state. The conservation equation (9) is solved for all participating species at a certain number of fixed points, which could be distributed along the space coordinate. No transformation of the space coordinate is used.

Since the program has been written primarily to evaluate experiments, the energy-conservation equation is not solved. Instead, the measured temperature profiles and the mass flow for experimental flames stabilized by a cooled burner are used as input parameters. The gradients of the species depend strongly on the temperature gradient mainly through the exponential dependence of rate constants on temperature.

For the numerical solution of the coupled system of ordinary equations an implicit first-order method was used with a linear least-square solution of the linear algebraic equations. This method was found to be stable. The methods for the solution of the equations of a stationary flat flame, and other aspects on computations of flat flames, have been thoroughly discussed by Smooke (1982).

For the calculations with only chemical kinetics, without any mass transport and diffusion, a part of this computer program was used.

Results and Discussion

Flame Experiments

Temperature and Perturbation. The flame temperature for an unperturbed flame is shown in Fig 1. Equation (1) was

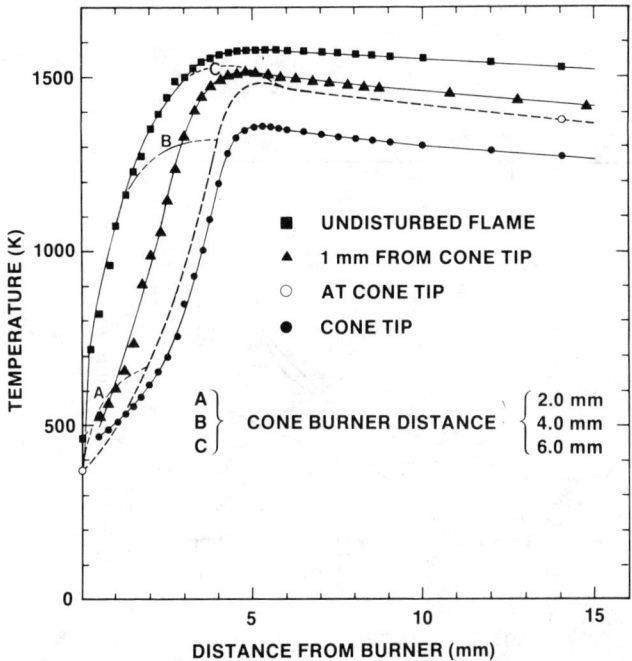

Fig. 1 Temperature profiles of a stoichiometric methanol/air flame at 13.3 kPa from thermocouple measurements.

used with $\varepsilon = 0.22$ and nitrogen values for λ, ρ, and η. The temperature at a fixed distance of 1 mm from the sampling cone along the flame was also recorded. The thermocouple is placed on the probe tip to measure the probe tip temperature. The gas temperature at the cone tip is shown by the broken line. This profile was determined from measurements of the inflow to the mass spectrometer, which is a function of temperature (Höglund and Rosengren 1983) and thermocouple measurements in the postcombustion zone. Due to heat conduction in the cone, the tip temperature close to the burner is higher than in the surrounding gas. Finally, the flame temperature profiles A, B, and C for the cone at the following fixed positions from the flameholder, 2.0, 4.0, and 6.0, respectively, are shown. These profiles are obtained from the temperature values recorded at different distances from the cone.

The temperature rise in the combustion zone is much lower when the cone is closer to the burner, which means that the flame development is delayed by the cone. The narrow CH_3 profile (see Fig. 2a) indicates that the disturbances of the flowfield around the probe tip are negli-

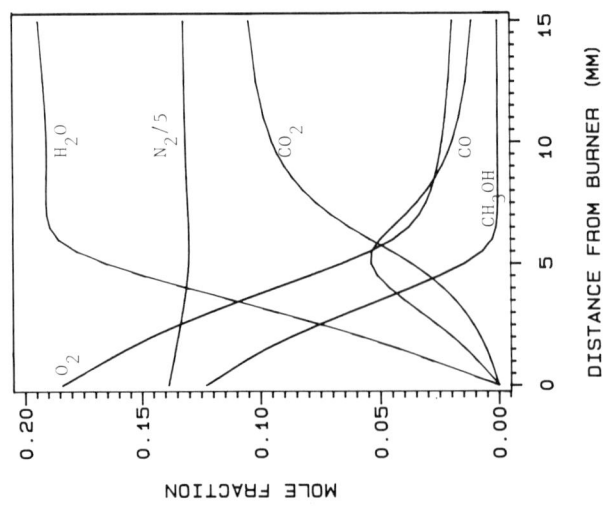

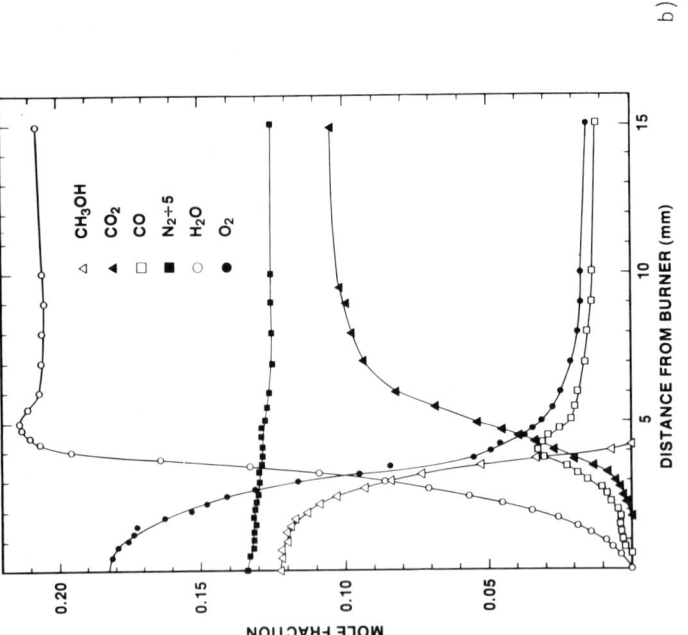

Fig. 2 Molar fractions of some major intermediates at a pressure of 13.3 kPa and a flame velocity of 0.25 m/s. a) Experimental measurements this study. b) Calculations with fundamental diffusion coefficients from Eq. (10), and a simplified Westbrook-Dryer mechanism with 18 species/56 reactions.

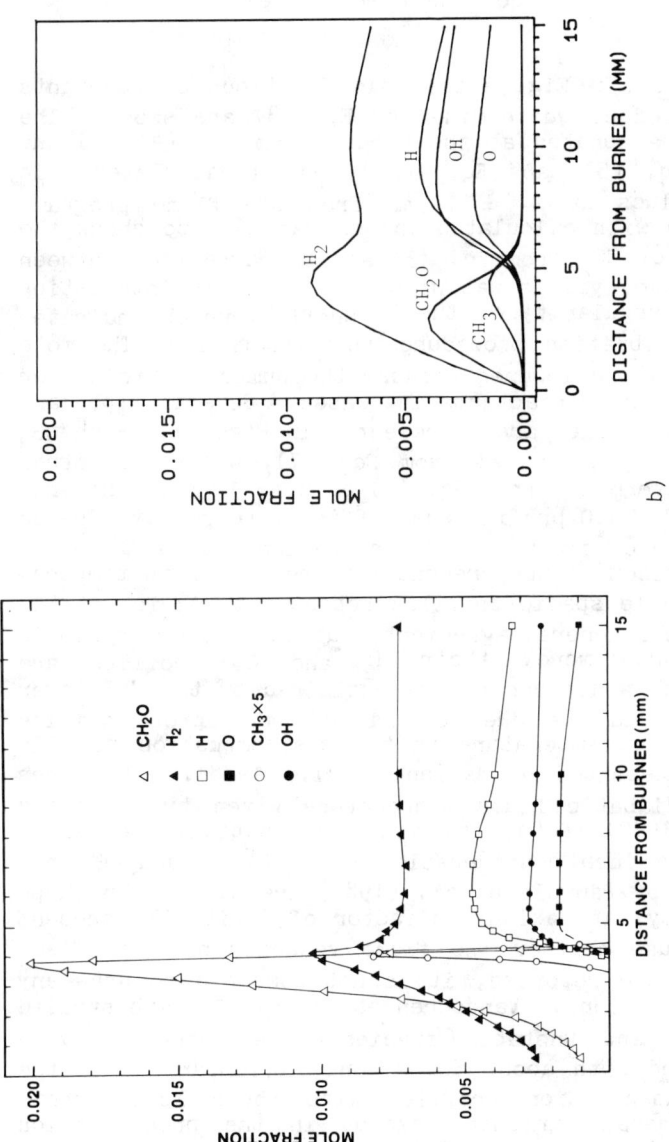

Fig. 3 Stoichiometric methanol/air flame. Molar fractions of major species at a pressure of 13.3 kPa and a flame velocity of 0.25 m/s. a) Experimental measurements this study. b) Calculations with fundamental diffusion coefficients from Eq. (10), and a simplified Westbrook-Dryer mechanism with 18 species/56 reactions.

gible (Hayhurst and Kittleson 1977). The major pertubation seem to be due to cooling of the flame by the cone, which reduces the amount of thermal energy (Biordi et al. 1974) and radicals (Stepowski et al. 1981) in the precombustion zone. Therefore, the flame is delayed.

Mole Fractions. In Fig 3a the mole fractions of reactants and major products determined by Eq. (8) are shown. The former are calibrated at the burner from Eq. (4) and the latter by Eq. (5), and Eq. (6) based on calculated mole fraction values at z_1 = 15 mm from the flame program. Equation (6) with calculated values was used to check the calibration of CO_2 from Eq. (5) and the agreement between the two sets of values was within 10%. It is interesting to note the regularity of the N_2 profile, which indicates that the calibration procedure works properly. The mole fraction of N_2 is lowered because the number of moles for certain amounts of reactants increases. Several important intermediate species are represented in Fig. 2a. Both CH_3 and CH_2O were calibrated from Eq. (7), while the other ones were determined from Eq. (6) at z_1 = 15 mm. The initial part of the O profile was difficult to resolve due to fragmentation of other species in the ion source.

A comparison of the present measurements with temperature and stable specie profiles reported by Akrich et al. (1978) shows a general agreement, but the H_2O/CO_2 ratio is larger. Furthermore, their H_2 and CO profiles are broader. The width can not be explained by the 25% lower pressure but may be due to a larger microprobe orifice diameter. The temperature in the postcombustion zone is lower than reported by Akrich et al. (1978). It agrees with the adiabatic flame temperature given by Westbrook and Dryer (1980) if the lower than adiabatic flame velocity is taken into consideration. The H, O, and OH profiles given by Pauwels et al. (1982) are similar in shape but differ by, at maximum, a factor of 2 with the present ones. Because neither the flame velocity nor the flame temperature are reported, it is not possible to draw any further conclusions. Vandooren et al. (1981) have studied both stable and unstable species in a methanol-oxygen flame diluted with about 50% argon. A comparison of the present concentration profiles with their measurements shows a general agreement except in the precombustion zone.

Comparisons with Other Flame Calculations. In addition to the evaluation of the experiments by computer modeling,

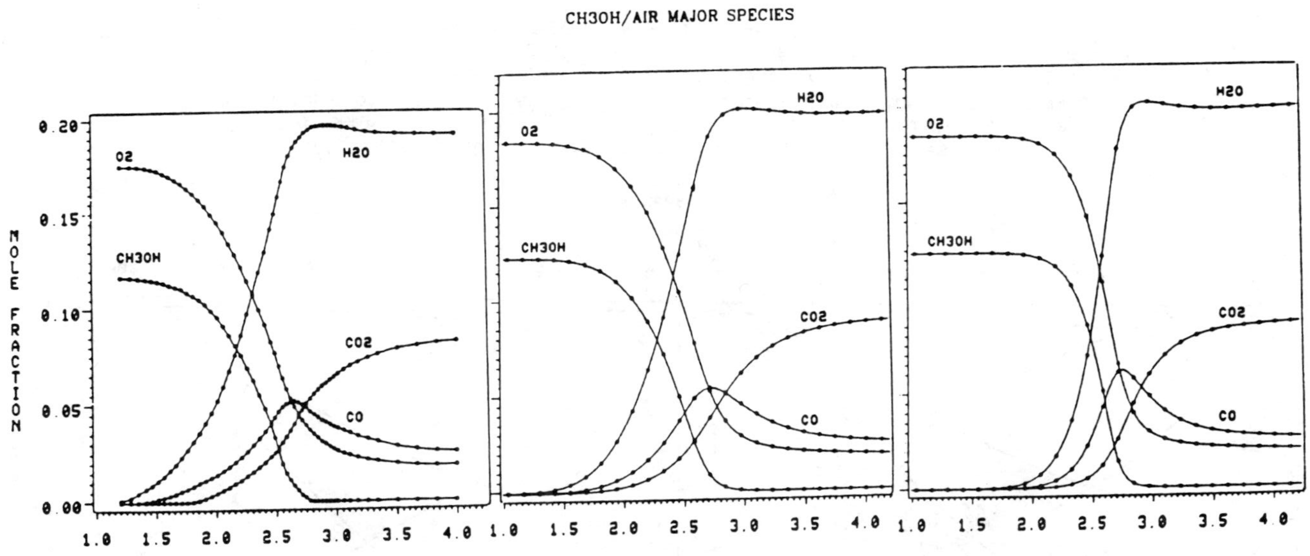

PREMIXED METHANOL-AIR FLAMES

Fig. 4 Stoichiometric methanol/air flame. Molar fractions of major species as a function of distance from an ideal burner. The pressure is 10 kPa and the flame velocity is 0.5 m/s. a) Calculations by Westbrook and Dryer (1980). b) Our calculations with empirical diffusion coefficients from Eq. (11). c) Our calculations with fundamental diffusion coefficients from Eq. (10).

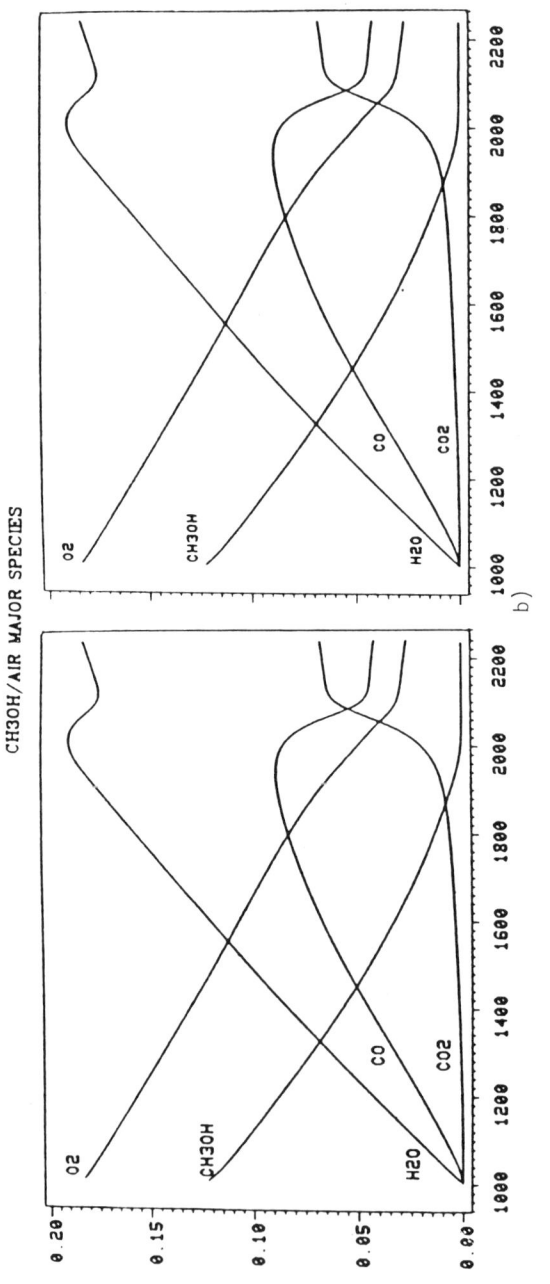

Fig. 5 Calculated molar fractions of major species as a function of temperature at a pressure of 10 kPa and an ignition temperature of 1000 K. a) Complete Westbrook-Dryer kinetic mechanism with 26 species/84 reactions. b) Subset of the Westbrook-Dryer kinetic mechanism with 18 species/ 56 reactions, where all hydrocarbons except CH_4 and CH_3 are excluded.

separate flame computations were made. The calculations for a pressure of 10 kPa and a flame velocity of 0.5 m/s are compared to the corresponding calculations of Westbrook and Dryer (1980). The temperature profile from their study is used in the calculations. The calculations were started with a stoichiometric composition of methanol and air at all grid points. A uniform grid of 40 points was used, in the computational interval from 0 to 5 cm, with about 12 points in the flame zone. Figure 4 shows the results for the major species with diffusion coefficients according to Eqs. (10) and (11), respectively. The calculations agree well with those of Westbrook and Dryer (1980), when empirical diffusion coefficients given by Eq. (11) are used. If the fundamental diffusion coefficients equation (10) are used, the specie profiles in comparison are compressed. For other species, the results were analogous. The reason is that the fundamental diffusion coefficients is three times smaller at 1000 K than the empirical ones.

To determine the predominant reaction paths, a subsystem of the complete mechanism was tested. A natural simplification is to exclude the species C_2H_6, C_2H_5, C_2H_4, C_2H_3, C_2H_2, C_2H, and CH and corresponding reactions. In Fig. 5, the results are shown for the complete mechanism 26 species/84 reactions and the simplified mechanism 18 species/56 reactions. The calculations describe the time evolution (temperature-evolution) of the species mole fractions at a constant pressure 10 kPa and a constant enthalpy without diffusion. As can be seen from Fig. 5, the major species profiles are identical for both the mechanisms within the reproducibility of the plots. The same good agreement was also found for the major intermediate species CH_2O, OH, O, H_2, and H. For CH_4 and CH_3, the two mechanisms of chemical kinetics gave profiles differing about 15%. These results were confirmed by an independent sensitivity analysis. Inclusion of the effects of diffusion in the calculation for both the 18 species and the 26 species mechanisms for the same conditions resulted in only small differences. Also, Akrich et al. (1978) in their experimental study of low-pressure methanol/air flames could not detect any CH_4 or C_2H_6 in lean or stoichiometric flames. The simplified mechanism 18 species and 56 reactions is a good approximation to the complete model, for stoichiometric composition and at a low pressure 10 kPa.

Computer Modeling of Flame Experiments. Figures 2 and 3 show experimental and computed molar fractions for a stoichiometric methanol/air flame at 13.3 kPa and a flame velocity of 0.25 m/s. The flame structure has been computed for the same conditions as the experiments. The measured flame velocity 0.25 m/s and undisturbed temperature profile (Fig. 1) have been used as input parameters. For these computations, the diffusion was included with diffusion coefficients modeled by Eq. (10), and the simplified kinetic mechanism (18 species and 56 reactions) was used. The initial starting profiles were those calculated in the absence of diffusion. This calculation was stopped at 5 cm with the corresponding values used as the hot boundary conditions. A nonuniform grid with 26 points was used, about ten of which were placed in the flame zone.

As can be seen from the Figs. 2 and 3 the experimental and the computed for each of the species are of about the same form. The discrepancies are within the ranges of the experimental and computational uncertainties. The agreement is better for CH_3OH, O_2, CO_2, and H_2O as expected, as these species are easier both to measure and model. The measured maximum of CH_2O is ten times higher than the its calculated maximum. This difference can hardly be explained by uncertainties in the experimental calibration. Possibly the reaction mechanism related to CH_2O formation/consumption needs to be revised. Specie profiles from the experiments are steeper than the corresponding computed profiles. Although the measured undisturbed temperature profiles have been used in the computations, the calculations have also been performed with nearly linear changes in the temperature profile (see Fig. 1), to model the varying degrees of cooling caused by the sampling cone. The specie profiles shifted toward (or from) the burner as the temperature increased (or decreased), but only minor changes in the gradients of the profiles or the peak values were evident. Other nonlinear perturbations of the flat flame can cause the experimental profiles to have steeper gradients than the calculated profiles (Stepowski et al. 1981). A flame/cone sampling system can be seen as a family of stationary flames with different temperature profiles and boundary conditions. Thus the assumed nearly one-dimensional behavior in the experimental flat flame sampling cone systems is probably an overidealization.

Conclusions

The computer program developed for this work can reproduce the computational results of Westbrook and Dryer (1980) to within a few percent. For the stoichiometric

conditions studied here, only CH_4 and CH_3 of the hydrocarbon species were found to be important.
The calculations agreed well with experiments in this study at 13.3 kPa when fundamental diffusion coefficients, which are three times smaller at 1000 K than the empirical ones, are used in flame model. An exception was CH_2O, for which the measured maximum was ten times higher than the computed maximum. The experimental profiles were steeper than the corresponding computed profiles. This discrepancy cannot be explained by simple thermal disturbances of the flame by the sampling cone.

Acknowledgements

The authors would like to thank Dr Bengt Hakberg and Professor Thure Högberg for helpful discussions and consistent support of this work. They are also grateful to Lennart Karlsson and Stellan Rosengren, who assisted with the computer graphics. This work was financially supported by the Swedish National Board for Technological Development and by AB Volvo.

References

Akrich, R., Vovelle, C., and Delbourgo, R. (1978) Flame properties and combustion mechanisms of methanol/air under reduced pressure. Combust. Flame 32, 171-179.

Biordi, J. C., Lazarra, C. P., and Papp, J. F. (1974) Molecular beam mass spectrometry applied to determining the kinetics of reactions in flames. I. Empirical characterization of flame perturbation by molecular beam sampling probes. Combust. Flame 23, 73-82

Fristrom, R. M. and Westenberg, A. A. (1965) Flame Structure. McGraw-Hill Inc, New York.

Hayhurst, A. N. and Kittleson, D. B. (1977) Mass spectrometric sampling of ions from atmospheric pressure flames. III. Boundary layer and other cooling of the sample. Combust. Flame 28, 137-143

Höglund, A, and Rosengren, L. G. (1983) A new sensitive molecular beam mass spectrometer. Int. J. Mass Spectrom. (in press).

Höglund, A., Rosengren, L. G., and Christenson, B. (1981) Experimental system for studies of premixed flat flames. Volvo Report LM 57729, AB Volvo, Göteborg, Sweden.

Kaskan, W. E. (1957) The dependence of flame temperature on mass burning velocity. Sixth Symposium (International) on Combustion, pp. 134-142. The combustion Institute, Pittsburgh, Pa.

Olsson, J. O. and Andersson, L. L. (1982) The structure of premixed laminar methanol/air flames. II. Computational results. Volvo report LM 57735, AB Volvo, Göteborg, Sweden.

Pauwels, J. F., Carlier, M., and Sochet, L. R. (1982) Analysis by gas-phase electron spin resonance of H, O, OH and Halogen atoms in flames. J. Phys. Chem. 86, 4330-4335

Smooke, M. D. (1982) Solution of burner-stabilized premixed laminar flames by boundary value methods. J. Comput. Phys. 48, 72-105.

Stepowski, D., Puechberty, D., and Cottereau, M. J. (1981) Use of a laser-induced fluorescence of OH to study the perturbation of a flame by a probe. 18th Symposium (International) on Combustion, pp. 1567-1573. The Combustion Institute, Pittsburgh, Pa.

Vandooren, J., Balakhine, V. P., and Van Tiggelen, P. J. (1981) Mass spectrometric investigation of the structure of methanol flames. Arch. Combust. 1, 229-242

Warnatz, J. (1979) The structure of freely propagating burner-stabilized flames in the H_2-CO-O_2 system. Ber. Bunsenges. Phys. Chem. 83, 950-957

Westbrook, C. K. and Dryer, F. L. (1980) Prediction of laminar flame properties of methanol/air mixtures. Combust. Flame 37, 171-192.

Westbrook C. K., and Dryer, F. L. (1981) Simplified mechanism for the oxidation of hydrocarbon fuels in flames. Combust. Sci. Technol. 27, 31-43.

Burning Velocities of Ethanol-Air and Ethanol-Water-Air Mixtures

Ömer L. Gülder*
National Research Council of Canada
Ottawa, Ontario, Canada

Abstract

Laminar burning velocities of ethanol and ethanol-water mixtures burning in air have been determined as a function of equivalence ratio, water concentration, unburned mixture temperature and pressure. Measurements were made during the constant pressure period in a constant volume spherical bomb with a radius of 0.162 m. A density correction scheme which accounts for presence of preheat layer was used for the calculation of burning velocities from the measured flame growth rates. Maximum burning rates were found to occur at an equivalence ratio of approximately 1.08, independent of unburned mixture temperature and water content of the mixture. Pressure, temperature, equivalence ratio, and water content dependence of burning velocity are represented by empirical and semi-empirical (modified Semenov) relations. Obtained results suggest that the effect of water addition on burning velocity is mostly thermal rather than chemical kinetic. The range of experimental observations covered equivalence ratios from 0.7 to 1.4. Unburned mixture temperature was varied from 350 to 600 K and pressure from 100 to 800 kPa. Water content of ethanol was varied from 0 to 50% by mass.

Nomenclature

A = constant in Eq. (5)
B = constant in Eq. (9)
C_w = mole fraction of water vapor in unburned mixture
C_p = specific heat at constant pressure
E_a = apparent activation energy

Presented at the 9th ICODERS, Poitiers, France, July 3-8, 1983. Copyright © 1984 by Ö. L. Gülder. Published by the American Institute of Aeronautics and Astronautics with permission.
*Research Associate, Division of Mechanical Engineering.

H	= enthalpy of combustion
I	= density correction factor
K	= constant in Eq. (15)
m	= pressure exponent
N	= ratio of number of mols of products to that of reactants
n	= temperature exponent
P	= pressure
R°	= universal gas constant
r	= distance along radial axis
r_b	= flame radius
S_u	= laminar burning velocity
S_g	= gas velocity
T	= temperature
t	= time
$\dot{w}$	= reaction rate
z	= pressure exponent
α	= constant in Eq. (9)
β	= constant in Eq. (9)
δ	= preheat zone thickness
λ	= thermal conductivity
ρ	= density
ϕ	= fuel-air equivalence ratio

Subscripts

ad	= adiabatic
b	= burned
i	= 1,2,...q
q	= number of computational segments in preheat layer
o	= reference condition
u	= unburned

Introduction

For analysis and performance predictions of various combustion engines, fueled with gaseous and liquid fuels, laminar burning velocity data are of prime importance. The majority of turbulent combustion models require a knowledge of laminar burning velocity of the fuel-air mixture as a function of pressure, temperature and mixture strength. Also, reliable experimental data are needed in order to test and calibrate thermokinetic combustion models which have been quite successful for combustion predictions of simple hydrocarbon fuels (See Tsatsaronis 1978; Warnatz 1981; Westbrook and Dryer 1981).

Although alcohol fuel applications have recently received much attention, fundamental experimental data on combustion characteristics of alcohol fuels are quite scarce and available information is limited to methanol-air mixtures. For ethanol-air mixtures, maximum burning velocity at 100°C was reported by Sachsse and Bartholome (1949) and recently a more detailed study was reported (Gülder 1983a).

The use of water as an ancillary combustion control technique has received much interest since the beginning of this century. The observed effects in internal combustion engines were eliminated detonation and pre-ignition and internal cooling. The major motivation in the last two decades has been the reduction of combustion generated nitric oxide emission by water addition. A short review of the subject was reported by Dryer (1977).

Alcohol-water mixtures as fuels were considered in order to decrease nitric oxide emissions in spark ignition engines (See Johnson 1978; Browning and Pefley 1978; Gülder 1980) and to boost the octane quality of the alcohol fuel for application in high compression ratio engines (Gülder 1981) in order to achieve higher conversion efficiencies.

Very limited information about the effect of water addition on combustion rates of hydrocarbon fuels exists. It has been long known that trace amounts of water vapor accelerates the carbon monoxide combustion. However, laminar methanol-air (Koda et al. 1982) and hydrogen-air (Liu and MacFarlane 1983) flames are inhibited by water vapor addition. In this study the laminar burning velocities of ethanol-air and ethanol-water-air mixtures, and their dependence on water concentration, mixture strength, unburned mixture temperature and pressure were determined using the constant volume spherical bomb technique.

Apparatus and Procedure

A spherical high-pressure vessel with an internal diameter of 0.325 m, was used for the measurements. Ionization probes were used to measure the spatial velocity of the flame during the prepressure period of combustion in the vessel. The description of the experimental setup has been reported in a previous paper (Gülder 1983a).

The burning velocity can be calculated from the measured flame growth rate by employing the following relation

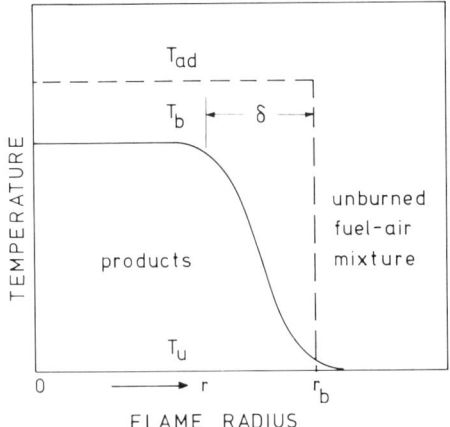

Fig. 1 Radial temperature distribution in a spherical kernel during constant pressure combustion period.

$$S_u = \frac{\bar{\rho}_b}{\bar{\rho}_u} \frac{dr_b}{dt} \qquad (1)$$

which is valid only for the prepressure period of combustion. dr_b/dt, $\bar{\rho}_u$, and $\bar{\rho}_b$ are the measured flame growth rate, unburned mixture density, and mean burned gas density, respectively. The main difficulty in employing Eq. (1) is the evaluation of the mean burned gas density. This problem arises due to presence of the preheat zone which has a considerable effect at low pressures and temperatures. The radial temperature distribution in a spherical kernel during the constant pressure combustion period, in which the increase in temperature resulting from the compression of the burned gases can be neglected, is shown in Fig. 1. If the burned gas within radius r_b is assumed to be at the adiabatic flame temperature, then the corresponding value of $\bar{\rho}_b$ is less than the actual one and Eq. (1) underestimates the burning velocity.

Assuming ideal gas behavior, Eq. (1) can be put into the following form

$$S_u = \frac{T_u}{T_{ad}} \frac{I}{N} \frac{dr_b}{dt} \qquad (2)$$

where

$$I = \bar{\rho}_b/\rho_{ad} \qquad (3)$$

and ρ_{ad} is the burned gas density calculated using the adiabatic flame temperature T_{ad}. N is the ratio of the number of moles of products to that of reactants.

Procedures were suggested for the estimation of I for constant volume bomb measurements (See Andrews and Bradley 1972; Garforth and Rallis 1978; Gülder 1983a; Gülder 1983b). The analysis of Andrews and Bradley (1972) yielded a density correction factor of nearly 1.2 for a stoichiometric methane-air flame propagating spherically at a flame radius of 25 mm with P = 1 atm and T_u = 300 K, using a flame front thickness value of 1.1 mm. Rallis and Garforth (1978) reported that the correction factor is approximately 1.05 for the same conditions. A much simpler scheme (Gülder 1983a) yielded a correction factor of 1.13 ± 0.05 for stoichiometric methanol-air flame at atmospheric conditions.

In this work a more rigorous scheme has been employed for density correction. For a steady one-dimensional planar flame, the energy equation for the preheating zone can be expressed as follows

$$\frac{d}{dx}\left(\lambda \frac{dT}{dx}\right) = S_u \rho_u \frac{d(C_p T)}{dx} + H \dot{w} \qquad (4)$$

where λ is the thermal conductivity, T the temperature, C_p the specific heat, H the enthalpy of combustion, and $\dot{w}$ the reaction rate. Integrating this expression, assuming that the reaction rate is negligible and the transport properties are constant, gives the temperature profile as (Bradley 1972),

$$T = T_u + A \exp\left(\frac{S_u \rho_u C_p}{\lambda} x\right) \qquad (5)$$

In order to allow for the variation of thermal conductivity and specific heat with temperature, the preheat zone can be divided into a number of elemental thicknesses and in each strip appropriate mean values of λ and C_p can be used. Then the sum of elemental thicknesses yields the preheat zone thickness δ as follows

$$\delta = \sum_{i=1}^{q-1} \frac{\lambda}{S_u \rho_u C_p} \ln\left(\frac{T_i - T_u}{T_{i+1} - T_u}\right) \qquad (6)$$

In this expression $T_1 = T_u + 5°C$, $T_q = T_b$ and q is the number of layers with appropriate mean temperatures within the preheat zone.

Knowing the temperature profile, the mean burned gas density is given by

$$\bar{\rho}_b = \frac{3}{r_b^3} \int_0^{r_b} \rho_b(r) \, r^2 \, dr \qquad (7)$$

Using Eqs. (2), (3), (6), and (7), the burning velocity can be evaluated from the measured flame growth rate by a simple iterative scheme.

Experimental burning velocities reported in this work were based on the measurements taken at a flame diameter of 0.05 m. Corresponding cut-off pressure rise is less than 1% of the initial mixture pressure. The range of experimental observations covered equivalence ratios from 0.7 to 1.4. Unburned mixture temperature was varied from 400 to 600 K and the range of pressure covers from atmospheric to 800 kPa. Water content of ethanol was varied from 0 to 50% by mass.

The adiabatic combustion temperatures were computed using an equilibrium computer program which considers 14 species in the products. Thermodynamic and transport properties of ethanol vapor, steam, air, and combustion products are either taken from standard tables (JANNAF 1971; Mayhew and Rogers 1974) or estimated by the methods recommended by Reid et al. (1977).

Results and Discussion

Density Correction

In the procedure of density correction, heat exchange between the mixture and the ignition electrodes and ionization probes was assumed to be negligible. Adiabatic combution temperatures necessary for data reduction, for the ethanol-water mixtures at 400 K and 100 kPa, are shown in Fig. 2.

In order to show the importance of density correction, computed values of preheat layer thickness and correction factors were plotted in Fig. 3. At 400 K and atmospheric pressure the correction factor is about 1.1, with increasing pressure (and temperature) it decreases rapidly and becomes negligible near 800 kPa. The measured flame front thicknesses of common fuel hydrocarbons (See Janisch 1971; Andrews and Bradley 1972; Dixon-Lewis and Wilson 1951; Fristrom and Westenberg 1957) and those computed by detailed thermokinetic models (Westbrook and Dryer 1979; Westbrook and Dryer 1981) are of the order of 1-2 mm at room temperature and atmospheric pressure. Our results at 400 K, when reduced to 300 K, are comparable to measured values reported in literature.

Laminar Burning Velocities

In order to assure that our experimental apparatus and density correction scheme give the correct result,

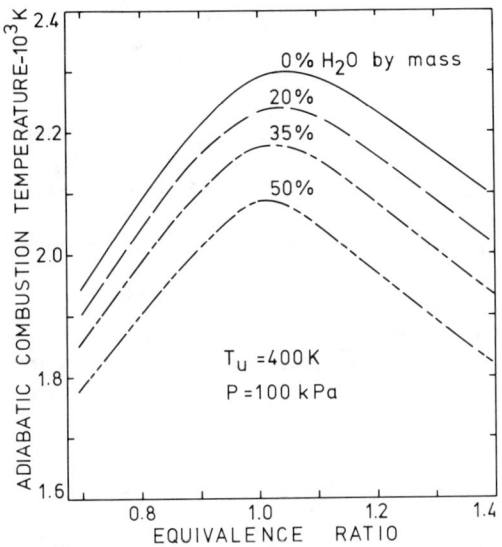

Fig. 2 Adiabatic combustion temperatures of ethanol-water-air mixtures at 100 kPa pressure and T_u = 400 K.

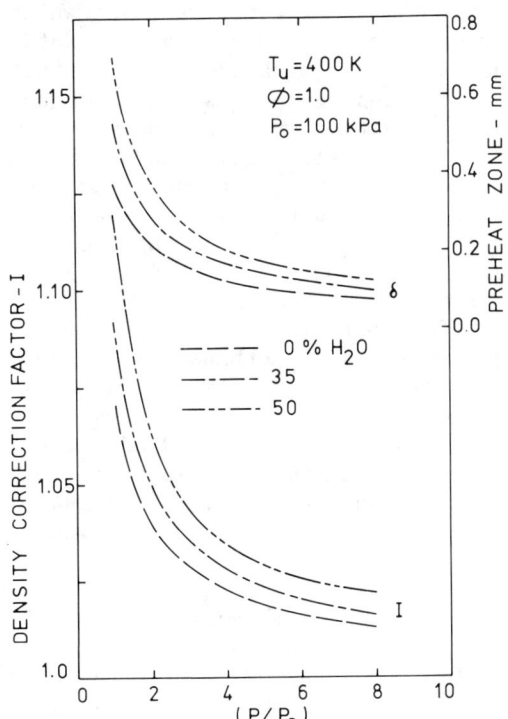

Fig. 3 Variation of preheat zone thickness and density correction factors with pressure.

preliminary measurements were made with methane-air mixtures on which an appreciable amount of experimental data exist. Results obtained at 300 K and 100 kPa are compared to recently reported burning velocity data of methane-air mixtures in Fig. 4. The agreement is very good and this confirms the validity of the method used.

Variation of laminar burning velocities of ethanol-air mixtures with equivalence ratio and initial mixture temperature is shown in Fig. 5. Maximum burning rates were obtained at an equivalence ratio of approximately 1.08, independent of the unburned mixture temperature.

Measured flame growth rates dr_b/dt and corresponding laminar burning velocities for ethanol-air flames are shown in Fig. 6. S_g is the gas velocity due to expansion of the burned gases, i.e.,

$$S_u = \frac{dr_b}{dt} - S_g \qquad (8)$$

Figure 7 shows the effect of water concentration on burning velocity at different equivalence ratios. Water percentages are given as mass fractions of water in water-ethanol blends. Increasing the amount of water in the mixture reduced the burning velocity as expected. Water concentration did not change the equivalence ratio at which maximum burning rates were attained. However, as the water content was increased the range of equivalence ratio within which the mixture could be ignited was narrowed on the lean side (Fig. 7). A similar observation was reported for methanol-water mixtures (Hirano et al. 1981) that with increasing water content the range of equivalence ratio for stable flames on a circular nozzle burner was narrowed. Present experiments were limited to $\phi \leqslant 1.4$ on the rich side and no ignition problem was experienced at $\phi = 1.4$ with increasing water content.

The obtained results, for stoichiometric mixtures, were compared to burning velocities of methanol-water and hydrogen-water mixtures. In Fig. 8, normalized velocities were plotted as a function of the mole fraction of water vapor in a fuel-water-air mixture. The observed effect of water on the burning velocity of ethanol is comparable to that in methanol combustion (Hirano et al. 1981) and that in hydrogen combustion (Liu and MacFarlane 1983). However, the inhibitory effect of water vapor on hydrogen flames is not as high as those of water vapor on alcohols.

The laminar burning velocities of ethanol and ethanol-water mixtures burning in air can be represented as a function of equivalence ratio and water concentration at 400 K and 100 kPa by means of a fitted equation. The

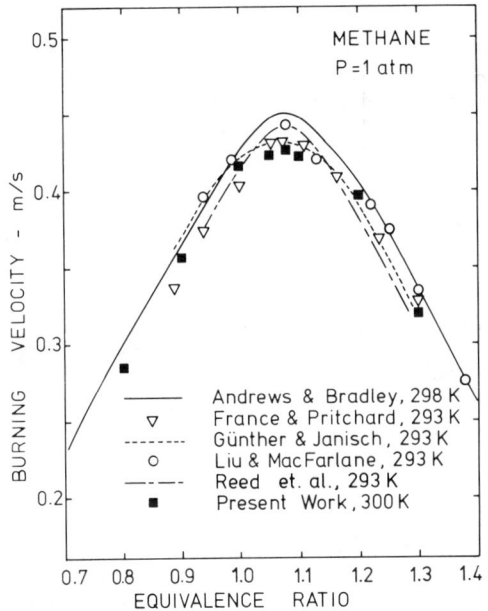

Fig. 4 Comparison of burning velocity of methane mixtures.

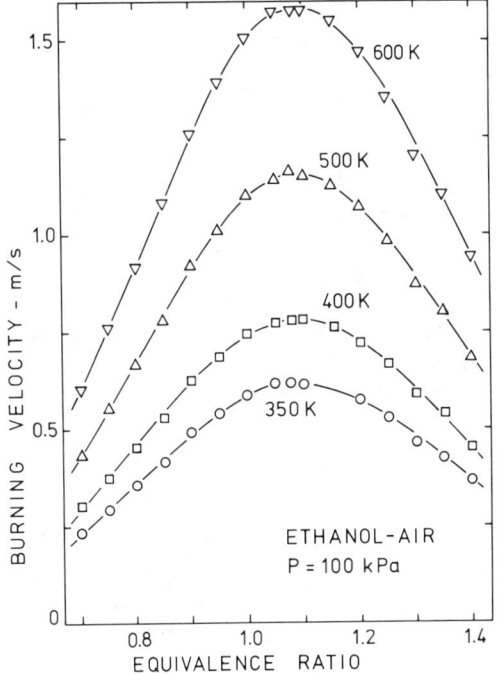

Fig. 5 Burning velocities of ethanol-air mixtures at 100 kPa pressure as function of equivalence ratio and unburned mixture temperature.

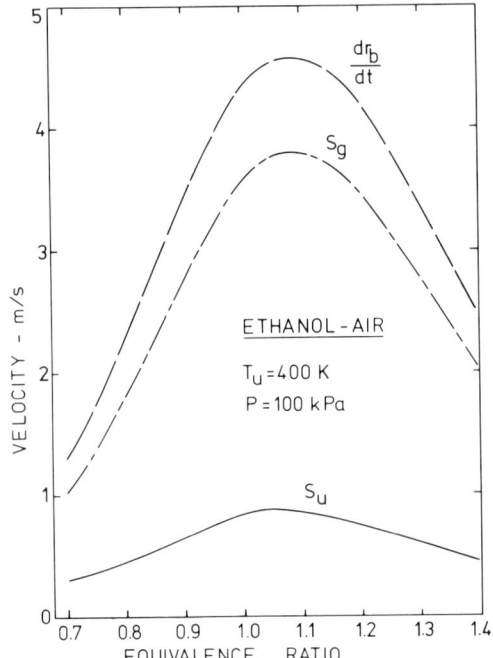

Fig. 6 Flame growth rate, gas velocity, and laminar burning velocity.

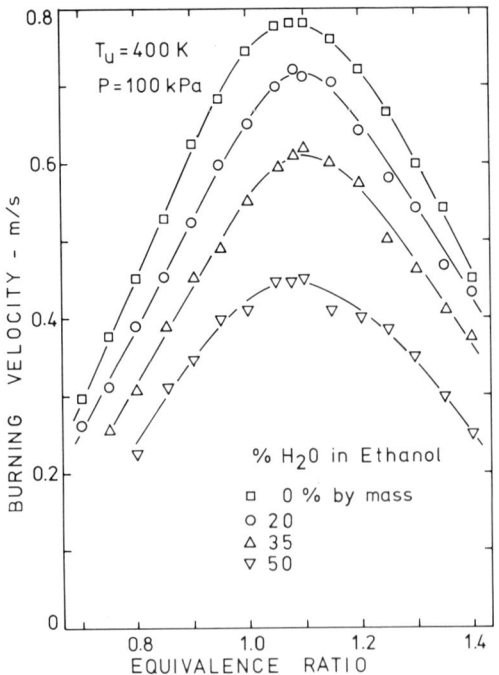

Fig. 7 Burning velocities of ethanol-water-air mixtures as a function of equivalence ratio.

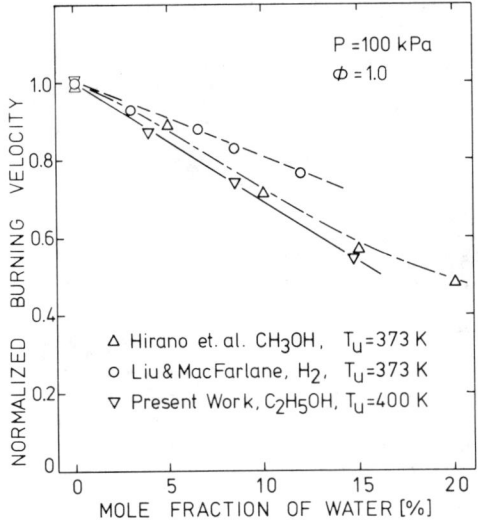

Fig. 8 Effect of water vapor concentration on laminar burning velocities of methanol, hydrogen, and ethanol.

following estimates the burning velocities with small deviations:

$$S_{uo}(\phi) = F \cdot B \cdot \phi^{\alpha} \cdot \exp\left[-\beta(\phi - 1.075)^2\right] \quad (9)$$

where B is 0.77 m/s for T_{uo} = 400 K, and

$$F = 1.0 - 3.125 \, C_w \quad (10)$$

where C_w is the mole fraction of the water in the stoichiometric unburned mixture. The values of α and β are 0.25 and 6.34, respectively. The maximum difference between experimental data and those calculated by Eq. (9) is less than 4%.

Dependence of burning velocities of ethanol and ethanol-water mixture on unburned mixture temperatures is illustrated in Fig. 9, for stoichiometric conditions. Temperature dependence of burning velocity can be represented by the following power law

$$S_u(T, \phi) = S_{uo}(\phi) \left(T_u/T_o\right)^n \quad (11)$$

The value of n was found to be 1.75 independent of the amount of water in the mixture.

The only experimental value reported in literature for the burning velocity of ethanol, to the author's knowledge, is 0.7 m/s measured at T_u = 100°C and ϕ = 1.09 (Sachsse and Bartholome 1949). This value agrees well with our measured velocity of 0.68 m/s at T_u = 375 K and ϕ = 1.0 (Fig. 9).

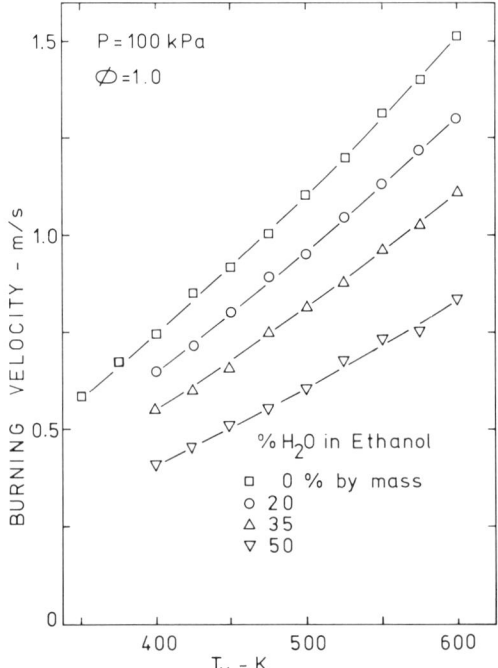

Fig. 9 Burning velocity as a function of unburned mixture temperature at 100 kPa pressure, stoichiometric mixture.

The variation of burning velocities with pressure at T_u = 400 K and ϕ = 1.0 is shown in Fig. 10. The data points, when plotted on logarithmic scales, indicate a pressure dependence in the form below

$$S_u(P,\phi) = S_{uo}(\phi)\left(P/P_o\right)^m \qquad (12)$$

where m is -0.17 for ethanol and increases, as the water content is increased, to -0.13 for 50% water by mass in ethanol. m can be represented by the following linear relation

$$m = 0.28\ C_w - 0.17 \qquad (13)$$

These pressure exponents are comparable to those of methanol and other common fuel hydrocarbons (Metghalchi and Keck 1982; Gülder 1983a) except that for methane.

Chemical kinetic effects of water addition to hydrocarbon fuels have not been extensively investigated. Cullis and Newitt (1956) found no effect of water vapor addition to ethanol on the length of the induction period of ethanol oxidation at 295°C. Dryer (1977) proposed that

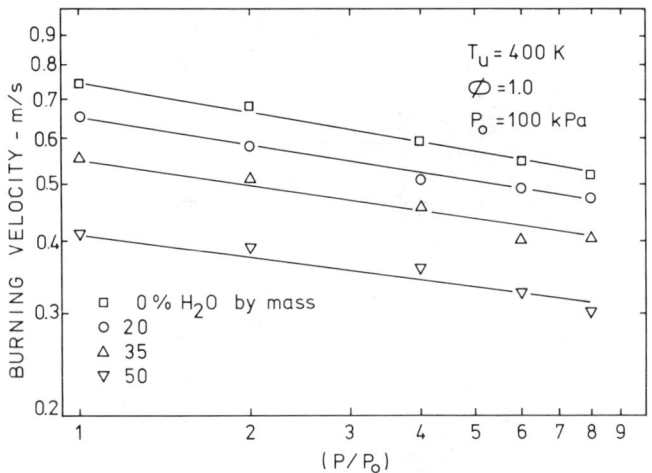

Fig. 10 Variation of burning velocities with pressure at T_u = 400 K, stoichiometric mixture.

water vapor probably exerts some chemical inhibition on hydrocarbon autoignition kinetics other than through decreasing the system temperature. A possible mechanism was also suggested by Eberius et al. (1969) that water vapor promotes the hydrogen-oxygen radical recombination reactions. Thus, lower concentrations of hydrogen atom would reduce the rate of chain branching through the reaction

$$H + O_2 \rightarrow OH + O \tag{14}$$

which would result in longer induction times. Using the elementary kinetic mechanism of Bowman (1975) for methane oxidation, Dryer (1977) predicted that 10% volume air nitrogen replaced by water vapor, maximum hydroxyl radical concentration increases by 30% and oxygen atom concentration decreases by 22%. Maximum hydrogen atom concentration was found to be only weakly affected and overall oxidation rate of methane was slightly decreased.

In order to predict the dominant mode of effect of water on burning velocity, i.e., thermal or chemical kinetic, it would be helpful to represent the present data by modified Semenov equation derived from Semenov's thermal model (Dugger 1952):

$$S_u(T,P,\phi) = K \cdot (P/P_o)^z \left[f(T) \cdot N^2 \cdot \exp\left(-E_a/R^\circ T_{ad}\right) \right]^{\frac{1}{2}} \tag{15}$$

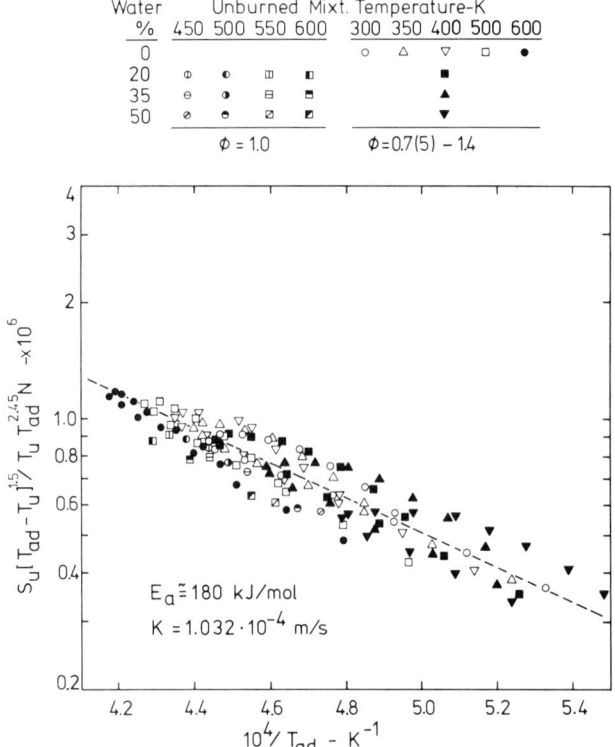

Fig. 11 Correlation of data for the determination of constants in modified Semenov equation.

where
$$f(T) = T_u^2 T_{ad}^{4.9}/(T_{ad} - T_u)^3 \tag{16}$$

For ethanol-air mixtures, apparent activation energy was found as 180 MJ/kmole, z as −0.24 and the value of K as $1.032 \cdot 10^{-4}$ m/s. Equation (15) with the same set of constants was found to describe the burning velocities of ethanol-water mixtures as well, with more or less the same deviations (Fig. 11). However, slightly increased values of pressure exponent z are required with increasing water content for better estimations (z increases with increasing water content and reaches −0.18 for $C_w = 0.144$). Although Eqs. (9-13) predict the burning velocities with better accuracy than Eq. (15) does, the latter shows that the presence of water in the mixture causes mostly thermal effects. Water seems to act as a heat sink and a diluent, as a result it lowers the combustion temperature which yields lower burning rates. This finding is in agreement with the conclusion of Koda et al. (1982) for methanol-water mixtures.

Summary and Conclusions

The laminar burning velocities of ethanol and ethanol-water mixtures burning in air were determined over the range of $\phi = 0.7 - 1.4$, and at various temperatures and pressures.

1) Pressure, temperature, equivalence ratio, and water content dependence of burning velocities is described by empirical and semi-empirical equations.

2) The semi-empirical relation Eq. (15) suggests that the effect of water on burning velocity is mainly thermal. Water lowers the combustion temperature resulting in reduced burning rates.

3) Small changes in pressure exponents of empirical and semi-empirical relations were observed as the water content was varied. With the present results it is not possible to argue on the presence of any chemical interactions due to existence of primary water. To search for such chemical kinetic inhibition effects detailed thermokinetic modelling studies are necessary.

4) As the water content is increased the range of equivalence ratio within which the mixture can be ignited is narrowed on the lean side.

References

Andrews, G. E. and Bradley, D. (1972) The burning velocity of methane-air mixtures. Combustion and Flame 19, 275-288.

Bowman, C. T. (1975) Non-equilibrium radical concentrations in shock-initiated methane oxidation. 15th Symposium (International) on Combustion, pp. 869-882, The Combustion Institute, Pittsburgh, Pa.

Bradley, J. N. (1972) Flame and Combustion Phenomena. Chapman-Hall, London, pp. 39-40.

Browning, L. H. and Pefley, R. K. (1978) Predicted methanal-water fueled SI engine performance and emissions. 2nd Intl. Symp. Alcohol Fuels Technology, pp. 6.4(1)-6.4(6), DOE, Washington, D.C.

Cullis, C. F. and Newitt, E. J. (1956) The gaseous oxidation of aliphatic alcohols. I. Ethyl alcohol; the products formed in the early stages. Proc. Royal Soc. A237, 530-542.

Dixon-Lewis, G. amd Wilson, M. J. G. (1951) A method for the measurement of the temperature distribution in the inner core of a bunson flame. Trans. Faraday Soc. 46, 1106-1114.

Dryer, F. L. (1977) Water addition to pratical combustion systems-concepts and applications. 16th Symposium (International) on

Combustion, pp. 279-295, The Combustion Institute, Pittsburgh, Pa.

Dugger, G. L. (1952) Effect of initial mixture temperature on flame speed of methane-air, propane-air, and ethylene-air mixtures. NACA Report 1061.

France, D. H. and Pritchard, R. (1976) Laminar burning velocity measurements using a laser-doppler anemometer. J. Inst. Fuel 49, 79-82.

Eberius, Von H., Hoyerman, K., and Wagner, H. Cg. (1969) Zur Reaktion $H + H + H_2O \rightarrow H_2 + H_2O$. Ber. Bunsenges. Physik. Chem. 73, 962-966.

Fristrom, R. M. and Westenberg, A. A. (1957) Flame zone studies IV- Microstructure and material transport in a laminar propane-air flame front. Combustion and Flame 1, 217-228.

Garforth, A. M. and Rallis, C. J. (1978) Laminar burning velocity of stoichiometric methane-air: Pressure and temperature dependence. Combustion and Flame 31, 53-68.

Gülder, Ö. L. (1980) Performance and exhaust emissions of a multicylinder SI engine fueled with methanol and gasoline. Proc. 15th IECEC, pp. 698-703, Vol. 1, AIAA, New York.

Gülder, Ö. L. (1981) Effect of compression ratio on the performance of a SI engine fuel with methanol and ethanol. Proc. 16th IECEC, pp. 1160-1165, Vol. 2, ASME, New York.

Gülder, Ö. L. (1983a) Laminar burning velocities of methanol, ethanol, and isooctane-air mixtures. 19th Symposium (International) on Combustion, pp. 275-281, The Combustion Institute, Pittsburgh, Pa.

Gülder, Ö. L. (1983b) Laminar burning velocities of methanol, isooctane and isooctane/methanol blends. Combust. Sci. Technol. 33, 179-192.

Günther, R. and Janisch, G. (1972) Measurements of burning velocity in a flat flame front. Combustion and Flame 19, 49-53.

Hirano, M., Oda, K., Hirano, T., and Akita, K. (1981) Burning velocities of methanol-air-water gaseous mixtures. Combustion and Flame 40, 341-343.

JANNAF Thermochemical Tables (1971) 2nd Edition, NSRDS-NBS 37.

Janisch, G. (1971) Geschwindigkeits- und Temperaturverteilung in einer laminaren Flammenfront. Chemie- Ing.- Techn. 43, 561-565.

Johnson, R. T. (1978) A comparison of gasoline, methanol and a methanol/water blend as spark ignition engine fuels. 2nd Intl. Symp. Alcohol Fuels Technology, pp. 4.3(1)-4.3(7), DOE, Washington, D.C.

Koda, S., Oda, K., Hirano, H., Hirano, T., and Akita, K. (1982) Burning characteristics of methanol-water-air mixtures in a constant volume combustion vessel. Combustion and Flame 46, 17-28.

Liu, D. D. S. and MacFarlane, R. (1983) Laminar burning velocities of hydrogen-air and hydrogen-air-steam flames. Combustion and Flame 49, 59-71.

Mayhew, Y. R. and Rogers, G. F. C. (1974) Thermodynamic and Transport Properties of Fluids, Basil Blackwell, Oxford.

Metghalchi, M. and Keck, J. C. (1982) Burning velocities of air with methanol, isooctane, and indolene at high pressure and temperature. Combustion and Flame 48, 191-210.

Reed, S. B., Mineur, J., and McNauchton, J. P. (1971) The effect on the burning velocity of methane of vitiation of combustion air. J. Inst. Fuel 44, 149-155.

Reid, R. C., Prausnitz, J. M., and Sherwood, T. K. (1977) The Properties of Gases and Liquids, 3rd Ed., McGraw-Hill, New York.

Sachsse, Von H. and Bartholome, E. (1949) Beitrage zur Frage der Flammengeschwindigkeit. Z. Elektrochem. 53, 183-190.

Tsatsaronis, G. (1978) Prediction of propagating laminar flames in methane-oxygen-nitrogen mixtures. Combustion and Flame 33, 217-239.

Warnatz, J. (1981) Flame velocity and structure of laminar hydrocarbon-air flames. Combustion in Reactive Systems: AIAA Progress in Astronautics and Aeronautics. (Edited by J. R. Bowen, A. K. Oppenheim, N. Manson, and R. I. Soloukhin), Vol. 76, pp. 501-521, AIAA, New York.

Westbrook, C. K. and Dryer, F. L. (1979) Modeling of flame properties of methanol. 3rd Intl. Sypm. Alcohol Fuels Technology, University of Santa Clara, Vol. 1, Asilomar, California.

Westbrook, C. K. and Dryer, F. L. (1981) Chemical kinetics and modeling of combustion processes. 18th Symposium (International) on Combustion, pp. 749-767, The Combustion Institute, Pittsburgh, Pa.

Computer Modeling Study of Acetylene-Oxygen Ignition and Flames Using a Truncated Reaction Mechanism

Soon-Muk Hwang* and William C. Gardiner Jr.†
University of Texas, Austin, Texas
and
Jürgen Warnatz‡
University of Heidelberg, Federal Republic of Germany

Abstract

The ignition and flame propagation processes in stoichiometric to lean acetylene-oxygen mixtures were investigated by computer modeling using detailed reaction mechanisms. It was found that a branched chain with H, O, OH, and CH_2 as chain centers provides good agreement with ignition and flame profiles using rate coefficient expressions derived from a variety of combustion and noncombustion experiments. The areas where disagreement still exists consist of the forms and magnitudes of OH profiles in flames and shock wave initiation, the relative amounts of CO and CO_2 produced in the reaction zones, and the intensity of CH* emission computed to be produced in shock-initiated combustion. These discrepancies suggest that significant amounts of C_2H are produced from the reactions of O and OH with C_2H_2 and that H atoms are not the only chain centers produced in the reaction of CH_2 with O_2. Adding such elementary reactions would not improve the match to exponential growth constant and ignition delay data significantly beyond that achieved already. It is to be expected, however, that the remaining discrepancies with profiles can then be removed and that the heat release rate during the primary reaction zone can be accounted for quantitatively.

Presented at the 9th ICODERS, Poitiers, France, July 3-8, 1983. Copyright © American Institute of Aeronautics and Astronautics, Inc., 1984. All rights reserved.
*Research Associate, Department of Chemistry.
†Professor, Department of Chemistry.
‡Professor, Department of Applied Physical Chemistry.

Introduction

Detailed modeling of combustion reactions using sets of elementary reactions that are presumed to contain all of the steps that contribute to the combustion process has been attempted for many fuels. Only in the case of H_2-O_2 and H_2-O_2-CO combustion has a level of confidence been attained such that detailed numerical modeling is as accurate as laboratory experiments. For hydrocarbon combustion, however, there are in general so many elementary reactions, with unknown or poorly known rate coefficient expressions, that combustion experiments cannot be modeled at the present time with comparable confidence. For studying the gasdynamics of the combustion process this is an unsatisfactory situation for two reasons. First, the fuels that one is most interested in are just the ones for which the combustion mechanisms are complex; second, in hydrocarbon combustion a major fraction of the reaction energy is released in the main reaction zone by the formation of C-O bonds, in contrast to the H_2-O_2 case where the main energy release occurs later by slow termolecular reactions. This circumstance accounts for most of the different detonation and flame stability behavior in hydrocarbon combustion.

Acetylene combustion appears to offer the best chance to develop a detailed hydrocarbon combustion mechanism that will be both useful and accurate. The complexities of CH_3-radical chemistry and the addition reactions of olefins are absent, and there is a broad range of combustion and noncombustion data suitable for modeling (Westbrook and Dryer 1981; Olson and Gardiner 1980). In the present study we set a goal of critically testing a minimal set of elementary reactions that can account for the main features observed in flame and ignition experiments. To avoid the complexities arising from polyacetylene formation and oxidation, we restricted the modeling to stoichiometric to lean combustion. Also, where minor species are known to be produced in the main chain reactions without contributing to the progress of ignition, their further reactions were ignored even though in the real world they are consumed as final equilibrium is approached.

The reaction mechanism resulting from this study proved to account more or less quantitatively for all of the available flame and ignition data. In addition, since it also included C_2H_2 pyrolysis reactions and H_2-O_2-CO reactions as subsets, it is in accord with the available experimental data on these reactions as well. The main limitation it has is in ignoring the elementary reactions which become important in rich combustion and hence as steps in the process of soot formation. There prove to be small but definite disagreements as well with some of the flame and ignition profiles, suggesting that elementary reactions may have been overlooked or, more likely, that reaction channels have been incorrectly assigned.

The reader's attention is called to a recent modeling study of C_2H_2 combustion by Miller et al.(1982) in which questions related to those discussed herein are also addressed.

Methods

The numerical procedures used for steady flame modeling were described previously, (Warnatz 1981) as were those used to model shock-initiated ignition (Gardiner et al. 1981). The constraint used for flame modeling was burner-stabilization at the unburned gas flow speed used in low-pressure laboratory flames by Fristrom et al. (1959), Westenberg and Fristrom (1965), and by Eberius et al. (1973). For shock-initiated combustion, steady flow with laminar wall boundary-layer formation was used for modeling incident shock wave experiments and constant-density reaction was used for reflected shock wave experiments.

The set of elementary reactions and rate coefficients was assembled by selecting from a critical survey of combustion reactions (Warnatz 1984) those that were expected to be important in stoichiometric and lean combustion of C_2H_2. Reactions that proved to be unimportant contributors to the initiation, branching, or equilibration processes under any conditions studied were eliminated as the modeling progressed. No adjustments were made to the rate coefficient expressions for the H_2-O_2-CO reactions or the C_2H_2 pyrolysis reactions, and only small adjustments were made to a few of the C/H/O reactions in the course of testing the model, as described in the next section. The reaction mechanism given in Table 1 differs from the one described in a previous study of the shock-initiated ignition process (Hidaka et al. 1984) in the inclusion of HO_2 and H_2O_2 reactions and of reactions of CH*; the former were for the purpose of completing the description of the colder parts of flames and the latter to permit comparison with CH* emission profiles in shock-initiated combustion.

Results

The mechanism given in Table 1 leads to the flame profiles shown together with the experimental data in Fig. 1. It can be seen that the agreement with the major species

Table 1 Reaction mechanism and rate coefficient expressions[a]

Reaction	A	m	E_A		Reference
1 $C_2H_2+M \rightarrow C_2H+H+M$	4.2+16		447	**	Warnatz 1984
2 $C_2H_2+C_2H_2 \rightarrow C_4H_3+H$	2.0+12		192	*	Tanzawa 1980
3 $C_2H+H_2 \rightarrow H+C_2H_2$	8.0+12		11	*	Warnatz 1984
4 $C_2H_2+C_2H \rightarrow C_4H_2+H$	3.5+13				Warnatz 1984
5 $C_2H_2+O \rightarrow CH_2+CO$	4.1+08	1.5	7		Warnatz 1984
6 $C_2H_2+O \rightarrow C_2HO+H$	4.3+14		51		Warnatz 1984
7 $C_2H_2+OH \rightarrow C_2H_2O+H$	4.0+14		78		This work
8 $C_2H_2O+M \rightarrow CO+CH_2+M$	3.6+15		248	*	Warnatz 1984
9 $C_2H_2O+OH \rightarrow C_2HO+H_2O$	1.0+13		11		Estimated
10 $C_2H_2O+H \rightarrow C_2HO+H_2$	3.0+13		36		Estimated
11 $C_2H+O_2 \rightarrow C_2HO+O$	5.0+13		6		Warnatz 1984
12 $C_2HO+H \rightarrow CH_2+CO$	3.0+13				Warnatz 1984
13 $CH_2+O_2 \rightarrow CO_2+H+H$	1.3+13		6		Warnatz 1984
14 $C_2H+O_2 \rightarrow CO_2+CHEX$	5.0+14		105		Estimated
15 $CHEX+M \rightarrow CH+M$	4.0+10	0.5			Estimated
16 $CHEX+O_2 \rightarrow CH+O_2$	2.4+12	0.5			Estimated
17 $CHEX \rightarrow CH$	1.9+06				Becker 1980
18 $CH+O_2 \rightarrow CO+OH$	2.0+13				Warnatz 1984
19 $CH+O \rightarrow CO+H$	4.0+13				Warnatz 1984
20 $H+O_2 \rightarrow OH+O$	7.8+15	-0.6	70	***	Warnatz 1984
21 $O+H_2 \rightarrow OH+H$	1.5+07	2.0	32	***	Warnatz 1984
22 $OH+H_2 \rightarrow H_2O+H$	1.0+08	1.6	14	***	Warnatz 1984
23 $OH+OH \rightarrow H_2O+O$	1.5+09	1.1			Warnatz 1984
24 $H+O_2+M \rightarrow HO_2+M$	7.0+17	-0.8			Warnatz 1984
25 $HO_2+H \rightarrow OH+OH$	1.5+14		4	*	Warnatz 1984
26 $HO_2+H \rightarrow H_2+O_2$	2.5+13		3	*	Warnatz 1984
27 $HO_2+OH \rightarrow H_2O+O_2$	2.0+13				Warnatz 1984
28 $HO_2+HO_2 \rightarrow H_2O_2+O_2$	2.0+12				Warnatz 1984
29 $H_2O_2+M \rightarrow OH+OH+M$	1.1+17		190		Estimated
30 $H_2+M \rightarrow H+H+M$	2.2+14		402	***	Warnatz 1984
31 $CO+OH \rightarrow CO_2+H$	4.4+06	1.5	-3	***	Warnatz 1984

[a]Units are mole/cm^3, s, and kJ. Rate coefficients are expressed in the form k = $AT^m\exp(-E_A/RT)$. Literature citations are given by Warnatz (1984). This set has the $H_2/CO/O_2$ reactions needed for both ignition and flame propagation chemistry. Reactions oxidizing C_2H_2O, C_2HO, C_4H_2, and C_4H_3 are ignored; it can be assumed that such reactions may be important in the ignition processes of rich mixtures. Some of the rate coefficient expressions cannot be extrapolated below 1000 K. The asterisk denotes a Guide Michelin rating of the confidence that may be placed in the accuracy of the rate coefficient expressions. For reaction 13, alternate product channels were considered in the modeling. (See text.)

and temperature profiles is very good, the slight discrepancies with CO and CO_2 being smaller than the lack of carbon balance in the flame data. The OH profile, on the other hand, while it does have the observed form, is computed to be about a factor of 2 below the experimental data.

The rate coefficient expressions for the chain reactions of Table 1 differ from those used in a previous study of exponential growth constants and ignition delays

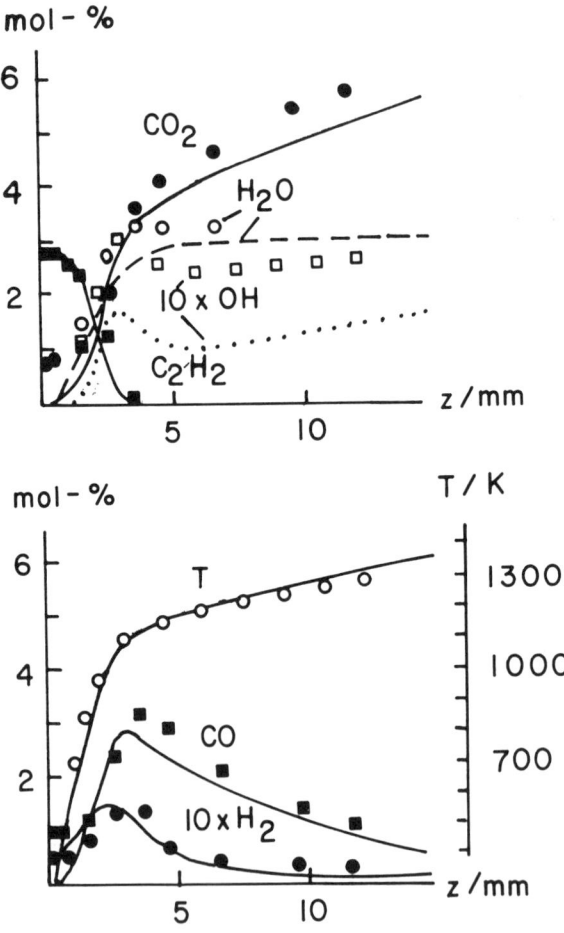

Fig. 1 Mole fraction and temperature profiles in a burner-stabilized 3% C_2H_2 - 97% O_2 flame at P = 0.1 bar. Points: measurements by Eberius et al. (1973). Lines: computed using the Table 1 mechanism.

(Hidaka et al. 1984) only in the adoption here of an updated expression for the rate coefficient of $H + O_2 = OH + O$. While the Table 1 expression is distinctly different from the one used before, the resulting rate coefficient itself differs only at temperatures well below those used in the shock tube experiments. The match to the exponential growth constant and ignition delay data for the Table 1 mechanism is therefore the same as for the previous mechanism. Similarly, the same qualitative features of the CO, CO_2, and OH profiles that appeared before to be different than observed are still present in the profiles computed with the Table 1 mechanism. We now consider the computed effects of altering the Table 1 mechanism so as to adjust these profiles in a direction that would bring them closer to the experimental ones.

The CO and CO_2 profiles, and hence the rate of heat release in the reaction, are strongly dependent upon the reaction channels chosen for the $CH_2 + O_2$ reaction. While the profiles of rich flames (Warnatz 1981) indicate that two H-atoms are produced, it is possible that some fraction of

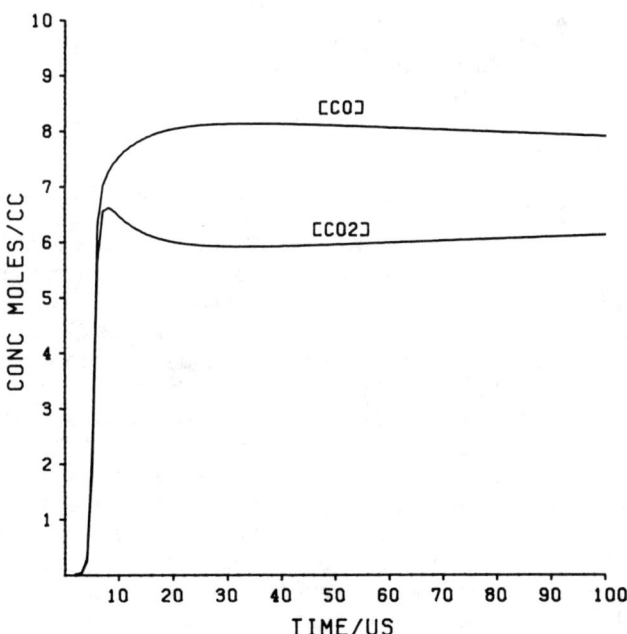

Fig. 2 Computed CO and CO_2 profiles for an incident shock experiment in a 1% C_2H_2, 3.75% O_2, 95.25% Ar mixture with P_1 = 5.33 kPa and T_2 = 2175 K (Jachimowski 1977). Products of the $CH_2 + O_2$ reaction assumed to be CO_2 + H + H. Units of ordinate 10^{-8} mole/cm^3.

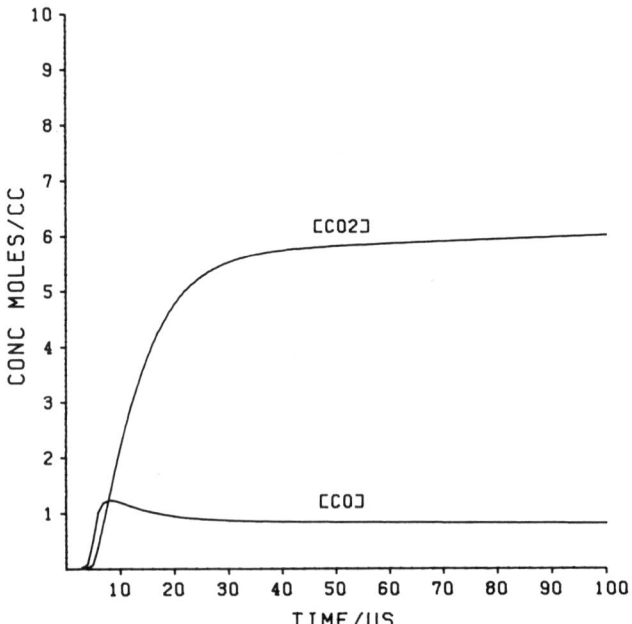

Fig. 3 Computed CO and CO_2 profiles for same experiment as in Fig. 2 with the products of the $CH_2 + O_2$ reaction assumed to be CO + OH + H. Units of ordinate 10^{-7} mole/cm^3 for CO and 10^{-8} mole/cm^3 for CO_2.

reaction leads to other products, such as CO + OH + H. In Figs. 2 and 3 the behavior of the CO_2 and CO profiles can be seen to be drastically affected by replacement of one channel by the other. The experiments (Jachimowski 1977) show that CO and CO_2 appear at the end of the induction period as step functions, after which there is a slow rise in the CO_2 profile and a slow decrease in the CO profile, attributable to the CO + OH reaction. Figures 2 and 3 show that the two assumptions about the experiments - exclusive reaction into one or the other of the two $CH_2 + O_2$ reaction channels - appear to bracket the leads to experimentally unobserved short peaks and CO_2 profiles that are either too steep or too slow. Adjustment of the relative amounts of the two reaction channels to bring these profiles into agreement with experiment was not attempted for the reasons given below in connection with the failure of the mechanism to account for the observed CH* yield of the reaction.

In Figs. 4 and 5 profiles computed for conditions used in laser Schlieren experiments are shown, again with two extreme assumptions about the reaction channels of $CH_2 + O_2$. It can be seen that the magnitude of the laser Schlieren

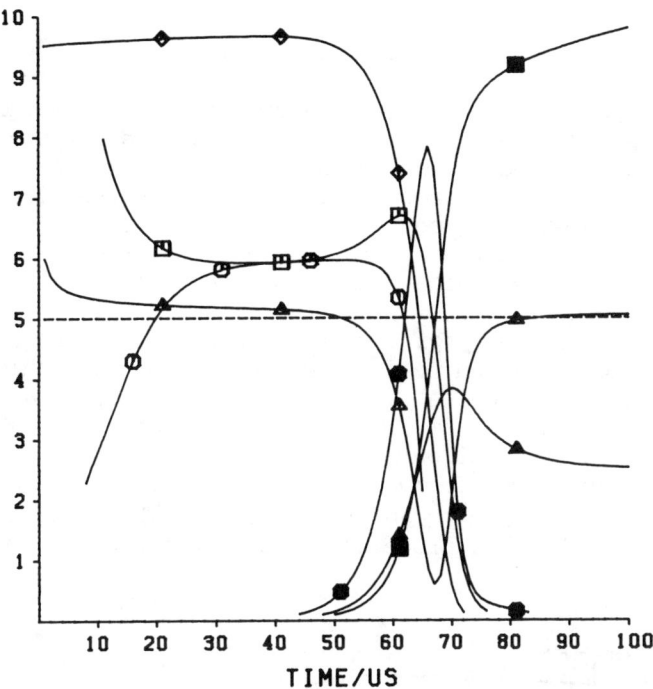

Fig. 4 Computed reaction progress for shock-initiated combustion of a 0.5% C_2H_2, 5.0% O_2, 94.5% Ar mixture in an incident shock wave with P_1 = 1.33 kPa and T_2 = 1700 K (Hidaka et al. 1984). Products of the $CH_2 + O_2$ reaction assumed to be CO_2 + H + H. The units of the ordinates are 10^{-9} mole/cm^3 for C_2H_2, CO_2, and OH, 10^{-16} mole/cm^3 for CH*, ug/cm^4 for dρ/dx, and 10^{-10} dm^3 mole-1s-1 for α/[O2] (Gutman and Matsuda 1970) and α*/[O2] (Matsuda et al. 1972). Symbols: □ α/[O2], O α*/[O2], △ dρ/dx, ◇ C_2H_2, ■ CO_2, ● CH*, ▲ OH.

signal is computed to change by a factor of 2. The reason for the difference can be seen immediately in the computed CO_2 profile, which rises steeply at the end of the induction period in Fig. 4 and only slowly in Fig. 5.

Also shown in Figs. 4 and 5 are computed CH* emission profiles based upon the assumptions that CH* is formed in the reaction $C_2H + O_2 = CO_2 + CH*$ and that C_2H is formed in the reaction $H + C_2H_2 = H_2 + C_2H$ only. The equality of the CH* and CO exponential growth constants is shown to persist over the exponential growth range, as found experimentally (Gutman and Matsuda 1970). Assuming that the $C_2H + O_2$ reaction proceeds at the maximum rate permitted by its activation energy (Matsuda et al. 1972) then provides maximum CH* concentration profiles that can be compared with experiments (Amrich 1967).

Discussion

The good agreement between computed and observed flame profiles, exponential growth constants, and ignition delays suggests that the Table 1 mechanism and rate coefficients may be close to a correct description of near-stoichiometric to lean acetylene combustion. On the other hand, the discrepancies between computed and observed CO and CO_2 profiles clearly indicate that an important feature is not yet included.

If only these profiles are considered, then agreement between computation and experiment can be achieved by dividing the channels of $CH_2 + O_2$ into $CO + OH + H$ and $CO_2 + H + H$ as a fitted function of temperature. This was not done for the following reason. The intensity of CH* chemiluminescence from acetylene combustion in shock waves can be reduced to a quantum yield of CH* production by direct calibration procedures that are certainly correct as

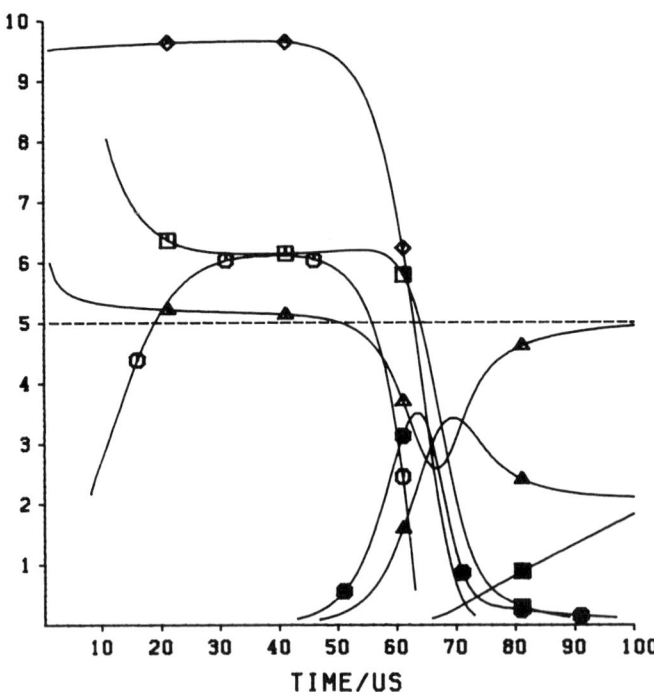

Fig. 5 Computed reaction progress for the same experimental conditions as in Fig. 4. Products of the $CH_2 + O_2$ reaction assumed to be $CO + OH + H$. Symbols and units on ordinate as in Fig. 4.

to order of magnitude and probably even better (Amrich 1967). Furthermore, there are strong arguments identifying the reaction of C_2H with O_2 as the source of CH^* under these conditions (Matsuda et al. 1972) and a host of consistent measurements of the radiative lifetime of CH^* (Becker et al. 1980 and references cited therein). A maximum value for the computable CH^* emission intensity can readily be found from the computed C_2H profiles, an assumed upper limit rate constant for $C_2H + O_2$ and the radiative lifetime. For conditions used in the laser Schlieren experiments, for example, the computed CH^* concentrations are in the range of 10^{-15} mole/cm^3, while the quantum yield measurements indicate a range of 10^{-11} mole/cm^3. The discrepancy of four orders of magnitude can not be explained in any other way than as an error in the C_2H profile.

In the Table 1 mechanism C_2H is produced only in the reaction of H atoms with C_2H_2, a relatively minor channel for both species. It is not known from the experiments of Löhr and Roth (1981) whether O atoms abstract from C_2H_2 in the temperature range considered here. Another candidate reaction capable of producing C_2H is attack on C_2H_2 by OH. The production of ketene and H assumed for the Table 1 mechanism is supported essentially by low-temperature experiments only, where the pressure dependence of the rate coefficient suggests that an addition reaction takes place. It is therefore quite likely that the $O + C_2H_2$ and $OH + C_2H_2$ reactions at high temperature may have different reaction channels, and abstraction could well be the major ones (Warnatz 1984).

If this is the case, then the discrepancy with the chemiluminescence quantum yield is removed, but only at the cost of revising the computed course of reaction during the induction period. Instead of the relatively unreactive ketene and H atoms, the OH chain center is transferred to C_2H, which can then undergo a variety of presumably fast elementary reactions with O_2 as well as its known reaction with C_2H_2 to form C_4H_2 and H. Reaction with O_2 could, for example, produce primarily CO and CHO, the latter leading primarily to H atoms through dissociation and to some HO_2 formation through atom transfer; or it could form CO_2 and ground-state CH. In either case the flow of the chain reaction is essentially changed, and if CH is formed there is another chain center identity that has to be characterized by experiment.

In addition to these aspects of the CO and CO_2 profiles there remains the facts that computed OH profiles show experimentally unobserved transient peaks and appear earlier than observed in the experiments of Stubbeman and Gardiner (1964). The difficulty with the CO and CO_2 profiles may indeed be coupled to that of the OH profiles, and all of the discrepancies may disappear together when the last parts of the induction zone mechanism are found.

It would appear that there are two experimental routes to clarifying these remaining problems with the acetylene combustion model. The first is to use the traditional shock tube techniques of laser Schlieren measurements of heat-release profiles together with i.r. chemiluminescence measurements of CO and/or CO_2 profiles to achieve simultaneous measures of the main reaction routes in the induction period. The second is to search for ground-state CH under similar conditions. Unless an independent source of information on the high-temperature rate constants of OH and CH reactions in these systems can be devised, however, the coupling between the CH_2, C_2H, and CH chemistry would seem to be too strong to permit rate coefficient expressions for their reactions to be found by combustion modeling alone.

Conclusions

Flame and ignition profiles in acetylene combustion can be understood in terms of a reaction mechanism containing a small number of elementary reactions, the essence of which is a branched chain with H, O, OH, and CH_2 as chain centers. Some remaining discrepancies between computation and experiment may be resolvable by adding reactions of C_2H and CH to the mechanism.

Acknowledgments

Acknowledgment is made to the donors of the Petroleum Research Fund, administered by the American Chemical Society, for partial support of this work. Additional support was provided by the U.S. Army Research Office and the Robert A. Welch Foundation.

References

Amrich, M. J. Jr. (1967) Quantum yields of OH and CH chemiluminescence from the combustion of acetylene in shock waves. M.A. Thesis, University of Texas, Austin, Texas.

Becker, K. H., Brenig, H. H., and Tatarczyk, T. (1980) Lifetime measurement on electronically excited CH radicals. Chem. Phys. Lett. 71, 242-245.

Eberius, K. H., Hoyermann, K., and Wagner, H. Gg. (1973) Structure of lean acetylene-oxygen flames. 14th Symposium (International) on Combustion, pp. 147-156. The Combustion Institute, Pittsburgh, Pa.

Fristrom, R. M., Avery, W. H., and Grunfelder, C. (1959) Reactions of simple hydrocarbons in flame fronts-microstructure of C_2 hydrocarbon-oxygen flames. 7th Symposium (International) on Combustion, pp. 304-310. The Combustion Institute, Pittsburgh, Pa.

Gardiner, W. C. Jr. and Olson, D. B. (1980) Chemical kinetics of high temperature combustion. Ann. Rev. Phys. Chem. 31, 377-399.

Gardiner, W. C. Jr., Walker, B. F., and Wakefield, C. B. (1981) Mathematical methods for modeling chemical reactions in shock waves. Shock Waves in Chemistry, (edited by A. Lifshitz), pp. 319-374. Marcel Dekker, New York.

Gutman, D. and Matsuda, S. (1970) Shock-tube study of the acetylene-oxygen reaction. I. CH chemiluminescence and CO production during the induction period. J. Chem. Phys. 52, 4122-4132.

Hidaka, Y., Eubank, C. S., Gardiner, W. C. Jr., and Hwang, S.-M. (1984) Shock tube and modeling study of acetylene oxidation. J. Phys. Chem., 88, 1006-1012.

Jachimowski, C. J. (1977) An experimental and analytical study of acetylene and ethylene oxidation behind shock waves. Combust. Flame 29, 55-66.

Löhr, R. and Roth, P. (1981) Shock tube measurements of the reaction behaviour of acetylene with O-atoms. Ber. Bunsenges. Phys. Chem. 85, 153-158.

Matsuda, S., Slagle, I. R., Fife, D. J., Marquart, J. R., and Gutman, D. (1972) Shock-tube study of the acetylene-oxygen reaction. IV. Kinetic study of CH, C_2, and continuum chemiluminescence during the induction period. J. Chem. Phys. 57, 5277-5285.

Miller, J. A., Mitchell, R. E., Smooke, M. D., and Kee, R. J. (1982) Toward a comprehensive chemical kinetics mechanism for the oxidation of acetylene: comparison of model predictions with results from flame and shock tube experiments. 20th Symposium (International) on Combustion, pp. 181-196. The Combustion Institute, Pittsburgh, Pa.

Stubbeman, R. F. and Gardiner, W. C. Jr. (1964) Shock tube study of the acetylene-oxygen reaction. J. Phys. Chem. 68, 3169-3176.

Warnatz, J. (1981) The structure of laminar alkane, alkene, and acetylene flames. 18th Symposium (International) on Combustion, pp. 369-384. The Combustion Institute, Pittsburgh, Pa.

Warnatz, J. (1984) Survey of rate coefficients in the C/H/O system. Combustion Chemistry, (edited by W. C. Gardiner Jr.), Springer-Verlag, New York, (in press).

Westbrook, C. K. and Dryer, F. L. (1981) Chemical kinetics and modeling of combustion processes. 18th Symposium (International) on Combustion, pp. 749-767. The Combustion Institute, Pittsburgh, Pa.

Westenberg, A. A. and Fristrom, R. M. (1965) H and O atom profiles measured by ESR in C_2 hydrocarbon-O_2 flames. 10th Symposium (International) on Combustion, pp. 473-478. The Combustion Institute, Pittsburgh, Pa.

Prediction of Laminar Flame Properties of Propane-Air Mixtures

Charles K. Westbrook* and William J. Pitz†
Lawrence Livermore National Laboratory, University of California
Livermore, California

Abstract

A numerical model including a detailed chemical kinetic reaction mechanism is used to study laminar flame propagation in propane-air mixtures. The effects of variations in pressure and fuel-oxidizer equivalence ratio are examined. Propane-air flames are compared with methane-air, methanol-air, and ethylene-air laminar flames. Quenching of propane-air flames in thermal boundary layers is examined, and the results are compared with previous studies of flame wall quenching for methane-air and methanol-air mixtures, to assess the influence of fuel molecule size on unburned hydrocarbon emissions from internal combustion engines. Finally, the addition of small fractions of propane to natural gas and its influence on laminar flame burning velocity is studied.

Introduction

Propane is an important combustion fuel and is often used in laboratory studies of oxidation processes in internal combustion engines, detonations, and other environments. Unlike hydrocarbon fuels with simpler structures such as methane or ethane, the thermochemical and combustion properties of propane are similar in many ways to those of more complex practical fuels.

Relatively few detailed modeling studies of propane and propylene combustion have appeared (Warnatz 1979, 1981, 1983; Cathonnet et al. 1981; McLain and Jachimowski

---

Presented at the 9th ICODERS, Poitiers, France, July 3-8, 1983. Copyright © by the American Institute of Aeronautics and Astronautics, Inc.,1984. All rights reserved.
*Senior Physicist, Theoretical Physics Division.
†Physicist, Theoretical Physics Division.

1977; Hautman et al. 1981; Edelson and Allara 1980; Layokun and Slater 1979; Lifshitz and Frenklach 1975; Burcat 1975). Most of these have treated only pyrolysis, and only Warnatz (1979, 1981,1983) has examined propane oxidation in laminar flames. In each of these previous oxidation and pyrolysis studies, the detailed reaction mechanism was developed specifically for only one type of combustion environment. Therefore chemical species and elementary reactions not important for that application were generally discarded.

The present modeling study represents a continuation in the development of "comprehensive" reaction mechanisms (Westbrook and Dryer 1981a) for the combustion of hydrocarbon fuels. In this approach, mechanisms are intended to be simultaneously applicable to a wide variety of combustion regimes. Therefore, reactions and species which are important in any relevant application are retained. When this is done, the resulting mechanism can be used to predict combustion phenomena over wide ranges of operating conditions. The development and validation of this type of reaction mechanism for propane oxidation (Westbrook and Pitz 1983) included comparisons between numerical predictions and experimental data from shock tubes and the turbulent flow reactor, and the resulting mechanism has been used to simulate propane oxidation in detonation waves (Westbrook et al. 1983a). In the present paper this mechanism is used, together with a numerical model for laminar flame propagation, to predict variations in laminar flame properties over wide ranges of pressure, temperature, and equivalence ratio.

Numerical Model

The numerical model used for these calculations was developed originally by Lund (1978) and has been employed extensively in the analysis of laminar flames (Westbrook 1980b, 1983; Westbrook and Dryer 1980, 1981b; Westbrook et al. 1980, 1983b). The equations of conservation of mass, momentum, and energy are solved together with the chemical species conservation equations including the kinetics terms, in a planar, one-dimensional geometry. Thermodynamic data and approximate transport coefficients for the chemical species have been validated in earlier studies and are used here without further modification.

The differential equations are solved in finite-difference form and are solved simultaneously, using a block tridiagonal matrix inversion technique. The model is fully time-dependent, so the steady propagation of a laminar flame is treated as the time-asymptotic solution of an unsteady problem. Each flame model consists of an unburned gas mixture with a specified composition and temperature, a burned gas in which the composition and temperature depend on the adiabatic flame properties of the fuel-oxidizer mixture, and a flame region between them. An initial guess is made of the spatial variations of the species concentrations and temperature through the flame, and the numerical model then computes the time evolution of the flame. A moving grid system enables additional zones to be concentrated in the flame region where the gradients are largest. The flame propagates into the unburned gas region at a rate that gradually approaches a steady value. The approach to that steady propagation condition can be quite rapid if the initial guess of the flame profiles is reasonably accurate. At this steady state the laminar burning velocity is defined conventionally as the rate of flame propagation relative to the velocity of the unburned gas.

Chemical Kinetic Mechanism

The oxidation of propane is described by the detailed kinetic reaction mechanism summarized in Table 1, in which the forward and reverse reaction rate coefficients are given. The development of this mechanism and its validation through comparison with experimental data is described in another paper (Westbrook and Pitz 1983). The mechanism was shown to reproduce experimental results for propane oxidation in shock tubes and plug flow reactors over wide ranges of pressure, temperature, and fuel-oxidizer equivalence ratio ϕ. In addition, the same reaction mechanism describes propane oxidation in detonation waves (Westbrook et al. 1983a). Finally, the propane pyrolysis submechanism is nearly identical to that developed by Hautman et al. (1981) to describe flow reactor propane pyrolysis. This mechanism has been built sequentially upon earlier mechanisms for methane (Westbrook et al. 1977; Westbrook 1979), methanol (Westbrook and Dryer 1979), and ethylene (Westbrook et al. 1983c) oxidation and simultaneously describes oxidation of these fuels in shock tubes, flow reactors, laminar flames, and detonations, in addition to the present results for propane.

Like the simpler hydrocarbon fuels studied previously, the first step in the combustion of the propane molecule involves abstraction of single H atoms, primarily by radical species including particularly H, O, and OH. However, in contrast with simpler fuels, the radical species thus formed, the propyl radical C_3H_7, exists in two distinct forms, the isopropyl radical iC_3H_7 and

Table 1 Fuel oxidation mechanism. Reaction rates in cm^3-mole-s-kcal units, $k = AT^n \exp(-E_a/RT)$

	Reaction	Forward rate			Reverse rate		
		log A	n	E_a	log A	n	E_a
1.	$H+O_2 \rightarrow O+OH$	14.27	0	16.79	13.17	0	0.68
2.	$H_2+O \rightarrow H+OH$	10.26	1	8.90	9.92	1	6.95
3.	$H_2O+O \rightarrow OH+OH$	13.53	0	18.35	12.50	0	1.10
4.	$H_2O+H \rightarrow H_2+OH$	13.98	0	20.30	13.34	0	5.15
5.	$H_2O_2+OH \rightarrow H_2O+HO_2$	13.00	0	1.80	13.45	0	32.79
6.	$H_2O+M \rightarrow H+OH+M$	16.34	0	105.00	23.15	-2	0.00
7.	$H+O_2+M \rightarrow HO_2+M$	15.22	0	-1.00	15.36	0	45.90
8.	$HO_2+O \rightarrow OH+O_2$	13.70	0	1.00	13.81	0	56.61
9.	$HO_2+H \rightarrow OH+OH$	14.40	0	1.90	13.08	0	40.10
10.	$HO_2+H \rightarrow H_2+O_2$	13.40	0	0.70	13.74	0	57.80
11.	$HO_2+OH \rightarrow H_2O+O_2$	13.70	0	1.00	14.80	0	73.86
12.	$H_2O_2+O_2 \rightarrow HO_2+HO_2$	13.60	0	42.64	13.00	0	1.00
13.	$H_2O_2+M \rightarrow OH+OH+M$	17.08	0	45.50	14.96	0	-5.07
14.	$H_2O_2+H \rightarrow HO_2+H_2$	12.23	0	3.75	11.86	0	18.70
15.	$O+H+M \rightarrow OH+M$	16.00	0	0.00	19.90	-1	103.72
16.	$O_2+M \rightarrow O+O+M$	15.71	0	115.00	15.67	-0.28	0.00
17.	$H_2+M \rightarrow H+H+M$	14.34	0	96.00	15.48	0	0.00
18.	$CO+OH \rightarrow CO_2+H$	7.11	1.3	-0.77	9.15	1.3	21.58
19.	$CO+HO_2 \rightarrow CO_2+OH$	14.18	0	23.65	15.23	0	85.50
20.	$CO+O+M \rightarrow CO_2+M$	15.77	0	4.10	21.74	-1	131.78
21.	$CO_2+O \rightarrow CO+O_2$	12.44	0	43.83	11.50	0	37.60
22.	$HCO+OH \rightarrow CO+H_2O$	14.00	0	0.00	15.45	0	105.15
23.	$HCO+M \rightarrow H+CO+M$	14.16	0	19.00	11.70	1	1.55
24.	$HCO+H \rightarrow CO+H_2$	14.30	0	0.00	15.12	0	90.00
25.	$HCO+O \rightarrow CO+OH$	14.00	0	0.00	14.46	0	87.90
26.	$HCO+HO_2 \rightarrow CH_2O+O_2$	14.00	0	3.00	15.56	0	46.04
27.	$HCO+O_2 \rightarrow CO+HO_2$	12.60	0	7.00	12.95	0	39.29

(Table continued on next page)

Table 1 (continued) Fuel oxidation mechanism. Reaction rates in cm^3-mole-s-kcal units, $k=AT^n \exp(-E_a/RT)$

Reaction	Forward rate			Reverse rate		
	log A	n	E_a	log A	n	E_a
28. $CH_2O+M \rightarrow HCO+H+M$	16.52	0	81.00	11.15	1	-11.77
29. $CH_2O+OH \rightarrow HCO+H_2O$	12.88	0	0.17	12.41	0	29.99
30. $CH_2O+H \rightarrow HCO+H_2$	14.52	0	10.50	13.42	0	25.17
31. $CH_2O+O \rightarrow HCO+OH$	13.70	0	4.60	12.24	0	17.17
32. $CH_2O+HO_2 \rightarrow HCO+H_2O_2$	12.00	0	8.00	11.04	0	6.59
33. $CH_4+M \rightarrow CH_3+H+M$	17.15	0	88.40	11.45	1	-19.52
34. $CH_4+H \rightarrow CH_3+H_2$	14.10	0	11.90	12.68	0	11.43
35. $CH_4+OH \rightarrow CH_3+H_2O$	3.54	3.08	2.00	2.76	3.08	16.68
36. $CH_4+O \rightarrow CH_3+OH$	13.20	0	9.20	11.43	0	6.64
37. $CH_4+HO_2 \rightarrow CH_3+H_2O_2$	13.30	0	18.00	12.02	0	1.45
38. $CH_3+HO_2 \rightarrow CH_3O+OH$	13.51	0	0.00	10.00	0	0.00
39. $CH_3+OH \rightarrow CH_2O+H_2$	12.60	0	0.00	14.08	0	71.73
40. $CH_3+O \rightarrow CH_2O+H$	14.11	0	2.00	15.23	0	71.63
41. $CH_3+O_2 \rightarrow CH_3O+O$	13.68	0	29.00	14.48	0	0.73
42. $CH_2O+CH_3 \rightarrow CH_4+HCO$	10.00	0.5	6.00	10.32	0.5	21.14
43. $CH_3+HCO \rightarrow CH_4+CO$	11.48	0.5	0.00	13.71	0.5	90.47
44. $CH_3+HO_2 \rightarrow CH_4+O_2$	12.00	0	0.40	13.88	0	58.59
45. $CH_3O+M \rightarrow CH_2O+H+M$	13.70	0	21.00	9.00	1	-2.56
46. $CH_3O+O_2 \rightarrow CH_2O+HO_2$	12.00	0	6.00	11.11	0	32.17
47. $C_2H_6 \rightarrow CH_3+CH_3$	19.35	-1	88.31	12.95	0	0.00
48. $C_2H_6+CH_3 \rightarrow C_2H_5+CH_4$	-0.26	4	8.28	10.48	0	12.50
49. $C_2H_6+H \rightarrow C_2H_5+H_2$	2.73	3.5	5.20	2.99	3.5	27.32
50. $C_2H_6+OH \rightarrow C_2H_5+H_2O$	9.94	1.05	1.81	10.23	1.05	20.94
51. $C_2H_6+O \rightarrow C_2H_5+OH$	13.40	0	6.36	12.66	0	11.23
52. $C_2H_5+M \rightarrow C_2H_4+H+M$	15.30	0	30.00	10.62	0	-11.03
53. $C_2H_5+O_2 \rightarrow C_2H_4+HO_2$	12.00	0	5.00	11.12	0	13.70
54. $C_2H_4+C_2H_4 \rightarrow C_2H_5+C_2H_3$	14.70	0	64.70	14.17	0	-2.61

(Table continued on next page)

Table 1 (continued) Fuel oxidation mechanism. Reaction rates in cm^3-mole-s-kcal units, $k = AT^n \exp(-E_a/RT)$

Reaction	Forward rate log A	n	E_a	Reverse rate log A	n	E_a
55. $C_2H_4 + M \rightarrow C_2H_2 + H_2$	16.97	0	77.20	12.66	1	36.52
56. $C_2H_4 + M \rightarrow C_2H_3 + H + M$	18.80	0	108.72	17.30	0	0.00
57. $C_2H_4 + O \rightarrow CH_3 + HCO$	12.52	0	1.13	11.20	0	31.18
58. $C_2H_4 + O \rightarrow CH_2O + CH_2$	13.40	0	5.00	12.48	0	15.68
59. $C_2H_4 + H \rightarrow C_2H_3 + H_2$	7.18	2	6.00	6.24	2	5.11
60. $C_2H_4 + OH \rightarrow C_2H_3 + H_2O$	12.68	0	1.23	12.08	0	14.00
61. $C_2H_4 + OH \rightarrow CH_3 + CH_2O$	12.30	0	0.96	11.78	0	16.48
62. $C_2H_3 + M \rightarrow C_2H_2 + H + M$	14.90	0	31.50	11.09	1	-10.36
63. $C_2H_3 + O_2 \rightarrow C_2H_2 + HO_2$	12.00	0	10.00	12.00	0	17.87
64. $C_2H_2 + M \rightarrow C_2H + H + M$	14.00	0	114.00	9.04	1	0.77
65. $C_2H_2 + O_2 \rightarrow HCO + HCO$	12.60	0	28.00	11.00	0	63.65
66. $C_2H_2 + H \rightarrow C_2H + H_2$	14.30	0	19.00	13.62	0	13.21
67. $C_2H_2 + OH \rightarrow C_2H + H_2O$	12.78	0	7.00	12.73	0	16.36
68. $C_2H_2 + OH \rightarrow CH_2CO + H$	11.51	0	0.20	12.50	0	20.87
69. $C_2H_2 + O \rightarrow C_2H + OH$	15.51	-0.6	17.00	14.47	-0.6	0.91
70. $C_2H_2 + O \rightarrow CH_2 + CO$	13.83	0	4.00	13.10	0	54.67
71. $C_2H + O_2 \rightarrow HCO + CO$	13.00	0	7.00	12.93	0	138.40
72. $C_2H + O \rightarrow CO + CH$	13.70	0	0.00	13.50	0	59.43
73. $CH_2 + O_2 \rightarrow HCO + OH$	14.00	0	3.70	13.61	0	76.58
74. $CH_2 + O \rightarrow CH + OH$	11.28	0.68	25.00	10.77	0.68	25.93
75. $CH_2 + H \rightarrow CH + H_2$	11.43	0.67	25.70	11.28	0.67	28.72
76. $CH_2 + OH \rightarrow CH + H_2O$	11.43	0.67	25.70	11.91	0.67	43.88
77. $CH + O_2 \rightarrow CO + OH$	11.13	0.67	25.70	11.71	0.67	185.60
78. $CH + O_2 \rightarrow HCO + O$	13.00	0	0.00	13.13	0	71.95
79. $CH_3OH + M \rightarrow CH_3 + OH + M$	18.48	0	80.00	13.16	1	-10.98
80. $CH_3OH + OH \rightarrow CH_2OH + H_2O$	12.60	0	2.00	7.27	1.66	25.31
81. $CH_3OH + O \rightarrow CH_2OH + OH$	12.23	0	2.29	5.90	1.66	8.35

(Table continued on next page)

Table 1 (continued) Fuel oxidation mechanism. Reaction rates in cm^3-mole-s-kcal units, $k=AT^n \exp(-E_a/RT)$

Reaction	Forward rate			Reverse rate		
	log A	n	E_a	log A	n	E_a
82. $CH_3OH+H \rightarrow CH_2OH+H_2$	13.48	0	7.00	7.51	1.66	15.16
83. $CH_3OH+H \rightarrow CH_3+H_2O$	12.72	0	5.34	12.32	0	36.95
84. $CH_3OH+CH_3 \rightarrow CH_2OH+CH_4$	11.26	0	9.80	6.70	1.66	18.43
85. $CH_3OH+HO_2 \rightarrow CH_2OH+H_2O_2$	12.80	0	19.36	7.00	1.66	11.44
86. $CH_2OH+M \rightarrow CH_2O+H+M$	13.40	0	29.00	16.69	-0.66	7.58
87. $CH_2OH+O_2 \rightarrow CH_2O+HO_2$	12.00	0	6.00	17.94	-1.66	28.32
88. $C_2H_3+C_2H_4 \rightarrow C_4H_6+H$	12.00	0	7.30	13.00	0	4.70
89. $C_2H_2+C_2H_2 \rightarrow C_4H_3+H$	13.00	0	45.00	13.18	0	0.00
90. $C_4H_3+M \rightarrow C_4H_2+H+M$	16.00	0	60.00	11.92	1	2.54
91. $C_2H_2+C_2H \rightarrow C_4H_2+H$	13.60	0	0.00	14.65	0	0.55
92. $C_4H_2+M \rightarrow C_4H+H+M$	17.54	0	80.00	12.30	1.0	-16.40
93. $C_2H_3+H \rightarrow C_2H_2+H_2$	13.30	0	2.50	13.12	0	68.08
94. $C_3H_8 \rightarrow CH_3+C_2H_5$	16.23	0	84.84	10.18	1	-0.32
95. $CH_3+C_3H_8 \rightarrow CH_4+iC_3H_7$	15.04	0	25.14	15.64	0	32.12
96. $CH_3+C_3H_8 \rightarrow CH_4+nC_3H_7$	15.04	0	25.14	15.64	0	32.12
97. $H+C_3H_8 \rightarrow H_2+iC_3H_7$	6.94	2	5.00	12.89	0	15.87
98. $H+C_3H_8 \rightarrow H_2+nC_3H_7$	7.75	2	7.70	12.96	0	14.46
99. $iC_3H_7 \rightarrow H+C_3H_6$	13.80	0	36.90	13.00	0	1.50
100. $iC_3H_7 \rightarrow CH_3+C_2H_4$	10.30	0	29.50	4.66	1	4.29
101. $nC_3H_7 \rightarrow CH_3+C_2H_4$	13.98	0	31.00	8.34	1	5.79
102. $nC_3H_7 \rightarrow H+C_3H_6$	14.10	0	37.00	13.00	0	1.50
103. $iC_3H_7+C_3H_8 \rightarrow nC_3H_7+C_3H_8$	10.48	0	12.90	10.48	0	12.90
104. $C_2H_3+C_3H_8 \rightarrow C_2H_4+iC_3H_7$	11.00	0	10.40	11.12	0	17.80
105. $C_2H_3+C_3H_8 \rightarrow C_2H_4+nC_3H_7$	11.00	0	10.40	11.12	0	17.80
106. $C_2H_5+C_3H_8 \rightarrow C_2H_6+iC_3H_7$	11.00	0	10.40	10.56	0	9.93
107. $C_2H_5+C_3H_8 \rightarrow C_2H_6+nC_3H_7$	11.00	0	10.40	10.56	0	9.93
108. $C_3H_8+O \rightarrow iC_3H_7+OH$	6.70	2	3.00	5.52	2	7.41

(Table continued on next page)

Table 1 (continued) Fuel oxidation mechanism. Reaction rates in cm^3-mole-s-kcal units, $k = AT^n \exp(-E_a/RT)$

Reaction	Forward rate log A	n	E_a	Reverse rate log A	n	E_a
109. $C_3H_8+O \rightarrow nC_3H_7+OH$	6.70	2	3.00	5.52	2	7.41
110. $C_3H_8+OH \rightarrow iC_3H_7+H_2O$	8.68	1.4	0.85	8.93	1.25	22.37
111. $C_3H_8+OH \rightarrow nC_3H_7+H_2O$	8.76	1.4	0.85	9.01	1.25	22.37
112. $C_3H_8+HO_2 \rightarrow iC_3H_7+H_2O_2$	12.70	0	18.00	12.01	0	8.43
113. $C_3H_8+HO_2 \rightarrow iC_3H_7+H_2O_2$	12.70	0	18.00	12.01	0	8.43
114. $C_3H_6+O \rightarrow C_2H_4+CH_2O$	13.77	0	5.00	13.76	0	86.67
115. $iC_3H_7+O_2 \rightarrow C_3H_6+HO_2$	12.00	0	5.00	11.30	0	17.48
116. $nC_3H_7+O_2 \rightarrow C_3H_6+HO_2$	12.00	0	5.00	11.30	0	17.48
117. $C_3H_8+O_2 \rightarrow iC_3H_7+HO_2$	13.60	0	47.50	12.31	0	0.00
118. $C_3H_8+O_2 \rightarrow nC_3H_7+HO_2$	13.60	0	47.50	12.31	0	0.00
119. $C_3H_6+OH \rightarrow C_2H_5+CH_2O$	12.90	0	0.00	13.66	0	17.35
120. $C_3H_6+O \rightarrow C_2H_5+HCO$	12.55	0	0.00	11.85	0	29.92
121. $C_3H_6+OH \rightarrow CH_3+CH_3CHO$	11.54	0	0.00	11.44	0	20.40
122. $C_3H_6+O \rightarrow CH_3+CH_3CO$	13.07	0	0.60	12.25	0	38.37
123. $CH_3CHO+H \rightarrow CH_3CO+H_2$	13.60	0	4.20	13.25	0	23.67
124. $CH_3CHO+OH \rightarrow CH_3CO+H_2O$	13.00	0	0.00	13.28	0	36.62
125. $CH_3CHO+O \rightarrow CH_3CO+OH$	12.70	0	1.79	12.00	0	19.16
126. $CH_3CHO+CH_3 \rightarrow CH_3CO+CH_4$	12.23	0	8.43	13.48	0	28.00
127. $CH_3CHO+HO_2 \rightarrow CH_3CO+H_2O_2$	12.23	0	10.70	12.00	0	14.10
128. $CH_3CHO \rightarrow CH_3+HCO$	15.85	0	81.78	9.58	1	0.00
129. $CH_3CHO+O_2 \rightarrow CH_3CO+HO_2$	13.30	0.5	42.20	7.00	0.5	4.00
130. $CH_3CO \rightarrow CH_3+CO$	13.48	0	17.24	11.20	0	5.97
131. $C_3H_6+H \rightarrow C_3H_5+H_2$	12.70	0	1.50	12.18	0	17.70
132. $C_3H_6+CH_3 \rightarrow C_3H_5+CH_4$	10.95	0	8.50	11.87	0	25.18
133. $C_3H_6+C_2H_5 \rightarrow C_3H_5+C_2H_6$	11.00	0	9.20	5.00	0	56.77
134. $C_3H_6+OH \rightarrow C_3H_5+H_2O$	12.60	0	0.00	7.18	0	69.69
135. $C_3H_8+C_3H_5 \rightarrow iC_3H_7+C_3H_6$	11.60	0	16.20	11.30	0	6.50

(Table continued on next page)

Table 1 (continued) Fuel oxidation mechanism. Reaction rates in cm^3-mole-s-kcal units, $k=AT^n \exp(-E_a/RT)$

	Forward rate			Reverse rate		
Reaction	log A	n	E_a	log A	n	E_a
136. $C_3H_8+C_3H_5 \rightarrow iC_3H_7+C_3H_6$	11.60	0	16.20	11.30	0	6.50
137. $C_3H_5 \rightarrow C_3H_4+H$	13.60	0	70.00	8.00	1	0.00
138. $C_3H_5+O_2 \rightarrow C_3H_4+HO_2$	11.78	0	10.00	11.08	0	10.00
139. $_1C_4H_8 \rightarrow C_3H_5+CH_3$	19.18	-1	73.40	13.13	0	0.00
140. $_1C_4H_8 \rightarrow C_2H_3+C_2H_5$	19.00	-1	96.77	12.95	0	0.00
141. $_1C_4H_8+O \rightarrow CH_3CHO+C_2H_4$	13.11	0	0.85	12.32	0	85.10
142. $_1C_4H_8+O \rightarrow CH_3CO+C_2H_5$	13.11	0	0.85	12.37	0	38.15
143. $_1C_4H_8+OH \rightarrow CH_3CHO+C_2H_5$	13.00	0	0.00	12.97	0	19.93
144. $_1C_4H_8+OH \rightarrow CH_3CO+C_2H_6$	13.00	0	0.00	12.99	0	32.43
145. $C_3H_4+O \rightarrow CH_2O+C_2H_2$	12.00	0	0.00	12.03	0	81.73
146. $C_3H_4+O \rightarrow HCO+C_2H_3$	12.00	0	0.00	10.47	0	30.82
147. $C_3H_4+OH \rightarrow CH_2O+C_2H_3$	12.00	0	0.00	11.93	0	18.25
148. $C_3H_4+OH \rightarrow HCO+C_2H_4$	12.00	0	0.00	11.77	0	33.81
149. $C_3H_6 \rightarrow C_3H_5+H$	13.00	0	78.00	11.00	0	0.00
150. $C_2H_2+O \rightarrow HCCO+H$	4.55	2.7	1.39	2.70	2.7	12.79
151. $CH_2CO+H \rightarrow CH_3+CO$	13.04	0	3.40	12.38	0	40.20
152. $CH_2CO+O \rightarrow HCO+HCO$	13.00	0	2.40	11.54	0	33.50
153. $CH_2CO+OH \rightarrow CH_2O+HCO$	13.45	0	0.00	13.44	0	18.50
154. $CH_2CO+M \rightarrow CH_2+CO+M$	16.30	0	60.00	10.66	0	0.00
155. $CH_2CO+O \rightarrow HCCO+OH$	13.70	0	8.00	10.86	0	8.00
156. $CH_2CO+OH \rightarrow HCCO+H_2O$	12.88	0	3.00	11.03	0	11.00
157. $CH_2CO+H \rightarrow HCCO+H_2$	13.88	0	8.00	11.39	0	8.00
158. $HCCO+OH \rightarrow HCO+HCO$	13.00	0	0.00	13.68	0	40.36
159. $HCCO+H \rightarrow CH_2+CO$	13.70	0	0.00	13.82	0	39.26
160. $HCCO+O \rightarrow HCO+CO$	13.53	0	2.00	13.92	0	128.26
161. $C_3H_6 \rightarrow C_2H_3+CH_3$	15.80	0	85.80	10.00	1	0.00
162. $C_3H_5+H \rightarrow C_3H_4+H_2$	13.00	0	0.00	13.00	0	40.00
163. $C_3H_5+CH_3 \rightarrow C_3H_4+CH_4$	12.00	0	0.00	13.00	0	40.00

the normal propyl radical nC_3H_7. The rates of H atom abstraction from propane do not depend strongly on which site the H atom occupies, but the effects of the subsequent reactions of these propyl radicals are very important. The nC_3H_7 radical decomposes by two paths

$$nC_3H_7 \rightarrow CH_3 + C_2H_4 \quad (101)$$

$$nC_3H_7 \rightarrow H + C_3H_6 \quad (102)$$

The rate of Reaction 101, producing methyl radicals and ethylene, is somewhat greater than the path producing H atoms and propylene. The opposite trend is observed for the iC_3H_7 radicals

$$iC_3H_7 \rightarrow H + C_3H_6 \quad (99)$$

$$iC_3H_7 \rightarrow CH_3 + C_2H_4 \quad (100)$$

The rate of Reaction 100 is negligibly small, so effectively all of the iC_3H_7 radicals produce H atoms. The subsequent fate of the products of propyl radical decomposition reactions is responsible for the importance of the distinction between the two isomeric forms. When H atoms are produced, a significant fraction then react with O_2 molecules

$$H + O_2 \rightarrow O + OH \quad (1)$$

in what is the principal chain branching reaction in this and most other hydrocarbon oxidation mechanisms. In contrast, the methyl radicals produced by the nC_3H_7 radical decomposition are relatively unreactive. A significant fraction recombine (Warnatz 1979,1981; Westbrook et al. 1977) to produce ethane, methane, and methanol, effectively reducing the size of the radical pool rather than enlarging it by means of Reaction 1. Therefore, the production of iC_3H_7 radicals accelerates the overall rate of combustion of propane while the production of nC_3H_7 radicals actually inhibits the overall process.

The reactions of propylene are not as well established as those of propane. In the construction of the present mechanism two separate paths have been adopted. The first consists of continued abstraction of H atoms to produce allyl (C_3H_5) radicals, while the second path involves addition reactions with O and OH which lead to breaking of C-C bonds. The allyl radical path yields C_3H_4 (no distinction is made between allene and propyne in this mechanism) through further decomposition, or butene by means of recombination with methyl radicals

$$C_3H_5 + CH_3 \rightarrow C_4H_8 \qquad (139)$$

The relatively small amounts of C_3H_4 are then oxidized by means of addition reactions with O and OH, followed by fragmentation to produce ethylene and formaldehyde, together with radical species. The product distributions for addition reactions of propylene are taken from several sources (Warnatz 1979,1981; McLain and Jachimowski 1977) and are somewhat parallel to analogous reactions of ethylene (Westbrook et al. 1983c). However, further work is needed to establish both the product distributions and rates of many of these reactions.

Flame Models

A series of laminar flame model calculations was carried out in which both the pressure and the fuel-air equivalence ratio were varied. In the next section a reference model is described indetail, a stoichiometric C_3H_8-air mixture at atmospheric pressure and an unburned gas temperature T_u of 300 K. This flame model will be compared with similar flames in which the fuel is CH_4, CH_3OH, and C_2H_4. Subsequent sections will describe the effects of variations in equivalence ratio and pressure.

Reference Case

The basic reference model for a stoichiometric mixture of propane-air predicts a laminar burning velocity of 41 cm/s, in excellent agreement with available experimental results (Singer 1953; Gilbert 1957; Gibbs and Calcote 1959; Kuehl 1962; Metghalchi and Keck 1980) and previous modeling predictions (Warnatz 1979, 1981, 1983). Other experimental results which are also consistent with these are summarized by Dugger et al. (1959). Temperature and concentration profiles for fuel and CO are plotted in Fig. 1 as functions of distance from an arbitrary flame position. This flame position is defined here as the point in the flame at which the rate of heat release from chemical reactions attains its maximum value. Also shown for comparison in Fig. 2 are the same quantities from analogous methane-air, methanol-air, and ethylene-air laminar flames. The fuel concentrations have each been normalized by dividing by the mole fraction of fuel in the unburned fuel-air mixture, while the CO concentrations

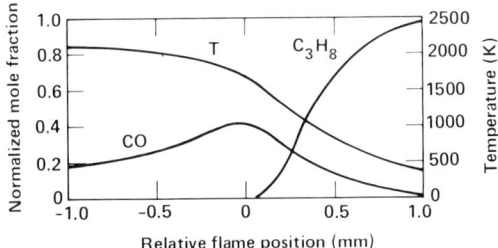

Fig. 1 Spatial profiles of temperature and normalized C_3H_8 and CO species concentrations, in reference propane-air laminar flame.

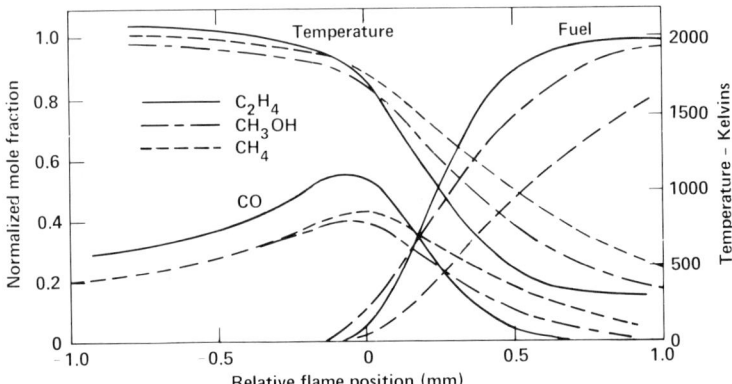

Fig. 2 Spatial profiles of temperature and normalized fuel and CO species concentrations for stoichiometric reference flames in methane-air, methanol-air, and ethylene-air mixtures.

have been normalized by dividing by the mole fraction of carbon atoms in the unreacted fuel-air mixture. From these figures it can be seen that the propane results fall within the same ranges as the other hydrocarbon fuels. The thickness of the flame region in the propane-air model is somewhat greater than for the ethylene-air model and is closest to the methanol-air results. Similarly, the CO curve in the propane-air reference case most closely resembles the methanol-air results. The laminar burning velocity of the methanol-air (44 cm/s) is also closest to that of propane-air (70 cm/s for ethylene-air and 36 cm/s for methane-air), so these computations indicate that the flame thickness and CO level both scale smoothly with the burning velocity.

The sequential fragmentation of the propane fuel is indicated in Figs. 3-7, showing the concentration profiles

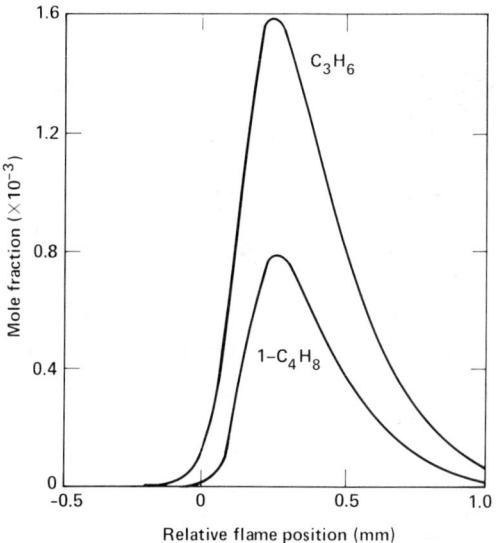

Fig. 3 Spatial profiles for C_3 and C_4 olefin species concentrations in reference propane-air flame.

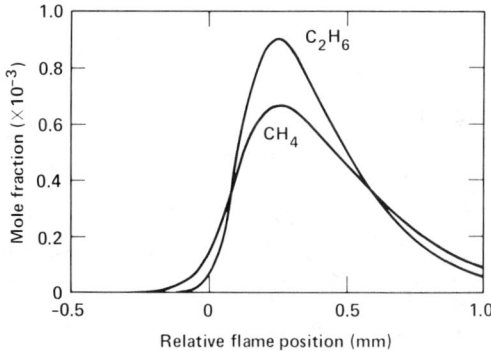

Fig. 4 Spatial profiles for intermediate alkane species concentrations in reference propane-air flame.

of the major intermediate species in the reference flame. In Fig. 3 the C_3 and C_4 olefin species are plotted, and from the positions at which these reach their peak values it can be seen that these species are formed very early in the flame. The saturated smaller alkanes CH_4 and C_2H_6 are shown in Fig. 4, both peaking early in the flame. Both species are produced in large quantities by radical recombination reactions including

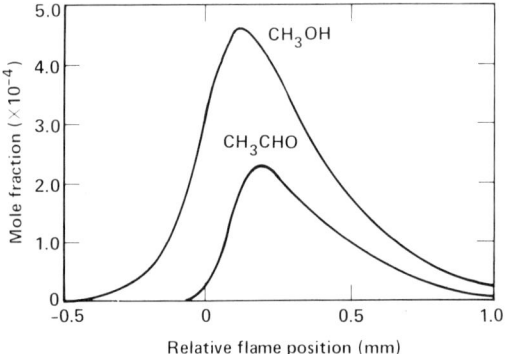

Fig. 5 Spatial profiles for methanol and acetaldehyde concentrations in reference propane-air flame.

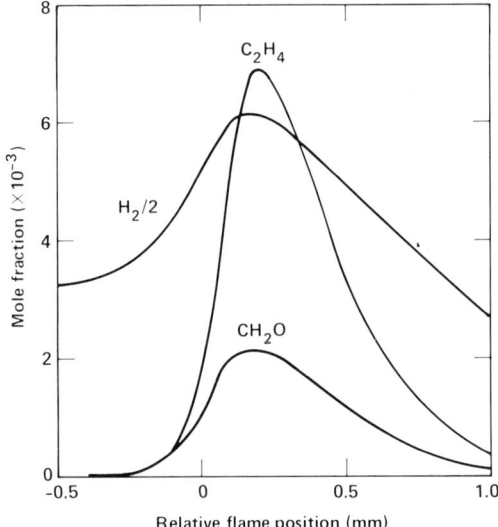

Fig. 6 Spatial profiles for major intermediate species concentrations in reference propane flame.

$$CH_3 + CH_3 \rightarrow C_2H_6 \tag{47}$$

$$CH_3 + H + M \rightarrow CH_4 + M \tag{33}$$

The methyl radicals are produced by several important reactions, among them

$$nC_3H_7 \rightarrow CH_3 + C_2H_4 \tag{101}$$

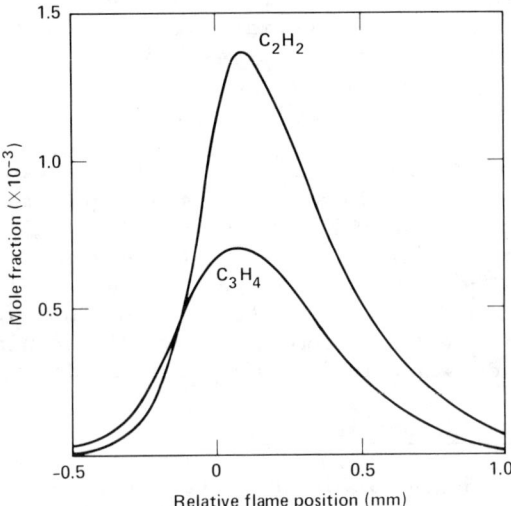

Fig. 7 Spatial profiles for C_2H_2 and C_3H_4 concentrations in reference propane-air flame.

$$C_3H_6 + OH \rightarrow CH_3 + CH_3CHO \quad (121)$$

$$C_3H_6 + O \rightarrow CH_3 + CH_3CO \quad (122)$$

The fact that most of the methyl radicals are products of nC_3H_7 and C_3H_6 consumption is consistent with the observation that the peak methane and ethane concentrations follow that of C_3H_6 very closely. Since butene is also produced by reactions of methyl radicals

$$CH_3 + C_3H_5 \rightarrow 1C_4H_8 \quad (139)$$

the concentrations of all of the species in Figs. 3 and 4 are closely interconnected.

Two other associated species are the oxygenated hydrocarbons CH_3OH and CH_3CHO, shown in Fig. 5. Methanol is produced primarily by recombination reactions of methyl and hydroxyl radicals

$$CH_3 + OH + M \rightarrow CH_3OH + M \quad (79)$$

while acetaldehyde is produced by

$$C_3H_6 + OH \rightarrow CH_3CHO + CH_3 \quad (121)$$

The major stable intermediate species C_2H_4, CH_2O, and H_2 are plotted in Fig. 6. Their peak concentrations occur slightly later in the flame than those in Figs. 3-5, consistent with the sequential overall oxidation of propane. Following these species in this sequence are C_2H_2 and C_3H_4, shown in Fig. 7.

Effectively all of the hydrocarbon species, except the least saturated species C_2H_2 and C_3H_4, are consumed before the temperature in the flame reaches 1500 K, at a relative flame location of about -0.1 mm. The peak CO concentration and the onset of CO oxidation coincide with the final disappearance of these hydrocarbon intermediates. The presence of C_2H_2 and C_3H_4, both of which have relatively low rates of reaction with H atoms and therefore do not compete with the chain branching Reaction 1, does not inhibit CO and H_2 oxidation.

Effects of Variations in Pressure

The dependence of burning velocity on pressure was determined for stoichiometric propane-air mixtures in a series of model computations. The results are summarized in Fig. 8, together with computed results for other fuel-air mixtures. In all of the cases examined, an increase in the pressure resulted in a decrease in the burning velocity. As previously observed for CH_4-air (Smith and Agnew 1957; Tsatsaronis 1978; Westbrook and Dryer 1980; Westbrook et al. 1983b), CH_3OH-air (Westbrook and Dryer 1980), and C_2H_4-air (Westbrook et al. 1983b) laminar flames, the curve of laminar burning velocity vs pressure on a logarithmic scale like that of Fig. 8 shows two separate regimes, one for pressure below atmospheric and the other above about 4 atmospheres. In each pressure regime the curves are relatively straight, indicating a simple power law variation

$$S_u = A\ P^n$$

with different values of the pressure exponent n for each regime. For the present propane-air flames, the computed low-pressure exponent is n ≃ -0.14, falling to n ≃ -0.19 at elevated pressures. Gilbert (1957) measured the variation in burning velocity with pressure for propane-air mixtures at pressures below atmospheric (50 < P < 760 Torr), with results that agree very closely with the computed values in Fig. 8. Gilbert then determined a value for the pressure exponent n, based on estimates of the influence of flame quenching in the long ducts being used, arriving at a value of n ≃ -0.215. However, the analysis in Gilbert's paper presupposed that the pressure

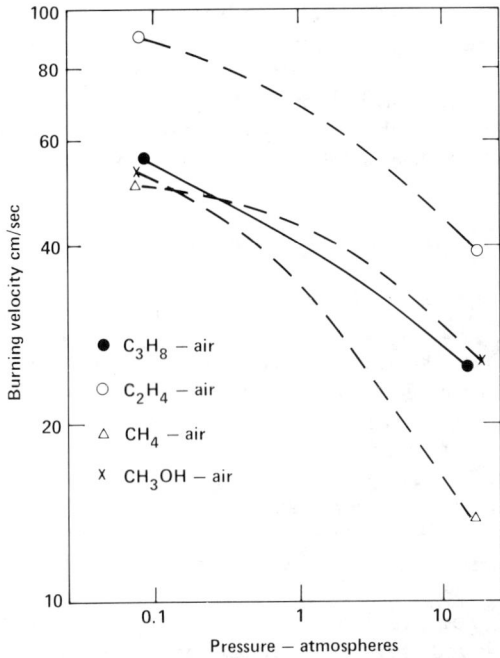

Fig. 8 Variation in laminar burning velocity with pressure for indicated stoichiometric fuel-air mixtures, with T_u = 300 K.

exponent would be a constant, while the present kinetic modeling results indicate that n is a function of pressure. Gilbert measured a burning velocity of about 60 cm/s at 0.1 atm, and 40 cm/s at atmospheric pressure, both in excellent agreement with the predictions of the present flame model. Kuehl (1962) observed a value for n of approximately -0.2 and Agnew and Graiff (1961) a value of -0.11, both for low-pressure propane-air flames. For pressures primarily above atmospheric (0.4 < P < 40 atm), Metghalchi and Keck (1980) found a value of n = -0.17, while Agnew and Graiff (1961) found n = -0.28. Only the study of Agnew and Graiff covered a pressure range sufficiently wide to observe the variation in pressure exponent n with pressure for propane-air mixtures.

For other fuel-air mixtures, high-pressure values of n include -0.5 for methane-air (Tsatsaronis 1978; Andrews and Bradley 1972) and a large amount of data for methanol, ethanol, ethylene, and iso-octane (Westbrook and Dryer 1980; Westbrook et al. 1983b; Gulder 1983; Metghalchi and Keck 1982) with -0.17 > n > -0.22, so the present computed values of the pressure exponent for propane-air are consistent both with available experimental results and with related results for other fuel-air mixtures except methane-air.

The effect of variations of pressure on the burning velocity of hydrocarbon-air mixtures has been explained (Westbrook and Dryer 1980) to result from a competition between Reactions 1 and 7

$$H + O_2 \rightarrow O + OH \qquad (1)$$

$$H + O_2 + M \rightarrow HO_2 + M \qquad (7)$$

As the pressure is increased, the rate of the pressure-dependent Reaction 7 increases much more rapidly than that of Reaction 1. This competition becomes important for pressures above atmospheric. When Reaction 7 is omitted from the reaction mechanism, the curved region in Fig. 8 disappears and the low-pressure exponent $n \simeq -0.14$ also describes the high-pressure computed results. Reactions 1 and 7 are common to all hydrocarbon-air flames, so the curved transition region in Fig. 8 occurs at about the same pressure for all four of the fuel-air mixtures and should be observed for most other hydrocarbon-air mixtures as well.

Effects of Variations in Equivalence Ratio

The predicted variation of burning velocity with equivalence ratio was computed at 0.1-, 1.0-, and 10.0-atm pressure, and the results are summarized in Fig. 9. All of the results show the same overall shape, with a maximum occurring slightly on the rich side of stoichiometric. Both the maximum burning velocity of approximately 44 cm/s and the equivalence ratio $\phi \simeq 1.1$ at which the maximum occurs agree well with available data (Warnatz 1979, 1981, 1983; Singer 1953; Gibbs and Calcote 1959; Kuehl 1962; Metghalchi and Keck 1980) for the atmospheric pressure case. However, similar data for the other pressures could not be found for comparison.

Extrapolation of the atmospheric pressure curve yields a value of $\phi_L \simeq 0.5$ for the lean limit of flammability, also in good agreement with experimental data (Dugger et al. 1959). However, extrapolation of the rich portion of the curve is more difficult for two principal reasons. First, although not shown in Fig. 9, the model was used to compute burning velocities for propane-air mixtures with $\phi > 2$. For these very rich mixtures the model predicts a very slowly decreasing burning velocity,

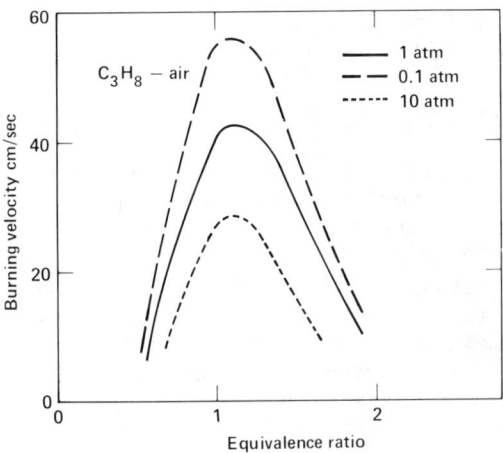

Fig. 9 Variation in laminar burning velocity with equivalence ratio for propane-air mixtures at indicated pressures, with T_u = 300 K.

falling to $S_u \simeq 5$ cm/s at $\phi = 3$. The numerical model is adiabatic, so it will continue to predict flame propagation at any fuel-air equivalence ratio. However, previous experience with the model has suggested that a predicted burning velocity less than 5 cm/s will correspond to a nonflammable mixture. Therefore, the present results point to a numerical prediction of $\phi_R \simeq 3$ for atmospheric pressure propane-air flames, in reasonably good agreement with available experimental data (Dugger et al. 1959) giving a rich limit of $\phi_R \simeq 2.8$. In addition to the variation in burning velocity with pressure at a given value of ϕ, another effect of variation in pressure is a modification of the rich and lean limits to flammability. As shown in Fig. 9, the curves become gradually narrower as the pressure is increased, thereby increasing the lean limit and reducing the rich limit.

The second principal reason why the prediction of the rich limit is more difficult than the lean limit is that the reaction mechanism gradually becomes more uncertain as the equivalence ratio increases. The importance of individual chemical species and elementary reactions changes substantially with equivalence ratio, and for rich propane-air mixtures these include reactions of C_3H_6, C_3H_5, C_3H_4, and C_4H_8. Reactions of these species in Table 1 are considerably more uncertain than most of the rest of the mechanism. Furthermore, other species and reactions not included in the mechanism such as C_4H_{10} can be expected to become important as ϕ

increases, and their neglect will introduce some errors into the computed flame models.

Laminar Flame Wall Quenching

In an earlier paper (Westbrook et al. 1980) a numerical study of laminar flame quenching at a cold surface was carried out, using both methane and methanol as fuels. The physical model assumes the end-wall configuration in which the flame propagates directly toward the wall which is held at a fixed temperature. The progress of the flame towards the wall is arrested, at a distance which agrees well with experimental observations (Daniel 1957), trapping a layer of unburned fuel between the flame and the wall in a quench layer. However, the model predicts that, subsequent to the quenching of the flame, the fuel and other partially oxidized hydrocarbons in this quench layer diffuse out of this layer into the hot product gases and are completely burned, on a time scale which is short when compared to the available residence time in a typical internal combustion engine chamber.

One of the limitations of this earlier study was that both of the fuels considered were molecules with low molecular weights and correspondingly large molecular diffusivities. Since the mechanism for the consumption of the quench layer involved diffusion of the fuel, the possibility remained that a heavier fuel molecule such as propane might not be transported rapidly enough into the hot region and oxidized within the available residence time. Sloane and Schoene (1983) recently addressed this question, using methane as the fuel but assigning the fuel molecule an artificially enhanced molecular weight and reduced diffusivity, concluding that even with a reduced rate of diffusion, the fuel was still consumed very rapidly. In the present work, the mechanism in Table 1 was used to repeat earlier computations (Westbrook et al. 1980), adding a computation for a propane-air mixture. The methane-air and methanol-air model calculations were repeated and compared with the propane-air results. The overall conclusion of this numerical analysis is that the mechanism of flame quenching at the wall, followed by rapid diffusion and consumption of the fuel, is exactly the same for propane-air as for methane-air and methanol-air mixtures. The details of the three model calculations vary slightly; the thickness of the propane-air flame is somewhat less than the others, so the flame stagnates a little closer to the wall. As a result the peak heat-transfer rate to the wall is slightly higher, and the distance through which the quench layer contents have to diffuse to reach the hot product region is slightly smaller than for the other two fuels. However, although the molecular weight of propane is nearly three times greater than that of methane, there is no significant change in the overall features of the quenching event. Based on the additional results for propane-air, there is even less reason to believe that flame quenching in thermal boundary layers is responsible for a major fraction of the unburned hydrocarbon emissions from internal combustion engines.

Effects of Propane on Natural Gas Combustion

Although natural gas consists primarily of methane, small fractions of ethane and propane are also present. In a recent experimental program dealing with liquefied natural gas (LNG) spills (Koopman et al. 1982), the methane fractions varied from 95% down to about 80%, with nearly all of the remainder consisting of ethane and propane. Most typically, about 90% of the fuel is methane, 6-9% is ethane, and the small remainder is propane. These fractions will vary significantly, depending in large part on the original source of the natural gas and upon its method and time of storage. Previous studies (Westbrook 1979, 1980a) indicated that the presence of ethane in the natural gas had a profound effect on the ignition and detonation properties of natural gas, but very little effect on the laminar burning velocity or flammability limits. Subsequent analysis of the effects of propane on the ignition and detonation properties has been presented (Westbrook et al. 1983a), and the same type of analysis has been carried out in the present study to determine the influence of small amounts of propane on the laminar flame properties of natural gas.

Three stoichiometric fuel-air mixtures were considered, containing 90% CH_4/10% C_2H_6, 90% CH_4/8% C_2H_6/2% C_3H_8, and 90% CH_4/10% C_3H_8. Although some of the intermediate hydrocarbon species concentrations varied among the three flame models, the burning velocity, flame thickness, adiabatic flame temperature, and radical species profiles were almost exactly the same in each case. This result means that, for modeling purposes, it is probably adequate to assume that the flame properties of natural gas are essentially the same as those of methane. This certainly does not suggest that for modeling other combustion properties, these minor constituents in the fuel can be neglected.

Summary

A detailed kinetic reaction mechanism for the oxidation of propane has been used to study laminar flame propagation under a wide range of conditions. Model predictions of burning velocity and flammability limits agree well with available experimental data. The same reaction mechanism has been shown (Westbrook and Pitz 1983) to reproduce experimental results under shock tube, detonation, and turbulent flow reactor conditions, and represents a reliable description of the key features of high-temperature propane oxidation. Subsequent applications of this laminar flame model to examine propane oxidation in thermal boundary layers and in natural gas flames have been described.

Acknowledgments

This work was performed under the auspices of the U.S. Department of Energy by the Lawrence Livermore National Laboratory under Contract No. W-7405-ENG-48.

References

Agnew, J. T. and Graiff, L. B. (1961) The pressure dependence of laminar burning velocity by the spherical bomb method. Comb. Flame 5, 209.

Andrews, G. E., and Bradley, D. (1972) The burning velocity of methane-air mixtures. Comb. Flame 19, 275.

Burcat, A. (1975) Cracking of propylene in a shock tube. Fuel 54, 87.

Cathonnet, M., Boettner, J. C., and James, H. (1981) Experimental study and numerical modeling of high temperature oxidation of propane and n-butane. Eighteenth Symposium (International) on Combustion, p. 903. The Combustion Institute, Pittsburgh, Pa.

Daniel, W. A. (1957) Flame quenching at the walls of an internal combustion engine. Sixth Symposium (International) on Combustion, p. 886. Reinhold, New York.

Dugger, G. L., Simon, D. M., and Gerstein, M. (1959) Laminar flame propagation. Chap. 4 in NACA Report 1300, Basic Considerations in the Combustion of Hydrocarbon Fuels in Air.

Edelson, D. and Allara, D. L. (1980) A computational analysis of the alkane pyrolysis mechanism: Sensitivity analysis of individual reaction steps. Int. J. Chem. Kinet. 12, 605.

Gibbs, G. J. and Calcote, H. F. (1959) Effect of molecular structure on burning velocity. J. Chem. Eng. Data 4, 226.

Gilbert, M. (1957) The influence of pressure on flame speed. Sixth Symposium (International) on Combustion, p. 74. Reinhold, New York.

Gulder, O. L. (1983) Laminar burning velocities of methanol, ethanol, and isooctane-air mixtures. Nineteenth Symposium (International) on Combustion, The Combustion Institute, Pittsburgh, Pa.

Hautman, D. J., Santoro, R. J., Dryer, F. L., and Glassman, I. (1981) An overall and detailed kinetic study of the pyrolysis of propane. Int. J. Chem. Kinet. 14, 149.

Koopman, R. P., Cederwall, R. T., Ermak, D. L., Goldwire, H. C., Hogan, W. J., McClure, J. W., McRae, T. G., Morgan, D. L., Rodean, H. C., and Shinn, J. H. (1982) Analysis of Burro series 40 m^3 LNG spill experiments. J. Hazardous Mat. 6, 43.

Kuehl, D. K. (1962). Laminar burning velocities of propane-air mixtures. Eighth Symposium (International) on Combustion, p. 510. The Combustion Institute, Pittsburgh, Pa.

Layokun, S. K. and Slater, D. H. (1979) Mechanism and kinetics of propane pyrolysis. Ind. Eng. Chem. Process Des. Dev. 18, 232.

Lifshitz, A. and Frenklach, M. (1975) Mechanism of the high temperature decomposition of propane. J. Phys. Chem. 79, 686.

Lund, C. M. (1978) HCT-A general computer program for calculating time dependent phenomena involving one-dimensional hydrodynamics, transport, and detailed chemical kinetics. University of California Lawrence Livermore National Laboratory Report UCRL-52504.

McLain, A. G. and Jachimowski, C. J. (1977) Chemical kinetic modeling of propane oxidation behind shock waves. NASA TN D-8501.

Metghalchi, M. and Keck, J. C. (1980) Laminar burning velocity of propane-air mixtures at high temperature and pressure. Comb. Flame 38, 143.

Metghalchi, M. and Keck, J. C. (1982) Burning velocities of mixtures of air with methanol, isooctane, and indolene at high pressure and temperature. Comb. Flame 48, 191.

Singer, J. M. (1953). Burning velocity measurements on slot burners; comparison with cylindrical burner determinations. Fourth Symposium (International) on Combustion, p. 352. Williams and Wilkins, Baltimore, Md.

Sloane, T. M. and Schoene, A. Y. (1983) Computational studies of end-wall flame quenching at low pressure: The effects of heterogeneous radical recombination and crevices. Comb. Flame 49, 109.

Smith, D. and Agnew, J. T. (1957) The effect of pressure on the laminar burning velocity of methane-oxygen-nitrogen mixtures. Sixth Symposium (International) on Combustion, p. 83. Reinhold, New York.

Tsatsaronis, G. (1978) Prediction of propagating laminar flames in methane, oxygen, nitrogen mixtures. Comb. Flame 33, 217.

Warnatz, J. (1979) Flame velocity and structure of laminar hydrocarbon-air flames. Combustion in Reactive Systems: AIAA Progress in Astopmaitocs amd Aeronautics, (edited by J. R. Bowen, N. Manson, A. K. Oppenheim, and R. I. Solorkhin, Vol. 76, pp. 501-521. AIAA, New York.

Warnatz, J. (1981) The structure of laminar alkane-, alkene-, and acetylene flames. Eighteenth Symposium (International) on Combustion, p. 369, The Combustion Institute, Pittsburgh, Pa.

Warnatz, J. (1983) The mechanism of high temperature combustion of propane and butane. Comb. Sci. Tech. (in press).

Westbrook, C. K., Creighton, J., Lund, C., and Dryer, F. L. (1977) A numerical model of chemical kinetics of combustion in a turbulent flow reactor. J. Phys. Chem. 81, 2542.

Westbrook, C. K. (1979) An analytical study of the shock tube ignition of mixtures of methane and ethane. Comb. Sci. Tech. 20, 5.

Westbrook, C. K. and Dryer, F. L. (1979) A comprehensive mechanism for methanol oxidation. Comb. Sci. Tech. 20, 125.

Westbrook, C. K. (1980a) Modeling of laminar flames in mixtures of vaporized liquefied natural gas (LNG) and air. University of California Lawrence Livermore National Laboratory Report UCID-18540.

Westbrook, C. K. (1980b) Inhibition of laminar methane-air and methanol-air flames by hydrogen bromide. Comb. Sci. Tech. 23, 191.

Westbrook, C. K. and Dryer, F. L. (1980) Prediction of laminar flame properties of methanol-air mixtures. Comb. Flame 37, 171.

Westbrook, C. K., Adamczyk, A. A., and Lavoie, G. A. (1980) A numerical study of laminar flame wall quenching. Comb. Flame 40, 81.

Westbrook, C. K., and Dryer, F. L. (1981a) Chemical kinetics and the modeling of combustion processes. Eighteenth Symposium (International) on Combustion, p. 749. The Combustion Institute, Pittsburgh, Pa.

Westbrook, C. K. and Dryer, F. L. (1981b) Simplified reaction mechanisms for the oxidation of hydrocarbon fuels in flames. Comb. Sci. Tech. 27, 31.

Westbrook, C. K., and Pitz, W. J. (1983) A comprehensive chemical kinetic reaction mechanism for oxidation and pyrolysis of propane and propene (submitted for publication).

Westbrook, C. K., Pitz, W. J., and Urtiew, P. A. (1983a) Chemical kinetics of propane oxidation in gaseous detonations (submitted for publication).

Westbrook, C. K., Dryer, F. L., and Schug, K. P. (1983b) Numerical modeling of ethylene oxidation in laminar flames. Comb. Flame (in press).

Westbrook, C. K., Dryer, F. L., and Schug, K. P. (1983c) A comprehensive mechanism for the pyrolysis and oxidation of ethylene. Nineteenth Symposium (International) on Combustion, p. 153. The Combustion Institute, Pittsburgh, Pa.

Westbrook, C. K. (1983) Numerical modeling of flame inhibition by CF_3Br. Comb. Sci. Tech. (in press).

Homogeneity and Propagation of Autoignited Cool and Blue Flames

Yasuhiko Ohta*
Nagoya Institute of Technology, Nagoya, Japan
and
Hitoshi Takahashi†
Meijo University, Nagoya, Japan

Abstract

High-speed schlieren photography has been used to study the autoignition processes of a compressed mixture of diethyl-ether and air or n-heptane and oxygen-argon. Cool and blue flames were each clearly isolated from the subsequent flames, especially from the hot flame explosion, with a special single-cycle motored engine. In general, the inflammation of cool flame occurs almost uniformly throughout the mixture, originating from the flame kernels counted by the hundred. The transit time of these flames for each run has been estimated to be 5 ms; that means 0.2 m/s as a flame speed. On the other hand, only a few sources are established for the onset of the blue flame; then they propagate and pass across the chamber. Approximation of the flame speed of the blue flame front indicates a value of 2.5 m/s, which results in requiring 20 ms to cover the whole volume. The reason for the nonuniformity of the blue flame appearance may be due to localized temperature gradients in the charge accelerated by the post-cool-flame reactions from the fine gradients in the mixture induced by the entrainment of quench layers off the walls during the compression process. The experimental results described in this paper show that the flame that gives the inhomogeneity of the flame onsets is not the cool flame but the blue flame, the secondary low-temperature flame.

Presented at the 9th ICODERS, Poitiers, France, July 3-8, 1983. Copyright © American Institute of Aeronautics and Astronautics, Inc., 1984. All rights reserved.
*Lecturer, Department of Mechanical Engineering.
†Professor, Department of Mechanical Engineering.

Aims and Background

High-speed schlieren photography has been used to examine the autoignition processes of fuel-air mixtures in the engine cylinder, from low-temperature flames to knock combustion. Miller and Olsen (1943) and Osterstrom (1947) have discovered flame mottlings prior to the occurrence of knock. Male (1949) has shown the autoignited flame with n-heptane as fuel, separated from the spark-initiated normal flame propagation. The dark areas observed in the center of the end gas zone were denoted as "disturbances" and postulated to be cool flames by Lewis and von Elbe (1961). The propagation of preknock flames through an engine combustion chamber, followed by autoignition, was observed by Ball (1955).

Visualization has also been performed in rapid compression machines. Schlieren photographs of autoignition of isooctane revealed a series of "concentric rings" followed by two "prominent centers of activity" spreading through the volume (Livengood and Leary 1951).

Although blue flame (secondary low-temperature flame) is often confused with cool flame, there has been no clear distinction between cool and blue flames. Blue flames often appear overlapped with hot flames, particularly under high-pressure conditions (Ohta and Takahashi 1983). The locality of onsets (or inhomogeneity of autoignition) and the propagation of hot flame may be well known, but the behavior of low-temperature flames (and their propagation velocities) have not been investigated. To obtain a clear understanding of the characteristics of low-temperature flames, it is necessary to distinctly isolate each from the subsequent flames, especially from the hot flame explosion, in the engine cylinder. The purpose of this paper is to photograph the two separate and different phenomena: the first appearance of autoignition termed cool flame and a reaction directly before the knock termed blue flame and to obtain a new interpetation about their appearance and propagation.

Experimental Details

A motorcycle engine (Yamaha: 2J2) was modified so that photography of the entire volume of gas within the disk-shaped combustion chamber along the cylinder axis could be observed. The engine has a single cylinder with 87-mm bore and 84-mm stroke. The compression ratio was 9.1:1. Sectional views of the combustion chamber of the engine are shown in Fig. 1. A transparent flat cylinder head was made of

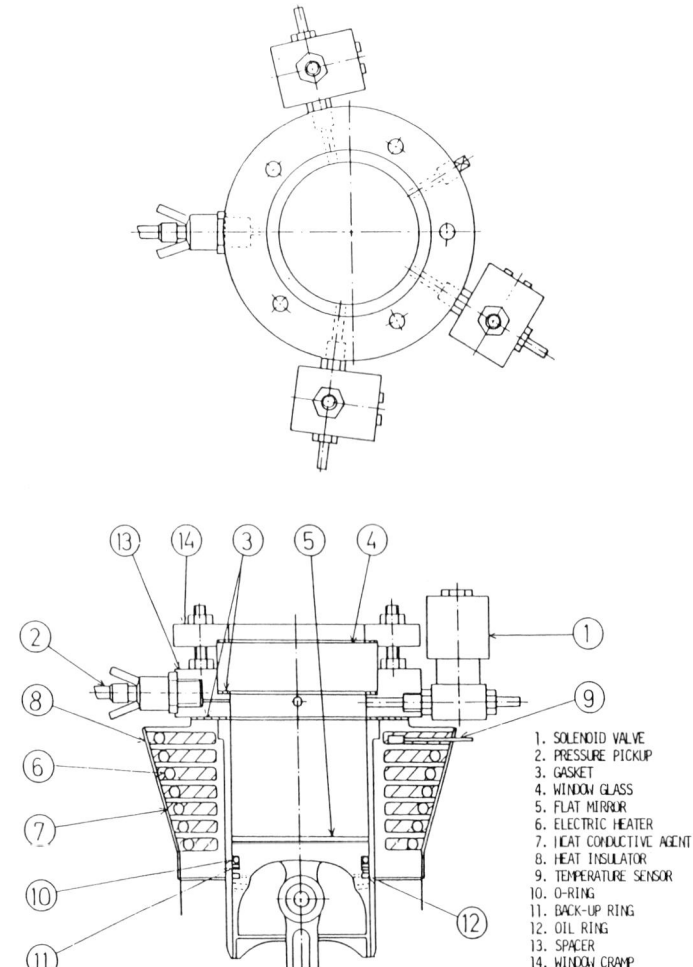

Fig. 1 Simplified longitudinal section and top view of engine.

1. SOLENOID VALVE
2. PRESSURE PICKUP
3. GASKET
4. WINDOW GLASS
5. FLAT MIRROR
6. ELECTRIC HEATER
7. HEAT CONDUCTIVE AGENT
8. HEAT INSULATOR
9. TEMPERATURE SENSOR
10. O-RING
11. BACK-UP RING
12. OIL RING
13. SPACER
14. WINDOW CRAMP

Pyrex glass, and a flat-surface mirror was fastened to the top of the piston by an adhesive agent. Openings around its wall provided access for a strain-gage type pickup (Kyowa: PE-30KF) for pressure measurements and also mixture passages to the solenoid valves. A pulse generator, mounted over the flywheel, produces one pulse every 6 crank angle degrees. Two photomultiplier tubes (Hamamatsu TV: 931A), which viewed the chamber through color glass filters (Toshiba: V-42 (blue, 320-510 nm) and R-62 (red, 620- nm)), were used to measure radiation intensities of blue and red colors from the flames.

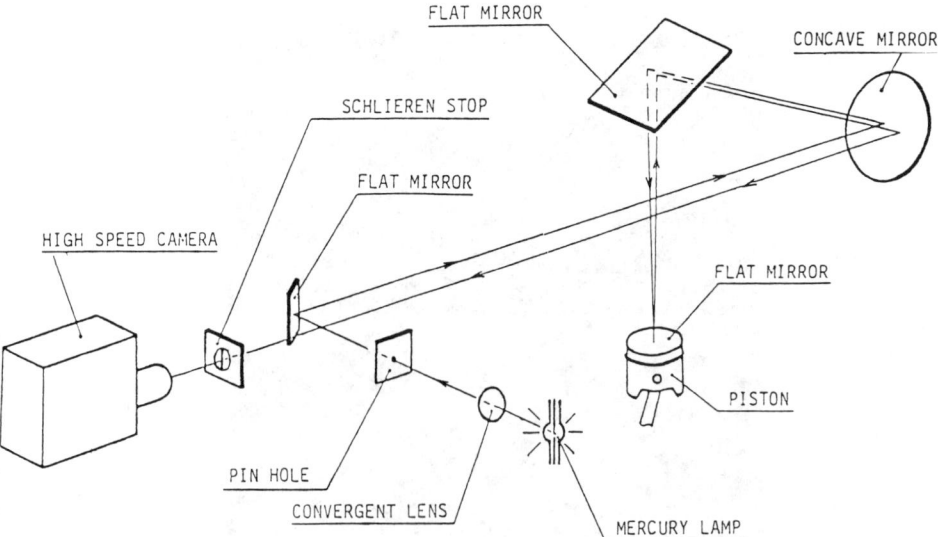

Fig. 2 Diagrammatic sketch of schlieren optical system.

A diagrammatic sketch of the schlieren optical system is shown in Fig. 2. It consisted of a spherical and a flat mirror, a point light source, and a knife edge. Schlieren photographs were taken with a 16-mm high speed camera (Himac 16HS), which was operated at about 4000 frames per second with a 1/5 speed shutter. Timing and crank angle markers on each side of the film permitted the film and pressure data to be synchronized.

The engine was driven to a standard revolution rate of 240 rpm by a small electrical motor whose torque was just sufficient to move the piston to the top dead center of the power cycle. As a consequence, angular velocity of the cranking in the compression stroke was low, and that in the expansion stroke, high. This cycle causes the onset of the cool flame in the compression stroke, onset of the blue flame near and after the top dead center, and suppression of the hot flame by the fast expansion. Compression was used to autoignite a single premixed charge. Spark-initiated runs were photographed separately as references with the spark plug located at the center of the combustion chamber.

The fuels were diethyl-ether or n-heptane, and oxidizers were air or a mixture of oxygen and argon. The equivalence ratio was 0.6. The mixture was well mixed in the storage tank. The mixture reservoir and engine cylinder were connected through three solenoid valves and were electrically heated to keep the initial mixture and chamber wall temperatures at $40°C$.

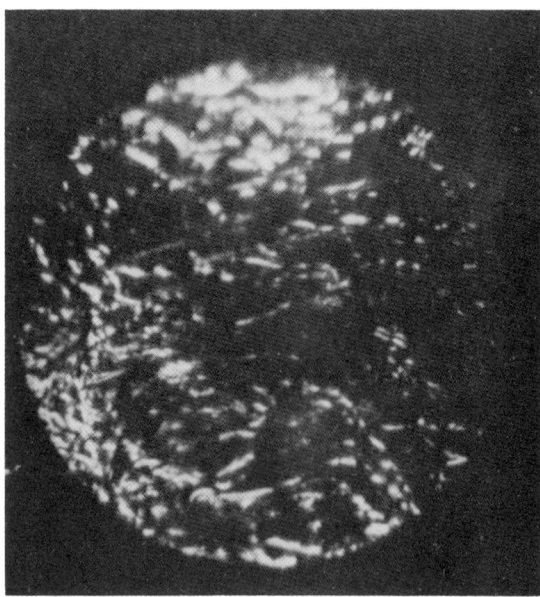

Fig. 3 Schlieren photograph of marble-like structure at 20 deg bTDC during compression stroke. Engine speed: 240 rpm, compression ratio: 9.1, mixture: diethyl-ether-air, equivalence ratio: 0.6, initial temperature: 40°C.

Results and Discussion

Autoignition may be considered to occur in three stages: 1) The cool flame stage is gently exothermic and produces a weak pale-blue radiation believed to be due to excited formaldehyde. 2) The blue flame stage is exothermic and produces a bright-blue (not yellow or red) radiation. 3) The hot flame stage is highly exothermic and generally rapid, produces intense yellow radiation, and is responsible for audible knock vibration in engines.

The gas motion at the beginning of the compression stroke is difficult to discern, because the density gradients in the bulk gas are too small. As the piston rises during the compression stroke, a marble-like structure develops and can be observed in schlieren photographs at relatively high schlieren sensitivity (see Fig. 3). This structure is assumed to be indicative of the scale of the turbulence in the cylinder. During the compression stroke there was little evidence of swirl.

Scraping off and rolling up of the low-temperature boundary layer on the cylinder wall by the piston movement

AUTOIGNITED COOL AND BLUE FLAMES

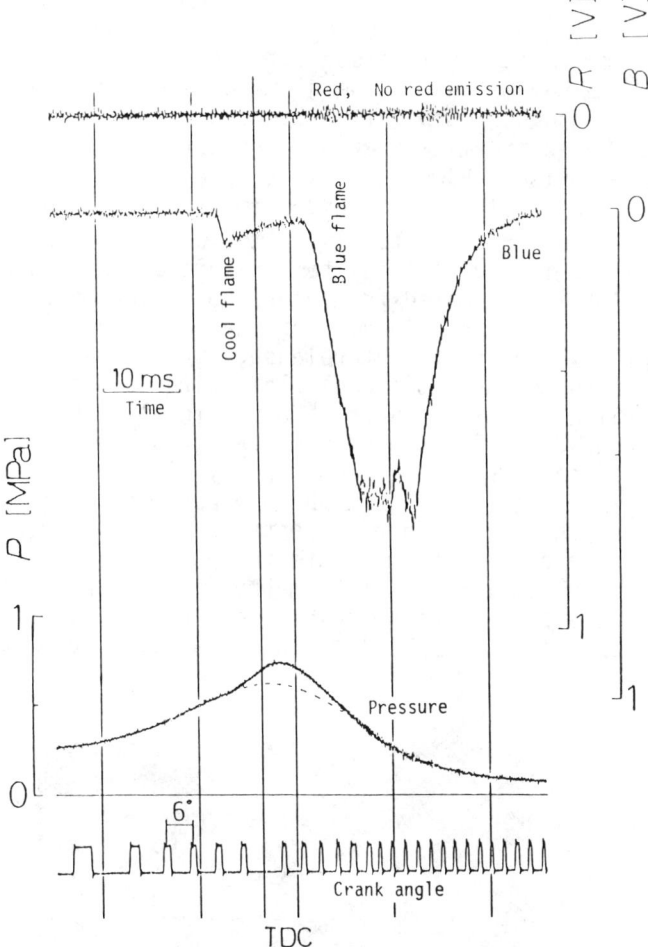

Fig. 4 Simultaneous pressure and luminosity records of the auto-ignited process. B: luminosity through blue glass filter, R: luminosity through red glass filter, engine speed: 240 rpm, compression ratio: 9.1, mixture: diethyl-ether-air, equivalence ratio: 0.6, initial temperature: 40°C.

may produce fine temperature gradients in the mixture during the compression stroke, which result in the marble-like structure.

Simultaneous pressure and correlated blue and red luminosity records are shown in Fig. 4. The absence of red light emission indicates the absence of hot flames behind the low-temperature flames. The onset and propagation of the pure low-temperature flames is evident from the blue light emission.

When the piston is 10 deg before top dead center (bTDC), the mottling is distributed almost uniformly over the entire volume and can be distinguished from the background of rougher marble-like structure (see Fig. 5). At this time, exothermic reactions are sufficient to produce a pressure rise. Luminosity is scarcely detectable at the time of the mottlings, at which time the pressure starts to increase over that due to piston motion as indicated by the broken line in Fig. 4. The temperature of the charge at this point is estimated from pressure measurements and the gas law to be 598 K. In view of the weak blue luminosity and the low levels of heat release, it is likely that combustion at this stage is a cool flame with hundreds of kernels. Similar results were obtained in all runs regardless of whether diethyl-ether or n-heptane was used as a fuel.

The concentric rings reported by Livengood and Leary (1951) were not observed in these experiments. When the schlieren photographs are projected as motion pictures, the cool flames appear to propagate for extremely short distances with locally radial symmetry (i.e., concentricity). Each of the cool flame kernels enveped the mixture in a distance of the order of 1 mm in a transit time of about 5

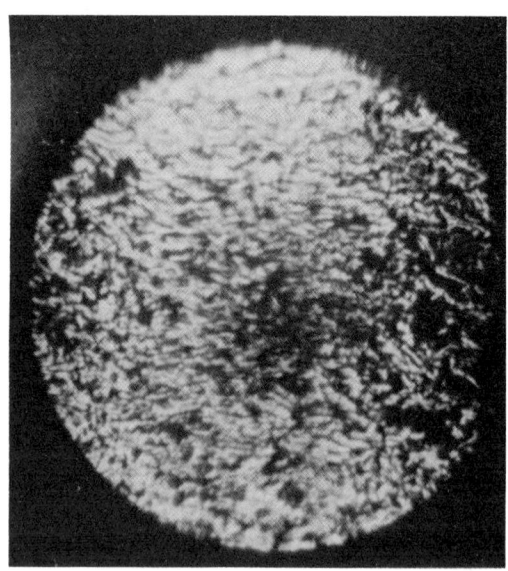

Fig. 5 Schlieren photograph of cool flame appearance at 10 deg bTDC during compression stroke. Engine speed: 240 rpm, compression ratio: 9.1, mixture: diethyl-ether-air, equivalence ratio: 0.6, initial temperature: 40°C.

ms. The cool flame speed of 0.2 m/s is about one-tenth that of normal spark-initiated flame in the chamber under similar conditions. The limits on schlieren sensitivity were set by the necessity to determine the onset of cool flames.

As the compression cycle proceeds, the marble-like structure begins to show considerable distortion. After the piston passes through top dead center, the second stage of mottlings, which is characterized by a few source areas and in the schlieren photographs appears as localized sequential dark portions, originates near the center of the chamber. The flame followed a path different from that of the cool flames. Figure 6 includes schlieren photographs and explanatory sketches of flame onset (al, bl, cl) and propagation. The flame propagation appears similar to gathering of cumuli (from al and bl to a2 and b2), and the products of the cool flame in the chamber are enveloped by these new low-temperature flame. As these secondary low-temperature flame reac-

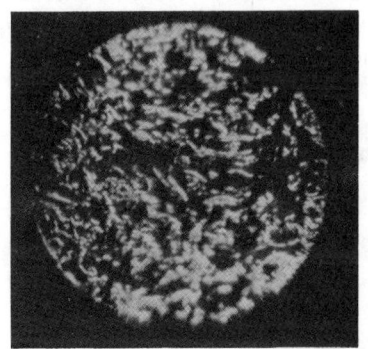
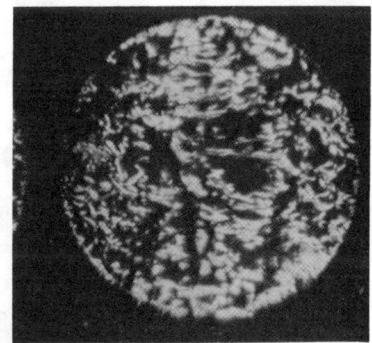
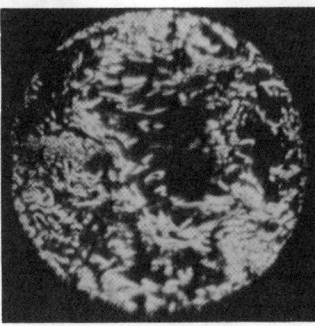

Fig. 6 Schlieren photographs of typical processes of onset and propagation of blue flames. a) 10 deg aTDC during expansion stroke, b) 15 deg aTDC during expansion stroke, c) 25 deg aTDC during expansion stroke. Engine speed: 240 rpm, compression ratio: 9.1, mixture: diethyl-ether-air, equivalence ratio: 0.6, initial temperature: 40ºC.

tions reduced the temperature gradient, the inner structure of the flame front was lost in the mottlings (a3 and b3). Considerable pressure was being developed in opposition to the continuous expansion as the piston descended. The bright-blue emission corresponding to the pressure trace for autoignition is shown in Fig. 4. From the slope and obtained peak value of the pressure-time records, it is evident that the all the heat of combustion of the fuel would not have been released up to the end of the propagation of these flames. The estimated average temperature of the gas was 855 K at 10 deg after top dead center (Fig. 6a). These secondary flames are believed to be blue flames rather than second cool flames. The speed of the blue flame front is approximately 2.5 m/s, which is comparable to that of the normal spark-initiated flame propagation. However, the flame apparently passes across the combustion chamber rather slowly, with initiation at 5 deg after top dead center (aTDC) and completion by about 35 deg aTDC, which requires nearly 20 ms. The key features observable from these photographs (and more discernible from the movies) are the non-uniformity of the onset of the blue flames in the geometric

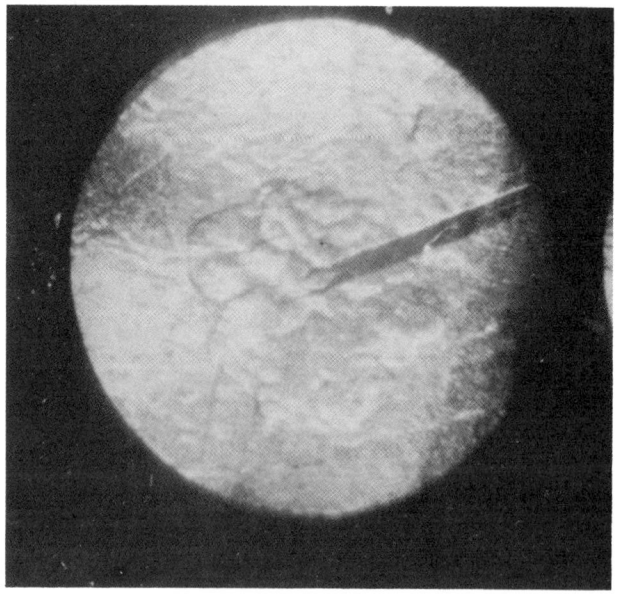

Fig. 7 Schlieren photograph of spark initiated flame at 45 deg bTDC, during compression stroke. Engine speed: 240 rpm, compression ratio: 9.1, mixture: diethyl-ether-air, equivalence ratio: 0.6, initial temperature: 40°C.

sense and their persistence, notwithstanding their flame speed, 10-15 times as much as that of cool flames.

The reason for this nonuniformity of the blue flame appearance is not definitely clear. The nonuniformity may be due to many localized temperature gradients, which could be caused in part by the post-cool-flame reactions in the mixture induced by the entrainment of quench layers off the walls during the compression process.

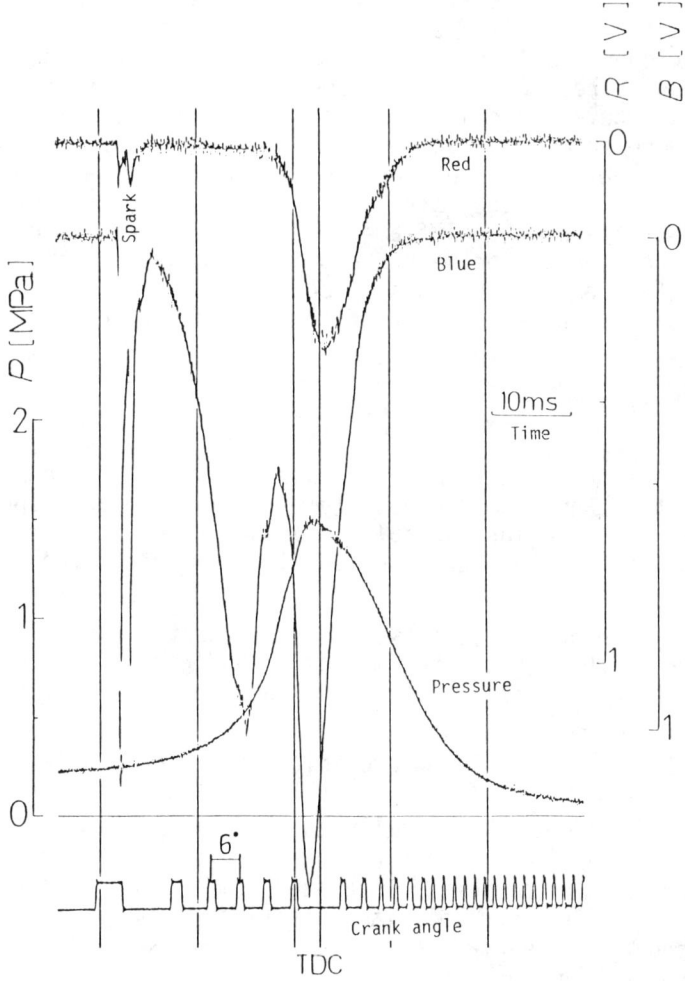

Fig. 8 Simultaneous pressure and luminosity records of the spark initiated process. B: luminosity through blue glass filter, R: luminosity through red glass filter, engine speed: 240 rpm, compression ratio: 9.1, mixture: diethyl-ether-air, equivalence ratio: 0.6, initial temperature: 40°C.

The "disturbance" noted by Miller and Olsen (1943) and "prominent centers of activity" noted by Livengood and Leary (1951) are thought to be blue flames followed by hot flames. These experimental results, when considered in the context of those of other investigators using motored engines (Male 1949; Ball 1955) or rapid compression machine (Livengood and Leary 1951), show that the flame that produces nonuniformity of flame onset is the blue flame not the cool flame.

Spark-initiated flame propagation of the same mixture under the same initial conditions is shown in Fig. 7. Flame fronts are recognizable as thin flame sheets distorted but almost spherically developed. Simultaneous pressure and luminosity records of this flame are illustrated in Fig. 8. Pressure development and blue and red light emissions are much more intense because of the development of hot flames.

In the case of autoignited blue flame propagation, it proceeds through the "stressed" (Ball 1955) charge influenced by the post-cool-flame and pre-blue-flame reactions so that the boundary between the background and the blue flame or post-blue flame regions is not so much a thin flame frontal area as the border of the temperature difference.

Conclusions

Cool and blue flames were successfully isolated from each other and from the hot flame explosion with a special single-cycle engine. Simultaneous pressure and correlated luminosity records were used to distinguish the cool and blue flames.

1) In autoignition, cool flame onset occurs almost uniformly throughout the mixture at hundreds of kernels. The flame speed of the cool flame has been estimated to be 0.2 m/s.

2) The blue flames are initiated at only a few sites in the cylinder and propagate throughout the chamber. The speed of blue flame is estimated to be 2.5 m/s.

3) The reason for the nonuniformity of the blue flame appearance may be due to localized temperature gradients caused by the post-cool-flame reactions in the mixture induced by the entrainment of quench layers off the walls during the compression process.

4) The experimental results described in this paper show that the flame that gives the inhomogeneity of the flame onsets is not the cool flame but the blue flame, the secondary low temperature flame.

References

Ball, G. A. (1955) Photographic studies of cool flames and knock in an engine. Fifth Symposium (International) on Combustion, pp.366-372. Reinhold, New York.

Lewis, B. and von Elbe, G. (1961) Combustion, Flames and Explosions of Gases, pp.187-191. Academic Press, New York.

Livengood, J. C. and Leary, W. A. (1951) Autoignition by rapid compression. Ind. Eng. Chem. 43, pp.2797-2805.

Male, T. (1949) Photographs at 500,000 frames per second of combustion and detonation in a reciprocating engine. Third Symposium on Combustion and Flame and Explosion Phenomena, pp.721-726. Williams and Wilkins, Baltimore.

Miller, C. D. and Olsen, H. L. (1943) Identification of knock in NACA high-speed photographs of combustion in a spark-ignition engine. NACA Report, 761.

Ohta, Y. and Takahashi, H. (1983) Temperature and pressure effects in cool and blue flames. Flames, Lasers, and Reactive Systems, AIAA Progress in Astronautics and Aeronautics, (edited by J. R. Bowen, N. Manson, A. K. Oppenheim, and R. I. Soloukhin), Vol.88, pp.38-56. AIAA, New York.

Osterstrom, G. E. (1947) Knocking combustion observed in a spark-ignition engine with simultaneous direct and schlieren high-speed motion pictures and pressure Records. NACA Report, 897.

Stability of Solid Propellant Combustion Subject to Nonplanar Perturbations

F. J. Higuera* and A. Liñán†
Escuela Técnica Superior de Ingenieros Aeronáuticos
Madrid, Spain

Abstract

An analysis is given of the intrinsic stability of the heterogeneous deflagration of solid propellants subject to nonplanar perturbations when the process can be modeled by a pyrolysis reaction at the interface followed by a gas reaction. The quasisteady and quasiplanar approximations are used in the analysis of the gas-phase process. The stability limit to planar perturbations lies within the unstable domain of nonplanar perturbations. Traveling waves of nonzero wavenumber are found when crossing the stability limit.

Introduction

The analysis of the transient response of solid propellants plays an important role in the stability of solid propellant rockets (Williams et al., 1969). In particular, there is an extensive literature on the evaluation of the admittance function that relates the changes in the solid propellant regression rate to the frequency and amplitude of the pressure perturbations (Culick 1968; Barrère 1970). In most of these studies, the solid-phase reactions are modeled by a surface pyrolisis reaction, and there are additional reactions in the gas phase, where the combustion process is assumed to be quasisteady. This last assumption, which results from the small value of the gas-to-solid density ratio, was introduced by Zel'dovich (1942; 1964) and used in a more general form by Novozhilov (1965). It is implicitly consid-

---

Presented at the 9th ICODERS, Poitiers, France, July 3-8, 1983. Copyright © 1984 by the American Institute of Aeronautics and Astronautics, Inc., All rights reserved.
*Assistant Professor, Department of Fluid Mechanics.
†Professor, Department of Fluid Mechanics.

ered in these analyses that the gas-phase quasisteady state is stable.

If, in addition, nonplanar effects are also neglected in the analysis of the gas-phase combustion processes, it is possible to obtain the following two relations:

$$m = m(\phi, p) \qquad T_s = T_s(\phi, p) \tag{1}$$

for the surface mass burning rate m and the surface temperature T_s in terms of the ambient pressure p and the heat flux ϕ from the interface to the inert solid. These relations can be obtained from experimental data, under steady-state conditions, for different pressures and initial temperatures of the solid; or they can be calculated theoretically for different models of the interface and gas-phase processes. In particular, in the model of Denison and Baum (1961), combustion in the gas-phase is approximated by a one-step Arrhenius reaction, and the pyrolisis of the solid, which occurs at the interface, is assumed to follow an Arrhenius law.

$$m = B \exp(-E_s/RT_s) \tag{2}$$

with an activation energy E_s. The mass burning rate in the gas-phase is given by the following relation:

$$m = C\, p^{n/2} \exp(-E/2RT_f) \tag{3}$$

where C is a constant including gas and reaction parameters, and n and E are the order and activation energy of the reaction. The final combustion temperature, T_f, can be related to the surface temperature and the heat flux from the gas to the interface; thus, Eq. (3) together with the pyrolisis law (2) are equivalent to Eq. (1). Williams (1973) used an asymptotic treatment of the gas-phase equations in the large activation energy limit to justify Eq. (3). A review of some others models of gas-phase combustion can be found in Culick (1968), Williams et al. (1969), and Barrère (1970). In particular, the KTSS model (Krier et al. 1968), originated at Princeton University, deserves special mention because it has been widely used; see, for example, De Luca (1981).

It is obvious that nonplanar effects are neglected in most of the transient analysis because they are one-dimensional in character. In this work, a stability analysis is reported for the combustion of homogeneous solid propellants subject to nonplanar perturbations for a constant gas pressure. The thicknesses of the thermal layers in the gas and solid (δ_g and δ_s) are obtained from the balance between the convective transport and the conduction normal to the interface. The ratio is $\delta_g/\delta_s \sim (\lambda_g/c_p)/(\lambda_s/c_s)$, with the gas conductivity λ_g evaluated at the flame temperature. It does not depend on the relative magnitude of the temperature increments in the phases and can be assumed to be small due to the moderately small value of the gas-to-solid conductivity ratio λ_g/λ_s. In the energy equation for the gas-phase, the effect of the transverse heat conduction is of the order $(\delta_g/\delta_s)^2$ relative to the conduction normal to the interface; therefore, in the study of the stability of the solid propellant combustion, relations (1), with unsteady and nonplanar effects in the solid heat conduction zone, can be used.

Formulation

The heat conduction equation, in a reference frame moving with the steady-state recession velocity, describes the temperature distribution in the solid. Under steady-state conditions, the surface x=0 coincides with the gas-solid interface, the solid lying at x < 0. Within the solid, T satisfies the equation

$$\rho c_s \frac{\partial T}{\partial t} + m_0 c_s \frac{\partial T}{\partial x} = \lambda \nabla^2 T \qquad (4)$$

where ρ, c_s, and λ are the density, specific heat, and thermal conductivity of the solid, respectively, and are assumed to be constant; m_0 is the steady-state mass burning rate. The boundary condition at a large distance from the surface is given in terms of the initial temperature of the solid T_∞ by

$$x \to -\infty: \quad T = T_\infty \qquad (5)$$

Equation (4) should be solved with the boundary condition (5) and additional conditions at the unknown propellant surface $x = X(y,z,t)$, provided by the relations (1) coming from the quasisteady and quasiplanar analysis of the pyrolysis and gas-phase combustion zones. For a given gas pressure p, it is possible to derive from Eq. (1) the following relation:

$$m = m_s(T_s, p) \qquad (6)$$

giving the mass burning rate as a function of the surface temperature. The surface regression rate m/ρ is related to

$X(y,z,t)$ by

$$\frac{m}{\rho} = \frac{m_o/\rho - X_t}{(1+X_y^2+X_z^2)^{1/2}} \quad (7)$$

It is also more convenient to replace the other relation (1) by

$$m = m_f(T_f, p) \quad (8)$$

where the final combustion temperature T_f is a function of the surface temperature, and the heat flux $\phi = \lambda(\partial T/\partial n)_s$ from the interface to the inert solid. T_f is given by

$$mc_p T_f = mc_s T_s - \lambda\left(\frac{\partial T}{\partial n}\right)_s + mQ_{ef} \quad (9)$$

where c_p is the specific heat of the gas, and Q_{ef} is the net heat released per unit mass in the processes of gas-phase combustion and surface pyrolysis. Equation (9) is an energy balance between the interface, in the solid side, and the gas phase downstream the combustion zone. Equations (6), (8), and (9) are equivalent to relations (1). When the surface pyrolysis and the gas-phase combustion are modeled by Arrhenius reactions, as done by Denison and Baum (1961), relations (6) and (8) take the form of Eqs. (2) and (3).

The steady surface mass flux can be obtained from the global energy balance in the stationary case:

$$c_p T_{f_o} = c_s T_\infty + Q_{ef} \quad (10)$$

The subscript o means stationary values. Equation (10) shows that the final combustion temperature depends only on the initial solid temperature. The value of T_{f_o}, which results from Eq. (10), can be used with Eq. (8) to find m_o, and T_{s_o} may be determined from this result and the surface decomposition law (6). The temperature profile inside the solid, $x < 0$, is given by

$$T_o = T_\infty + (T_{s_o} - T_\infty) \exp(x m_o c_s/\lambda) \quad (11)$$

Equations (4-6), (8), and (9) are written in nondimensional form on the introduction of

$$\mu = m/m_o \qquad \theta = (T - T_\infty)/(T_{s_o} - T_\infty)$$

$$\xi = x m_o c_s/\lambda \qquad \tau = t m_o^2 c_s/\lambda\rho \quad (12)$$

where $\lambda/m_0 c_s$ is the thickness of the solid conduction zone, and $\lambda \rho/m_0^2 c_s$ the time of residence in that zone. The transversal coordinates and the interface equation $x = X(y,z,t)$ are nondimensionalized by the length $\lambda/m_0 c_s$. The result is

$$\frac{\partial \theta}{\partial \tau} + \frac{\partial \theta}{\partial \xi} = \nabla^2 \theta \tag{13}$$

$$\xi \to -\infty: \quad \theta = 0 \tag{14}$$

$$\xi = \Xi: \quad \theta = \theta_s \tag{15}$$

$$\xi = \Xi: \quad \left.\frac{\partial \theta}{\partial n}\right|_s - \mu \theta_s = -\mu \frac{c_p}{c_s} (\theta_f - \theta_{f_0}) \tag{16}$$

where $\Xi = X m_0 c_s/\lambda$ and $\mu = (1-\Xi_\tau)/(1+\Xi_\eta^2+\Xi_\zeta^2)^{1/2}$. When the Denison-Baum model is used for Eqs. (6) and (8), Eqs. (15) and (16) take the form

$$\theta_s = \left[1 + \frac{\omega}{\gamma-1} \ln \mu\right] / (1 - \omega \ln \mu) \tag{17}$$

$$\left.\frac{\partial \theta}{\partial n}\right|_s - \mu \theta_s = \frac{-\varepsilon_g \mu \ln \mu}{1 - \omega_g \ln \mu} \tag{18}$$

with

$$\omega = RT_{s_0}/E_s \qquad \gamma = T_{s_0}/T_\infty$$

$$\omega_g = \frac{2RT_{f_0}}{E} \qquad \varepsilon_g = \frac{c_p}{c_s} \frac{2RT_{f_0}^2}{E(T_{s_0}-T_\infty)} \tag{19}$$

Linear Stability

The steady-state solution in nondimensional form is

$$\Xi_0 = 0 \qquad \theta_0 = \exp \xi \tag{20}$$

for $\xi < 0$. To this a small perturbation periodic in the transversal direction η and exponential in time τ is added:

$$\theta = \theta_0 + \sigma \theta'(\xi) \exp(\Omega \tau + ik\eta)$$

$$\Xi = \sigma \Xi' \exp(\Omega \tau + ik\eta) \tag{21}$$

with $\sigma \ll 1$. For the periodic disturbance (21) and small σ Eqs. (13-16) are linearized to

$$\Omega\theta' + \frac{d\theta'}{d\xi} = \frac{d^2\theta'}{d\xi^2} - k^2\theta' \qquad \varepsilon < 0 \qquad (22)$$

$$\xi \to -\infty: \quad \theta' = 0 \qquad (23)$$

$$\xi = 0: \quad \theta' + \Xi' = -\varepsilon_s \Omega \Xi' \qquad (24)$$

$$\xi = 0: \quad \frac{d\theta'}{d\xi}\bigg|_0 - \theta' + \Omega\Xi' = \varepsilon_g \Omega \Xi' \qquad (25)$$

where

$$\varepsilon_s = m_o \bigg/ (T_{s_0} - T_\infty) \left[\frac{\partial m_s}{\partial T_s}\right]_{T_{s_0}} \qquad (26)$$

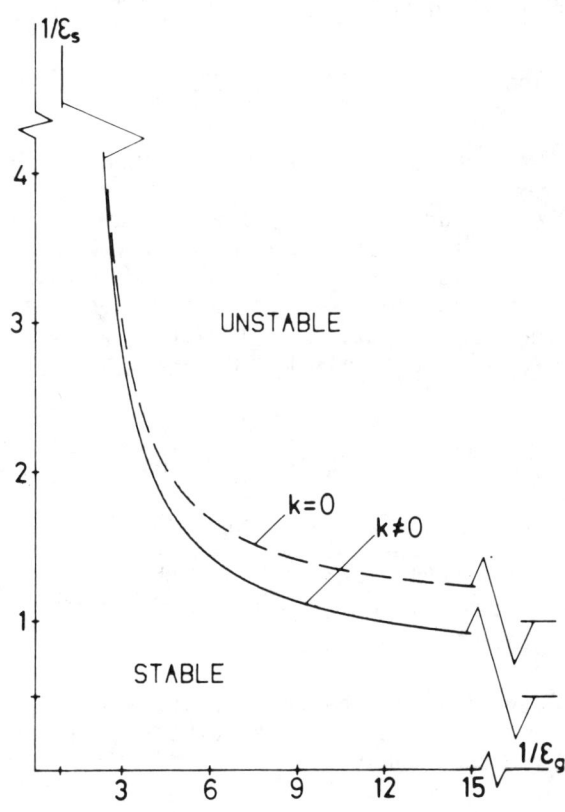

Fig.1 Stability limit for nonplanar perturbations, Eq.(31). The dashed line corresponds to planar perturbations, Eq.(33).

$$\varepsilon_g = \frac{c_p}{c_s} \left[m_0 \bigg/ (T_{s_0} - T_\infty) \left(\frac{\partial m_f}{\partial T_f} \right)_{T_{f_0}} \right] \quad (27)$$

For the Denison-Baum model, ε_g is given by Eq. (19), and

$$\varepsilon_s = \gamma \omega / (\gamma - 1) = RT_{s_0}^2 / E_s (T_{s_0} - T_\infty) \quad (28)$$

is the inverse of the parameter A of Denison and Baum (1961). The Zel'dovich (1942; 1964) model, of constant surface temperature, is recovered for $\varepsilon_s = 0$. The case $\varepsilon_g = 0$ corresponds to a constant flame temperature model. Solutions of the homogeneous system equations (22-25), exist only if the following dispersion relation

$$\varepsilon_s^2 \Omega^3 + \left[\varepsilon_s^2 k^2 + (1+\varepsilon_g)\varepsilon_s - (1-\varepsilon_g)^2 \right] \Omega^2 + (2\varepsilon_s k^2 + \varepsilon_g) \Omega + k^2 = 0 \quad (29)$$

between Ω and k is satisfied.

Analysis of the Dispersion Relation

For fixed ε_g and ε_s, there are wavenumbers k_n for which the perturbations do not decay or amplify due to the vanishing of the real part of Ω. For $\Omega = i\Omega_n$ and $k = k_n$, Eq. (29) yields, in addition to the trivial solution $\Omega_n = 0$, $k_n = 0$ (which reflects an invariance of the problem against shifts in the ξ axis), solutions with $k_n \neq 0$, given by

$$\Omega_n = \pm \varepsilon_s^{-1} \left(\varepsilon_g + 2\varepsilon_s k_n^2 \right)^{1/2} \qquad k_n^4 + a k_n^2 + b = 0 \quad (30)$$

where

$$a = \frac{1+3\varepsilon_g}{2\varepsilon_s} - \frac{(1-\varepsilon_g)^2}{\varepsilon_s^2} \qquad b = \frac{\varepsilon_g(1+\varepsilon_g)}{2\varepsilon_s^2} - \frac{\varepsilon_g(1-\varepsilon_g)^2}{2\varepsilon_s^3}$$

The stability limit in the $(\varepsilon_g, \varepsilon_s)$ plane is obtained from the requirement that the two solutions of Eq. (30) for k_n^2 coincide. The requirement is satisfied for

$$\varepsilon_s = 2(1 - \sqrt{\varepsilon_g})^2 \qquad \varepsilon_g < 1 \quad (31)$$

and the corresponding values of k and Ω are

$$k_c = \varepsilon_g^{1/4}/2(1-\sqrt{\varepsilon_g})^{1/2} \qquad \Omega_c = \pm i\varepsilon_g^{1/4}/2(1-\sqrt{\varepsilon_g})^2 \qquad (32)$$

The critical wavelength goes to zero and the frequency diverges when $\varepsilon_g \to 1^-$. For $\varepsilon_g \ll 1$, corresponding to large activation energy of the gas-phase reaction, $k_c \sim \varepsilon_g^{1/4}$, and this result provides additional justification to the quasi-planar assumption in the analysis of the gas-phase response.

Equation (29) also has solutions with $k = 0$ that give the response of the system to planar perturbations. The stability limit in this case is given by

$$\varepsilon_s = (1-\varepsilon_g)^2/(1+\varepsilon_g) \qquad \varepsilon_g < 1 \qquad (33)$$

and then

$$\Omega = \pm i\sqrt{\varepsilon_g}\,(1+\varepsilon_g)/(1-\varepsilon_g)^2 \qquad (34)$$

which coincides with the result of Denison and Baum. Both stability limits are shown in Fig. 1. It can be seen that a part of the stability domain to planar perturbations is really unstable when $k \ne 0$.

The case $\varepsilon_g \ll 1$ is specially interesting, because the effective value of the activation energy for the gas-phase reaction is usually large. In this case, an asymptotic expansion of the dispersion relation can be given near the stability limit. When $\varepsilon_g = 0$, the stability limit is $\varepsilon_s = 2$, and k_c and Ω_c are zero. If α and K are defined by $\varepsilon_s = 2(1+\delta^2\alpha)$ and $k = \delta K$ and introduced in Eq. (29) with the expansion

$$\Omega = \delta\Omega_1 + \delta^2\Omega_2 + \ldots \qquad (35)$$

where $\delta^2 \ll 1$ is a scale for the distance to the stability limit, the following result is obtained

$$\Omega = \pm i\delta K \mp i\delta^3 K(\alpha+2K^2) - 2\delta^4 K^2(\alpha+4K^2) + \ldots \qquad (36)$$

When $\varepsilon_g \ne 0$, the distinguished limit corresponds to $\delta = \varepsilon_g^{1/4}$. If an expansion similar to Eq. (35), with $\varepsilon_s = 2(1+\varepsilon_g^{1/2}\alpha)$ and $k = \varepsilon_g^{1/4}K$, is introduced in Eq. (29), and terms of like order in ε_g are identified, the dispersion relation becomes

$$\Omega = \pm i\varepsilon_g^{1/4}K \mp i\varepsilon_g^{3/4}K(\alpha+2K^2) - \varepsilon_g\left[1/2 + 2K^2(\alpha+4K^2)\right] + \ldots \qquad (37)$$

The real part of Ω (of order ε_g) becomes positive for values of the bifurcation parameter $\alpha < -2$. The stability limit is given by $\alpha = -2$, in agreement with Eq. (31) for small ε_g. The third root of Eq. (29), which is not included in Eq. (37), is negative, $\Omega \to -1/4$ for small k. The expansion equation (37) breaks down for small values of K. For $k/\varepsilon_g = K_1 < 1/2$, in first approximation,

$$2\Omega/\varepsilon_g = -1 \pm \sqrt{1-4K^2} \qquad (38)$$

gives the dispersion relation. The results equations (37) and (38) have been shown schematically in Fig. 2. The limiting case $\varepsilon_s \to 0$ leads to the constant surface temperature model. For $\varepsilon_s \ll 1$, the dispersion relation near the stability limit can be written

$$\Omega = \pm \frac{2i}{\nu^2} \pm \frac{8i(K^2-2)}{\nu}$$

$$+ \left[-\frac{K^2(K^2-1)}{2} - 2\beta + \frac{1}{2} \mp i\left(-\frac{K^4}{4} + \frac{11K^2}{2} - 2\beta - \frac{3}{4}\right)\right] + \ldots \qquad (39)$$

in terms of $\nu = 1-\varepsilon_g$, $K = \nu^{1/2}k$, and $\varepsilon_s = \nu^2/2(1+\nu/2+\beta\nu^2)$, where β is a bifurcation parameter of order unity, equal to

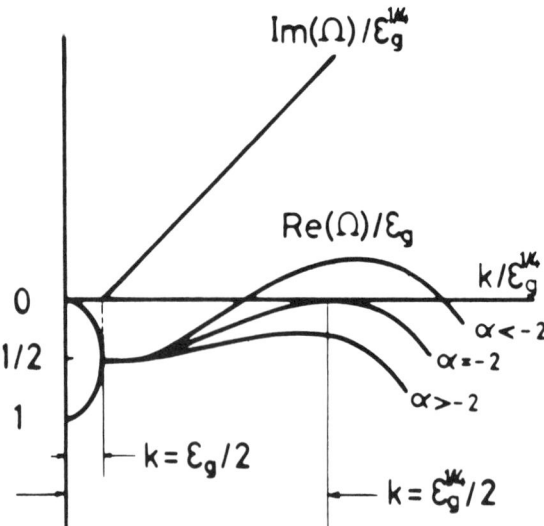

Fig.2 Sketch of the dispersion relation near the stability limit for $\varepsilon_g \ll 1$.

1/16 at the stability limit. The real part of Ω becomes positive for values of $\beta < 1/16$.

Conclusions

The analysis of the intrinsic stability of the heterogeneous deflagration of solid propellants subject to nonplanar perturbations has been carried out for a surface pyrolysis reaction and a gas-phase combustion process that follows the quasisteady and quasiplanar approximations. For the planar case, Culick (1968) has shown that there are only two parameters in the linear stability analysis, ε_s^{-1} and ε_g^{-1}, characterizing the temperature sensitivity of the pyrolysis and gas-phase reactions. The present work shows that this is also true for the linear stability analysis for the nonplanar case. The Denison and Baum stability limit, shown in Fig. 2, is recovered in the present analysis for planar perturbations, $k = 0$. However, this limit lies in the unstable domain for nonplanar perturbations. At the true stability limit, both k_c and the imaginary part Ω_c of Ω are different from zero; so that the appearance of traveling waves, of wavenumber k_c, and frequency Ω_c may be expected in the vicinity of the stability limit. The analysis can obviously be used to describe the stability for the simplified model in which only a pyrolysis reaction at the solid-gas interface is considered and the heat flux from the gas to the solid is neglected. In this case, T_f should be replaced by T_s in Eq. (9), and Eq. (8) does not appear in the analysis. The dispersion relation for this model is also given by Eq. (29) if ε_g is replaced by $(c_p/c_s)\varepsilon_s$. The stability limit can be written for the simplified model: $\varepsilon_s = 2/(1 + \sqrt{2c_p/c_s})^2$.

Special attention has been paid to the limit $\varepsilon_g \to 0$, corresponding to a large activation energy of the gas-phase reaction. In this limit, the characteristic transversal length and response time become large compared with the thickness of the transport zone in the solid and the residence time in this zone, respectively. In the stability limit, traveling waves appear whose nondimensional frequency is of order $\varepsilon_g^{1/4}$. At distances of order $\varepsilon_g^{1/2}$ from the stability limit, the nondimensional time of decay or amplification of the perturbations is of order ε_g^{-1}. The imaginary part of Ω vanishes outside a region around the stability limit whose width goes to zero with ε_g. The disparity of time and space scales appearing in the dispersion relation will be used to simplify the analysis of the nonlinear response near the stability limit of the steady solution. In the planar case, the bifurcated solution when $\varepsilon_g \ll 1$ has been analyzed by Crespo and Kindelán (1978).

It is difficult to observe directly the traveling waves predicted by the theory and therefore clearly determine their relevance in physical systems, because the intrinsic response of the solid propellant often appears intermingled with processes occurring in others parts of the systems. In cylindrical samples, the traveling waves become spinning waves, and these were first pointed out by Merzhanov et al. (1973) and analyzed by Sivashinsky (1981) within the framework of a model of gasless combustion.

References

Barrère, M. (1970) Transient response of solid propellant combustion. Astronaut. Acta 15 (5,6), 633-640.

Crespo, A. and Kindelán, M. (1978) Unsteady burning of solid propellant. AFOSR 76-3049, Air Force Office of Scientific Research, Bolling AFB, D.C. 20332, USA.

Culick, F. E. C. (1968) A review of calculations for unsteady burning of solid propellants. AIAA J. 6 (12), 2241-2255.

De Luca, L. (1981) Nonlinear burning stability theory of heterogeneous thin flames. Proceedings of the XVIII International Symposium on Combustion, University of Waterloo, Canada, August 1980, pp 1439-1450.

Denison, R. and Baum, E. (1961) A simplified model of unstable burning in solid propellants. ARS J. 31 (8), 1112-1122.

Krier, H., T'ien, J. S., Sirignano, W. A., and Summerfield, M. (1968) Nonsteady burning phenomena of solid propellants: Theory and experiments. AIAA J. 6 (2), 278-285.

Merzhanov, A. G., Filonenko, A. K., and Borovinskaya, I. P. (1973) New phenomena in combustion of condensed systems. Proc. Acad. Sci. USSR Phys. Chem. Sci. Sec. 208 (892), 122-125.

Novozhilov, B. V. (1965) Stability criterion for the steady state of gunpowder combustion. Zh. Prikl. Mekh. Tekhn. Fiz. 4, 157

Sivashinsky, G. I. (1981) On spinning propagation of combustion waves SIAM J. Appl. Math. 40 (3), 432-438.

Williams, F. A., Barrère, M., and Huang, N.C. (1969) Fundamental aspects of solid propellants rockets. AGARDograph 116, Technivision Services, Slough, England.

Williams, F. A. (1973) Quasi-steady gas-phase flame theory in unsteady burning of a homogeneous solid propellant. AIAA J. 11 (9), 1328-1330.

Zel'dovich, Ya. B. (1942) Theory of the combustion of gunpowders and explosive substances. Zh. Eksp. Teor. Fiz. 12 (11), 498-524.

Zel'dovich, Ya. B. (1964) Rate of combustion of gunpowders under varying pressure. Zh. Prikl. Mekh. Tekhn. Fiz. 5 (3), 126-130.

Chapter II. Diffusion Flames

Laminar Diffusion Flames with Cylindrical Symmetry, Arbitrary Values of Diffusion Coefficients and Inlet Velocities, and Chemical Reactions in the Approach Streams

S. S. Penner,* M. Y. Bahadori, † and E. M. Kennedy‡
University of California, San Diego, La Jolla, California

Abstract

The analysis of Penner and Sherman 1947 is used for heat flow in composite cylinders, together with the methodology for defining diffusion-flame boundaries (Burke and Schumann 1928) in the treatment of laminar, cylindrically symmetric diffusion flames to develop a simplified procedure for the analytical solution of coupled diffusion and laminar decomposition flames with cylindrical symmetry. Careful selection of dimensionless variables, in a manner that is essentially different from that used by Burke and Schumann, is required in the analysis. Presented are solutions for diffusion flames with arbitrary values of diffusion coefficients and inlet flow velocities and also comparisons between calculated and observed data for both underventilated and overventilated diffusion flames. Agreement with experimental data is generally satisfactory. The generalized analytical solution presented in this paper reduces smoothly to the classical solution as the

---

Presented at the 9th ICODERS, Poitiers, France, July 3-8, 1983. Copyright © American Institute of Aeronautics and Astronautics, Inc., 1984. All rights reserved.
  *Professor of Engineering Physics and Director, Energy Center, Department of Applied Mechanics and Engineering Sciences.
  †Research Assistant, Energy Center and Department of Applied Mechanics and Engineering Sciences.
  ‡Shell Graduate Aid and Research Assistant, Energy Center and Department of Applied Mechanics and Engineering Sciences.

flow velocities and diffusion coefficients of the fuel and oxidizer streams become equal. Numerical, finite-difference solutions provide exact fits to these generalized analytical solutions. The Shvab-Zel'dovich procedure, which was used long ago in the description of methane-oxygen diffusion flames by Penner 1959, is modified and shown to yield good agreement for temperature and concentration profiles with experimental data and numerical studies on methane-air flames. The methodology presented here is readily extended to include chemical reactions in fuel and oxidizer approach flows. The solution of this problem may prove to be useful in the description of heterogeneous solid-propellant combustion.

Nomenclature

A_{1j}, A_{2j}, B_{2j}	= constants
d	= diameter of the inner cylinder
d'	= diameter of the outer cylinder
D_K	= diffusion coefficient of species K
$D_{I,K} = k_K(d'/2)/u_K$	= Damköhler number of species K
F	= fuel
h	= flame height
I	= inert material
J_0, J_1	= Bessel functions of the first kind of orders zero and one, respectively
k	= reaction rate
K, K'	= species
K_1	= $1/(ScRe)_F \equiv D_F/(u_F d'/2)$
K_2	= $1/(ScRe)_O \equiv D_O/(u_O d'/2)$
$\dot{m}_F$, $\dot{m}_O$	= total mass flow rates of fuel and oxidizer, respectively
N_0, N_1	= Bessel functions of the second kind of orders zero and one, respectively
O	= oxidizer
P, P'	= products
r	= radial distance
$Re_K = u_K(d'/2)/\nu_K$	= Reynolds number of species K
$Sc_K = \nu_K/D_K$	= Schmidt number of species K
T	= temperature
$\bar{T}$	= effective temperature at which the diffusion coefficient is evaluated

LAMINAR DIFFUSION FLAMES 263

T^*	=	temperature rise under standard conditions
T_0	=	initial temperature of the reactants in the approach flow
T_f	=	adiabatic flame temperature
u_K	=	flow speed for species K
$\dot{V}$	=	volume flow rate
w_K	=	chemical source function for species K
W_K	=	molecular weight of species K
Y	=	variable defined in Eqs. (1) and (18)
Y_1, Y_2	=	variables defined in Eqs. (2) and (29)
Y_K	=	mass fraction of species K
Y_K^*	=	mass fraction of species K entering into the overall chemical change
Y_X	=	variable defined in Eq. (11)
z	=	downstream distance
Z_j	=	functions defined in the text in terms of Bessel functions
$\eta = z/(d'/2)$	=	nondimensional axial distance
$\lambda_{1j}, \lambda_{2j}$	=	eigenvalues resulting from separation of variables of the species conservation equations for the fuel and oxidizer, respectively
μ	=	$(K_2/K_1)^{1/2} \equiv (D_O u_F/D_F u_O)^{1/2}$
$\xi = r/(d'/2)$	=	nondimensional radial distance
ξ_f	=	nondimensional radial location of a hypothetical cylindrical diffusion flame
ρ	=	density
ν	=	stoichiometric ratio of g of fuel consumed per g of oxidizer at the flame front
ν_K	=	kinematic viscosity of species K
ϕ	=	$\nu(\rho u)_O/(\rho u)_F$

Subscripts

c	=	critical condition for infinite flame height

calc	=	calculated
f	=	flame
F	=	fuel
o	=	initial condition
ov	=	overventilated
obs	=	observed
O	=	oxidizer
P, P'	=	products
uv	=	underventilated

The Burke-Schumann Solution of Diffusion Flames and its Generalization for Arbitrary Flow Speeds and Transport Properties

The classical analysis of Burke and Schumann (1928) on laminar diffusion flames shows qualitative successes on comparisons with experimental data. There is magic in this paper because the diffusion problem is solved without proper disposition of combustion products. At every plane downstream, the radial diffusion problem is solved for the initial approach flows as though nothing happened either upstream or downstream. These physical assumptions allow the construction of a different analytical solution (Penner et al. 1983) based on the description of diffusion flames formed in an infinite set of semi-infinite, concentric, cylindrical shells and using a methodology developed many years ago by Penner and Sherman (1947).

In order to generalize the Burke-Schumann analysis, a dimensional error must first be corrected. Burke and Schumann used effectively the dimensionless mass fraction (see Penner 1957a)

$$Y = \begin{cases} Y_F + \nu Y_{O,o} & \text{for } 0 \leq \xi < \xi_f \quad (1a) \\ -\nu Y_O + \nu Y_{O,o} & \text{for } \xi_f < \xi < 1 \quad (1b) \end{cases}$$

where F and O identify fuel and oxidizer; the subscript o refers to inlet conditions; ν represents the stoichiometric ratio of g of fuel consumed per g of oxidizer at the flame front; ξ is a dimensionless radial coordinate; and ξ_f denotes the dimensionless radial distance at the flame front. Thus, write

$$Y_1 = Y_F + \phi Y_{O,o} \quad \text{for } 0 \leq \xi < \xi_f \quad (2a)$$

$$Y_2 = -\phi Y_O + \phi Y_{O,o} \quad \text{for } \xi_f < \xi < 1 \quad (2b)$$

where

$$\phi = \nu(\rho u)_O / (\rho u)_F \quad (3)$$

ρ is the density, and u represents the flow speed.

In Eqs. (1) and (2), Y_F is in g of fuel per g of mixture containing fuel (i.e., the mass fraction of fuel in the fuel flow) while $Y_{O,o}$ is the initial value of g of oxidizer per g of mixture containing oxidizer (i.e., the mass fraction of oxidizer in the approach flow). It is now apparent that $\nu Y_{O,o}$ has the physical dimension of (g of fuel/g of oxidizer)$_\text{stoichiometric mixture}$ × (g of oxidizer/g of oxidizer mixture), which is evidently not dimensionally homogeneous with Y_F. On the other hand, the dimensions of $\phi Y_{O,o}$ are

$$\left[\left(\frac{\text{g of fuel}}{\text{g of oxidizer}} \right)_{\substack{\text{stoichiometric} \\ \text{mixture}}} \times \frac{\text{g of oxidizer mixture}}{\text{g of fuel mixture}} \right]$$

$$\times \frac{\text{g of oxidizer}}{\text{g of oxidizer mixture}}$$

which is dimensionally homogeneous with Y_F. Hence, the definition given in Eqs. (2) is required for flows with $(\rho u)_O/(\rho u)_F \neq 1$ while the definition of Eqs. (1) is acceptable if we mean by ν not ν but ϕ, which is numerically equivalent to ν for $(\rho u)_O/(\rho u)_F = 1$. It was concluded that Burke and Schumann possessed remarkable insight (which they did) when they wrote ν for ϕ and used dimensionally inhomogeneous quantities, which are numerically equivalent to dimensionally homogeneous quantities for their special problem.

There is no known analytical procedure for effecting the solution of partial differential equations when the eigenvalues are continuous functions of the coordinates. A general solution of this type to the problem is not being sought. Instead, being sought is a particular solution that satisfies the diffusion-flame boundary and interface conditions (with fuel and oxidizer mass fractions decreasing downstream as the result of reactions at the flame front), as well as the following fundamental property: the solution boundary is defined by the <u>initial</u> conditions and by the <u>local radial diffusion</u> properties. For problems of this class, an analytical solution may be obtained by proceeding as follows. First, separate the variables in the partial differential equations in the normal manner and find solutions at each value of the dimensionless height (η). The solutions must also satisfy all flame boundary and interface conditions. The resulting concentration profile and flame-boundary equations will then represent physically observable properties.

In support of the validity of this approach, the following results are cited (Penner et al. 1983). (1) For the homogeneous medium, the solution presented here reduces smoothly to the known solution of the Burke-Schumann problem. (2) For the vertical flame, the solution requires the known stoichiometric inlet mass flowrate ratios. (3) The physically required conditions for overventilated and underventilated flames are clearly met. (4) Direct, numerical, finite-difference calculations yield results that are identical with our analytical relations. (This fact allows us to view the analytical solutions as happy guesses, which represent correctly the results of direct numerical solutions of the problem under discussion.) (5) Comparisons of calculated and observed flame heights and flame shapes are in acceptable agreement, including those overventilated hydrocarbon-air flames for which the Burke-Schumann solution does not yield real (observable) flame boundaries. Furthermore, the effects of different velocities in the fuel and oxidizer streams appear to be approximately accounted for.

The Penner-Sherman Methodology Applied to Diffusion Flames

The method of solution used in the work of Penner and Sherman (1947) applies, with minor modifications, to the case in which a discontinuity occurs at $\eta = 0$ (see Fig. 1). All of the steps of the original analysis (1947) may be followed, except that the (variable) interface parameter ξ_f appears in place of d/d' in the interface continuity and gradient conditions. As in the analysis of Burke and Schumann (1928), the flame location is defined by (1) the arrival of stoichiometric mass flux ratios of fuel and oxidizer, (2) vanishingly small mass fractions of fuel and oxidizer, and (3) the solution of the radial transport equation without any reference to reaction processes that have occurred upstream or may occur downstream, except insofar as the mass fractions of fuel and oxidizer are reduced by reactions at the flame front. It is hypothesis (3) that allows us to solve the problem parametrically by representing the complete solution as a series of independent solutions for different values of ξ_f with $\eta > 0$, the applicable value of ξ_f being determined by utilizing the conditions (1) and (2). [Normally, there will be a discontinuity at $\eta = 0$ because $\xi_f \neq d/d'$ (see Fig. 1) at the injection plane.] As in the analysis of Burke and Schumann (1928), the fuel and oxidizer flow speeds remain constant because no account is taken of the production of reaction products formed at the flame front (assumption 3) and heat transfer from the flame front is not considered.

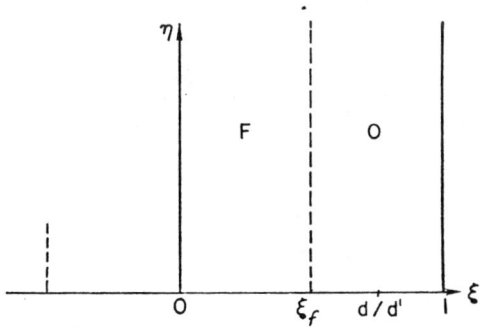

Fig. 1 Schematic diagram of concentric cylinders with fuel F entering at $\eta = 0$ for $0 \leq \xi < d/d'$ and oxidizer O for $d/d' < \xi < 1$. Here, $d/2$ is the radius of the inner cylinder and $d'/2$ that of the outer cylinder. The dimensionless coordinate ξ is defined as the ratio of radial distance (r) to $d'/2$. The dimensionless vertical coordinate η is defined as the downstream distance (z) measured in units of $d'/2$. The dimensionless distance ξ_f represents the location of a hypothetical cylindrical diffusion flame above the inlet port at $\eta = 0$. The precise values of ξ_f are ultimately determined as a function of η (see the text for details).

For the cylindrically symmetric flow of Fig. 1, inlet fuel flow occurs for $0 \leq \xi < d/d'$ and inlet oxidizer flow for $d/d' < \xi < 1$. With radial diffusion, the conservation equations for fuel and oxidizer above the burner port become, respectively,

$$\frac{\partial Y_F}{\partial \eta} = \frac{1}{(Sc\,Re)_F} \left(\frac{1}{\xi} \frac{\partial Y_F}{\partial \xi} + \frac{\partial^2 Y_F}{\partial \xi^2} \right) \quad \text{for } 0 \leq \xi < \xi_f \quad (4)$$

$$\frac{\partial Y_O}{\partial \eta} = \frac{1}{(Sc\,Re)_O} \left(\frac{1}{\xi} \frac{\partial Y_O}{\partial \xi} + \frac{\partial^2 Y_O}{\partial \xi^2} \right) \quad \text{for } \xi_f < \xi < 1 \quad (5)$$

where $Sc_K = \nu_K/D_K$ is a Schmidt number and $Re_K = u_K(d'/2)/\nu_K$ is a Reynolds number; u_F and u_O represent, respectively, the uniform (and constant downstream) entrance velocities in the fuel and oxidizer streams at $\eta = 0$. Using Eqs. (2), the continuity condition for mass fraction at the diffusion-flame boundary is $(Y_1)_{\xi_f^-, \eta} = (Y_2)_{\xi_f^+, \eta}$. The reacting species are delivered in stoichiometric proportions at the interface if

$$(\rho D)_F \left(\frac{\partial Y_1}{\partial \xi} \right)_{\xi_f^-, \eta} = (\rho D)_O \frac{(\rho u)_F}{(\rho u)_O} \left(\frac{\partial Y_2}{\partial \xi} \right)_{\xi_f^+, \eta}$$

We define $\mu = (K_2/K_1)^{1/2} \equiv (D_O u_F / D_F u_O)^{1/2}$, $K_1 \equiv 1/(Sc\,Re)_F$, $K_2 \equiv 1/(Sc\,Re)_O$. The method of Penner and Sherman (1947) then yields the results

$$K_1 \lambda_1^2 = K_2 \lambda_2^2$$

$$\frac{Y(\xi, \eta)}{(Y_{F,o} + \phi Y_{O,o})} = \left(\frac{d}{d'}\right)^2 + \sum_{j=2}^{\infty} \frac{\left\{ \left[\left(\frac{d}{d'}\right) J_1\left(\mu \lambda_{2j} \frac{d}{d'}\right) \right] / \mu \lambda_{2j} \right\}}{\left\{ \frac{1}{2} [A_{2j} J_0(\lambda_{2j}) + B_{2j} N_0(\lambda_{2j})]^2 \right\}} \cdots$$

$$\cdots \frac{\times Z_j(\lambda_j \xi)\exp(-K_2 \lambda_{2j}^2 \eta)}{+\frac{1}{2}\xi_f^2[(\mu^2-1)/\mu^2][J_1(\mu\lambda_{2j}\xi_f)]^2}\Bigg\}$$

$$= \begin{cases} Y_1(\xi,\eta)/(Y_{F,o}+\phi Y_{O,o}), & 0 \le \xi < \xi_f \\ Y_2(\xi,\eta)/(Y_{F,o}+\phi Y_{O,o}), & \xi_f < \xi < 1 \end{cases} \quad (6)$$

and

$$Z_j(\lambda_j \xi) = \begin{cases} Z_{1j}(\mu\lambda_{2j}\xi) = A_{1j}J_0(\mu\lambda_{2j}\xi), & 0 \le \xi < \xi_f \\ Z_{2j}(\lambda_{2j}\xi) = A_{2j}J_0(\lambda_{2j}\xi) \\ \qquad + B_{2j}N_0(\lambda_{2j}\xi), & \xi_f < \xi < 1 \end{cases}$$

A_{1j}, A_{2j}, and B_{2j} are constants; J_0 and J_1 are Bessel functions of the first kind of orders zero and one, respectively; and N_0 is a Bessel function of the second kind of order zero. The location of the flame boundary ($\xi = \xi_f$, $Y_F = Y_O = 0$) is

$$2\left(\frac{d}{d'}\right)\sum_{j=2}^{\infty} \frac{\left\{\left[J_1\left(\mu\lambda_{2j}\frac{d}{d'}\right)\right][J_0(\mu\lambda_{2j}\xi_f)]\right.}{(\mu\lambda_{2j})\{[A_{2j}J_0(\lambda_{2j})+B_{2j}N_0(\lambda_{2j})]^2} \cdots$$

$$\cdots \frac{\times \exp(-K_2\lambda_{2j}^2 \eta)}{+\xi_f^2[(\mu^2-1)/\mu^2][J_1(\mu\lambda_{2j}\xi_f)]^2}\Bigg\}$$

$$= \frac{\phi Y_{O,o}}{(Y_{F,o}+\phi Y_{O,o})} - \left(\frac{d}{d'}\right)^2 \quad (7)$$

Equation (7) is the general solution and may be used to obtain flame heights and flame shapes for arbitrary values of diffusion coefficients and flow velocities. Penner et al. 1983 verified by direct, numerical, finite-difference calculations that Eqs. (6) and (7) are correct analytical solutions of the problem posed. For $\mu = 1$ (i.e., $\phi = \nu$), Eqs. (6) and (7) reduce to the Burke-Schumann results.

The flame heights for overventilated (η_{ov}) and underventilated (η_{uv}) flames are obtained, respectively, by setting $\xi_f = 0$ or $\xi_f = 1$. The results are

$$\sum_{j=2}^{\infty} \frac{J_1\left(\mu\lambda_{2j}\frac{d}{d'}\right)\exp(-K_2\lambda_{2j}^2\eta_{ov})}{(\mu\lambda_{2j})[J_0(\lambda_{2j})]^2}$$

$$= \frac{\phi Y_{O,o}}{(Y_{F,o} + \phi Y_{O,o})\left(2\frac{d}{d'}\right)} - \left(\frac{d}{2d'}\right) \qquad (8)$$

and

$$\sum_{j=2}^{\infty} \frac{J_1\left(\mu\lambda_{2j}\frac{d}{d'}\right)\exp(-K_2\lambda_{2j}^2\eta_{uv})}{(\mu\lambda_{2j})J_0(\mu\lambda_{2j})}$$

$$= \frac{\phi Y_{O,o}}{(Y_{F,o} + \phi Y_{O,o})\left(2\frac{d}{d'}\right)} - \left(\frac{d}{2d'}\right) \qquad (9)$$

where the λ_{2j} are roots of $J_1(\lambda_{2j}) = 0$ and $J_1(\mu\lambda_{2j}) = 0$ for Eqs. (8) and (9), respectively.

When the mass flow-rate ratio of fuel to oxidizer is ν, the flame height must tend to infinity; for this condition, $\phi \equiv \phi_c$ is defined where

$$\phi_c = Y_{F,o} d^2 / Y_{O,o} (d'^2 - d^2) \qquad (10)$$

For $\phi/\phi_c > 1$, $\dot{m}_{F,o}/\dot{m}_{O,o} < \nu$ and the flame is overventilated; for $\phi/\phi_c < 1$, $\dot{m}_{F,o}/\dot{m}_{O,o} > \nu$ and the flame is underventilated.

Numerical Calculations and Comparisons with Experimental Data

Experimental data that are suitable for application of our theoretical relations have been published by Burke and Schumann (1928), Barr (1953), and Mitchell et al. (1980). These measurements include large and small ratios of u_F/u_O for which the Burke-Schumann model is fundamentally inapplicable, although it was used by Burke and Schumann after applying heuristic modifications to the input data.

Using as adjustable parameter the diffusion coefficient for a methane-oxygen flame at 600 K and scaling temperatures for other fuels in accord with the known heat release (Penner et al. 1983), the comparisons of calculated and observed flame heights are constructed as shown in Tables 1-3, which show generally acceptable agreement. Diffusion flame boundaries and the special case of H_2-air diffusion flames are discussed in a previous work Penner et al. (1983). Figures 2 and 3 show comparisons of calculated and observed flame boundaries for an underventilated and an overventilated diffusion flame, respectively. In Fig. 4, the calculated mass-

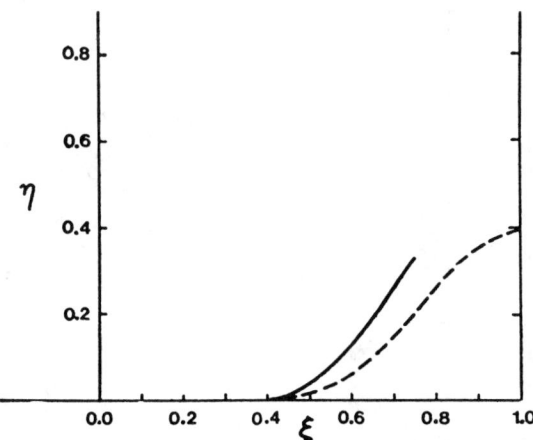

Fig. 2 Comparisons of calculated (the dashed curve was constructed by using the binary diffusion coefficient of fuel in nitrogen) and observed (solid curve) C_4H_{10}-air diffusion flames for Barr's (1953) experiment (e). The following conditions apply: d = 0.9 cm, d' = 2.20 cm, $\dot{V}_F$ = 10 cm^3/s, $\dot{V}_O$ = 15 cm^3/s, u_F = 15.72 cm/s, and u_O = 4.74 cm/s. Diffusion coefficients were evaluated at 1295 K; see Penner et al. (1983).

Table 1 Summary of calculated [from Eq. (8)] and observed heights of overventilated diffusion flames [The binary diffusion coefficients for oxygen and fuel in nitrogen were evaluated at 600 K. The blue-green flame boundary was used for the data given in the work of Mitchell et al. (1980). This table has been reproduced from Penner et al. (1983).]

Reference		Barr (1953)			Burke and Schumann (1928)[a]
		Fig. 3 (a)	Fig. 3 (b)	Fig. 1 (c)	Table I (e)
Experimental conditions					
Reactant mixture	Fuel	C_4H_{10}	C_4H_{10}	C_4H_{10}	CH_4
	Oxidizer	Air	Air	Air	Air
Initial mass fractions	$Y_{F,o}$	1	1	1	1
	$Y_{O,o}$	0.233	0.233	0.233	0.233
Dimensions	d, cm	0.90	0.90	0.90	0.953
	d', cm	2.20	2.20	2.20	2.54
	d/d'	0.409	0.409	0.409	0.375
Volume flow rates	$\dot{V}_F$, cm^3/s	1	1	1.2	15.7
	$\dot{V}_O$, cm^3/s	100	50	100	252
Inlet flow velocities	u_F, cm/s	1.57	1.57	1.89	22.1
	u_O, cm/s	31.6	15.8	31.6	57.8
	u_O/u_F	20.1	10.1	16.7	2.62
Observed (obs) flame heights	$h_{ov,obs}$, cm	4.90	6.80	6.10	20.6
	$\eta_{ov,obs}$	4.45	6.18	5.55	16.2
Calculated (calc) flame heights	$h_{ov,calc}$, cm	5.85	4.83	6.58	22.0
	$\eta_{ov,calc}$	5.32	4.39	5.98	17.31
$100 \times (\eta_{ov,calc} - \eta_{ov,obs})/\eta_{ov,obs}$, %		20	-29	8	7

(Table continued next page)

Table 1 (cont.) Summary of calculated [from Eq. (8)] and observed heights of overventilated flames.

Reference		Burke and Schumann (1928)[a]			
Experimental conditions		Table III (i)	Table III (ii)	Table III (iii)	Table III (iv)
Reactant mixture	Fuel	CH_4	CH_4	CH_4	CH_4
	Oxidizer	Air	Air	Air	Air
Initial mass fractions	$Y_{F,o}$	1	1	1	1
	$Y_{O,o}$	0.233	0.233	0.233	0.233
Dimensions	d, cm	0.953	0.953	0.953	0.953
	d', cm	2.54	2.54	2.54	2.54
	d/d'	0.375	0.375	0.375	0.375
Volume flow rates	$\dot{V}_F$, cm^3/s	2.99	5.90	7.86	10.4
	$\dot{V}_O$, cm^3/s	55.1	106	142	187
Inlet flow velocities	u_F, cm/s	4.20	8.28	11.1	14.6
	u_O, cm/s	12.6	24.4	32.5	43.0
	u_O/u_F	3.01	2.95	2.94	2.95
Observed (obs) flame heights	$h_{ov,obs}$, cm	4.39[b]	8.56	11.4	14.8
	$\eta_{ov,obs}$	3.46	6.74	8.94	11.6
Calculated (calc) flame heights	$h_{ov,calc}$, cm	4.32	8.37	11.2	14.7
	$\eta_{ov,calc}$	3.40	6.59	8.81	11.6
$100 \times (\eta_{ov,calc} - \eta_{ov,obs})/\eta_{ov,obs}$, %		-2	-2	-1	0

(Table continued on next page)

Table 1 (cont.) Summary of calculated [from Eq. (8)] and observed heights of overventilated flames.

Reference		Burke and Schumann (1928)[a]			Mitchell et al. (1980)
Experimental conditions		Table III (v)	Table III (vi)	Table III (vii)	
Reactant mixture	Fuel	CH_4	CH_4	CH_4	CH_4
	Oxidizer	Air	Air	Air	Air
Initial mass fractions	$Y_{F,o}$	1	1	1	1
	$Y_{O,o}$	0.233	0.233	0.233	0.233
Dimensions	d, cm	0.953	0.953	0.953	1.27
	d', cm	2.54	2.54	2.54	5.08
	d/d'	0.375	0.375	0.375	0.250
Volume flow rates	$\dot{V}_F$, cm^3/s	12.9	16.2	17.9	5.70
	$\dot{V}_O$, cm^3/s	232	292	323	188
Inlet flow velocities	u_F, cm/s	18.1	22.7	25.2	4.50
	u_O, cm/s	53.3	67.0	74.2	9.88
	u_O/u_F	2.94	2.95	2.95	2.20
Observed (obs) flame heights	$h_{ov,obs}$, cm	18.4	22.9	25.2	6.80
	$\eta_{ov,obs}$	14.5	18.0	19.8	2.68
Calculated (calc) flame heights	$h_{ov,calc}$, cm	18.3	23.0	25.6	7.52
	$\eta_{ov,calc}$	14.4	18.1	20.1	2.96
$100 \times (\eta_{ov,calc} - \eta_{ov,obs})/\eta_{ov,obs}$, %		-1	1	2	10

[a]The tube diameters for the experiments in Table III of Burke and Schumann (1928) are not known. However, comparison with the data of Table I(e) of this work indicates that the inner and outer cylinders must have radii of 3/16 and 1/2 in., respectively.
[b]This value corresponds to 1.73 in. rather than the apparently misprinted value 1.23 in. given in the work of Burke and Schumann (1928).

Table 2 Calculated and observed (Burke and Schumann 1928) flame heights for CO-air flames with air flowing through the inner cylinder [The binary diffusion coefficients of CO in N_2 and O_2 in N_2 were evaluated at 600 K. This table has been reproduced from Penner et al. (1983).]

Experiment	Fuel	Oxidizer	$Y_{O,o}$	$Y_{CO,o}$	d, cm	d', cm	d/d'	$\dot{V}_{air}$, cm^3/s	$\dot{V}_{CO}$, cm^3/s
VIII(c)	CO	Air	0.233	1	2.13	3.20	0.667	25.6	29.5
VIII(d)	CO	Air	0.233	1	2.13	3.20	0.667	43.7	49.21

Experiment	u_{air}, cm/s	u_{CO}, cm/s	$h_{ov,obs}$ cm	$\eta_{ov,obs}$	$h_{ov,calc}$ cm	$\eta_{ov,calc}$	$100\times(\eta_{ov,calc}-\eta_{ov,obs})/\eta_{ov,obs}$, %
VIII(c)	7.15	6.61	2.59	1.62	2.27	1.42	-12
VIII(d)	12.2	11.0	4.45	2.78	3.73	2.33	-16

Table 3 Summary of calculated [from Eq. (9)] and observed heights of underventilated hydrocarbon and CO diffusion flames burning with air [In these calculations, binary diffusion coefficients were used for O_2 in N_2 and for fuel in N_2 at appropriate temperatures (see Penner et al. 1983).]

Reference		Table I (a)	Burke and Schumann (1928)				Table VI (a)
Experiment in corresponding reference			Table I (b)	Table I (c)	Table I (d)		
Reactant mixture	Fuel Oxidizer	CH_4 Air	City gas Air	CH_4 Air	CH_4 Air		CH_4 Air
Initial mass- fractions	$Y_{F,o}$ $Y_{O,o}$	1 0.233	1 0.233	1 0.233	1 0.233		1 0.233
Dimensions	d, cm d', cm d/d'	1.27 3.18 0.4	1.27 3.18 0.4	0.953 1.91 0.5	1.59 3.18 0.5		0.953 2.54 0.375
Volume flow rates	$\dot{V}_F$, cm^3/s $\dot{V}_O$, cm^3/s	15.7 47.2	15.7 31.5	15.7 47.2	15.7 47.2		15.7 31.5
Inlet flow velocities	u_F, cm/s u_O, cm/s u_O/u_F	12.4 7.10 0.572	12.4 4.73 0.381	22.1 22.1 1	7.95 7.95 1		22.1 7.23 0.327
Observed flame heights	$h_{uv,obs}$, cm $\eta_{uv,obs}$	2.21 1.39	1.45 0.91	2.24 2.35	2.13 1.34		1.98 1.56
Calculated flame heights	$h_{uv,calc}$, cm $\eta_{uv,calc}$	2.87 1.81	1.30^a 0.82^a	1.63 1.71	1.60 1.01		2.69 2.12
$100 \times (\eta_{uv,calc} - \eta_{uv,obs})/\eta_{uv,obs}$, %		30	-10	-27	-25		36

(Table continued on next page)

Table 3 (cont.) Summary of calculated [from Eq. (9)] and observed heights of underventilated hydrocarbon and CO diffusion flames burning with air.

Reference		Table IV (3)	Burke and Schumann (1928)			
Experiment in corresponding reference			Table II (b)	Table VIII (a)	Table VIII (b)
Reactant mixture	Fuel	C_2H_6	CO	CO	CO
	Oxidizer	Air	Air	Air	Air
Initial mass-fractions	$Y_{F,o}$	0.233	0.233	0.233	0.233
	$Y_{O,o}$	1.27	2.13	2.13	2.13
Dimensions	d, cm	3.175	3.20	3.20	3.20
	d', cm	0.4	0.667	0.667	0.667
	d/d'	15.7	197	23.6	47.2
Volume flow rates	$\dot{V}_F$, cm^3/s	47.2	175	33.0	62.9
	$\dot{V}_O$, cm^3/s	12.4	55.0	6.60	13.2
Inlet flow velocities	u_F, cm/s	7.10	39.1	7.39	14.1
	u_O, cm/s	0.572	0.710	1.12	1.07
	u_O/u_F	1.57	17.7	2.34	4.45
Observed flame heights	$h_{uv,obs}$, cm	0.99	11.0	1.46	2.78
	$\eta_{uv,obs}$	1.76	21.8[b]	4.42[b]	8.27[b]
Calculated flame heights	$h_{uv,calc}$, cm	1.11	13.6[b]	2.76[b]	5.17[b]
	$\eta_{uv,calc}$				
$100 \times (\eta_{uv,calc} - \eta_{uv,obs})/\eta_{uv,obs}$, %		12	24[b]	89[b]	86[b]

[a] For city gas, we used the arithmetic mean of the binary diffusion coefficients of H_2 in N_2 and of H_2 in CH_4.
[b] Agreement between calculated and observed underventilated CO-air flames is greatly improved by using $T_{CO,uv}$ = 500 K instead of the prescribed value of 405 K (see Penner et al. 1983).

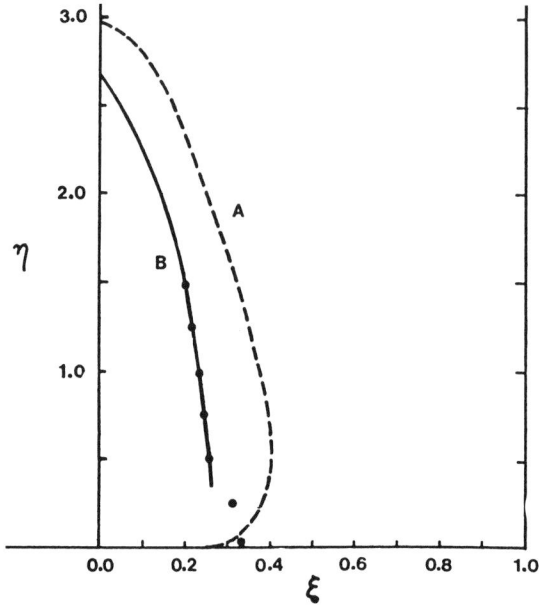

Fig. 3 Comparison of calculated (curve A) and observed (curve B) flame boundaries. The observed flame boundary refers to blue-green radiation from a methane-air flame and has been transcribed from Fig. 6 of Mitchell et al. (1980). The curve A was calculated according to Eq. (7) with $\mu = 0.642$ and $K_2 = 0.027$ (corresponding to D_O evaluated at 600 K); reproduced from Penner et al. (1983).

fraction profiles of fuel and oxidizer are compared with the results of a numerical solution of the conservation equations by Mitchell et al. (1980).

Approximate Estimations of Concentration and Temperature Profiles

Mitchell et al. (1980) calculated and observed concentration, temperature and velocity profiles for an over-ventilated methane-air diffusion flame. The use of similarity analysis§ for a Lewis number of unity permits facile estimations of temperature and concentration profiles for the major product species if one concentration profile is known. This application of the Shvab-Zel'dovich§ procedure is due to Penner (1959) and yields results that are in

---

§See Zel'dovich 1949 or 1950; Shvab 1944 or 1948; Penner and and Mullins 1959.

excellent agreement with the calculations and in acceptable agreement with the measurements published in the work of Mitchell et al. (1980). It may be shown (Penner et al. 1983) that

with
$$Y_X = \frac{Y_X^*}{Y_O^* Y_F^*} \frac{Y_{F,o} Y_{O,o} - Y_F Y_{O,o} - Y_{F,o} Y_O}{(Y_{F,o}/Y_F^*) + (Y_{O,o}/Y_O^*)}$$

$$Y_X = Y_P, \; Y_{P'}, \text{ and } (T - T_0) \quad \quad (11)$$

for the schematic bimolecular reaction $I + F + O \to P + P' + I$. The flame front (subscript f) is again identified by the condition $Y_F = Y_O = 0$ so that

for
$$Y_{X,f} = Y_X^* \frac{(Y_{F,o}/Y_F^*)(Y_{O,o}/Y_O^*)}{(Y_{F,o}/Y_F^*) + (Y_{O,o}/Y_O^*)}$$

$$Y_{X,f} = Y_{P,f}, \; Y_{P',f} \text{ or } (T_f - T_0)$$

where $Y_F^* = 5.506 \times 10^{-2}$ (F = CH$_4$), $Y_O^* = 0.2202$ (O = O$_2$), $Y_P^* = 0.1514$ (P = CO$_2$), $Y_{P'}^* = 0.1239$ (P' = H$_2$O), and $T_f^* = 1860$ K for the methane-air flame (Penner et al. 1983).

The numerical values at the flame front agree within 5% with computer calculations of Mitchell et al. (1980). Thus, the approximation of a Lewis number equal to unity and use of the Shvab-Zel'dovich procedure do not lead to significant errors as long as the Burke-Schumann flame-sheet model and the corresponding single-step reaction mechanism are employed.

Equation (11) is a scaling relation, which yields excellent approximations to the composition and temperature profiles for all species if data are available for one species (see Figs. 5 and 6, reproduced from Penner et al. 1983).

Coupled Decomposition and Diffusion Flames

Equal First-Order Decomposition-Reaction Rates, Flow Velocities, and Diffusion Coefficients in the Fuel and Oxidizer Streams

Using the notation of Penner (1957b), the conservation equation for species K' in the presence of chemical reactions, in a cylindrically symmetric flow, is

$$\frac{\partial Y_{K'}}{\partial z} = \frac{w_{K'}}{\rho u} + \frac{D_{K'}}{u}\left(\frac{1}{r}\frac{\partial Y_{K'}}{\partial r} + \frac{\partial^2 Y_{K'}}{\partial r^2} + \frac{\partial^2 Y_{K'}}{\partial z^2}\right) \quad (12)$$

The axial diffusion term $\partial^2 Y_{K'}/\partial z^2$ is included in the analysis since its effect may not be negligibly small for short flames. For the unimolecular decomposition reaction $K \xrightarrow{k_K} 2K'$ [where $K' = F'$ or O' is assumed to be a nonreactive decomposition product formed from fuel (F) and oxidizer (O), respectively]$^\pi$

$$\frac{d(K')}{dt} = 2k_K(K) = 2k_K(\rho Y_K/W_K)$$

or

$$w_{K'} = W_{K'}\frac{d(K')}{dt} = W_{K'}\left(\frac{2k_K \rho Y_K}{W_K}\right) \quad (13)$$

With $W_K = 2W_{K'}$ and $Y_K = 1 - Y_{K'}$, Eq. (13) yields

$$w_{K'} = k_K \rho Y_K = k_K \rho (1 - Y_{K'}) \quad (14)$$

[e.g., in g of K' produced/(cm^3-s)]. Substitution of Eq. (14) into Eq. (12) gives

$$\frac{\partial Y_{K'}}{\partial z} = \frac{k_K(1 - Y_{K'})}{u} + \frac{D_{K'}}{u}\left(\frac{1}{r}\frac{\partial Y_{K'}}{\partial r} + \frac{\partial^2 Y_{K'}}{\partial r^2} + \frac{\partial^2 Y_{K'}}{\partial z^2}\right) \quad (15)$$

Using $Y_K = 1 - Y_{K'}$ in Eq. (15) and replacing $D_{K'}$ by D_K and u by u_K leads to

$$\frac{\partial Y_K}{\partial z} = -\frac{k_K}{u_K}Y_K + \frac{D_K}{u_K}\left(\frac{1}{r}\frac{\partial Y_K}{\partial r} + \frac{\partial^2 Y_K}{\partial r^2} + \frac{\partial^2 Y_K}{\partial z^2}\right) \quad (16)$$

$^\pi$This reaction model is oversimplified. In general, we expect to encounter reactions between F' and O or O', between O' and F or F', as well as between F and O.

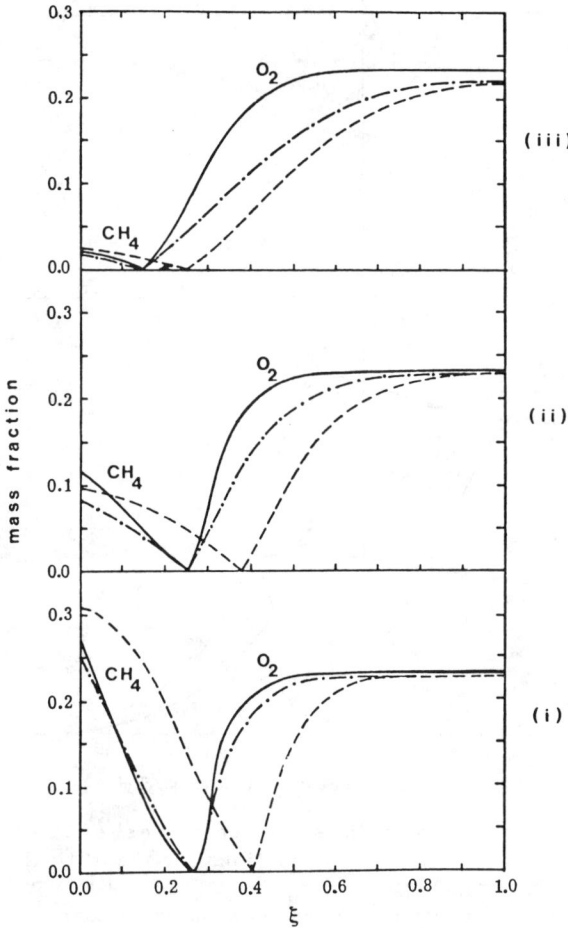

Fig. 4 Mass fraction profiles (dashed curves) of CH_4 and O_2, calculated from Eq. (6) for the overventilated methane-air diffusion flame of Mitchell et al. (1980) (see Table 1 and Fig. 3) at (1) 1.2 cm, (2) 2.4 cm, and (3) 5.0 cm above the burner port. Diffusion coefficients employed in Eq. (6) were evaluated at 600 K (see Penner et al. 1983) for diffusion of oxygen in nitrogen and of methane in nitrogen. The solid curves are given by Mitchell et al. (1980) and were obtained by numerical solution of the governing equations. The dot-dash curves are the calculated curves after allowing for differences in flame-front location; reproduced from Penner et al. (1983).

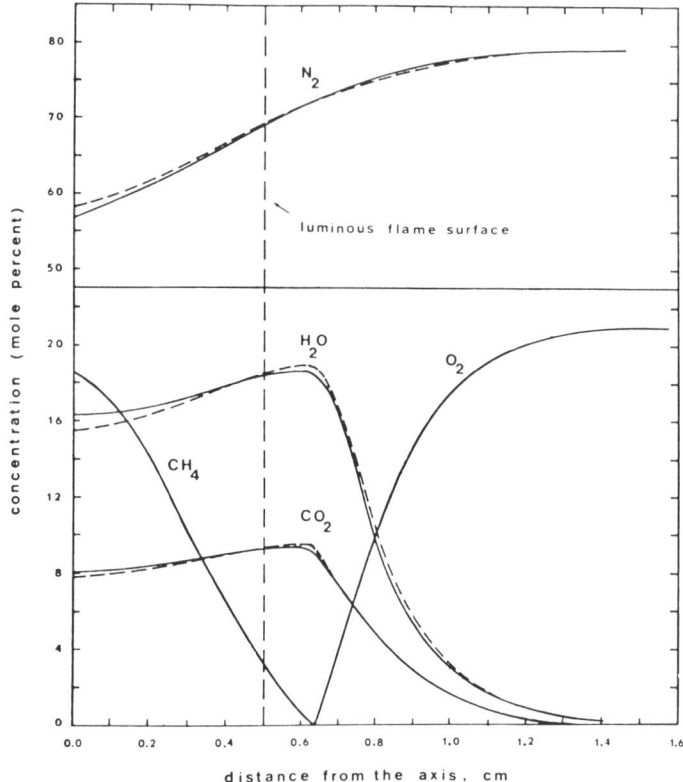

Fig. 5 Mole fraction profiles (dashed curves) of CO_2 and H_2O, calculated from the similarity relations for theoretical values (see Mitchell et al. 1980) on CH_4 and O_2, at a distance of 2.4 cm above the burner port, for an overventilated methane-air diffusion flame. The solid curves are given by Mitchell et al. (1980) and were obtained by numerical solution of the governing equations; reproduced from Penner et al. (1983).

With $K = F$, Eq. (16) applies for $0 \leq \xi < \xi_f$ and, with $K = O$, it applies for $\xi_f < \xi < 1$.

First, define the Damköhler numbers $D_{I,K} = k_K(d'/2)/u_K$ with $K = F$ or O. Next, use the simplifying assumptions

$$\left. \begin{array}{c} D_{I,F} = D_{I,O} = D_I \\ (Sc\,Re)_F = (Sc\,Re)_O = Sc\,Re \end{array} \right\} \quad (17)$$

hence, a single variable mass fraction over the entire radial region may be defined by

$$Y = \begin{cases} Y_F & \text{for } 0 \leq \xi < \xi_f, \quad \eta > 0 \\ -\phi Y_O & \text{for } \xi_f < \xi < 1, \quad \eta > 0 \end{cases} \quad (18)$$

The problem now reduces to the following single governing equation

$$\frac{\partial Y}{\partial \eta} = -D_I Y + \frac{1}{Sc\,Re}\left(\frac{1}{\xi}\frac{\partial Y}{\partial \xi} + \frac{\partial^2 Y}{\partial \xi^2} + \frac{\partial^2 Y}{\partial \eta^2}\right) \quad (19)$$

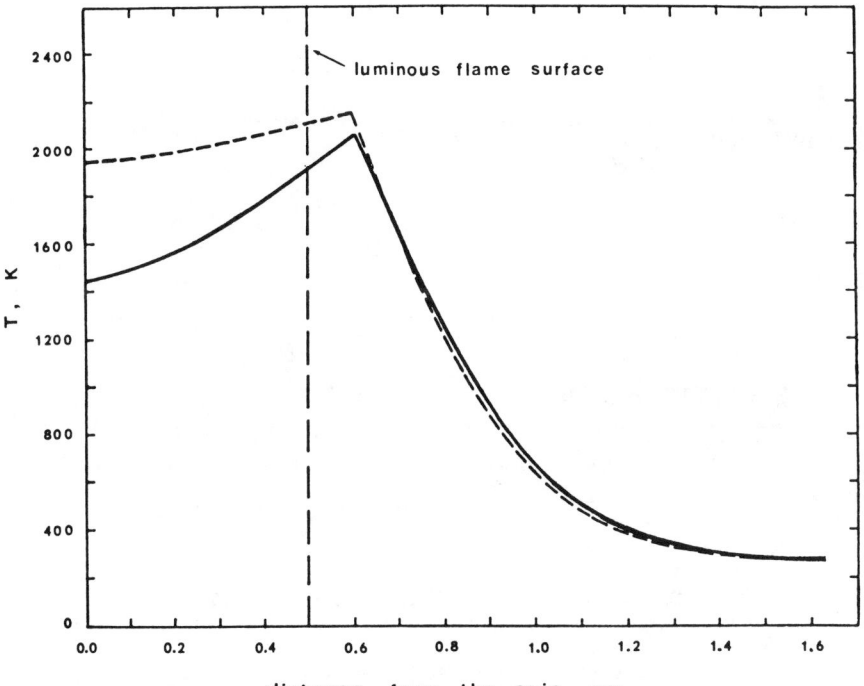

Fig. 6 The temperature profile (dashed curve), calculated from the similarity relations for theoretical values (see Mitchell et al. 1980) on CH_4 and O_2, at a distance of 2.4 cm above the burner port for an overventilated methane-air diffusion flame. The solid curves are given by Mitchell et al. (1980) and were obtained by numerical solution of the governing equations; reproduced from Penner et al. (1983).

over the region $0 \leq \xi < 1$. In view of Eq. (18), the initial conditions ($\eta = 0$) are

$$Y = Y_{F,o} \quad \text{for} \quad 0 \leq \xi < d/d' \tag{20}$$

$$Y = -\phi Y_{O,o} \quad \text{for} \quad d/d' < \xi < 1 \tag{21}$$

The continuity condition for mass fraction at the flame boundary is

$$(Y)_{\xi_f^-, \eta} = (Y)_{\xi_f^+, \eta} \tag{22}$$

The stoichiometry condition at the flame boundary is

$$\left(\frac{\partial Y}{\partial \xi}\right)_{\xi_f^-, \eta} = \left(\frac{\partial Y}{\partial \xi}\right)_{\xi_f^+, \eta} \tag{23}$$

where $(Sc\,Re)_F = (Sc\,Re)_O$ was used. The symmetry and boundary conditions are

$$(\partial Y / \partial \xi)_{0, \eta} = 0 \tag{24}$$

$$(\partial Y / \partial \xi)_{1^-, \eta} = 0 \tag{25}$$

Equation (19) may be solved by using separation of variables and the methods of solution of either Burke and Schumann (1928) or Penner and Sherman (1947). Using separation of variables

$$\frac{Y(\xi, \eta)}{(Y_{F,o} + \phi Y_{O,o})} = \left[\left(\frac{d}{d'}\right)^2 - \frac{\phi Y_{O,o}}{(Y_{F,o} + \phi Y_{O,o})}\right]$$

$$\times \exp\left\{\frac{Sc\,Re}{2}\left[1 - \sqrt{1 + \frac{4Sc\,Re\,D_I}{(Sc\,Re)^2}}\right]\eta\right\}$$

$$+ 2\left(\frac{d}{d'}\right)\sum_{j=2}^{\infty} \cdots$$

$$\frac{J_1\left(\lambda_j \frac{d}{d'}\right) J_0(\lambda_j \xi) \exp\left\{\frac{Sc\,Re}{2}\left[1-\sqrt{1+\frac{4(\lambda_j^2+D_I Sc\,Re)}{(Sc\,Re)^2}}\right]\eta\right\}}{\lambda_j [J_0(\lambda_j)]^2} \quad (26)$$

was found. The location of the flame boundary is obtained for $\xi = \xi_f$, where $Y = 0$, as

$$\left[\frac{\phi Y_{O,o}}{(Y_{F,o}+\phi Y_{O,o})}-\left(\frac{d}{d'}\right)^2\right] \exp\left\{\frac{Sc\,Re}{2}\left[1-\sqrt{1+\frac{4Sc\,Re\,D_I}{(Sc\,Re)^2}}\right]\eta\right\}$$

$$= 2\left(\frac{d}{d'}\right) \sum_{j=2}^{\infty} \cdots$$

$$\frac{J_1\left(\lambda_j \frac{d}{d'}\right) J_0(\lambda_j \xi_f) \exp\left\{\frac{Sc\,Re}{2}\left[1-\sqrt{1+\frac{4(\lambda_j^2+D_I Sc\,Re)}{(Sc\,Re)^2}}\right]\eta\right\}}{\lambda_j [J_0(\lambda_j)]^2} \quad (27)$$

As the chemical reaction rate goes to zero, $D_I \to 0$ and Eqs. (26) and (27) reduce to the Burke-Schumann problem with the term $\partial^2 Y_K'/\partial z^2$ included in Eq. (12). The Burke-Schumann result is obtained from Eqs. (26) and (27) for $D_I \to 0$ and $4\lambda_j^2/(Sc\,Re)^2 \ll 1$, which correspond to the absence of chemical reaction and deletion of the term $\partial^2 Y_K'/\partial z^2$ in Eq. (12), respectively.

The heights for overventilated (η_{ov}) and underventilated (η_{uv}) flames are obtained by setting $\xi_f = 0$ or $\xi_f = 1$, respectively, in Eq. (27). The results are

$$\left[\frac{\phi Y_{O,o}}{(Y_{F,o}+\phi Y_{O,o})}-\left(\frac{d}{d'}\right)^2\right] \exp\left\{\frac{Sc\,Re}{2}\left[1-\sqrt{1+\frac{4D_I}{Sc\,Re}}\right]\binom{\eta_{ov}}{\text{or}}{\eta_{uv}}\right\}$$

$$= 2\left(\frac{d}{d'}\right) \sum_{j=2}^{\infty} \cdots$$

$$\frac{J_1\left(\lambda_j \frac{d}{d'}\right) \begin{pmatrix} 1 \\ \text{or} \\ J_0(\lambda_j) \end{pmatrix} \exp\left\{\frac{Sc\,Re}{2}\left[1-\sqrt{1+\frac{4(\lambda_j^2 + D_I Sc\,Re)}{(Sc\,Re)^2}}\right]\right\} \begin{pmatrix} \eta_{ov} \\ \text{or} \\ \eta_{uv} \end{pmatrix}}{\lambda_j [J_0(\lambda_j)]^2} \tag{28}$$

Arbitrary Decomposition Rates, Diffusion Coefficients, and Flow Velocities in the Fuel and Oxidizer Streams

Again the chemical decomposition model of the previous section was used with $F \xrightarrow{k_F} 2F'$, $O \xrightarrow{k_O} 2O'$, where F' and O' are nonreactive while F and O form a diffusion flame, the location of which shall now be determined. Using the definitions,

$$Y_1 = Y_F \text{ for } 0 \leq \xi < \xi_f, \quad Y_2 = -\phi Y_O \text{ for } \xi_f < \xi < 1 \tag{29}$$

the governing equations are

$$\frac{\partial Y_i}{\partial \eta} = -D_{I,i} Y_i + K_i \left(\frac{1}{\xi}\frac{\partial Y_i}{\partial \xi} + \frac{\partial^2 Y_i}{\partial \xi^2} + \frac{\partial^2 Y_i}{\partial \eta^2}\right)$$

$$i = 1 \text{ for } 0 \leq \xi < \xi_f, \quad i = 2 \text{ for } \xi_f < \xi < 1 \tag{30}$$

where $D_{I,1} \equiv D_{I,F} = k_F(d'/2)/u_F$, $D_{I,2} \equiv D_{I,O} = k_O(d'/2)/u_O$, $K_1 = 1/(Sc\,Re)_F$, $K_2 = 1/(Sc\,Re)_O$. The interface continuity conditions are

$$(Y_1)_{\xi_f^-, \eta} = (Y_2)_{\xi_f^+, \eta} \tag{31}$$

$$\left(\frac{\partial Y_1}{\partial \xi}\right)_{\xi_f^-, \eta} = \mu^2 \left(\frac{\partial Y_2}{\partial \xi}\right)_{\xi_f^+, \eta} \tag{32}$$

where $\mu = (D_O u_F / D_F u_O)^{1/2}$. The symmetry and boundary conditions become

$$(\partial Y_1 / \partial \xi)_{0,\eta} = 0, \quad (\partial Y_2 / \partial \xi)_{1^-,\eta} = 0 \tag{33}$$

The initial conditions are

$$Y_1(\xi, 0) = Y_{F,0} \quad \text{for} \quad 0 \leq \xi < d/d',$$

$$Y_2(\xi, 0) = -\phi Y_{O,o} \quad \text{for} \quad d/d' < \xi < 1 \tag{34}$$

The Penner-Sherman methodology may be shown to yield

$$Y(\xi, \eta) = \sum_{j=1}^{\infty} \frac{\left\{\left[Y_{F,o} + \phi Y_{O,o}\left(\frac{\lambda_{1j}}{\mu \lambda_{2j}}\right)^2\right]\left(\frac{d}{d'}\right)\frac{J_1\left(\lambda_{1j}\frac{d}{d'}\right)}{\lambda_{1j}}\right.}{\left\{\frac{1}{2}[A_{2j}J_0(\lambda_{2j}) + B_{2j}N_0(\lambda_{2j})]^2\right.}$$

$$\cdots \frac{\left. \times Z_j(\lambda_j \xi)\exp\left\{\frac{1}{2K_2}\left[1 - \sqrt{1 + 4K_2^2\left(\lambda_{2j}^2 + \frac{D_{I,2}}{K_2}\right)}\right]\eta\right\}\right\}}{\left. + \frac{1}{2}\xi_f^2\left[1 - \left(\frac{\lambda_{1j}}{\mu^2 \lambda_{2j}}\right)^2\right][J_1(\lambda_{1j}\xi_f)]^2\right\}}$$

$$= \begin{cases} Y_F & \text{for } 0 \leq \xi < \xi_f \\ -\phi Y_O & \text{for } \xi_f < \xi < 1 \end{cases} \tag{35}$$

with $Z_j(\lambda_j \xi)$ given by

$$Z_j(\lambda_j \xi) = \begin{cases} Z_{1j}(\lambda_{1j}\xi) = A_{1j}J_0(\lambda_{1j}\xi), & 0 \leq \xi < \xi_f \\ Z_{2j}(\lambda_{2j}\xi) = A_{2j}J_0(\lambda_{2j}\xi) \\ \quad + B_{2j}N_0(\lambda_{2j}\xi), & \xi_f < \xi < 1 \end{cases}$$

It should be noted that only $D_{I,2}$ appears in the solution explicitly; however, a relation involving $D_{I,1}$ must be satisfied between λ_{1j} and λ_{2j} and is given in Eq. (36). Here, the coefficients $A_1 = A_{1j}$, $A_2 = A_{2j}$, and $B_2 = B_{2j}$ are the solutions of the following expressions for particular values of λ_{1j} and λ_{2j}:

$$A_2 J_1(\lambda_2) + B_2 N_1(\lambda_2) = 0 \quad \text{for} \quad \lambda_2 \neq 0$$

$$A_1 J_0(\lambda_1 \xi_f) \exp\{①\,\eta\} = [A_2 J_0(\lambda_2 \xi_f)$$

$$+ B_2 N_0(\lambda_2 \xi_f)] \exp\{②\,\eta\}$$

$$\lambda_1 A_1 J_1(\lambda_1 \xi_f) \exp\{①\,\eta\} = \mu^2 \lambda_2 [A_2 J_1(\lambda_2 \xi_f)$$

$$+ B_2 N_1(\lambda_2 \xi_f)] \exp\{②\,\eta\}$$

where ① and ② denote, respectively, the values of

$$\frac{1}{2K_i}\left[1 - \sqrt{1 + 4K_i^2\left(\lambda_i^2 + \frac{D_{I,i}}{K_i}\right)}\right]$$

with $i = 1$ for $0 \leq \xi < \xi_f$ and $i = 2$ for $\xi_f < \xi < 1$. This analysis shows that the following relation must hold between the eigenvalues λ_{1j} and λ_{2j}:

$$(1/K_1)\left\{1 - \sqrt{1 + 4K_1^2[\lambda_{1j}^2 + (D_{I,1}/K_1)]}\right\}$$

$$= (1/K_2)\left\{1 - \sqrt{1 + 4K_2^2[\lambda_{2j}^2 + (D_{I,2}/K_2)]}\right\} \quad (36)$$

The eigenvalues λ_{2j} and λ_{1j} are obtained from

$$\begin{vmatrix} 0 & J_1(\lambda_{2j}) & N_1(\lambda_{2j}) \\ J_0(\lambda_{1j}\xi_f) & J_0(\lambda_{2j}\xi_f) & N_0(\lambda_{2j}\xi_f) \\ \lambda_{1j} J_1(\lambda_{1j}\xi_f) & \mu^2 \lambda_{2j} J_1(\lambda_{2j}\xi_f) & \mu^2 \lambda_{2j} N_1(\lambda_{2j}\xi_f) \end{vmatrix}$$

$$= \lambda_{1j} J_1(\lambda_{1j}\xi_f) [J_1(\lambda_{2j}) N_0(\lambda_{2j}\xi_f) - J_0(\lambda_{2j}\xi_f) N_1(\lambda_{2j})]$$

$$- \mu^2 \lambda_{2j} J_0(\lambda_{1j}\xi_f) [J_1(\lambda_{2j}) N_1(\lambda_{2j}\xi_f)$$

$$- J_1(\lambda_{2j}\xi_f) N_1(\lambda_{2j})] = 0 \qquad (37)$$

The zeros of Eq. (37) may be found by plotting the determinant as a function of λ_{2j} after using Eq. (36) to find λ_{1j} in terms of λ_{2j}.

The flame location is obtained by setting $Y_F = Y_O = 0$ and $\xi = \xi_f$ in Eq. (35) and using $[Z_{1j}(\lambda_{1j}\xi)]_{\xi_f^-} = [Z_{2j}(\lambda_{2j}\xi)]_{\xi_f^+}$. It is found that

$$\sum_{j=1}^{\infty} \left\{ \frac{\left[Y_{F,o} + \phi Y_{O,o}\left(\frac{\lambda_{1j}}{\mu\lambda_{2j}}\right)^2\right] J_1\left(\lambda_{1j}\frac{d}{d'}\right) J_0(\lambda_{1j}\xi_f)}{\lambda_{1j}\left\{[A_{2j} J_0(\lambda_{2j}) + B_{2j} N_0(\lambda_{2j})]^2\right.} \right.$$

$$\left. \cdots \frac{\times \exp\left\{\frac{1}{2K_2}\left[1 - \sqrt{1 + 4K_2^2\left(\lambda_{2j}^2 + \frac{D_{I,2}}{K_2}\right)}\right]\eta\right\}}{\left. + \xi_f^2 [1 - (\lambda_{1j}/\mu^2\lambda_{2j})^2][J_1(\lambda_{1j}\xi_f)]^2\right\}} = 0 \qquad (38)$$

which reduces to Eq. (7) for $D_{I,2} = 0$, $4K_2^2 \lambda_{2j}^2 \ll 1$, and $\lambda_{1j} = \mu\lambda_{2j}$ after properly re-evaluating the Fourier-Bessel coefficients.

Generalization to More Complex Chemical Processes and Applications to Solid Propellants

The procedure which has been described may be applied for more realistic chemical changes than unimolecular decompositions. Separation of the radially- and axially-dependent components of the governing equations is generally not possible. However, numerical solution of the governing equations, subject to the specified initial and boundary conditions, may be effected without difficulty (compare the Appendix of Penner et al. 1983).

An interesting and important application of this methodology relates to development of solutions for composite solid-propellant burning. Since introduction of the granular diffusion flame model by Summerfield and his associates, [For a review of the early work of Summerfield et al., as well as for the discussion of the sandwich burner model of Nachbar and Johnson, see Penner (1962).], it has become customary to formulate theories of composite solid-propellant burning in terms of mixed models involving coupled diffusion processes and chemical reactions. The more recent studies originated with Hermance (1966) and have been pursued especially by Beckstead and his associates (see Beckstead 1981). All of these analyses of composite propellant burning utilize the classical laminar diffusion flame model of Burke and Schumann for the estimation of flame heights or flame stand-off distances. The approach has remained largely heuristic in the manner of Summerfield, although improved statistical techniques have been introduced for both time and space averaging over heterogeneous particles. Ignition delay times are typically used to characterize reaction rates. Because of the heuristic introduction of estimates for important flame parameters, it appears inappropriate to label the recent developments as theories. The last rigorous attempt at constructing a theory from first principles dates back to the sandwich burner model of Nachbar and Johnson, although this approach has not been fruitful for extension to statistical distributions of oxidizer particles in a fuel matrix. A heuristic approach has been followed because no one has as yet succeeded in constructing a tractable, preferably analytical model in which diffusion and reaction processes are coupled. Using our results on diffusion flames, the following applications should be possible:

1) The generalized analytical procedure for describing laminar diffusion flames should be extended to arbitrary chemical reaction processes occurring either in the oxidizer stream or in the fuel stream or in both.

2) The coupled laminar diffusion and reaction flame is solved by utilizing the special assumptions of the

Burke-Schumann model to relate the gas flames (with reactions in the oxidizer stream or in the fuel stream or in both) to the diffusion flame. The result should be a representation of all aspects of these mixed diffusion-reaction flames. In particular, diffusion-flame shapes and flame heights can be determined as functions of the chemical processes occurring in the fuel and/or oxidizer streams.

3) Utilizing the results of the analysis described under 2, in conjunction with refined statistical procedures that are related to those used in recently published studies on composite propellant burning, an improved theoretical model may then be developed for composite propellant burning.

References

Barr, J. (1953) Diffusion flames. Fourth Symposium (International) on Combustion, pp. 765-771. Williams and Wilkins, Baltimore, Md.

Beckstead, M. W. (1981) A model for solid propellant combustion. Eighteenth Symposium (International) on Combustion, pp. 175-185. The Combustion Institute, Pittsburgh, Pa.; (see also earlier publications cited in this article).

Burke, S. P. and Schumann, T. E. W. (1928) Diffusion flames. Ind. Eng. Chem. 20, 998-1004.

Hermance, C. E. (1966) A model of composite propellant combustion including surface heterogeneity and heat generation. AIAA J. 4, 1629-1637.

Mitchell, R. E., Sarofim, A. F., and Clomburg, L. A. (1980) Experimental and numerical investigation of confined laminar diffusion flames. Combustion and Flame 37, 227-244.

Penner, S. S. (1957) Chemistry Problems in Jet Propulsion, (a) pp. 273-275, and (b) pp. 235-238. Pergamon Press, New York.

Penner, S. S. (1962) Chemical Rocket Propulsion and Combustion Research, Chap. 4. Gordon and Breach, New York.

Penner, S. S., Bahadori, M. Y., and Kennedy, E. M. (1983) Laminar diffusion flames with cylindrical symmetry and arbitrary values of diffusion coefficients and inlet velocities.

Energy Center, University of California, San Diego, La Jolla, June 1983; see also the Ph.D. Thesis of M. Y. Bahadori, University of California, San Diego, June 1984.

Penner, S. S. and Mullins, B. P. (1959) *Explosions, Detonations, Flammability and Ignition*, AGARDograph No. 31, Part I, pp. 9-15. Pergamon Press, New York.

Penner, S. S. and Sherman, S. (1947) Heat flow through composite cylinders. *J. Chem. Phys.* 15, 569-574.

Shvab, V. A. (1944) Thesis, Leningrad; (1948) *Gos. Energ. izd.*, Moscow-Leningrad.

Zel'dovich, Y. B. (1949) On the theory of combustion of initially unmixed systems. *Zhur. tekhn. fiz.* 19, 1199-1210; (1950) English translation in *N.A.C.A. Technical Memorandum No. 1296*, Washington, D.C.

Transition and Transport in the Initial Region of a Turbulent Diffusion Flame

A. R. Masri,* S. H. Stårner,† and R. W. Bilger‡
The University of Sydney, Sydney, Australia

Abstract

The initial region of a diffusion flame of 2 H_2/N_2 fuel issuing from a round jet into a coflowing stream has been studied at Reynolds numbers from 5000 to 30,000 using shadowgraphs and LDV. It is found that in addition to the (velocity) shear layer there is a scalar layer whose maximum gradient lies completely outside the shear layer. The shear layer undergoes transition by means of the classical Kelvin-Helmholtz instabilities which roll up, pair, and produce three-dimensional turbulence. On the other hand, a completely separate transition of this outer scalar layer can occur without disturbing the transitional process in the inner shear layer. This outer layer transition is thought to be caused by transition of the boundary layer on the outside of the nozzle. Measurements by LDV, although hampered by thermophoretic effects on the seed, confirm that the velocity turbulence is confined to the inner layer when the outer appears laminar. It is also found that the Reynolds stress in the "coherent structures" region closely follows the mean velocity gradient. The results are found to be insensitive to the N_2/H_2 ratio of the fuel and similar phenomena can be seen in published photographs of hydrocarbon diffusion flames. It is concluded that molecular transport

---

Presented at the 9th ICODERS, Poitiers, France, July 3-8, 1983. Copyright © American Institute of Aeronautics and Astronautics, Inc., 1984. All rights reserved.
  *Postgraduate Research Student, Department of Mechanical Engineering.
  †Research Fellow, Department of Mechanical Engineering.
  ‡Professor of Mechanical Engineering, Department of Mechanical Engineering.

effects are of great importance in the initial region even at high initial Reynolds number.

Introduction

The initial region, say up to ten diameters, of burning fuel jets is a region which is interesting for several reasons: It is here that the flame is stabilized either at the nozzle lip or as a lifted flame, the mechanism of this stabilization as yet not being clear; it is here that transition from laminar to turbulent flow occurs with a Reynolds number that apparently depends on fuel type (Hottel and Hawthorne, 1949); and it is also here that the "coherent structures" of much contemporary interest occur most distinctly.

Although experimental evidence (Chigier and Yule 1979 a,b) suggests that the coherent structures control the development of the initial region of the flame, little is known about the mechanism of transition and the role of such structures in turbulent transport.

Only recently has attention been drawn to the fact that there exist two concentric regimes near the nozzle in jet diffusion flames, an inner cold fuel jet surrounded by a flame envelope. Takahashi et al. (1982) in working with buoyant H_2 and H_2/N_2 flames into still air show that the transition distances to turbulence onset for these two layers are not equal, and argue that the outer laminar layer is broken down by the spread of turbulent jet fluid once the jet transition has occurred. In this paper we study jet flames in a co-flowing airstream and demonstrate that the transition to turbulence can occur independently for each layer.

Mean velocity, mean temperature, turbulence, and shear stress data are obtained and presented together with shadowgraphs of the far upstream flame region.

Experimental

In the present study, fuel jets of mixed hydrogen and nitrogen issued horizontally from a contoured and thin-lipped nozzle into a coflowing stream of air. The results have been obtained for mixtures with a hydrogen to nitrogen ratio of 2 to 1 by volume. The nozzle diameter $\bar{D_j}$ was 15.2 mm and the mean jet velocity $\bar{u_j}$ was varied from 8.6 to 51.9 m/s giving jet Reynolds numbers of 5000 - 30,000 (overbars denote time-averaging). The external flow velocity $\bar{u_e}$ was varied from 1.7 to 17.3 m/s and was adjusted so that jet to freestream velocity ratios of 10, 5, and 2 to 1 were obtained. The internal profile of the nozzle was such that a thin laminar boundary layer was produced at the end of a 10 to 1 overall contraction. The external boundary was relatively thick and changed from laminar to turbulent over the velocity range quoted above. The wind tunnel is that used by Kent and Bilger (1973) and has a freestream turbulence level of about 0.2%.

Visualization of the flow was produced by a shadowgraph system using a 150 mm diameter spherical mirror of 1524 mm focal length, a "Strobotac" light source with a flash duration of 3 μs, a ground glass screen, and a Mamiya RB67 single lens reflex camera. LDV measurements were made using the two-color system described earlier (Stårner 1983). To avoid noise problems with the Bragg Cell shift system, the two colors were used at 0- and 45- deg alignment to the flow direction and instantaneous u (axial) and v (transverse) velocities deduced by computer processing of the records from the two separate counter-processors. Both the fuel stream and the near-nozzle part of the airstream were seeded with fine (0.3 μm) Al_2O_3 and MgO particles, respectively.

Results and Discussion

Figures 1 and 2 show shadowgraphs of flames at a velocity ratio of 5/1 and Reynolds numbers of 10,000 and 20,000, respectively. It is seen that the initial region of both flames is characterized by a double layer structure.

The fuel issuing from the nozzle produces a laminar shear layer with a natural instability which grows to produce vortex rings further downstream. These vortex rings grow, pair, and break up into three-dimensional turbulence towards the end of the potential core. This description is

Table 1 Vortex frequency and convection velocity for the flames of Figs. 1 and 2

Re	x/D_j	$\bar{u}_j$ (m/s)	$\bar{u}_e$ (m/s)	Strouhal number St	Vortex frequency f	$\dfrac{\bar{u}_c - \bar{u}_e}{\bar{u}_j - \bar{u}_e}$
10,000	4.9	17.3	3.5	1.2	850	0.45
10,000	7.3	17.3	3.5	0.59	400	0.38
20,000	3.0	34.6	6.9	2.2	3000	0.62
20,000	4.7	34.6	6.9	1.1	1500	0.7
20,000	6.0	34.6	6.9	0.64	870	0.6

similar to that given by List (1982) for axisymmetric turbulent jets.

The first vortex pairing occurs at x/D = 6 to 7 in the flame of Fig. 1 and at x/D = 4 in that of Fig. 2. Power spectra of the streamwise velocity component, taken for both flames at various radial and axial positions (for full details, see Masri (1982), give the vortex frequency f, taken as the frequency of the peak spectral density. The results shown in Table 1 indicate that vortex pairing is indeed taking place.

The Strouhal number shown in Table 1 is defined as $St = f\, D_j / \tfrac{1}{2}(\bar{u}_j + \bar{u}_e)$. By measurements of the distance ℓ between adjacent vortices on the shadowgraphs, an estimate of the convection velocity $\bar{u}_c = f\, \ell$ has been obtained. Yule et al. (1981) found that for a rich premixed flame of Re = 10,000 at x/D = 8, $(\bar{u}_c - \bar{u}_e)/(\bar{u}_o - \bar{u}_e) = 0.3$, where $\bar{u}_o$ is the centerline mean velocity. Their Strouhal number was about 0.1, an order of magnitude lower than for the present results. The normalized convection velocities of Table 1 are comparable to the values 0.6-0.7 which have been measured by Hussain (1980) in the shear layers of cold round jets, using hot wire anemometers.

Shadowgraphs of the full range of flames with Reynolds numbers from 5000 to 30,000 show that the position of the first vortex pairing is shifted downstream by a decrease in the jet velocity $\bar{u}_j$. The trend is consistent over the whole range of flames.

The external layer, which we will call the scalar layer, remains laminar up to x/D = 16 in Fig. 1 and x/D - 5 in Fig.

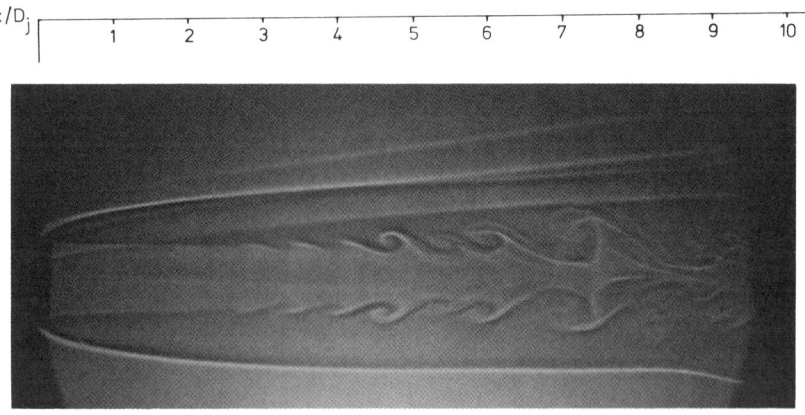

Fig. 1 Shadowgraph of $2H_2/N_2$ diffusion flame, Re = 10,000, $\bar{u}_j/\bar{u}_e = 5/1$, $\bar{u}_j = 17.3$ m/s, $\bar{u}_e = 3.5$ m/s.

TRANSITION AND TRANSPORT IN DIFFUSION FLAMES   297

2. It is clear from these figures, and from many other shadowgraphs not shown here, that the scalar layer will remain laminar until broken up by the more rapidly spreading turbulent inner layer. By contrast, the flame of Fig. 3 (Re = 20,000, $\bar{u}_j/\bar{u}_e$ = 2/1) exhibits a separate, early transition of the scalar layer. The only difference between the flames of Figs. 2 and 3 is in the external velocity; 6.9 m/s in Fig. 2 and 17.3 m/s in Fig. 3. It appears that the increased external velocity causes transition of the external nozzle boundary layer. The turbulence that develops shows S-shaped streaks and lacks circumferential coherence.

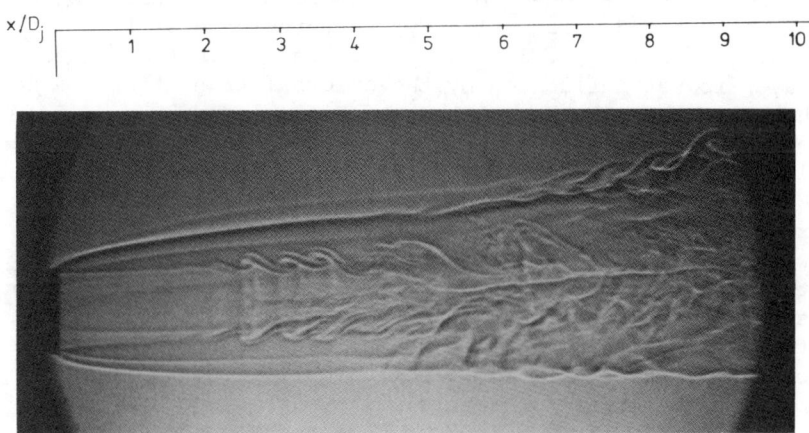

Fig. 2 Shadowgraph of $2H_2/N_2$ diffusion flame, Re = 20,000, $\bar{u}_j/\bar{u}_e$ = 5/1, $\bar{u}_j$ = 34.6 m/s, $\bar{u}_e$ = 6.9 m/s.

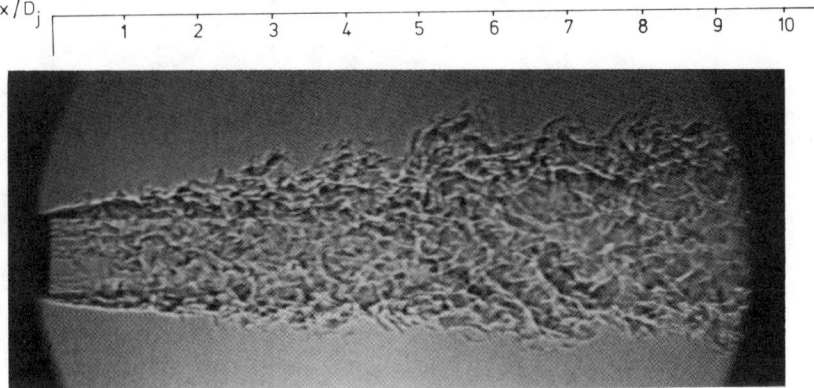

Fig. 3 Shadowgraph of $2H_2/N_2$ diffusion flame, Re = 20,000, $\bar{u}_j/\bar{u}_e$ = 2/1, $\bar{u}_j$ = 34.6 m/s, $\bar{u}_e$ = 17.3 m/s.

Figures 4 and 5 show the velocity and turbulence profiles of the flames in Figs. 1, 2, and 3 at x/D = 4.9, 4.0, and 4.7, respectively. The radial coordinate is normalized by the nozzle radius a. The radial position of the edge of the scalar layer, as marked on the velocity profiles of Fig. 4, indicates that the edge of this layer is advected downstream at the external velocity $\bar{u}_e$. The turbulence profiles of Fig. 5 show that the velocity turbulence is confined to the inner (velocity) layer when the outer layer remains laminar. The scatter in the data points at the outer radial positions is due to LDV fringe bias error and to thermophoretic effects on the seeding density. The sketched lines represent the expected profiles where these sources of error are eliminated. The velocity and turbulence profiles of the flame shown in Fig. 3 at x/D = 4.0 reveal that the shear layer is not influenced by the external transition and behaves in the same way as that of Figs. 1 and 2.

Temperature profiles were obtained using an uncoated 13% Rh-Pt/Pt thermocouple with a bead diameter of 0.8 mm. Figure 6 shows results for the flames of Figs. 1 and 3 at x/D_j = 4.9 and 2.0, respectively. The results are presented without correction for any of the errors associated with thermocouple contamination or radiation losses. Figure 6 shows that the luminous flame lies slightly outside the mean temperature maximum. This is believed to be due to the catalytic effects of the free radicals. The radial distance between the mean temperature maximum which occurs at an equivalence ratio $\phi \simeq 1.07$ and the reaction zone which occurs at ϕ = 1.0 is estimated to be approximately 0.8 mm

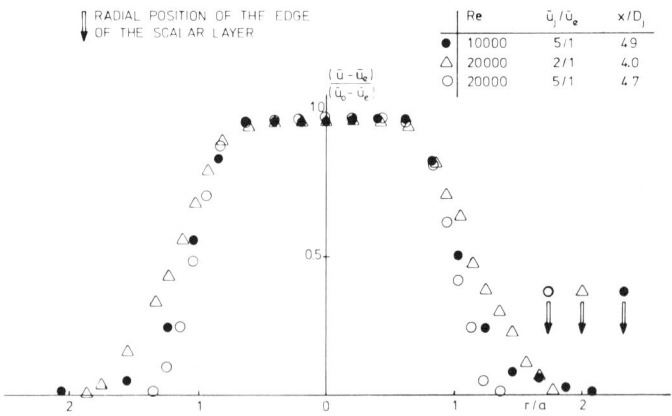

Fig. 4 Radial profiles of normalized mean excess velocity $(\bar{u} - \bar{u}_e)/(\bar{u}_o - \bar{u}_e)$. The radial coordinate is normalized by the jet radius a.

for $x/D_j < 12$. Further out radially, the edge of the thermal layer coincides with the edge of the scalar layer as determined from the shadowgraphs. This is as expected, since the lines on the shadowgraphs represent the rate of change of density gradient. Figures 7 and 8 relate the mean temperature, mean velocity, turbulence intensity, and shear stress profiles to the shadowgraph pictures of the flames of Figs. 1 and 3 at $x/D = 4.9$ and 2.0, respectively. It is obvious that maximum turbulence and shear stress occur in the coherent structures region, and that the peak of the thermal layer lies well outside the shear layer.

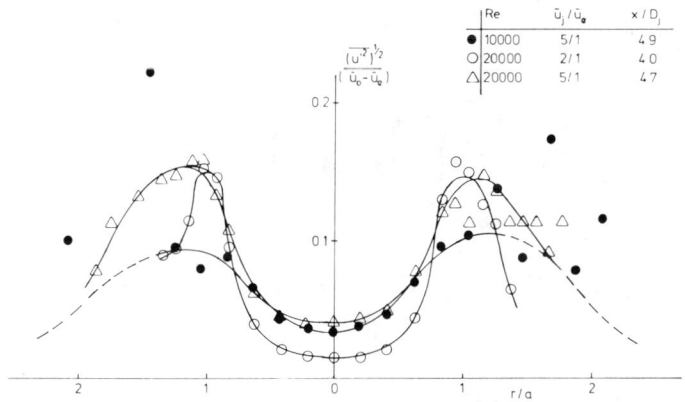

Fig. 5 Radial turbulence intensity profiles $(\overline{u'^2})^{\frac{1}{2}}/(\bar{u}_o - \bar{u}_e)$.

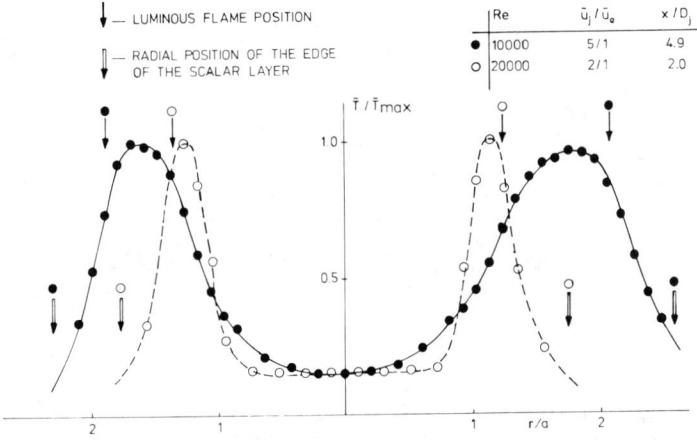

Fig. 6 Normalized radial temperature profiles $\bar{T}/\bar{T}_{max}$.

Figure 9 shows the shear stress $\overline{u'v'}$ and the corresponding velocity gradient $\partial \overline{u}/\partial r$, both normalized by their maximum values, for the flame of Fig. 2 (Re = 20,000, $\overline{u}_j/\overline{u}_e$ = 5/1). The behavior exhibited by $(\overline{u'v'})/(\overline{u'v'})_{max}$ and $-(\partial \overline{u}/\partial r)/|\partial \overline{u}/\partial r|_{max}$ is quite similar except towards the scalar layer where the LDA measurements lack the accuracy obtained in the shear layer. This is an indication that the concept of gradient diffusion in turbulence modeling may yield reasonable results even this far upstream, although no sound theoretical basis exists for its application to the early development of turbulence.

The foregoing indicates that the initial region of the turbulent jet diffusion flame contains two layers which

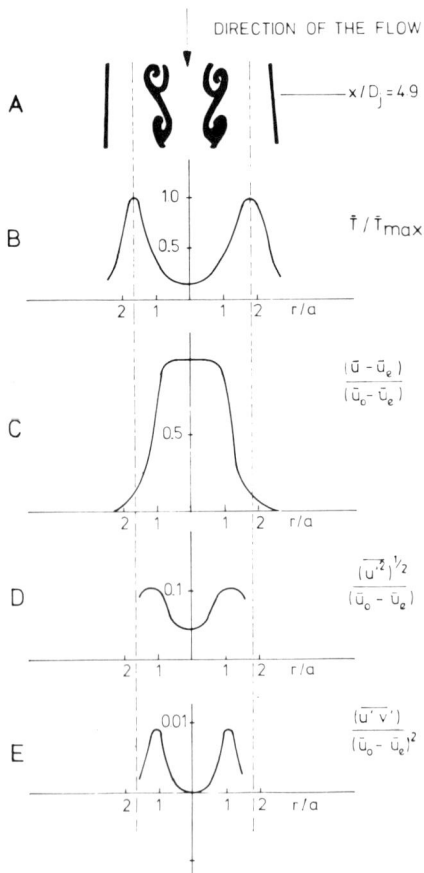

Fig. 7a) Re = 10,000, $\overline{u}_j/\overline{u}_e$ = 5/1, x/D = 4.9; b) radial normalized temperature profiles; c) radial profiles of normalized mean excess velocity; d) radial turbulence intensity profiles; and e) radial normalized shear stress profiles.

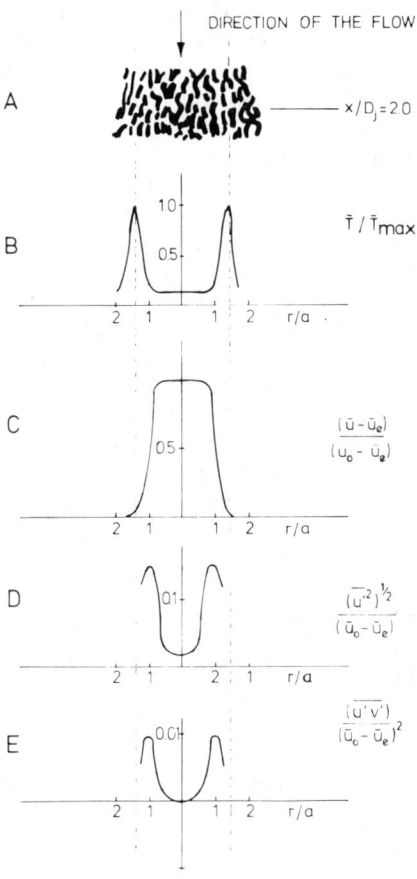

Fig. 8a) Re = 20,000, $\bar{u}_j/\bar{u}_e$ = 2/1, x/D = 2.0; b) radial normalized temperature profiles; c) radial profiles of normalized mean excess velocity; d) radial turbulence intensity profiles; and e) radial normalized shear stress profiles.

undergo transition by means of separate instability mechanisms. The radial species diffusion in the initial region of these flames is clearly much larger than the momentum diffusion, leading to the double layer structure; a feature not seen in cold or slightly heated jets. The laminar fuel element diffusivity in a hydrogen flame rises to over 20 cm²/s in the hottest part of the flame, and the Schmidt number for molecular hydrogen in nitrogen is of order 0.1. For hydrocarbon flames, such as methane-air and acetylene-air, the corresponding figures are around 8 cm²/s and 0.5, so one would expect a less pronounced double layer structure. However, shadowgraphs of an acetylene flame (C_2H_2/

$2N_2$, Re = 10,000, $\bar{u}_j/\bar{u}_e$ = 2/1) studied in this work show only minor differences from the hydrogen flame of Fig. 1. The reader may also consult Wohl et al. (1949) for a shadowgraph of a city gas diffusion flame showing the same features.

A contributing reason for the high ratio of species to momentum diffusion near the nozzle lip in these flames is that this is where the species gradient is steepest, while the coflowing stream produces a boundary layer with a positive velocity gradient which in part cancels the negative $\partial\bar{u}/\partial r$ of the fuel jet, hence reducing the radial momentum diffusion in the near wake of the nozzle lip.

The jet velocity profile at the nozzle will obviously affect the momentum diffusion in the first few diameters. In the present work we have a near-laminar "top-hat" profile produced by the 10/1 contraction just upstream of the nozzle. The fully developed turbulent pipe flow profile often used in diffusion flame experiments would certainly result in different initial momentum transfer, and could be expected to influence the jet transition. There are also other unanswered questions in this complex topic, such as the influences of nozzle diameter and boundary-layer thickness, which we have omitted from this early study, but which should prove worthwhile subjects for more detailed work.

Conclusions

Regardless of the type of fuel used, the initial region of a turbulent jet diffusion flame is found to have a double

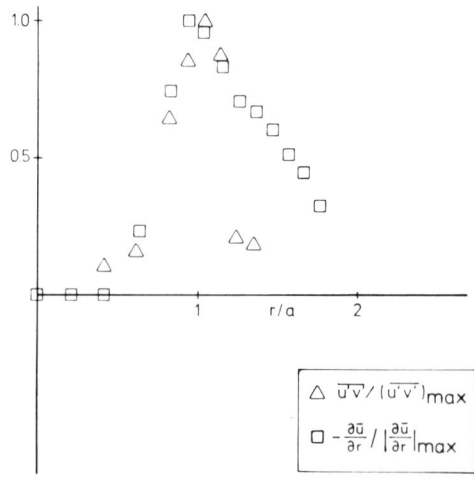

Fig. 9 Radial profile of Reynolds stress; Re = 20,000, $\bar{u}_j/\bar{u}_e$ = 5/1, x/D_j = 4.7.

layer structure. As with cold jets, the inner (velocity) layer becomes turbulent by the formation of ring vortices which pair, break up, and lose coherence. The outer (scalar) layer is formed by high molecular diffusivities and steep gradients of the fuel species combined with the momentum deficit in the immediate wake of the nozzle lip. For high external air velocity, the scalar layer independently becomes turbulent close to the nozzle, apparently by a transition of the external boundary layer; this process does not show the clear vortex features of the velocity shear layer. Velocity and temperature profiles show that the reaction zone is well outside the shear layer.

Measured shear stresses in the ring-vortex region have radial profiles similar to those of the mean axial velocity gradient, implying that gradient transport modeling has some experimental, if not theoretical, rationale even in this region.

Acknowledgment

This work has been supported by a grant from the Australian Research Grants Scheme.

References

Chigier, N.A. and Yule, A.J. (1979a) The physical structure of turbulent flames. AIAA Pper 79-0217, Seventeenth Aerospace Sciences Meeting, New Orleans, La.

Chigier, N.A. and Yule, A.J. (1979b) The structure of eddies in turbulent flames-I, Project SQUID. Technical Report US-1-PU Purdue University, West Lafayette, Ind.

Hottel, H.C. and Hawthorne, W.R. (1949) Diffusion in laminar flame jets. Third Symposium on Combustion, Flame and Explosion Phenomena, 254-265. Williams and Wilkins, Baltimore, Md.

Hussain, A.K.M.F. (1980) Coherent structures and studies of perturbated and unperturbated jets. The Role of Coherent Structures in Modeling Turbulence and Mixing, Lecture Notes in Physics No. 136, 252-291. Springer-Verlag, New York.

Kent, J.H. and Bilger, R.W. (1973) Turbulent diffusion flames. Fourteenth Symposium (International) on Combustion, 615-625. The Combustion Institute, Pittsburgh, Pa.

List, E.J. (1982) Turbulent jets and plumes. Annual Review of Fluid Mechanics 14, 189-212.

Masri, A.R. (1982) The structure of the initial region of turbulent diffusion flames. B.E. Thesis, The University of Sydney, Australia.

Stårner, S.H. (1983) Joint measurements of radial velocity and scalars in a turbulent diffusion flame. <u>Combust. Sci. and Technol.</u> 30, 145-170.

Takahashi, F., Mizomoto, M., and Ikai, S. (1982) Transition from laminar to turbulent free jet diffusion flames. <u>Combustion and Flame</u> 48, 85-95.

Wohl, K. Gazley, C., and Kapp, N. (1949) Diffusion flames. <u>Third Symposium on Combustion, Flames and Explosion Phenomena</u>, P. 288. Williams & Wilkins, Baltimore, Md.

Yule, A J., Ventura, J. M. P., and Chigier, N. A. (1981) On large eddy structure and turbulent mixing in flames. <u>Seventh Biennial Symposium on Turbulence</u>, University of Missouri-Rolla, Mo.

Predicted Structure of Stretched and Unstretched Methane-Air Diffusion Flames

S. K. Liew* and J. B. Moss†
Cranfield Institute of Technology, Bedford, England
and
K.N.C. Bray‡
University of Southampton, Southampton, England

Abstract

The incorporation of realistic chemical kinetics into the prediction of turbulent nonpremixed hydrocarbon-air combustion is a crucial element in much of the present model development. The interpretation of turbulent burning as being comprised of an ensemble of laminar flamelets, which in themselves present more tractable analytical problems, has gained a wide currency. As part of such an approach, this paper reports numerical solutions for both unstretched and stretched nonpremixed, laminar, methane-air flames in a time-dependent, one-dimensional configuration. The unstretched flame structure, supported by a chemical kinetic mechanism of 38 reactions, is substantially time-independent and compares favorably with detailed measurements in the literature. While these solutions can be readily incorporated into a turbulent flowfield calculation employing the conserved scalar approach, many practical combustors may involve such severe stretching of the flame zone that the local structure is significantly influenced, perhaps to the point of extinction. Therefore time-dependent stretching of the flame is also introduced which permits these alternative thermochemical structures to be investigated. The effect of local stretching is characterized by the property χ ($\equiv 2D(\partial \xi/\partial x_k)^2$), whose time-averaged value in the turbulent flame is the rate of scalar dissipation. A criterion, based on χ, is identified which distinguishes critical changes in flame structure. This is compared with

---

Presented at the 9th ICODERS, Poitiers, France, July 3-8, 1983. Copyright © American Institute of Aeronautics and Astronautics, Inc. 1984. All rights reserved.
*Research Fellow, School of Mechanical Engineering.
†Professor, School of Mechanical Engineering.
‡Professor, Department of Aeronautics and Astronautics.

extinction criteria derived from alternative approaches and its incorporation into turbulent flowfield calculation is outlined.

Introduction

The incorporation of realistic chemical kinetics into the prediction of turbulent nonpremixed hydrocarbon-air combustion is a crucial element in much of the present model development. The interpretation of turbulent burning as being comprised of an ensemble of laminar flamelets, which in themselves present more tractable analytical problems, has gained a wide currency. Liew et al.(1981) have shown, for example, that dramatic improvements in the prediction of CO levels in turbulent open nonpremixed methane-air flames result from the introduction of an undisturbed laminar flamelet as the microscopic element in the ensemble, rather than the commonly encountered assumption of complete chemical equilibrium. Such an approach also encourages an interest in the structure of strained laminar flames and in conditions perhaps leading eventually to extinction. These features might be expected to play an important role in circumstances of turbulent burning less benign than the open flame.

The most common method for the prediction of stretched laminar flames involves analytical techniques based on large activation energy asymptotic expansions. This technique produces analytical results for the entire range of Damkohler number, leading to the well-known S-curve (Linán 1974). An important criticism of these techniques applied to hydrocarbon flames is that the mathematical complexity involved restricts the chemistry description to a single-step irreversible reaction.

Peters and Williams (1982) have recently reported an extinction criterion derived from the measurements of Tsuji and Yamaoka (1970). Some uncertainty arises in this extinction criterion through their assumption of constant flame properties and, more importantly, indications in the experiment of significant heat loss to the fuel-sphere in the counterflow configuration employed.

This paper will describe a different strategy to that outlined above. It will be assumed that the effects of stretch are consequent on the decrease of flame thickness and the time-dependent structure of a one-dimensional flamelet in a simple straining flow will be investigated. The formulation in part parallels that of Eickhoff and Grethe (1981), although the interpretation of the results in terms of a local scalar dissipation rate will be emphasized and a more extensive chemical reaction mechanism to describe methane-air combustion will be employed. The implications of incorporating the structure of a stretched laminar flame into models of turbulent nonpremixed flames are briefly addressed. These strained flame solutions are used to identify a criterion for modifying the turbulent flame model based on undisturbed flamelets, and this criterion is contrasted to the conventional extinction criterion from the steady-state solutions which are described by the S-curve for T_{max} vs Damkohler Number.

Formulation of Problem

In premixed flame studies, Klimov (1963) has shown that the one-dimensional, time-dependent flame structure under stretch $\gamma(t)$ may be described by the equation :

$$\frac{\partial \phi}{\partial t} - \gamma(t) \frac{x \partial \phi}{\partial x} + \frac{u \partial \phi}{\partial x} - \frac{\partial}{\partial x}\left(\frac{\nu \partial \phi}{\partial x}\right) = \frac{1}{t_c} F(\phi) \qquad (1)$$

where ϕ can be viewed either as a reactant mass fraction or as a nondimensional temperature and $F(\phi)$ is a nondimensional reaction rate function, ν is the diffusivity, and t_c is a characteristic chemical time.

Williams (1975) recognized that a similar formulation is applicable to nonpremixed flames. In this context Eq.(1) reduces simply to a form of the diffusive-reactive equation if the convective flux is neglected. This prescription reproduces the moving observer feature of the ESCIMO model proposed by Spalding (1978). ESCIMO envisages the combustion domain as being occupied by interleaved folds which are formed at different locations within the fuel-air layer.

The general form of the governing equation employed in ESCIMO and also used here is

$$\frac{\partial \phi}{\partial t} = \psi^{-2} \frac{\partial}{\partial \omega}\left(\rho \, \Gamma_\phi \frac{\partial \phi}{\partial \omega}\right) + d_\phi \qquad (2)$$

where ψ is a quantity, akin to the stream function, which is a function of time and ω is the nondimensional stream function; Γ_ϕ is the transport coefficient and d_ϕ is the source term.

Spalding (1978) supposed that the stretch rate R is constant throughout the life of the fold such that ψ varies with time in accordance with

$$\psi = \psi_0 \, e^{-Rt} \qquad (3)$$

where ψ_0 is the initial value of ψ.

Stretching of the flame is thus attained by a time-dependent decrease of the thickness y of the interface of the fold. The appropriate concentration and energy equations can be written in a form similar to Eq.(2) and so describe the transient one-dimensional laminar diffusion flame.

The transport properties for the various species are calculated from the appropriate Enskog-Chapman expressions (Hirschfelder et al. 1967). The mixture conductivity is obtained from Wilke's semiempirical formula (Reid et al.

Table 1 The chemical reaction mechanism

1.	M	+ CH_4		=	CH_3	+ H	+ M
2.	M	+ CHO		=	CO	+ H	+ M
3.	M	+ CO_2		=	CO	+ O	+ M
4.		CH_4	+ CH_2	=	CH_3	+ CH_3	
5.		CH_4	+ H	=	CH_3	+ H_2	
6.		CH_4	+ OH	=	CH_3	+ H_2O	
7.		CH_4	+ O	=	CH_3	+ OH	
8.		CH_3	+ OH	=	CH_2	+ H_2O	
9.		CH_3	+ O	=	CH_2O	+ H	
10.		CH_2	+ H	=	CH	+ H_2	
11.		CH_2	+ OH	=	CH	+ H_2O	
12.		CH_2	+ OH	=	CH_3	+ O	
13.		CH_2	+ OH	=	CH_2O	+ H	
14.		CH_2	+ H_2	=	CH_3	+ H	
15.		CH_2	+ O	=	CHO	+ H	
16.		CH_2	+ O_2	=	CH_2O	+ O	
17.		CH	+ CO_2	=	CHO	+ CO	
18.		CH	+ OH	=	CHO	+ H	
19.		CH	+ O	=	CO	+ H	
20.		CH	+ O_2	=	CHO	+ O	
21.		CH_2O	+ H	=	CHO	+ H_2	
22.		CH_2O	+ OH	=	CHO	+ H_2O	
23.		CH_2O	+ O	=	CHO	+ OH	
24.		CHO	+ HO_2	=	CH_2O	+ O_2	
25.		CHO	+ O	=	CO	+ OH	
26.		CO	+ OH	=	CO_2	+ H	
27.		CO	+ O_2	=	CO_2	+ O	
28.	M	+ H_2		=	H	+ H	+ M
29.	M	+ OH	+ O	=	HO_2		+ M
30.	M	+ H	+ O_2	=	HO_2		+ M
31.	M	+ H_2O		=	OH	+ H	+ M
32.		H	+ OH	=	H_2	+ O	
33.		H	+ HO_2	=	OH	+ OH	
34.		OH	+ H_2	=	H	+ H_2O	
35.		OH	+ OH	=	H_2O	+ O	
36.		OH	+ O	=	H	+ O_2	
37.		HO_2	+ H	=	O_2	+ H_2	
38.		HO_2	+ O	=	OH	+ O_2	

Table 2 Values of A_j, N_j, and T_j for forward and reverse reactions

	A_j Forward rate	N_j	T_j	A_{-j} Backward rate	N_{-j}	T_{-j}
1.	1.000E+14	0	4.465E+04	2.900E+05	1.000E+00	-9.344E+03
2.	2.500E+17	-1.500E+00	8.485E+03	1.100E+11	-5.000E-01	-6.040E+02
3.	3.800E+27	-4.000E+00	6.769E+04	3.800E+18	-3.000E+00	3.116E+03
4.	1.300E+09	7.000E-01	1.010E+04	2.300E+05	7.000E-01	1.223E+04
5.	5.000E+07	1.000E+00	5.051E+03	1.400E+06	1.000E+00	4.455E+03
6.	3.000E+10	0	2.525E+03	4.100E+09	0	9.713E+03
7.	1.900E+11	0	5.909E+03	2.400E+09	0	4.288E+03
8.	2.000E+08	7.000E-01	1.010E+03	1.500E+08	7.000E-01	6.069E+03
9.	2.600E+11	0	1.010E+03	4.100E+12	0	3.551E+04
10.	3.200E+08	7.000E-01	2.525E+03	1.800E+08	7.000E-01	3.806E+03
11.	5.000E+08	5.000E-01	3.030E+03	1.400E+09	5.000E-01	1.209E+04
12.	5.000E+08	5.000E-01	3.030E+03	7.400E+09	5.000E-01	6.781E+03
13.	1.000E+10	0	2.525E+03	2.300E+12	0	4.078E+04
14.	3.000E+09	0	3.535E+03	2.000E+10	0	6.260E+03
15.	5.000E+08	5.000E-01	2.020E+03	2.900E+09	5.000E-01	4.782E+04
16.	5.000E+08	5.000E-01	3.535E+03	7.200E+09	5.000E-01	3.334E+04
17.	1.000E+07	5.000E-01	3.030E+03	9.500E+05	5.000E-01	3.637E+04
18.	1.000E+07	5.000E-01	5.051E+03	2.300E+08	5.000E-01	5.059E+04
19.	5.000E+08	5.000E-01	0	2.300E+10	5.000E-01	8.882E+04

(Table continued on next page)

Table 2 (continued) Values of A_j, N_j, and T_j for forward and reverse reactions

	A_j	Forward rate N_j	T_j	A_{-j}	Backward rate N_{-j}	T_{-j}
20.	5.000E+08	5.000E-01	3.030E+03	6.900E+08	5.000E-01	4.012E+04
21.	1.300E+07	1.000E+00	1.616E+03	6.900E+05	1.000E+00	1.019E+04
22.	3.200E+07	1.000E+00	0	8.500E+06	1.000E+00	1.635E+04
23.	2.000E+08	1.000E+00	2.222E+03	5.000E+06	1.000E+00	9.768E+03
24.	1.000E+11	0	1.515E+03	4.300E+12	0	2.183E+04
25.	6.300E+08	1.000E+00	2.525E+02	1.300E+09	1.000E+00	4.353E+04
26.	1.500E+04	1.300E+00	-3.864E+02	3.600E+06	1.300E+00	1.182E+04
27.	3.200E+09	0	2.525E+04	4.600E+10	0	2.900E+04
28.	2.100E+12	7.000E-02	5.244E+04	2.200E+05	1.070E+00	-9.556E+02
29.	1.600E+14	-1.500E+00	0	1.000E+22	-2.500E+00	3.296E+04
30.	3.000E+09	0	-5.051E+02	1.200E+16	-1.000E+00	2.400E+04
31.	2.300E+21	-2.000E+00	6.192E+04	4.900E+13	-1.000E+00	7.409E+02
32.	2.700E+14	-9.400E-01	7.419E+03	6.000E+14	-9.400E-01	8.444E+03
33.	2.500E+11	0	9.596E+02	1.600E+14	0	2.037E+04
34.	2.500E+10	0	2.626E+03	1.200E+11	0	1.041E+04
35.	6.000E+09	0	5.051E+02	6.500E+10	0	9.314E+03
36.	6.300E+08	5.000E-01	0	1.000E+10	5.000E-01	8.450E+01
37.	2.500E+10	0	3.535E+02	5.900E+10	0	2.924E+04
38.	5.000E+10	0	5.051E+02	5.300E+10	0	2.837E+04

1977). The mixture viscosity is calculated similarly. Thermodynamic properties are computed from JANNAF thermochemical tables (1971) and are expressed as polynomials in temperature.

The gas-phase reaction mechanism for mixtures of methane-oxygen-nitrogen reported by Hahn and Wendt (1981) has been slightly modified for the present study. The mechanism adopted here contains 38 elementary reactions for the 16 species with nitrogen considered inert (Tables 1,2). The only species and reactions that are in the original mechanism of Hahn and Wendt and which are ignored here involve the element nitrogen. Comparisons between experiment and prediction indicated that the best fit is found with the rate coefficients reported by Hahn and Wendt.

The numerical computation used here is a modified version of the CREK routine proposed by Pratt (1977). A set of Newton-Raphson correction equations is solved simultaneously and iteratively at each grid node until all the Newton-

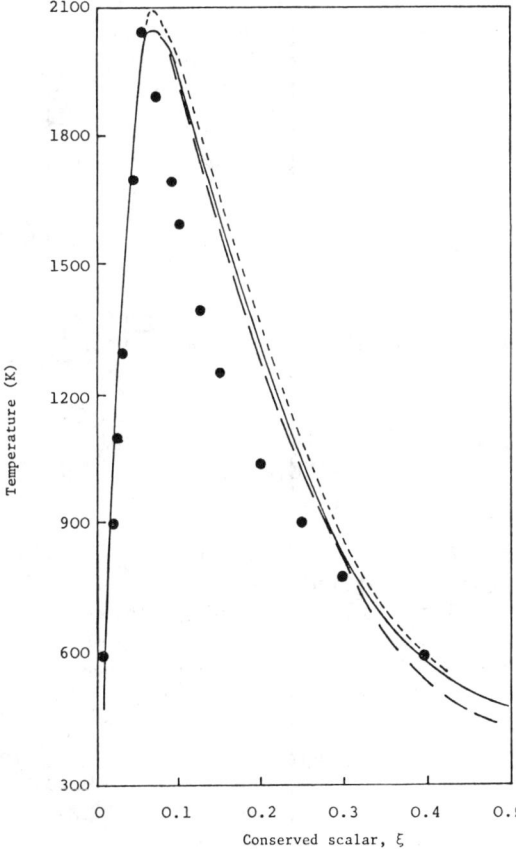

Fig. 1 Comparison between experimental steady-state temperature profile(● - Mitchell et al. 1980) and predictions of present model at different elapsed times, (– – –1, ——1.5, and ----2.5 ms).

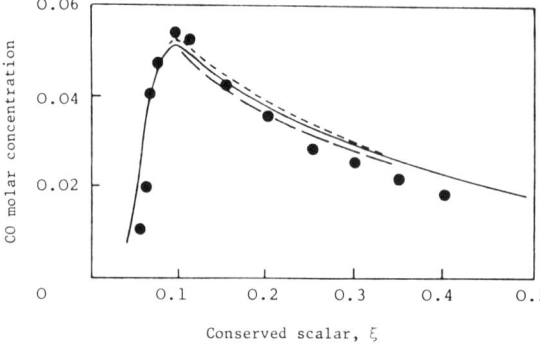

Fig. 2 Comparison between experimental steady-state CO concentration(● - Mitchell et al. 1980) and model predictions at different elasped times, (– – –1, ——1.5, and ----2.5 ms).

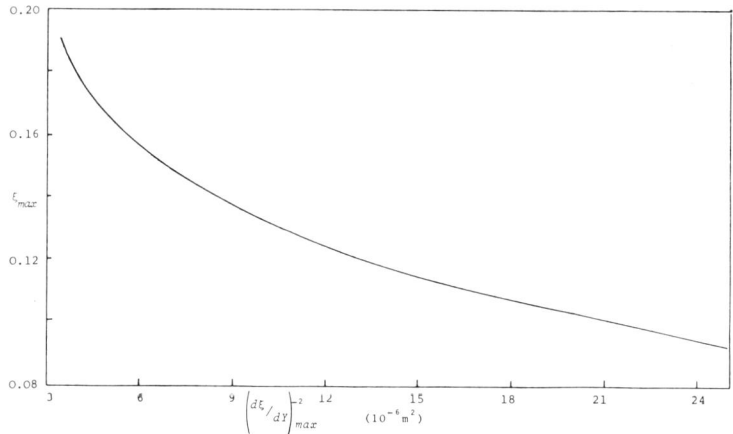

Fig.3 Conserved scalar as a function of the inverse of its gradient squared and evaluated at the maximum temperature.

Raphson correction variables are equal to some preset values. This ensures that all the species and enthalpy equations are satisfied point-by-point. In the present calculations a total of 40 radial nodes were used. These were found to be adequate to resolve the steepest gradients for any scalar in both undisturbed and stretched flames. The global time scale, $t = 10^{-5}$s, was used as the largest allowable time step for the desired accuracy. Further details of the numerical technique are reported elsewhere (Liew 1983).

The Unstretched Flame

In order to validate the predictions of undisturbed flame structure, comparisons are made with the measurements of Mitchell et al. (1980). The local mixture fraction (ξ) is selected as the coordinate against which to plot the results since this should be substantially independent of flame geometry and position in the flame.

Differential diffusion effects lead to slight differences in the fuel-lean domain between the mixture fractions constructed from the C, O, H, and N atomic elements. With the possible exception of H_2-air flames this effect is usually of minor importance and the average value of the mixture fraction based on the four chemical elements is adequate.

The calculations have been performed over a total time interval of 3 ms. Further marching in time is unnecessary since the domain of present interest lies around the early regions of the evolving flame. It is envisaged that the predicted flame structure will attain chemical equilibrium eventually as spatial gradients become small.

Figures 1 and 2 show the distributions of temperature and CO concentration. The predicted profiles are taken after elapsed times of 1, 1.5, and 2.5 ms. Good agreement in both peak height and width of profile can be observed between numerical predictions and experimental data. Other major reaction products and reactants which are not shown here are also in good agreement with experimental data.

The reaction zone concentrations of OH, H, and O are all broadly similar, falling in the range 2×10^{-3} - 8×10^{-3} mole fraction while that of CH_3, CH_2O, CHO, CH_2, and CH fall below 10^{-4} mole fraction. There are no experimental data available for comparison on these trace species. However, the orders of magnitude are consistent with the predictions using the NASA equilibrium program (Gordon and McBride 1971) and are also in agreement with the perturbation analysis of Melvin and Moss (1975).

The computed temperature and concentration profiles are found to be independent of time and provide an alternative to the measured profiles as a representation of the instantaneous thermochemical state in the turbulent nonpremixed flame of the earlier study (Liew et al. 1981).

The Stretched Laminar Flame

Calculations of stretched flame structure have been performed for various stretching rates in the range $1500 < R < 2000$ (s^{-1}). From these predictions it is found that the scalar profiles in conserved scalar space do not differ appreciably with R over this range. Before describing these results further, it is important to identify, as part of a wider prediction strategy, a suitable parameter

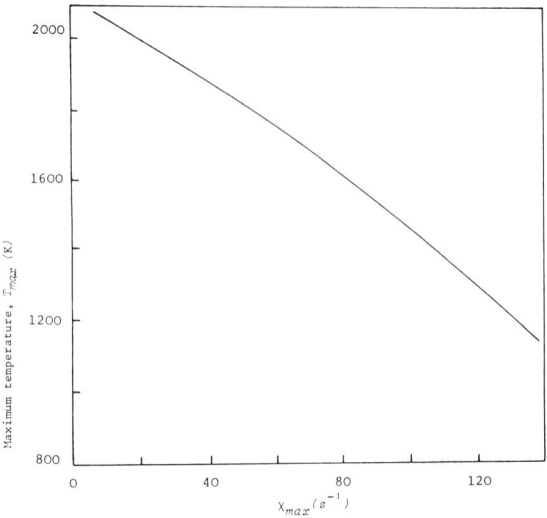

Fig.4 Peak temperature variation with local scalar dissipation rate.

to link the local value of flame stretch in the present calculations to some feature characterizing turbulent mixing. This parameter is evidently most usefully defined in terms of the local properties of the flame zone, since the key features of the turbulent flame are envisaged as depending critically on the interaction between chemistry and hydrodynamic stretch.

The above requirement is satisfied by defining the variation in terms of the instantaneous scalar dissipation rate which can be written as

$$\chi = 2D \left(\frac{\partial \xi}{\partial x_k}\right)^2 \quad (4)$$

where D is the mixture mass diffusion coefficient.

Using the large activation energy asymptotics approach to laminar flame structure, in mixture fraction space, Liñán (1974) and Peters (1980) have demonstrated an increase in reactant intermixing and displacement of the reaction zone which increases with stretch. The present numerical predictions, employing multichemistry, confirm these observations. Fig.3 shows the variation of $(d\xi/dy)^{-2}{}_{max}$, where the subscript max denotes evaluation at the maximum temperature. For weakly stretched flames $((d\xi/dy)^{-2}{}_{max} >> 1)$ ξ_{max} approaches asymptotically the undisturbed value (cf the stoichiometric value for equal diffusivities $\xi_s = 0.055$) while with increasing stretch, approaching the extinction regime discussed later, a significant increase in ξ_{max} is observed.

If the mixture mass diffusion coefficient in Eq.(4) is described approximately by the binary diffusion coefficient of CH_4 in excess N_2, the scalar dissipation rate χ can be evaluated. Fig. 4 shows the variation of peak temperature with χ_{max}. It is clear that the present time-varying flamelet prediction will produce a unique solution given the parabolic form of the describing equation and therefore will not reproduce the unstable burning branch which features in the multiple-valued S-curve. This must lead to some ambiguity in the precise identification of an extinction criterion analogous to the "knee" of the S-curve. Figs.5-7 illustrate profiles of temperature, CO and reactant concentrations, respectively, at three different stretching rates. The experimental data of Tsuji and Yamaoka (1971) exhibit qualitatively similar behavior in

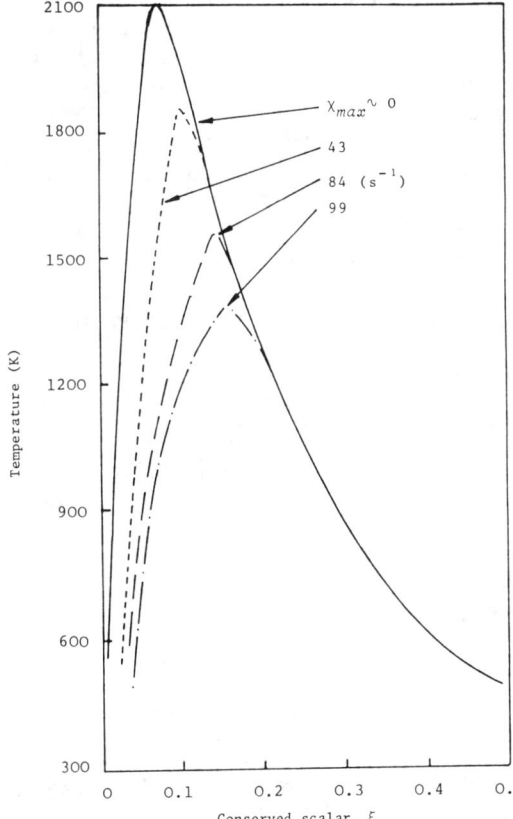

Fig. 5 Temperature profiles in conserved scalar space for different values of local scalar dissipation rate at elapsed time 2.5 ms.

respect of the influence of stretching rate on flame structure.

Extinction and Flame Stretch Criteria

While it is not the objective of this study to investigate laminar flame extinction, some aspects of the present calculations find parallels in extinction studies. Although stable solutions were only found for values of $\chi_{max} < 130$ s^{-1}, and a corresponding peak temperature of $T_{max} = 1200$ K, clearly this breakdown of the numerical solution cannot be confidently interpreted as extinction. The present solutions do however suggest that the limiting stretch must be of order $\chi_{max} \sim 100$ s^{-1}. Peters and Williams (1982), using the experimental data of Tsuji and Yamaoka (1970), deduced a value of χ at extinction of only 5 s^{-1}. The substantial difference between these values may be attributed both to the nonadiabatic nature of the experiment and also to their evaluation of the critical scalar dissipation rate at the fixed stoichiometric value and not at T_{max} which varies in mixture fraction space with stretch (see Fig.3). Their criterion is sensitive to this stoichiometric condition and evaluating χ at T_{max} (and ξ_{max}) in the same framework leads to a value of 32 s^{-1}. Given the inevitable uncertainty in the identification of precise extinction conditions these values are then broadly comparable.

It is clearly impractical in a turbulent flowfield calculation to average over all the different flame structures which arise as a result of variations in stretching rate. The present approach identifies a critical value of χ

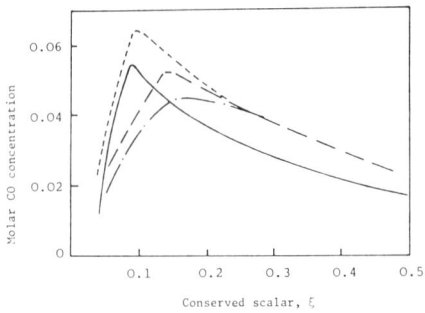

Fig.6 Variation of CO concentration profiles at different values of local scalar dissipation rate, (——— $\chi_{max} \sim 0$, ---- 43, - - - 84, - · - 99 s^{-1} at elapsed time 2.5 ms.

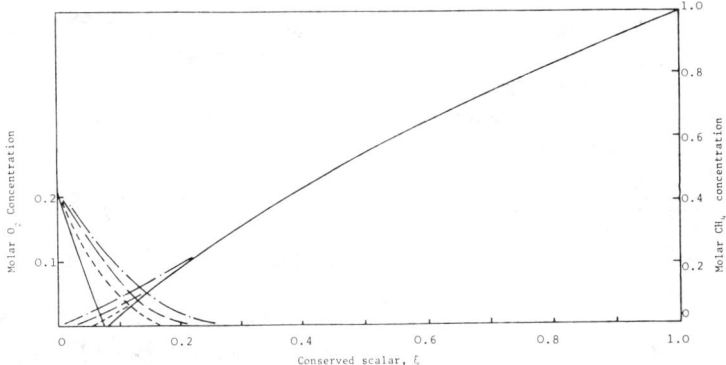

Fig.7 Variation of reactant concentrations at different values of local scalar dissipation rate, (—— $\chi_{max} \sim 0$, ---- 43, - - - 84, - · - 99 s^{-1} after elapsed time 2.5 ms.

as an approximate criterion which, when reached, implies that the undisturbed flamelet model (Liew et al.1981) is no longer a suitable representation of the local flame structure. A second model must then be invoked to provide an alternative microscopic description. In the extinction limit, the alternative forms might then be undisturbed flamelet burning and nonreactive mixing.

In less extreme circumstances, a limited range of stretched flame solutions might be incorporated. By way of illustration drawn from the present calculations, the undisturbed flamelet model may be supposed inappropriate for values of $\chi_{max} > 50$ s^{-1} and hence, from Fig.4, for those flamelets having peak temperatures, $T_{max} < 1800$ K. The flame structure predicted at $\chi_{max} = 50$ s^{-1} may then be used as the required alternative thermochemical model. In this way, several different criteria may be identified to effectively discretize the microscopic time-varying behavior under flamelet burning.

Present limitations on space do not permit a more detailed description of how these features are to be incorporated into a complete statistical model of turbulent burning. This aspect will be considered in a separate publication (see Liew 1983).

Conclusions

Numerical solutions of both stretched and unstretched nonpremixed laminar methane-air flames in a time-dependent, one-dimensional configuration are reported. The unstretched flame structure, supported by a realistic chemical kinetic mechanism, compares very favorably with detailed experiments reported in the literature. The structure of stretched flames plausibly reproduces important features of a limited range of comparable experiments.

The present study permits identification of an approximate criterion for the validity of the undisturbed flamelet representation of the local thermochemical state in a turbulent flame. A strategy is outlined by which stretched flamelets can be incorporated in a turbulent ensemble.

Acknowledgments

This work is supported by the Procurement Executive, Ministry of Defence (United Kingdom). The authors gratefully acknowledge access to computational facilities at Imperial College, London, provided by Professor D. Spalding.

References

Eickhoff, H. and Grethe, K. (1981) Investigations on a reaction model for turbulent diffusion flames, <u>Third International Symposium on Turbulent Shear Flow</u>, University of California, Davis, Ca.

Gordon, S. and McBride, B. (1971) Computer program for calculation of complex chemical equilibrium compositions, rocket performance, incident and reflected shocks, and Chapman-Jouguet detonations. NASA SP-273.

Hahn, W. and Wendt, J. (1981) NO_x formation in flat, laminar, opposed jet methane diffusion flames. <u>Eighteenth Symposium (International) on Combustion</u>, pp. 121-131. The Combustion Institute, Pittsburgh, Pa.

Hirschfelder, J., Curtis, C., and Bird, R. (1967) <u>Molecular Theory of Gases and Liquids</u>. John Wiley and Sons, New York.

Klimov, A. (1963) Laminar flame in a turbulent flow, <u>Zhur. Prikl. Mekh. I Tekhn. Fiz.</u> 3, 49-58.

Liew, S., Bray, K., and Moss, J. (1981) A flamelet model of turbulent non-premixed combustion. <u>Combustion Science and Technology</u> 27, 69-73.

Linãn, A. (1974) The asymptotic structure of counter-flow diffusion flames for large activation energies. <u>Acta Astronautica</u> 1, 1007-1039.

Melvin, A. and Moss, J. (1975) Structure in methane-oxygen diffusion flames. <u>Fifteenth Symposium (International) on Combustion</u>, pp. 625-636. The Combustion Institute, Pittsburgh, Pa.

Mitchell, R., Sarofin, A., and Clomburg, L. (1980) Experimental and numerical investigation of confined laminar diffusion flames. <u>Combustion and Flame</u> 37, 337-244.

Peters, N. (1980) Local quenching due to flame stretch and non-premixed turbulent combustion. Spring meeting, Western State Section, The Combustion Institute.

Peters, N. and Williams, F. (1982) Lift-off characteristics of turbulent jet diffusion flames. Aerospace Sciences Meeting, Orlando, Fla.

Pratt, D. (1977) Calculation of chemically reacting flows with complex chemistry. Studies in Convection, Vol. 2, (edited by B. E. Launder), Academic Press, New York.

Reid, R., Prausnitz, J., and Sherwood, T. (1977) The Properties of Gases and Liquids, 3rd Edition, McGraw-Hill Book Co., New York.

Spalding, D. (1978) Chemical reactions in turbulent fluids. Proceedings of Levich Birthday Conference, Advance Publications, Vol. 1, pp. 321-338.

Stull, D. and Prophet, H. (1971) JANNAF Thermochemical Tables. 2nd Edition, NSRDS-NBS37, U.S. Department of Commerce, Washington, D.C.

Tsuji, H. and Yamaoka, I. (1970) Structure analysis of counterflow diffusion flames in the forward stagnation region of a porous cylinder. Thirteenth Symposium (International) on Combustion, pp. 723-731. The Combustion Institute, Pittsburgh, Pa.

Williams, F. (1975) Recent advances in theoretical descriptions of turbulent diffusion flames. Turbulent Mixing in Non-Reactive and Reactive Flows, (edited by S. N. Murthy), Plenum Press, New York.

An Experimental Study of Turbulent Jet Diffusion Flames

O. K. Sønju*
Norwegian Institute of Technology, Trondheim, Norway
and
J. Hustad†
The Foundation for Scientific and Industrial Research
at the Norwegian Institute of Technology, Trondheim, Norway

Abstract

Experiments have been performed on turbulent flames of methane and propane, issuing from jets into quiescent air at atmospheric pressure and temperature for both small-scale laboratory flames and larger flames during field tests. Release tube diameters were between 2 and 80 mm, and jet exit Mach numbers were between 0.03 and unity, resulting in calculated heat release rates of up to 7.5 MW and, in observed flame heights, up to 8 m. Measurements included mean temperature profiles, mean concentration profiles for O_2, CO, and CO_2, radiative energy flux profiles, and flame heights and diameters and liftoff heights. Flame blowout was also studied under certain conditions with propane. Average flame heights and diameters were found to scale approximately with the Froude number (which varied between 80 and 6×10^5), and the nondimensional heights and diameters for propane flames were found to be larger than those for methane flames by approximately 30 and 60%, respectively. Correlations of liftoff heights were obtained in terms of strain rates. The average radiation flux from the flame was found to increase linearly with the flame height, when the flame shape was approximated by a cylinder. The radiation fluxes for propane flames were found to be approximately twice as large as those for methane flames. An estimated 8 to 30% of the heat release was radiated depending on the conditions.

Presented at the 9th ICODERS, Poitiers, France, July 3-8, 1983. Copyright © American Institute of Aeronautics and Astronautics, Inc. 1984. All rights reserved.
*Professor, Division of Heat and Combustion Engineering, Department of Mechanical Engineering.
†Research Engineer.

Nomenclature

A	= constant in Froude number scaling
C	= constant in liftoff characteristics, s
D	= average flame diameter, m
d	= nozzle exit diameter, mm
Fr	= Froude number = u^2/gd
F_{1-2}	= shape factor between flame and radiation sensor
g	= gravity acceleration, m/s^2
H	= average flame height from nozzle exit to visible flame tip, m
H*	= flame radiation height, m
h	= flame liftoff height, m
m	= exponent in Froude number scaling
q"	= thermal radiation flux, W/m^2
T	= flame temperature, K
u	= nozzle exit velocity, m/s
x	= radial distance from flame centerline, m
y	= distance along flame centerline from the nozzle exit, m
y*	= distance along the centerline of the visible flame zone = y - h, m
σ	= Stefan-Boltzmann constant, W/m^2K^4
ε	= flame emissivity

Introduction

This paper reports an investigation of large turbulent diffusion flames for gaseous fuels released from straight tubes at up to sonic velocities at atmospheric pressure. The goals have been to establish approximate correlations and scaling laws for these flames, to compare with results available in the literature, and to contribute to understanding of flame structure and liftoff mechanisms. This study is part of a program started in 1979 to assess fire hazards and damage potential connected with industrial gas releases, and gas and oil/gas releases and blowout situations related to offshore oil and gas exploration. Two reports by the authors describing this work are given in the list of references.

Investigations of various aspects of turbulent diffusion flames resulting from jet releases into quiescent air have been published by many authors over the last 30 to 40 years. However, no complete and unified treatment of the various characteristics of such flames have been found in the literature. Further, no data for large turbulent jet diffusion flames of the type of interest here have been published previously.

The present work extends the range of data for average flame height and liftoff height significantly for propane and methane flames and checks flame blowout velocities for tube diameters up to 16 mm. A simple, approximate correlation for the prediction of the thermal radiation from the flames is derived from the approximation of the flame as a radiating cylindrical surface with an average flame diameter and radiation shape factors. In addition, detailed average temperature and specie-concentration contour maps in laboratory size flames have been measured to establish more detailed knowledge of the flame structure.

Regarding studies of flame height, the recent works of Suris et al. (1977) and Becker and Liang (1978) are relevant here. Suris and co-workers suggest an explicit correlation of height with a Froude number based on the exit diameter, while Becker and Liang employ an implicit formulation with a Richardson number based on the flame height. In the present work, a direct scaling with Froude number is chosen. No unified definition of the flame height is presented in the literature. As the flame fluctuates considerably with time, an average value will be used here.

A recent theoretical study of liftoff characteristics for turbulent jet diffusion flames has been published by Peters and Williams (1983). They recommend the correlation of liftoff heights with strain rates in the turbulent flow and compare theoretical predictions with data for methane flames. In the present work, the parameter ranges have been extended considerably, and comparisons for both propane and methane flames are included. Blowout limits have been studied by Kalgathi (1981) and by Annushkin and Sverdlov (1979), among others.

As far as thermal radiation from these flames is concerned, only measurements of small laboratory-scale flames appear to be available in the literature. Recent studies of interest here include publications by Markstein (1976) and (1978), Tien and Lee (1982), and Yuen and Tien (1976). However, the main emphasis in these references is on local raditaion fluxes and associated temperatures and emissivities, rather than the

JET DIFFUSION FLAMES

global time-averaged radiation fluxes that are measured in the present work. Flame temperature measurements were taken earlier by Becker and Yamazaki (1978).

The experimental apparatus, the test procedures, the test results, and the proposed correlations and scaling laws are discussed in more detail in the following sections of the paper.

Experimental Apparatus

Two different test rigs were used in this work, namely, one for small flames in the laboratory and another for large flames at an outdoor fire-testing area. A schematic of the field test apparatus is shown in Fig. 1. The laboratory test apparatus has a similar layout. The maximum fuel flow rate and thermal heat-release rate for propane (using 20 standard bottles manifolded together) were approximately 160 grams/s and 7.5 MW, respectively, and for methane, approximately 30 grams/s and 1.6 MW, respectively.

The flow rate, temperature, and pressure of the fuel are measured as shown in Fig. 1. The flame geometry is measured from photographs of the flame. The flame height, flame diameter, and flame liftoff heights are obtained from the averages of the data measured from five photographs of the flame taken over

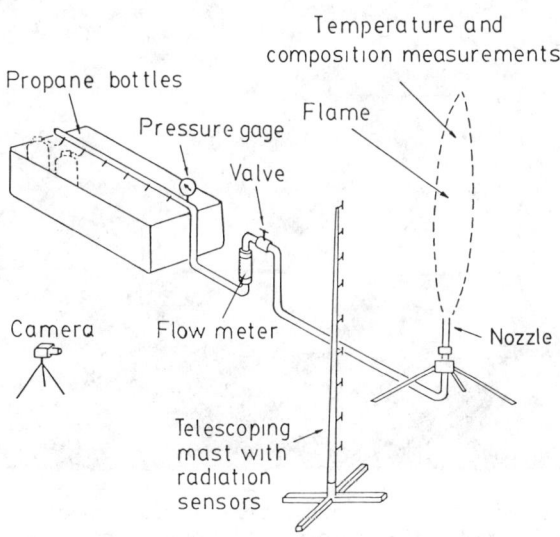

Fig. 1 Schematic of fuel supply and flame test setup.

a period of several seconds. Each photograph is shot with an exposure of 2 ms. Movies are also taken of some of the flames. Platina-type (Pt/Pt-10% Rh) thermocouples are used to measure an average flame temperature at each location. The thermal radiation fluxes from the flames are measured with special thermopile sensors mounted on telescoping masts. Wide-angle sensors made by Sensors Inc. are used so that the total thermal radiation from the whole flame can be measured at the various sensor locations. The sensor surfaces are placed on what correspond to concentric cylindrical surfaces at various distances from the flame centerline. A mast for flame temperature measurements and a telescoping mast for thermal-radiation-flux measurements are shown together with a typical large turbulent propane flame in Fig. 2.

The chemical composition of the flame was measured in laboratory flames for both propane and methane. Continuous analyzers were used to measure the O_2-, CO-, and CO_2-concentrations by traversing across the flame at various heights above the nozzle exit. To measure CO and CO_2, Beckman infrared analyzers were used, and O_2-concentrations were measured by a Servomex continuous analyzer. The diameter of the water-cooled sampling probe was 2 mm, and the measurements were made on a dry basis.

Fig. 2 Photograph showing typical propane flame, telescoping masts for radiation sensors, and temperature probes.

The experimental results to be reported were obtained with circular nozzles having diameters of 2.3, 3.2, 5, 10, 16, 20, 33, 40, and 80 mm. All nozzles were straight tubes with length-to-diameter ratios greater than 50, discharging vertically upward.

Flame Temperatures and O_2-, CO-, CO_2-Concentrations

Average flame temperatures have been measured at various locations in both laboratory and field tests. In a few laboratory tests, traverses across the flame at various heights from the nozzle exit have been made, and temperature-contour plots have been constructed as shown for a methane flame in Fig. 3. It is seen that the maximum average temperature is approximately 1650 °C. The maximum average centerline temperature in the same methane flame is approximately 1550 °C, i.e., the highest average temperature was found to occur off the centerline and at a relative low height. For a propane flame, the measured maximum centerline temperature was approximately 1300 °C at the same total heat-release rate as for the methane flame. Becker and Yamazaki (1978) measured maximum centerline temperature from

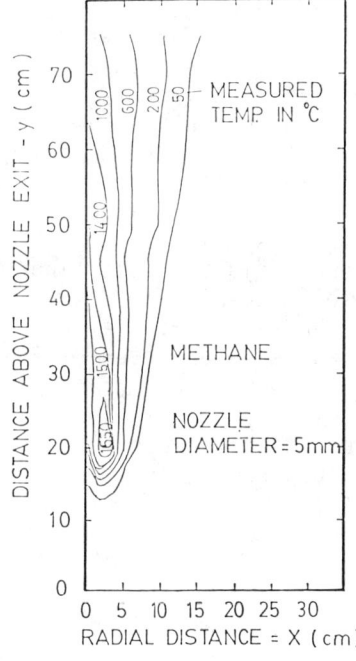

Fig. 3 Typical measured temperature contours for methane flames.

1250 to 1700 °C as the nozzle velocity increased for propane flames with a nozzle diameter of approximately 4.5 mm. The present measurement of 1300 °C is about 150 to 200 °C below the corresponding value obtained by Becker and Yamazaki.

In a few of the field tests, the temperatures have been measured at three positions along the flame centerline to obtain approximate maximum average centerline temperatures and their locations. Both propane and methane have been used, and the nozzle exit Mach numbers have been varied from about 0.1 to 0.5 with heat releases of from 0.4 to 2.2 MW at a nozzle diameter of 16 mm. The measured maximum centerline temperatures for the propane flames varied from 800 to 1200 °C and were observed at about 60% of the flame height. For the methane flames, the maximum average temperatures varied from about 1000 to 1150 °C and were located at about 50% of the flame height. The temperature was observed to increase with increasing heat-release rate.

The measured maximum temperatures for the large-scale flames are from 100 to 500 °C below the measurements for the small flames. These are average temperatures rather than instantaneous peak temperatures, and the large flames fluctuate considerably. Further measuremens including intermediate size flames must be made to establish a complete set of temperature data.

Average chemical compositions within the flames have been measured in some of the laboratory tests. Traverses across the flames at various heights above the nozzle exit were made, and some typical results for O_2-, CO-, CO_2-concentrations for a methane flame are shown in Fig. 4. It is seen that the average oxygen content in the flame is fairly high, especially near the center of the flame at the lower height, indicating an entrainment of air in the liftoff region. The oxygen content along the centerline decreases with height. The CO_2-content in the flame increases by a factor of 2 between the two y locations. CO- and CO_2-concentrations of 4 to 5% by volume are measured at $y = 50$ cm. These results appear reasonable; however, no results have been found in the literature for methane under comparable conditions that can be used for comparison.

Flame Geometry

To describe the flame geometry, the spatial and temporal averages of the flame diameter and the flame height are used. The flame is in this way approximated by a cylinder of height H which is measured from the nozzle exit to the top of the visible flame section, and diameter D which is the average over the visible flame height. Tests were run with the various nozzle exit diameters d from low exit velocities u of 5 m/s to high velocities of 200-250 m/s. For the large nozzles, the maximum velocities were, however, much smaller because of limiting capacities of the fuel supplies. The Froude number (Fr) based on the exit conditions of the nozzle varied from approximately 80 to 6×10^5 and 5×10^2 to 1×10^5 for the propane and methane flames, respectively. The data for the flame height and flame diameter divided by the nozzle diameter d are plotted in Fig. 5 for propane flames and for nozzle diameters larger than 10 mm.

To fit this data, a simple direct scaling with a Froude number suggested for the flame height by Suris et al.(1977) is used. Their scaling law is of the form

$$H/d = A \times Fr^m$$

where $Fr = u^2/gd$, in which A and m are constants. For Froude numbers larger than 3×10^4, Suris and coworkers

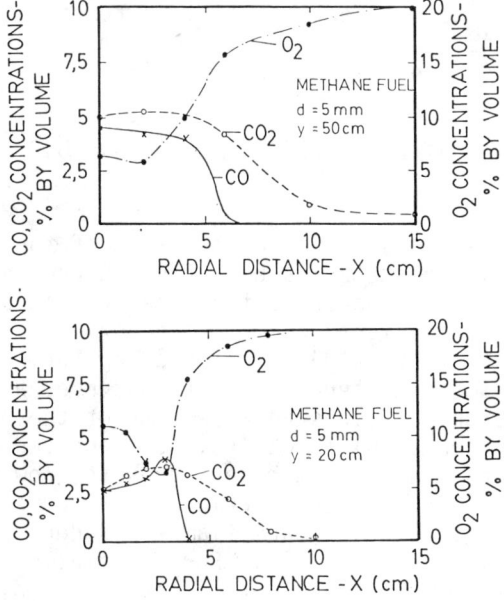

Fig. 4 O_2-, CO-, CO_2-concentrations for a methane flame for different heights above the nozzle exit.

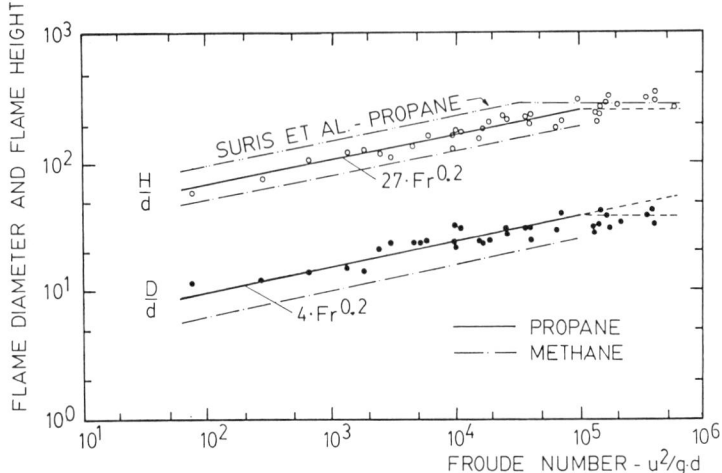

Fig. 5 Flame height/nozzle diameter and flame diameter/nozzle diameter as functions of Froude number for propane and methane flames.

indicate that the exponent m becomes zero and the flame height becomes independent of the Froude number.

They find that m=0.2 and A=40 for propane and m=0.2 and A=29 for methane. Although Suris et al. do not, however, report flame diameters, the flame diameter data can be correlated by a similar function. The correlations for flame height and diameter of the propane flames with the present data (Fig. 5) are:

$$H/d = 27 \times Fr^{0.2} \qquad D/d = 4 \times Fr^{0.2}$$

These correlations indicate the ratio of the flame height to its diameter is 6.75 for the propane flames. The present data indicates that the transition to constant flame height occurs at approximately $Fr=10^5$ rather than at $Fr=3 \times 10^4$ as observed by Suris et al. For the flame-diameter data, dotted curves are shown for the large Fr values because some of the data for tests with rectangular nozzles, not reported here, indicate that the diameter continues to grow (Hustad and Sønju 1983). The correlation due to Suris et al., shown in Fig. 5 for propane flames, is derived from data for small flames only. For the smallest nozzles (2.3 to 5 mm) in the present laboratory study, the non-dimensional flame heights, not shown in Fig. 5, were greater than those of the field study but generally fall between the two correlation curves (Hustad and Sønju 1983). Only the results for the larger flames (greater than 1 m in height) are plotted here, since they are of main interest in this work.

The following correlations for the data on flame height and diameter for large methane flames are also shown in Fig. 5:

$$H/d = 21 \times Fr^{0.2} \qquad D/d = 2.5 \times Fr^{0.2}$$

These correlations indicate that the ratio of flame height to diameter is 8.4. There are fewer tests for methane than for propane; these results are discussed in more detail by Hustad and Sønju (1983).

When all of the data for both small and large nozzles for both propane and methane are taken into consideration, the scatter in the data fall within +45% and -25% of the suggested correlations.

Liftoff Heights and Flame Blowout

The liftoff height h is defined as the distance along the centerline from the nozzle exit to the plane at which the flame is initiated and stabilized. The measurements were done both with laboratory scale flames and flames in the field. The nozzle diameters were varied from 2.3 to 80 mm for propane and from 2.3 to 5 mm for methane. In laboratory scale flames, the liftoff heights were measured from the same photographs that provided the flame height and diameter. In the field tests, the liftoff heights were mostly measured with a yardstick directly on the flames, because it was often difficult to make accurate measurements on the photographs. Appreciable differences in the results occur, depending on whether the heights to the blue or yellow zones are used. The heights to the blue flames are used in this work, yellow heights sometimes are appreciably larger.

Peters and Williams (1983) have developed the following functional dependence of liftoff heights on strain rates for the turbulent flows:

$$h/d = C(d/u)^{-1}$$

where C is a constant with the dimension of time. Peters and Williams recommend, based on analysis of measurements on methane flames that:

$$h/d = 3.6 \times 10^{-3} (d/u)^{-1}$$

The observed h/d and d/u, plotted for propane flames and methane flames in Fig. 6, agree quite well with their results.

Flame blowout is another important characteristic of turbulent jet diffusion flames. The blowout limit marks the nozzle exit velocity beyond which the flame connot be stabilized in the mixing region. A few tests were made to measure such limits for propane for comparison with the results of Kalgathi (1981) and Annushkin and Sverdlov (1979). Kalgathi found that blowout for propane flames occured only for nozzle diameters less than about 17 mm, while Annushkin and Sverdlov indicated that blowout occurs only for diameters less than 2 to 3 mm. In the present tests with propane, flame blowout was obtained for nozzles with diameters of from 4 to 10 mm at nozzle exit velocities

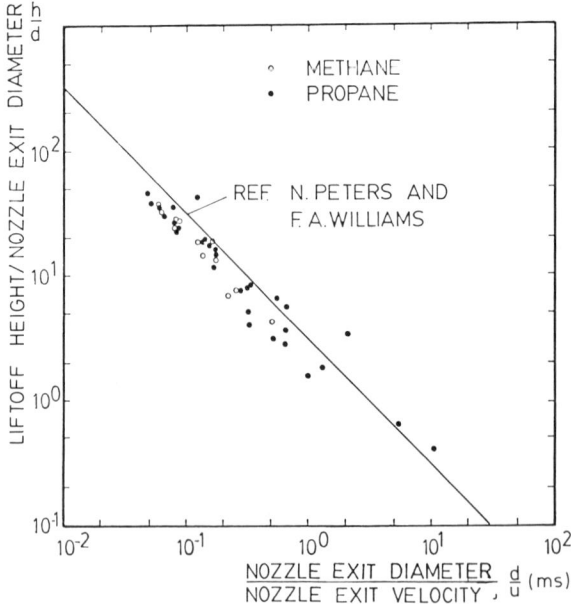

Fig. 6   Liftoff characteristics.

of about 120 and 250 m/s, respectively. These results agreed quite closely with the results of Kalgathi. For a 16-mm nozzle, the flame was stable even at sonic velocity. Higher velocities were not achieved in these experiments.

Flame Radiation

Thermal radiation fluxes were measured for propane and methane flames over the ranges of nozzle diameter, nozzle exit velocity, and Froude number given earlier. The corresponding heat-release rates for the field tests are 0.6 to 7.5 MW for propane and 0.4 to 1.6 MW for methane, while the tests in the laboratory were in the kilowatt range. For data correlation, thermal radiation is assumed to occur from a cylinder whose dimensions are the flame radiation height, H*=H-h, and the flame diameter D. The effective intensity of the radiation from the flame surface will be determined from the experiments.

The radiation flux is:

$$\dot{q}'' = F_{1-2} <\varepsilon\sigma T^4> \quad (W/m^2)$$

where $\dot{q}''$ is the radiation flux received in W/m^2; F_{1-2} is the shape factor between the flame and the sensor; σ is the Stefan-Boltzmann constant in W/m^2K^4; ε is the flame emissivity; T is the radiation temperature K; and $<\varepsilon\sigma T^4>$ is the average effective value of this product in W/m^2.

The shape factor F_{1-2} can be calculated by standard formulas for radiation from a cylinder (area 1) to a unit area on another concentric cylindrical surface (area 2) (Schlünder 1977). The values of $<\varepsilon\sigma T^4>$, calculated for both propane and methane flames from the measured radiation fluxes and the predicted F_{1-2} values, are plotted for circular nozzles in Fig. 7 as a function of the flame radiation height H*. The following straight lines fit the results, for propane:

$$<\varepsilon\sigma T^4> = 1 + 0.78 \, H^*$$

and for methane:

$$<\varepsilon\sigma T^4> = 0.5 + 0.39 \, H^*$$

These correlations suggest that the radiation flux from the surface increases with size even beyond H* = 7 m event though it is expected that thermal radiation should taper off as H* increases. As there is considerable scatter in the data; more data, especially at larger flame heights, are required. For methane, the experimental data are not shown, but these results are discussed by Hustad and Sønju (1983). Although the small flame data are also not shown, they are in general agreement with the large flame data, but tend to fall somewhat above the curve. All results of the present investigation fall within ±60% of the curve except for two points.

A theoretical radial shape factor distribution for the radiating cylinder and sensor geometry is plotted in Fig. 8 for a cylinder with a constant ratio of flame height-to-diameter of 6.7, as obtained earlier for the propane flames and at the half-height of the cylinder where the radiation is highest. The experimentally determined shape factors, plotted in Fig. 8, were calculated from the measured radiation fluxes and the $<\varepsilon\sigma T^4>$ values given in Fig. 7. Lines showing deviations of +50% and -33% are drawn. Quite good agreement exists between the experimental and theoretical shape factors. For the methane flames, the same procedure is used as for the propane flames, but in this case a constant height-to-diameter ratio of 8.4 is used. The data for the methane flames fit quite well with the theoretical curve, as shown in Fig. 9.

The maximum radiation fluxes from propane and methane flames at any distance from the flame cen-

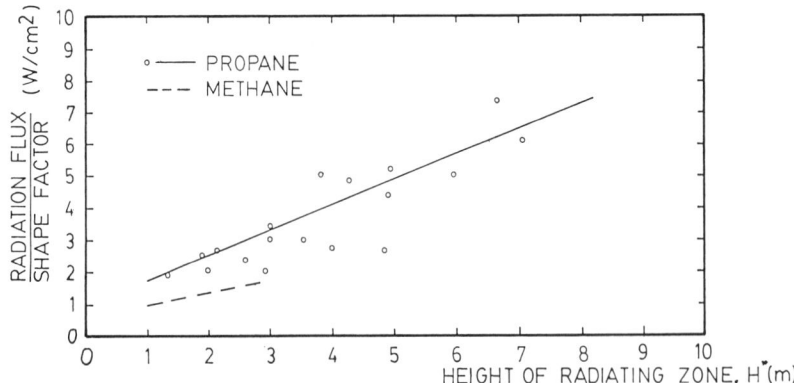

Fig. 7   Radiation flux/shape factor as a function of height of the radiating zone.

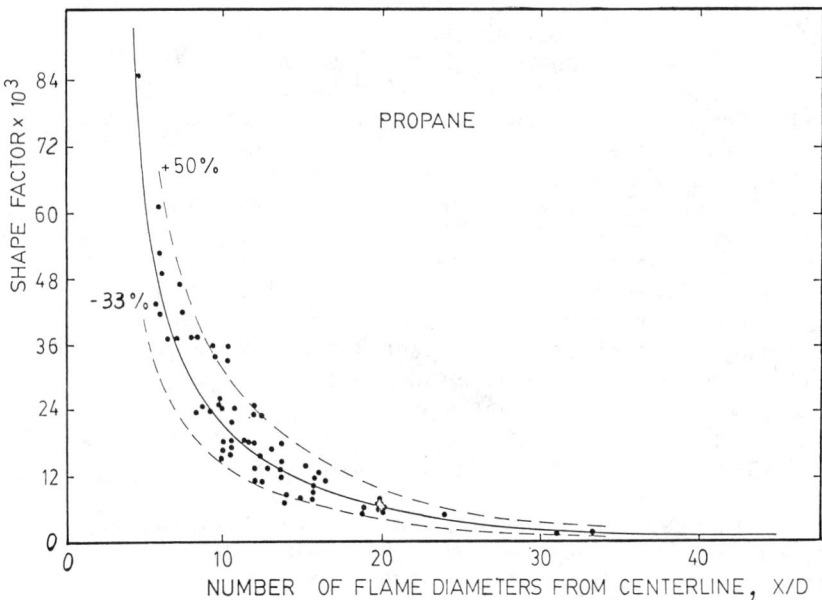

Fig. 8 Shape factor as a function of number of flame diameters from centerline for propane flames.

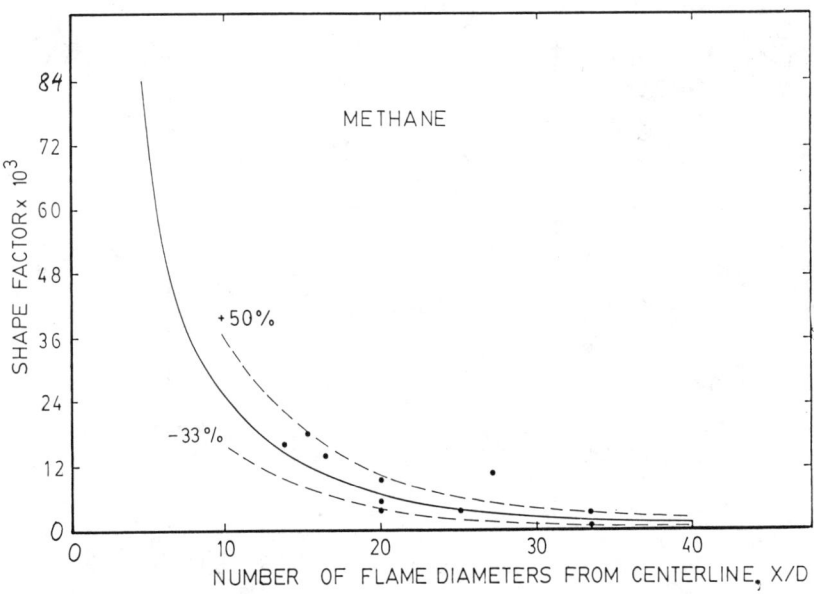

Fig. 9 Shape factor as a function of number of flame diameters from centerline for methane flames.

terline can now be calculated. To find the radiation flux at locations above or below the location for the maximum radiation flux, one must determine the vertical radiation flux distributions. Again, the shape factors can be calculated from well-known formulas (Schlünder 1977), and sample plots of the normalized radiation flux for radial distances of 7 and 20 flame diameters are shown in Figs. 10 and 11. Data for two propane flames and one methane flame are shown for comparison in Figs. 10 and 11, respectively. The data show quite good agreement with the calculated values. Comparisons for many tests of both small and large flames have been made with good results (Hustad and Sønju 1983).

Discussions and Conclusions

This research has helped to improve knowledge of turbulent-jet diffusion flames oriented vertically and reveals systematic dependences on scale.

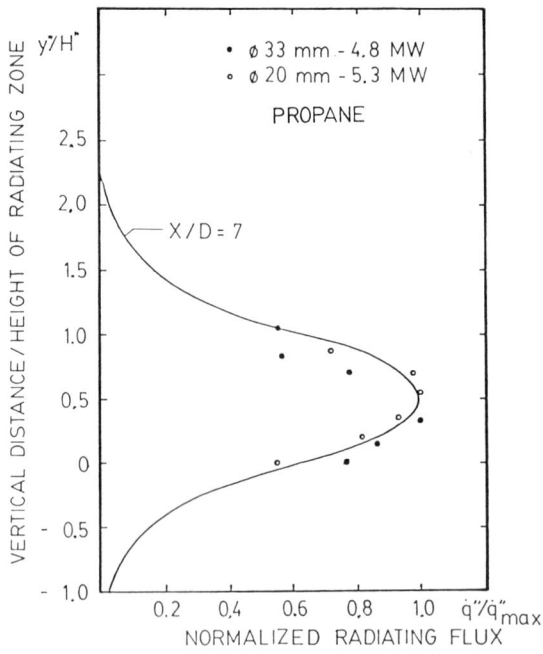

Fig.10 Normalized vertical flame radiation profiles for propane flames at a distance of seven flame diameters from the centerline.

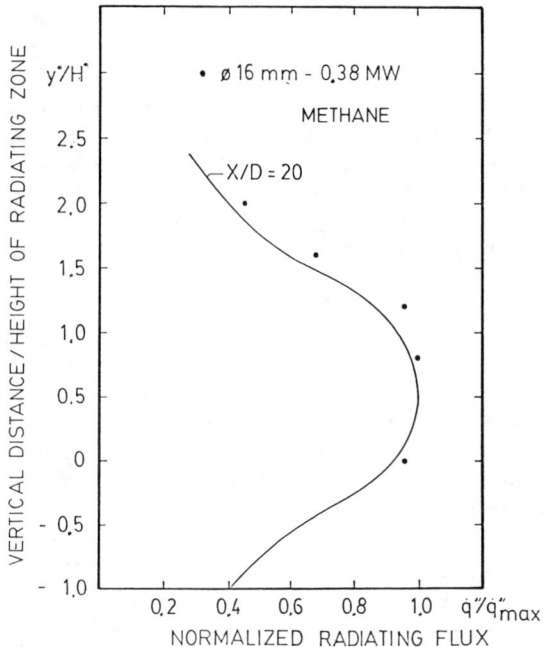

Fig.11 Normalized vertical flame radiation profile for a methane flame at a distance of 20 flame diameters from the centerline.

Thermocouple measurements of average temperatures in flames are difficult to perform, and the results are influenced to some extent by thermocouple properties, such as bead diameters. Inaccuracies associated with thermocouple response times are expected to be largest in the smaller flames, at the higher exit velocities, and nearer to the tube exit, because characteristic fluctuation times and length scales are smaller in these cases. Nevertheless, systematic trends can be identified from the data. The observation of highest temperatures at lower heights in the smaller flames and the observed increase in the average flame temperatures with increasing exit velocity suggest that the trends observed are real and are not due to limitations related to the thermocouples. The off-axis peak at the lower heights is consistent with off-axis average location of the reaction zone. The lower peak values at higher elevations are consistent with influences of radiant losses in the upstream regions, which lower the overall temperature levels downstream, and with larger-scale fluctuations at the higher elevations, which bring larger fractions of cooler gas to the thermocouple. These same effects can cause the observed increase in the maximum average centerline temperature as jet velocity increases and the lower temperature levels generally found in the larger flames. The lower average temperatures of the large-scale propane flames, in comparison with the large-scale methane flames at the same heat-release rates, may be due to greater radiative heat-loss rates in the propane flames.

The off-axis peak of the average CO-concentration at the lower elevation is consistent with chemical production of CO in the flame zone. The relatively steep decrease of CO-concentrations with increasing radius, in comparison with the more gradual decrease of CO_2-concentrations, is consistent with the burn-up of CO in the outer wings of the turbulent flame. The surprisingly high on-axis concentrations of O_2 (on the order of 5%) at the lower heights must reflect mixing near the tube exit prior to establishment of the flame. These large values do not necessarily represent mixing at a molecular scale, because the observations may result from an average of samplings of fuel and air pockets. The subsequent decrease in the centerline oxygen concentration with increasing height is consistent with its consumption in the combustion process.

The exponent of the Froude number in the flame-height correlations is consistent with that obtained in previous studies over most of the range of Froude number tested and thereby supports the physical mechanisms in the literature that produce this exponent. However, a previously unrecognized systematic variation of the factor of proportionality with scale has been found. At a given Froude number, the ratio of the flame height to the exit diameter seems to tend to decrease with increasing scale for both propane and methane. Because of uncertainties in flame-height definitions, especially for the large-scale flames, which were observed to exhibit large fluctuations in flame height, the extent to which this systematic variation is a function of measurement methods is unclear at present. However, it must be concluded that careful study is needed to define flame heights of large-scale flames.

It was found that an average flame diameter scales with the Froude number in the same way as the average flame heights. Thus, the major geometrical factors of the flame exhibit a Froude number scaling over the range of conditions tested, within a reasonable accuracy. There may be some dependences of these scalings on the exit geometry, but in another study where rectangular exit geometries (Hustad and Sønju 1983) were used, reasonable agreement was also obtained when the equivalent circular diameter was used in the scaling. The ratio of the average flame height to the average flame diameter was found to be constant but to depend on the fuel and to be larger for methane than for propane. The difference might be due to different stoichiometry or to different radiative properties that effect the experimental definitions of heights and diameters.

The measured liftoff heights correlate well with the theory of Peters and Williams (1983) over more than two order of magnitude in the ratio of liftoff height to nozzle diameter. This correlation is achieved for the large-scale flames only if the blue height (the distance from the nozzle exit to the onset of blue radiation) is employed. The yellow height does not correlate in this way for the large flames. The yellow is an indication of soot, which is not present at the lower elevations in the highly lifted flames; yellow heights and blue heights may differ by a factor of 2. The correlation thus is seen to apply to the exothermic blue flame and not to the complicated finite-rate chemistry of soot production. The correlation is consistent with the theory but does not prove the theory, because it might also be consistent with other theories that have not been tested with the data. If the theory is correct, then the agreement between the propane and methane data suggests that the overall reaction rates at flame temperature for these two fuels are approximately the same.

Exit velocities obtained for blowout were consistent with earlier data of Kalgathi (1981). However, the ability to achieve blowout varies with experiments of different investigators. This suggests that blowout may depend relatively strongly on the experimental setup and on the exit geometry. Further study of blowout conditions is needed.

Relatively extensive measurements were completed of magnitudes of radiant energy fluxes from these flames and of the dependences of the fluxes on radial and vertical positions of the receiver. Over a wide range of parameters it was established that for purposes of calculation, the flame may be characterized in terms of shape factors for a radiating cylinder of an effective radiation height given by the difference between the flame height and the liftoff height (or stabilization height) at which point radiation begins. This provides a relatively simple approach to the estimation of radiant fluxes of practical interest, if the flame heights and diameters and the radiant fluxes emitted by the flame are available. Correlations for each of these last three quantities have been obtained for propane and methane flames. The last exhibits a linear increase of the emitted flux with the height of the emitting zone, over the entire range of conditions of the large-scale tests performed. This would require a corresponding increase in emissivities and/or emitting temperatures.

In the literature, constant emitting temperatures and corresponding emissivities for these fuels are suggested. Attempts to estimate emissivities from the data with constant temperatures were unsuccessful in that satisfactory agreement could not be obtained between predicted emissivities and those available in the literature. Conversely, use of literature emissivities to predict emitting temperatures produced unacceptably low results. Therefore, it seems best, at present, to characterize these large flames in terms of the average radiant flux emitted. This flux can not continue to grow linearly but must level off for sufficiently large values of the height of the radiating zone; however, the heights needed to achieve this may be appreciably greater than 7 m. It may be concluded that further research is needed on radiative properties of large-scale turbulent diffusion flames.

Acknowledgments

The authors would like to acknowledge the important contributions, especially in the early phases of this work, of several students and engineers at The Norwegian Institute of Technology and at The Foundation for Scientific and Industrial Research. Furthermore, we appreciate the many helpful discussions with F.A. Williams at Princeton University. This work was supported in part by The Royal Norwegian Council for Scientific and Industrial Research.

References

Annuskin, Y.M. and Sverdlov, E.D. (1979) Stability of submerged diffusion flames in subsonic and underexpanded supersonic gas-fuel streams. Combust. Explos. Shock Waves USSR 14 (5), 597.

Becker, H.A. and Liang, D. (1978) Visible length of vertical free turbulent diffusion flames. Combust. Flame 32 (2), 115-137.

Becker, H.A. and Yamazaki, S. (1978) Entrainment, momentum flux and temperature in vertical free turbulent diffusion flames. Combust. Flame 33 (2), 123-149.

Hustad, J. and Sønju, O. K. (1983) Radiation and size scaling for large leakages of gas and oil/gas mixtures. SINTEF Report STF15 A83036, SINTEF, Trondheim, Norway.

Kalgathi, G.T. (1981) Blowout stability of gaseous jet diffusion flames. Combust. Sci. Technol 26 (5,6), 233-239.

Markstein, G.H. (1976) Radiative energy transfer from turbulent diffusion flames. Combust. Flame 27 (1), 51-63.

Markstein, G.H (1976) Scaling of radiative characteristics of turbulent diffusion flames. 16th Symposium (International) on Combustion, pp. 1407-1419. Massachusetts Institute of Technology, Cambridge, Massachusetts.

Peters, N. and Williams, F.A. (1983) Lift-off characteristics of turbulent jet diffusion flames. AIAA 21 (3), 423-429.

Schlünder, E.U. (1977) VDI-Wärmeatlas. VDI-Verlag GmbH, Düsseldorf, West Germany.

Suris, A.L., Flankin, E.V., and Shorin, S.N. (1977) Length of free diffusion flames. Combust. Explos.Shock Waves 13 (4), 459-462.

Tien, C.L. and Lee, S.C. (1982) Flame Radiation. Prog. Energy Combust. Sci. 8 (1), 41-59.

Yuen, W. W. and Tien, C. L. (1976) A simple calculation scheme for the luminous-flame emissivity. 16th Symposium (International) on Combustion, pp. 1481-1487. Massachusetts Institute of Technology, Cambridge, Massachusetts.

Chapter III. Turbulent Combustion

On Sound Sources in Turbulent Combustion

N. Kidin* and V. Librovich†
Institute for Problems in Mechanics, Moscow, USSR
and
J. Roberts‡ and M. Vuillermoz§
Polytechnic of the South Bank, London, United Kingdom

Abstract

Sound emission from turbulent gaseous combustion has been successfully described macroscopically in terms of a random distribution of elementary monopole sources. This paper presents an extension to the currently accepted model of sound generation by these elementary sources which takes particular account of the rate of volume evolution at the terminal stage in the burning of small pockets of a combustible mixture. The model is based on hydrodynamic relationships applied to the elementary volume of combustible mixture which are solved using first-order perturbation methods. Sound emission is then predicted from an estimate of the characteristic time of chemical reaction. An experimental arrangement is described which has enabled the acoustic pressure from such small isolated burning pockets of burning fuel to be measured. Predictions from the theoretical model are shown to be in reasonable accord with the experimental findings.

Nomenclature

a = bubble radius, m
c = velocity of sound, m/s
C_p = specific heat (constant pressure), J kg^{-1} K^{-1}

Presented at the 9th ICODERS, Poitiers, France, July 3-8, 1983. Copyright © American Institute of Aeronautics and Astronautics, Inc., 1984. All rights reserved.
*Senior Research Worker, Thermogasdynamics Laboratory.
†Deputy Director.
‡Principal Lecturer, Institute of Environmental Science and Technology.
§Senior Lecturer, Department of Physical Science and Technology.

C_v = specific heat (constant volume), J kg^{-1} K^{-1}
D = diffusion coefficient, m^2 s^{-1}
E = activation energy, J kg^{-1}
K = rate constant, s^{-1}
n = order of reaction
p = acoustic pressure, Pa
P = ambient pressure, Pa
Q = heat of reaction, J kg^{-1}
R = gas constant for unit mass, J kg^{-1} K^{-1}
S = surface area, m^2
T = temperature, K
u = flame velocity, m s^{-1}
v = particle velocity, m s^{-1}
V = volume, m^3
α = concentration, kg m^{-3}
γ = specific heat ratio
ν = symmetry parameter
ρ = density, kg m^{-3}
τ = characteristic time, s
Φ = source term, W m^{-3}
χ = thermal diffusivity, m^2 s^{-1}

Introduction

Problems associated with acoustic instability of many working combustion processes and the increased noise levels of modern high intensity combustion systems continue to stimulate research into the mechanisms of sound emission from turbulent combustion.

In general, sound emission from any combustion process stems from a rapid variation in the rate of volume evolution of the combustion products. A variety of mechanisms might contribute to the causes of such variations, the significance of which will depend on the nature and scale of the processes involved. In the case of turbulent combustion it is the variation in the local burning rate in the flame region which is believed to provide the major source of sound and which may be described macroscopically in terms of a distribution of monopole sources (Bragg 1963; Smith and Kilham 1963). In simple terms the flame front becomes distorted into complex shapes by the turbulence to form many small zones of intense chemical reaction. It is the increased reaction rates within these zones which then provides the source of sound emission (Fig.1). The random nature of the process accounts for the broadband acoustic spectrum characterizing this type of combustion.

A model for describing sound generation by these individual elementary sources has been proposed (Thomas and Williams 1961; Hurle, Price, Sugden, and Thomas 1968) where the necessary changing rate of volume evolution comes directly from the changing geometry of the elementary volume as the fuel is consumed at an initial rate associated with the local laminar flame velocity. However, a further stage in this process may be reached when the temperature of the as yet unburnt fuel and oxidant is raised sufficiently for it to undergo a small thermal explosion and thus generate an additional terminating sound pulse. The experimental verification of the Thomas and Williams model used relatively large initial volumes of gas, confined by soap films, which were at room temperature at the moment of ignition. Such conditions might well have caused the sound wave generated in the first stage of the process to overshadow the small nonlinear effect at the termination of burning. In a real turbulent flame the pockets would be surrounded in the main by the recirculating combustion products and would, when they commence burning, already be at a relatively high temperature. The critical volume at which the explosive phase occurs could then be significantly larger than for the case of the initially cold gas and it may thus make a contribution to the total sound emission which has been hitherto unrecognized.

The purpose of this paper is to describe an experimental arrangement to investigate the sound output from such small gaseous sources and to develop a model in a form which could be experimentally tested.

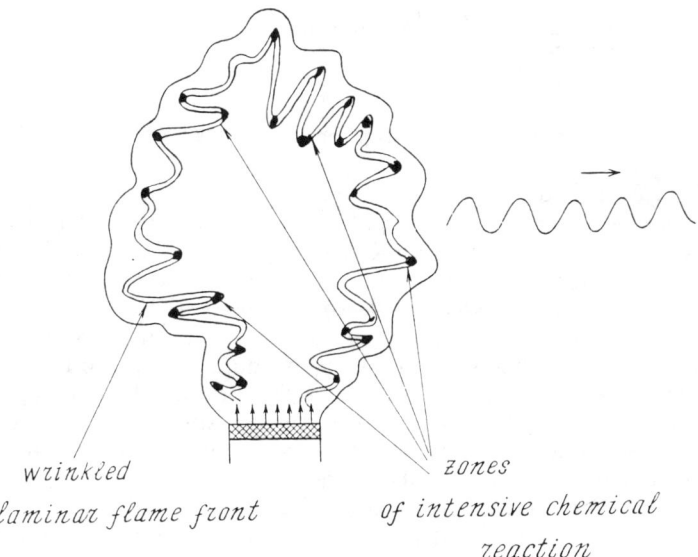

Fig. 1 Sound emission by turbulent flames.

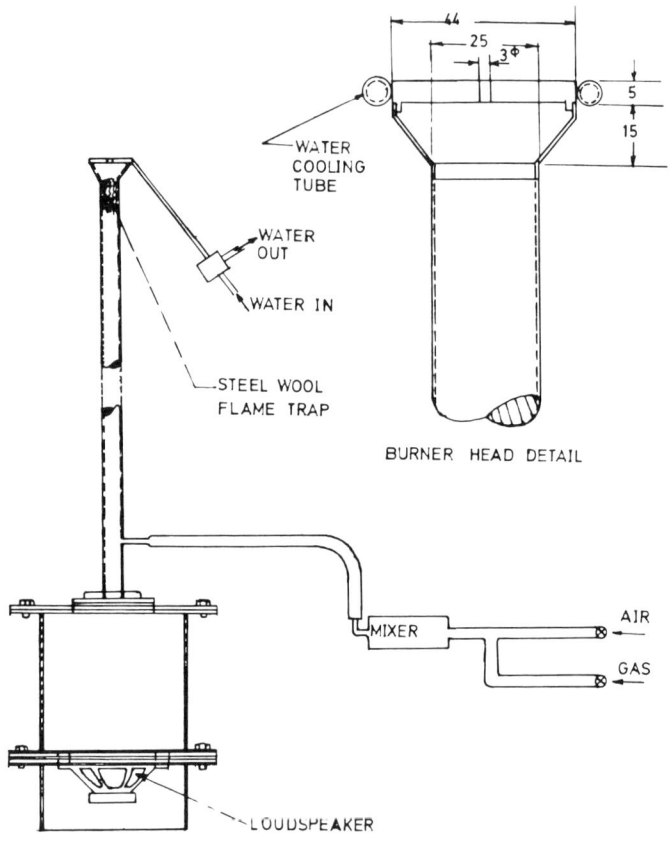

Fig. 2 Burner system.

Experimental Arrangements and Major Findings

The experimental configuration is shown in Figs. 2 and 3. The burner was supplied with a 1:6 natural gas-air mixture with a total flow rate of 0.7 litre/min to give a distinct conical flame front. A loudspeaker at the base of the supply tube was used to impose a sinusodial acoustic perturbation on the velocity of the gas-air mixture at the flame. This changing velocity produced a periodic variation in the flame profile giving it a mushroom shaped appearance with the maximum distance between the base of the flame and the head a function of the acoustic signal amplitude. During the cycle the stalk lengthened and its radius decreased (as the flame front propagated into the remaining combustible mixture) until it suddenly underwent a catastrophic collapse and disappeared leaving the head as a burning bubble of gas. Similarly, the bubble radius decreased until it also underwent a catastrophic collapse. It has been shown by the authors that the collapse of the neck and bubble coincided precisely with the generation of pulses of sound (Smith el al. 1982).

A stroboscopic optical system with variable time delay was used to observe the flame shape. The projected "shadow" of the flame was photographed throughout one complete cycle at 100 Hz. A time history of the acoustic pressure output from the flame was recorded and a typical plot is shown in Fig. 4. The first peak represents the collapse of the flame neck and the second the collapse of the bubble. The maximum acoustic pressure pulse at the moment of collapse of the neck (and bubble detachment) as a function of the corresponding length of the cylindrical neck is shown in Fig. 5. The least-squares fit line through the data points has a slope of 17 Pa/m. Also shown, in Fig. 6 is a plot of maximum acoustic pressure recorded at the collapse of the bubble against the bubble diameter (the latter estimated from the shadowgraphs).

Hydrodynamic Model

The following model has been developed in an attempt to account for these experimental findings. The elementary volume of burning gas is taken as a hydrodynamic system with a steadily decreasing volume and surface area. The normal conservation equations will be

$$\frac{\partial \rho}{\partial t} + \text{div}(\rho \vec{v}) = 0 \qquad (1)$$

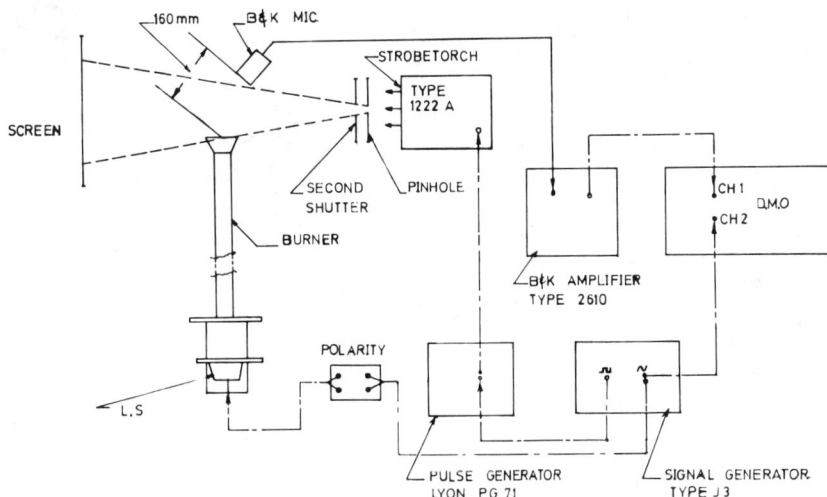

Fig. 3 Instrumentation arrangement.

$$\frac{\partial \vec{v}}{\partial t} + \rho(\vec{v}\nabla)\vec{v} = -\text{grad } P \qquad (2)$$

$$\frac{\partial}{\partial t}\left(\frac{\rho v^2}{2} + \rho C_v T\right) = -\text{div}\left[\rho\vec{v}\left(\frac{v^2}{2} + \frac{P}{\rho} + C_v T\right)\right] + \Phi \qquad (3)$$

The quantity Φ represents the heat release rate per unit volume due to chemical reactions and is assumed to take the simple Arrhenius form.

$$\Phi = Q\alpha^n K(T) \exp(-E/RT) \qquad (4)$$

Equations (1-3) may be integrated over the volume using the Gauss-Ostrogradsky theorem to give rate equations in terms of spatial averages. Mass flow rate $\vec{m}$ replaces $\rho\vec{v}$ and a time-dependent ratio of surface area to volume A(t) is introduced. The time-dependent source term Φ is averaged over the internal volume of the gas to exclude the surface burning. This is justified as the surface volume of burning gas is negligible compared to the whole volume. The rate

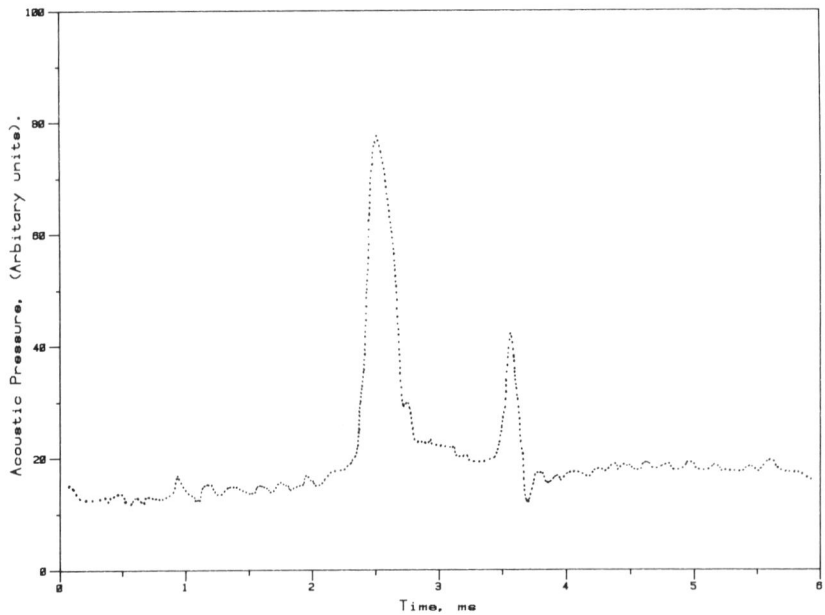

Fig. 4 Digital memory oscilloscope record.

equations then become

$$\frac{d\bar{\rho}}{dt} = -m_s \frac{S}{V} + (\rho_s - \bar{\rho})\frac{1}{V}\frac{dV}{dt} \qquad (5)$$

$$\frac{d\bar{m}}{dt} = -\left(\frac{m^2}{\rho} + P\right)_s \frac{S}{V} + (m_s - \bar{m})\frac{1}{V}\frac{dV}{dt} \qquad (6)$$

$$\frac{d}{dt}\left[\overline{\left(\frac{m^2}{2\rho}\right)} + \overline{\rho C_v T}\right] = -\left[m\left(\frac{m^2}{2\rho^2} + C_p T\right)\right]_s \frac{S}{V}$$

$$+ \left[\left(\frac{m^2}{2\rho} + \rho C_v T\right)_s - \overline{\left(\frac{m^2}{2\rho}\right)} - \overline{\rho C_v T}\right]\frac{1}{V}\frac{dV}{dt} + \bar{\Phi} \qquad (7)$$

where the overbars indicate the spatial average over the volume and the subscripts refer to quantities evaluated at the surface in front of the flame. Equations (5-7) may be linearized by considering first-order perturbation of the parameters at a point corresponding to the start of the explosive phase and solved for the variables using straightforward algebraic techniques. This results in the following linear approximation for temperature perturbation

$$\frac{dT'}{dt} - \frac{\nu+1}{\gamma\nu+1}\frac{\Phi'}{\rho_1 C_v} = \frac{\gamma\nu}{\gamma\nu+1}\frac{1}{A_1^2 c_1^2}\left(\frac{d^3 T'}{dt^3} - \frac{1}{\rho_1 C_v}\frac{d^2\Phi'}{dt^2}\right) \qquad (8)$$

where ν is a symmetry parameter which takes the value 2 for a cylinder and 3 for a sphere and perturbation terms are primed.

The quantity $1/(c_1 A_1)$ is a measure of the time taken for a pressure disturbance at the surface to cross to the centre of the volume of gas, the so-called acoustic characteristic time. If this time is very small compared to the characteristic time of development of the chemical reaction the right-hand side of Eq. (8) may be neglected so that

$$\frac{dT'}{dt} - \frac{\nu+1}{\gamma\nu+1}\frac{\Phi'}{\rho_1 C_v} = 0 \qquad (9)$$

The contrary condition of the chemical reaction proceeding at a rate faster than the speed of sound is not pursued at this stage.

The Acoustic Pressure

The sound pressure at distance r from a small monopole source of volume V is given by

$$p = \frac{\rho_0}{4\pi r} \frac{d^2 V}{dt^2} \qquad (10)$$

If, as before, we consider a small perturbation of the mass flow rate when the surface area is S_1, the rate of change of volume is given by

$$\frac{dV}{dt} = \frac{S_1 m'}{\rho_1}$$

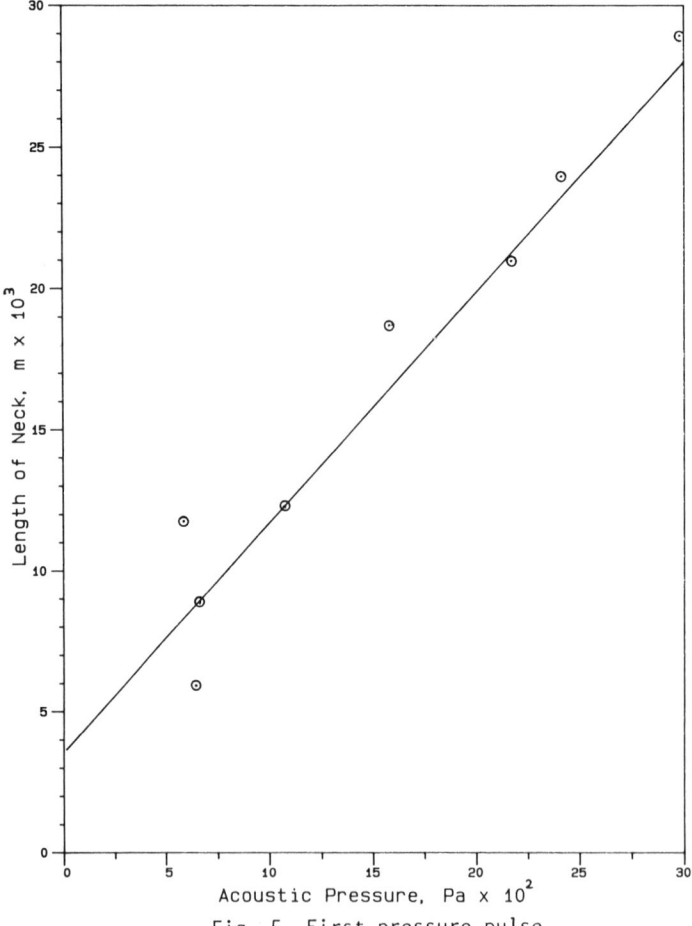

Fig. 5 First pressure pulse.

so that

$$p = \frac{1}{2} \frac{\rho_o}{\rho_1} \frac{S_1}{4\pi r} \frac{dm'}{dt} \quad (11)$$

The factor of 1/2 is introduced to account for the acoustical impedance matching between the source and the surrounding medium. The internal acoustical impedance has been shown to be the complex conjugate of the external impedance (Vuillermoz and Roberts 1984). Combining the linearized

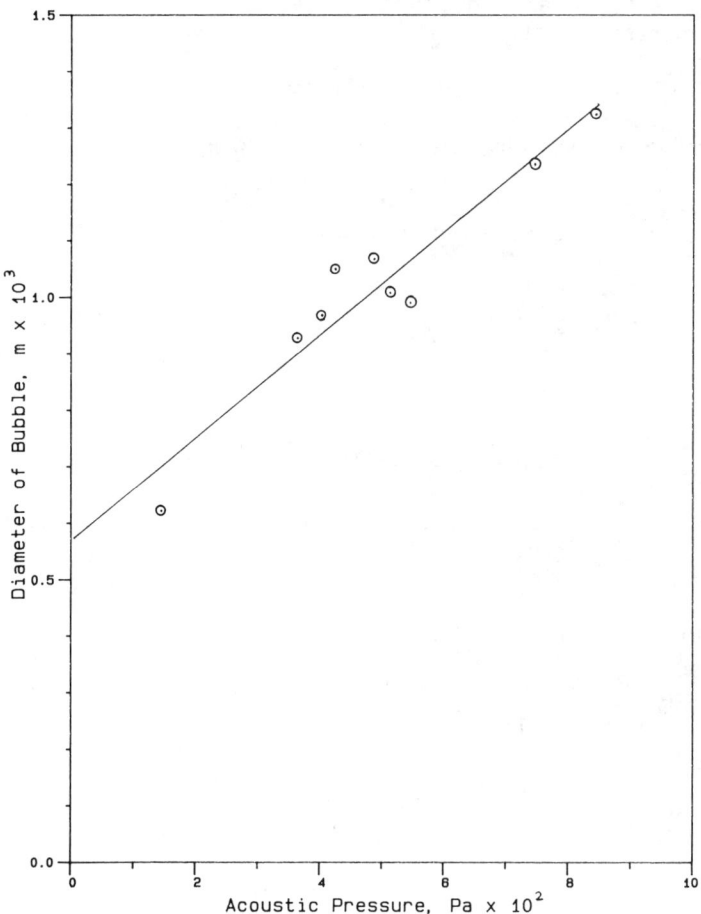

Fig. 6 Second pressure pulse.

forms of Eqs. (5-7) with Eq. (11) it may readily be shown that

$$p = \frac{PV_1}{8\pi r(\gamma-1)T_1}\left(\frac{1}{\rho_1 C_v}\frac{d\phi'}{dt} - \frac{d^2 T}{dt^2}\right) \quad (12)$$

and using Eq. (9) we obtain

$$p = \frac{\nu\gamma_o(\gamma-1)}{\nu\gamma + 1}\frac{V_1}{8\pi r c_o^2}\frac{d\phi'}{dt}$$

$$= \frac{\nu\gamma_o(\gamma-1)}{\nu + 1}\frac{V_1 \rho_1 C_v}{8\pi r c_o^2}\frac{d^2 T'}{dt^2} \quad (13)$$

where c_o and γ_o are the speed of sound and the specific heat ratio, respectively, for the surrounding medium.

The Source Term

Power to sustain the explosive phase is assumed to come solely from the chemical reaction. Power per unit volume expressed in the Arrhenius form [see Eq. (4)] may be linearized as before at about, say, the burning temperature of the flame to give

$$\phi' = Q\alpha_1^n K(T_1)\exp\left[\frac{E^*}{RT_1}\left(\frac{T'}{T_1} - 1\right)\right] \quad (14)$$

E^* is the effective mean activation energy which includes the temperature dependence of K, so that

$$\frac{d\phi'}{dt} = \frac{E^*}{RT_1^2}\phi'\frac{dT}{dt} \quad (15)$$

and again using Eq. (9) we obtain

$$\frac{d\phi'}{dt} = \frac{E^*}{RT_1^2}\frac{\gamma(\nu+1)}{\gamma\nu + 1}\frac{1}{\rho_1 C_p}(\phi')^2 \quad (16)$$

Expressing the rate of chemical reaction in terms of the charateristic time τ of the reaction at temperature T_1 we have

$$\frac{d\phi'}{dt} = \frac{\nu+1}{\gamma\nu+1}\frac{\rho_1 C_v RT_1^2}{E^*}\frac{e^{2\theta}}{\tau^2} \quad (17)$$

where

$$\tau = \rho_1 C_v RT_1^2 \exp(E^*/RT_1) / E^* Q\alpha_1^n K(T_1)$$

and

$$\theta = E^* T'/RT_1^2$$

If Eq. (17) is used with Eq. (13) we obtain finally the time developing amplitude of the acoustic pressure pulse as

$$p(t) = \frac{\nu(\nu+1)}{(\gamma\nu+1)^2} \frac{V_1 RT_1 \rho_o}{8\pi r E^* \tau^2} \left[1 - \frac{(\nu+1)t}{(\gamma\nu+1)\tau}\right]^{-2} \quad (18)$$

This equation does not include the consequences of the finite nature of the combustible mixture, so it can only describe the leading edge of the pressure pulse and not its decay.

Pressure Maximum

In order to obtain an estimate of the pressure maximum of the acoustic pulse we need an estimate of the characteristic time of the chemical reaction which is obtained in the following way. The equivalent thickness of the preheated layer for the laminar flame front h_1 is taken as

$$h_1 = \chi/u_n$$

Where χ is the heat diffusivity coefficient and u_n the normal flame velocity. The thickness of the zone of chemical reaction h_2 in the flame will be less than this by a factor of RT_1/E so that

$$h_2 = \chi RT_1/u_n E$$

The characteristic time of chemical reaction τ will be of the order of the time particles spend in this zone. Particles cross the zone (of mean temperature T_1) with an average speed u_b, thus

$$\tau \approx \frac{h_2}{u_b} = \frac{h_2 T_o}{u_n T_1} = \frac{\chi RT_o}{u_n^2 E} \quad (19)$$

For the exploding cylinder we then postulate that the pressure maximum will occur soon after the radius has been reduced to a value of, at most, the order of the thickness of the preheated layer h_1. Under these circumstances the temperature of the gas will be close to the temperature of the flame, and θ from Eq. (17) will approach zero. From Eq. (12), and with these assumptions, we obtain

$$p_{max} = \frac{\nu(\nu+1)\rho_o u_n^2 \, T_1 \, E^* \, L}{(\gamma\nu+1)^2 \, 8\pi r \, RT_o^2} \quad (20)$$

where L is the length of the cylindrical neck.

Discussion

For the experiments described the only fuel gas used was methane with varying gas-air ratios. For a primary mixture of fuel gas-to-air of 1:6, we may assume $E/RT_o \approx 50$ and $u_n \approx 0.4$ m/s. With these values Eq. (20) predicts that at the position of the microphone (0.16 m from the flame), the maximum of the pressure pulse on collapse of the cylindrical stalk will be 17 Pa/m; this compares well with the experimental value.

To use this method for investigating the contribution and significance of various physical and chemical processes in the formation of thermal explosions of spherical bubbles of combustible mixtures, it is necessary to rewrite Eq.(10) as

$$p = \frac{\rho}{2r} \frac{d}{dt}\left(a^2 \frac{da}{dt}\right) \quad (21)$$

where a is the radius of the bubble. If the flame speed varies according to the Markstein expression (Zel'dovich et al. 1980), we have

$$-\frac{da}{dt} = u_n\left(1 + \frac{\mu}{a}\right) \quad (22)$$

where, as before, u_n is the normal burning velocity and μ is the Markstein constant. Combining Eqs. (21) and (22) we obtain

$$p = \frac{\rho}{2r} u_n^2 \, a\left(2 + \frac{\mu}{a}\right)\left(1 + \frac{\mu}{a}\right) \quad (23)$$

This last equation reduces to that of Thomas and Williams (1966) for bubbles of large radius when $\mu/a \ll 1$. However, for a bubble of small radius near the end of its combustion, when $\mu/a > 1$, the predicted acoustic pressure will be substantially higher.

We may obtain an estimate for μ from the expression given in (Zel'dovich et al. 1980)

$$\mu = \left[\frac{\chi-D}{\chi}\frac{E(T_b-T_o)}{2RT_b^2} + \frac{D}{\chi}\right]\frac{\chi}{u_n} \quad (24)$$

which shows that for the Lewis number (D/χ) of unity, the factor μ/a will play a role only at a small bubble radius. However, for Lewis numbers not equal to unity, the ratio of the Markstein length to the thickness of the preheat layer may be as large as 10; thus μ/a would play a significant role at a bubble radius much greater that the thickness of the preheat layer.

The determination of both the radius at which bubbles undergo the thermal explosion described and the accompanying acoustical pressure pulse, for different gases, different air-fuel ratios and fuel additives may thus provide a useful method for investigating the interdependence of important combustion parameters such as Lewis number, flame velocity, and Markstein length.

Acknowledgements

This work forms part of the collaborative research carried out under the terms of an Agreement between the Institute for Problems in Mechanics, Moscow and the Polytechnic of the South Bank, London.

References

Bragg, S.L. (1963) Combustion noise. J. Inst. Fuel 36, 12-16.

Hurle, I., Price, R., Sugden, T., and Thomas, A. (1968) Sound emission from turbulent premixed flames. Proc. R. Soc. London, Ser. A: 303, 409-427.

Smith, J., Vuillermoz, M., and Roberts, J. (1983).A possible source of non-linearities in pyro-acoustic amplification. Acoustics Letters 5, 66-69.

Smith, T. J. B. and Kilham, J. K. (1963) Noise generation by open turbulent flames. JASA 35, 715-724.

Thomas, A. and Williams, G. (1966). Sound emission from spark-ignited bubbles of combustible gas. Proc. R. Soc. London, Ser. A: 294, 449-466.

Vuillermoz, M. and Roberts, J. (1984) Impedance matching for small gaseous sound sources. (in preparation).

Zel'dovich, Ya. B., Barenblatt, G. I., Librovich, V., and Makhviladze, G. M. (1980). The Mathematical Theory of Combustion and Explosion. pp. 404-407. Moscow, Nayka.

Comparisons of Experimental and Computed Length Scales and Velocities in Turbulent Combustion

A. Y. Abdalla,* D. Bradley,† S. B. Chin,‡ and C. Lam§
University of Leeds, Leeds, United Kingdom

Abstract

Experimental values are presented of turbulent length scales obtained in traverses through the reaction zone in a jet-stirred conical reactor. The values were based on fluctuating ionization measurements with electrostatic probes and on fluctuating temperatures with thermocouples compensated for thermal inertia. Space-time correlations were derived from the linked computer. Time scales were found from the area under correlation curves. Gas velocities were obtained from twin electrostatic probes, a known distance apart, from a derivation of the time at which the value of the space-time correlation coefficient for the two signals was a maximum. Multiplication of ionization and temperature autocorrelation times by the measured velocities gave the corresponding length scales. In addition, further values of length scales were derived from the cross correlation of ion current signals with variable probe separation distance. Gas velocities and the various length scales are compared with those predicted by a mathematical model with a k-ε model of turbulence. Agreement is better in the region of the central jet than close to the wall. Some of the limitations in the k-ε model are discussed in relation to the experiments.

Presented at the 9th ICODERS, Poitiers, France, July 3-8, 1983. Copyright © American Institute of Aeronautics and Astronautics, Inc., 1984. All rights reserved.
*Research Fellow, Mechanical Engineering Department.
†Professor, Mechanical Engineering Department.
‡Rolls-Royce Research Fellow, Mechanical Engineering Department.
§Research Student, Mechanical Engineering Department.

Introduction

Turbulent length scales, although important in characterizing turbulence, have not been as extensively measured as other parameters in turbulent flames. Furthermore, knowledge of their values can aid in the assessment of the validity of mathematical models. For example, in the k-ε model, the expression for eddy viscosity is in terms of both k, the turbulent kinetic energy, and ε, the rate of dissipation of that energy. A length scale only becomes explicit in the expression for ε. Thus, not only root mean square turbulent velocities, but also length scales should be measured as part of any examination of the validity of such a model. Any such assessment, however, requires compatibility between the definition of the scale and the associated numerical constants in the model (Abdalla et al. 1981).

The present work describes measurements of length scales, based upon ionization and temperature, as well as of velocities inside a jet-stirred reactor. Because of limited access, velocities could not be measured with laser velocimetry.

Experimental Technique

Measurements were made inside a previously described (Abdalla et al. 1981) conical, jet-stirred reactor, shown in Fig. 1. Results are reported for a methane-air premixture, of equivalence ratio, $\phi = 0.84$, with mean entry velocities of 60 and 130 ms^{-1}. Space-time correlations were obtained from signals from electrostatic probes and thermocouples, inserted through ports A and B, and these yielded gas velocities and turbulent length scales. Data collection, storage and processing were based upon a VAX 11-780 computer.

The double electrostatic probes for velocity measurements comprised two identical Pt - 40% Rh wires of 0.1-mm diam with spherical tips of 0.15-mm diam. One probe was aligned behind the other in the direction of the computed stream lines (Abdalla et al. 1981). For the single probe measurements, the basic construction was the same as for the double sensor. These wires were insulated within an alumina tube up to 2 mm from the sphere, whilst a water jacket tube around the alumina maintained its insulation at high temperature. The nozzle, electrically grounded, acting as the other electrode.

An electronic control unit, with a bandwidth of 15 kHz, provided both a variable stabilized dc biasing voltage to

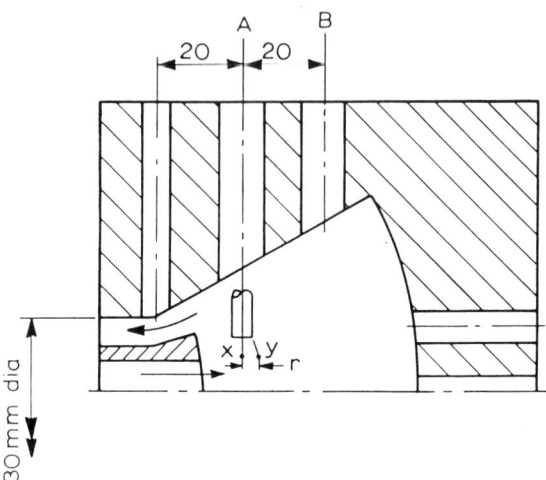

Fig. 1 Reactor and double electrostatic probe.

the probe in the range of ±12 V and also the required amplification. The analog signals from the probe were amplified close to the limits (±10 V) of the 14-bits analog-to-digital converter (ADC) for better resolution. The signals were digitized by the ADC and fed into the VAX 11-780 computer via a direct memory access unit DR-11B interface. A total of 16,384 point values were sampled at time intervals of 33 µs and stored on a magnetic tape for subsequent processing. The probe was biased at -3 V and the nozzle served as the other, Earthed electrode. In order to avoid any Earth loop, all hardware was connected to one common Earthing point. Details of the fluctuating temperature measurement techniques, including electronic compensation for thermal inertia, have been described by Abdalla et al. (1983).

The cross-correlation, or space-time correlation coefficient, $R_{xy}(r,\tau)$ describes the similarity of two random signals $x(t)$ and $y(t)$ from two sensors at a distance apart r as a function of the time shift τ between the signals. When r=0, $y(t) = x(t)$ and the cross correlation becomes the autocorrelation coefficient $R_{xx}(0,\tau)$ of $x(t)$.

Experimental Results

Convection Velocity

Taylor's hypothesis, that at low turbulence levels gradually distorting eddies possess a convection velocity (though not necessarily equal to the mean velocity), has been used previously with space-time correlations of velocity fluctuations in studies of turbulent shear flows (Wills 1964; Favre 1965). This technique also has been applied to signals from thermocouples and electrostatic probes (Cox 1976; Cox and Chitty 1980) and from laser Schlieren light and thermocouples (Lewis and Moss 1979) in flame studies. Computed turbulence intensities were in the region of 10% and the same technique appeared valid in the present circumstances. The maximum value of the space-time correlation coefficient occurs when the time delay τ_m is such that both signals are recording the passage of the same eddy. So, if the sensing probes are at a distance r apart, the convection velocity U_c is defined as the ratio r/τ_m; where τ_m satisfies

$$\frac{\partial R_{xy}(r,\tau)}{\partial \tau} = 0 \qquad (1)$$

This technique was applied to signals from the double electrostatic probe. To obtain a sufficient degree of correlation the convection time between the sensors should not be greater than the eddy lifetime. For the present conditions this led to r < (4.4 x the integral length scale). Separation distances of either 2.5 or 5.0 mm were selected on this basis. Typical cross-correlation coefficients, $R_{xy}(r,\tau)$ are plotted against time delay τ in Fig. 2. In Fig. 3 the derived convection velocities are shown together with the local mean velocities, computed on the basis of the mathematical model previously described (Abdalla et al. 1981). The principal error in the measured velocity arises more from that in the measurement of r than in the definition of the peak value of the correlation coefficient. The associated error in the velocity is in the region of ± 5%.

Ion and Temperature Length Scales

A characteristic time scale τ_a can be defined as the area under the autocorrelation curve, namely

$$\tau_a = \int_0^\infty R_{xx}(0,\tau) \, d\tau \qquad (2)$$

The correlograms always showed a monotonically decreasing function with time, with positive correlation, close to exponential decay. Therefore, a convenient definition of τ_a is that time in which the autocorrelation coefficient falls to 1/e of its initial value. For saturation ion current signals to present an autocorrelation time based upon ion

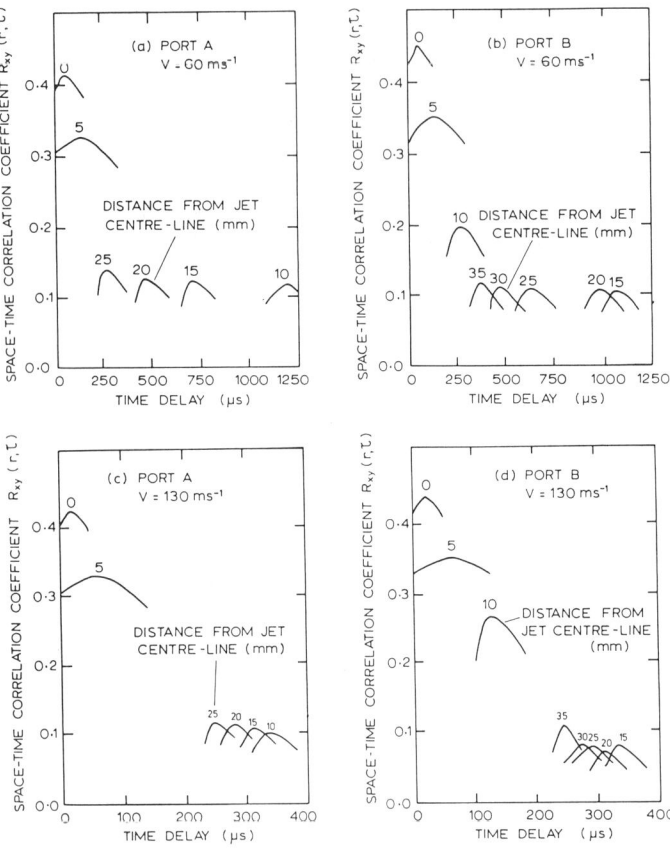

Fig. 2 Cross-correlation coefficient of ion current: r = 5 mm, except at centerline where r = 2.5 mm.

density it is necessary to allow for the nonlinearity of the probe current response characteristic with respect to this density. Therefore, a correction factor to the ion current characteristic time scale has to be applied. Page et al. (1975) argued that since it was generally accepted that the probe current was proportional to the 3/4 power of the ion density, the characteristic time scale could be defined as that time at which the autocorrelation coefficient fell to $(1/e)^{0.75}$ or 0.472. Such time scales were multiplied by the measured convection velocities to give the ion length scales L_{ia} shown in Fig. 4. Also shown in this figure are ion length scales based upon the computed velocities.

Another, "two point" ion length scale L_{ic} was estimated from the cross correlation with zero time delay between the

EXPERIMENTAL LENGTH SCALES

fluctuating ion current signals from two sensors at a variable distance r apart. The two point ion length scale L_{ic} is defined as the area under the space-correlation coefficent curve, namely

$$L_{ic} = \int_0^\infty R_{xy}(r,0) \, dr \qquad (3)$$

The nonlinearity of the ion current response was corrected on the same basis as for the single probe, and the value of L_{ic} was taken as the distance at which R_{xy} fell to a value of 0.472. The values so obtained along traverses are shown in Fig. 4.

Autocorrelation coefficients derived from fluctuating temperature signals were also obtained. Corresponding

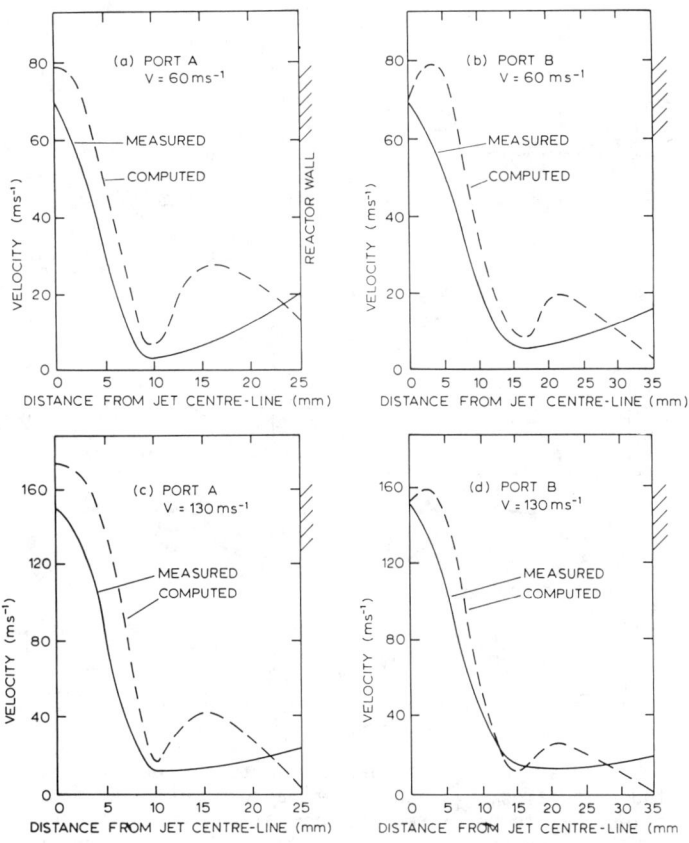

Fig. 3 Comparisons of measured convection velocity and computed local mean velocity.

thermal length scales L_{Ta} were estimated using the same method as for autocorrelated ion current signals, except that the characteristic time scale was taken to be that at which the autocorrelation coefficients had fallen to 1/e. These values also are shown in Fig. 4, based upon both measured and computed velocities.

The mathematical model has been discussed elsewhere (Abdalla et al. 1981, 1983). The flow modeling employed rests upon the k-ε model. This has largely been derived on the basis of isotropic, incompressible, thin shear flows without chemical reaction. Nevertheless, it has been used

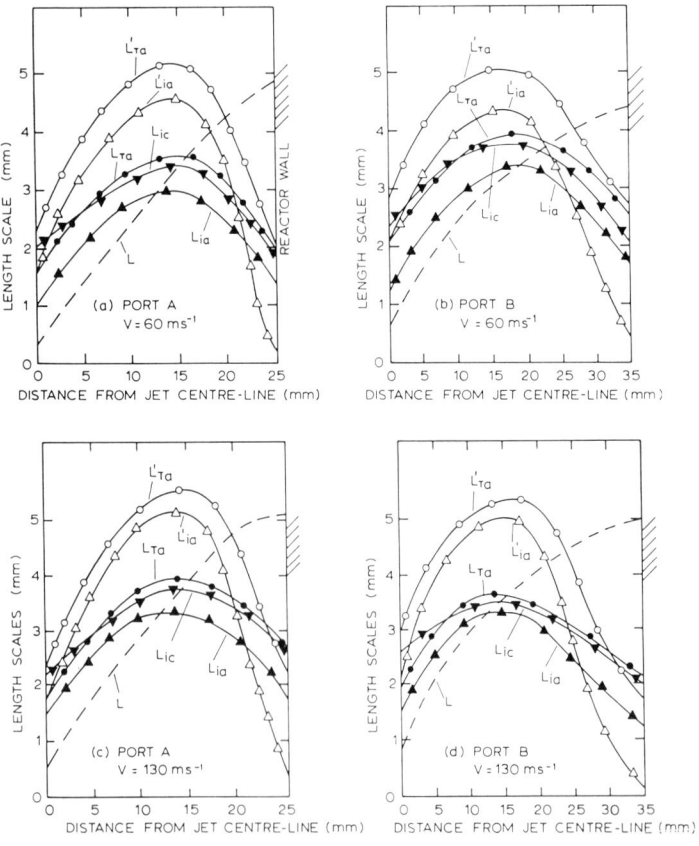

Fig. 4 Length scales: L'_{Ta} temperature autocorrelations from computed velocities, L_{ia} ion current autocorrelations from computed velocities, L_{Ta} temperature autocorrelations from measured velocities, L_{ic} ion current cross correlations, L_{ia} ion current autocorrelations from measured velocities, and L integral length scale (k-ε model).

widely in combustion models and the present experimental work was designed to give some indication of its general validity in this field. A dissipative length scale is associated with the k-ε model (Bilger and Beck 1975) and it is more desirable for purposes of experimental comparisons to recover the integral length scale L. Abdalla et al. (1981) have shown this is done in

$$L = C_D k^{1.5} \varepsilon^{-1} \tag{4}$$

with a value of 0.202 assigned to C_D.

Because the present flow pattern, with its round jet flow and recirculation was rather different from those for which the k-ε model was developed, the values of the other constants were computer optimized to the cold flow measurements in a reactor of Hankinson (1975). This yielded values of $C_{\varepsilon 1}$ and $C_{\varepsilon 2}$ of 1.48 and 1.98, respectively. The wall boundary conditions were matched to the two adjacent computational nodes within the fluid by a quadratic extrapolation procedure. The values of L yielded by the model with these modifications are shown by the broken curves in Fig. 4.

Discussion

Figure 3 shows the profiles of computed local mean flow velocity and measured velocity. The possibility arises that the correlation technique for velocity measurement inherently might give values that are too low. There is some evidence for this not only in the present work, but also in studies of isothermal jets (Willis 1964). Certainly, the measured velocities are generally below those computed. The elevation of the computed above the measured velocity is greatest in the recirculation zone. This probably reflects the well-known deficiencies of the k-ε model in such zones. At less than 3 mm from the wall the computed velocities are lower than the measured convection velocities. This suggests an underestimation of velocity by the k-ε model, possibly arising from its inherent assumption of isotropic diffusivity. The cold flow hot wire velocity measurements of Hankinson (1975) in the cone also revealed such a tendency.

The space-time correlation measurement technique is most suitable when the mean flow is unambiguous. For the jet-stirred reactor, particularly in the recirculation zone, a probe with four sensors would be required to measure the components of velocity along three orthogonal axes. However, geometrical limitations precluded this approach. Figure 4 shows the computed length scales are in best agreement with the measured ion and thermal length scales near the jet and in the recirculation zone. The discrepancy increases as the wall is approached, again suggesting deficiencies in the model in this region. The model assumption of isotropic diffusivity is clearly incorrect as the wall is approached.

On the basis of the autocorrelations and the measured convection velocities, the derived ion length scales L_{ia} are everywhere smaller than the thermal scales L_{Ta}. Similar results have been observed in turbulent round jet diffusion flames (Yanagi and Mimura 1981). Although there is a fair agreement between L_{ia} and L_{Ta} in the jet region and reaction zone, the discrepancy becomes greater in the recirculation zone and also near the wall. The smaller length scales obtained from the ion current fluctuation beyond the reaction zone might result partially from the recombination of ions and the ensuing more uniform ionic composition after combustion. The profiles of L_{ia} and those of the "two point" ion length scales L_{ic} based upon the cross correlation show some degree of similarity. Autocorrelation length scales based upon the computed velocities are, in general, higher and indicated by L'_a. In any comparison of these experimental length scales with those computed on the basis of the k-ε approach L it must be remembered that the latter are essentially based upon velocity fluctuations. There is no fundamental reason why they should be identical.

Recently, there has emerged some experimental evidence, from both diffusion (Takagi et al. 1981) and premixed (Yanagi and Mimura 1981) flames, that indicates a degree of correlation between temperature and velocity fluctuations. It has also been shown that in the reaction zone of premixed flames, the measured integral length scale L, derived from velocity fluctuations, is approximately twice the thermal length scale derived from temperature fluctuations.

It is instructive, however tentatively, to discuss how measured values of length scales might give rise to changes in the values assigned to the numerical constants in the k-ε model. In the present k-ε model, the integral length scale was computed from Eq. (4). A value of 1.0 for C_D has been used by other workers (Jones and Whitelaw 1982), but in those cases L is not the integral length scale. The eddy viscosity is given by

$$\mu_t = C_D C_\mu \rho k^2/\varepsilon \qquad (5)$$

in which ρ is the density and the value of the constant C_μ had been optimized previously at 0.36 (Abdalla et al. 1981).

Table 1 Values of C_D and C_μ

	C_D	C_μ	$C_D C_\mu$
Jones and Whitelaw (1982)	1.00	0.09	0.09
Abdalla et al. (1981)	0.202	0.36	0.072
Revised values	0.27	0.27	0.072

Values of C_D and C_μ in these references are compared in Table 1.

The isothermal, hot wire results of Hankinson (1975) in a conical reactor suggest the autocorrelation technique gives a velocity length scale L_x appropriate to the axial direction of the flow. For the flame brush around the central jet, it is the length scale normal to the jet flow that is important in the flame transport processes, and Hinze (1975) has suggested that this length scale is $L_x/2$. Reference to Fig. 4 does indeed show, at the jet axis, computed values of L that are approximately half those of the two autocorrelation length scales.

On the other hand, if velocity length scales normal to the jet were to be taken as twice the thermal scales in that direction, as suggested by some workers, to obtain a value of L from that of L_{Ta} the measured value of the latter should be doubled on this count and halved to give a value normal to the jet. Thus L should be equal to L_{Ta}. In the reaction zone, about 5 mm from the jet centerline, the present work gives L approximately equal to 0.75 L_{Ta}. Equation (4) shows that to make L equal to L_{Ta}, the value of C_D should be 1/0.75 of the original values. This revised value of 0.27 also is listed in Table 1. To maintain the same value of μ_t, Eq. (5) suggests the product $C_D C_\mu$ should be unchanged and this determines the revised value of C_μ. This exercise is not intended basically to "improve" the model by alterations to the values of some of the constants. Clearly, it has more serious, fundamental, limitations that must be addressed for combustion in recirculating flows.

Acknowledgments

Support is acknowledged from the Science and Engineering Research Council, Leeds University and Rolls-Royce Ltd. (Aero Division).

References

Abdalla, A. Y., Ali, B. B., Bradley, D., and Chin, S. B. (1981) Stratified combustion in recirculating flow. Combust. Flame 43, 131-143.

Abdalla, A. Y., Bradley, D., Chin, S. B., and Lam, C. (1983) Temperature fluctuations in a jet-stirred reactor and modeling implications. Nineteenth Symposium (International) on Combustion, pp. 477-486. The Combustion Institute, Pittsburgh, Pa.

Bilger, R. W. and Beck, R. E. (1975) Further experiments on turbulent jet diffusion flames. Fifteenth Symposium (International) on Combustion, pp. 541-552. The Combustion Institute, Pittsburgh, Pa.

Cox, G. (1976) Some measurements of fire turbulence. Fire and Materials, 1, 116-122.

Cox, G. and Chitty, R. (1980) A study in the deterministic properties of unbounded fire plumes. Combust. Flame 39, 191-209.

Favre, A. (1965) Review on space-time correlations in turbulent fluids. J. Appl. Mech. 32E, 241-257.

Hankinson, G. (1975) Combustion in recirculating flow. Ph.D. Thesis, Leeds University, Leeds, United Kingdom.

Hinze, J. O. (1975) Turbulence p. 185. McGraw-Hill, New York.

Jones, W. P. and Whitelaw, J. H. (1982) Calculation methods for reacting turbulent flows: a review. Combust. Flame 48, 1-26.

Lewis, K. J. and Moss, J. B. (1979) Time-resolved scalar measurements in a confined turbulent premixed flame. Seventeenth Symposium (International) on Combustion, pp. 267-277. The Combustion Institute, Pittsburgh, Pa.

Page, F. M., Roberts, W. G., and Williams, H. (1975) An experimental study of interaction of chemical kinetic effects and turbulent flow in flames. Fifteenth Symposium (International) on Combustion, The Combustion Institute, pp. 617-624. Pittsburgh, Pa.

Takagi, T., Shin, H. D., and Ishio, A. (1981) A study on the structure of turbulent diffusion flames: properties of fluctuations of velocity, temperature, and ion concentration. Combust. Flame 41, 261-271.

Wills, J. A. B. (1964) On convection velocities in turbulent shear flows. J. Fluid Mech. 20, 417-432.

Yanagi, T. and Mimura, Y. (1981) Velocity-temperature correlation in premixed flame. Eighteenth Symposium (International) on Combustion, The Combustion Institute, pp. 1031-1039. Pittsburgh, Pa.

Flow Rate and Equivalence Ratio Influences on the Thermal Field of a Turbulent Cool Flame

I. Gökalp*
Centre National de la Recherche Scientifique, Orléans, France
and
N. Zarrad,† G.M.L. Dumas,† and R. I. Ben Aïm†
Université Pierre et Marie Curie, Paris, France

Abstract

The aim of this work is to show that turbulent cool flames (TCF) may be used to investigate low-Damköhler-number (slow chemistry) turbulent flames and to obtain information about the influence of turbulence on chemical reactions. In the present work, a premixed heptane-air cool flame is stabilized in a horizontal conico-cylindrical reactor that is placed in an oven. Extensive measurements of the mean and fluctuating thermal field of the TCF have been made. Fluctuating temperatures are measured by electronically compensated 25- m-diam thermocouples. The global behavior of the TCF at various flow rates and equivalence ratios was determined from the mean temperature axial profiles. The variations of the turbulent propagation velocity of the TCF, with the turbulence level and the equivalence ratio, are similar to those of normal turbulent flames. Fluctuating temperature axial profiles are parabolic, with a maximum always situated at the half-width of the flame front. This maximum temperature fluctuation is not influenced by flow rate but increases with increasing equivalence ratio. The observed normalized temperature fluctuation agrees well with that predicted by the Bray-Moss unified model, with a uniform shape for the burning mode function $f(c)$. The measured temperature probability density fuctions

---

Presented at the 9th ICODERS, Poitiers, France, July 3-8, 1983. Copyright © 1984 by the American Institute of Aeronautics and Astronautics, Inc. All rights reserved.
*Centre de Recherches sur la Chimie de la Combustion et des Hautes Températures.
†Laboratoire de Chimie Générale.

show large proportions of intermediary states in the main reaction zone of the TCF.

Introduction

Theoretical models in turbulent flames are mainly devoted to the wrinkled laminar flames for which the fast chemistry hypothesis is valid, and extensive studies of the turbulence structure have been performed for such flames. The alternative model, i.e., the distributed flame front, was anticipated in the pioneering work of Damköhler (1940). Although some evidence suggests that this model describes some practical combustion problems, the distributed flame front is very difficult to realize in laboratory flames. Consequently, experimental information about its structure is almost nonexistent. To analyze turbulent combustion in terms of interactions between chemical reactions and turbulence, experimental data for the main reaction zone of a turbulent flame are needed, since in this region this interaction is the more important.

A thorough investigation of this zone, by the usual measurements techniques is only possible with distributed flame fronts. The unified models-such as that of Bray and Moss (1977)-use physical ideas (coexistence of burnt, unburnt, and burning gas pockets) that are more consistent with the extended flame front configuration. These models are, up to now, compared with measurements made for the "upper limiting case" (wrinkled laminar flames) with the assumption that the flame front flapping to and fro past the observation point produces a similar result to that produced by the passage of discernible burnt and unburnt gas pockets.

Experimental information for extended turbulent flame front configurations with slow (or finite) chemistry should enrich the understanding of the turbulent interactions. Furthermore, as recently noted by Jones and Whitelaw (1982) finite chemistry considerations are important in ignition and extinction phenomena and also for calculations of pollutant formation. Since 1978, the turbulent cool flame (TCF), a distributed flame front of low exothermicity, have been studied (Gökalp 1981). Cool flames of hydrocarbons have been extensively studied in the laboratoire de Chimie Générale in Paris, either in a static system (Ben-Aim 1960) or in a laminar flow (Dumas et al. 1973). Cool flames are characterized by a very incomplete combustion (with the production of about 60 species), a very weak light emission, a weak exothermicity, and, in a dynamic system at atmospheric pressure, a laminar front thickness of about 1 cm. Several physical simplifications for the experimental investigation of the turbulent combustion arise from these peculiar features. The most important is a thick flame front that permits precise turbulence measurements in the main reaction zone, without the intermittency effects due to the wrinkling of the flame front. The low exothermicity of the TCF permits the use of 1) fine wires (resistance thermometer or thermocouple) for thermal field measurements, and 2) nickel, chrome-nickel-alloy thermocouples without any coating since there are minimal catalytic effects. Because of the TCF low exothermicity, comparisons are possible with nonreactive flows having a similar mean temperature field and similar dynamic parameters as the reacting flow. Fluctuating temperature field comparisons between the TCF and a strongly heated airflow presenting an axially homogeneous temperature field are reported in Gökalp et al. (1981). For turbulent field modifications with strong heating, see also Gökalp and Lasek 1979, Gökalp et al. 1982, and Gökalp 1982. Comparisons between the TCF and a heated airflow presenting an axial temperature gradient similar to the one of the TCF are in progress in our group; some preliminary results are given in Zarrad (1982). Weak exothermicity of the TCF diminishes the dilatation effect, which is an important element of the turbulence balance in normal flames (Bray and Libby 1976). As flame stabilization for the TCF is obtained by auto-ignition in a conical reactor, the complications introduced by the presence of flameholders or pilot flames are avoided.

Previous measurements on the thermal structure of the TCF were presented in Gökalp et al. (1981) with a particular emphasis on the characteristics of the fluctuating temperature spectrum in different regions of the TCF. When compared with similar measurements in heated nonreactive flows, these results have shown the specific influence of the chemical reactions on the thermal field fine structure. The small eddies (high-frequency components) in a heated nonreactive airflow become less and less energetic with increasing mean temperature, while, on the contrary, they become more energetic in the flame for a comparable mean temperature level. Although spectral measurements exist in the literature, the probabilistic analysis of the turbulent combustion is now largely preferred to the spectral one. However, a thorough understanding of this complex phenomenon requires the use of the physical information arising from both of these approaches. The present work is part of on-going investigation of the TCF. While the combustion apparatus is the same as in Gökalp (1981), the fluctuating temperature measurements are performed with an electronically compensated fine thermo-

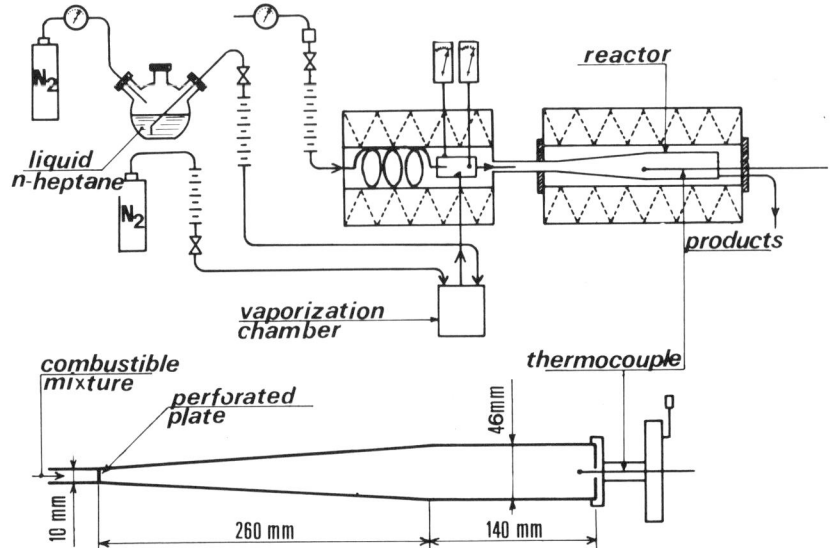

Fig. 1 Schematic representation of the combustion apparatus.

couple instead of the previously used digitally compensated fine-wire technique. The mean and the fluctuating temperatures and the associated probability density functions are measured in different regions of TCF for various values of the flow rate and the equivalence ratio.

Experimental Considerations

The heptane-air premixed cool flame is stabilized by auto-ignition in a horizontal adiabatic flow reactor placed in a regulated oven (Fig. 1). Liquid normal heptane of 99% purity is admitted to a vaporization chamber that is maintained at the fixed temperature of 150°C. Compressed air is filtered and preheated to 340°C in an oven that also contains a device to mix air and vaporized heptane. The mixture is then fed at atmospheric pressure to the axisymmetric quartz reactor. The conical section has an inlet diameter of 1 cm, a total cone angle of 7 deg, and a length of 26 cm. The cylindrical sections diameter is 4.6 cm and the total reactor length L is 40 cm. A pefforated plate, with 35 holes of a diameter of 0.8 mm each, is placed at the conical inlet section and serves as a turbulence grid. Its position defines the origin of the space coordinate X along the reactor axis. The cold flow dynamic field characteristics are obtained by the hot-wire

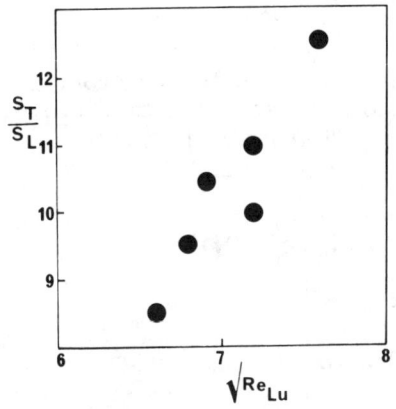

Fig. 2 Correlation of the TCF S_T/S_L data.

anemometry technique. The mean and the fluctuating temperature fields in the TCF are determined with 25-μm-diameter nickel-chrome-nickel alloy thermocouples. The electric welding technique gives a bead diameter sligthly greater than the wire diameter. The thermocouple is welded on a probe support made of 200-μm wires maintained with 5-mm spacing.

For temperature fluctuation measurements, the frequency response of the thermocouple is enhanced with a compensating electronic circuit (Lockwood and Moneib 1980). The response curves of the thermocouple have been tested for different time constants. The mean time constant in the flame is calculated theoretically from the heat balance equation of a thin cylindrical wire with negligible radiation and conduction effects and with local values of the flow parameters. A correction due to the presence of the bead is also applied. The final adjustments of the mean time constant are made with a knowledge of the measured temperature probability density functions(PDFs) in the flame. It is found that the time constant is practically invariant throughout the flame, with a mean value of 20 ms. The instantaneous temperature signal is tape recorded and then processed on a frequency and probability analyzer.

Characterization of the TCF

The TCF corresponds to the Damköhler lower limit, that is, the turbulence length scales are small in comparison with the laminar thickness. Measurements in laminar cool flames for variuos flow parameters and equivalence ratios (Gökalp 1981) have shown a laminar flame thickness e_L of about 1 cm at atmospheric pressure and a laminar burning velocity S_L of about 0.1 ms^{-1}. When these laminar flame parameters are compared with the axial characteristics of the ambient temperature dynamic field in the reactor, the following ratios calculated at the main stabilization region of the TCF (16 cm X 25 cm) for a flow rate of 1.03x10^{-3} m^3 s^{-1}, are obtained:

$$3200 < Re < 5600 \quad 0.14 < L_u/e_L < 0.17$$

$$3 < u'/S_L < 6 \quad 2.5 \text{ ms} < t_t = L_u/u' < 5 \text{ ms}$$

$$t_c = e_L/S_L = 100 \text{ ms} \quad 0.025 < Da = t_t/t_c < 0.05$$

where Re is the Reynolds number based on the mean axial velocity and the reactor diameter; L_u is the longitudinal integral length scale; u' is the rms axial velocity fluctuation; t_t is a characteristic turbulent time; t_c is a characteristic chemical time; and Da is the Damköhler number. The TCF is then characterized by a chemical time greater then the turbulence time, and is located at the top left-hand corner of the Bray diagram for turbulent flame propagation regimes (Bray 1980).

As Damköhler suggested, the principal effect of the turbulence in this limiting case is to enhance the transport processes within the flame. With some restricted hypotheses it may be shown that the ratio of turbulent to laminar burning velocities S_T/S_L is proportional to $(R_L)^{0.5}$, where R_L is the turbulent Reynolds number based on the longitudinal integral scale. Figure 2 shows the correlation of the TCF S_T/S_L data (for three values of the equivalence ratio and two values of the flow rate) with $(R_L)^{0.5}$. Stationary flame propagation velocities are calculated, at each stabilization section, from the room temperature values of the flow parameters. The good correlation observed on Fig. 2 confirms the distributed nature of the TCF.

Description of the Thermal Structure of the TCF

Figure 3 presents the general shape of the axial temperature profile in the TCF. Three distinct regions are clearly identified:
Region 1 corresponds to the slow oxidization reactions. The slight decrease of the temperature below the initial temperature of the reactor in the very beginning of this region is due to the endothermic reactions usually encountered in the first stages of low temperature oxidization (Ben-Aim 1960).

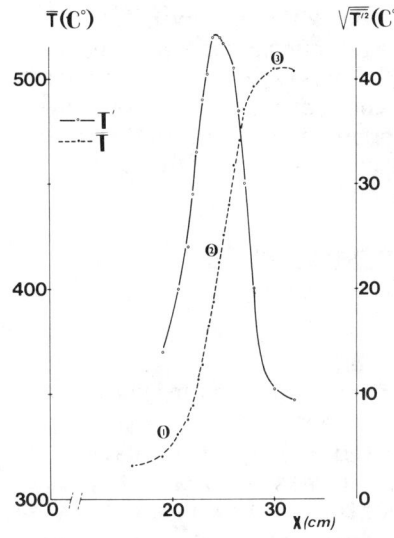

Fig. 3 Mean and fluctuating temperature profiles for a TCF where Φ = 0.5 and Q = 1800 liter/h.

Region 2 corresponds to a quasilinear temperature gradient of the order 25°C/cm. This zone, where the main combustion reactions occur, gives an idea of the flame front thickness which is about 6 cm.

Region 3 corresponds to the post-combustion zone, where the maximum temperature is 510°C. The cooling of the burnt gases by the walls gives rise to a slight decrease of the mean temperature with increasing distance.

The temperature fluctuations axial profile with a compensation level of 20 ms (Fig. 3) is parabolic with a maximum fluctuation of 43°C located at the middle of the flame front. The compensation level used for the thermocouple influences the observed magnitude of the maximum but does not affect its location. Typical instantaneous temperature signals, corresponding to the three regions of the TCF, are presented on Fig. 4. In region 1, narrow positive temperature spikes are observed; they indicate the existence of ocasionally burnt gas pockets in a low-temperature medium. In region 2, positive and negative temperature fluctuations occur with appearently equal probability. In region 3, narrow and high negative amplitude temperature spikes appear, indicating the existence of ocasionally unburnt mixture pockets in a high-temperature medium. The temperature PDFs corresponding to the instantaneous signals shown in Fig. 4 are presented in Fig. 5. In region 1, the PDF is a narrow quasi-Gaussian, with a positive tail corresponding to the burnt pockets of low probability. In the post-combustion zone, where the mean temperature increase is maximum, the PDF is asymmetrical, with a high-amplitude negative tail corresponding to the low-probability unburnt pockets; the most probable temperature is higher than the mean. In the intense reaction zone (region 2), where the maximum temperature fluctuation is observed, the PDF is bimodal with an important intermediary state corresponding to the burning mode.

Effect of Flow Rate and Equivalence Ratio on Mean Temperature

The influence of volumetric flow rate Q on the TCF mean temperature axial profiles is shown on Fig. 6. The superior limit of the flow rate is imposed by the stabilization conditions of the TCF. The mean temperature profiles are quite similar to those of normal flames (see, for instance, Chen and Churchill 1972) and with increased flow rate, the TCF is shifted toward the exit section of the reactor (X = L). The maximum mean temperature decreases with increased flow rate because of increased heat losses by the walls due to the thickening of the flame front. The decrease of the mean temperature in the post-combustion zone is less important when the flow rate increases, decreasing the cooling time of the hot gases. As shown on Fig. 7, increased equivalence ratios at constant flow rate shift the TCF

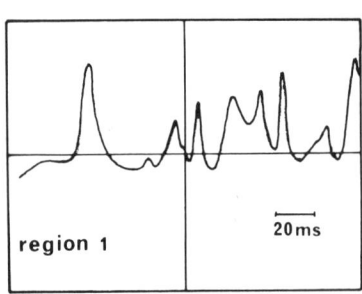
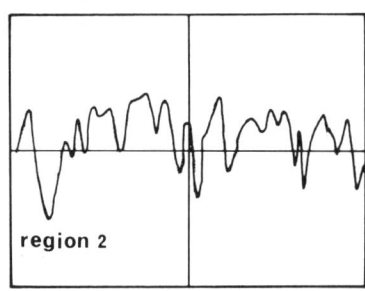
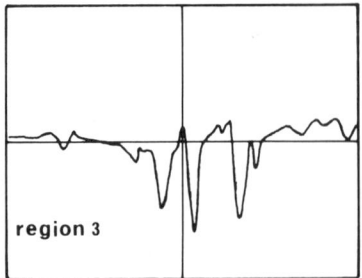

Fig. 4 Instantaneous temperature signals in the three regions of a TCF where ϕ = 0.5, Q = 1800 liter/h, τ = 20 ms.

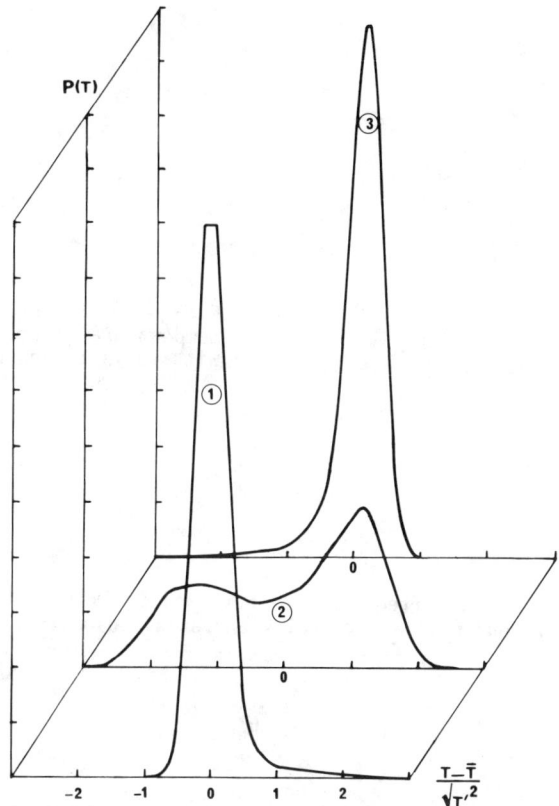

Fig. 5 Temperature probability density functions in the three regions of a TCF where $\Phi = 0.5$, $Q = 1800$ liter/h, $\tau = 20$ ms.

toward the inlet section of the reactor (X = 0). This result is also observed in laminar cool flames (Sahetchian et al. 1980). Simultaneously, the maximum mean temperature increases with increasing equivalence ratio.

Effect of Flow Rate and Equivalence Ratio on the Gross Characteristics of the TCF

The effect of flow rate and equivalence ratio on the following three properties of the TCF: 1) the flame propagation velocity S_T, 2) the magnitude of the mean temperature axial gradient $d\bar{T}/dX$, and 3) the thickness of the flame front e_T may be deduced from the behavior of the axial temperature increase profiles. The propagation velocities of the TCF are deduced from the continuty equation written for each stabilization section. Table 1 gives S_T values for two

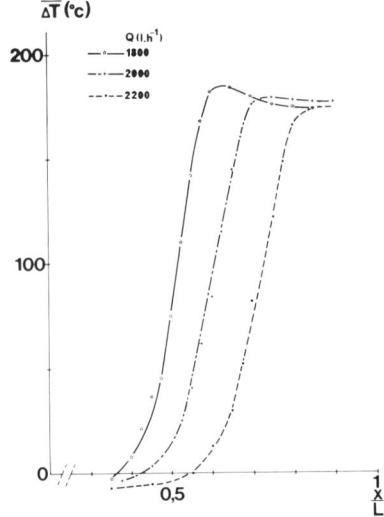

Fig. 6 Evolution of the mean temperature increase axial profiles with varying flow rate, $\Phi = 0.5$.

Table 1 Effect of flow rate and equivalence ratio on turbulent flame velocity and axial velocity fluctuation

Q (liter/h)	1800		2000	
Φ	S_T m s^{-1}	u' m s^{-1}	S_T m s^{-1}	u' m s^{-1}
0.5	1.00	(0.52)	0.86	(0.43)
0.7	1.10	(0.52)	0.95	(0.46)
0.8	1.26	(0.58)	1.05	(0.48)

flow rate and three equivalence ratio values. The rms axial velocity fluctuations (shown in parentheses) are indicative of the trend behavior of the turbulence energy but do not give the true turbulence level in the flame for the following reasons. First, they correspond to a higher flow rate, and second, they are measured at ambient temperature in an inert airflow.

Two main observations may be noted: 1) S_T decreases with increased flow rate at constant equivalence ratio. This result may seem anomalous, as it is well established that increases in the flow rate increase the level of turbulence and, hence, the propagation velocity. The TCF data are, however, not inconsistent since the turbulence energy in the reactor decreases with distance from the perforated

Table 2 Effect of flow rate and equivalence ratio on mean temperature axial gradient

Q (liter/h)	1800	2000	2200
ϕ	dT/dx C cm^{-1}	dT/dx C cm^{-1}	dT/dx C cm^{-1}
0.5	32.5	28.2	22.7
0.7	41.0	31.2	26.3
0.8	46.5	39.8	27.0

plate situated at the inlet section (Zarrad 1982). As the TCF shifts toward the exit section with increased flow rate, the reaction zone is stabilized in a region where the turbulence energy is more attenuated and S_T is observed to decrease. 2) S_T increases with increased equivalence ratio at constant flow rate. As the TCF shifts toward the inlet section, with increased equivalence ratio, the reaction zone is stabilized in a region of higher turbulence energy. These observations are in good agreement with the results obtained in normal flames (see, for instance, Dandekar and Gouldin 1982).

The magnitude of the mean temperature axial gradient is calculated from the linear part of the temperature increase profiles. Values of dT/dX, for three flow rates and three equivalence ratios are reported in Table 2.

Two definitions of the TCF thickness e_T were used. The first is based on the axial mean temperature profiles and equates e_T with $\Delta T_{max} (dX/dT)_{max}$. The second definition is calculated from the axial temperature fluctuations profiles (see the next section) where e_T is equated with axial distance between two points where the temperature fluctuation is the half of its maximum value.

As shown on Table 3, the e_T values given by these two definitions are quite comparable and indicate that, for constant equivalence ratio, e_T increases with increased flow rate and that, for constant flow rate, it decreases with increased equivalence ratio.

The Effect of Flow Rate and Equivalence Ratio on the Fluctuating Temperature Profiles and Discussion of the TCF Thermal Structure

Figures 8 and 9 show the axial behavior of the temperature fluctuations T' in the TCF. For both cases, the same

Table 3 Effect of flow rate and equivalence ratio on the TCF thickness

Q (liter/h)	1800		2000		2200	
	e_{T_1}	e_{T_2}	e_{T_1}	e_{T_2}	e_{T_1}	e_{T_2}
ϕ	cm	cm	cm	cm	cm	cm
0.5	5.60	5.68	6.30	6.10	7.57	7.47
0.7	4.73	5.26	6.06	5.89	....	....
0.8	4.41	4.63	4.87	5.47	7.00	6.73

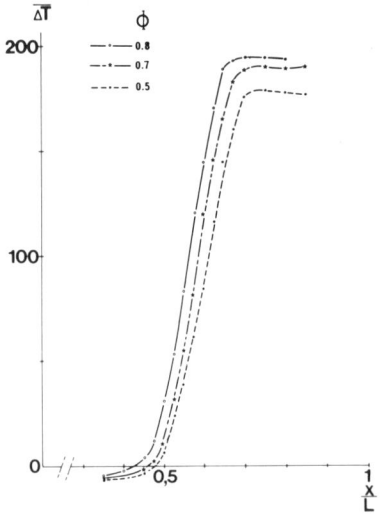

Fig. 7 Evolution of the mean temperature increase axial profiles with varying equivalence ratio, Q = 2000 liter/h.

mean compensation level of 20 ms is applied. Although the maximum temperature fluctuation is not influenced by the increased flow rate, it increases with increased equivalence ratio. Examination of the T' profiles and the mean temperature increase profiles indicates that the temperature fluctuation maximum is always located at the middle of the flame front for which $\Delta T = (\Delta T)_{max}/2$.

Figures 10 and 11 show the axial behavior of the normalized temperature fluctuations. The maximum thermal turbulence intensity $T'_{max}/\Delta T_{max}$ increases sligthly with increased flow rate because ΔT_{max} decreases when the flow rate increases (see Fig. 7). This observation suggests that the thermal turbulence intensity is more dependent on the chemical reaction rates than on the total heat release.

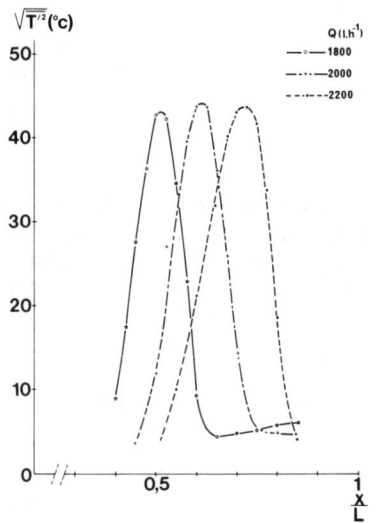

Fig. 8 Evolution of the temperature fluctuation axial profiles with varying flow rate; $\phi = 0.5$, $\tau = 20$ ms.

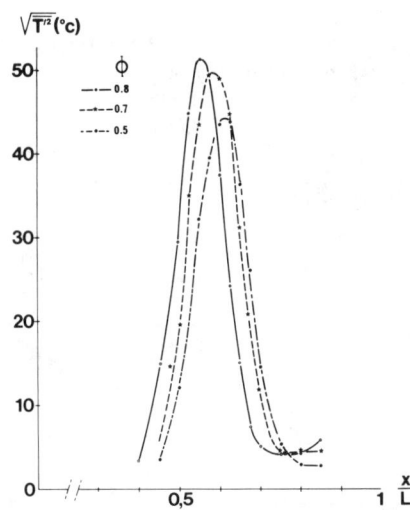

Fig. 9 Evolution of the temperature fluctuation axial profiles with varying equivalence ratio; $Q = 2000$ liter/h, $\tau = 20$ ms.

Normalized temperature measurements provide the basis for the discussion of the structure of the turbulent premixed flames. Following Damköhler, the ratio of a turbulence time t_t to a chemical time t_c is the starting point. The latter is proportional to e_l/S_l and the former is proportional to an energetic eddy time L_u/u'. The Damköhler upper limit is approached with normal flames for which the chemistry is fast, and consequently, t_t/t_c is infinite.

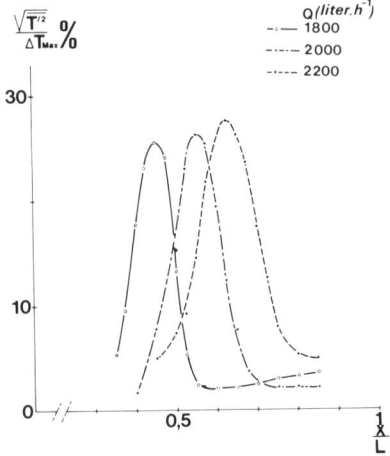

Fig. 10 Evolution of the temperature fluctuation intensity axial profiles with varying flow rate; $\Phi = 0.8$, $\tau = 20$ ms.

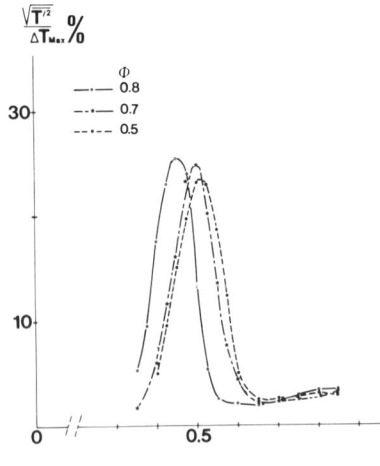

Fig. 11 Evolution of the temperature fluctuation intensity axial profiles with varying equivalence ratio; $Q = 2000$ liter/h, $\tau = 20$ ms.

The temperature field measurements of Yoshida and Günther (1980) are representative of this limiting case. Their measured temperature PDFs are effectively bimodal, with negligibly small intermediary states. The observed dependence of $T'/\Delta T_{max}$ on the temperature progress varible agrees well with the theoretical upper limit curve derived by Lewis and Moss (1979) for a temperature PDF modeled by two Dirac peaks.

To move from this upper limit, t_t may be reduced or t_c may be increased. Generally it is easier to reduce t_t, either by a reduction in the integral scale or by an increase in the velocity fluctuation. For instance, Bill et

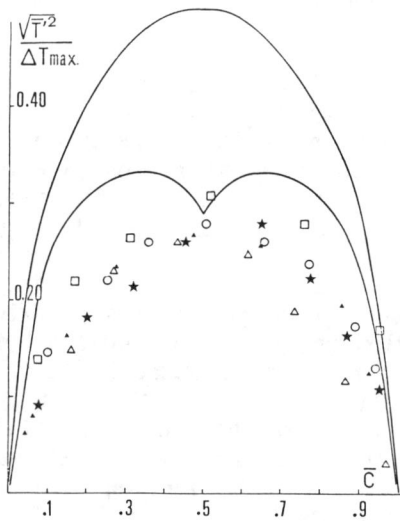

Fig. 12 Correlation of the temperature fluctuation intensity with the temperature progress variable. □ Q = 1800 liter/h, ϕ = 0.5; ○ Q = 2000 liter/h, ϕ = 0.5; △ Q = 2200 liter/h, ϕ = 0.5; ▲ Q = 2000 liter/h, ϕ = 0.7; ★ Q = 2000 liter/h, ϕ = 0.8.

al. (1982) show intermediary state proportions of about 30% for a propane-air burner flame. This observation is attributed to an insufficiently large integral length scale in comparison with the laminar flame thickness. In a jet stirred reactor, Abdalla et al. (1982) diminished t_t by an increase of velocity fluctuations which resulted from increased jet mean velocity. This does not mean that the dynamic field is composed of such small eddies located in the intense reaction zone, so that a distributed flame arises.

It is difficult to obtain an extended flame front without a modification of the laminar flame stucture. One of the major advantages of the TCF is that its laminar thickness for normal laboratory conditions is about 1 cm and its laminar propagation velocity is about 10 cm/s so that the chemical time is about 0.1 s. As the large eddy turbulence time varies between 2.5 and 5 ms, t_t/t_c varies between 0.025 and 0.05. Figure 12 shows, for various flow rates and equivalence ratios, the dependence of $T'/\Delta T_{max}$ on a mean temperature progress variable $c = \Delta T/\Delta T_{max}$. The experimental points are situated well below the thin flame limit curve, and the maxima of $T'/\Delta T_{max}$ varies between 0.23 and 0.30. The theoretical curve for a uniform temperature PDF in the unified model of Bray and Moss (1977) is also shown on Fig.12. This model predicts that $(T'/_\Delta T_{max})_{max}$ is 0.32.

Conclusion

The principal aim of this work was not to investigate the specific behavior of cool flames in a turbulent regime. But for reasons explained in the introduction, the cool flame is only used as an experimental support in order to investigate turbulent flames near the low Damköhler number limit. The thermal field measurements reported in this paper show that the TCF really constitutes an extended stabilized flame front of slow chemistry and low exothermicity. But, in spite of these particular features, the global behavior of the TCF is not different from the normal turbulent flames. Indeed, it has been shown that the propagation velocity and the mean temperature profiles of the TCF vary similarly to normal turbulent flames with varying flow rate, equivalence ratio, or turbulence level.

We think therefore, that the TCF is an appropriate experimental support for acquiring information about, on the one hand, low-Damköhler-number turbulent flames and, on the other hand, the turbulence chemical reactions interaction. Indeed, the study of the behavior of temperature fluctuations and temperature PDFs in the TCF shows clearly the existence of intermediary states corresponding to burning gas products that were observed with great difficulty in normal turbulent flames.

References

Abdalla, A. Y., Bradley, D., Chin, S. B. and Lam, C. (1982) Temperature fluctuations in a jet-stirred reactor and modelling implications. 19th International Symposium on Combustion, p. 495. The Combustion Institute, Pittsburgh, Pa.

Ben-Aïm, R. I. (1960) Théorie des explosions. Applications au cas des flammes froides du pentane normal. Université de Paris.

Bill, R.G., Namer, I. and Talbot, L. (1982) Density fluctuations of flames in grid-induced turbulence. Combust. Flame 44 (1-3), 277-285.

Bray, K. N. C. and Libby, P. A. (1976) Interaction effects in turbulent premixed flames. Phys. Fluids 19 (4), 1687 - 1701.

Bray, K. N. C. and Moss, J. B. (1977) A unified statistical model of the premixed turbulent flame. Acta Astronaut. 4, 291-320.

Bray, K. N. C. (1980) Turbulent flows with premixed reactants. Turbulent Reacting Flows in Turbulent Reacting Flows. (Edited by P. A. Libby and F. A. Williams), p. 130. Springer Verlag, Berlin.

Chen, J. L. P. and Churchill, S. W. (1972). Stabilization of flames in refractory tubes. Combust. Flame 18 (1), 37-48.

Damköhler, G. Z. (1940) Z. Electrochem. 46, 601-626. English Translation in the effect of turbulence on the flame velocity in gas mixtures (1947) NACA TN 1112.

Dandekar, K. V. and Gouldin (1981) Temperature and velocity measurements in premixed turbulent flames. AIAA paper 81-0179, Saint-Louis, Mo.

Dumas, G. M. L., Heiss A. and Ben-Aïm, R. I. (1973) Autoinflammation en régime dynamique de mélanges d'hexane normal et d'air. C. R. Acad. Sci. Paris 277 C, 1097-1100.

Gökalp, I. and Lasek, A. (1979) Fluctuations de temperature dans un écoulement à température élevée. Int. J. Heat Mass Transfer 22 (9), 1309-1317.

Gökalp, I. (1981) Sur les problèmes d'interaction entre une flamme et un écoulement turbulent. Flamme froide et écoulement à température élevée. Thèse de Doctorat ès-Sciences, Université Paris VI.

Gökalp, I., Gougat, P., Lasek, A. and Martin, F. (1982) Some characteristics of turbulent flows at elevated temperatures. Structure of Turbulence in Heat and Mass Transfer, (Edited by Z. P. Zariç). pp. 265-276, Hemisphere Publishing Corp. New York.

Gökalp, I. (1982) Thermal field structure in a turbulent flow laminarizing under strong heating. Proceedings of the 7th International Heat Transfer Conference. (Edited by U. Grigull, E. Hahne, K. Stephan and J. Straub). Volume 3, pp. 253-256, Hemisphere Publishing Corp. New York.

Jones, W. P. and Whiklaw, J. H. (1982) Calculation methods for reacting turbulent flows : A review. Combust. Flame, 48 (1), 1-26.

Lewis, J. K. and Moss, J. B. (1979) Time resolved scalar measurements in a confined turbulent premixed flame. 17th International Symposium on Combustion, p. 267. The Combustion Institute, Pittsburgh, Pa.

Lockwood, F. C. and Moneib, H.A. (1980) Fluctuating temperature measurements in a heated round free-jet. Combus. Sci. Technol 22 (1-2). 63-81.

Sahefchian, K. A., Heiss, A. and Dumas, G. M. L. (1980) Evolution des peroxydes moléculaires et radicalaires dans une flamme froide de n-heptane. J. Chimie Phys. 77 (6), 507-513.

Yoshida, A. and Gunther, R. (1980) Temperature and ionization measurements in turbulent premixed flames. AIAA Paper 80-0207, Pasadena, Calif.

Zarrad, N. (1982) Structure du champ thermique dans une flamme froide turbulente. Thèse de 3ème Cycle, Université Paris VI.

Turbulent Reacting Concentric Jets: Comparison Between pdf and Moment Calculations

P. Givi,* J. I. Ramos,† and W. A. Sirignano‡
Carnegie-Mellon University, Pittsburgh, Pennsylvania

Abstract

A pdf transport equation and a two-equation model of turbulence have been employed to study the extremely fast chemical reaction which occurs between ozone and nitric oxide in a two concentric jet geometrical configuration. The inner jet consists of nitrogen and traces of ozone; in the outer jet, nitrogen and traces of nitric oxide are used. The effects of unmixedness on the chemical reaction rate are studied by various approaches involving a joint pdf transport equation for the reactant concentrations, three eddy-break-up models, two second-order closure models, and a model which simply neglects the contribution of the concentration fluctuations to the chemical reaction rate. Calculations have been performed at different Damkohler numbers in order to establish the contribution of the unmixedness terms to the reaction rate. The effects of different inner-to-outer jet velocity ratios and time scales in the molecular mixing terms have also been investigated. The results indicate that the second-order closure models predict a lower reactant conversion rate, higher unmixedness, and a thinner reaction zone than the pdf model. If the contribution of the reactant concentration fluctuations to the mean reaction rate is neglected, the reactant conversion rate is larger than that predicted by the pdf model. Increasing the inner-to-outer jet

---

Presented at the 9th ICODERS, Poitiers, France, July 3-8, 1983. Copyright by the American Institute of Aeronautics and Astronautics,Inc.,1984. All rights reserved.
*Graduate Student, Department of Mechanical Engineering.
†Assistant Professor, Department of Mechanical Engineering.
‡G.T. Ladd Professor and Head, Dept. of Mechanical Engineering.

velocity ratio results in higher unmixedness and lower reactant conversion rates. An increase in the frequency of mixing results in lower unmixedness and higher reactant conversion rates. Three eddy-break-up models have also been studied. It has been found that there are no universal constants which can be applied at different Damkohler numbers and inner-to-outer jet velocity ratios thereby discrediting these models in the range of parameters investigated. Even for a fixed Damkohler number and inner-to-outer jet velocity ratio the values of these constants are found to be a function of space. The numerical results also indicate that the reactant conversion rates depend on turbulent mixing rates, molecular mixing rates, and chemical kinetic rates and that it does not seem reasonable to make any limiting assumption whereby the reactant conversion rates would depend only upon one or two of these basic rates.

Introduction

Turbulent diffusion flames have been the subject of numerous theoretical and experimental studies (Bilger 1976, 1980). Some of these studies have employed the eddy-break-up model of Spalding (1971b), a flame-sheet approximation (Bilger 1976), a chemical equilibrium assumption (Bilger 1980), a probabilistic model (Givi et al. 1983), or a second-order model (Fishburne and Varma 1977). The experimental work performed has generally involved complex chemical reaction mechanisms and temperature variations. In this paper we consider a two concentric axisymmetric jet configuration at atmospheric pressure and temperature. The inner jet discharges nitrogen and traces of ozone, while the outer jet discharges nitrogen and traces of nitric oxide into a stagnant atmosphere. The chemical reaction between the ozone and nitric oxide is extremely fast and accurately known but since traces of reactants are employed the flow is isothermal and incompressible (Givi et al. 1983). Thus we have selected a flow system in which the chemical reaction does not affect the flowfield. This configuration allows us to determine the effects of the turbulent flowfield on the rates of conversion of nitric oxide and ozone.

Under isothermal and isobaric flow conditions we can assess the influence of several chemical reaction models on the consumption of reactants. Because of the uncoupling between the hydrodynamics and chemistry, this configuration will allow us to establish the validity of some turbulence models for the chemical reaction rates. In this paper, we employ a two-equation turbulence model to account for the hydrodynamic flowfield and study the influence of different reaction rate models on the reactant consumption. The calculations have been performed by solving a transport equation for the joint pdf (probability density function) of three scalars which are related to the reactant and product concentrations. In this equation, the chemical reaction terms appear in closed form. However, the molecular mixing and the convection terms need to be modelled. We have employed Curl's model (Curl 1963) and two different molecular mixing frequencies. The first molecular mixing frequency only depends on the energy containing eddy frequency, i.e., the inverse of the characteristic eddy turnover time. In the second model, we have employed a molecular mixing frequency which depends on the molecular mass diffusivity (or equivalently the laminar Schmidt number and the molecular kinematic viscosity) and the dissipation rate of turbulent kinetic energy. Calculations performed with these two time scales are compared in order to assess the influence of the mixing frequency on the reactant conversion rates.

The pdf calculations have been compared with those obtained by means of two second-order closure models. The comparisons have been made at different Damkohler numbers and for different shear layer strain levels. The validity of second-order closure models for nonpremixed flow situations is discussed. Calculations have also been performed with a mean reaction rate model in which the contribution of the reactant concentration fluctuations to the reaction rate is neglected. These calculations are compared with those obtained with the pdf transport equation in order to assess the effects of unmixedness on the reactant consumption.

The pdf calculations have also been used to determine the constants which appear in the reactant conversion terms in eddy-break-up models. Three of these models have been considered in the present study: Spalding's model (Spalding 1971b) assumes that the reaction rate is proportional to the gradients of the reactant mean concentrations; Magnussen and Hjertager (1977) consider that the reaction is controlled by the minimum of the reactant mean concentrations; and Ramos (1982) assumes that the reaction rate is proportional to the square root of the product of the reactant mean concentration gradients. The determination of these constants has been performed at different Damkohler numbers and inner-to-outer jet velocity ratios in order to assess the effects of mixing and reaction on eddy-break-up models.

In the next sections we first introduce the hydrodynamic equations, the chemical reaction, and the boundary conditions. We then present the pdf transport equation, the mean reaction rate model in which the contribution of the unmixedness terms to the reaction rate is neglected, and two second-order closure models. We then consider three eddy-break-up models and present the numerical results. A discussion of the results concludes the paper.

Hydrodynamic Equations

We consider the geometrical configuration shown in Fig. 1. This configuration consists of two concentric axisymmetric jets which discharge into a stagnant atmosphere and corresponds to an experiment to be developed at Aerochem Research Laboratories. The inner jet of 4-cm diam discharges nitrogen (N_2) and traces of ozone (O_3) at a velocity u_i. The inner jet radius is R_i. The outer jet of 30.48-cm diam discharges nitrogen and traces of nitric oxide (NO) at a velocity u_o and has a radius equal to R_o. Both jets and the stagnant atmosphere are assumed to be at a pressure and a temperature of 1 atm and 300 K, respectively. We introduce the axial and radial coordinates x and r as shown in Fig. 1 and write the conservation equations of mass, axial momentum u, turbulent kinetic energy k, and dissipation rate of turbulent kinetic energy ε, as

$$L(\theta) = S_\theta \qquad (1)$$

where

$$L(\theta) = \partial(u\theta)/\partial x + (1/r)\ \partial(rv\theta)/\partial r - (1/r)\ \partial(r\Gamma_\theta \partial\theta/\partial r)/\partial r \qquad (2)$$

where θ, Γ_θ, and S_θ denote variables, diffusion coefficients, and source terms, respectively. For the continuity

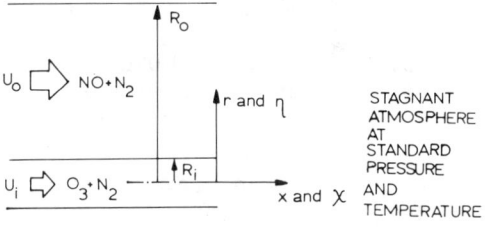

Fig. 1 Schematic of the axisymmetric two concentric jet configuration.

equation $\theta = 1$ and $\Gamma_\theta = S_\theta = 0$; for the axial momentum equation $\theta = u$, $\Gamma_\theta = \nu_t$, and $S_\theta = 0$, where ν_t is the turbulent kinematic viscosity; for the turbulent kinetic energy equation $\theta = k$, $\Gamma_\theta = \nu_t$, and $S_\theta = \nu_t (\partial u/\partial r)^2 - \varepsilon$; and for the dissipation rate of turbulent kinetic energy $\theta = \varepsilon$, $\Gamma_\theta = \nu_t/1.3$ and $S_\theta = (C_{\varepsilon 1} \nu_t (\partial u/\partial r)^2 - C_{\varepsilon 2} \varepsilon)\varepsilon/k$, where $C_{\varepsilon 2} = 1.90$.

In the above equations the radial velocity v was determined from the continuity equation while the value of ν_t is given, at high Reynolds numbers, by (Launder and Spalding 1972)

$$\nu_t = C_\mu k^2/\varepsilon \qquad (3)$$

where $C_\mu = 0.09$.

The value of $C_{\varepsilon 1}$ is known to be nonuniversal (Pope 1978). For example, Launder et al. (1973) proposed a $k\varepsilon_1$ model in which both C_μ and $C_{\varepsilon 1}$ are functions of the local jet half-width and the local axial derivative of the centerline velocity. However, in the calculations performed by Givi and Ramos (1983), it was found that by using a value for $C_{\varepsilon 1}$ equal to 1.52, while keeping $C_\mu = 0.09$ and $C_{\varepsilon 2} = 1.90$, the computed velocity profiles and spreading rate were in good agreement with the experimental data of Wygnanski and Fiedler (1969) for a single axisymmetric jet discharging air into a stagnant atmosphere. These values for C_μ, $C_{\varepsilon 1}$, and $C_{\varepsilon 2}$ may need to be modified to predict coaxial turbulent jets such as the ones described by Champagne and Wygnanski (1971). However, in the calculations reported here, we have employed the values of the constants which produce good agreement with the available experimental data for round free jets.

Chemical Reaction

We consider the well-known and fast chemical reaction between ozone and nitric oxide defined as

$$O_3 + NO \rightarrow NO_2 + O_2$$

where the reaction rate is $K_f = 9.28 \times 10^6$ m³/kg-mole/s at atmospheric pressure and temperature. For convenience we define species 1, 2, 3, and 4 as the ozone, nitric oxide, nitrogen dioxide, and oxygen, respectively.

The transport equations for the ensemble-averaged concentrations, i.e., $<c_j>$, where $< >$ denotes a long time or ensemble average, can be written as

$$L(<c_j>) = <w_j> \quad , \quad j = 1,2,3,4 \qquad (4)$$

where

$$\langle w_1 \rangle = \langle w_2 \rangle = -\langle w_3 \rangle = -\langle w_4 \rangle \tag{5}$$

and

$$\langle w_1 \rangle = -K_f (\langle c_1 \rangle \langle c_2 \rangle + \langle c'_1 c'_2 \rangle) \tag{6}$$

where w is the reaction rate, $\langle c \rangle$ is the mean concentration, and c' is the fluctuating concentration.

We introduce the following Shvab-Zeldovich variables

$$\phi_1 = (c_1 + c_3)/c_{1i} \tag{7}$$

$$\phi_2 = (c_2 + c_3)/c_{2i} \tag{8}$$

and

$$\phi_3 = c_3/c_{2i} \tag{9}$$

$$\alpha = c_{1i}/c_{2i} \tag{10}$$

$$\beta = c_{2i}/c_5 \tag{11}$$

$$\Omega = pK_f/RT \tag{12}$$

where c_{1i}, c_{2i}, and c_5 denote the initial concentration of ozone in the inner jet, the initial concentration of nitric oxide in the outer jet, and the concentration of nitrogen, respectively. Since only dilute mixtures are considered in this paper, the nitrogen concentration can be determined from the equation of state for an ideal gas whose pressure and temperature are p and T, respectively, by neglecting the other species concentrations. In this manner, the nitrogen concentrations in the inner and outer jets are identical. In Eq. (12), R is the universal gas constant. At atmospheric pressure and temperature $\Omega = 4 \times 10^5$/s.

Substitution of Eqs. (7-12) into Eq. (4) yields

$$L(\langle \phi_1 \rangle) = L(\langle \phi_2 \rangle) = 0 \tag{13}$$

and

$$L(\langle \phi_3 \rangle) = \langle w(\phi_1, \phi_2, \phi_3) \rangle = \beta\Omega\langle (\phi_2-\phi_3)(\alpha\phi_1-\phi_3) \rangle \tag{14}$$

where L is the linear operator defined by Eq. (2).

Equations (1), (13), and (14) can be cast in nondimensional form by introducing the mapping

$$(r, x) \to (\eta, \chi) \tag{15}$$

defined by

$$\eta = r/\delta(x) \tag{16}$$

and

$$\chi = \int_0^x dx/\delta(x) \tag{17}$$

where $\delta(x)$ is the jet half-width, i.e., the radial location at which the axial velocity is equal to a half of its centerline value. We also define the following nondimensional quantities

$$u^*(\eta,\chi) = u(r,x)/u_{CL}(x), \quad k^* = k/u_{CL}^2(x) \tag{18}$$

$$v^*(\eta,\chi) = v(r,x)/u_{CL}(x) - u^*(\eta,\chi) \, \eta \, d\delta/dx \tag{19}$$

$$\Gamma_\phi^* = \Gamma_\phi/u_{CL}(x)/\delta(x), \quad \varepsilon^* = \varepsilon\delta/u_{CL}^3(x) \tag{20}$$

where $u_{CL}(x)$ is the centerline velocity. The diffusion coefficient Γ_ϕ is assumed equal for all the species and can be written as

$$\Gamma_\phi = \nu_t/Sc_t \quad (\Gamma_\phi^* = 0.09 \, k^{*2}/\varepsilon^*/Sc_t) \tag{21}$$

where Sc_t is the turbulent Schmidt number which was determined by comparing the conserved scalar profile for a single jet with the experimental data of Chevray and Tutu (1978). In the calculations reported here $Sc_t = 0.80$. This value differs from the value of 0.71 obtained by Forstall and Shapiro (1950) and reported by Hinze (1975). When $Sc_t = 0.70$ was employed in the calculations, the experimental data of Chevray and Tutu (1978) were not accurately predicted with $C_\varepsilon = 1.52$.

Substitution of Eqs. (16-20) into Eqs. (13) and (14) yields

$$L^*(<\phi_1>) = L^*(<\phi_2>) = 0 \tag{22}$$

$$L^*(<\phi_3>) = Da <(\phi_2 - \phi_3)(\alpha \, \phi_1 - \phi_3)> \tag{23}$$

where Da is the local Damkohler number defined as

$$Da = \delta(x)\beta\Omega/u_{CL}(x) \qquad (24)$$

and

$$L^*(\theta) = u^*\partial\theta/\partial x + v^*\partial\theta/\partial\eta - (1/\eta)\partial(\eta\Gamma_\theta^*\partial\theta/\partial\eta)/\partial\eta \qquad (25)$$

Boundary Conditions

The following boundary conditions were used at the two concentric jet exit

$$k^* = 0.05, \quad \varepsilon^* = 0.02, \quad <\phi_1> = 1,$$

$$<\phi_2> = <\phi_3> = 0, \quad 0 \leq \eta \leq 1 \qquad (26)$$

$$k^* = 0.05, \quad \varepsilon^* = 0.02, \quad <\phi_1> = <\phi_3> = 0,$$

$$<\phi_2> = 1, \quad 1 \leq \eta \leq R_o/R_i \qquad (27)$$

$$k^* = 0.05, \quad \varepsilon^* = 0.02, \quad <\phi_1> = <\phi_2> = <\phi_3> = 0,$$

$$R_o/R_i \leq \eta \leq 8 \qquad (28)$$

where R_o and R_i denote the outer and inner jet radii, respectively. The initial jet half-width is equal to R_i.

At the jet boundary, i.e., at $\eta = 8$, we have employed

$$u^* = k^* = \varepsilon^* = <\phi_j> = 0, \quad j = 1,2,3 \qquad (29)$$

and at the centerline, i.e., $r = 0$,

$$\partial<\phi_j>/\partial\eta = \partial u^*/\partial\eta = \partial k^*/\partial\eta$$

$$= \partial\varepsilon^*/\partial\eta = v^* = 0, \quad j = 1,2,3 \qquad (30)$$

At any axial location different from the two concentric jet exit the values of k^*, ε^*, u^*, and concentrations were set to zero at $\eta = 8$ (c.f., compare with $R_o/R_i = 7.62$). This value, which represents the jet edge, was determined so that the numerical results are not affected by the location of the computational domain boundary. At the two concentric jet exit there is a discontinuity in the values of k^* and ε^* at the jet edge. This discontinuity is smoothed out downstream and does not affect the numerical results. Because of the nondimensionalization employed in this paper, the turbulent intensity of the outer jet at x=0 is very high, i.e., 0.05 $(u_i/u_0)^2$. However, when the outer jet turbulent intensity is set equal to 0.05, very small differences are observed in the mean axial velocity profiles. The differences in the k^* and ε^* values are at most 20 and 10%, respectively. The largest differences occur at the interjet shear layer. There are also differences in the outer jet velocity, turbulent kinetic energy, and dissipation rate profiles, but they do not seem to affect the flowfield except at the interjet shear layer.

Joint pdf Transport Equation

A complete statistical description of the scalar properties in a turbulent flowfield can be obtained from the joint probability density function $P[(\underline{\psi}, \eta, \chi)|\underline{\psi} = (\psi_1, \psi_2, \psi_3)]$ of the scalars ψ_1, ψ_2, and ψ_3, where $P(\psi_1, \psi_2, \psi_3)d\psi_1 d\psi_2 d\psi_3$ denotes the probability of finding $\underline{\psi}$ between $\underline{\psi}$ and $\underline{\psi} + d\underline{\psi}$. In a two-dimensional parabolic flowfield the normalized transport equation for P can be written (Pope 1979) as

$$u^* \partial P/\partial \chi + v^* \partial P/\partial \eta - \partial[Da\, P\, w(\psi_1, \psi_2, \psi_3)]/\partial \psi_3$$
$$=(1/\eta)\, \partial(\eta\, \Gamma_p^*\, \partial P/\partial \eta)/\partial \eta + \Omega_T^*\, E(\psi_1, \psi_2, \psi_3) \qquad (31)$$

where $\Gamma_p^* = \Gamma_p/u_0(x)/\delta(x)$, $\Gamma_p = \nu_t/Sc_t$, and $\Omega_T^* = \Omega_T\, \delta/u_0$. Ω_T is the molecular mixing frequency. The term E is the molecular mixing which has been modelled (Curl 1963) as

$$E(\psi_1, \psi_2, \psi_3) = 8 \int P(\underline{\psi} + \underline{\psi}^*)\, P(\underline{\psi} - \underline{\psi}^*)\, d\underline{\psi}^* - P(\underline{\psi}) \qquad (32)$$

where the integration is over the whole of $\underline{\psi}$ space.

It should be pointed out that Eq. (32), i.e., the molecular mixing term, has been modelled but that the reaction terms which appear in Eq. (31) are exact and do not need to be modelled. The transport terms have been modelled using a Boussinesq approximation. As indicated in Eq. (31) the molecular mixing term depends on the frequency Ω_T which needs to be specified. In the calculations reported here we have used two mixing frequencies. If it is assumed that the rate of molecular mixing is controlled by the energy containing eddy turnover time, i.e., by the large-scale eddies, then (Spalding 1971a)

$$\Omega_T = \varepsilon/k \qquad (33)$$

However, if the molecular mixing terms are assumed to be controlled by the dissipative, i.e., small-scale, eddies then the molecular mixing frequency is given by (Corrsin 1964)

$$1/\Omega_T = 1.25 \, \lambda \, / \, u' \, / \, (Sc \, (3-Sc^2))^{1/2} \qquad (34)$$

where Sc is the laminar Schmidt number, u' denotes a characteristic fluctuating velocity, and λ is the Taylor microscale which in homogeneous isotropic turbulence can be written as

$$\lambda^2 = 10\nu k/\varepsilon \qquad (35)$$

where ν is the molecular kinematic viscosity.

Equation (34) is different from that proposed by Janicka et al. (1979) in that it does not involve the dissipation rate of a passive scalar. In the calculations reported here, we have not employed Eq. (34). Instead, we have taken Ω_T as the ratio of a characteristic turbulent velocity, i.e., $\sqrt{2k/3}$, and the Taylor microscale for the concentration fluctuations λ_c, which is related to the dissipation rate of concentration fluctuations as (Corrsin 1964)

$$\varepsilon_c = 12D \, {c'}^2/\lambda_c^2 \qquad (36)$$

where ε_c and D are the dissipation rate of concentration fluctuations and the molecular mass diffusivity, and c' is a characteristic concentration fluctuation. Equation (36) is valid for homogeneous turbulence.

If an inertial subrange is assumed for the concentration spectrum then

$$\varepsilon_c = A \, u' \, {c'}^2/\ell \qquad (37)$$

where A is a constant of order one, u' is a characteristic velocity, and ℓ is a characteristic length of the energy containing eddies which is related to k and ε through (Hinze 1975)

$$\ell = Q \, {u'}^3/\varepsilon \qquad (38)$$

where Q is a constant of order one.

Also, we can take

$$u' = (2k/3)^{1/2} \quad (39)$$

Substitution of Eqs. (37-39) into Eq. (36) yields

$$\lambda_c = \left(8 \frac{Q}{A} D \frac{k}{\varepsilon}\right)^{1/2} \quad (40)$$

From Eq. (39) and (40) one can obtain

$$1/\Omega_T = \lambda_c/u' = \left(12 \frac{Q}{A} \frac{\nu}{\varepsilon} \frac{1}{Sc}\right)^{1/2} \quad (41)$$

which depends on the molecular diffusivity $D = \nu/Sc$ and the dissipation rate of turbulent kinetic energy. In the calculations reported here we have used $Sc = 0.73$ and $A = Q = 1$.

The solution of Eq. (31) yields the values of P from which the concentration mean values and second-order moments can be calculated. Equation (31) was solved by means of a Monte Carlo method (Pope 1981).

Mean Reaction Rate Model

In this model the mean reaction rate $<w_1>$ [Eq. (6)] has been modelled as

$$<w_1> = -K_f <c_1><c_2> \quad (42)$$

i.e., the unmixedness terms $<c'_1 c'_2>$ have been neglected. In Eq. (42) the instantaneous concentrations have been replaced by the mean concentrations. This approach has been employed in numerous turbulent combustion studies and will be evaluated here by comparing the reactant conversion rates obtained by employing Eq. (42) with those obtained by means of Eqs. (31) and (6).

A Second-Order Closure Model

An equation for the concentration correlations $<c'_i c'_j>$, $i,j = 1,2$, can be obtained from the transport equation

$$L(c_i) = w_i, \quad i = 1,2 \quad (43)$$

where c_i and w_i are the instantaneous concentration, i.e., $c_i = <c_i> + c'_i$, and w_i is the reaction rate, i.e.,

$$w_i = - K_f c_i c_j, \quad i,j = 1,2 \qquad (44)$$

A long time average of Eq. (43) can be subtracted from the combination of Eqs. (43) and (44) and the resulting equation can be multiplied by c'_j and averaged to yield

$$L(<c'_i c'_j>) = 2 \frac{\nu_t}{Sc_t} \frac{\partial <c_i>}{\partial r} \frac{\partial <c_j>}{\partial r} - \frac{2\nu}{Sc} <\frac{\partial c'_i}{\partial r} \frac{\partial c'_j}{\partial r}>$$

$$- K_f [<c_i> (<c'_i c'_j> + <c'_j c'_j>) + <c_j> (<c'_i c'_i>$$

$$+ <c'_i c'_j>) + <c'_i c'_i c'_j> + <c'_i c'_j c'_j>], \quad i,j = 1,2$$

$$(45)$$

where the laminar diffusion terms have been neglected on account of the large Reynolds number flows considered and where the Boussinesq approximation has been used to model the velocity-concentration correlations. Equation (45) was solved by means of the same finite-difference scheme as the one used to solve Eq. (42) and the hydrodynamic equations, i.e., Eq. (2). However, in solving Eq. (45) we have neglected the concentration triple correlations and introduced the following modelling (Ramos 1982)

$$- 2 \frac{\nu}{Sc} <\frac{\partial c'_i}{\partial r} \frac{\partial c'_j}{\partial r}> = - B \frac{\varepsilon}{k} <c'_i c'_j> \qquad (46)$$

and $2/Sc_t = F$, so that Eq. (45) can be written as

$$L(<c'_i c'_j>) = F \nu_t \frac{\partial <c_i>}{\partial r} \frac{\partial <c_j>}{\partial r} - B \frac{\varepsilon}{k} <c'_i c'_j>$$

$$- K_f [<c_i> (<c'_i c'_j> + <c'_j c'_j>)$$

$$+ <c_j> (<c'_i c'_i> + <c'_i c'_j>)] \qquad (47)$$

where F and B are constants equal to 2.857 and 2.0, respectively (Spalding 1971a).

Equation (47) was solved for $<c'_1 c'_2>$; simultaneously Eqs. (6) and (4) were solved to obtain the mean reactant concentrations. Equation (47) was also solved for $<c'_1 c'_1>$ and $<c'_2 c'_2>$ and the results were compared with those obtained from the pdf transport equation, i.e., Eq. (31). It should be pointed out that $<c'_1 c'_1>$ and $<c'_2 c'_2>$ appear in the transport equation for $<c'_1 c'_2>$, i.e., Eq. (47). Thus this second-order closure model requires the solution of three transport equations for $<c'_1 c'_1>$, $<c'_1 c'_2>$, and $<c'_2 c'_2>$ in order to calculate the mean reactant concentrations and the mean reaction rates. These calculations can be eliminated if we introduce the following simplified second-order closure model (Spalding 1971b; Ramos 1982).

A Simplified Second-Order Closure Model

Equation (47) can be simplified if we neglect the contribution of the chemical reaction terms, i.e., if we neglect the terms proportional to K_f. This simplification results in the following equation

$$L(<c'_i c'_j>) = F \nu_t \frac{\partial <c_i>}{\partial r} \frac{\partial <c_j>}{\partial r} - B \frac{\varepsilon}{k} <c'_i c'_j> \quad (48)$$

which yields the values of $<c'_1 c'_2>$. These values can then be substituted into Eqs. (6) and (4) to yield the mean reaction rate and mean reactant concentrations. This model was proposed by Ramos (1982) and in a different form by Spalding (1971a). Note that Eq. (48) was only solved for $<c'_1 c'_2>$. If instead of solving for $<c'_1 c'_2>$ we solve for $<c'_1 c'_1>$ we shall obtain Spalding expression for the concentration fluctuations (Spalding 1971a).

Equations (4) and (6) can be further simplified if somehow $<c'_1 c'_2>$ is related to the mean concentrations. This can be achieved in different ways as explained in the following sections.

Eddy-Break-Up Models

Three eddy-break-up models are presented and discussed in this section. The first one is due to Spalding (1971a, b) and is based on the idea that the reactant conversion rate depends on a typical frequency of the energy containing eddies and the gradient of the mean reactant concentration. Thus, the conversion rate is assumed to be proportional to the square root of the rate of dissipation of turbulent eddies. These eddies will react around their edges only when the species gradients are high. Spalding

(1971b) proposed that under mixing controlled conditions

$$<w_1> = - G\, k^{1/2} \min \left(\left|\frac{\partial <c_1>}{\partial r}\right|, \left|\frac{\partial <c_2>}{\partial r}\right| \right) \quad (49)$$

where G is a constant of order one. Note that by using Eq. (49), Eq. (4) can be solved directly to yield $<c_1>$ and $<c_2>$.

Magnussen and Hjertager (1977) have also introduced an eddy-break-up model in which the mean reactant conversion rate is related to the rate of dissipation of turbulent eddies and the mean reactant concentrations as

$$<\bar{w}_1> = - H\, \frac{\varepsilon}{k} \min \left(<\bar{c}_1>, \frac{<\bar{c}_2>}{s}, B\, \frac{<\bar{c}_p>}{1+s} \right) \quad (50)$$

where H is a constant of order one, s is the stoichiometric coefficient of the reaction, i.e., s = 48/30, overbars denote concentrations in mass (rather than in moles) per unit volume, c_p is the concentration of the reaction products, and B = 0.5 (Magnussen and Hjertager 1977).

Recently, Ramos (1982) introduced a new eddy-break-up model based on Eq. (48). In this model, the mean reaction rate is given by

$$<w_1> = - R\, \left(k\, \left| \frac{\partial <c_1>}{\partial r}\, \frac{\partial <c_2>}{\partial r} \right| \right)^{1/2} \quad (51)$$

where R is a constant of order one. Equation (51) can be obtained from Eq. (48) by setting the righthand side term of this equation equal to zero, i.e., by assuming equilibrium or that the production of concentration correlations is equal to its dissipation rate. Furthermore, the conversion rate is controlled by the dissipation rate of the concentration correlations.

It should be pointed out that the mean reaction rates of the eddy-break-up models defined by Eqs. (49-51) are only valid at high Damkohler numbers. One of the goals of this study is to determine the constants G, H, and R which appear in these models and establish if these constants are Damkohler number independent. In addition, the results of the eddy-break-up models will be compared with those obtained from the pdf transport equation in order to validate the numerical calculations. However, note that Eqs. (49), (50), and (51) are very dependent on the initial specification of the turbulent kinetic energy profile. An incorrect specification results in an incorrect concentration field which cannot be corrected in the downstream region as the length scale and turbulent intensity are modified by the flowfield. In addition, the reaction rates of Eqs. (49) and (51) are functions of the mean concentration gradients, which are initially very large. This may result in an extremely large reaction rate. Some authors (Ramos 1984) prescribe an initial turbulent kinetic energy profile so that the reaction rate in the initial region is not exceedingly large. Others (Spalding 1971b) prescribed a reaction rate which is the geometric mean of those given by Eqs. (49) and (42), i.e., at high Damkohler numbers, Eq. (49) dominates over Eq. (42).

The models described in the previous sections range from very sophisticated pdf equations, second-order closure schemes, and eddy-break-up approximations to mean reaction rate models and have been frequently employed in turbulent combustion calculations. The parabolic partial differential equations which govern these models have been solved by means of an implicit finite-difference scheme (Givi et al. 1983; Pope 1977) except for the pdf equation [Eq. (31)] which has been solved by means of the Monte Carlo method developed by Pope (1981).

Presentation and Discussion of Results

The calculations reported here were performed in an unequally spaced grid which consisted of 96 points in the η direction. In the calculations, the pdf was represented by an ensemble of 200 elements at each grid point. The calculations were performed with a step size equal to 0.0025δ in the axial direction. Other calculations were performed with smaller step sizes in order to verify that the results were grid-independent.

The statistical errors associated with the Monte Carlo technique are proportional to $N^{-1/2}$, where N is the number of elements which represent the probability density function at each grid point. It can be shown (Pope 1981) that the average of n computer runs with N number of elements is equivalent to a computer run with nN elements, and that such a run substantially reduces the statistical errors. In the calculations reported, the differences between two runs each with N=200 were at most 5%. Therefore, instead of making extensive calculations, we simply smoothed the concentration profiles obtained with the Monte Carlo method using available software. This software minimizes the standard deviation resulting from the 5% error. The smoothed results were compared with those obtained by averaging five computer runs, each with 200 elements and an excellent agreement was observed for all values of η. The smoothing procedure also resulted in more affordable calculations, but did not yield an exact zero normal gradient condition at $\eta = 0$ as shown in some of the figures presented in this section. Note though that the zero gradient constraint was applied in the calculations.

In Fig. 2 we show the mean axial velocity profile in m/s as a function of η at different axial locations. These axial locations have been nondimensionalized by the inner jet diameter, i.e., $D = 2R_i$. This figure shows the diffusion of axial momentum and was computed with $u_o = 5.25$ m/s and $u_i = 40$ m/s. In Fig. 3 we present the normalized turbulent kinetic energy profiles as a function of η at dif-

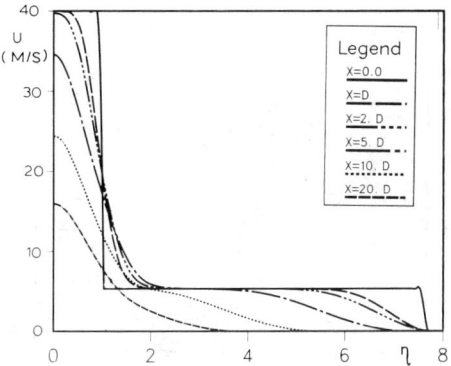

Fig. 2 The mean axial velocity profiles in m/s as a function of η.

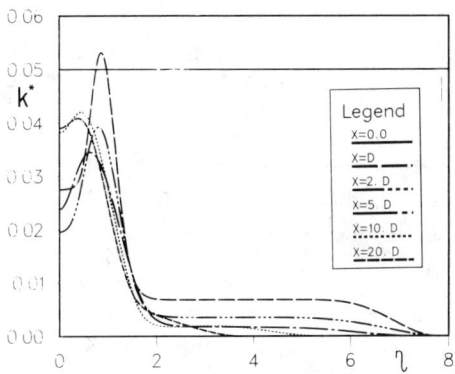

Fig. 3 The normalized turbulent kinetic energy profiles as a function of η.

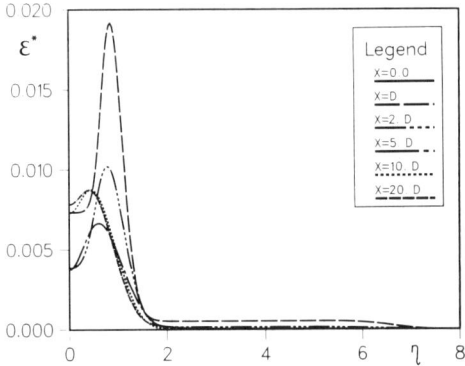

Fig. 4 The normalized profiles of the dissipation rate of turbulent kinetic energy as a function of η.

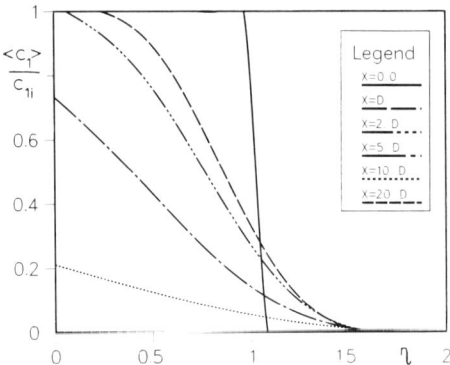

Fig. 5 The normalized mean concentration profiles of ozone calculated from the pdf transport equation as a function of η.

ferent axial locations. The figure indicates that turbulence is produced at the interjet shear layer but decays downstream. The profiles of turbulent energy dissipation are shown in Fig. 4 and indicate that the interjet shear layer is characterized by high levels of dissipation. This is clearly visible in the dissipation rate peak at $x/D = 1$. Downstream of this location the dissipation rate profiles decay because the production of turbulent kinetic energy decreases. Although the k/ε model employed in the calculations reported here is Reynolds number independent because the equations governing the transport of k and ε do not contain the laminar viscosity, it is convenient to define a theoretical Reynolds number based on the inner jet velocity and diameter. For the calculations shown in Figs. 2-4 this

Table 1 The jet half-width and centerline velocity values

x/D	δ (cm)	u_{CL} (m/s)
0	2	40.00
1	2.28	39.99
2	2.39	39.74
5	2.87	34.49
10	4.40	24.43
20	7.80	15.91

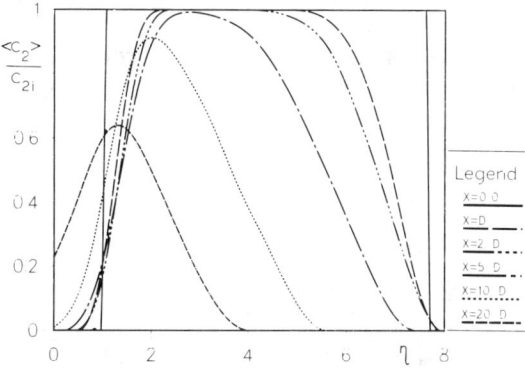

Fig. 6 The normalized mean concentration profiles of nitric oxide calculated from the pdf transport equation as a function of η.

Reynolds number is about 1×10^5. In order to have a complete description of the hydrodynamics, in Table 1 we show the jet half-width and the centerline velocity at several diameters downstream of the inner pipe exit. From the results presented in this table, one can deduce a characteristic diffusion time.

Unless otherwise indicated the initial (at x=0) Damkohler number is 10, and the initial concentrations of ozone and nitric oxide are $c_{1i}/c_5 = 0.1$ and $c_{2i}/c_5 = 0.05$, respectively, where c_5 is the nitrogen concentration which can be calculated from the equation of state. Also, unless otherwise stated, the results were obtained by the solution pdf transport [Eq. (31)] with molecular mixing rates controlled by the large-scale eddy time.

In Fig. 5 we present the ozone mean concentration profiles normalized by the inner jet ozone concentration as a function of η at different axial locations for the same flow conditions as those used in the calculations of Figs. 2-4. These profiles decrease monotonically with the axial distance from the jet exit. In Fig. 6 we present the mean concentration profiles of nitric oxide. These profiles are normalized by the initial concentration of the nitric oxide, and show that the reactant is consumed along the geometrical configuration centerline axis. The profiles of nitrogen dioxide normalized by the inner jet ozone concentration are presented in Fig. 7. Nitrogen dioxide is produced at the interjet shear layer where ozone and nitric oxide mix and react with each other. The nitrogen dioxide concentration increases with the distance from the jet exit but diffuses radially in such a manner that its largest magnitude no longer appears at the interjet shear layer but at the centerline after an axial distance of about x/D = 10.

In Fig. 8 we present the mean nitrogen dioxide concentration profiles computed with the second-order closure

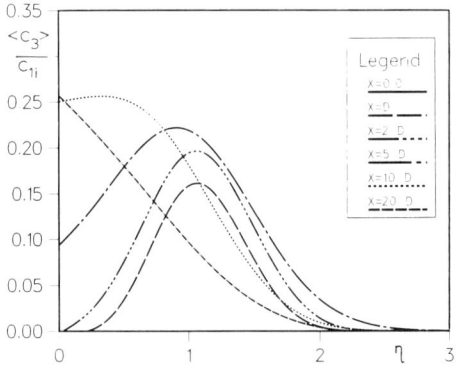

Fig. 7 The normalized mean concentration profiles of nitrogen dioxide calculated from the pdf transport equation as a function of η.

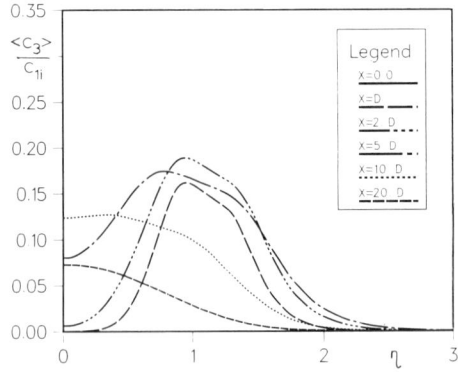

Fig. 8 The normalized mean concentration profiles of nitrogen dioxide calculated from the second-order closure model as a function of η.

model defined by Eq. (47). A comparison of Figs. 7 and 8 indicates that the second-order closure model underestimates the conversion rates of ozone and nitric oxide. This underestimation seems likely to be largely due to the neglect of the triple concentration correlations in the second-order closure model. It may also be due somewhat to the modeling of some of the terms which appear in the transport equations for the double correlations of the concentrations. The nitrogen dioxide profiles computed with the simplified second-order closure model are shown in Fig. 9 and are slightly smaller than those computed with the

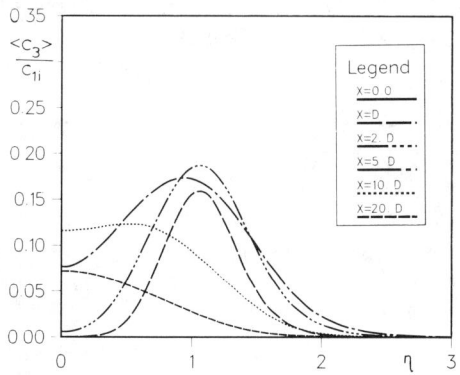

Fig. 9. The normalized mean concentration profiles of nitrogen dioxide calculated from the simplified second-order closure model as a function of η.

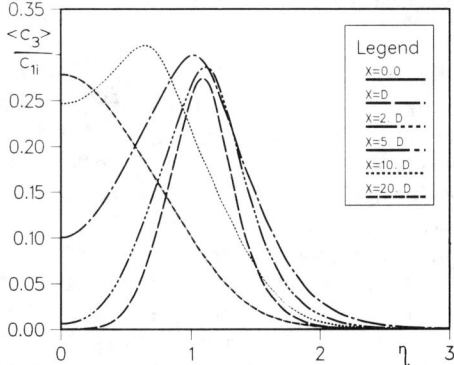

Fig. 10 The normalized mean concentration profiles of nitrogen dioxide calculated from the mean reaction rate model as a function of η.

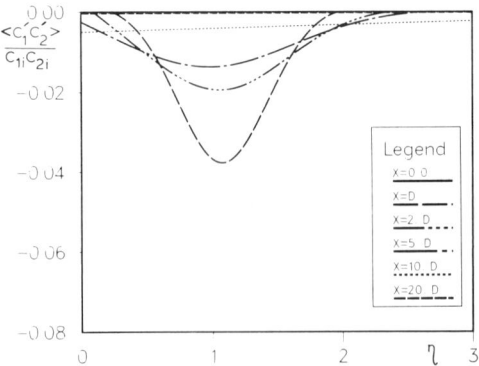

Fig. 11 The normalized unmixedness term profiles calculated from the pdf transport equation as a function of η.

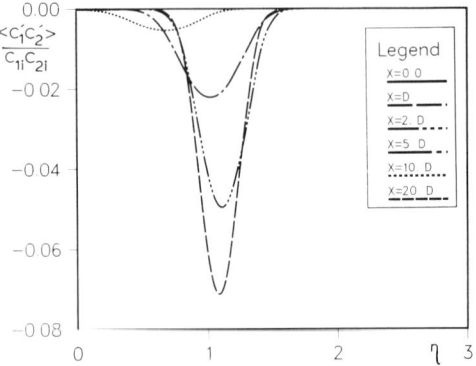

Fig. 12 The normalized unmixedness term profiles calculated from the second-order closure model as a function of η.

second-order closure model. Probably, this is due to the neglect of chemistry terms in Eq. (48). Thus, neglecting the concentration triple correlations and the chemistry terms in the reaction rates apparently results in lower reactant conversion rates.

The nitrogen dioxide profiles computed with the mean reaction rate model which neglects the contribution of the concentration fluctuations to the reaction rate are presented in Fig. 10. This figure indicates that product formation occurs at the interjet shear layer and is higher than that calculated from the pdf transport equation. This is not surprising since the unmixedness term $<c'_1 c'_2>$ has been neglected in the mean reaction rate model. This term is negative as shown in Fig. 11 which is based upon calculations from the pdf transport equation. Thus, a mean reaction rate model always overestimates the reactant conversion rates in a nonpremixed situation. In Fig. 11 the unmixedness term $<c'_1 c'_2>$ has been normalized by the product of the initial ozone and nitric oxide concentrations. The largest value of $<c'_1 c'_2>$ occurs at $x/D = 1$ and is about -0.04 of the product of the initial concentrations of ozone and nitric oxide. The profiles of $<c'_1 c'_2>$ normalized by the product of the initial concentrations of nitric oxide and ozone and computed with the second-order closure model are shown in Fig. 12. A comparison of Figs. 11 and 12 indicates that the second-order closure model overestimates the unmixedness terms computed from the pdf transport equation. The largest normalized peak of $<c'_1 c'_2>$ calculated from the second-order closure model is about -0.07, i.e., it is about twice the value of $<c'_1 c'_2>$ calculated from the pdf transport equation. This justifies, in part, that the reactant conversion rates predicted by the second-order closure model are smaller than those predicted by the pdf model. An interesting behavior of the $<c'_1 c'_2>$ profiles computed with both models is that they always are negative, i.e., for nonpremixed flows the unmixedness term $<c'_1 c'_2>$ is always negative. Another interesting feature shown in Figs. 11 and 12 is that the second-order closure model presents a much more narrow peak in the unmixedness term profiles than the pdf transport equation.

In Fig. 13 we show the normalized unmixedness term profiles calculated with the simplified second-order closure model. The magnitude of these profiles is larger than that computed from the pdf transport equation but is approxi-

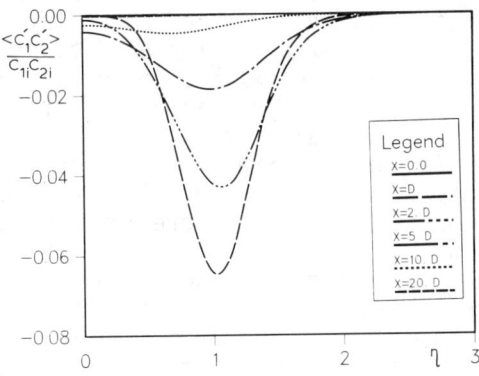

Fig. 13 The normalized unmixedness term profiles calculated from the simplified second-order closure model as a function of η.

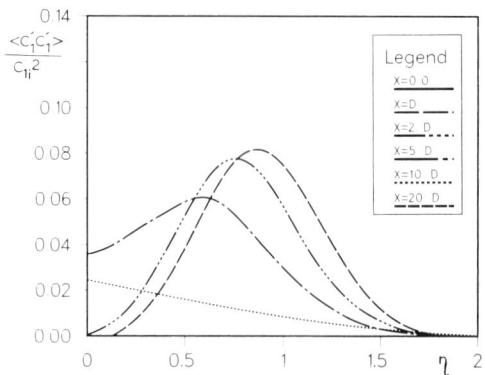

Fig. 14 The normalized $<c'_1 c'_1>$ profiles calculated from the pdf transport equation as a function of η.

mately equal to that computed with the second-order closure model. The main difference between the unmixedness term profiles computed with the second-order closure models is the width of the profiles. The width calculated with the simplified second-order closure model is similar to that of the pdf transport equation but broader than that of the second-order closure model. Since the unmixedness terms computed with the simplified second-order closure model are larger than those predicted by the pdf transport equation, the reactant conversion rates are larger in the simplified second-order closure model than in the pdf model (compare Figs. 7-9).

Figure 14 presents the $<c'_1 c'_1>$ correlation normalized by the square of the initial ozone concentration. This correlation has been calculated from the pdf transport equation and is always positive as it should be. Large $<c'_1 c'_1>$ values occur at the interjet shear layer but radially diffuse after an axial distance of about $x/D = 2$. The largest normalized $<c'_1 c'_1>$ value occurs at $\eta = 1$, i.e., at the interjet shear layer, and is about 0.08. The same correlation has been calculated with the second-order closure model and is shown in Fig. 15. The second-order closure model predicts a positive correlation which is much larger (by a factor of 2) than that predicted by the pdf transport equation. The largest $<c'_1 c'_1>$ value occurs at the interjet shear layer but decreases much faster, after $x/D = 2$, in the second-order closure model than in the pdf formulation. It should be pointed out that both the second-order closure model and the pdf transport equation results show the same qualitative trends; they only differ on the magnitude of the second-order correlations which are intimately related to the reactant conversion rates.

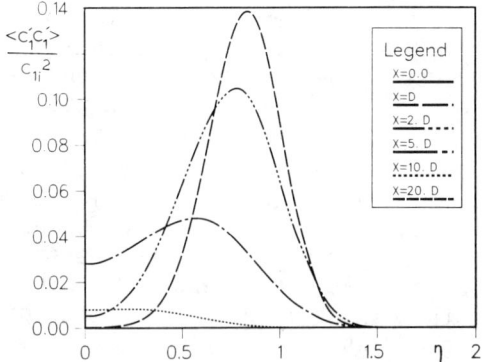

Fig. 15 The normalized $<c'_1 c'_1>$ profiles calculated from the second-order closure model as a function of η.

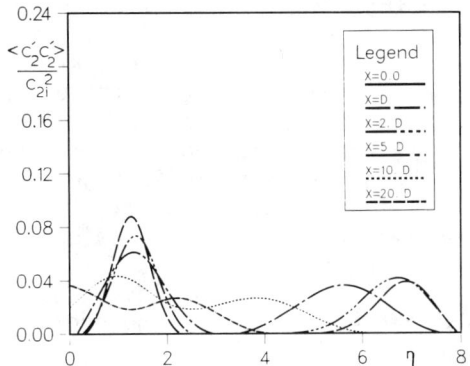

Fig. 16. The normalized $<c'_2 c'_2>$ profiles calculated from the pdf transport equation as a function of η.

The profiles of $<c'_2 c'_2>$ normalized by the square of the initial nitric oxide concentration are presented in Figs. 16 and 17. These profiles were calculated from the pdf transport equation and the second-order closure model, respectively. Figure 16 indicates that the $<c'_2 c'_2>$ profiles peak at the interjet shear layer and near the outer jet boundary. This is because most of the reactant conversion takes place at the shear layer. Furthermore, initially the outer jet is composed of nitric oxide only. These peaks diffuse radially when x/D increases. The profiles in Fig. 17 show the same qualitative trends as those obtained from the pdf transport equation; however, their magnitude is larger by about a factor of 2. An interesting feature of Figs. 14 and 15 and Figs. 16 and 17 is that the

$<c'_1 c'_1>$ and $<c'_2 c'_2>$ profiles at the interjet shear layer are broader for the pdf transport equation than for the second-order closure model. The same trends were observed in the $<c'_1 c'_2>$ profiles, i.e., Figs. 11 and 12.

The pdf calculations (and the other calculations) reported so far have been performed with a molecular mixing frequency equal to that of the energy containing eddies, i.e., Eq. (33). Calculations were also performed with the same initial and boundary conditions, and the same Damkohler and Reynolds numbers as those reported in the previous figures but where the mixing frequency was that of the smaller dissipating eddies and therefore had a higher value. These calculations are presented in Figs. 18 and 19 and show the $<c_3>$ and $<c'_1 c'_2>$ profiles normalized by the ozone initial concentration and the product of the initial concentrations of nitric oxide and ozone, respectively. It should be emphasized that Figs. 7 and 11 present the same type of profiles as those of Figs. 18 and 19. The earlier figures were calculated from the pdf transport equation with a mixing frequency equal to that of the energy containing eddies. Figures 7 and 18 and Figs. 11 and 19 present the same qualitative trends. However, the calculations based on the characteristic dissipating eddy frequency show larger reactant conversion rates, i.e., larger nitrogen dioxide concentrations, than those based on the characteristic energy containing eddy frequency. This is not surprising since the dissipating eddy frequency is larger than that of the energy containing eddies, and these frequencies are used to determine the rates of molecular mixing. Thus, one could expect that an increase of the molecular mixing terms would result in larger reactant con-

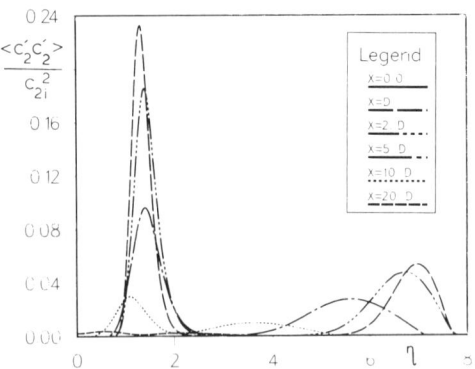

Fig. 17 The normalized $<c'_2 c'_2>$ calculated from the second-order closure model as a function of η.

version rates. Furthermore, since the same initial and boundary conditions were used in the calculations one can also expect that the unmixedness terms are smaller when the dissipating eddy frequency is used than when the energy containing eddy frequency is employed. This is clearly shown in Figs. 11 and 19. These figures indicate that the largest normalized $<c'_1 c'_2>$ value occurs at the interjet shear layer and is about -0.04 and -0.004 for the calculations performed with the pdf transport equation and the energy containing eddy and the dissipating eddy frequencies, respectively. Another interesting feature of Figs. 11 and 19 is that the normalized $<c'_1 c'_2>$ profiles are wider for the calculations performed with the energy containing eddy frequency than for those performed with the dissipating eddy frequency.

Both the mean reaction rate (deterministic) and the pdf models yield the same values of the Shvab-Zel'dovich variables defined in Eqs. (7) and (8) except for the statistical errors in the solution of the pdf equation. These errors are caused by the use of a finite number (200) of elements to represent the pdf at each grid point and are about 5%.

Calculations have also been performed with a Damkohler number equal to 10 and velocities $u_i = 20$ m/s and $u_0 = 5.25$ m/s in order to assess the influence of the shear layer intensity, i.e., velocity gradients, on the reactant conversion rates and $<c'_1 c'_2>$ profiles. These calculations have been performed with $c_{1i}/c_5 = 0.05$ and $c_{2j}/c_5 = 0.025$. Some sample results are shown in Figs. 20 and 21 and correspond to the normalized nitrogen dioxide concentrations and the normalized $<c'_1 c'_2>$ correlations. These profiles have been normalized by the initial ozone concentration and the product of the initial concentrations of ozone and nitric oxide, respectively. A comparison of Figs. 7 and 20 shows that increasing the inner-to-outer jet velocity ratio, the nitrogen dioxide concentration decreases, i.e., an increase in the inner-to-outer jet velocity ratio results in smaller reactant conversion rates. However, the decrease in the reactant conversion rate is less than about 1%. This seems to be due to the unmixedness terms shown in Figs. 11 and 21. These figures indicate that larger inner-to-outer jet velocity ratios result in higher unmixedness, i.e., larger $<c'_1 c'_2>$ values. It should be pointed out that the profiles shown in Figs. 7, 11, 20, and 21 are normalized and correspond to a Damkohler number of 10. In order to keep a fixed Damkohler number while changing the inner-to-outer jet velocity ratio it was necessary to change the initial concentrations of ozone and nitric oxide.

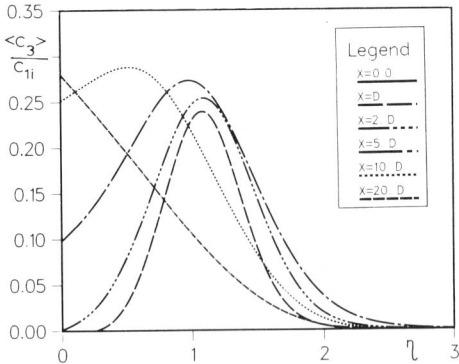

Fig. 18 The normalized mean concentration profiles of nitrogen dioxide calculated from the pdf transport equation with a dissipating eddy frequency as a function of η.

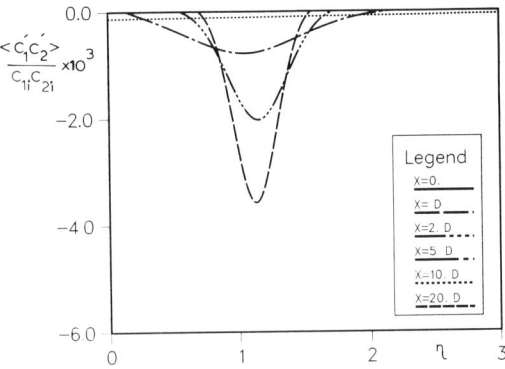

Fig. 19 The normalized values of the unmixedness terms calculated from the pdf transport equation with a dissipating eddy frequency as a function of η.

Therefore, as the velocity ratio is decreased, the concentrations decrease and the shear layer thickness increases resulting in a decrease in the concentration gradients and a decrease in the production of the concentration correlations. Evidently, as the initial concentrations of ozone and nitric oxide are increased the isothermal flow assumption may break down. For example, for u_i = 40 m/s and u_0 = 5.25 m/s a Damkohler number of 100 can only be reached when c_{2i} = 0.5c_5. These reactant concentrations are very large and, in general, for a Damkohler number of 100 the flow is not isothermal.

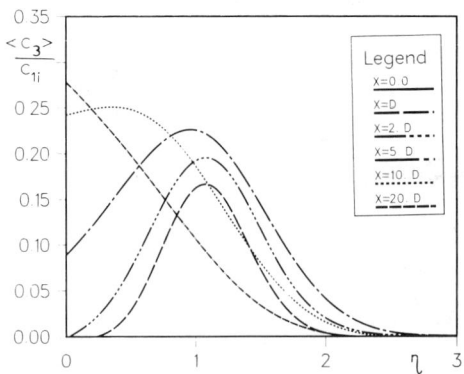

Fig. 20 The normalized mean concentration profiles of nitrogen dioxide calculated from the pdf transport equation with u_j = 20 m/s as a function of η.

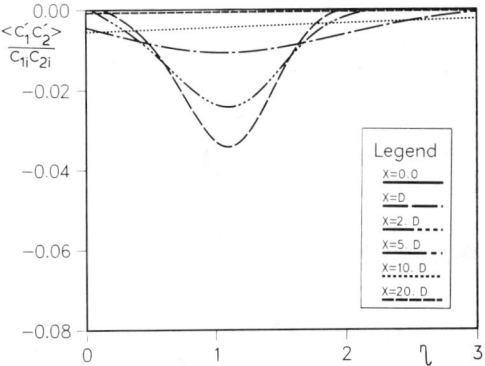

Fig. 21 The normalized unmixedness term profiles calculated from the pdf transport equation with u_j = 20 m/s as a function of η.

Calculations have also been performed with u_j = 40 m/s, u_o = 5.25 m/s, and a Damkohler number of 10 in order to determine the "constants" G^{-1}, H^{-1}, and R^{-1} which appear in the eddy-break-up models, i.e., Eqs. (49), (50), and (51). These "constants" were determined by equating Eqs. (49), (50), and (51) to Eq. (6), i.e., to the mean reaction rate obtained from the pdf transport equation which was solved using the energy containing eddy frequency. The reaction rate terms which appear in Eqs. (49), (50), and (51) were evaluated with the results obtained from the pdf transport equation.

The values of the "constants" G^{-1}, H^{-1}, and R^{-1} determined in the manner described before are presented in Figs.

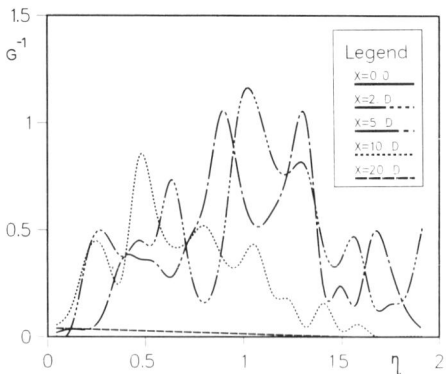

Fig. 22 The value of G^{-1} as a function of η.

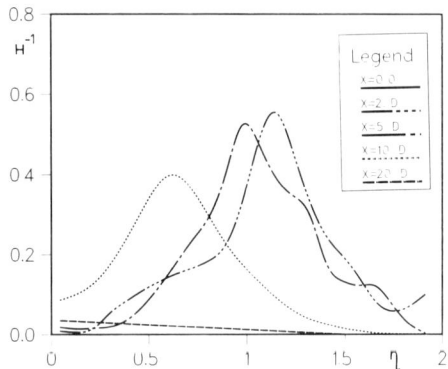

Fig. 23 The value of H^{-1} as a function of η.

22, 23, and 24. Figure 22 shows that the value of G^{-1} which appears in the eddy-break-up model defined by Eq. (49) (Spalding 1971b) is not a constant but oscillates in the radial direction. The profile of G^{-1} shows some peaks which are due to the use of the mean concentration gradients in Eq. (49). They are also due to the unmixedness terms which appear in the reaction rate evaluated from the pdf transport equation. The value of H^{-1} of the Magnussen and Hjertager model [Eq. (50)] does not present as many oscillations as that of G^{-1} in the Spalding model [Eq. (49)] as indicated in Fig. 23. Both the G^{-1} and H^{-1} values show a peak at the interjet shear layer where reactant consumption is most important. However, there is not a constant value of H^{-1} which can be employed throughout the whole jet flow both axially and radially. The value of R^{-1} which appears in the eddy-break-up model defined by Eq.

(51) shows similar oscillations as those presented in Fig. 22. However, the oscillation amplitudes are smaller. In addition, Fig. 24 indicates that the value of R^{-1} is closer to a constant than the values of G^{-1} and H^{-1}.

Similar calculations have been performed at different Damkohler numbers and inner-to-outer jet velocity ratios. These are shown in Table 2 and indicate that the values of G^{-1}, H^{-1}, and R^{-1} are Damkohler number and velocity ratio

Table 2 The values of G^{-1}, H^{-1}, and R^{-1} for the mixing controlled models

Flow conditions[a]	Eq. (49)	Eq. (50)	Eq. (51)
$Da = 0.2$ $u_i = 40$ m/s[b]	$0 \le G^{-1} \le 3$	$0 \le H^{-1} \le 3$	$0 \le R^{-1} \le 4$
$Da = 10$ $u_i = 40$ m/s[b]	$0 \le G^{-1} \le 1.2$	$0 \le H^{-1} \le 0.6$	$0 \le R^{-1} \le 4$
$Da = 10$ $u_i = 40$ m/s[c]	$0 \le G^{-1} \le 0.8$	$0 \le H^{-1} \le 0.25$	$0 \le R^{-1} \le 1.2$
$Da = 10$ $u_i = 20$ m/s[b]	$0 \le G^{-1} \le 1.2$	$0 \le H^{-1} \le 0.6$	$0 \le R^{-1} \le 4$
$Da = 100$ $u_i = 40$ m/s[b]	$0 \le G^{-1} \le 0.7$	$0 \le H^{-1} \le 0.3$	$0 \le R^{-1} \le 0.9$

[a]The outer jet velocity is $u_o = 5.25$ m/s for all the calculations presented in this table, and the Damkohler number is evaluated at $x = 0$. [b]The pdf transport equation was solved with a mixing frequency equal to that of the energy containing eddies. [c]The pdf transport equation was solved with a mixing frequency equal to that of the dissipating eddies.

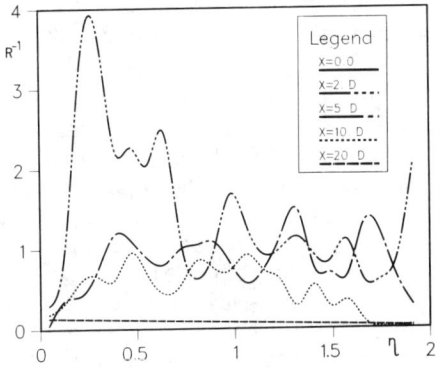

Fig. 24 The value of R^{-1} as a function of η.

Table 3 CPU times for different turbulent models

Model	CPU time[a]
Pdf model	3.5
Second-order closure model	2.0
Simplified second-order closure model	1.3
Mean reaction rate model	1.0

[a]Unity corresponds to 20 min of CPU time in a DEC-20 computer.

dependent. Thus, these values are not universal constants. Table 2 also shows the values of G^{-1}, H^{-1}, and R^{-1} for a Damkohler number equal to 0.2, i.e., when the reaction rate is controlled by the chemistry rather than the mixing. For this Damkohler number the values of G^{-1}, H^{-1}, and R^{-1} are higher than for larger Damkohler numbers. This is because the applicability of eddy-break-up models breaks down at low Damkohler numbers.

The calculations reported here were performed in a DEC-20 computer. The central processing unit times are shown in Table 3. The most expensive calculations correspond to those performed with the pdf transport equation. These are followed by the second-order closure model, the simplified second-order closure model, and the mean reaction rate model. The pdf calculations require more than three times the computer time required by the mean reaction rate model.

Summary and Conclusions

An axisymmetric two concentric jet configuration which consists of nitrogen and ozone traces in the inner jet, and nitrogen oxide traces in the outer jet has been studied at constant pressure and temperature. Under these conditions the flow is incompressible and the hydrodynamic field is decoupled from the chemistry. This is an ideal situation for determining the validity of several models for the reactant conversion rates.

Most of the calculations reported here were performed with a two-equation model of turbulence and a pdf transport equation in which the reaction terms appear in closed form and do not need to be modelled. The transport terms have been modelled by using a Boussinesq-type approximation. The Curl model has been used to model the molecular mixing terms. Two models have been employed for the mixing frequency. In the first model, this frequency corresponds to that of the energy containing eddies; while, in the second model the mixing frequency is that of the dissipating eddies. The calculations performed with these models indicate that, when the dissipating eddy frequency is employed in the mixing terms, there is a higher reactant conversion rate and smaller unmixedness than when the energy containing eddy frequency is used. This is to be expected since the dissipating eddy frequency is larger than that of the energy containing eddies. The use of a single frequency for molecular mixing is a consequence of the two-equation model of turbulence employed in the calculations if the energy containing eddy frequency is employed. However, the use of a single frequency for the dissipating eddies is consistent with the fact that at high Reynolds numbers the smallest turbulent scales are isotropic and only depend on the molecular kinematic viscosity and the dissipation rate of turbulent kinetic energy.

Calculations have also been performed with a second-order closure model in which the concentration triple correlations are neglected. These calculations show that the second-order closure model predicts a thinner reaction zone, higher unmixedness, and lower reactant conversion rates than the pdf model. Calculations have also been performed with a simplified second-order closure model in which the contribution of the chemistry terms to the transport of the unmixedness terms is neglected. This model only requires the solution of a transport equation for the unmixedness terms and transport equations for the mean reactant concentrations. The calculations based on the simplified second-order closure model predict unmixedness and reactant conversion rates comparable to that of the second-order closure model. However, the reaction zone thickness predicted by the simplified second-order closure model is about the same as that predicted by the pdf model.

The neglect of the unmixedness terms in the mean reaction rates increases the reactant conversion rates because the unmixedness terms are negative throughout the entire flowfield.

The ozone-ozone and nitric oxide-nitric oxide correlations predicted by the pdf model are smaller than those predicted by the second-order closure model. Both correlations peak at the interjet shear layer where the reaction takes place. The nitric oxide-nitric oxide correlations also present a peak at the jet edge.

The effect of the shear layer strain levels has been examined by decreasing the inner jet axial velocity. The calculations indicate that, at constant Damkohler number, larger shear levels decrease the reactant conversion rates and increase the unmixedness terms.

Three eddy-break-up models for the mean conversion rate expression have also been evaluated at different Damkohler numbers and inner-to-outer jet velocity ratios. The results of these evaluations show that there is no universal constant which can predict the reactant conversion rates at different velocity ratios and Damkohler numbers. The results also indicate that the value of these "constants" is a function of the axial distance.

The calculations reported here indicate that results for the conversion rate will vary as turbulent mixing rates (which depend upon velocity ratio), molecular mixing rate (which is related to the mixing frequency), or the chemical kinetic rate (which is represented through the Damkohler number) are varied. It does not seem reasonable therefore to make any limiting assumptions whereby the conversion rate would depend only upon one or two of those three basic rates. Note further that the Damkohler number range that was studied here would be similar to the range pertinent to typical hydrocarbon oxidation reactions. While the nonisothermal behavior will cause some differences in the quantitative results, it is doubtful that the qualitative behavior will be significantly different.

The pdf model has been taken as the basis for comparison since the chemical kinetic term is treated without modelling and the modelling of the turbulent transport term is not worse than for other models. The authors do have reservations, however, about the use of Curl's model for a description of the molecular mixing. One can only say that it can give the correct order of magnitude for the mixing rate if the frequency coefficient is correctly chosen. The details of the molecular mixing are simply poorly described by that model and a better model is needed in the future.

Experiments are needed to verify the calculations reported here.

Acknowledgments

The calculations reported here were performed under Grant No. CPE-8014661 from the National Science Foundation. Review comments by Professor R. W. Bilger were very useful in preparing the revision of this paper.

References

Bilger, R. W. (1976) Turbulent jet diffusion flames. Prog. Energy Combust. Sci. 1, 87-109.

Bilger, R. W. (1980) Turbulent flows with nonpremixed reactants. Turbulent Reacting Flows, (edited by P. A. Libby and F. A. Williams), pp. 65-113. Springer-Verlag, New York.

Champagne, F. H. and Wygnanski, I. J. (1971) An experimetnal investigation of coaxial turbulent jets. Int. J. Heat Mass Transfer 14, 1445-1464.

Chevray, R. and Tutu, N. K. (1978) Intermittency and preferential transport of heat in a round jet. J. Fluid Mech. 88, 133-160.

Corrsin, S. (1964) The isotropic turbulent mixer: Part II: Arbitrary Schmidt number. AIChE J. 10, 870-877.

Curl, R. L. (1963) Dispersed phase mixing: Part I: Theory and effects in simple reactors. AIChE J. 9, 175-181.

Fishburne, E. S. and Varma, A. K. (1979) Investigations of chemical reactions in a turbulent media. Acta Astronautica 6, 297-308.

Forstall, W. and Shapiro, A. H. (1950) Momentum and mass transfer in coaxial jets. J. Appl. Mech. 17, 399-408.

Givi, P. and Ramos, J. I. (1984) On the calculation of heat and momentum transport in a round jet. Int. Comm. Heat and Mass Transfer 14, 173-182.

Givi, P., Sirignano, W. A., and Pope, S. B. (1983) Probability calculations for turbulent key flows with mixing and reaction of NO and O_3. (submitted for publication).

Hinze, J. O. (1975) Turbulence. McGraw-Hill Publishing Co., New York, 790 p.

Janicka, J., Kolbe, W., and Kollman, W. (1979) Closure of the transport equation for the probability density function of turbulent scalar fields. J. Non-Equilib. Thermodyn. 4, 47-66.

Launder, B. E., Morse, A., Rodi, W., and Spalding, D. B. (1973) Prediction of free shear flows: A comparison of the performance of six turbulence models. Free Turbulent Shear Flows, Volume I-Conference Proceedings, NASA SP-321, 361-426.

Magnussen, B. F. and Hjertager, B. H. (1977) On mathematical modeling of turbulent combustion with special emphasis on soot formation and combustion. Sixteenth Symposium (International) on Combustion, pp. 719-729. The Combustion Institute, Pittsburgh, Pa.

Pope, S. B. (1977) A novel calculation procedure for free shear flows. Report FS/77/8, Department of Mechanical Engineering, Imperial College of Science and Technology, London, UK.

Pope, S. B. (1978) An explanation of the turbulent round-jet/plane-jet anomaly. AIAA J. 16, 279-281.

Pope, S. B. (1979) The statistical theory of turbulent flames. Phil. Trans. Royal Soc. London 291, 529-568.

Pope, S. B. (1981) A Monte Carlo method for the pdf equations of turbulent reactive flow. Comb. Science Techn. 25, 159-174.

Ramos, J. I. (1982) Unpublished manuscript on modeling of nonpremixed turbulent reactive flows.

Ramos, J. I. (1984) The numerical solutions of nonpremixed reactive flows in a swirl combustor model. To appear in Numerical Methods for Non-linear Problems, (edited by E. Onate, C. Taylor, E. Hinton, and D. R. J. Owen). Pineridge Press, Swansea, UK.

Spalding, D. B. (1971a) Concentration fluctuations in a round turbulent free jet. Chem. Eng. Science 26, 95-107.

Spalding, D. B. (1971b) Mixing chemical reations in steady confined turbulent flames. Thirteenth Symposium (International) on Combustion, pp. 649-657. The Combustion Institute Pittsburgh, Pa.

Tennekes, H. and Lumley, J. L. (1972) A First Course in Turbulence, The MIT Press, Cambridge, Mass., 300 pp.

Wygnanski, I. and Fiedler, H. (1969) Some measurements in the self-preserving jet. J. Fluid Mech. 38, 577-612.

Chapter IV. Constant Volume Combustion

Influence of Turbulent Motion on Spark Ignition

Peter S. Tromans* and Simon J. O'Connor†
Shell Research Ltd., Chester, Great Britain

Abstract

In this paper the effects of turbulent motion on the minimum spark energy required for the ignition of gaseous fuel-air mixtures are investigated. The regime in which the scale of the smallest eddies is larger than or nearly equal to the laminar flame thickness is considered; these circumstances are likely to be satisfied in many practical combustion systems. Though turbulent mixing may not be important in this regime, previous work has suggested that fluid motion may still have a strong influence on combustion through turbulent straining. A continuous flow combustion chamber with grid-induced turbulence has been used to perform ignition experiments: turbulence increases the minimum spark energy required for ignition. The data are well correlated by the root-mean-square turbulent strain rate. The influence of turbulence on ignition can only be described statistically: increasing turbulence intensity broadens the band of spark energies in which ignition is uncertain. It has been observed that other factors, such as the precise nature of the electric discharge, can have a strong influence on ignition energy.

Introduction

The spark ignition of premixed gases has been studied widely and, in recent years, the influence of turbulence has received particular attention (see Lewis and von Elbe 1951; de Soete 1971; Ballal and Lefebvre 1974, 1977; Akindele,

---

Presented at the 9th ICODERS, Poitiers, France, July 3-8, 1983. Copyright © 1984 by Shell Research Ltd. Published by the American Institute of Astronautics and Aeronautics with permission.
*Senior Scientist, Thornton Research Centre.
†Technician, Thornton Research Centre.

Bradley, Mak, and McMahon 1982). Much of the experimental work, that has been reported, has been limited either by the use of low pressures or by the failure to measure a sufficient number of parameters of the turbulent motion. However, the results indicate that minimum ignition energy is increased by the presence of turbulence in a way that depends on the intensity and length scales of the motion. A number of correlations and highly empirical theories have been used to explain these observations; the theories involve balances between heat generation and heat loss by conduction and turbulent mixing. The results show rough agreement with experiments.

The experiments described here are intended to overcome some of the above limitations. However, they are concentrated on the regime in which the laminar flame thickness is roughly equal to or smaller than the scale of the smallest eddies (the Kolmogorov length scale). These circumstances are likely to be satisfied in many practical combustion systems. In this regime turbulent mixing is not believed to be important; the dominant mechanism by which fluid motion influences combustion is turbulent straining (Tromans 1981). The strong influence strain fields have on laminar flame propagation has been established both analytically and experimentally (see Tromans 1981; Buckmaster 1978; Daneshyar et al. 1983; Law et al. 1981): straining inhibits flame propagation. The inhibition results from increased gradients in the flame front and convection along the front. The relevance of these processes to turbulent flames has been convincingly argued (see Tromans 1981; Klimov 1963; Libby and Williams 1982). Though no analysis has been made of the influence of strain fields on spark ignition, it seems likely that the influence should be strong and worthy of investigation.

Experimental Apparatus

The experiments described in this paper were carried out at room temperature and ambient pressure in a continuous flow combustion chamber. The combustion chamber forms the working section of a vertical, open-circuit, steel wind tunnel shown schematically in Fig. 1. Air enters at the base where it is thoroughly mixed with the fuel. The flow rates of methane used in these experiments were measured by means of Rotameters. The flow of fuel-air mixture is monitored by observing the pressure drop across an orifice plate situated between the mixture generator and the settling tank. In the 50-mm square working section the mixture flows first through a turbulence generating grid and then past the spark electrodes. The grid is a perforated plate. The size and position of the grid can be adjusted so that the parameters of the turbulence in the region of the electrodes can be varied to some extent independently of each other. Downstream of the working section the flammable mixture (or burned gas) is diluted with air and finally exhausted through a centrifugal fan.

Hot-wire anemometry was used to obtain turbulence intensity u', integral length scale L, Taylor microscale λ, Kolmogorov length scale η, and root mean square (r.m.s.) strain rate $S = [(du_1/dx_1)^2]^{1/2}$ of the turbulence in the region of the electrodes. The signal was collected and processed by means of a Disa 55D01 anemometer, a differentiator, a Disa 55D25 auxillary amplifier, and a Hewlett-Packard HP 3721A correlator. The measurements indicated that the turbulent flow was reasonably homogeneous at distances of 8.5 times the diameter of the grid perforations downstream of the grid (the closest position reached in the ignition experiments). The turbulence intensity varied from 0.4 to 1.4 m/s, the r.m.s. strain rate from 250 to 1600 s^{-1}, the Kolmogorov length scale from 60×10^{-3} to 130×10^{-3} mm, and the integral length scale from 2 to 7 mm.

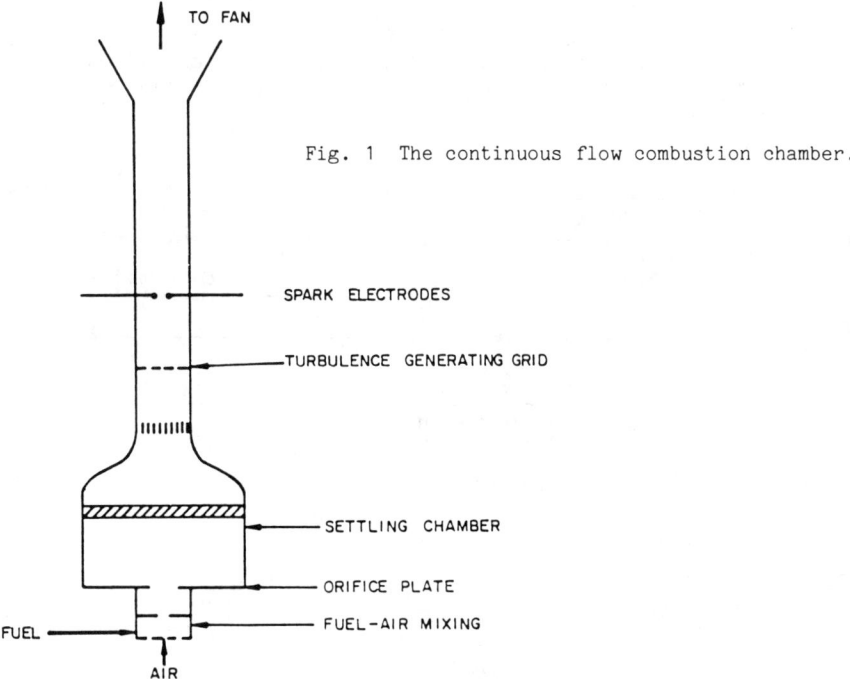

Fig. 1 The continuous flow combustion chamber.

The spark electrodes consist of 1.5-mm-diam steel ball bearings brazed onto 1.5-mm-diam brass rods. The spark-generating circuit was designed such that discharge energy and duration could be varied independently; the circuit is shown schematically in Fig. 2. The high voltage generator charges the 1-µF storage capacitor C_1. When the triggered spark gap T_1 is fired by the first trigger unit the storage capacitor is connected to the live electrode in the combustion chamber through the resistor R. The voltage at the live electrode rises until the gap between the electrodes breaks down and current flows. When this happens the pulse transformer P triggers the timer which, after some set interval, fires trigger unit 2 and, hence, the triggered spark gap T_2. Operation of T_2 short-circuits the electrodes in the combustion chamber and so terminates the discharge there. Thus, the spark duration is set by the timer while the energy can be adjusted by changing the resistor R and the voltage of the storage capacitor C_1.

The electrical energy dissipated in the spark gap is calculated from the readings of voltage and current obtained from a Tektronix P6015 high voltage probe and either a Tektronix P6021 current probe or a Pearson 411 current transformer. The energy consists of two components: a capacitive (breakdown) component arising from the discharge of the stray capacitance C_g of the spark gap; and a relatively long duration glow discharge arising from the partial discharge of the storage capacitor C_1 through the resistor R. The forms of the voltage and current traces are

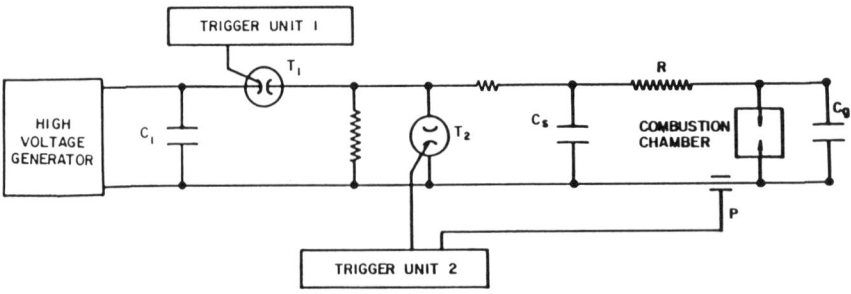

C_1 STORAGE CAPACITOR, 1µF
C_g STRAY CAPACITANCE AT ELECTRODES, 1.5 pF
C_s STRAY CAPACITANCE IN CIRCUIT
P PULSE TRANSFORMER
R LARGE RESISTOR
T_1, T_2 TRIGGERED SPARK GAPS

Fig. 2 The spark generating circuit.

shown diagrammatically in Fig. 3. The total discharge energy is given by

$$E = \frac{1}{2} C_g V_b^2 + \int_0^t V_g I_g \, dt$$

where V_b is the breakdown voltage and V_g and I_g are the glow voltage and current. Most of the energy dissipated in the gap is provided by the glow discharge. This type of discharge was used since it provides more consistent experimental results than a purely capacitive, breakdown spark and it is difficult to produce stable arc discharges of sufficiently low energy. The glow discharge is rarely used in ignition experiments because it usually involves dissipation of a large amount of energy in a small region adjacent to the negative electrode, where there is a large voltage drop. This criticism cannot be made of the present experiments: in

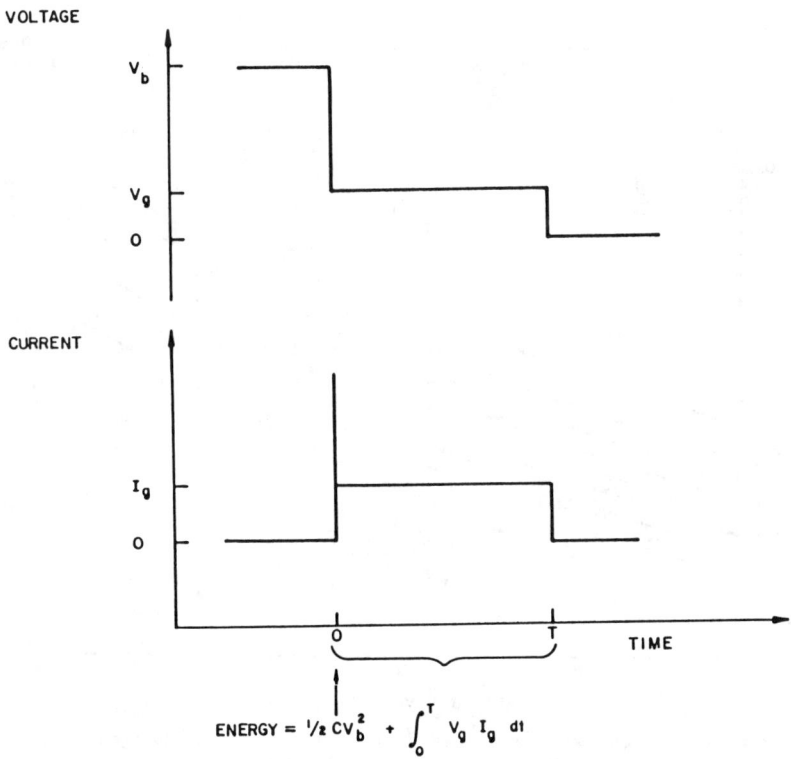

Fig. 3 Discharge voltage and current.

the high voltage, low current discharges used here the voltage gradient, and hence the energy dissipation, is fairly uniformly distributed along the spark gap (Gambling and Edels 1954).

Experimental Results

The experiments were performed with three grids at various distances upstream of the electrodes using methane-air mixtures of equivalence ratio 0.87. The mean velocity was 7.25 m/s. Both spark gap and duration were varied. The optimum settings were 2.5-3.5-mm gap and 50-100-μs duration; these values showed no dependence on fluid motion and were used to obtain all the results in this paper.

The graph in Fig. 4 shows the energies and breakdown voltages of discharges produced in a low turbulence flow. There is no sharp boundary between regions where ignition occurs and does not occur. This overlap of ignition and nonignition regions was noted by Lewis and von Elbe in their work with capacitive discharges at atmospheric and

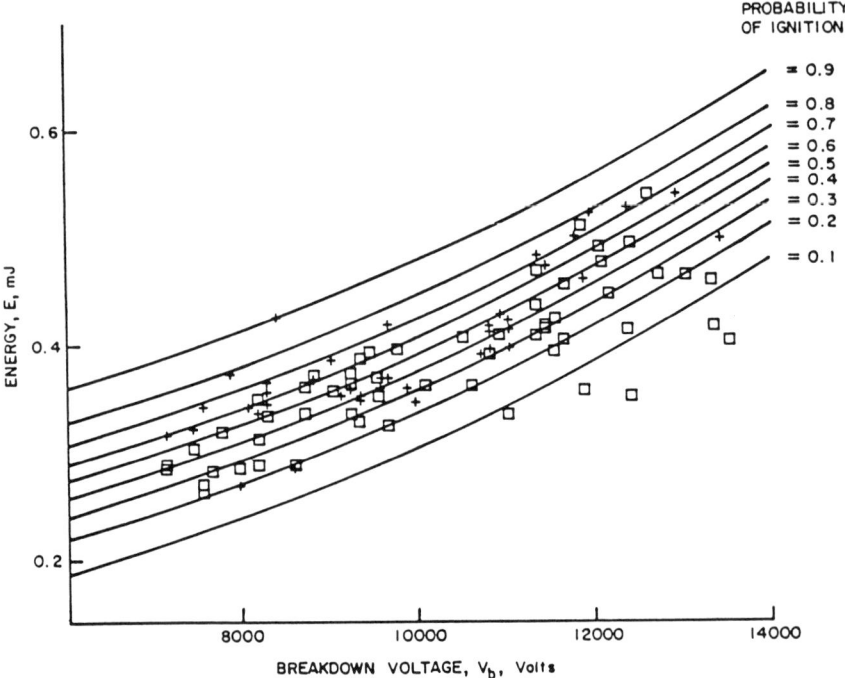

Fig. 4 The energies and breakdown voltages of discharges in low turbulence flow. ■ Ignition; □ Nonignition; $u' = 0.43$ m/s, $S = 260$ s^{-1}; $\eta = 0.120$ mm, $L = 5.5$ mm

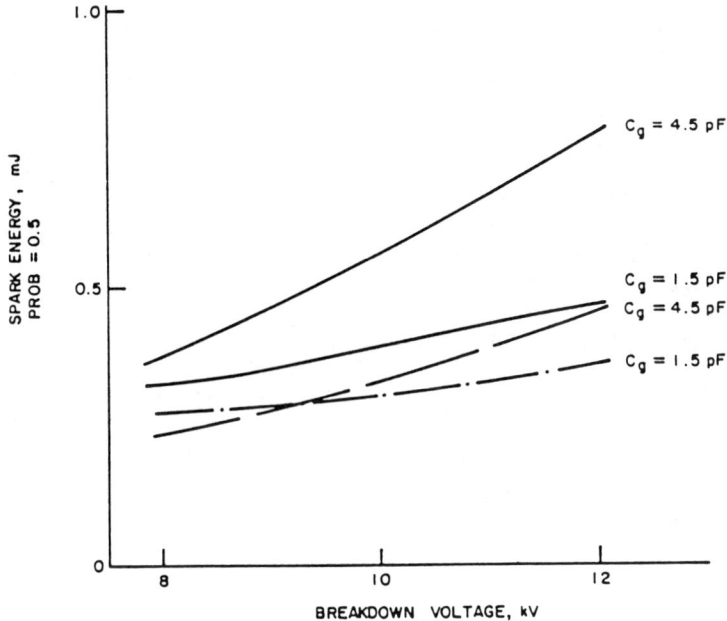

Fig. 5 The effect of spark gap capacitance on the energy and breakdown voltage for 50% probability of ignition.

higher pressures: they did not observe it at low pressures. Curves of constant probability of ignition are also shown in Fig. 4; these were fitted to the experimental data with the GENSTAT routines of an IBM 4341 computer. Figure 4 shows that the probability of ignition depends on the voltage at which the gap broke down as well as the spark energy. It seems likely that this phenomenon is associated with the capacitive component of discharge energy. To investigate this some experiments were performed with the spark gap capacitance C_g increased from 1.5 to 4.5 pF. The resultant curves, for which the probability of ignition is 0.5, are shown in Fig. 5. The lowering of capacitance reduces the sensitivity of ignition energy to breakdown voltage. Moreover, if the capacitive component of energy is totally neglected in the calculation of spark energy the sensitivity is further reduced. This suggests that, in the present experiments, the small amount of energy supplied by the stray capacitance is very ineffective as a contributor to the ignition process and may even have an inhibiting effect.

Plots of energy and breakdown voltage, similar to those shown in Fig. 4, have been obtained at various distances downstream of a variety of grids. Each plot involves a large number of individual measurements. However, to study the influence of turbulence the influence of breakdown voltage was removed by choosing one value of breakdown voltage V_b = 10 kV and replotting to produce graphs of probability of ignition against spark energy such as those in Fig. 6. One curve is for a relatively low intensity turbulent field: the other for a relatively high intensity one. The increase in turbulence intensity broadens the band of spark energies in which ignition is uncertain: in particular, it increases the spark energy required for ignition to be certain. To investigate possible relationships the spark energy for 50% probability of ignition has been plotted against various properties of the turbulent flow, such as r.m.s. strain rate, turbulence intensity, and eddy viscosity. The only parameter with which ignition energy was found to increase monotonically was the r.m.s. strain rate; a graph is shown in Fig. 7. The spark energies for 10 and 90% probability of ignition are also plotted in Fig. 7; it appears that the broadening of the band of energies within which ignition is uncertain can also be related to strain rate.

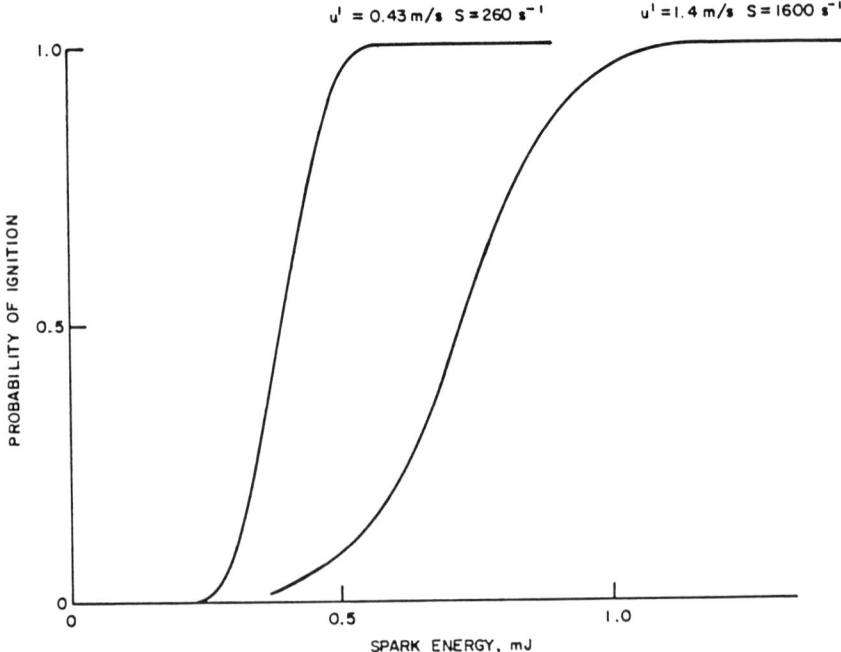

Fig. 6 The variation of probability of ignition with spark energy at a fixed breakdown voltage, V_b = 10 kV.

Discussion

At present there are no theories of spark ignition that can be used to compare with these experimental results or to reduce them to a more general form. However, the analyses of steady, strained laminar flames quoted in the introduction suggest that V/δ = (laminar burning velocity ÷ laminar flame thickness) might provide a suitable scaling with which to make the r.m.s. strain rate nondimensional. It seems reasonable to scale the ignition energy on its value at zero strain rate, obtained here by extrapolation. Thus, the 50% probability of ignition line shown in Fig. 7 is given by

$$(E/E_o) = 1 + 3.3 \ (S\delta/V) \qquad (1)$$

where E_o = 0.30 mJ, V = 0.28 m/s, and δ = 0.075 × 10^{-3} m.

Strain rate has not been used previously to correlate any spark ignition data. However, many of the results of Ballal and Lefebvre (1974) fall into the regime where laminar flame thickness is less than the Kolmogorov scale. The r.m.s. strain rate has been calculated for these data and plotted against minimum ignition energy in Fig. 8. The

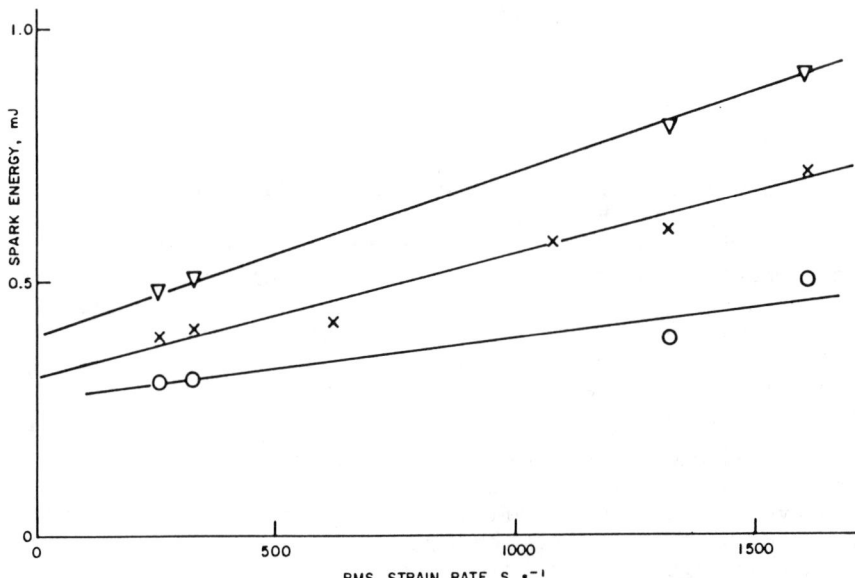

Fig. 7 Spark energy for a given probability of igniton plotted against R.M.S strain rate. Probability of ignition = 0.1, O; 0.5, ■ ; 0.9, △.

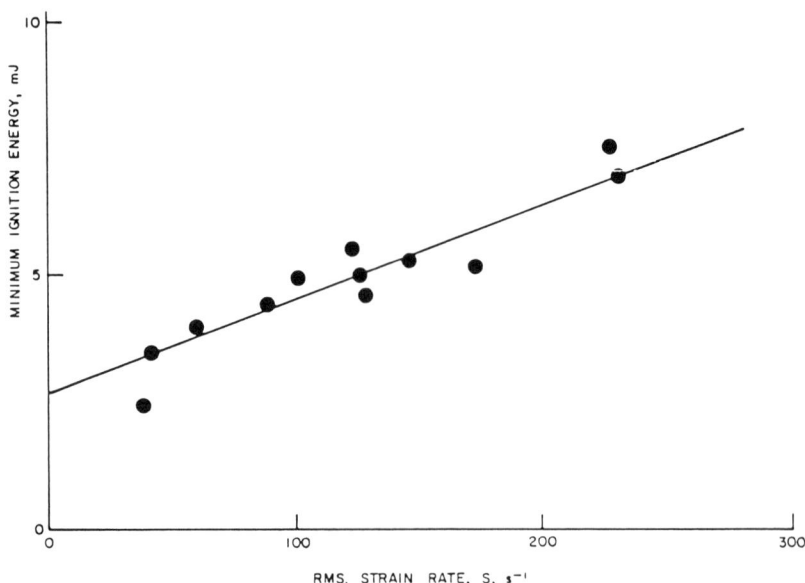

Fig. 8 Dependence of ignition energy on the strain rate parameter (calculated from the data of Ballal and Lefebvre 1974). Experimental conditions: propane-air mixture, equivalence ratio = 1.04, pressure = 0.17 atm.

line in Fig. 8 is given by

$$(E/E_o) = 1 + 12 \, (S\delta/V) \qquad (2)$$

where E_o = 2.7 mJ, V = 0.5 m/s, and δ = 0.29 × 10^{-3} m.

In the present work methane was the fuel, a glow discharge served as the ignition source, and the experiments were performed at atmospheric pressures: in Ballal and Lefebvre's investigation propane was the fuel, an arc discharge served as the ignition source and the experiments were performed at a pressure of 0.17 atm. In view of these differences and the various possible sources of systematic error in the results, the agreement between Eqs. (1) and (2) is encouraging.

A major difference between Ballal and Lefebvre's results and the present ones is that Ballal and Lefebvre observed no statistical effects: they found a sharp boundary between the regions in which ignition does and does not occur rather than a band of energies in which ignition is uncertain. This contrast can be attributed to the difference in pressures between the two experiments. At the low pressure of Ballal and Lefebvre's work diffusivities, and hence quenching distances, are large and a long spark gap is optimum; in fact, the spark gap is of order five times the integral length scale of the turbulent motion. Thus local fluctuations are averaged out and every identical spark produces the same result. In the present work at atmospheric pressure the quenching distance and optimum spark gap are much smaller, less than the intergral length scale. Because of the statistical nature of the turbulent field in the region of the spark, the result will not be the same at every attempt at ignition. Thus, there is a band of energies in which ignition is uncertain. This band broadens as the r.m.s. strain rate, and hence the amplitude of fluctuations in strain rate, increases.

Conclusions

The main conclusions of this work are:

1) Turbulence increases the spark energy required to achieve a given probability of ignition. The increase can be correlated with the r.m.s. strain rate.

2) The influence of turbulence on ignition can only be described statistically.

3) There are optimum settings for spark duration and spark gap. These seem insensitive to fluid motion.

4) The spark energy required to achieve a given probability of ignition depends on breakdown voltage in the present experiments.

References

Akindele, O. O., Bradley, D., Mak, P. W., and McMahon, M. (1982) Spark ignition of turbulent gases. Combustion and Flame 47, 129-155.

Ballal, D. R. and Lefebvre, A. H. (1974) The influence of flow parameters on minimum ignition energy and quenching distance. Fifteenth Symposium (International) on Combustion, pp.1473-1481. The Combustion Institute, Pittsburgh, Pa.

Ballal, D. R. and Lefebvre, A. H. (1977) Ignition and flame quenching in flowing gaseous mixtures. Proc. R. Soc. Lond. A 357, 163-181.

Buckmaster, J. D. (1978) The quenching of a deflagration wave held in front of a bluff body. Seventeenth Symposium (International) on Combustion, pp.835-842. The Combustion Institute, Pittsburgh, Pa.

Daneshyar, H., Ludford, G. S. S., Mendes-Lopes, J. M. C. and Tromans, P. S. (1983) The influence of straining on a premixed flame and its relevance to combustion in SI engines. Conference on Combustion in Engineering, pp.191-199. I. Mech. E, London.

De Soete, G. G (1971) The influence of isotropic turbulence on the critical ignition energy. Thirteenth Symposium (International) on Combustion, pp.735-743. The Combustion Institute, Pittsburgh, Pa.

Gambling, W. A. and Edels, H. (1954) The high pressure glow discharge in air. Brit. J. of Appl. Phys. 5, 36-39.

Klimov, A. M. (1963) Laminar flame in a turbulent flow. Zh. Prikl. Mekh. Tekn. 3, 49-58.

Law, C. K., Ishizuka, S., and Mizomoto, M. (1981) Lean limit extinction of propane/air mixtures in the stagnation-point flow. Eighteenth symposium (International) on Combustion, pp.1791-1798. The Combustion Institute, Pittsburgh, Pa.

Lewis, B. and von Elbe, G. (1951) Combustion, Flame and Explosion of Gases. Academic Press, New York.

Libby, P. A. and Williams, F. A. (1982) Structure of laminar flamelets in pre-mixed turbulent flames. Combust. Flame, 44, 287-303.

Tromans, P. S. (1981) The interaction between strain fields and flames - a possible source of combustion variations in spark ignition engines. Symp. on Fluid Mech. of Comb. Systems, pp.201-206. ASME, New York.

Vibratory Combustion Triggered by a Small Cavity in the Wall of a Constant Volume Combustion Chamber

A. Girard,* F. Fisson,† and J. C. Leyer‡
Université de Poitiers, Poitiers, France

Abstract

The reported experiments demonstrate that a cavity of small volume (20 to 700 mm^3) in the wall of a flat cylindrical closed combustion chamber (16x10^3 mm^3) is responsible for knock occurrence during the last phase of flame propagation. Flame Schlieren pictures and chamber pressure-time records indicate that the observed fast end-gas burning is triggered by the turbulence of a small unsteady jet issuing out of the small cavity as the flame front approaches the chamber wall. With a stoichiometric propane-oxygen mixture, diluted by argon, used through the experiments, knock is still enhanced by self-ignition phenomena that start within the cavity as the initial chamber pressure p_i exceeds a certain limit ($p_i \sim 0.8$ MPa). The main parameters that influence the process were found to be: i) the cavity diameter; ii) its location relative to the ignition spark; iii) the ratio γ of the specific heat of the mixture as far as self-ignition is concerned.

Introduction

The explosion of gaseous hydrocarbon-air mixtures in closed chambers is often accompanied by severe pressure oscillations appearing in the last phase of the combustion. Such oscillations are observed to occur preferentially with rich mixtures at high initial pressure. A detailed description and attempted explan-

---

Presented at the 9th ICODERS, Poitiers, France, July 3-8, 1983. Copyright American Institute of Aeronautics and Astronautics, Inc., 1984. All rights reserved.
 *Boursier du CNRS, Laboratorie d'Energétique et de Détonique, E.N.S.M.A.
 †Maître-Assistant, Laboratoire d'Energétique et de Détonique, E.N.S.M.A.
 ‡Professeur, Laboratoire d'Energétique et de Détonique, E.N.S.M.A.

ation of this phenomenon are given by Sokolik (1963) in the section of his book devoted to "knock in the absence of self-ignition ahead of the flame front". In fact, vibratory explosions in bombs have attracted the attention of many researchers because the similarities with knocking combustion in internal combustion engines (ICE) are striking (for instance, the excitation of the few first acoustic modes of the combustion chamber) and because experimental observations are more easily obtained in bombs than in running engines. However, the mean temperature reached in engines at the end of the compression stroke cannot be reproduced in closed chambers when they are filled with mixtures at varying initial pressures but generally at room temperature. As a consequence, results obtained in bombs cannot easily be applied to engine conditions.

Numerous causes of knock have been proposed (see, for instance, Starkman 1962). These involve either the effects of chemical reactions with self-ignition in the end-gas or physical effects due to amplification of some transverse acoustic modes of the engine chamber triggered by a sudden variation in the burning rate. This sudden variation may be the consequence--as observed by Leyer and Manson (1970) in a rectangular closed chamber--of the reduction in the burning area when the flame front reaches the walls after the initial spherical phase of the propagation. Turbulence may also cause acceleration of the flame and consequently promote vibratory combustion (see Maly and Ziegler 1982). However, in ICE, the evidence suggests that turbulence reduces the tendency to knock at high compression ratios (see, for instance, Ricardo 1931). Therefore, the actual role of turbulence in the onset of vibratory combustion is still now an open question.

The present paper reports the preliminary results of experiments carried out in a flat cylindrical chamber that demonstrate that localized turbulence (in the form of a high-velocity jet) enhances vibratory combustion of stoichiometric propane-oxygen-argon mixtures. This turbulence is produced by the fluid interactions with a cavity in the side wall of the chamber. The cavity is produced by drilling a small hole in the side wall.

The main features of the jet generation as the flame front approaches the cavity and the onset of vibrations are described for various values of the governing parameters of the phenomenon, which have been identified to be the cavity diameter and the pressure level in the chamber mainly.

When the initial chamber pressure p_i exceeds a certain limit ($p_i \sim 0.8$ MPa), a strong pressure peak occurs in the cavity and triggers high-pressure oscillations in the chamber. Although the origin of this sudden pressure pulse is not yet completely understood, it could be self-ignition or surface-ignition inside the cavity. Work is now in progress to clarify this point.

Experimental Conditions

Experiments were performed with a flat cylindrical (80 mm in diameter) chamber with two plane glass walls, separated by the distance h = 8 mm (see Fig. 1). A crescent-shaped acrylic plastic piece was inserted between the glass walls. In this piece, a hole (labeled C on Fig. 1) of diameter d_c in the range 1-6 mm was drilled to a depth of 25 mm. This mounting defined in the chamber a nearly cylindrical combustion space (diameter: $\phi_c \sim 50$ mm; volume: 16 cm^3) provided with the small cavity C. Through the glass windows, the complete flame front could be visualized by Schlieren cinematography (4000 frames per second) along its total path in the space and eventually in the cavity. Ignition was obtained by a wall-mounted spark plug. As shown in Fig. 1, two spark-cavity relative locations, referred to as C_1 and C_2 in the text, were used in experiments.

The stoichiometric mixture was prepared from pure propane (99.9 % of volume) and oxygen diluted by argon to obtain the volumetric composition $C_3H_8 + 5\,O_2$

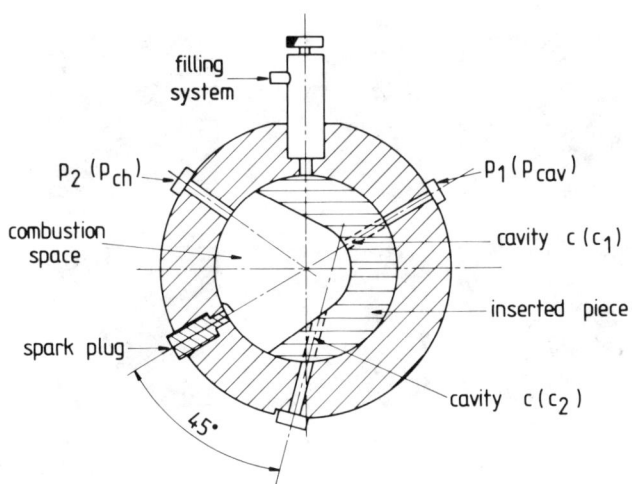

Fig. 1 Experimental mounting.

+ 18.8 A. The inert gas volume is the same as the volume of nitrogen in air. Argon was used to increase the specific heat ratio γ of the mixture from 1.4 to about 1.6, and hence to increase the adiabatic compression temperature for a given compression ratio. The mixture was forced through the slight gap (<< 0.2 mm) that exists between the crescent-shaped piece and the windows to fill the combustion space. The thickness of the gap was sufficiently small to quench the flame. The space and the cavity C were filled at room temperature T_i = 293 K and initial pressure p_i ranging from 0.1 to 0.9 MPa. Pressure-time variations in the combustion space (p_{ch}) and at the bottom of the cavity (p_{cav}) were recorded simultaneously by the transducers P_1 and P_2, respectively.

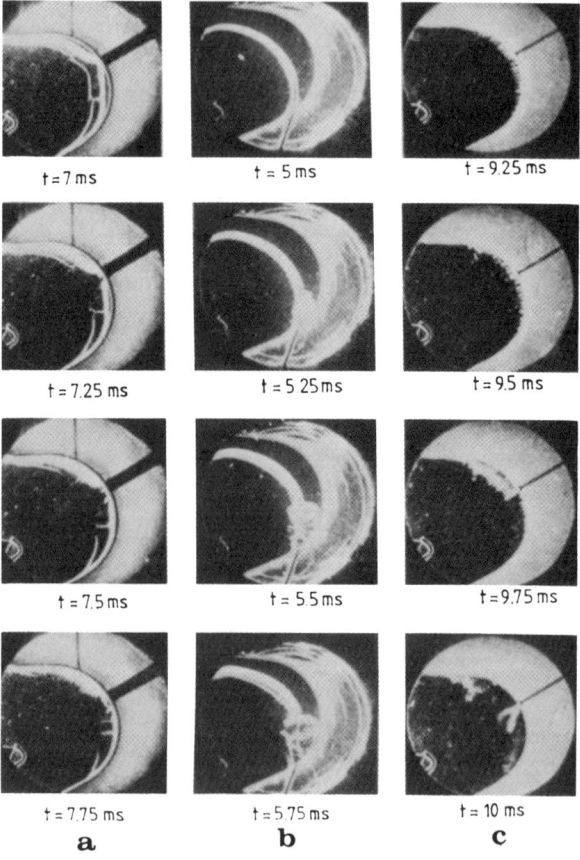

Fig. 2 Schlieren records of the flame propagation
p_i = 0.1 MPa; a) C_1, d_c = .5mm, b) C_2, d_c = 1mm, c) C_1, d_c = 1mm.

The initial experiments were performed without the cavity C but at the conditions just described. The time of flame propagation through the combustion space ranged from 9 to 11 ms as the initial pressure was increased, whereas the final pressure p_c was found in the interval $5 p_i < p_c < 6.0 p_i$. No pressure vibration was detectable on the pressure records. For a compression ratio of 5.2 and a value of $\gamma = 1.6$, the maximum adiabatic compressed end-gas temperature would be $T_{fmax} = 535$ K for an initial value $T_{fi} = 288$ K.

Results

The Flame Development

Figures 2 and 3 are typical examples of Schlieren records showing the flame front propagation, respectively, for low and high initial pressures p_i and several values of the cavity diameter d_c. It is evident from the records that the cavity is responsible for a pronounced disturbance of the approaching flame front. For large values of d_c ($d_c > 4$ mm; see Fig. 2a), the front seems to be entrained into the cavity. For small values of d_c ($d_c \sim 1$ mm), a localized turbulent jet is shown to issue from the cavity and to accelerate the flame front (Figs. 2b and 2c; Figs. 3b and 3c). The turbulent effect of the jet is more pronounced if the cavity has the location C_2. At higher pressure, the propagation is typically of the same nature, but secondary ignitions are observed whatever the value of d_c (see Fig. 3). For $d_c > 4$ mm, self-ignition points occur in the chamber, while for $d_c < 4$ mm, ignition points are located inside the cavity for small d_c.

For small diameter d_c, the flame development is highly influenced by the jet phenomenon for both locations C_1 and C_2 of the cavity at high pressure, and the end-gas burns fast. However, at low pressure and for the location C_1 of the cavity, the jet is produced just at the end of the propagation and the acceleration of burning is not so evident in the end-gas.

Pressure in the Chamber and in the Cavity

Pressure-time diagrams recorded in the chamber $p_{ch}(t)$ and at the bottom of the cavity $p_{cav}(t)$ provide additional information on the origin of the previously described turbulent jet (see Figs. 4 and 5). Here the pressure-time history for a low explosion pressure

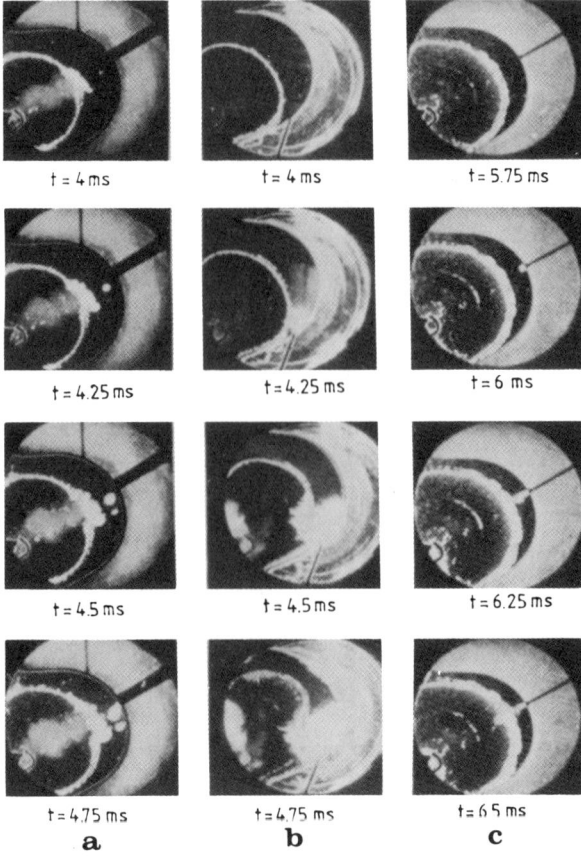

Fig. 3 Schlieren records of the flame propagation
p_i = 0.9 MPa; a) C_1, d_c = 5mm, b) C_2, d_c = 1mm, c) C_1, d_c = 1mm.

(p_i = 0.1 MPa) is first described and then the differences that have been noticed for a high explosion pressure (p_i = 0.9 MPa) are mentioned. It is evident from the records on Fig. 4 that the cavity location is mainly responsible for the discrepancies between the pressure history in the cavity and in the main chamber.

Pressure-time diagrams are identical before the instant labeled t_a or t_b, corresponding to the appearance of the jet in the main chamber (if d_c < 4 mm) or to the entrance of the flame in the cavity (if d_c > 4 mm), respectively. For $t > t_a$ or t_b, the pressure in the cavity is observed either to increase when the cavity is at the C_1 location (this fact seems to indicate that the gas, confined in the cavity is burning continuously), or to decrease when the cavity is at the C_2 location. In the latter case, cooling of the gas in the cavity during burning may be responsible for the observed decrease in pressure. For the relative cavity location C_2, the turbulent jet strongly accelerates the final phase of combustion in the main chamber and leads to an obvious increase in the ratio dp/dt and a decrease in the duration t_c of the combustion in the main chamber.

If observations deduced from Figs. 2 and 4 are correlated, it appears that the turbulent jet is formed by the combustion products confined in the cavity in which a flame front, ignited by the main front as it arrives at the edge of the hole, propagates. Finally it can also be noticed that combustion in small cavities is characterized by the appearance of vibratory phenomena that occur just before the jet appears in the main chamber.

For high initial pressure values (see Fig. 5), the phenomena described just above are complicated, due to the appearance of secondary ignition spots. Pressure-

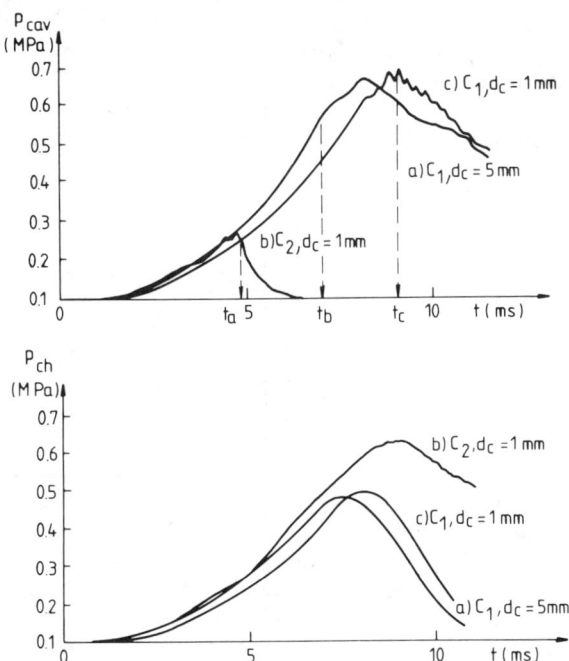

Fig. 4 Pressure variations recorded in the chamber $p_{ch}(t)$ and at the bottom of the cavity $p_{cav}(t)$; p_i = 0.1 MPa; a) C_1, d_c = 5mm, b) C_2, d_c = 1mm, c) C_1, d_c = 1mm.

time records characteristics depend not only on the relative cavity location in the main chamber, as mentioned previously, but also strongly on the cavity diameter d_c. For large values of the cavity diameter ($d_c \geq 4$ mm), secondary ignition spots occur ahead of the main flame front at time t_c and lead to an obvious increase in the final pressure in the chamber (and also to smaller values of the combustion time t_c) compared to those observed without the cavity. At this moment the unburned gas estimated temperature reaches about 500 K- a value that is probably very close to the self-ignition temperature. This temperature is estimated by Lewis and von Elbe (1961) to be T_{ig} = 580 K for a rich propane-air mixture (equivalence ratio r = 1.2) under a pressure of 1.35 MPa.

For small values of the diameter d_c ($d_c \leq 4$ mm), the pressure in the cavity exhibits oscillations--with a frequency f_c in the range 6-7 kHz--that start several tenths of a millisecond before the appearance of a jet in the chamber at time $t = t_a$ (see Fig. 5).

These first oscillations are followed by the occurrence in the cavity of a sudden pressure pulse, with a peak-to-peak amplitude of around 2 MPa. Simultaneously, chamber pressure oscillations begin to grow with a frequency in the range 8-12 kHz, corresponding approximately to the frequency F_c of the first transverse mode of the chamber:

$$F_c = 1.841 c / \pi \Psi_c$$

where c designs the sound speed in gases that are burning (650 m/s $\leq$ c $\leq$ 1000 m/s).

The reason for the pressure pulse occurrence in the cavity is not clear at the moment. One may think, first, of self-ignition inside the cavity, as observed in the chamber for $d_c \geq 4$ mm. However, the estimated isentropic compression temperature in the cavity, calculated at $t = t_a$, is in the range 400-450 K, and, due to large heat losses in a high surface to volume ratio space, the actual temperature is certainly lower. So it seems that self-ignition conditions cannot be reached in the cavity. However, the mean temperature in the cavity could be increased by the dissipation of acoustic waves of finite amplitude (of the order of 0.15 MPa peak-to-peak at high-frequency 6-7 kHz). This dissipation occurs in a rugged capillary tube. As a consequence, self- or surface-ignition occurrence in the cavity cannot be completely discarded. Further

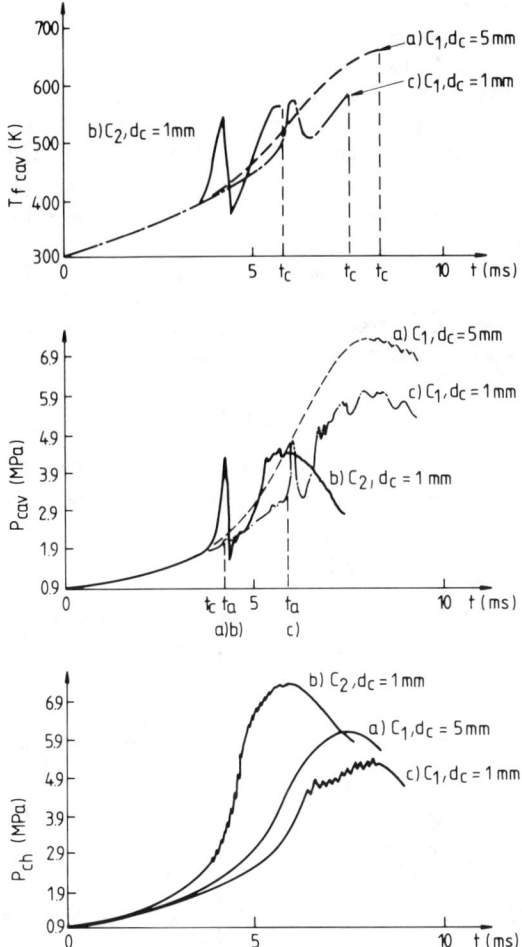

Fig. 5 Pressure variations recorded in the chamber $p_{ch}(t)$ and at the bottom of the cavity $p_{cav}(t)$; and calculated unburned mixture isentropic compression temperature in the cavity; $p_i = 0.9$ MPa; a) C_1, d_c = 5mm, b) C_2, d_c = 1mm, c) C_1, d_c = 1 mm.

experiments are needed to explain completely the observed phenomena.

Conclusion

It has been observed that a small cavity drilled in the wall of a closed combustion chamber may trigger vibratory combustion with characteristics comparable to knock.

The fundamental process that induces vibrations seems to be the appearance of a localized turbulence in the form of a turbulent high-velocity jet flowing from the cavity into the main chamber. This turbulence accelerates the burning of the last fraction of the mixture. When the pressure in the chamber is high ($p_i \simeq$ 0.8 MPa for the mixture used in the experiments), secondary ignitions make the end-gas combustion still more complicated.

The present study demonstrates again the importance of localized turbulences and localized acoustic perturbations to the onset of vibratory combustion that may resemble knock in internal combustion engines.

Acknowledgment

The authors are greatly indebted to C. Guerraud, Ingénieur CNRS, for his help in performing experiments.

References

Lewis, B. and von Elbe, G. (1961) Combustion Flame and Explosion of Gases, p. 178, Academic Press, New York.

Leyer, J. C. and Manson, N. (1970) Development of vibratory flame propagation in short closed tubes and vessels. 13th Symposium (International) on Combustion, pp. 551-558, The Combustion Institute, Pittsburgh, Pa.

Maly, R. and Ziegler, G. (1982) Thermal combustion modelling, theoretical and experimental investigation of the knocking process. SAE Technical Paper Series No. 820759.

Ricardo, H. R. (1931) The High Speed Internal-Combustion Engine, pp. 95-128. Blackie and Son Limited, London.

Sokolik, A. S. (1963) Self-ignition, flame and detonations in gases, pp. 438-442. Israel Program for Scientific Translations, Jerusalem.

Starkman, E. S. (1962) Reciprocating engine combustion research - a status report. 9th Symposium (International) on Combustion, pp. 1005-1011, The Combustion Institute, New York.

Direct Measurement of the Head-on Flame Quenching Distance in Closed Chambers

A. Girard* and J. C. Leyer†
Université de Poitiers, Poitiers, France

Abstract

An optical method is proposed as a means of monitoring the head-on approach of a flame front to a plane wall of a closed rectangular combustion chamber. This method depends on the analysis of, in a narrow band of wavelength $\lambda = (431 \pm 6.6)$ nm, the CH radical flame front emission at several distances d_w ($0 < d_w < 2$ mm) from a nonadiabatic wall. For the propagation of a well-reproducible flame in the chamber, recording the instant of appearance of the CH peak and the relative decay of its amplitude as d_w is decreased towards zero, gives both the flame trajectory and an estimate of the head-on quenching distance. Measurements are presented for lean methane-air flames at pressures between 0.1- 0.7 MPa and unburned gas temperatures of about 500 K. It is shown that the cold wall begins to influence the propagation on a distance equivalent to 10-18 times the thermal flame thickness ℓ. These results are compared to those values obtained from several different models recently proposed to describe the flame approach against a nonadiabatic cold wall.

Introduction

Fuel economy and reduction of exhausted pollutants are, today, necessary requirements for reciprocating engine

---

Presented at the 9th ICODERS, Poitiers, France, July 3-8, 1983. Copyright American Institute of Aeronautics and Astronautics, Inc. 1984. All rights reserved.

* Boursier du CNRS, Laboratoire d'Energétique et de Détonique, E.N.S.M.A.

† Professeur, Laboratoire d'Energétique et de Détonique, E.N.S.M.A.

design. Both requirements are achieved through increased compression ratio and combustion of lean mixtures in conventional or stratified engines. But decreased equivalence ratio increases combustion time, renders ignition more difficult, and misfires frequently occur. The flame front-wall interaction phenomena also become more important as the fuel content is reduced. Thermal transfer coupled with changes in chemical reaction mechanisms and rates close to the wall may result in incomplete combustion, which in turn may constitute a significant source of unburnt hydrocarbon emission in the exhaust.

The characteristics of flame quenching at the walls are not yet completely understood. The quenching distance (defined as the critical propagation distance d_c in narrow tubes or between parallel plates) and its dependence on temperature and pressure are well known and found in current textbooks on combustion (see, for instance, Fristrom and Westenberg 1965). Few experimental results on quenching distance values in the case of head-on approach (labeled d_f in the next) have been reported but the predictions of several models[‡], describing the flame front approach to a nonadiabatic wall recently proposed, vary by several orders of magnitude.

For these reasons, a method has been developed to estimate the head-on quenching distance d_f so that the more appropriate models may be selected from among those proposed. This paper reports an effort to define such a method based on the observation of the flame front CH emission, close to the wall in pure methane-air lean mixtures (equivalence ratio ϕ = 0.75 and 0.9) which burn in a rectangular chamber initially filled at room temperature ($T_i \sim 293$ K) under pressure p_i ranging from 0.025 to 0.1 MPa.

Experimental Method

The stainless steel rectangular chamber (V = 384 cm^3), schematically drawn on Fig. 1, is identical to the chamber described by Girard and Leyer (1980) in their study of stratified constant volume combustion. The chamber has two windows for observation of the total process of flame development by Schlieren photographs. Ignition of the lean pure methane-air mixtures was obtained by a spark in the center of the lower horizontal wall (electrode E, Fig. 1)

[‡] (See Kurkov and Mirsky 1969; Adamczyk and Lavoie 1978; Bush et al. 1980; Henningsen 1980; Ishikawa and Branch 1977; Hocks et al. 1981; Westbrook et al. 1981).

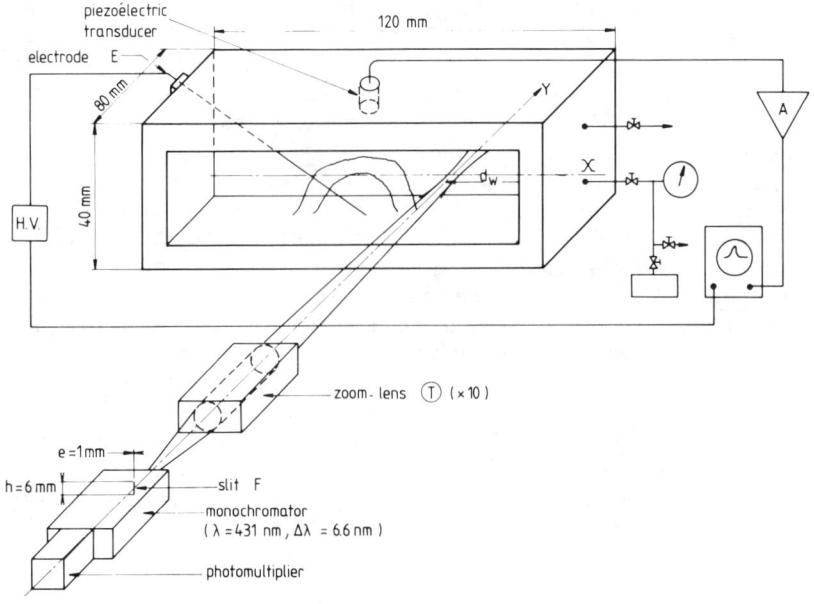

Fig. 1 Experimental arrangement.

of the chamber. A piezoelectric gage recorded the chamber pressure vs time as the flame propagates.

The optical system used to view the flame front approach from a side vertical wall of the chamber (wall area: 80 x 40 mm^2) is composed mainly of a lens system T which extends about 10 times the region viewed in the X.Y plane of the chamber and which focuses the image of this region on the slit F of a monochromator. Height and width of the slit were maintained constant (6 and 1 mm, respectively) during the experiments. The monochromator passed a wavelength band with $\Delta\lambda$ = 6.6 nm, centered on λ = 431 nm, which corresponds to one of the emission bands of the radical CH present in the reaction zone of hydrocarbon-air flames (see Gaydon 1957). Intensity of this radiation was monitored by a photomultiplier (see Fig. 1). During experiments, the distance d_w between the axis of the optical mounting and the chamber plane wall was varied by translation of the combustion chamber in the horizontal direction. Because the flame propagation was highly reproducible from run to run, the flame approach to the wall could be monitored by recording the instant of the CH emission start as the distance d_w was varied. Preliminary experiments indicated that the volume viewed by the photomultiplier was unchanged provided that the optical axis was $d_w \geq 0.25$ mm. Thus, with this condition fulfilled, the observed relative decay of the CH emission peak intensity as d_w was decreased from 2 to 0.25 mm was representative of the influence of the cold wall on the flame reaction zone structure.

On Fig. 2 are shown typical output signals of the photomultiplier corresponding to various distances d_w ($d_w = 0$ at the wall). On these records the time origin corresponds to the instant of spark ignition. The CH peak, corresponding to the flame front arrival at the distance d_w, is clearly distinguished on the records. Following the peak emission, the slow return of the signal to zero level is attributed to part of the continuous spectrum of CO oxidation, a well-known phenomenon which is called afterburning (see, for instance, Wohl and Welty 1955) and is responsible for burned gas "reillumination". From such records, the variations of the spatial flame velocity $u(t)$, in the distance range $2 > d_w > 0.25$ mm, were

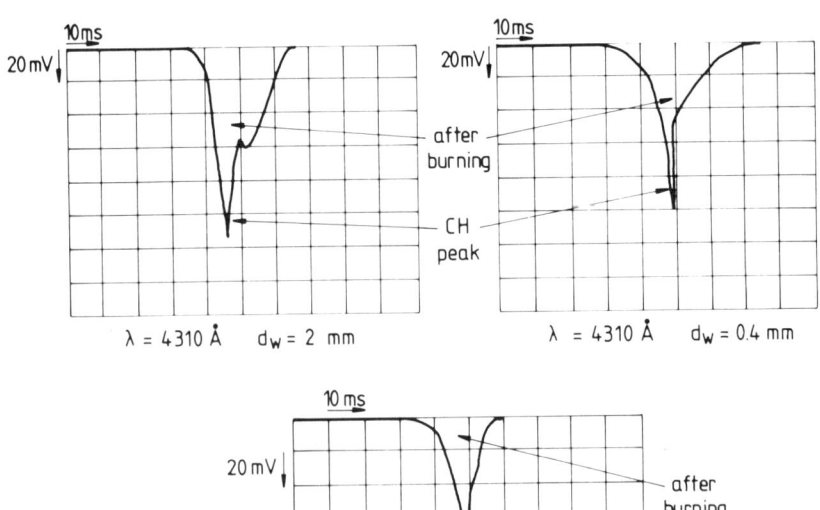

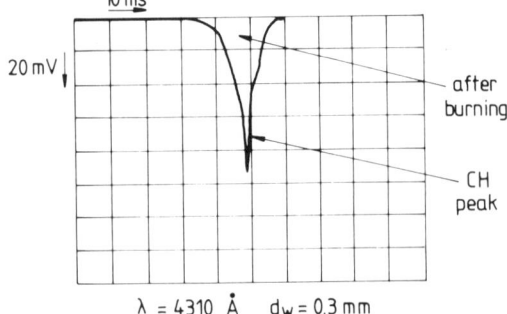

Fig. 2 Typical photomultiplier output signals for various distances d_w between the chamber wall and the mounting optical axis.

obtained. The decrease of the ratio $R = I/I_{max}$ of the local CH peak amplitude from its maximum value (measured for $d_w > 2$ mm) to its value at the wall was determined to evaluate the head-on quenching distance.

Results

Flame Path

To validate this method the flame trajectory obtained from Schlieren picture records (4000 frames/s) was shown to be in good agreement with that deduced from the CH emission method. Agreement was obtained for $d_w \geq \sim 1$ mm. For smaller d_w, the refractive index gradient is so great that the Schlieren method fails to give the flame position with sufficient accuracy. The CH emission method gives acceptable results for $d_w \geq \sim 0.25$ mm, at which the size of the volume viewed by the photomultiplier decreases. As a consequence, the spatial flame speed $u(t)$ could be calculated for $d_w \geq 0.3$ mm from the wall with acceptable accuracy. From these observations the distance d_w at which $u(t)$ equals the adiabatic normal velocity u_o^w may be defined. Close to the wall the spatial velocity and the burning gas velocity tend to be equivalent because the unburned gas velocity goes to zero at the wall. To calculate u_o under actual condition of temperature T_u and pressure p of the fresh mixture at the end of the flame propagation in the chamber for a given mixture ratio ϕ, the dependence given by Andrews and Bradley (1972) is used:

$$u_o(\phi, p, T_u) \text{ m/s} = u_o(\phi, p = 0.1 \text{ MPa}, T_u = 298 \text{ K})/$$
$$u_o(\phi=1, p=0.1 \text{ MPa}, T_u = 298 \text{ K}) \times 10^{-1.5}/\sqrt{p} \times \quad (1)$$
$$(10 + 0{,}000371 \, T_u^2)$$

where $u_o(\phi, p = 0.1 \text{ MPa}, T_u = 298 \text{ K})$ and $u_o(\phi = 1, p = 0.1 \text{ MPa}, T_u = 298 \text{ K})$ are the reference values recommended by Andrews and Bradley (1972).
In this relationship p is expressed in MPa. To estimate the temperature T_u, an isentropic compression ($\gamma = 1.4$) of the fresh mixture was assumed.

Head-on Quenching Distance

The head-on quenching distance was estimated by plotting the variation of the intensity ratio R vs the reduced distance $\delta = d_w/\ell$, where ℓ is the thermal thickness of the flame defined as:

$$\ell = a/u_o \tag{2}$$

in which a is the thermal diffusivity of the unburned gas again evaluated in the thermodynamic state corresponding just to the end of the combustion. As shown on the example of Fig. 3, for a mixture ratio $\phi = 0.75$ and an initial pressure range $0.025 < p_i < 0.1$ MPa (corresponding to final chamber pressure: $0.1 < p < 0.7$ MPa), the measured flame velocity becomes of the order of the adiabatic burning velocity u_o when the relative intensity ratio R decreases about to 95% of its maximum value. This observation leads to the conclusion that the effects of the wall are observable at this moment. As a result the criterion defining the head-on quenching distance d_f is:

$$d_w = d_f \quad \text{at} \quad R \lessapprox 0.95 \tag{3}$$

In the particular example of Fig. 3, the head-on quenching distance is of the order of $12.5\,\ell$. In general, for the experimental conditions studied, it has been found that the estimate of d_f was between 10 and 18 times the thermal flame thickness ℓ.

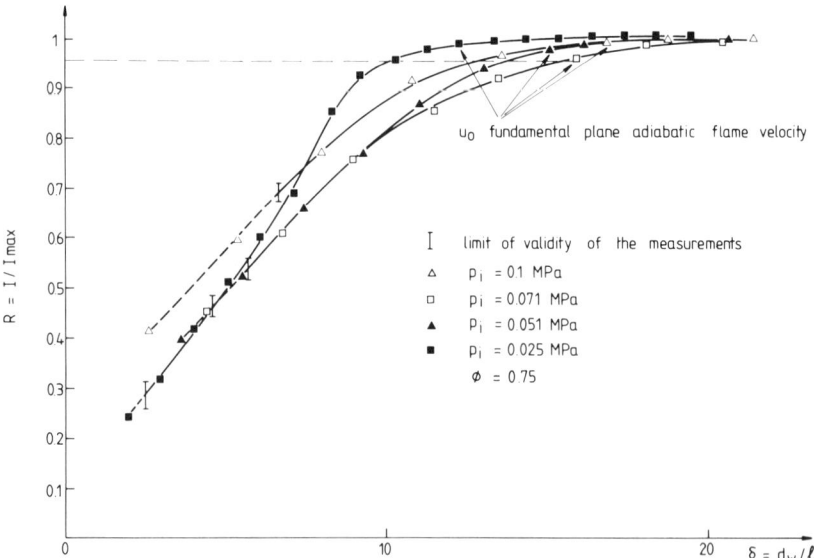

Fig. 3 Variation of CH intensity emission ratio R with the reduced flame-wall distance δ. Methane-air mixture $\phi = 0.75$ at initial pressure $0.025 < p_i < 0.1$ MPa.

Table 1   Key for Fig. 4

	References
	Number indicates results from models

(1)   Adamczyk and Lavoie (1978)
          — — — — $\phi = 1$ ☆ $\phi = 0.9$ ○ $\phi = 0.75$

(2)   Bush et al. (1980)
          —·— (a) $\phi = 0.9$ —··— (b) $\phi = 0.75$

(3)   Henningsen (1980)
          □ r = 0.9

(4)   Hocks et al. (1981)
          —··— (a) $\phi = 0.9$ —···— (b) $\phi = 0.75$

(5)   Ishikawa and Branch (1977)
          —···— (a) $\phi = 0.9$ —····— (b) $\phi = 0.75$
          Flame temperature 1880 K   Flame temperature 1660 K

(6)   Kurkov and Mirsky (1969)
          —·····— (a) $\phi = 0.9$ —······— (b) $\phi = 0.75$

(7)   Westbrook et al. (1981) CH_3OH + air
          —······— $\phi = 1$ ▽ $\phi = 0.9$ △ $\phi = 0.75$

Symbols for experimental results

● ✱   Daniel (1955) - C_3H_8-air; $\phi = 0.99$; $\phi = 0.7$

■   Gat and Kauffman (1980)
          $\phi = 0.965$

▼ ▲   Present work $\phi = 0.9$; $\phi = 0.75$
Curve   Extrapolation of our results
— (a)
— (b)   (a) $\phi = 0.9$ (b) $\phi = 0.75$

Discussion

These measurements of d_f were compared to experimental or theoretical results available in the literature (see the Fig. 4 and the key for this figure in Table 1). For the pressure range of this work (p $\leq$ 0.7 MPa) the measured results are in good agreement with those reported for a constant volume bomb by Gat and Kauffman (1980). The reported results are also in good agreement with those calculated on the basis of thermal losses by Ishikawa and Branch (1977).

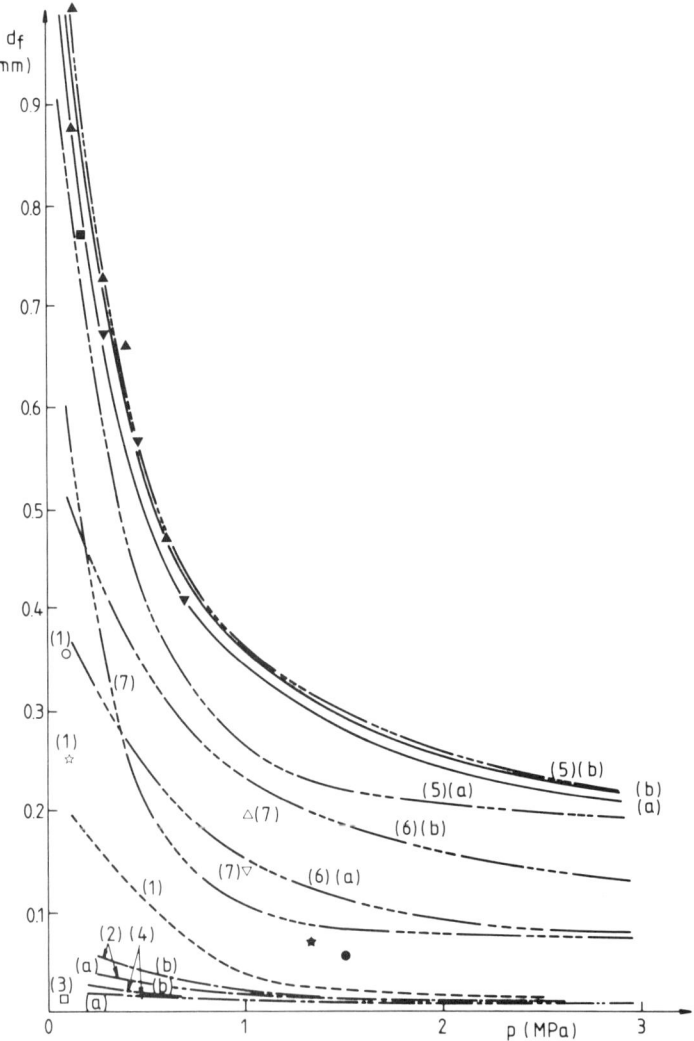

Fig. 4 Comparison of present measurements of the head-on quenching distance d_f with other experimental or calculated values of the literature (Table 1 for the key symbols).

Models which involve explicitly a given kinetic scheme of reaction predict somewhat lower values of the quenching distance (see Bush et al. 1980; Hocks et al. 1982, for instance). Models (particularly the model proposed by Westbrook et al. 1981) which use a complete kinetic scheme for methane predict intermediate values. As a consequence, a large dispersion is observed in the results displayed in Fig. 4. The observed discrepancies between the different models may be explained in part by the choice of the kinetic scheme in the model and the choice of the criterion to define the quenching distance. The quenching distance has been defined as that distance which: 1) separates the flame from the wall when the heat liberated by the reaction equals the heat transferred to the wall (Kurkov and Mirsky 1969; Ishikawa and Branch 1977), 2) is the closest approach at the maximum of the reaction rate (Adamczyk and Lavoie 1978; Bush et al. 1980; Hocks et al. 1981; Westbrook et al. 1981), or 3) is given by the defined fresh gas converted hydrocarbon profile distance from the wall (Henningsen 1977).

Conclusion

Observation of CH radical flame front emission has been used to follow the approach of a flame to a wall with a good accuracy up to a distance of 0.25 mm from the wall. The decrease in the relative intensity of CH radiation peak, is a measure of the characteristic distance at which the cold wall begins to influence the flame approach. The method provides an estimate of the head-on quenching distance. Models of wall flame interactions lead to numerical values of quenching distance which differ widely. The measurements are limited to quenching distances larger than 0.25 mm, or for methane air-mixtures to pressures below 1 MPa. The method should be refined to improve spatial resolution, if applications to reciprocating engine operation are contemplated.

Acknowledgment

The authors are greatly undebted to J. L. Brugier for his help in performing experiments.

References

Adamcyzk, A. A. and Lavoie, G. A. (1978) Laminar head on flame quenching, a theoretical study. SAE Technical Paper Series No. 780969.

Andrews, G. E. and Bradley, D. (1972) The burning velocity of methane-air mixtures. Combustion and Flame 19, pp. 275-288.

Bush, W. B., Fendell, F. E., and Fink, S. F. (1980) Effect of boundary thermal constraint on planar premixed flame-wall interaction. Combustion Science and Technology 24, pp. 53-70.

Daniel, W. A. (1955) Flame quenching at the wall of an internal combustion engine. 6th Symposium (International) on Combustion pp. 866-892. The Combustion Institute, New Haven, Connecticut.

Fristrom, R. M. and Westenberg, A. A. (1965) *Flame Structure*, pp. 27-28. McGraw-Hill Book Co., New York.

Gat, N. and Kauffman, C. W. (1980) The effect of exhaust gas recirculation and turbulence on the burning velocity, dead space thickness and minimum ignition energy in premixed methane-air combustion. *Combustion and Flame* 23, pp. 1-15.

Gaydon, A. G. (1957) *The Spectroscopy of Flames*, pp. 113-136. Chapman and Hall Ltd., London.

Girard, P. and Leyer, J. C. (1981) Constant volume combustion of uniform and stratified lean mixtures of propane and air. *Combustion in Reactive Systems: AIAA Progress in Astronautics and Aeronautics* (edited by J. R. Bowen, N. Manson, A. K. Oppenheim, R. I. Soloukhin), pp. 564-577. Vol 76, AIAA, New York.

Henningsen, S, (1977) A theoretical model for propagating and quenching of a one-dimensional laminar two reactions flame. SAE Technical Paper Series No. 800105.

Hocks, W., Peters, N., and Adomeit, G. (1981) Flame quenching in front of a cold wall under two-step kinetics. *Combustion and Flame* 41, pp. 157-170.

Ishikawa, N., and Branch, M. C. (1977) A simple model of transient thermal quenching. SAE Technical Paper Series No. 770648.

Kurkov, A. P. and Mirsky, W. (1969) An analysis of the mechanism of flame extinction by a cold wall. *12th Symposium (International) on Combustion* pp. 615-622. The Combustion Institute, Poitiers, France.

Westbrook, C. K., Adamcyzk, A. A., and Lavoie, G. A. (1981) A numerical study of laminar flame-wall quenching. *Combustion and Flame* 40, pp. 81-99.

Wohl, K. and Welty, F. (1955) Spectrophotometric traverses through flame fronts. *5th Symposium (International) on Combustion* pp. 746-753. The Combustion Institute, Pittsburgh, Pa.

Chapter V.   Spray Combustion

Timed Ignition of Explosives and Flammables from Desensitized Solutions

Melvin Gerstein* and P. Roy Choudhury*
University of Southern California, Los Angeles, California

Abstract

This paper is concerned with the evaporation of single drops of binary mixtures composed of an explosive solute in a solvent (ammonium azide in water and ozone in liquid oxygen) and a spontaneous flammable solute (white phosphorus) in carbon disulphide. The equations are general and may be applied to more complex systems. The work is easily expanded to groups of drops to simulate a spray and to sprays if a distribution function is known. The general concept using purely equilibrium evaporation and not the time dependent characteristics have been reported to explain anomalies in flash point determinations (Gerstein and Stine 1973). The problem has been studied for vapor phase ignition using aluminum alkyls (Gerstein, to be published). The study of controlled explosion and liquid-phase ignition is new to this study.

Introduction

Many explosives and spontaneously flammable substances can be insensitized by dilution in an appropriate solvent. The solutions of these reactive materials can be stored and handled in relative safety. Upon evaporation of the solvent, the material returns to its original condition.

On one hand, the evaporation of solvent can represent an explosion hazard. Materials in solution, being handled at temperatures above their explosion or ignition temperature may be rendered hazardous as the solvent evaporates. The problem is particularly acute if the solution is

---

Presented at the 9th ICODERS, Poitiers, France, July 3-8, 1983. Copyright © American Institute of Aeronautics and Astronautics, Inc., 1984. All rights reserved.
*Professor of Mechanical Engineering.

released as a spray, since the evaporation can occur very rapidly.

On the other hand, the timed release of a reactive solute in a spray can be used to provide a controlled, timed ignition or explosion source. The concentration of the solution, volatility of the solvent, and the characteristics of the environment control the time at which a flammable or explosive condition occurs.

This paper is concerned with the evaporation of single drops of binary mixtures composed of an explosive or spontaneously flammable solute and an inert solvent. Three combinations are investigated: 1) ammonium azide in water, 2) white phosphorus in carbon disulphide and, 3) ozone in liquid oxygen. The equations are general and may be applied to more complex systems. The work is easily expanded to groups of drops to simulate a spray and to sprays if a distribution function is known.

There are a large number of potential applications of the model

1) It can be used to analyze the behavior of explosive clouds prepared by atomization of a solution.

2) The concept may be used to analyze the hazard associated with the release of sprays from storage or reactor systems containing explosives or flammables in solution.

3) The concept of timed and progressive explosion or reaction may be used in connection with various models of shock wave amplification by energy addition or free radical production.

4) Programmed ignition can be used to increase the flame spread rate in high-speed combustion systems.

The general concept using purely equilibrium evaporation and not the time-dependent characteristics have been reported to explain anomolies in flash point determinations (Gerstein 1973). The problem has been studied for vapor phase ignition using aluminum alkyls (Gerstein, to be published). The study of controlled explosion and liquid-phase ignition is new to this study.

Analytical Model

A single droplet consisting of an explosive/incendiary species dissolved in a solvent is allowed to evaporate and thermally interact with the stagnant surroundings. The droplet has an initial injection velocity and its deceleration is assumed to be given by Stokes' drag. The evaporation rate depends upon the droplet velocity and is assumed to follow the Ranz-Marshall correlation equation.

For the ammonium azide-water solution, the droplet is injected into a heated environment and as long as the liquid phase remains the evaporation takes place at the saturation temperature of the mixture. Thus, a quasi-steady-state evaporation is assumed. The following equations describe the behavior of ammonium azide-water system.

x momentum

$$\frac{du}{dt} = -\frac{18\mu_g}{\rho_{az} D^2} u \left[\frac{1 + m(\rho_{az}/\rho_{H_2O})}{1 + m}\right] \quad (1)$$

Energy

$$(h_{fg})\frac{dm}{dt} = -\frac{6k}{\rho_{az}}\frac{Nu}{D^2}(T_\infty - T)(1 + m\frac{\rho_{az}}{\rho_{H_2O}}) \quad (2)$$

For simplicity, energy required to superheat the evaporating species is neglected.

Ideal mixture temperature

$$\frac{1}{T} = \frac{1}{T_s} + \frac{R}{h_{fg}} \ln\left[\frac{m}{m + (M_{H_2O}/M_{az})}\right] \quad (3)$$

Droplet location

$$\frac{dx}{dt} = u \quad (4)$$

From Eq. (2) diameter reduction rate

$$\frac{dD^2}{dt} = -\frac{4k}{\rho_{H_2O} h_{fg}} Nu(T_\infty - T) \quad (5)$$

where

D = droplet diameter
h_{fg} = enthalpy of vaporization
k = thermal conductivity
m = mass of H_2O/mass of azide
M = molecular weight
Nu = Nusselt number $2 + 0.6\ Re^{\frac{1}{2}} Pr^{1/3}$
Pr = Prandtl number
R = Universal gas constant
Re = Reynolds number
t = time
T = droplet temperature

T_s = saturation temperature of pure phase
u = droplet velocity
x = droplet location
ρ = density
μ = viscosity coefficient

This system of five equations consisting of five unknowns u, m, D, T, and x is solved by a fourth-order Runge-Kutta method.

Ammonium Azide-Water System

Dissolving ammonium azide in water reduces or eliminates the explosion hazard since the reactivity is reduced by dilution and the temperature limited by the boiling temperature of the solution. If a solution of ammonium azide in water is allowed to evaporate, however, solid ammonium azide can precipitate due to the loss of solvent and the boiling point of the solution rises. For the purposes of this study, ammonium azide is assumed to precipitate when $m < 0.5$ (Weast 1981-82) and explosion is assumed to occur when a temperature of 407 K is reached.

It is assumed that the solution sprays into an ambient atmosphere with a temperature above 407 K.

Ozone-Liquid Oxygen System

Dilute solutions of ozone in liquid oxygen are known to be stable. Mixtures containing 30% or less of ozone in liquid oxygen have been used safely in experimental rocket studies (Miller 1960). Although one way to make ozone safe is to dilute it with liquid oxygen, the potential advantages of liquid ozone as a rocket oxidant have been limited because ozone is easily detonated. A phase diagram for an ozone-liquid oxygen system is illustrated in Fig. 1 (Miller 1960).

At the normal boiling point of liquid oxygen 90 K and 1-atm pressure, ozone separates as a separate phase when a concentration of 25% by weight of O_3 in O_2 is reached. At this point it is assumed that the pure ozone phase will detonate.

It is assumed that the solution sprays into air at an ambient temperature of 300 K.

Examination of the process shows that the boiling temperature of a 25% O_3 in O_2 solution is almost the same as pure oxygen. As a result, Eq. (3) was eliminated from the calculation and the system temperature was assumed constant.

Phosphorus-Carbon Disulphide System

For the case of phosphorus-carbon disulfide solution, the ambient temperature is lower than the droplet temperature and a quasi-steady-state approximation can no longer be justified. Here the rate of change of droplet temperature is due to the following three coupled effects: 1) heat transfer from the droplet to the surroundings, 2) cooling of the droplet due to evaporation, and 3) temperature change due to the rate of change of concentration. The change of evaporation rate is estimated from the rate of change of vapor pressure at the drop surface. In addition to Eq. (1), (3), and (4) with appropriate density and molecular weight, the following two equations describe the behavior of phosphorus-carbon disulfide system:

Energy + mixture temperature

$$\frac{dT}{dt} = -\frac{6Nu}{D^2}\left[\frac{1}{8}\frac{\rho_{CS_2}}{\rho_P}\frac{\beta}{c_d}h_{fg}A + \frac{k(T-T_\infty)A}{c_d \rho_P} \\ -\frac{1}{8}\frac{RT^2}{h_{fg}}\beta\frac{\rho_{CS_2}}{\rho_P}\frac{M_{CS_2}/M_P}{m(m+\frac{M_{CS_2}}{M_P})}(1+m\frac{\rho_P}{\rho_{CS_2}})\right] \quad (6)$$

$A = \left[1 + m(\rho_P/\rho_{CS_2})\right]/(1+m) \qquad c_d = $ droplet specific heat

The second term of Eq. (6) is due to effect 1 and the first term is due to effect 2. The last term is due to effect 3 and is obtained by differentiating Eq. (3)

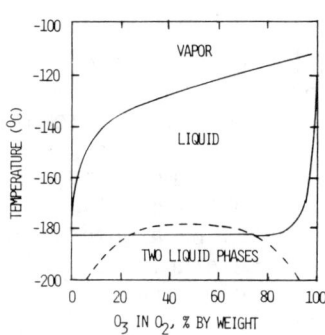

Fig. 1 Phase diagram of ozone-oxygen at 1 atm (Miller 1960).

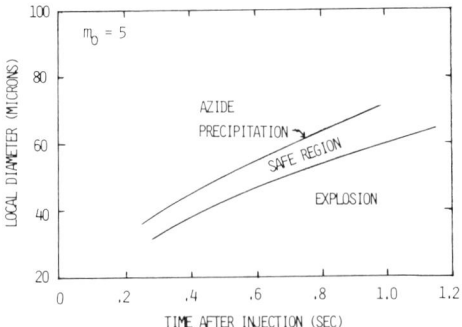

Fig. 2 Safe and danger zones for ammonium azide-water droplets. T_∞ = 450 K, initial droplet velocity = 250 cm/s, 1 atm, stagnant environment, m_0 = 5.

Diameter reduction rate

$$\frac{dD^2}{dt} = -\beta \frac{Nu}{2} \qquad (7)$$

Using the Clausius-Clapeyron equation, β can be expressed as

$$\beta = \beta_0 \frac{T_0}{T} \exp\left[\frac{h_{fg}}{R}\left(\frac{1}{T_0} - \frac{1}{T}\right)\right] \qquad (7a)$$

where β is the evaporation rate, m the mass of CS_2/mass of P, and subscript o the initial values.

Evaporation rate [from Eq. (7)]:

$$\frac{dm}{dt} = -\frac{3}{4} Nu \frac{\rho_{CS_2}}{\rho_p} \frac{\beta}{D^2}\left(1 + \frac{\rho_p m}{\rho_{CS_2}}\right) \qquad (8)$$

Equations (1), (4), (6), (7), and (8) consisting of 5 unknowns u, x, T, D, and m are solved by a fourth-order Runge-Kutta method. Phosphorus is assumed to precipitate at $m \leq 0.143$. If the droplet temperature at that point happens to exceed 317 K, the precipitate is expected to melt and coexist in liquid phase with CS_2. Under such a condition liquid phosphorus at the droplet surface will be exposed to air and burn. It is possible to control the droplet temperature at the point of precipitation by changing the initial concentration of CS_2. In other words, solid phosphorus can

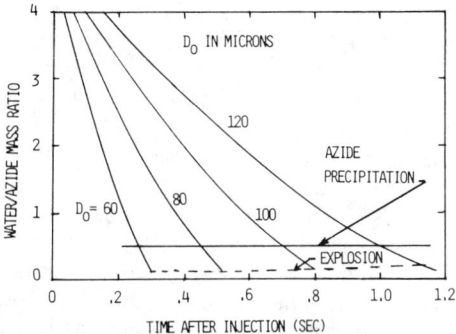

Fig. 3 Effect of water evaporation on mass ratio and drop size of ammonium azide-water system. Conditions are identical to those of Fig. 2.

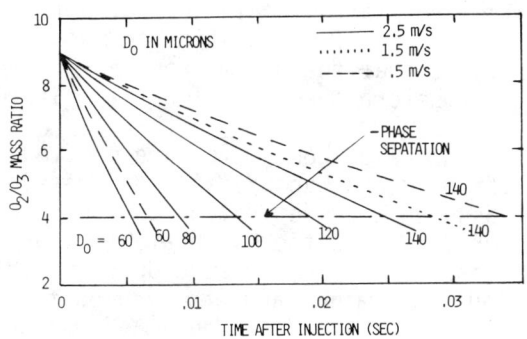

Fig. 4 Ozone-oxygen droplets with various initial velocities.

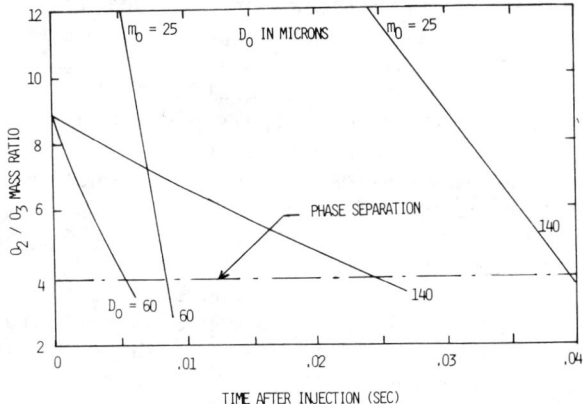

Fig. 5 Effect of initial mass ratio on ozone-oxygen system. Initial drop velocity = 250 cm/s.

be made to precipitate at a given distance from the injection point by a suitable choice of initial conditions.

Results

Figure 2 shows a result of calculations for an ammonium azide and water solution with an initial concentration of 5-gm water/gm of azide. Initial drop diameters of 60, 80, 100, and 120 μm were examined evaporating in a still atmosphere at 450 K. The drops were assumed to have an initial velocity of 250 cm/s. As the drop evaporates, water is removed from the liquid phase. A concentration is reached at which the ammonium azide begins to precipitate, but the drop temperature has not yet reached 407 K, the explosion temperature for ammonium azide.

At a later time, the drop temperature reaches 407 K and explosion is assumed to occur.

The results are plotted again in Fig. 3 to illustrate more clearly the effect of evaporation on mass ratio and of drop size. Note that the 60-μm drop has evaporated almost completely before the explosion temperature is reached.

Results for ozone-oxygen system are plotted in Fig. 4. Phase separation occurs at an oxygen/ozone mass ratio of 4.0 and it is assumed that detonation can occur when separation occurs. Due to the low heat of vaporization of liquid oxygen, the liquid temperature does not change appreciably during evaporation and was assumed to be near the boiling temperature of liquid oxygen. As with the case of ammonium azide-water solutions, small drops evaporate faster and lead to an unsafe condition more quickly. Since the velocity of the drop relative to the gas also affects evaporation rate, higher initial drop velocities (which may also be associated with smaller drop sizes) decreases the time to phase separation.

An initial oxygen/ozone mass ratio of 25 is compared to one of 4 in Fig. 5. Due to the rapid evaporation of liquid oxygen at normal ambient temperature, the increased dilution does not have as marked an effect on time to phase separation as one might expect.

White phosphorus ignites spontaneoulsy when exposed to air at room temperature. Solution in carbon disulphide can eliminate the ignition of the phosphorus. As the carbon disulphide evaporates, however, phosphorus separates from the solution either as a solid or a liquid, depending on the drop temperature (Weast 1981-82). The results for phosphorus-carbon disulphide solutions are shown in Fig. 6.

With an initial CS_2/P mass ratio of 0.3, precipitation and incendiary action occur very quickly since the drop

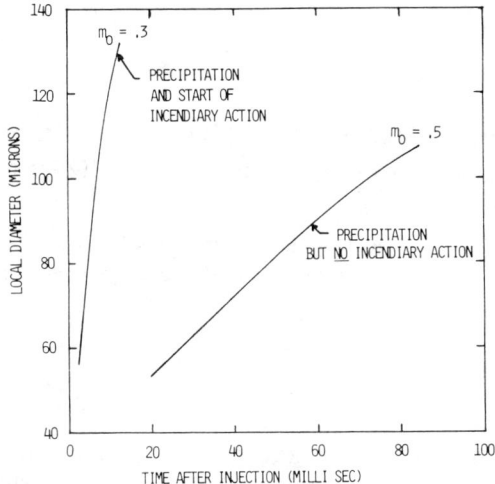

Fig. 6 Behavior of P-CS_2 droplets with two different initial mass ratios.

temperature also increases quickly. However, as the initial CS_2/P mass ratio is increased to 0.5, precipitation is delayed, the temperature of the drop rises more slowly and incendiary action does not start until all the CS_2 evaporates. Since the temperature at the precipitation line for $m_0 = 0.5$ is low, the phosphorus remains solid and is shielded from the oxidizing atmosphere; thus much longer time intervals are required for incendiary action.

References

Gerstein, M. and Stine, W. B. (1973) Anomalies in flash points of liquid mixtures. <u>I & EC</u> 12, 253.

Miller, R. O. (1960) Proceedings of the Propellant Thermodynamics and Handling Conference. Ohio State University; Special Report (U. S. Government) #12, Safety and handling of ozone-oxygen mixtures. 493-502.

Weast, R. C. (1981-82) <u>CRC Handbook of Chemistry and Physics</u>. CRC Press. Cleveland, Ohio.

Comparative Study of Droplet Heating and Vaporization at High Reynolds and Peclet Numbers

H. A. Dwyer* and B. R. Sanders†
Sandia National Laboratories, Livermore, California

Abstract

The heating and vaporization of liquid hydrocarbon fuels is a very complex process and requires considerable computational resources for its calculation in the Reynolds and Peclet number ranges of practical interest. This paper is concerned with detailed calculations of the unsteady heating and vaporization of liquid droplets for Reynolds numbers between 10 and 100, and Prandtl numbers between 10 and 100. The calculations are based on a numerical solution of the unsteady Navier-Stokes equations for both the liquid and gas phases. A vorticity-stream function formulation has been solved together with the energy and species equations, and the equations have been written in generalized nonorthogonal coordinates. At the liquid-gas interface conservation balances have been employed, and a form of the Clausius-Clapeyron relation was used. Results are presented for droplet heating with and without vaporization over a range of parameter space. A comparative study is presented of the effects of surface mass transport on droplet heat transfer and fluid dynamics. Additionally, isotherm and streamline distributions are presented for variations in Reynolds number and Prandtl number, where the product of these two parameters is held constant. The heat transport in the droplet for these constant Peclet number cases shows a very large sensitivity to liquid Prandtl number.

Presented at the 9th ICODERS, Poitiers, France, July 3-8, 1983. This paper is declared a work of the U.S. Government and therefore is in the public domain.
 *Consultant; also Professor, University of California, Davis, Calif.
 +Supervisor, Thermofluids Division.

Introduction

Many practical combustion devices utilize liquid sprays to rapidly introduce large quantities of fuel to an oxidizing environment. Such systems tend to be locally unsteady and the complicated processes of fuel droplet heating, vaporization and combustion play an important role in the overall performance of these devices. Due to the large number of fuel droplets in a spray, it is impractical to simultaneously model directly the details of individual droplet processes and the influence of droplets upon one another. This paper presents numerical solutions for the initial heating phase of a single vaporizing fuel droplet in a hot gas stream for cases of high Reynolds number and moderate Prandtl number, and for low Reynolds number and high Prandtl number. Solutions are based on finite difference formulations of the Navier-Stokes equations for both the gas phase surrounding the droplet and the liquid internal circulation driven by shear at the liquid-gas interface. This model of single droplet unsteady transport processes will serve as the basis of a complete single droplet model, which will include the effects of variable transport processes and gas phase combustion in future studies.

The dynamics of vaporizing fuel droplets has been the focus of many analytical and experimental research efforts during the past decade. However, what is known about the details of such a complex combination of physical processes is largely due to inference from experimental observations. The environment seen by a fuel droplet from the initial conditions of injection into hot oxidizing surroundings until evaporation is complete is so harsh that detailed measurements of droplet processes are nearly impossible. To gain a full understanding of droplet heating mechanisms one needs temporally-resolved velocity and temperature fields inside of the droplet, measurements which are not currently available. Vaporization rates can be measured through unsteady droplet size measurements, but again the competing physical processes governing thermal, mass, and momentum transport have not been sorted out for transient droplet heating and vaporization. The advancement in understanding of droplet dynamics in the near future will undoubtedly come from better measurement techniques to resolve flow-field structure around and inside the droplet and correlation of these measurements with detailed models of the important transport processes.

The droplet and spray combustion literature is rich with review articles summarizing the state of understanding of these important combustion problems. Three excellent reviews have appeared in the past year, which discuss in detail the attempts to improve droplet and spray models by adding more physics to describe unsteady heating, vaporization, and combustion processes. In his review of droplet vaporization and combustion topics, Law (1982) describes the classical d^2-law and its current limitations pertaining to convective flow, transient heating, fuel vapor accumulation, and variable transport properties. Sirignano (1983) in his comprehensive discussion of fuel droplet vaporization and spray combustion theory emphasizes analytical and computational contributions to the theory. In particular, he gives a physical discussion of droplet drag where vaporization is taken into account, a description of internal droplet circulatory flow due to shear stresses imposed at the gas-liquid interface where a differential velocity exists between the droplet and gas surroundings, and the effects of multicomponent fuels on the unsteadiness of droplet processes. In Chigier's (1982) review of spray combustion processes, he gives a broad description of experimental techniques applicable to sprays, and a discussion of the methods and implications of group combustion models for the structure and burning characteristics of liquid fuel sprays. From each of these reviews it is obvious that much work still remains to understand the dynamics of individual drops in transient environments, so that models of vaporization rates, drag coefficients, and other unsteady processes can be incorporated into comprehensive spray simulations.

It is apparent from these three review articles that a hierarchy of modeling efforts are currently being pursued, all with the ultimate goal of improving the predictive capability for spray combustion. The d^2-law, while the simplest in formulation, requires the most empirical input. Sirignano[‡] and co-workers have made great strides in matched boundary-layer analysis models, where simplifying assumptions are made in both the gas phase and internal liquid flowfields, but the essential physics of mass momentum and energy transport can still be considered deterministic rather than empirical. All spray combustion models which inherently treat large numbers of droplets must include models for droplet drag, heating, and vaporization rates. The basis for these models must be derived from experimental observation, or detailed modeling from which phenomenological models are constructed.

The objectives of the work described in this paper are to: 1) provide a detailed model of droplet dynamic pro-

---

[‡]Prakash and Sirignano (1978); Prakash and Sirignano (1980); Lerner, Homan, and Sirignano (1980).

cesses with which experimental droplet observations may be interpreted, 2) provide solutions with as few assumptions as is practical so that models of droplet dynamics with less detailed physics can be verified or constructed, and 3) provide a framework to study highly nonsteady droplet processes which are not amenable to asymptotic analysis. The following sections of this paper will describe the numerical results obtained in the solution of the transient Navier-Stokes equations, coupled to the two-phase boundary conditions at the liquid-gas interface. Results will be presented of unsteady droplet heating for cases of droplet Reynolds numbers of 10 and 100, and for liquid Prandtl numbers of 10 and 100. Comparisons are shown between the cases of droplet heating with and without vaporization, which isolates the effects of surface mass transfer. Comparisons are also given for cases of constant Peclet number, where the Peclet number is defined as the product of gas phase Reynolds number and liquid-phase Prandtl number. It will be shown that the Peclet number defined in this way is not a good scaling parameter, as there is a large sensitivity in the heating solutions to liquid Prandtl number, which is not fully reflected in the Peclet number.

Physical Parameters of the Problem

It is evident for hydrocarbon fuels that a wide range of transport properties and boiling points occur as a function of pressure and temperature. In this paper we have chosen a set of parameters which are of practical interest, but which do not cover a very wide range of possibilities. The conditions chosen are as follows:

Gas Pressure $\quad P = 5$ atm

Gas Temperature $\quad T = 1000$ K

Initial Liquid Temperature $\quad T_I = 400$ K

Average Viscosity Ratio $\quad \dfrac{\mu_\ell}{\mu_g} = 25$

Average Density Ratio $\quad \dfrac{\rho_\ell}{\rho_g} = 300$

Latent Heat of Vaporization of Liquid	$\dfrac{L}{C_p} = 106 \text{ K}$
Boiling Point of Liquid	$T_B = 573 \text{ K}$
Vapor Pressure of Liquid	$\ln\left(\dfrac{P}{17.5}\right) = 6.675 - \dfrac{4542}{T}$
Prandtl Number of Gas	$Pr_g = 1.0$
Prandtl Number of Liquid	$1.0 \leq Pr_\ell \leq 100$
Reynolds Number of Drop	$10 \leq Re_g = \dfrac{UD}{\nu_g} \leq 100$

The properties listed above have been held constant during the calculation and the results will be limited by these factors. In another paper the influence of variable liquid viscosity will be evaluated, as well as the influence of variable Reynolds number. Also, the present results give significant insight into the importance of these factors.

Basic Equations and Numerical Methods Employed

For the Reynolds and Prandtl numbers of interest a full Navier-Stokes treatment of the fluid mechanics is necessary, and a complete treatment of the equations of low speed mass transport is needed. In terms of cylindrical coordinates and vorticity-stream function formulation, the equations used in the study in compact vector form are:

$$\frac{\partial \vec{Q}}{\partial t} + \frac{\partial \vec{E}}{\partial r} + \frac{\partial \vec{F}}{\partial z} = \frac{\partial \vec{R}}{\partial r} + \frac{\partial \vec{S}}{\partial z} + \vec{H}$$

where

$$\vec{Q} = \begin{bmatrix} 0 \\ \rho r \omega \\ \rho r\, C_p\, T \\ \rho r\, Y_f \end{bmatrix} \qquad \vec{E} = \begin{bmatrix} 0 \\ \rho r v \omega \\ \rho r v\, C_p\, T \\ \rho r v\, Y_f \end{bmatrix} \qquad \vec{F} = \begin{bmatrix} 0 \\ \rho r u \omega \\ \rho r u\, C_p\, T \\ \rho r u\, Y_f \end{bmatrix}$$

$$\vec{R} = \begin{bmatrix} \frac{1}{r}\frac{\partial \psi}{\partial r} \\ r\frac{\partial}{\partial r}[\mu\omega] \\ r\, k\, \partial T/\partial r \\ \rho r\, D_f \frac{\partial Y_f}{\partial r} \end{bmatrix} \qquad \vec{S} = \begin{bmatrix} \frac{1}{r}\frac{\partial \psi}{\partial z} \\ \frac{\partial}{\partial z}[\mu\omega] \\ r\, k\, \partial T/\partial z \\ \rho r\, D_f \frac{\partial Y_f}{\partial z} \end{bmatrix} \qquad \vec{H} = \begin{bmatrix} -\omega \\ -\frac{\mu\omega}{r} + \rho v \omega \\ 0 \\ 0 \end{bmatrix}$$

The following notation has been employed: ψ = stream function, ω = vorticity, T = temperature, Y_f = fuel mass fraction in the gas, μ = viscosity, k = thermal conductivity, ρ = local gas density, and C_p = specific heat. Also, the dependent variables t, r, and z are the time, radial coordinate, and axial position, respectively.

The formulation of the equations can be made much more useful by transformation to generalized nonorthogonal coordinates which allow for arbitrarily shaped bodies as well as the change in sphere radius with time. The details of the transformed equations can be found in Dwyer et al. (1983).

At the interface between the liquid and gas phases of the droplet a special treatment of the dependent variables must be employed. The stream function and surface fuel density are determined by the mass transfer and Clausius-Clapeyron relationship, respectively. For the vorticity and energy equations a special spherical element is formed which is part liquid and part gas. It is essential that all terms in the transport equations be retained, since at various times in the unsteady analysis they can play a key role. Also, in the present work, the mass transport equation has only been solved in the gas, since the liquid phase is homogeneous.

In addition, the energy equation at the interface must be modified to include the influence of latent heat of vaporization. The amount of vaporization is determined from fuel mass conservation with diffusion which yields:

$$\rho D_f \frac{\partial Y_f}{\partial n}\bigg|_s = \rho v_s (Y_f - 1)$$

where ρv_s is the surface mass flux of fuel per unit area of surface and n is the coordinate normal to the interface. For the vorticity, energy, and fuel density equation an ADI numerical method was employed, while successive overrelaxation was used to calculate the stream function. Due to the lack of a time scale in the stream function equation and a well known sensitivity to the lack of convergence, an overall iteration procedure was applied to the entire system. This iteration procedure is primarily employed in order that the stream function and vorticity be coupled and updated, and is not particularly needed for the energy and species equations.

As has been mentioned previously, the exponential nature of the Clausius-Clapeyron equation can lead to serious instabilities at the liquid gas interface. These difficulties were considerably reduced by employing Newton's linearization method on the system and solving the interface equations simultaneously with the liquid and gas phases. However, as the boiling point of the liquid phase is approached, the problem is more severe and it is sometimes necessary to reduce the problem time step in order to control stability problems. This is obviously an area where new techniques and new modeling ideas can play a key role in the future.

Results

The computational grid used in the calculation of droplet heating and vaporization is shown in Fig. 1. Since the problem is symmetric, only the top half is shown. The liquid droplet in all the results is of unit radius with center at $x = 0$. The gas phase surrounding the liquid sphere is for $|x| > 1$. Only a small portion of this computa-

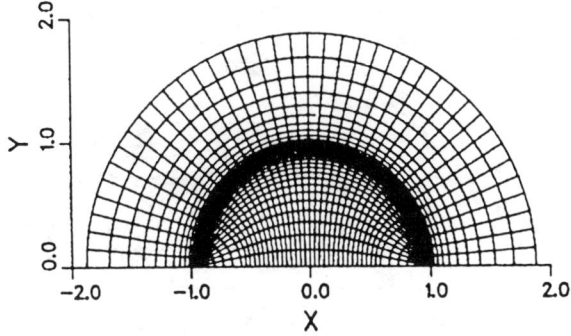

Fig.1 Nonorthogonal computational grid used in all computations. Droplet is of unit radius with center at x = 0. Gas flow is from left to right.

tional zone for the gas phase is shown here, however, the computations extend to many radii away from the drop. For all calculations which will be presented, the grid is fixed in time and the gas flow is from left to right. The dark area just inside the droplet surface is the result of so many grid lines placed in this area that the eye cannot resolve them individually. The details of the grid formation can be found in Dwyer et al. (1983).

Figs. 2 through 6 show the development of the isotherms and streamlines inside and outside of a droplet at a Reynolds number of 100 and a liquid Prandtl number of 10. The results are given for a nondimensional gas time scale defined by $\tau_g = t\nu_g/R^2$ where t is time, R is droplet radius and ν_g is gas kinematic viscosity. The relationship between the liquid and gas time scales can be computed with the liquid parameters for this problem, and is $\tau_\ell = \tau_g/120$. Although the time scale for transport in the gas phase is much shorter than for the liquid phase, the entire problem remains unsteady due to the unsteady surface mass transfer rate.

Fig. 2 shows the comparison between the solutions for no vaporization (upper figure) and with vaporization (lower figure) at an early time after the droplet has been immersed in the hot gas stream. At this early time the isotherms are very similar, as significant surface mass transport due to vaporization has not yet begun. The 410 K isotherm is being convected across the droplet, but a significant fraction of the liquid is still at the initial temperature of 400 K. The isotherms shown are separated by 100 K in the gas phase and 20 K in the liquid. The skewness of the isotherms persists throughout the droplet heating, primarily due to the separated flow in the gas phase

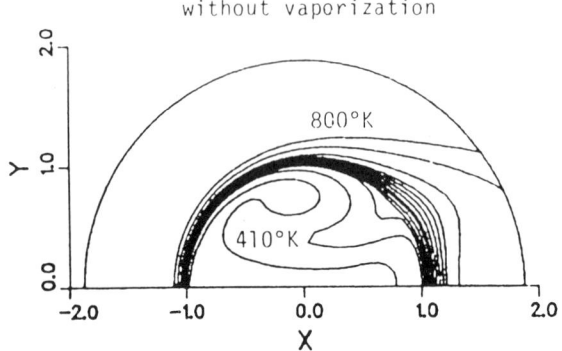

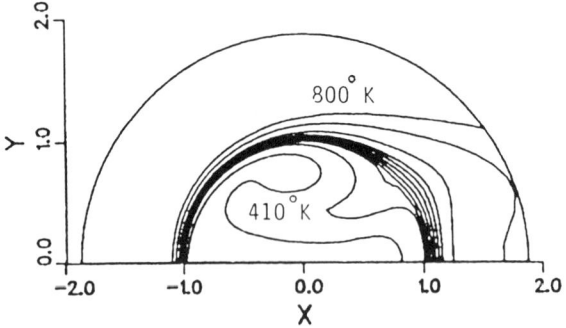

Fig.2 Isotherm distributions. Re = 100, τ_g = 1.909

approximately 45 degrees from the rear stagnation point. The fluid molecules near the surface are heated as they are convected from the front stagnation point at the left, until they reach the gas-phase separation point. There the molecules have a fairly long residence period, and then abruptly break away and flow across the droplet interior. Fig. 3 shows the two cases without and with vaporization at a later time. Here the fluid has recirculated several times inside the droplet leaving the isotherm pattern shown. The upper figure without vaporization shows a smaller closed streamline at 450 K, compared to the lower half with vaporization. The difference in heat transfer is attributed to the surface mass flux which lowers the overall heat transfer. Fig. 4 shows a late time result for the va-

DROPLET HEATING AND VAPORIZATION 473

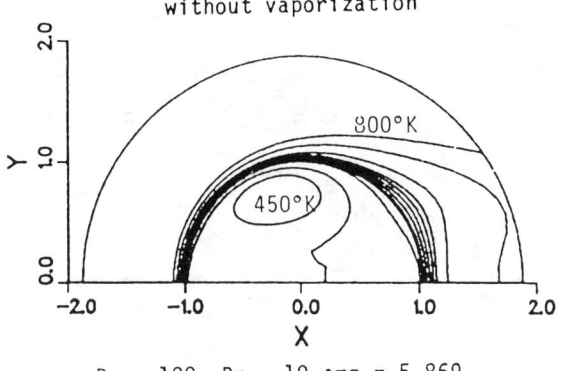

Re = 100   Pr = 10 τ_g = 5.869

Tmin = 430°K $\Delta T\ell$ = 20°K ΔT_g = 100°K

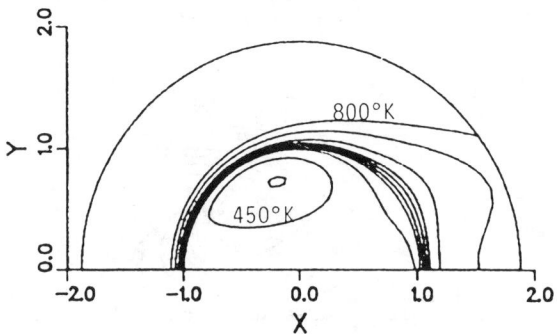

Fig.3   Isotherm distributions.   Re = 100, τ_g = 5.869

Re = 100   Pr = 10 τ_g = 13.789

Tmin = 490°K $\Delta T\ell$ = 20°K ΔT_g = 100°K

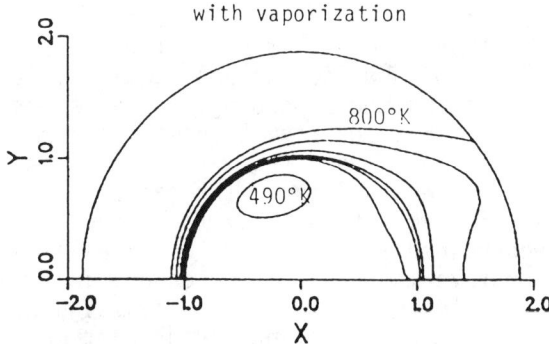

Fig.4   Isotherm distribution.   Re = 100, τ_g = 13.789

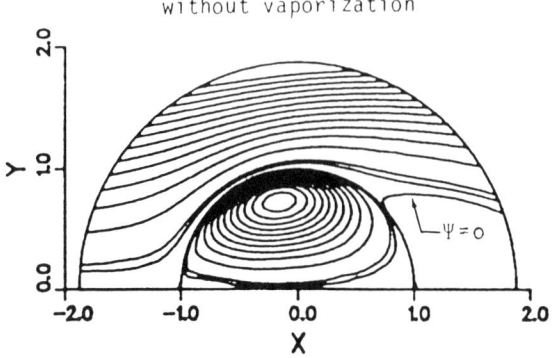

Re = 100 Pr = 10 τ_g = 1.909

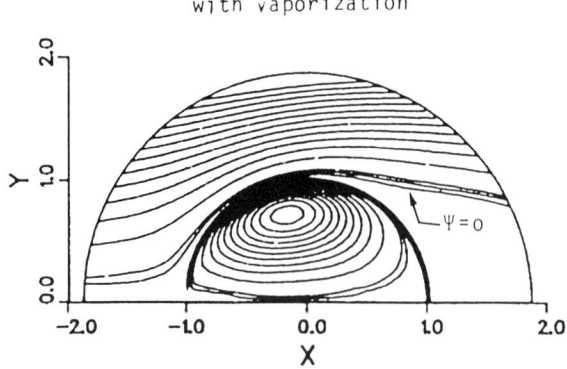

Fig.5 Streamline distributions. Re = 100, τ_g = 1.909

porization case only. At this time the droplet would have lost approximately 18% of its initial mass. There is still a relatively large volume of cold liquid and a distinctly nonsymmetric streamline pattern from front to back.

Figs. 5 and 6 show the streamline distribution at early and late times corresponding to the isotherm solutions. The upper half of Fig. 5 shows the streamlines without vaporization. Here the dividing streamline between liquid and gas, $\psi = 0$, splits at the gas separation point. For the lower figure with vaporization, the $\psi = 0$ streamline has been pushed away from the original liquid surface due to evaporative mass flux. The change in flowfield due to mass flux at the droplet surface will significantly change both the droplet heating and droplet drag by reduc-

DROPLET HEATING AND VAPORIZATION

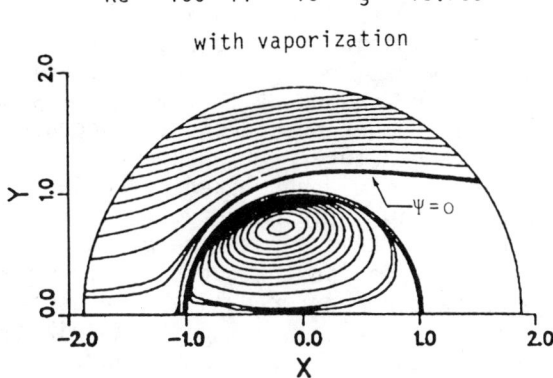

Fig.6   Streamline distribution.   Re = 100, τ_g = 13.789

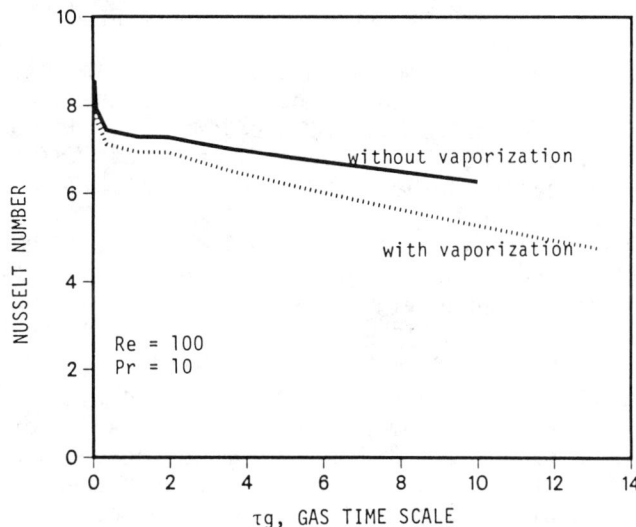

Fig.7   Nusselt number variation as a function of gas time scale.

ing, respectively, the thermal gradients and shear gradients. Fig. 6 shows the late-time streamline pattern for the case of vaporization. Without vaporization, a steady-state gas phase is established in a short time, τ_g = 2, and is represented by Fig. 5a. With vaporization there is never a steady state.

Fig. 7 shows the comparison of dimensionless heat transfer with and without vaporization for the cases of Re = 100 and Pr = 10. Here the Nusselt number is defined by $N_u = QD/kA\Delta T$, where Q = surface heat transfer, D = droplet diameter, A = surface area, k = thermal conductivity, and ΔT = the initial temperature difference between the droplet and the gas (600 K for these cases). The initial high values were caused by the impulsive start of the flow. The Nusselt number then decreases because of the rise in surface temperature. At a time of $\tau_g = 2$ one complete circulation of fluid has taken place inside the droplet. As the hot fluid is transported from the rear to the front of the droplet, this surface temperature suddenly rises, causing a plateau in the Nusselt number. Subsequently, the Nusselt number gradually decreases due to droplet heating and, for the case with vaporization, the decrease is partly due to surface mass flux.

The above results for Re = 100 and Pr = 10 are representative of actual hydrocarbon droplets when first introduced into a hot environment by fuel injection. These conditions lead to a Peclet number of 1000 inside the droplet, based on the Peclet number defined as the product of Reynolds number and Prandtl number. To demonstrate the difference in the solutions with constant Peclet number, we have lowered the Reynolds number and increased the Prandtl number by a factor of 10. At the lower Reynolds number the inertia of the gas phase is greatly reduced, while this increase in Prandtl number causes less conductive heat transfer inside the drop.

The results in Figs. 8 through 12 are for the case of Re = 10 and Pr = 100. Fig. 8 shows the early isotherm distribution; the upper figure is for the case without vaporization and the lower is with vaporization. As in the previous case of higher Reynolds number, there is not much difference with and without vaporization at this early stage. However, in comparison with the results of Fig. 2, there is a tremendous difference in isotherm distribution. It is evident that at the lower Reynolds number the internal liquid circulation is drastically reduced, thereby reducing the effectiveness of circulation as a droplet heating mechanism. With this low Reynolds number/moderate Prandtl number case, the molecules at the surface of the drop heat up and flow towards the rear stagnation point and then slowly penetrate the cold liquid core of the drop. Fig. 9 shows the same results at three times later in the calculation. Contrasted to the results of Fig. 3, one can see large differences. Fig. 10 shows the streamline distribution at the time corresponding to Fig. 8. There is a major difference between the gas phase flow for the case with and without vaporization. Fig. 10b shows the surface streamline being pushed away from the droplet surface,

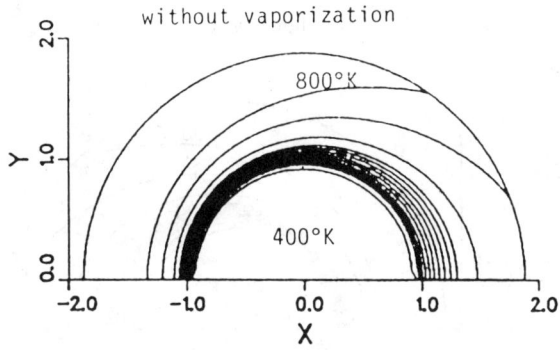

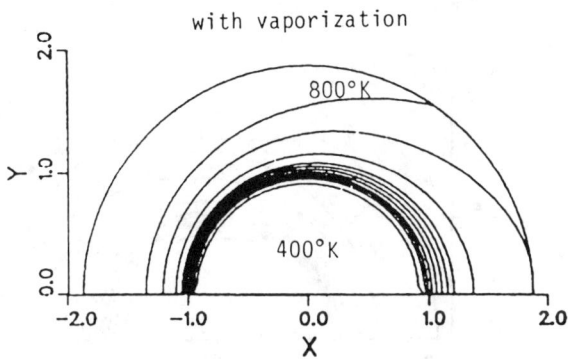

Fig. 8 Isotherm distributions.   Re = 10, τ_g = 1.909

which (as in the higher Reynolds number case) will reduce the overall heat transfer and drag. Fig. 10a shows a streamline pattern, which is essentially the steady state solution without vaporization. Fig. 11 shows the streamline at a later time for the case of vaporization and, here, the surface streamlines have been further displaced.

A comparison of Nusselt number variation with time is shown in Fig. 12 for both cases with and without vaporization. Even though the surface temperatures are higher for these cases than for the results shown in Fig. 7, the overall heat transfer is less due to less mass circulation inside the liquid region. Although the lower and higher Reynolds number cases have the same overall liquid Peclet number, the results are not similar and cannot be scaled with Peclet number.

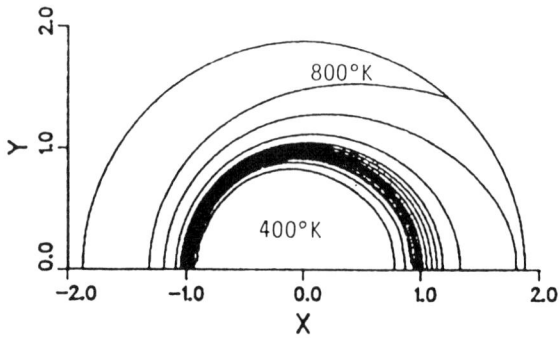

Re = 10 Pr = 100 τg = 5.869

Tmin = 400°K $\Delta T\ell$ = 20°K ΔTg = 100°K

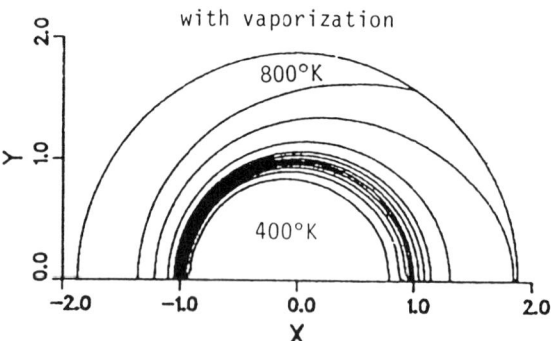

Fig. 9 Isotherm distributions. Re = 10, τ_g = 5.869

Fig. 13 shows the comparison of dimensionless mass flux with time for the cases of Re = 100, Pr = 10 and Re = 10, Pr = 100. The quantities plotted are the total surface mass fluxes normalized by free stream parameters, and they agree in magnitude with the results presented by Prakash and Sirignano (1980). The droplet lifetime would be lower for the lower Reynolds number case even though the dimensionless mass transfer is greater. If one calculates the actual surface flux for the same size droplet for the two cases, the higher Reynolds number case gives the larger mass flux.

DROPLET HEATING AND VAPORIZATION

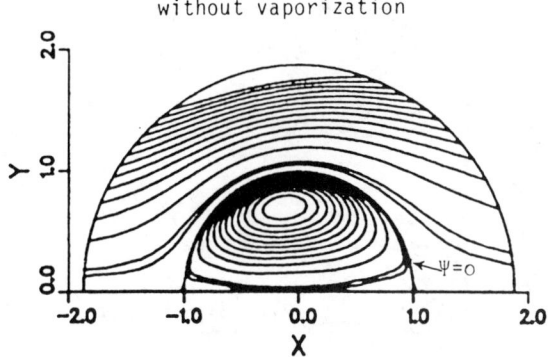

Fig.10 Streamline distributions.   Re = 10, τ_g = 1.909

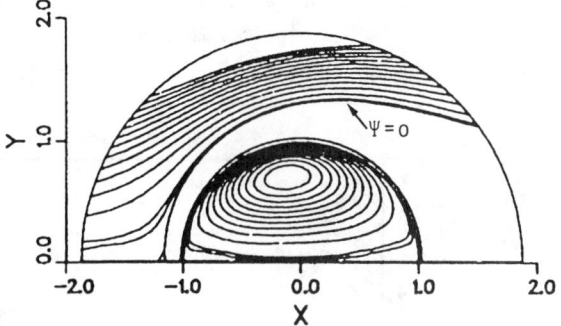

Fig.11 Streamline distribution.   Re = 10, τ_g = 9.829

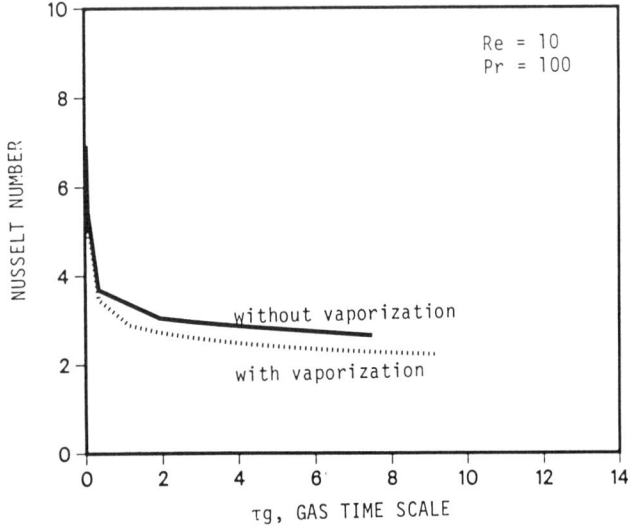

Fig.12 Nusselt number variation as a function of gas time scale.

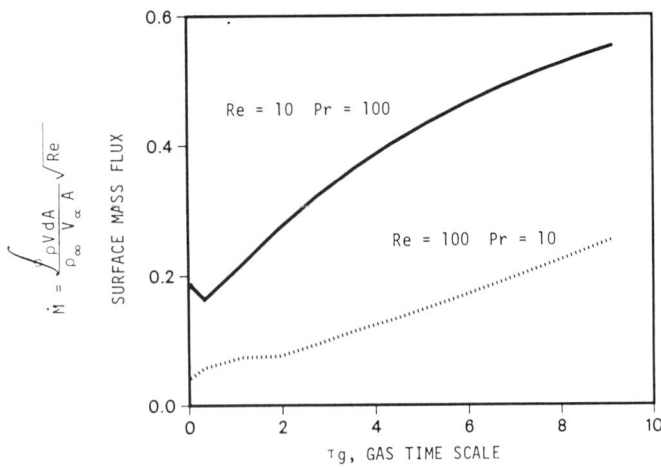

Fig.13 Vaporization mass flux variation as a function of gas time scale.

Fig. 14 shows the surface temperature normalized by the freestream gas-phase temperature, T_∞, for the high and low Reynolds number cases with vaporization. The results are plotted as a function of angular position in radians, as measured from the front stagnation point to the rear stagnation point. The shapes of the two curves are a result of the internal droplet heat transfer by both convec-

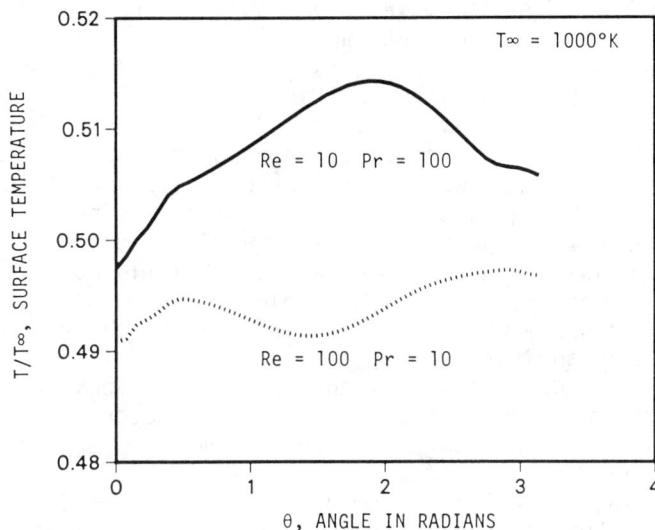

Fig.14 Surface temperature with vaporization as a function of angle measured from front stagnation point.

tion and conduction. For the Re = 100 cases, the temperature rises as fluid molecules travel from the front stagnation region towards the back. At approximately 45 degrees from the front stagnation point the temperature begins to drop again because of the reduced internal temperature gradients. The temperature rises again as the rear stagnation region is approached due to the separated and reversed flow in this region. For the Re = 10 case, the absence of a separated wake region gives a significantly different pattern. Even though the absolute temperature differences across the droplet are small, they have a large effect on the surface mass flux due to the exponential behavior of the Clausius-Clapeyron relationship.

Conclusions

The results presented show a highly unsteady behavior, even for these cases of constant droplet diameter, physical properties, and relative freestream velocity. The reduction of droplet velocity due to drag and the reduction in size due to vaporization will add additional unsteadiness to the droplet histories. However, the dominant feature for unsteadiness will be the dependency of liquid phase transport properties on temperature. It is possible for the viscosity of hydrocarbon fuels to vary over several orders of magnitude between 300 K and boiling point temperatures. The subject of variable transport properties will be treated in a subsequent paper.

For the two different sets of Reynolds and Prandtl number results summarized in this paper, a number of conclusions can be drawn.

1) The unsteady heating and vaporization of droplets in the ranges computed show the need for full unsteady solutions of the equations of motion. The results can be very sensitive to pressure, which has large consequences on the gas-phase Reynolds number while not influencing the liquid conditions. Hence, any solution which has a time-varying pressure would be very difficult to characterize by quasisteady analysis.

2) The droplet heat transport shows a very large sensitivity to Prandtl number, which is not described by the Peclet number. A change in Prandtl number causes significant variation in surface temperature as well as radical changes in internal isotherm distribution. For all cases computed, the isotherms and surface temperature distributions are asymmetric.

3) Surface mass flux due to vaporization gives a much larger gas-phase boundary-layer displacement thickness at lower Reynolds numbers than at higher Reynolds numbers. This increased displacement thickness at low Reynolds numbers is due to the lack of momentum in the gas phase to penetrate the vaporizing mass flux near the droplet surface. The heat transfer and droplet drag are complicated functions of the gas-phase flowfield, and will require a great deal of computational effort to resolve over a wide parametric range.

4) If one estimates droplet drag, it appears that a droplet starting at Re = 100 will lose half its velocity by the time 20% of the liquid has been vaporized. Even though the droplet diameter change is small for 20% mass reduction, the change due to drag makes constant average Reynolds number a poor assumption for droplet models.

5) The isotherm distribution difference between cases of no vaporization and with vaporization is not large, but vaporization has a large effect on the gas-phase flow and the resulting drag. Drag models must be established which account for the effects on the gas flow, not just due to surface regression and surface liquid velocity.

The calculations presented here are a small part of our overall effort to add the effects of variable transport properties, surface regression, deceleration, and multicomponent fuels to our numerical model. Detailed calculations such as these will aid in the interpretation of droplet experimental results and extrapolation to practical combustion situations. In addition, with these detailed results it will be possible to reduce the empiricism which must be included in spray transport equations for droplet heatup, vaporization rate, and drag.

References

Chigier, N. (1983) Group Combustion Models and Laser Diagnostic Methods in Sprays: A Review. Combustion and Flame, 51(2), 127-139.

Dwyer, H. A., Kee, R. J., Barr, P. K. and Sanders, B. R. (1983) Transient Droplet Heating at High Peclet Number. J. Fluids Engineering 105, 83-88.

Law, C. K. (1982) Recent Advances in Droplet Vaporization and Combustion. Progress in Energy and Combustion Science 8(3), 169-199.

Lerner, S. L., Homan, H. S., and Sirignano, W. A. (1980) Multicomponent Droplet Vaporization at High Reynolds Numbers: size, composition and trajectory histories. AICLE Meeting, Chicago, Ill.

Prakash, S. and Sirignano, W. A. (1978) Liquid Fuel Droplet Heating with Internal Circulation. Int. J. Heat and Mass Transfer, 21, 885-895.

Prakash, S. and Sirignano, W. A. (1980) Theory of Convective Droplet Vaporization with Unsteady Heat Transfer in the Circulating Liquid Phase. Int. J. Heat and Mass Transfer, 23, 253-268.

Sirignano, W. A. (1983) Fuel Vaporization and Spray Combustion Theory; Report Carnegie-Mellon University, Pittsburgh, Pa. (preliminary version).

Comparisons of Computed and Measured Dense Spray Jets

L. Martinelli* and F. V. Bracco†
Princeton University, Princeton, New Jersey
and
R. D. Reitz‡
General Motors Technical Center, Warren, Michigan

Abstract

Steady spray jets are considered from single cylindrical orifices under conditions of direct fuel injection in internal combustion engines, but at room temperature. Computations were made with a two-dimensional unsteady model that uses atomization results as nozzle exit boundary conditions, a k-ε submodel for gas turbulence, and a stochastic algorithm to compute drop events, including collisions and coalescence. Centerline velocity decay, spray width, average drop velocity distributions and standard deviation and skewness and flatness of the drop velocity distribution are compared at several axial and radial locations with corresponding laser Doppler velocimetry data from sprays at three different conditions. Agreement is very good with mean quantities, but the computed standard deviation of the drop velocity distribution (i.e., the drop velocity fluctuation) is generally smaller than the measured one. These nonvaporizing sprays achieve a structure that is very similar to that of fully developed incompressible jets, but 5-10 times further downstream. After development, the drops have little influence on the gas, but the gas controls the motion of the drops.

Presented at the 9th ICODERS, Poitiers, France, July 3-8, 1983. Copyright American Institute of Aeronautics and Astronautics, Inc., 1984. All rights reserved.
*Graduate Student, Department of Mechanical and Aerospace Engineering.
†Professor, Department of Mechanical and Aerospace Engineering.
‡Research Engineer, Department of Fluid Mechanics, Research Laboratories.

Nomenclature

C_1 = turbulence model constant = 1.5
C_2 = turbulence model constant = 1.9
C_3 = turbulence model constant = -1.0
C_4 = turbulence model constant = 0.09
C_a = drop size constant = 0.86
C_d = drag coefficient
c_p = gas specific heat at constant pressure
C_θ = nozzle spray angle constant = 5.2
d = nozzle diameter
d^* = numerical nozzle diameter at θ = 0.9
D = molecular diffusivity
D_t = turbulence diffusivity
f = drop number distribution, $f(\underline{x}, \underline{v}, r, t)$
F_{v_x} = flatness, $\overline{v_x'^4} / \overline{v_x'^2}^2$
$\underline{g}$ = gravity acceleration
h_g = enthalpy of the gas
$\underline{\underline{I}}$ = unit tensor
k = turbulence kinetic energy
L = nozzle length
p = pressure
Pr = Prandtl number, Pr_k = 1.0, Pr_ε = 1.3
r = droplet radius; radial coordinate
$r_{0.5,g}$ = spray half-radius using half the average gas centerline velocity, $r_{0.5,g}(x,t)$
$r_{0.5,\ell}$ = spray half-radius using half the average drop centerline velocity, $r_{0.5,\ell}(x,t)$
R = universal gas constant
Re_r = drop Reynolds number = $2\rho_g |\underline{u} + \underline{u}' - \underline{v}| r / \mu_g$
Re_j = liquid jet Reynolds number = $\rho_\ell \overline{U} d / \mu_\ell$
SMR = Sauter mean radius
S_{v_x} = skewness, $\overline{v_x'^3} / \overline{v_x'^2}^{3/2}$
t = time
T_g = temperature of the gas (and of mixture)
$\underline{u}$ = gas velocity vector of components $u_x(x,r,t)$ and $u_r(x,r,t)$
$\overline{U}$ = mass mean injection velocity

U^*	=	numerical uniform axial velocity of gas and drops at $\theta = 0.9$
$\underline{v}$	=	drop velocity vector of components $v_x(x,r,t)$ and $v_r(x,r,t)$
$\underline{\dot{v}}$	=	drop acceleration
W	=	molecular weight
We_j	=	jet Weber number = $\rho_\ell \overline{U}^2 d/\sigma$
$\underline{x}$	=	space coordinate
x	=	axial coordinate
δ	=	Dirac delta function
Δ	=	numerical increment
ε	=	rate of dissipation of turbulence kinetic energy
η_c	=	coalescence efficiency
ν_{ab}	=	collision frequency
θ	=	gas volume fraction; spray angle
λ_t	=	turbulence length scale
μ_g	=	gas viscosity
μ_ℓ	=	liquid viscosity
σ	=	surface tension
σ_{ab}	=	transition probability function
σ_{v_x}	=	standard deviation of the fluctuation of the drop axial velocity, $\sigma_{v_x}(x,r,t)$
ρ_g	=	mass of gas per unit volume of gas
$\hat{\rho}_g$	=	mass of gas per unit volume of mixture = $\theta \rho_g$
ρ_ℓ	=	mass of liquid per unit volume of liquid
$\underline{\underline{\tau}}_L$	=	turbulence stress tensor
τ_c	=	drop-turbulence correlation time

Superscripts

$()'$	=	fluctuation
$(\overline{})$	=	mean

Introduction

The family of sprays considered in this work is characterized by small length scales, high velocities, and complete or very high opacity. Because of the latter property, these sprays are loosely termed dense. Typically, the nozzle is a straight cylindrical hole 100-300 μm in diameter, the field of interest is 10 cm, the injection velocity is 100 m/s, the average drop radius is 10-30 μm, and, without evaporation, the light transmissivity through the core is still negligible hundreds of nozzle diameter downstream. An example is shown in Fig. 1. They are used primarily in Diesel and stratified charge engines in which fuel injection, vaporization, mixing, and combustion occur in small chambers and must be completed in less than 10 ms.

In such applications, neither the injection pressure nor the chamber gas are steady during injection. The comparisons discussed in this paper are with drop velocity data taken 4-10 cm from the nozzle in non vaporizing, room temperature, steady sprays (Wu et al. 1984). No spatially resolved, accurate measurements within such sprays closer to the nozzle or under transient conditions are available. Although similar comparisons do not seem to have been previously reported, recent reviews of closely related subjects are available (Faeth 1983; Sirignano 1983). The model is outlined first and then the comparisons are discussed.

The Model

The equations of the model for the transient and steady state of non evaporating and evaporating dense sprays are those of O'Rourke and Bracco (1980). In the application of

Fig. 1 Typical high velocity spray (n-hexane, $\bar{U}$ = 222 m/s, ρ_ℓ/ρ_g = 73, d = 0.343 mm, L = 1.72 mm).

this paper, droplet vaporization and temperature effects are negligible and the corresponding terms have been dropped from the more general equations, thus obtaining the following:

Spray equation

$$\frac{\partial f}{\partial t} + \nabla \cdot (f\underline{v}) + \nabla_{\underline{v}} \cdot (f\underline{\dot{v}}) = \frac{1}{2} \iint_{ab} \nu_{ab} \{\sigma_{ab} - \delta(r-r_a)\delta(\underline{v}-\underline{v}_a) - \delta(r-r_b)\delta(\underline{v}-\underline{v}_b)\} \, dr_a \, d\underline{v}_a \, dr_b \, d\underline{v}_b \quad (1)$$

Gas mass equation

$$\frac{\partial \hat{\rho}_g}{\partial t} + \nabla \cdot (\hat{\rho}_g \underline{u}) = 0 \quad (2)$$

Gas momentum equation

$$\frac{\partial \hat{\rho}_g \underline{u}}{\partial t} + \nabla \cdot (\hat{\rho}_g \underline{u}\,\underline{u}) + \nabla p = \nabla \cdot \underline{\underline{\tau}}_t - \iint \frac{4}{3}\pi r^3 \underline{\dot{v}} \rho_\ell f \, dr \, d\underline{v} \quad (3)$$

Gas energy equation

$$\frac{\partial \hat{\rho}_g h_g}{\partial t} + \nabla \cdot (\hat{\rho}_g h_g \underline{u}) = \theta(\frac{\partial p}{\partial t} + \underline{u} \cdot \nabla p) + \nabla \cdot (\hat{\rho}_g c_p D_t \nabla T_g) - \iint f \rho_\ell \frac{4}{3}\pi r^3 \underline{\dot{v}}_a \cdot (\underline{u} + \underline{u}' - \underline{v}) \, dr \, d\underline{v} + \underline{\underline{\tau}}_t : \nabla \underline{u} \quad (4)$$

where

$$\underline{\underline{\tau}}_t = \hat{\rho}_g D_t (\nabla \underline{u} + \nabla \underline{u}^T - \frac{2}{3} \nabla \cdot \underline{u} \, \underline{\underline{I}})$$

Turbulence kinetic energy

$$\frac{\partial \hat{\rho}_g k}{\partial t} + \nabla \cdot (\hat{\rho}_g \underline{u} k) = \nabla \cdot (\frac{\hat{\rho}_g D_t}{Pr_k} \nabla k) + G - \hat{\rho}_g \varepsilon \quad (5)$$

Rate of turbulent energy dissipation

$$\frac{\partial \hat{\rho}_g \varepsilon}{\partial t} + \nabla \cdot (\hat{\rho}_g \underline{u} \varepsilon) = \nabla \cdot (\frac{\hat{\rho}_g D_t}{Pr_\varepsilon} \nabla \varepsilon) + \frac{\varepsilon}{k}(C_1 G - C_2 \hat{\rho}_g \varepsilon) + C_3 \hat{\rho}_g \varepsilon \nabla \cdot \underline{u} \quad (6)$$

where

$$G = D_t \{ 2 [(\frac{\partial u_x}{\partial x})^2 + (\frac{\partial u_r}{\partial r})^2 + (\frac{u_r}{r})^2] + (\frac{\partial u_x}{\partial r} + \frac{\partial u_r}{\partial x})^2 \}$$

$$D_t = D + C_4 k^2/\varepsilon$$

Momentum exchange rate and state equations

$$\dot{\underline{v}}_a = \frac{3}{8} \frac{\rho_g}{\rho_\ell} \frac{|\underline{u} + \underline{u}' - \underline{v}|}{r} (\underline{u} + \underline{u}' - \underline{v}) C_d$$

$$\dot{\underline{v}} = \dot{\underline{v}}_a - (\nabla p)/\rho_\ell - g \tag{7}$$

$$p = \rho_g R T_g / W \tag{8}$$

$$h_g = c_p T_g \tag{9}$$

Equation (1) is Williams' (1962) spray equation plus a term that accounts for drop collisions and coalescence (the integral on the right-hand side). The collision frequency between drops with subscript a and those with subscript b is

$$\nu_{ab} = f_a f_b \pi (r_a + r_b)^2 |\underline{v}_a - \underline{v}_b| \tag{10}$$

In Eq. (1), the portion of the integrand within the brackets gives the sources (given by the transition probability function σ_{ab}) and the sinks (given by the delta functions δ) of drops of velocity $\underline{v}$ and radius r due to collisions between drops of classes a and b.

The transition probability function determines whether the outcome of a collision is coalescence or separation. Its mathematical expression is given by O'Rourke (1981). The criterion for drop separation after collision is that the rotational energy of the coalesced drop pair exceeds the surface energy required to reform the original drops from the coalesced pair. For the resulting coalescence efficiency η_c, which is the probability of coalescence given that collison has occurred, O'Rourke and Bracco (1980) give the expression

$$\eta_c = \min (2.4 \, g(\xi)/We_c, \, 1.0) \tag{11}$$

where $We_c = \rho_\ell |\underline{v}_a - \underline{v}_b|^2 r_a/\sigma$, $r_a \leqslant r_b$, $\xi = r_b/r_a$

and $$g(\xi) = \xi^3 - 2.4\xi^2 + 2.7\xi$$

In the present application, most of the colliding drops have radii of similar magnitudes so that We_c is important, the coalescence efficiency is generally < 1.0, and drop reseparation (grazing collision) is significant.

In the gas-phase momentum and energy conservation equations, the integrals on the right-hand sides represent the exchange functions. They are the sum over all drops at point $\underline{x}$ and time t of the rate of momentum and energy exchanges between each drop and the gas.

The drop acceleration $\underline{\dot{v}}$, given by Eq. (7), has a contribution due to aerodynamic drag and one due to the mean pressure gradient that has been shown to be important in some applications (O'Rourke 1981). In Eq. (7) there is the drag coefficient C_d. After a detailed survey of experimental and theoretical studies, O'Rourke and Bracco (1980) proposed the following correlations to account for the effect of the gas volume fraction θ:

$$C_d(\theta, Re_r) = 24 (\theta^{-2.65} + Re_r^{2/3}\theta^{-1.78}/6)/Re_r \qquad (12)$$

The effect of turbulence on the gas phase is accounted for by the terms involving D_t in Eqs. (3) and (4), where D_t is the turbulent diffusivity found from the turbulent kinetic energy and its rate of dissipation as shown in Eqs. (5) and (6). The turbulence effects on the drops are calculated by adding to the mean gas velocity $\underline{u}$ a fluctuating component $\underline{u}'$ when computing the aerodynamic drag force. $\underline{u}'$ is chosen randomly from an isotropic Gaussian distribution with mean square deviation 2/3k, where k is the turbulence kinetic energy. For each drop, after a turbulent correlation time τ_c, a new value of $\underline{u}'$ is chosen and τ_c is given by

$$\lambda_t = \int_t^{t+\tau_c} |\underline{v}(t') - \underline{u}(t')| dt' \qquad (13)$$

where $\underline{u}(t')$ is the mean gas velocity at the drop position at time t' and λ_t is the eddy size

$$\lambda_t = C_4^{3/4} k^{3/2}/\varepsilon \qquad (14)$$

A detailed derivation and discussion of the equations of the dense spray model is given by O'Rourke (1981).

All the computations were initiated at that axial location of the spray where the gas volume fraction θ is approximately 0.9 (eight nozzle diameters downstream of the nozzle in our case). At that axial location, gas and liquid are given equal axial velocity U* computed by using conservation of momentum. U* is somewhat lower than the experimentally measured mass mean liquid injection velocity U and the equivalent numerical nozzle diameter d* is somewhat larger than the actual nozzle diameter because of the divergence of the spray. The procedure to compute U* and d* is explained by O'Rourke (1981).

The initial spray angle and mean size of the drops was computed using correlations proposed by Reitz and Bracco (1982) for the atomization process. They were able to show that under the condition of $(\rho_\ell/\rho_g)(Re_j/We_j)^2 \gg 1$, the measured initial spray angle θ is correlated well by

$$\tan \theta/2 = (1/C_\theta)[4\pi(\rho_g/\rho_\ell)^{1/2} \sqrt{3}/6] \qquad (15)$$

where the proportionality constant C_θ depends on the geometry of the nozzle. Reitz and Bracco argued that the corresponding initial mean drop size should be correlated by

$$SMR = C_a [4\pi(\sigma/\rho_g \bar{U}^2)(3/2)] \qquad (16)$$

where the constant C_a is independent of the nozzle geometry and of order one. Note that in Eq. (16) the initial mean drop size is predicted to decrease when the chamber gas density increases. However, the computed downstream mean drop size is found to increase with increasing gas density due to collisions and recombinations and in agreement with measurements (Kuo and Bracco 1982b).

The computer code LDEF (Lagrangian Drop, Eulerian Fluid) developed by O'Rourke (1981) was employed with some modifications. This code incorporates the stochastic parcel method of Dukowicz (1980). The two-dimensional unsteady deterministic Eulerian equations for the gas and the stochastic Lagrangian equations for the drops are solved fully coupled with an implicit pressure iteration technique. In the present study, the k-ε model of turbulence was also included to describe the gas turbulence. The initial value of k in the injection cell was $k = 0.03 U^{*2}$ and ε was found from Eq. (14) with $\lambda_t = 0.2d$. The initial values of k and ε in the rest of the domain were set equal to small values and the resulting diffusivity was nearly the laminar one.

In the computational mesh, the cells were smaller near the nozzle exit where $\Delta y = 0.05$ cm and $\Delta x = 0.1$ cm. Away from the nozzle exit, the size of the cells increases in both the axial and radial direction with corresponding expansion factors of 4 and 7%. A total of 44 cells in the axial direction and 26 cells in the radial direction were used, giving the computational domain a length of 9.5 cm and width of 3.2 cm.

The top and right boundaries were treated as open boundaries, the left boundary as a solid wall, and the drops and gas injected from the cell of the left boundary next to the axis of symmetry. The condition on the top boundary allows for free entrainment from the surroundings. The pressures at the top and right boundaries were assumed to be uniform and equal to the ambient value to simulate a spray in a semi-infinite gas.

At any given time, the numerical solution of Eqs. (1-16) gives a distribution of the drop sizes and, for a given drop size, of the drop velocity in each elemental volume (practically, the numerical cell volume) around any point in the two-dimensional space. Also obtained are the instantaneous properties of the gas within the same volume. Even in steady state, the computed quantities fluctuate in time as they do in actual sprays, because of the stochastic nature of the atomization process reflected in the stochastic technique used to solve the spray equation. To compare with the measurements, many events are computed and then averaged, after which the fluctuating components of the various quantities are evaluated. Equivalently, one can let a steady spray run and sample the computations at appropriate time intervals and then again calculate the averages and fluctuating components. The latter was the technique used to obtain the results reported here.

The accuracy of the numerical solution of the equations is checked indirectly by reducing spatial and temporal increments and increasing the particle injection rate until the results become acceptably insensitive to them (O'Rourke 1981; Kuo 1982). The accuracy of the model is checked by comparisons with measured data. Thus, comparisons were made by O'Rourke and Bracco (1980) of the computed tip penetration rate and downstream drop size distribution with those measured by Hiroyasu and Kadota (1974) in one Diesel-type injection; by O'Rourke (1981) of the drop size and velocity distributions with those measured by Groeneweg (1967) at several locations within a spray from a swirl atomizer; and by Kuo and Bracco (1982b) with more of the penetration and drop size data of Hiroyasu and Kadota (1974) and with traditional spray tip penetration rate correlations.

Details on the model, the method of solution, comparisons with measured data, and computed structure of dense sprays are given by O'Rourke (1981) and Kuo (1982).

Comparison and Discussion

For clarity and future reference, it is useful first to give a brief description of the structure that sprays and other "compressible" jets must attain far from the injector.

In a steady nonvaporizing spray, the mass flow rate of the drops is the same at all axial locations but that of the entrained gas increases almost linearly with increasing distance from the nozzle. A station is reached where the mass and the momentum of the drops are neglegible with respect to those of the entrained gas.

Downstream of this station, it is entrained gas that entrains more gas, as in "incompressible" jets. Fully developed incompressible jets exhibit self-preserving velocity profiles and relationships between fluctuating and mean components that can be used to express the total axial momentum in terms of the centerline velocity. Then, momentum conservation gives

$$C_\ell \rho_\ell \bar{v}^2(0,0,\infty)\pi d^2/4$$
$$= C_g \rho_g \bar{u}^2(x,0,\infty)\pi r_{0.5,g}^2$$
$$= C_g \rho_g \bar{u}^2(x,0,\infty)\pi C^2(x-x_o)^2$$

or

$$\frac{\bar{u}(x,0,\infty)}{\bar{v}(0,0,\infty)} = \left(\frac{C_\ell}{C^2 C_g 4}\right)^{1/2} \frac{d(\rho_\ell/\rho_g)^{1/2}}{x-x_o} \quad (17)$$

where ℓ is any injected fluid and g any ambient fluid; C_ℓ is defined by the injection velocity and velocity fluctuation profiles and C_g by the fully developed velocity and velocity fluctuation profiles; and C relates the half-radius (i.e., half-the-width at half-the-depth) to the distance from the virtual origin x_o. Equation 17 shows that, at sufficient distance from the nozzle, the ratio of the centerline velocity to the injection velocity becomes a function only of $x/d(\rho_\ell/\rho_g)^{1/2}$. This is true for all jets that have the same injection velocity profiles. If not, to the same injection centerline velocity there correspond different momenta and C_ℓ is different. Figure 2 shows this relationship and the departure from the single line for large values of $x/d(\rho_\ell/\rho_g)^{1/2}$ must be due to the different C_ℓ and to experimental inaccuracies. The solid line of Fig. 2 is obtained with $C_\ell = 1$, $C_g = 0.846$, $C = 0.087$, in which case the coefficient of Eq. (17) is 6.3. Wu et al.(1984) give $C_g = 0.846 \pm 2.9\%$ and $C = 0.0868 \pm 7.9\%$. For $C_\ell = 1$, the coefficient of Eq. (17) then is 6.3±9%, which is in general agreement with those in the literature; e.g., Capp and George (1982) give 5.8.

Thus at sufficient distance from the injector, the axial scale of all steady jets becomes $d(\rho_\ell/\rho_g)^{1/2}$. Hinze (1975) gives references for the origin of this scale.

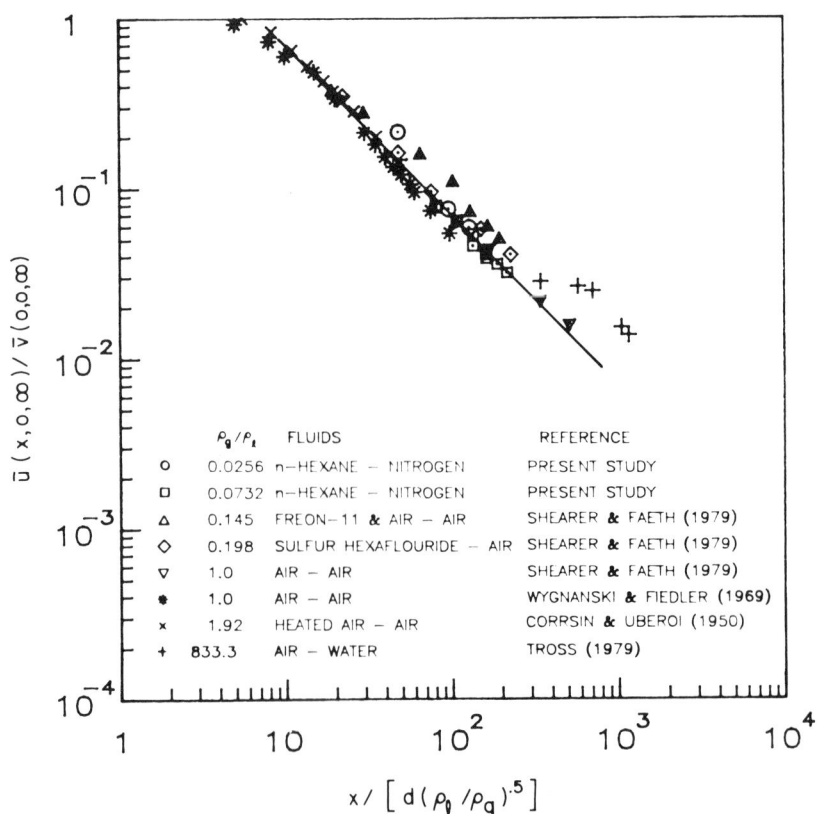

Fig. 2   Universal far field scaling for steady jets (Wu et al. 1984).

In the same axial region and in the zone of maximum radial gradient of the mean velocity, the turbulence intensity is about proportional to the centerline mean velocity $|\underline{u}'| \propto u(x,0,\infty)$ and the turbulence scale is proportional to the jet (width that is proportional to the) axial distance $\ell \propto (x-x_0)$ so that the turbulent diffusivity becomes proportional to centerline velocity and axial distance: $D_t \propto \ell |\underline{u}'| \propto (x-x_0) u(x,0,\infty)$. Then, using Eq. (17), we get

$$D_t \propto (\rho_\ell/\rho_g)^{1/2} \frac{d}{2} \bar{v} (0,0,\infty) \qquad (18)$$

Thus for all "compressible" jets at sufficient distance from the injector, the diffusivity varies as the square root of the density ratio (Kleinstein 1964). Notice that the value of the proportionality constant changes depending on the injection velocity profile. From our computations, we find that its most probable value is 0.016 $(\pi C_\ell)^{1/2}$.

Comparisons are now made between the results obtained with the model described in the previous section and the laser Doppler velocimetry (LDV) drop velocity measurements of Wu et al. (1984). They measured the distribution function of the axial and radial components of the drop velocity at various radial and axial locations within steady sprays. The conditions of their experiment are given in Table 1 and computations were made for cases A-C. The three cases differ in gas-liquid density ratio and in injection velocity. Notice that the measurements were taken within 300-800 nozzle diameters from the nozzle exit.

Using only their experimental data, Wu et al. (1984) concluded that these sprays approach the above-mentioned fully developed incompressible jet limit at x = 300d. Thus in the figures, the corresponding quantities measured in steady incompressible jets by Wygnanski and Fiedler (1969)

Table 1 Spray conditions[a] (Wu et al. 1984)

Series	P_g, MPa	ρ_g/ρ_ℓ	Δp, MPa	$\bar{U}$, m/s	Nozzle,d(μm) −L/d	X/d
A	1.48	0.0256	11.0	127	127-4	600,800
B	4.24	0.0732	11.0	127	127-4	300,400,500,600
C	4.24	0.0732	26.2	194	127-4	400,500,600
D	4.24	0.0732	11.0	149	76-4	300,600,700,800
E	1.48	0.0256	11.0	125	76-1	300

[a]Liquid: n-hexane, ρ_ℓ = 665 kg/m^3, μ_ℓ = 3.2x10^4 N. s/m^2, σ_ℓ = 1.84x10^2 N/m; gas: nitrogen; room temperature.

are given for reference and comparison. Wygnanski and Fiedler measured the gas velocity in air jets using hot wire anemometry (HWA), Wu et al. measured the drop velocities in sprays by LDV, and we computed both gas and drop velocities in the same sprays. In our computations, the assumption was not made that the gas and drop velocities are the same and the distinction is maintained in the discussion of the results. However, similarities exist, as will become apparent.

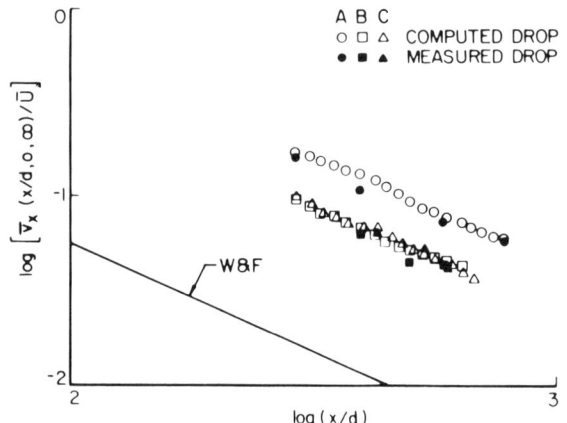

Fig. 3 Measured (Wu et al. 1984) and computed mean centerline drop velocity.

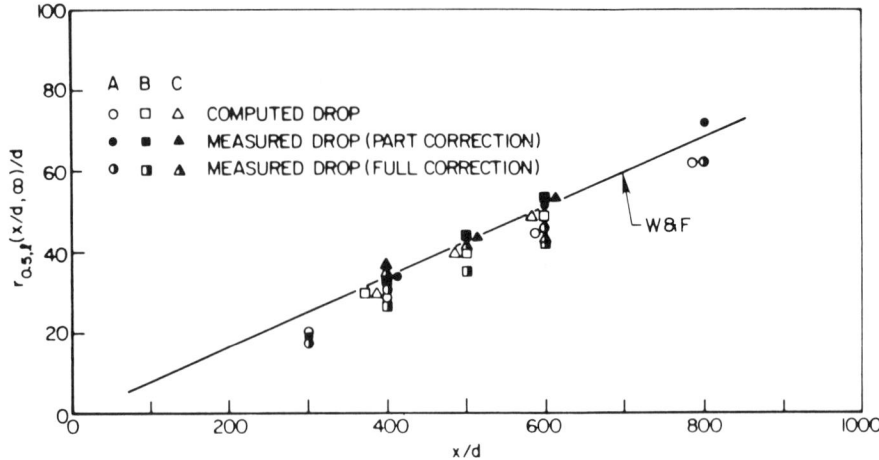

Fig. 4 Measured (Wu et al. 1984) and computed jet half radius based on the mean drop velocity.

First the mean quantities and then the fluctuations are compared.

For applications, the two most important parameters are the steady-state centerline velocity and the angle of the spray. For constant pressure injections into unconfined quiescent gases, the tip velocity is about 70% of the steady-state centerline velocity (Whitehouse and Sareen 1974; Kuo and Bracco 1982a) and behind the head vortex the spray quickly attains its steady configuration, including its steady angle. Thus, the ability to reproduce steady centerline velocity and angle also implies the ability to predict the global transient behavior. Figures 3 and 4 show good agreement between the computed and measured values. In Fig. 4 $r_{0.5,\ell}$ is the half-radius as determined by the drop velocity. The most likely value of this half-radius is between the fully and the partially corrected data, where the computed values fall. The corresponding half angle is 4.95 deg ± 0.025 deg as compared with 4.95 deg ± 0.04 deg for incompressible jets (Wu et al. 1984).

Agreement of centerline velocity and jet width does not necessarily imply agreement over the entire cross section. However, Fig. 5 shows that such agreement does exist. The mean value of the axial component of the drop velocity is reproduced adequately for all cases and axial and radial locations. This figure and an examination of the data also show that the computed average axial drop velocity is only a few percents greater than the computed average axial gas velocity. (The stochastic computation of the drop leaves an indetermination about the exact values of all drop quantities that brings about large uncertainties about small differences.) Finally, Fig. 6 shows that at the axial locations of the comparisons both computations and measurements indicate that self-preserving profiles have been reached by the mean axial velocities of drops and gas and that these profiles are indistinguishable from that of incompressible jets. (In Fig. 6 also notice for future reference that the measurements show drops at larger radial distances than the computations.)

Having recovered the structure of incompressible jets as far as the mean quantities are concerned, we can compare the predicted diffusivity with the constant diffusivity that is sufficient to characterize the global behavior of fully developed turbulent jets. Figure 7 shows that the computed diffusivity varies radially and axially but for x ⩾ 300d and for the region with the sharpest mean velocity gradients, it tends to the constant value of Schlichting (1979) after scaling it by the factor $(\rho_\ell/\rho_g)^{1/2}$; i.e. in Eq. (18), the proportionality constant is 0.0285 for $C_\ell = 1$ and with $v(0,0,\infty)$ replaced by $\bar{U}$.

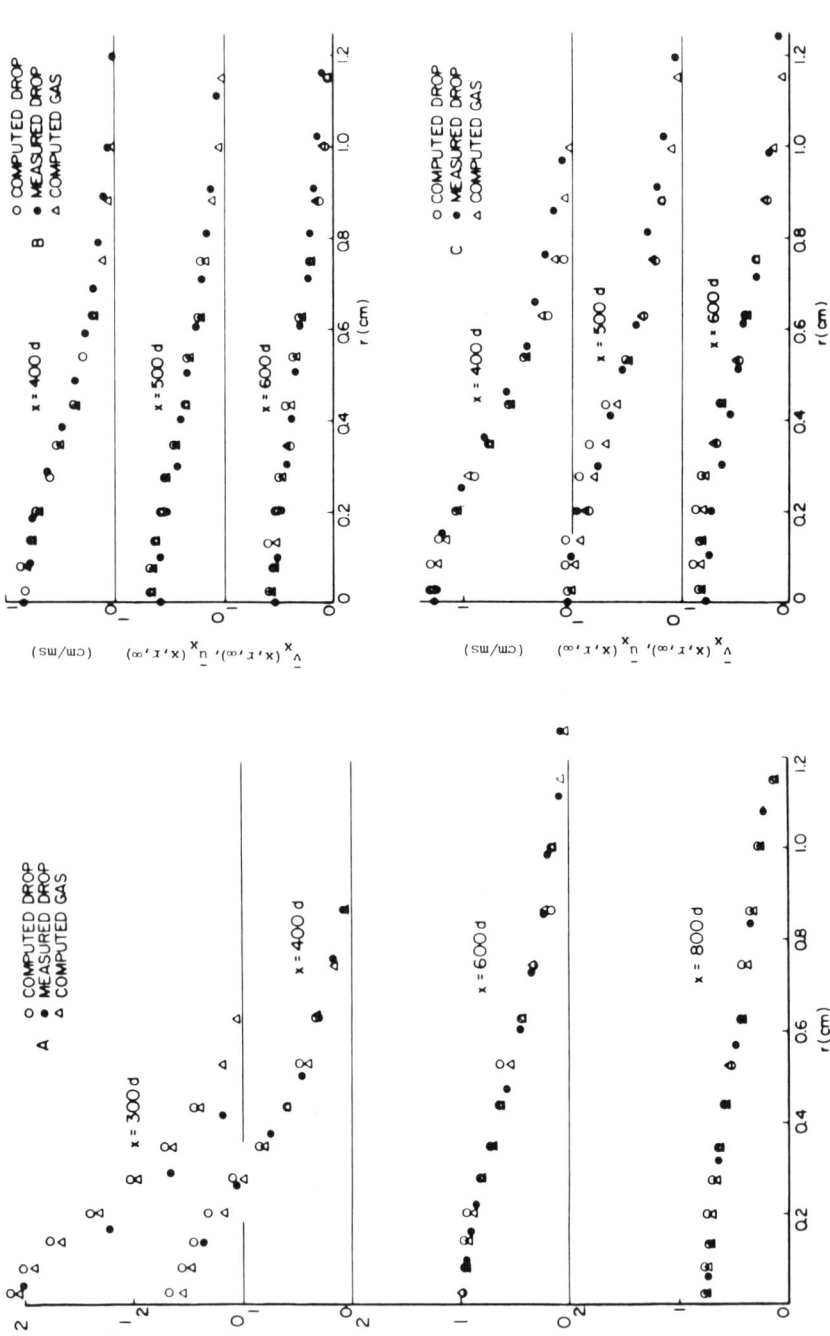

Fig. 5 Measured (Wu et al. 1984) and computed mean axial drop velocity and computed mean axial gas velocity for sprays A, B, and C of Table 1.

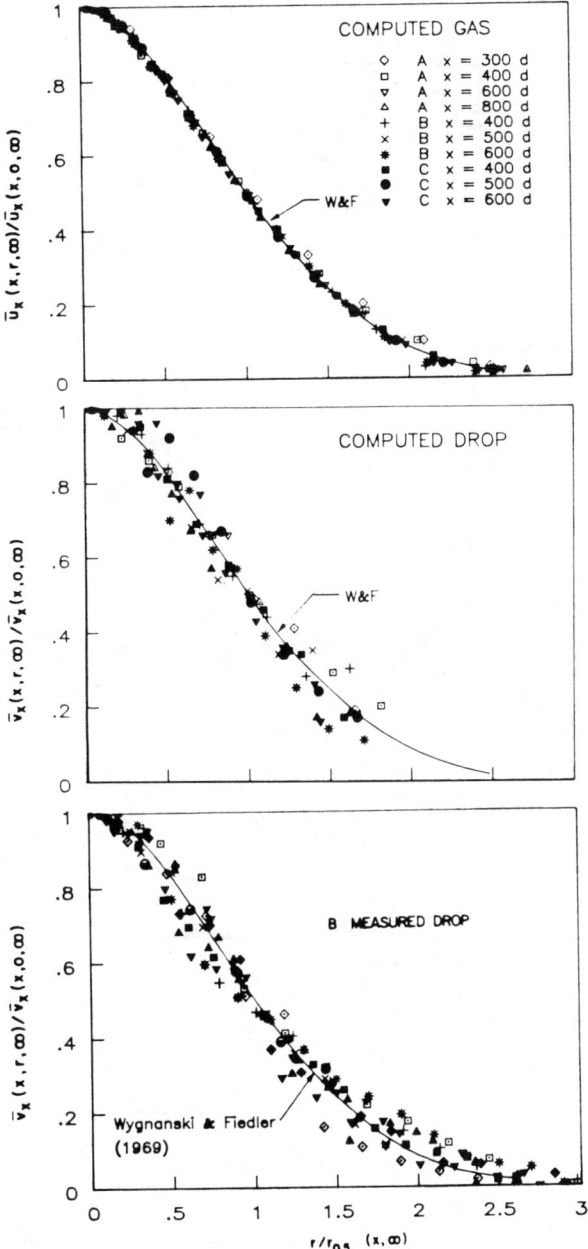

Fig. 6 Measured (Wu et al. 1984) and computed self preserving profiles of mean axial drop and gas velocities for sprays A-C of Table 1.

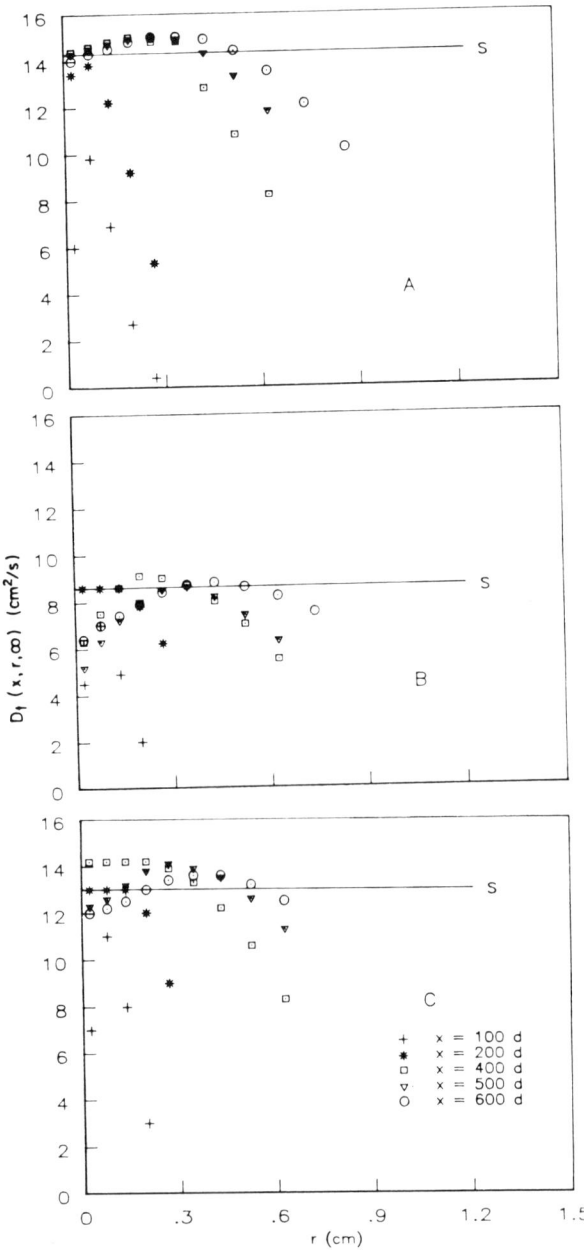

Fig. 7 Computed radial and axial distribution of the gas diffusivity for sprays A-C of Table 1.

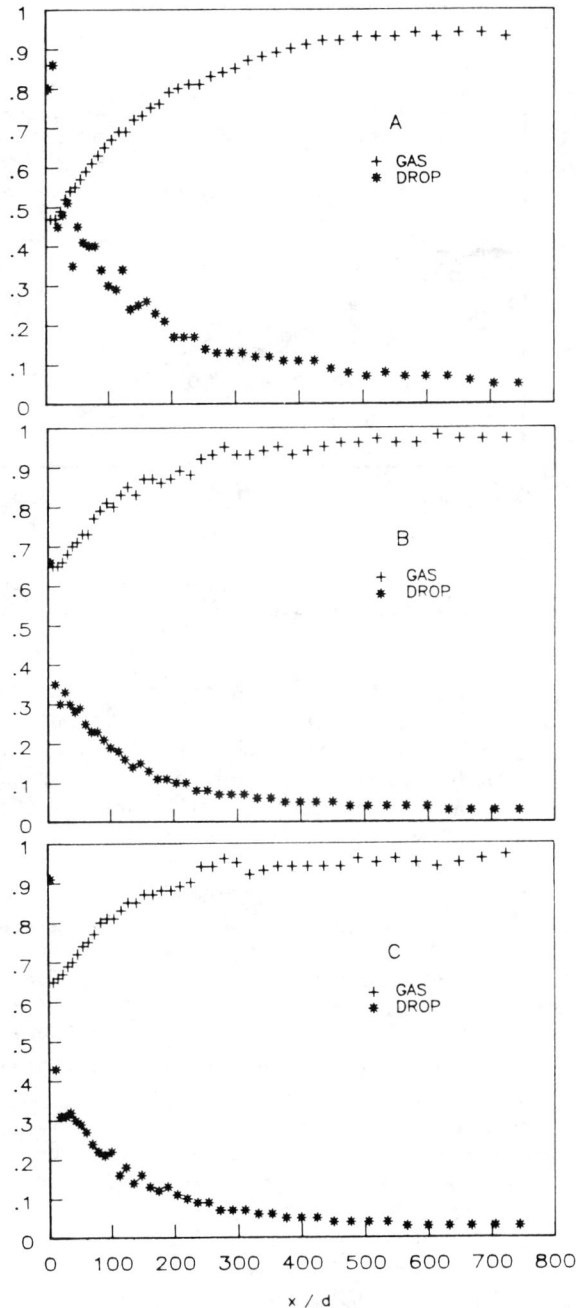

Fig. 8 Computed gas and liquid fractions of the axial momentum for sprays A-C of Table 1.

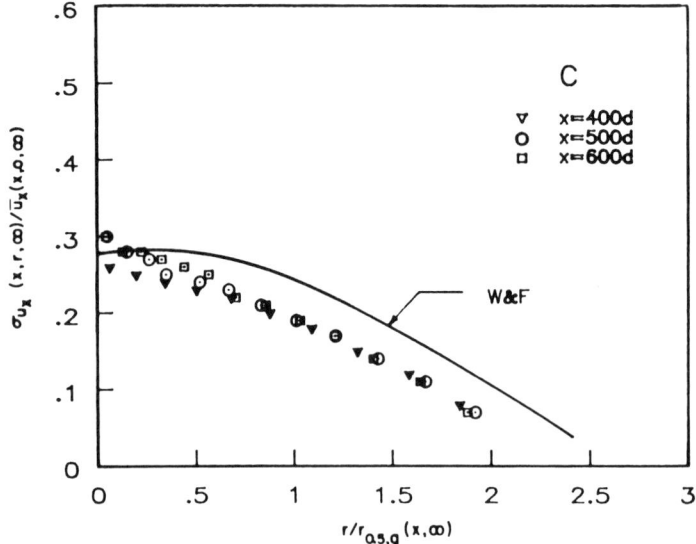

Fig. 9 Computed normalized turbulence intensity of the gas at various radial and axial locations for Case C of Table 1.

Thus, all indications are that for these sprays and $x > 300d$ the mentioned limit, in which the entrained gas entrains more gas and controls the structure of the jet, practically has been reached. Indeed for $x > 300d$, no less than 87% of the axial momentum is with the gas (Fig. 8).

Parenthetically, it is noted that the magnitude of the jet diffusivities of Fig. 7 is similar to that predicted for the air in the combustion chamber of Diesel engines (Grasso and Bracco 1983). It suggests that engine sprays may also be directly affected by the turbulence of the charge.

Having considered average quantities, we can now examine the fluctuations. Figure 9 shows the computed normalized gas turbulence intensity for one of the three sprays. It is seen to have reached its fully developed limit and to agree reasonably well with that measured by Wygnanski and Fiedler (1969). Presently we will point out that classical HWA turbulence intensity measurements in incompressible round jets are not necessarily very accurate, particularly at the edge of the spray where they tend to underestimate the turbulence intensity.

Figures 10-12 show the computed relative fluctuation amplitude (normalized standard deviation), skewness, and flatness of the axial component of the drop velocity at various axial and radial locations for the three sprays. Figure 13 plots the measured data of Wu et al. (1984) for

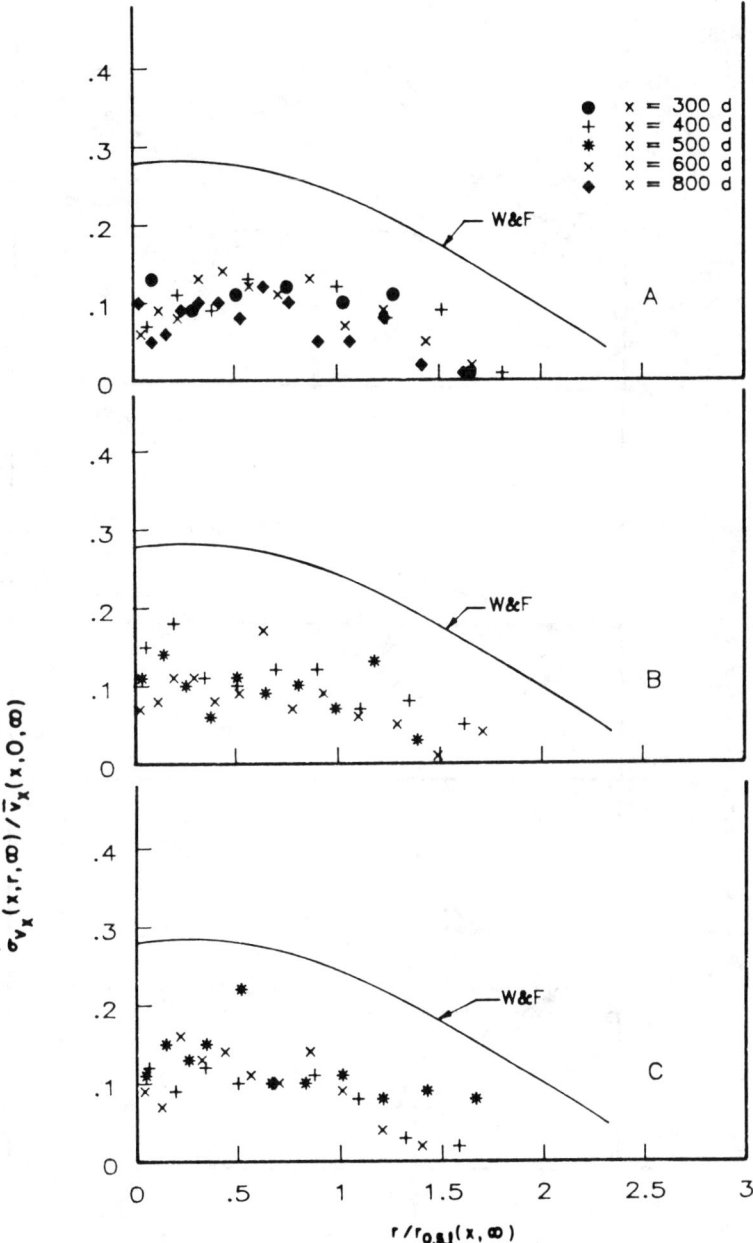

Fig. 10 Computed normalized standard deviation of the axial component of the drop velocity for sprays A-C of Table 1.

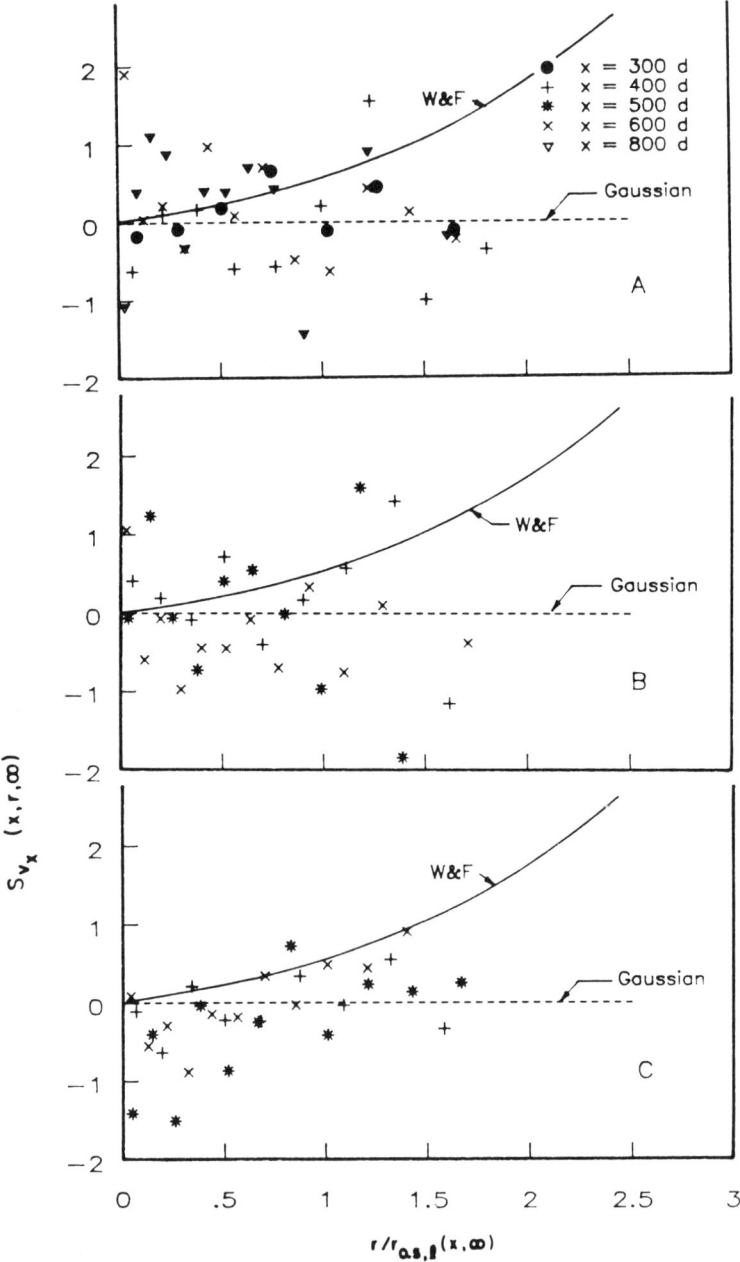

Fig. 11 Computed skewness of the axial component of the drop velocity for sprays A-C of Table 1.

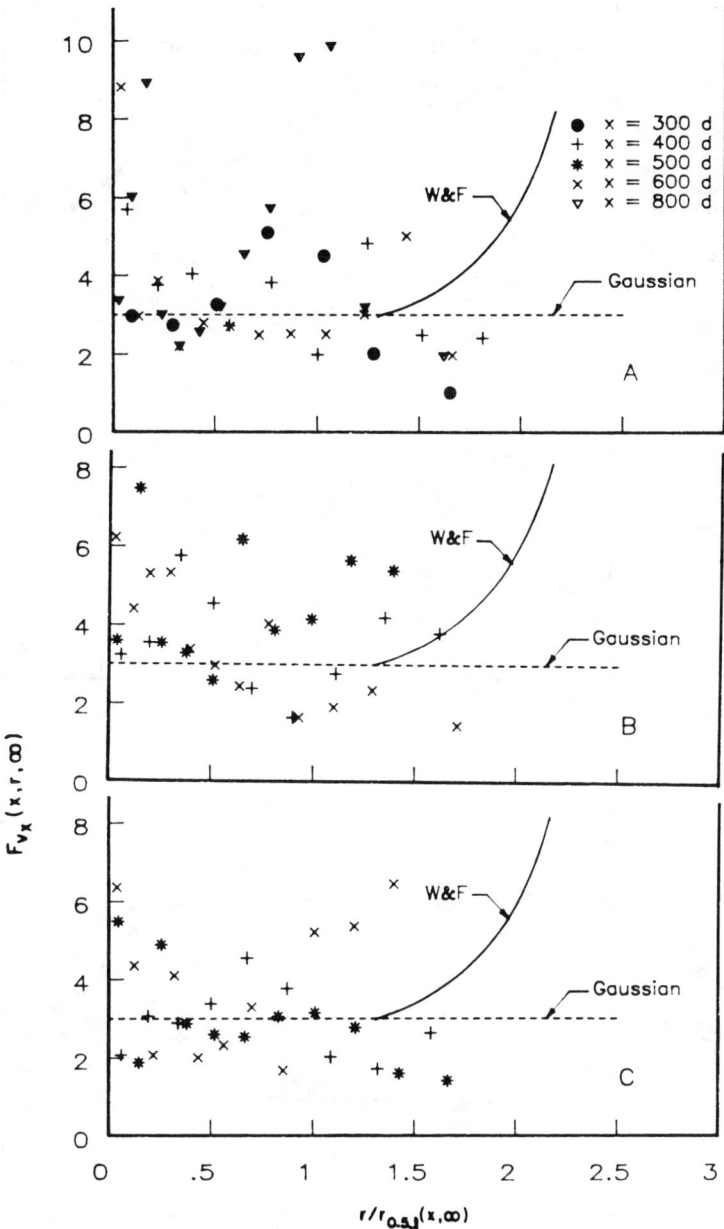

Fig. 12 Computed flatness of the axial component of the drop velocity for sprays A-C of Table 1.

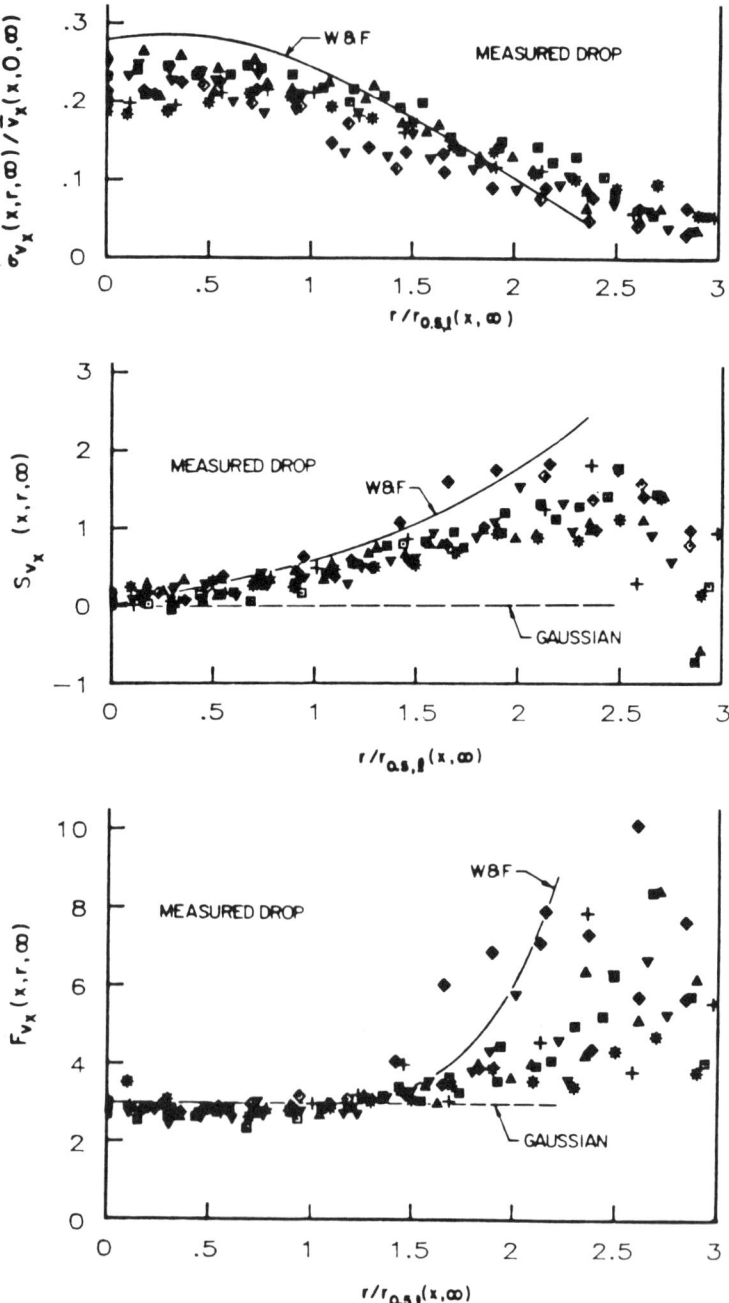

Fig. 13 Measured (Wu et al. 1984) normalized standard deviation, skewness, and flatness of the axial component of the drop velocity for sprays A-C of Table 1.

the same quantities. Also shown are the corresponding fluid quantities measured by Wygnanski and Fiedler (1969) in incompressible jets and those pertaining to Gaussian distributions. Both computed and measured data show considerable scatter, but the amplitude of the computed drop velocity fluctuation is about one half of that measured. However, the shape of the two distributions appear to be the same. Also, whereas the measured and computed mean drop quantities were shown to agree with those of incompressible jets, the measured drop fluctuations appear to disagree.

Wu et al. (1984) considered the influence of directional error in HWA and velocity biasing error in LDV and concluded that both classical HWA data and recent LDV data can be affected by large errors outside the half radius, particularly the fluctuations. When the possible sign and magnitude of the errors is considered, a tendency to better agreement is found. The accuracy of classical HWA measurements of fluctuations in incompressible jets had already been questioned (List 1982). We are inclined to conclude that the disagreement between the measured drop data and the incompressible jet data is due largely to experimental errors and that the amplitude of the fluctuation of the drop velocity should be equal to or smaller than that of the gas in the region where the gas dominates.

We are still left with the disagreement between the computed and the measured drop fluctuation amplitude. In the computations the scattering of the drop by the gas turbulence is determined by Eq. (13). A drop is assumed to feel the velocity of an eddy for a time that equals the residence time of the drop in the eddy. If the correlation time τ_c is much shorter than the drop relaxation time, $\tau_d \simeq \rho_\ell \, r^2/4.5 \, \mu_g$, then the drops cannot respond to the gas velocity fluctuations, exhibit smaller fluctuations and remain closer to the axis of the spray. This trend was checked with a computation of spray C in which the turbulent component of the gas velocity felt by the drop was changed at each Δt of the numerical integration, i.e., τ_c was reduced to the minimum value that can be computed (normally $\Delta t/\tau_c < 10^{-3}$). As expected, the standard deviation of the drop velocity decreased and became a half of that computed with Eq. (13) (collisions, whether or not followed by coalescence, scatter the drops and put a floor on σ_{v_x}). Also the drops did remain even closer to the axis than in Fig. 6. Thus, both trends indicate that it is necessary to account for the scattering effect of turbulence and that the coupling must be increased beyond that given by the present model.

At the other extreme, when the correlation time is much greater than the drop relaxation time, the drops follow the eddies and σ_{v_x} tends to σ_{u_x}. That is, the drops behave as the seeding particles for LDV turbulence measurements. This limit is approached as the distance from the injector increases, since τ_c continues to increase as x^2 whereas τ_d levels off when collisions become infrequent.

In the computations the ratio of the correlation time to the drop relaxation time was in the range 0.1-10, depending on drop size, location, and time of sampling. Therefore, σ_{v_x} should respond to adjustments in the correlation time and better agreement in both drop fluctuation amplitude and drop population at the edge of the spray should be possible.

As a whole, when the comparisons of mean quantities and shapes of distributions at the various locations and for the three sprays are considered, one can conclude that the far field is represented adequately. Unfortunately satsifactory

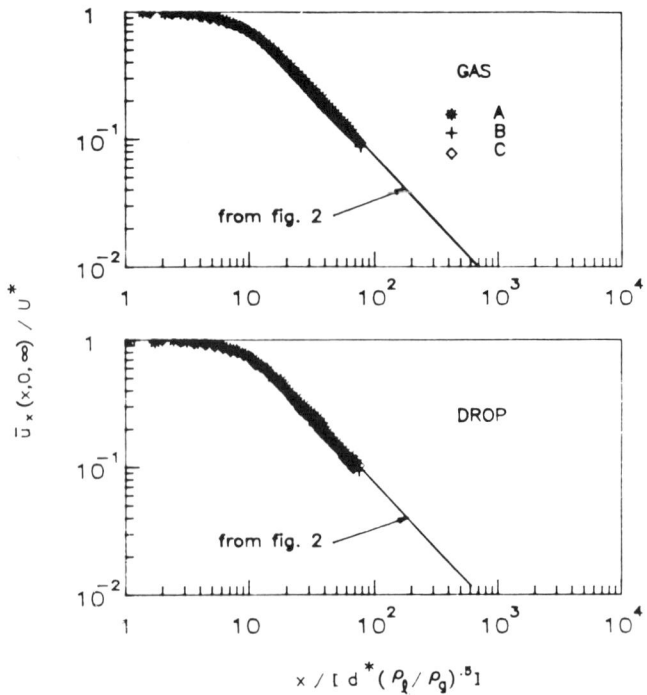

Fig. 14 Computed universal far-field scaling for gas and drops of sprays A-C of Table 1.

comparisons for x ≥ 300d assure only that the model is not in gross error for 0 ≤ x ≤ 300d. Thus we will comment only briefly on what the model suggests about the structure of this inner region.

Figure 8 shows that a large fraction of the total axial momentum is transferred very rapidly by the drops to the gas, but Fig. 14 shows that the centerline velocity of the drops remains nearly constant up to $x \approx 7\ d^*(\rho_\ell/\rho_g)1/2$ and then starts decaying and rapidly approaches the limit of Eq. (17). In Fig. 14 the centerline gas velocity appears to be identical to the centerline drop velocity. Actually, at x = 100d the mean axial drop velocity is about 10% larger than the mean axial gas velocity over the entire gross section and for all three sprays. But the difference gets progressively smaller until at x = 300d the liquid velocity is only a few percentage points higher than the gas velocity, as already stated and shown in Fig. 5.

Finally, we should point out that while our model attempts to account for the effect of turbulence on the drop motion, it does not account for direct effects of the drops on the gas motion [Eqs. (5) and (6) do not include any terms due to the presence of drops]. Such effects should be important in the development region (Elghobashi and Abou-Arab 1983), but are neglegible in the far field of our comparisons because the mass of the drops is small in comparison to that of the entrained gas. Even any retarding effect on the jet development is lost at sufficient distance from the nozzle.

Conclusions

Narrow full-cone high-velocity nonvaporizing isothermal sprays from steady injections into semi-infinite initially-quiescent gas were considered. The liquid-to-gas density ratio was 39 and 13.7.

Due to the rapid entrainment, the structure of these sprays becomes dominated by the ambient gas within distances of the order of hundreds of nozzle diameters. The entrained gas becomes a fully developed incompressible turbulent jet and is influenced little by the presence of the drops. The drops adjust their motion to that of the gas and local equilibrium is approached selectively.

In the fully developed region, centerline velocity and tip penetration rate are determined by the injection momentum and the gas density. To reproduce them, a simple model based on a constant diffusivity is sufficient. The diffusivity is that of incompressible jets multiplied by $(\rho_\ell/\rho_g)^{1/2}$. This is because differences in the development process and length (i.e. in the virtual origin) become negligible at sufficient distance from the nozzle. However, care must be exercised in extrapolating these results to transient jets. Thus, for example, if the ambient gas or the injection velocity are unsteady the details of the development region can be important, depending on the ratios of various characteristic times. The model used reproduced the steady state through the computation of an impulsively started transient.

The model reproduced well the measured mean axial velocity of the drops at all locations and conditions. It also predicted correctly that within the range of the measurements mean drop and gas velocities are very nearly equal. This had been inferred directly from the measurements of the drop velocities. However, the computed amplitude of the drop velocity fluctuations was a factor of two smaller than the measured one. Nonetheless, the model was helpful in clarifying relationships between drop velocity fluctuations, gas velocity fluctuations, drop scattering by drop collisions, and location of the drops within the spray.

Unfortunately, successful reproduction of the drop far field constitutes only a weak test of the validity of the model as far as the development region and transient sprays are concerned since the main requirements to achieve it are good numerical accuracy, correct steady global far field gas diffusivity, and a valid estimate of the order of magnitude of the scattering of drops by drop collisions and gas turbulence. But spatially resolved accurate measurements in the development region and during transient operations for our type of sprays are not available and very difficult to make. Thus stricter tests of the model for such region and conditions will have to wait for the availability of appropriate experimental data.

Acknowledgments

Support for this work was provided by the U.S. Department of Energy (Contract DE-AC-04-81AL16338), the U.S. Army Research Office (DAAG29-81-K-0135), General Motors, Komatsu, and Cummins Engine. These results were presented at the 17th DISC Meeting, Los Alamos National Laboratory, N. Mex., March 1983.

References

Capp, S. P. and George, W. K. (1982) Measurements in an axisymmetric jet using a two-color LDA and bust processing. International Symposium on Applied L.D.A. to Fluid Mechanics, Lisbon, Portugal.

Dukowicz, J. K. (1980) A particle-fluid numerical model for liquid sprays. J. Comp. Phys. 33 (2), 229-253.

Elghobashi, S. E. and Abou-Arab, T. W. (1983) A two-equation turbulence model for two-phase flows. Phys. Fluids 26 (4), 931-938.

Faeth, G. M. (1983) Evaporation and combustion of sprays. Prog. Energy Combust. Sci., 9 (1/2) 1-76.

Grasso, F. and Bracco, F. V. (1983) Sensitivity of chamber turbulence to intake flows in axisymmetric reciprocating engines. AIAA J. 21 (4), 637-640.

Groeneweg, J. F. (1967) The statistical description of a spray in terms of drop velocity, size, and location. Ph.D. Thesis, University of Wisconsin, Madison, Wisc.

Hinze, J. O. (1975) Turbulence, McGraw Hill Book Co., New York.

Hiroyasu, H. and Kadota, T. (1974) Fuel droplet size distribution in Diesel combustion chamber. SAE Paper 740715.

Kleinstein, G. (1964) Mixing in turbulent axially symmetric free jets. J. Spacecr. Rockets 1 (4), 403-408.

Kuo, T.-W. and Bracco, F. V. (1982a) On the scaling of transient laminar, turbulent, and spray jets. SAE Paper 820038.

Kuo, T.-W. and Bracco, F. V. (1982b) Computations of drop sizes in pulsating sprays and of liquid core length in vaporizing sprays. SAE Paper 820133.

Kuo, T.-W. (1982) On the scaling of transient laminar, turbulent, and spray jets. Ph.D. Thesis 1538-T, Dept. of Mechanical and Aerospace Engineering, Princeton University, Princeton, N. J.

List, E. J. (1982) Turbulent jets and plumes. Ann. Rev. Fluid Mech. 14, 189-212.

O'Rourke, P. J. and Bracco, F. V. (1980) Modeling of drop interactions in thick Sprays and comparison with experiments. Institution of Mechanical Engineers, Pub. ISBN 0 85298 4693, 101-116.

O'Rourke, P. J. (1981) Collective drop effects on vaporizing liquid sprays. Ph.D. Thesis 1532-T, Dept. of Mechanical and Aerospace Engineering, Princeton University, Princeton, N. J.

Reitz, R. D. and Bracco, F. V. (1982) Mechanism of atomization of a liquid jet. Phys. Fluids, 25 (10) 1730-1742.

Schlichting, H. (1979) Boundary-layer theory, McGraw-Hill Book Co., New York.

Sirignano, W. A. (1983) Fuel droplet vaporization and spray combustion theory, Prog. Energy Combust. Sci., 9(4) 291-322.

Whitehouse, N. D. and Sareen, B. K. (1974) Prediction of heat release in a quiescent chamber Diesel engine allowing for fuel/air mixing, SAE Paper 740084.

Williams, F. A. (1962) Progress in spray-combustion analysis, Eighth International Symposium on Combustion, pp. 50-69, Williams & Wilkins Co., Baltimore.

Wu, K.-J., Coghe, A., Santavicca, D. A., and Bracco, F. V. (1984) LDV measurements of drop velocity in Diesel-type sprays, AIAA J. (to appear).

Wygnanski, I. and Fiedler, H. (1969) Some measurements in the self-preserving jet, J. Fluid Mech., 38, (3), 577-612.

A Study of the Motion of Vaporizing Droplets in a Turbulent Flow

A. A. Mostafa* and S. E. Elghobashi†
University of California, Irvine, California

Abstract

A two-equation turbulence model for predicting isothermal steady two-phase flow including phase changes is presented. A set of equations describes the conservation of mass, momentum of each phase, concentration, kinetic energy of turbulence and its dissipation rate for the carrier fluid. Closure of the time-mean equations is achieved by modeling the existing turbulent correlations. Predictions of a turbulent axisymmetric jet laden with vaporizing droplets are discussed.

Introduction

Accurate prediction of spray combustion is rendered extremely difficult due to the complex physical and chemical phenomena encountered in this two-phase flow process. The interaction between droplets and the turbulent fluid, turbulence effects on chemical reaction and heat transfer (and hence on droplet vaporization) are just a few examples of the complexity. In order to understand the nature of these interactions, coordinated experimental and theoretical studies need to be performed in a stepwise manner thus isolating the phenomenon to be investigated.

Experimental (Yule et al. 1982) and analytical (Labowski and Rosner 1976) results show that in most practical situations spray burning, rather than being

---

Presented at the 9th ICODERS, Poitiers, France, July 3-8, 1983. Copyright American Institute of Aeronautics and Astronautics, Inc., 1984. All rights reserved.
* Research Assistant, Mechanical Engineering Department.
† Associate Professor, Mechanical Engineering Department.

controlled by "single droplet burning," is a droplet cloud process in which the majority of droplets vaporize in groups and reaction occurs at the vapor-air interface surrounding the clouds; similar to a gas diffusion flame. Thus modeling of spray burning requires modeling of droplet vaporization and droplet-turbulence interaction under conditions of heat, momentum, and mass transfer, between the droplets and the surrounding gas, which do not generally involve regions of reaction close to the droplet surface. Also vaporization of liquid sprays is an important process in its own right, even when the spray is not burning. In particular the premixed-pre-vaporized gas turbine combustor requires fuel sprays to be completely vaporized and mixed with air before reaction occurs in the combustion chamber (Yule et al. 1982).

The structure of sprays and other particle laden flows is generally influenced by turbulent dispersion of the discrete phase. The nature of the interaction between the discrete phase and the carrier fluid is rather complex and not well understood at present. There have been, however, several attempts to predict the behavior of solid particle laden turbulent flows under very limited conditions (see Saffman 1962; Owen 1969; Abramovich 1970).

Danon et al. (1977) proposed a turbulence model for two-phase jet flows without phase change. Their model failed to predict the flow of a two-phase jet. Agreement with experiment was achieved via arbitrary corrections to the kinetic energy equation.

Genchev and Karpuzov (1980) added to the single-phase equations of the turbulence kinetic energy and length scale a sink term to simulate the dispersed phase effects. They, however, ignored many significant correlations which appear in the rigorously derived equations, to be presented, of the turbulence kinetic energy and its dissipation rate (or length scale). Again the model was not successful in predicting the two-phase pipe flow.

The purpose of this paper is to: 1) present a set of transport equations describing the conservation of mass, momentum of each phase, concentration, kinetic energy of turbulence and its dissipation rate for the carrier fluid. These equations account for the evaporation (and thus the existence of a size distribution) of droplets and their effect on turbulence, 2) present a second-order closure model of turbulence which accounts for the interaction between the liquid droplets and the carrier fluid, 3) test the model for a turbulent axisymmetric jet laden with vaporizing droplets.

Conservation Equations

Equations of Motion

The following assumptions are made in deriving the governing equations:

1) Both the carrier fluid and dispersed phase behave macroscopically as a continuum, but only the carrier fluid behaves microscopically as a continuum. This means that the volume averaged equations are based on a control volume larger than the droplet spacing but much smaller than the characteristic volume of the flow system. Mutual exclusion of the phases is ensured, and chemical reactions do not occur.

2) The volume fraction of the dispersed phase is such that no collisions occur between the droplets. This assumption renders the equations valid only for dilute suspension.

3) Droplets of different sizes constitute different phases. This is from the point of view of "continuum" mechanics of a cloud of droplets, apart from the obvious definition of a multiphase system (mixture of phases of liquid droplet and gas) (Soo 1967). Therefore, the continuous droplet-size distribution will be divided into n intervals; d^k is the average diameter for droplets in the kth diameter range. If d^S and d^L are the smallest and largest droplet diameters, then the sizes are ordered as follows

$$d^S = d^n < d^{n-1} \text{---} < d^1 = d^L \tag{1}$$

Thus, n different diameter ranges constitute correspondingly n dispersed phases and the evaporated mass with the surrounding gas constitute the carrier phase.

4) Material properties for the different phases are constant.

The instantaneous, volume averaged momentum equations, in Cartesian tensor notations, of the carrier (lighter) phase are (Mostafa 1984)

$$(\rho_1 \Phi_1 U_i)_{,t} + (\rho_1 \Phi_1 U_i U_j)_{,j} = -\Phi_1 P_{,i} - \sum_k (F^k \Phi^k (U_i - V_i^k) + \dot{m}^k \Phi^k V_i^k) + (\mu_1 \Phi_1 (U_{i,j} + U_{j,i}))_{,j} \tag{2}$$

The corresponding equations for the kth dispersed phase are

$$(\rho_2 \Phi^k V_i^k)_{,t} + (\rho_2 \Phi^k V_i^k V_j^k)_{,j} = -\Phi^k P_{,i} + F^k \Phi^k (U_i - V_i^k)$$

$$- \dot{m}^k \Phi^k V_i^k + \mu_2 \Phi^k (V_{i,j}^k - V_{j,i}^k)_{,j} + g_i \Phi^k (\rho_2 - \rho_1) \tag{3}$$

The continuity equation for the kth dispersed phase is

$$(\rho_2 \Phi^k)_{,t} + (\rho_2 \Phi^k V_i^k)_{,i} = -\dot{m}^k \Phi^k \tag{4}$$

The global continuity is

$$\Phi_1 + \sum_k \Phi^k = 1 \tag{5}$$

Using the continuity equations for the different phases, Eqs. (2) and (3) can be written as

$$\rho_1 \Phi_1 U_{i,t} + \rho_1 \Phi_1 U_j U_{i,j} = -\Phi_1 P_{,i}$$

$$- \sum_k \Phi^k (F^k + \dot{m}^k)(U_i - V_i^k) + \mu_1 (\Phi_1 (U_{i,j} + U_{j,i}))_{,j} \tag{6}$$

$$\rho_2 \Phi^k V_{i,t}^k + \rho_2 \Phi^k V_j^k V_{i,j}^k = -\Phi^k P_{,i} + F^k \Phi^k (U_i - V_i^k)$$

$$+ g_i \Phi^k (\rho_2 - \rho_1) + \mu_2 (\Phi^k (V_{i,j}^k + V_{j,i}^k))_{,j} \tag{7}$$

In the equations above and throughout the paper the partial derivatives are represented by a subscript consisting of a comma and an index; e.g., $(\ )_{,t}$

$$\equiv \frac{\partial(\ )}{\partial t}, \quad U_{i,j} \equiv \frac{\partial U_i}{\partial x_j}, \quad U_{i,k\ell} \equiv \frac{\partial^2 U_i}{\partial x_k \partial x_\ell}$$

The subscripts 1 and 2 denote, respectively, the carrier fluid and dispersed phase; the superscript k denotes the kth dispersed phase; U_i is the velocity component of the carrier fluid; V_i^k is the velocity component of the droplets in the kth diameter range; ρ and μ are the material density and viscosity; P is the pressure; Φ is the volume fraction; g_i is the gravitational acceleration in the i direction; F is the interphase friction coefficient, and $\dot{m}$ is the evaporation rate per droplet volume.

The equations of motion for the mean flow are obtained from the instantaneous ones by performing the conventional Reynolds decomposition and averaging of Eqs. (2) and (3). This yields for the carrier fluid

$$\rho_1(\Phi_1 U_i U_j)_{,j} = -\Phi_1 P_{,i} - \overline{\phi_1 P_{,i}} - \sum_k \overline{\phi^k F^k}(U_i - V_i^k) + \sum_k \overline{\dot{m}^k \phi^k v_i^k}$$

$$- \sum_k F^k \overline{\phi^k(u_i - v_i^k)} + \sum_k \overline{\dot{m}^k \phi^k v_i^k} + \mu_1(\Phi_1(U_{i,j} + U_{j,i})$$

$$+ \overline{\phi_1(u_{i,j} + u_{j,i})})_{,j} - \rho_1(\Phi_1 \overline{u_i u_j} + U_i \overline{\phi_1 u_j} + U_j \overline{\phi_1 u_i} + \overline{\phi_1 u_i u_j})_{,j}$$

(8)

and for the kth phase

$$\rho_2(\Phi^k V_i^k V_j^k)_{,j} = -\Phi^k P_{,i} - \overline{\phi^k P_{,i}} + F^k \phi^k(U_i - V_i^k) - \overline{\dot{m}^k \phi^k v_i^k}$$

$$+ F^k \overline{\phi^k(u_i - v_i^k)} - \overline{\dot{m}^k \phi^k v_i^k} + \mu_2(\phi^k(V_{i,j}^k + V_{j,i}^k)$$

$$+ \overline{\phi^k(v_{i,j}^k + v_{j,i}^k)})_{,j} - \rho_2(\Phi^k \overline{v_i^k v_j^k} + V_i^k \overline{\phi^k v_j^k}$$

$$+ V_j^k \overline{\phi^k v_i^k} + \overline{\phi^k v_i^k v_j^k})_{,j} + g_i \phi^k(\rho_2 - \rho_1) \qquad (9)$$

The mean continuity equation of the kth phase is

$$\rho_2(\Phi^k V_i^k)_{,i} + \rho_2(\overline{\phi^k v_i^k})_{,i} = -\dot{m}^k \phi^k \qquad (10)$$

The mean global continuity is

$$\Phi_1 + \sum_k \Phi^k = 1 \qquad (11)$$

In Eqs. (8-11) the overbars indicate Reynolds averaged correlations.

Turbulence Kinetic Energy Equation (k)

The equation governing the mean kinetic energy ($k = \frac{1}{2}\overline{u_i u_i}$) of turbulence is obtained from the instantaneous momentum equation of the carrier fluid (6) following the same procedure as in the work of Elghobashi and Abou-Arab (1983).

The resulting k equation reads:

$$(\rho_1 \phi_1 U_\ell (\overline{u_i u_i}/2),_\ell) =$$

Convection

$$(-\rho_1 \phi_1 U_{i,\ell} \overline{u_i u_\ell} - \rho_1 U_{i,\ell} \overline{\phi_1 u_i u_\ell})$$

(I) Production

$$-(\overline{\phi_1 P,_i u_i} + \overline{P,_i \phi_1 u_i} - \overline{\phi_1 u_i P,_i} + \rho_1 \overline{\phi_1 u_i u_\ell u_{i,\ell}} + \rho_1 \overline{\phi_1 u_i u_\ell u_{i,\ell}})$$

(II) Turbulent Diffusion

$$-(\rho_1 U_\ell \overline{u_i \phi_1 u_{i,\ell}} + \rho_1 U_\ell U_{i,\ell} \overline{\phi_1 u_i})$$

(III) Production and Transfer

$$+ \sum_k (F^k + \dot{m}^k)(\overline{\phi^k u_i (V_i^k - U_i)} + \overline{\phi^k u_i (v_i^k - u_i)} + \overline{\phi^k u_i (v_i^k - u_i)})$$

(IV) Extra Dissipation (ε')

$$+ \mu_1 \overline{u_i ((u_{i,\ell} + u_{\ell,i}) \phi_1)},_\ell$$

(V) Viscous Diffusion and Dissipation

$$+ (\mu_1 \overline{u_i (\phi_1 (U_{i,\ell} + U_{\ell,i}))}),_\ell + \mu_1 \overline{u_i (\phi_1 (u_{i,\ell} + u_{\ell,i}))},_\ell$$

(VI) Extra Viscous Diffusion and Dissipation   (12)

Dissipation Rate Equation (ε)

The exact equation for the dissipation rate per unit volume $(\varepsilon = \nu_1 \overline{(u_{i,j} u_{i,j})})$ is derived by differentiating the instantaneous equation (6) with respect to x_j, then multiplying throughout by $\nu_1 u_{i,j}$, and finally time averaging. The exact equation of ε thus obtained is

$$(\rho_1 \phi_1 U_\ell \varepsilon,_\ell) =$$

Convection

$$(-2\mu_1 \overline{u_{i,j} u_{i,\ell}} (\phi_1 U_\ell),_j - 2\mu_1 \overline{u_{i,j} u_{\ell,j}} \phi_1 U_{i,\ell} - 2\mu_1 \overline{(\phi_1 u_\ell),_j u_{i,j}} U_{i,\ell}$$

(I) Production by the mean motion

$$-2\mu_1 \overline{u_{i,j} u_{i,\ell} (\phi_1 U_\ell)}_{,j} -2\mu_1 \overline{\phi_1 u_\ell u_{i,j} U_{i,\ell j}} -2\mu_1 \overline{u_{i,j} u_\ell (\phi_1 U_{i,\ell})}_{,j})$$

$$+ (-2\mu_1 \overline{u_{i,j} u_{i,\ell} (u_\ell \phi_1)}_{,j} - 2\mu_1 \overline{u_{i,j} u_{i,\ell} (u_\ell \phi_1)}_{,j})$$

(II) Production by Self-Stretching of Vortex Tubes

$$+ (-2\mu_1 \overline{\phi_1 u_{i,j} u_{i,\ell j} u_\ell} - 2\mu_1 \overline{\phi_1 u_{i,j} u_{i,\ell j} u_\ell})$$

(III) Turbulent Diffusion

$$+(-2\mu_1 \overline{\phi_1 u_{i,j} (U_\ell U_{i,\ell})}_{,j} -2\mu_1 \overline{u_{i,j} \phi_{1,j} U_\ell U_{i,\ell}} -2\mu_1 U_\ell \overline{\phi_1 u_{i,j} u_{i,\ell j}})$$

(IV) Production and Transfer

$$+(-2\nu_1 \overline{u_{i,j}(P_{,i}\phi_1)}_{,j} -2\nu_1 \overline{u_{i,j}(\phi_1 P_{,i})}_{,j} -2\nu_1 \overline{u_{i,j}(\phi_1 P_{,i})}_{,j})$$

(V) Spatial Transport by Pressure (fluctuation and mean)

$$+ -2\nu_1 \sum_k (F^k + \dot{m}^k) \overline{(u_{i,j}(\phi^k(U_i - V_i^k))}_{,j} + \overline{u_{i,j}((u_i - v_i^k)\phi^k)}_{,j}$$

$$+ \overline{u_{i,j}(\phi^k(u_i - v_i^k))}_{,j})$$

(VI) Extra Dissipation

$$+ 2\mu_1 \nu_1 \overline{u_{i,j}((u_{i,\ell} + u_{\ell,i})\phi_1)}_{,j\ell}$$

(VII) Viscous Diffusion and Destruction

$$+(2\mu_1 \nu_1 \overline{u_{i,j}(\phi_1(U_{i,\ell} + U_{\ell,i}))}_{,j\ell}$$

$$+ 2\nu_1 \mu_1 \overline{u_{i,j}(\phi_1(u_{i,\ell} + u_{\ell,i}))}_{,j\ell}) \quad (13)$$

(VIII) Extra Viscous Diffusion and Destruction

The Concentration Equation

The concentration C is defined as the ratio of the evaporated mass within a control volume to the mass of the carrier phase in the same volume. The instantaneous, volume averaged concentration equation (or the mean temperature equation) for the evaporating material is

$$(\rho_1 \Phi_1 C)_{,t} + (\rho_1 \Phi_1 U_j C)_{,j} = (\rho_1 \delta \Phi_1 C_{,j})_{,j} + \sum_k \phi_m^{k \cdot k} \qquad (14)$$

where δ is the molecular mass diffusivity of the evaporating material in air. The equation of the mean concentration is obtained by the Reynolds decomposition and time averaging.

$$(\rho_1 \Phi_1 U_j C)_{,j} = \left(\rho_1 \delta(\Phi_1 C_{,j} + \overline{\phi_1 c_{,j}})\right)_{,j} + \sum_k \phi_m^{k \cdot k}$$

$$- \rho_1 (\overline{\phi_1 c} U_j + C \overline{\phi_1 u_j} + \Phi_1 \overline{u_j c} + \overline{\phi_1 u_j c})_{,j} \qquad (15)$$

Modeled Equations

Modeling Assumputions

1) The turbulent diffusion by pressure correlations will be assumed to be absorbed in the turbulent diffusion by the turbulence quantities (see Daly and Harlow 1970; Lumley 1978; Hanjalic and Launder 1972).
2) All fourth order correlations such as $\overline{\phi_1 u_i u_i u_{i,\ell}}$, $\overline{u_{i,j} u_i (u_\ell \phi_1)_{,j}}$, and $\overline{\phi_1 u_{i,j} u_{i,\ell} u_\ell}$ will be neglected due to the dilutness assumption of the suspension.
3) The viscous diffusion will be neglected due to its small magnitude as compared to the turbulent diffusion in the free shear flows.
4) The correlation of two scalars such as $\overline{\phi_1 c}$ or $\overline{u_i \phi_1 c}$ will be neglected due to two reasons: 1) the lack of the experimental data (Lumley 1978) and 2) its relatively small values.

Closure of the Momentum and Concentration Equations

Elghobashi and Abou-Arab (1983) discuss in details the modeling of the different correlations that appeared in their two-equation turbulence model. Their model was derived for a two-phase flow without phase change and uniform size particles. Here, the modeling of the turbulent correlations needed to close the momentum and concentration equations with the previous assumptions will be presented.

$$\overline{\phi_1 u_i} = - (\nu_t / \sigma_\phi) \Phi_{1,i} \qquad (16)$$

where ν_t is the kinematic eddy viscosity ($= c_\mu k^2 / \varepsilon$), σ_ϕ the turbulent Schmidt number of the scalar quanity ϕ (of order unity), and c_μ a constant of value 0.09.

$$\overline{u_i u_j} = -\nu_t(U_{i,j} + U_{j,i}) + \frac{2}{3}\delta_{ij}k \qquad (17)$$

$$\overline{\phi_1 u_i u_j} = -c_{\phi 5}(k/\varepsilon)\left(\overline{u_i u_\ell}(\overline{u_j \phi_1})_{,\ell} + \overline{u_j u_\ell}(\overline{u_i \phi_1})_{,\ell}\right) \qquad (18)$$

where $c_{\phi 5}$ is a constant of value 0.1 (Launder 1976). For the concentration equation c will replace ϕ in Eqs. (16) and (18).

The correlations for the dispersed phase equations of motion are modeled similarly except that ν_t is replaced by ν_p^k, where ν_p^k is the droplet turbulent diffusivity. The details of the calculation of ν_p^k is given in the work of Elghobashi et al. (1983).

Closure of the Turbulence Kinetic Energy Equation

The exact equation of the turbulence kinetic energy k for the carrier fluid is given by Eq. (12). The terms are grouped according to their physical contribution to the conservation of k. The modeling of the turbulent correlations appearing in the k equation will be presented in this section.

The turbulent diffusion correlation $\overline{u_i u_\ell u_{i,\ell}}$ can be written as

$$[\overline{u_\ell (1/2\, u_i^2)}]_{,\ell}$$

which modeled as

$$[(\nu_t/\sigma_k)k_{,\ell}]_{,\ell}$$

where σ_k is an empirical diffusion constant of order one.

The extra viscous diffusion and dissipation [group VI in Eq. (12)] and the correlation $\overline{u_i \phi u_{i,\ell}}$ (the first term in group III) will be neglected due to its relatively small magnitude as compared to the other similar terms (see Launder 1976; Launder et al. 1975).

$$\overline{u_i(v_i^k - u_i)} = -(k/3)\left[1 - \int_0^\infty ((\Omega_1 - \Omega_R)/\Omega_2)\, F(\omega)\, d\omega\right] \qquad (19)$$

where ω is the harmonic frequency of turbulence and $F(\omega)$ is the Lagragian energy spectrum function of the carrier phase. Ω_1, Ω_2, and Ω_R are functions of the carrier and dispersed phases properties, the droplet diameter, and the harmonic frequency. They are given by

Chao (1964) as

$$\Omega_1 = (\omega/\alpha)^2 + \sqrt{6}\ (\omega/\alpha)^{3/2} + 3(\omega/\alpha) + \sqrt{6}\ (\omega/\alpha)^{1/2} + 1$$

$$\Omega_2 = \beta^{-2}(\omega/\alpha)^2 + \sqrt{6}\ \beta^{-1}(\omega/\alpha)^{3/2} + 3(\omega/\alpha) + \sqrt{6}\ (\omega/\alpha)^{1/2} + 1$$

$$\Omega_R = ((1-\beta)\ \omega/\alpha\beta)^2 \qquad (20)$$

$$\alpha = 12\ \nu_1/(d^k)^2$$

$$\beta = 3\rho_1/(2\rho_2 + \rho_1) \qquad (21)$$

$$F(\omega) = \frac{2}{\pi}\ \frac{T_L}{1+\omega^2 T_L^2}\ \text{where}\ \omega$$

ranges from 1 to 10^4 (Sec^{-1}) and T_L is the local Lagrangian integral time scale given by $T_L = (5/12)k/\varepsilon$

$$\overline{\phi^k u_i(v_i^k - u_i)} = -c_{\phi 5}(\frac{k}{\varepsilon})\overline{(v_i^k - u_i)u_i}\left((u_\ell \phi^k)_{,\ell} + ((v_\ell^k - u_\ell)\phi^k)_{,\ell}\right)$$

$$= -c_{\phi 5}(\frac{k}{\varepsilon})\overline{(v_i^k - u_i)u_i}\ \overline{(v_\ell^k \phi^k)}_{,\ell} \qquad (22)$$

$$\overline{\nu u_i((u_{i,\ell} + u_{\ell,i})\Phi_1)}_{,\ell} \simeq -\Phi_1 \varepsilon \qquad (23)$$

Closure of the Turbulence Energy Dissipation Rate Equation

The exact equation of the dissipation rate of turbulence energy ε for the carrier fluid is given by Eq. (13). The terms are also grouped similar to the k equation. Tennekes and Lumley (1972) based on an order-of-magnitude analysis, argued that the terms involving mean strain rates [Groups (I) and (IV)] are negligible at high Reynolds number. The production of ε by stretching of the vortex tubes can be modeled as

$$-2\nu_1 \overline{u_{i,j} u_{i,\ell}(u_\ell \Phi_1)}_{,j} = c_{\varepsilon 1} G_k \varepsilon/k \qquad (24)$$

where G_k is the total production of k [Group (I) and (III) in Eq. (12)], and $c_{\varepsilon 1}$ is a constant of value 1.43.

$$2\nu_1 \overline{u_{i,j} u_{i,\ell j} u_\ell} = -((\nu_t/\sigma_\varepsilon)\varepsilon_{,\ell})_{,\ell} \qquad (25)$$

At high turbulence Reynolds number the main viscous destruction is the dominant one [groups (VI) and (VIII) are neglected] and can be modeled as

$$-2\nu_1^2 \overline{u_{i,j}(u_{i,\ell} + u_{\ell,i})\Phi_1)_{,j\ell}} \approx -c_{\epsilon 2}\Phi_1 \epsilon^2/k \quad (26)$$

where the value of the constant $c_{\epsilon 2}$ is 1.92.

The extra dissipation due to the relative velocity between the phases [group (VI) in Eq. (13)] will be modeled as one term which is given by $-c_{\epsilon 3}\epsilon'\frac{\epsilon}{k}$ where ϵ' [term IV in Eq. (12)] is the extra dissipation of k. The constant $c_{\epsilon 3}$ was optimized by Elghobashi et al. (1983) for a two-phase jet flow. The value of this constant is 1.0.

Interphase Property Transfer

The nature of droplets-gas flow is primarily dependent on the transfer of mass, momentum, and energy between phases. The following section discusses some of the basic features of mass and momentum transfer between the phases and provides empirical equations to quantify each.

Mass Transfer

The mass transfer source term appearing in Eqs. (2-15) is given by

$$\dot{m}^k = \left(12\delta\rho/(d^k)^2\right) \text{Ln}(1+B)\text{Sh}^k \quad (27)$$

where B is the transfer number and Sh^k is the Sherwood number given by Eq. (30).

Here, evaporation occurs due to the concentration gradient; thus the transfer number is given by

$$B = (C_L - C)/(1-C_L) \quad (28)$$

where C, C_L are the concentrations of the evaporating material at the freestream conditions and at the droplet surface, respectively. C is obtained from the solution of the concentration transport Eq. (14) and C_L is obtained from Clausius-Clapeyron expression; which reads

$$X_v = \frac{P_o}{P} \exp\left(W_v L/R_o(1/T_B - 1/T_L)\right) \quad (29)$$

$$C_L = X_v W_v / X_v W_v + (1-X_v)W_a$$

where P_o and P are the atmospheric pressure and the partial pressure of the evaporating material at the droplet surface, respectively, W_v and X_v are the molecular weight and the molecular fraction of the evaporating material, respectively; T_B is the boiling temperature of the evaporating material; and R_o is the universal gas constant. L is the latent heat of vaporization/unit mass. T_L is the temperature at the droplet surface. W_a is the air molecular weight.

The Sherwood number in Eq. (27) is given by the empirical formula of Ranz and Marshall (1952)

$$Sh^k = \dot{m}^k/\pi d^k \delta(C_L - C) = 2 + 0.55 \, Re^{k^{1/2}} Sc^{1/3} \qquad (30)$$

where $Sc = \nu_t/\delta$ is the Schmidt number.

Momentum Transfer

As noted in the governing equations, momentum exchange between phases is accomplished by transfer of mass through a velocity difference or aerodynamic forces. Having already treated mass transfer, the focus here will be on the aerodynamic drag.

In the governing equations set, the drag force is expressed in terms of the interphase friction coefficient [$F^k = 18 \mu_1/(d^k)^2$ for Stokes' flow around a spherical droplet of diameter d^k]. In general F^k is given by

$$F^k = (3/4d^k)\rho_1 C_D^k \, |\vec{U} - \vec{V}| \qquad (31)$$

The drag coefficient C_D^k is primarily a function of the Reynolds number based on the relative velocity but may also depend on evaporation rate.

Droplet evaporation tends to reduce the friction drag coefficient only at very high evaporation rates (such as for droplets burning in pure oxidizing atmospheres). Also, the internal circulation of the liquid droplet decreases the boundary-layer thickness of the exterior flow and reduces the drag coefficient. But the effect is again very small. Sirignano (1983) in his critical review of the theory of droplet vaporization, discussed the drag coefficient of a vaporizing droplet. A balance exists between the decrease in friction drag and the increase in pressure drag due to the blowing effects associated with vaporization (Sirignano 1983). Thus the droplet drag coefficient is almost the same as that for a solid sphere of the same diameter. In the present case (low evaporation rate of a spherical liquid droplet in gas) the drag coefficient can be described by the standard experimental drag curve of a sphere in a steady motion. Clift et al. (1978) formulated this curve by

$$C_D^k = (24/Re^k)\left[1+0.135\ (Re^k)^{0.82-0.05w}\right] \quad 0.01 < Re^k \leqslant 20 \tag{32}$$

$$C_D^k = (24/Re^k)\left[1+0.1935(Re^k)^{0.6305}\right] \quad 20 < Re^k \leqslant 200 \tag{33}$$

where $w = Log_{10}\ Re^k$ and the droplet Reynolds number is calculated from

$$Re^k = \rho_1\ |\vec{u} - \vec{v}^k|\ d^k/\mu_1$$

Vaporizing Droplets In A Turbulent Round Jet

As a validation of the presented model, both the mean values and the turbulence quantities of the different phases will be predicted for the case of a turbulent axisymmetric jet laden with vaporizing droplets. In the first section the modeled equations in cylinderical coordinates will be presented. In section two, the sequence of solving the governing equations will be discussed. The results and discussions will be presented in the third section.

Modeled Equations in Cylindrical Coordinates

For a round jet, using the boundary-layer approximation with the previous modeled correlations, the modeled euquations in cylinderical coordinates can be written as follows

Momentum equations for the carrier fluid are as follows

$$\rho_1 \Phi_1 U_z U_{z,z} + \rho_1 \Phi_1 U_r U_{z,r} = -\Phi_1 P_{,z} - \sum_k \Phi^k (F^k + \dot{m}^k)(U_z - V_z^k)$$

$$+ \frac{1}{r}(\Phi_1 r \mu_t U_{z,r})_{,r} + c_{m1}\ \rho_1\ U_{z,r}(\frac{\nu_t}{\sigma_\phi}\Phi_{1,r})$$

$$+ c_{\phi 5}\ \frac{1}{r}\ (\frac{k}{\varepsilon}\ r\mu_t U_{z,r})_{,r}(\frac{\nu_t}{\sigma_\phi}\Phi_{1,r})_{,r}$$

$$+ c_{\phi 5}\ \frac{k}{\varepsilon}\ \mu_t U_{z,r}(\frac{\nu_t}{\sigma_\phi}\Phi_{1,r})_{,rr} \tag{34}$$

where $c_{m1} = 0.4$ and for the kth phase

$$\rho_2 \phi^k V_z^k V_{z,z}^k + \rho_2 \phi^k V_r^k V_{z,r}^k = - \phi^k P_{,z} + F^k \phi^k (U_z - V_z^k)$$

$$+ \frac{1}{r}(\phi^k r \mu_p^k V_{z,r}^k)_{,r} + c_{m1} \rho_2 V_{z,r}^k (\frac{\nu_p^k}{\sigma_\phi} \phi_{,r}^k)$$

$$+ c_{\phi 5} \frac{1}{r} (\frac{k}{\varepsilon} r \mu_p^k V_{z,r}^k)_{,r} (\frac{\nu_p^k}{\sigma_\phi} \phi_{,r}^k)_{,r}$$

$$+ c_{\phi 5} \frac{k}{\varepsilon} \mu_p^k V_{z,r}^k (\frac{\nu_p^k}{\sigma_\phi} \phi_{,r}^k)_{,rr} + (\rho_2 - \rho_1) g \phi^k \quad (35)$$

The mean continuity equation of the kth phase is

$$\rho_2 (\phi^k V_z^k)_{,z} + \rho_2 (\phi^k V_r^k)_{,r} - \rho_2 (\frac{\nu_p^k}{\sigma_\phi} \phi_{,z}^k)_{,z}$$

$$- \rho_2 (\frac{\nu_p^k}{\sigma_\phi} \phi_{,r}^k)_{,r} = - \dot{m}^k \phi^k \quad (36)$$

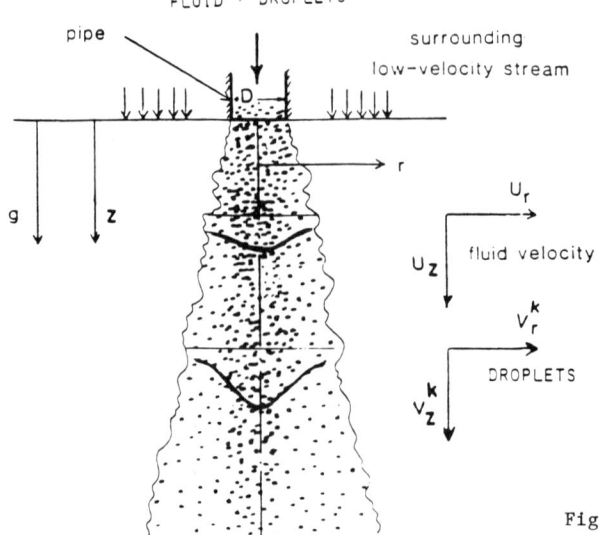

Fig. 1 Flow schematic.

The mean global continuity is

$$\Phi_1 + \sum_k \Phi^k = 1 \tag{37}$$

The concentration equation (C) is

$$\rho_1 \Phi_1 U_z C_{,z} + \rho_1 \Phi_1 U_r C_{,r} = \frac{1}{r}(\rho_1 r \Phi_1 \frac{\nu_t}{\sigma_c} C_{,r})_{,r}$$

$$+ \sum_k \Phi^k \dot{m}^k (1-C) + \rho_1 C_{,r}(\frac{\nu_t}{\sigma_\phi} \Phi_{1,r}) \tag{38}$$

where σ_c is a constant of value 0.7.

The turbulence kinetic energy equation (k)

$$\rho_1 \Phi_1 U_z k_{,z} + \rho_1 \Phi_1 U_r k_{,r} = \rho_1 \Phi_1 \nu_t U_{z,r} U_{z,r}$$

$$- \frac{4}{3} \rho_1 c_{\phi 5} (\frac{\nu_t}{c_\mu})(\frac{\nu_t}{\sigma_\phi} \Phi_{1,r})_{,r} U_{r,r}$$

$$+ \rho_1 c_{\phi 5} (\frac{k}{\varepsilon}) \nu_t (\frac{\nu_t}{\sigma_\phi} \Phi_{1,r})_{,r} U_{z,r} U_{z,r}$$

$$- \sum_k k \Phi^k (F^k + \dot{m}^k)(1 - \int_0^\infty (\frac{\Omega_1 - \Omega_R}{\Omega_2}) F(\omega) d\omega)$$

$$+ \sum_k (F^k + \dot{m}^k)((U_r - V_r^k)(\frac{\nu_p^k}{\sigma_\phi} \Phi^k_{,r})$$

$$- c_{\phi 5} (\frac{\nu_t}{c_\mu})(\frac{\nu_p^k}{\sigma_\phi} \Phi^k_{,r})_{,r} (1 - \int_0^\infty (\frac{\Omega_1 - \Omega_R}{\Omega_2}) F(\omega) d\omega))$$

$$+ \frac{1}{r}(\rho_1 \Phi_1 \frac{\nu_t}{\sigma_k} r k_{,r})_{,r} - \rho_1 \Phi_1 \varepsilon \tag{39}$$

The dissipation rate equation (ε) is

$$\rho_1 \Phi_1 U_z \varepsilon_{,z} + \rho_1 \Phi_1 U_r \varepsilon_{,r} = c_{\varepsilon 1} \frac{\varepsilon}{k} \rho_1 \Phi_1 \left(\nu_t U_{z,r} U_{z,r} \right.$$

$$- \frac{4}{3} \frac{c_{\phi 5}}{\Phi_1} (\frac{\nu_t}{c_\mu})(\frac{\nu_t}{\sigma_\phi} \Phi_{1,r})_{,r} U_{r,r}$$

$$+ \frac{c_{\phi 5}}{\Phi_1}(\frac{k}{\varepsilon}) \; \nu_t (\frac{\nu_t}{\sigma_\phi} \Phi_{1,r})_{,r} U_{z,r} U_{z,r} \Big)$$

$$- c_{\varepsilon 3} \frac{\varepsilon}{k} \left\{ \sum_k k\phi^k (F^k + \dot{m}^k) \left(1 - \int_o^\infty (\frac{\Omega_1 - \Omega_R}{\Omega_2}) F(\omega) d\omega \right) \right.$$

$$+ \sum_k (F^k + \dot{m}^k) \left((U_r - V_r^k)(\frac{\nu_p^k}{\sigma_\phi} \Phi^k_{,r}) \right.$$

$$\left. - c_{\phi 5}(\frac{\nu_t}{c_\mu})(\frac{\nu_p^k}{\sigma_\phi} \Phi^k_{,r})_{,r} (1 - \int_o^\infty (\frac{\Omega_1 - \Omega_R}{\Omega_2}) F(\omega) d\omega) \right) \Big\}$$

$$+ \frac{1}{r} (\rho_1 r \Phi_1 \frac{\nu_t}{\sigma_\varepsilon} \varepsilon_{,r})_{,r} - c_{\varepsilon 2} \rho_1 \Phi_1 \frac{\varepsilon^2}{k} \qquad (40)$$

Numerical Solution Procedure

The marching finite-difference solution procedure employed in this work is that developed and described in detail by Spalding (1979).

The coordinates of the expanding finite-difference grid are z and ψ, the stream function based on the gas-phase properties, defined as $\psi = \int_o^r \rho_1 \Phi_1 U_z r dr$. The steps followed to obtain the solution at a given axial location are:

1) Guess the downstream Φ_1^*, distribution (as the upstream values).
2) Solve for U_z downstream.
3) Solve for the k and ε, obtain r's and solve for U_r.
4) Obtain $p(r)$ from the carrier-phase lateral momentum equation.
5) Calculate the local mass evaporation rates (source terms), thus the droplets diameter distribution and the classification of the kth phases can be obtained.

MOTION OF VAPORIZING DROPLETS   529

6) Solve for downstream V_z^k, V_r^k, and Φ^k, and get Φ_1^*.

7) Adjust the calculated Φ_1 with the guessed one Φ_1^*.

8) Make corrections and repeat steps 1-7 until the solution converges before marching to the next station.

The Flow Considered

Figure 1 shows a sketch of the two-phase turbulent jet considered in this work. Air carrying a uniform size Methanol droplets of diameter 50 μm issues vertically downwards from a cylindrical pipe of diameter (D = 0.02 m). Qualitative comparison will be made with the flows of a turbulent jet laden with solid particles and a single-phase jet; those two flows have been predicted by Elghobashi et al. (1983). The initial and boundary conditions are those employed in that reference. Here the material density of the droplets is equal to that of liquid Methanol (810 kg m^{-3}) whereas in the previous work it was that of Silica (2990 kg m^{-3}).

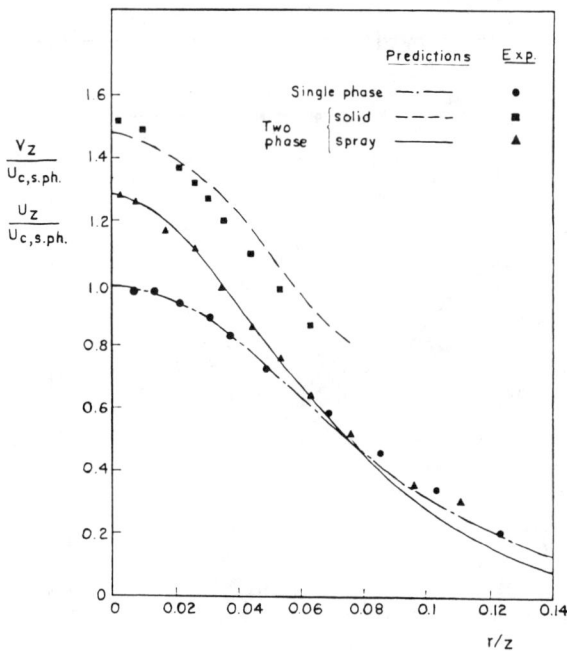

Fig. 2 Comparison of the predicted velocity profiles of a jet laden with solid particles by the present model with the experimental data at z/D = 20.

Results And Discussion

Since no experimental data exists at present for the flow of turbulent gaseous jet laden with uniform size evaporating droplets (at the pipe exit), the capability of the mathematical model is first shown by predicting a similar flow laden with solid spherical particles (Elghobashi et al. 1983) instead of droplets. Figure 2 (which is Fig. 5 of that reference) compares the predicted with the measured profiles of the mean axial velocities of the two phases at $z/D = 20$ and displays the mean velocity profiles of a turbulent single-phase jet. It is seen that the agreement is good. Comparison of other predicted quantities with experiment is given in the above mentioned reference.

Presented in the following are the predicted distributions of the mean velocities, volume fraction of both the carrier fluid and the liquid droplets, turbulence intensity and shear stress of the carrier fluid, the concentration of the evaporating material in the carrier phase, the droplet diameter, and the jet spreding rate.

Figure 3 shows the radial profiles of the mean axial velocities of the carrier phase, for both the solid-laden

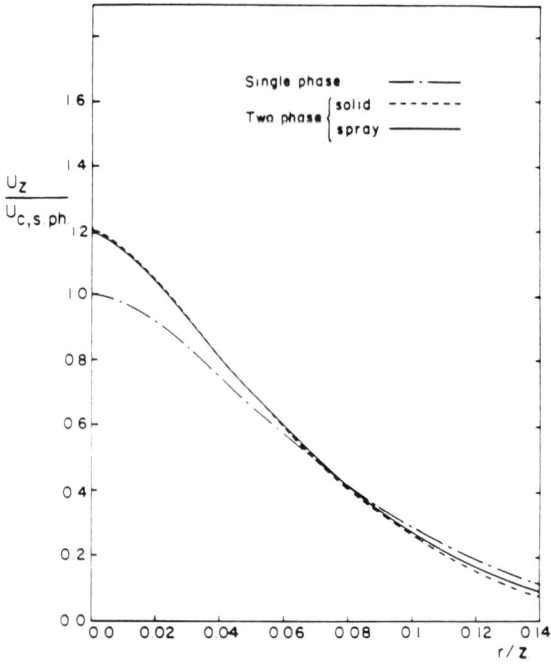

Fig. 3 Fluid mean axial velocity at $z/D = 10$.

and droplet-laden cases, normalized by the centerline velocity of the single-phase jet, $U_{c,s.ph}$ at $z/D = 10$. The mean velocity profiles of the carrier phase in both cases are nearly identical since close to the pipe exit little or no vaporization takes place. The mean velocity profiles for the carrier phase and the dispersed phase at $z/D = 20$ is shown in Fig. 4. Although the velocity of the carrier phase is nearly equal to that of the single-phase in the outer region, it is still closer to that of the carrier phase in the solid-laden case in the inner region. This is explained by the fact that droplets are confined to the inner region because their radial turbulent diffusivity is less than that of the carrier fluid; furthermore they are subjected to a net inward drag force by the fluid. Due to this confinement the droplets centerline velocity decays, with distance, at a slower rate than that of the fluid and thus they become a source of momentum to the fluid before equilibrium eventually takes place. The reason that the fluid centerline velocity in the liquid droplets case is less than that in the solid particles case is that vaporization reduces the

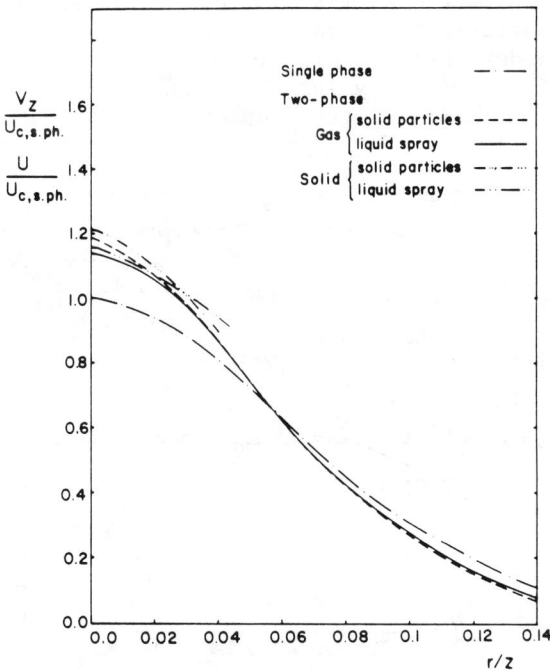

Fig. 4 Normalized mean velocity profiles at $z/D = 20$.

droplets diameter and thus, the momentum exchange in the former case is less than that in the latter.

The effect of the parital or complete droplet evaporation is reflected on the velocity distribution of the dispersed phase itself. The smaller the droplet diameter is the less the relative mean velocity ($U_z - V_z$) and the fluid and the higher the turbulent diffusivity of the dispersed phase is.

To understand the nature of the turbulent interaction between the carrier fluid and the vaporizing droplets, the main features of this type of flow are summarized:

1) The total mass of the dispersed phase continuously decreases and so does the volume fraction. Due to the reduction of the volume fraction, the momentum exchange terms (mean and/or fluctuating) are reduced.

2) The velocity of the evaporating material, as it leaves the droplet surface, is different from that of the carrier fluid. Thus there is an additional momentum transfer which depends on the evaporation rate and the relative velocity.

3) The momentum exchange coefficient is inversely proportional to the droplet diameter with an exponent ranging from 2 to 1.3 (for a Reynolds number less than 100). Thus, as the diameter is reduced the momentum exchange coefficient increases.

In the developing region (i.e., near the pipe exit) the turbulence intensity and the shear stress of the fluid are nearly the same for the droplet-laden and particle-

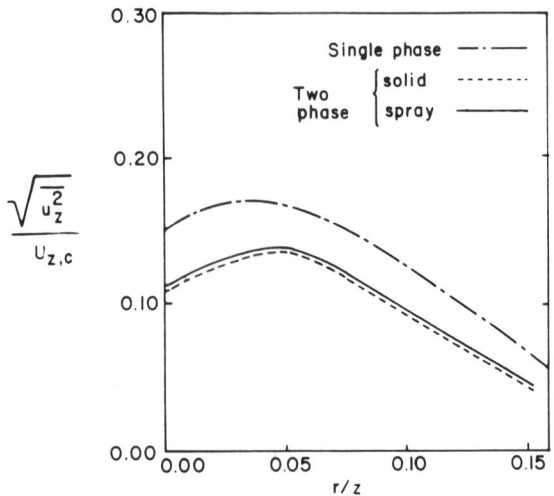

Fig. 5 Turbulence intensity profile at $z/D = 10$.

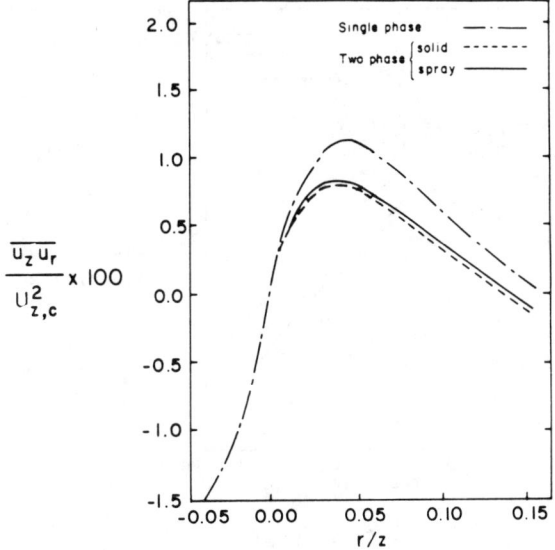

Fig. 6 Turbulent shear stress profile at $z/D = 10$.

laden cases (Figs. 5 and 6). Farther downstream, the effect of vaporization becomes more pronounced and the values of these turbulence quantities for both cases display significant differences (Figs. 7 and 8).

The volume fraction centerline decay for both the solid particles and the spray is shown in Fig. 9. It can be drawn from this figure that a complete evaporation occurs at a distance of $z/D \simeq 25$ after which the flow behaves as that of a single-phase jet.

Figure 10 shows the radial profile of the vapor concentration normalized by the centerline value at $z/D = 10$ and 20. Due to the continuous air entrainment by the jet and the turbulent diffusion of the vapor, the concentration of the evaporating material in the carrier fluid at $z/D = 20$ is less than the corresponding values at $z/D = 10$, although most of the vaporization process takes place between the two stations ($C_{,c}$ at $z/D = 20$ is 14% than that at $z/D = 10$).

The rate of evaporation is a function of the transfer number which is almost constant over the jet cross section and the droplet Reynolds number is maximum in the outer region and minimum at the centerline. So the rate of evaporation is maximum in the outer region, hence the minimum droplet diameter. This explains the radial distribution of the droplet diameter at the various sections as shown in Fig. 11. Also displayed is the

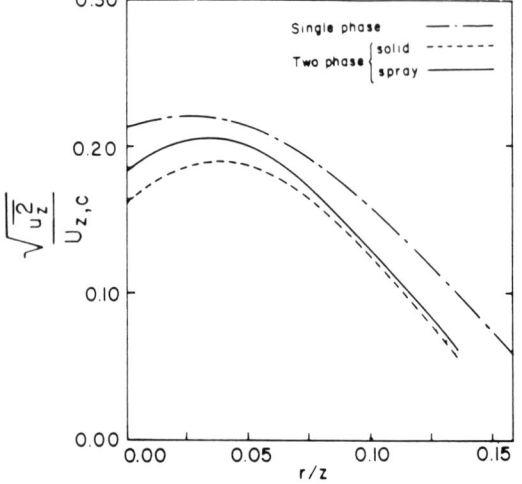

Fig. 7 Turbulence intensity at $z/D = 20$.

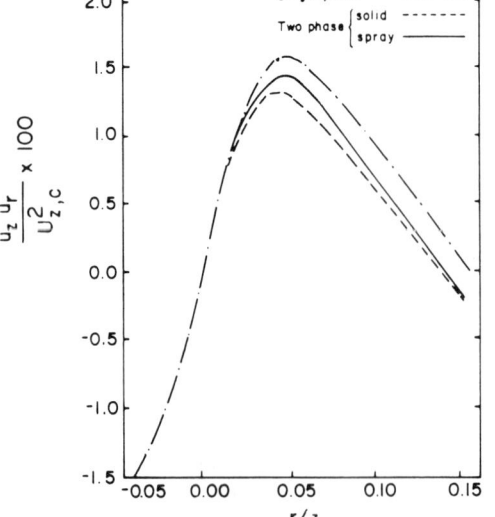

Fig. 8 Turbulent shear stress profile at $z/D = 20$.

monotonic reduction in droplet diameters with distance downstream accompanied by a narrowing of the extent of the region in which droplets exist until they eventually vanish.

Figure 12 shows the effect of the evaporating spray on the spreading rate of the jet by comparing the different $Y_{1/2} \sim z$ distributions, where $Y_{1/2}$ is the radius at which the fluid mean axial velocity is half that at the centerline. While for a turbulent single-phase jet the value of the slope ($dY_{1/2}/dz$) is constant ($\simeq 0.08$), that for a two-phase jet is a function of the dispersed phase properties such as droplet diameter, density and mass loading ratio. This dependence was discussed in the work of Elghobashi et al. (1983). In the developing region the spreading rate of the spray case is almost as that for the solid particles case with the same material density. As

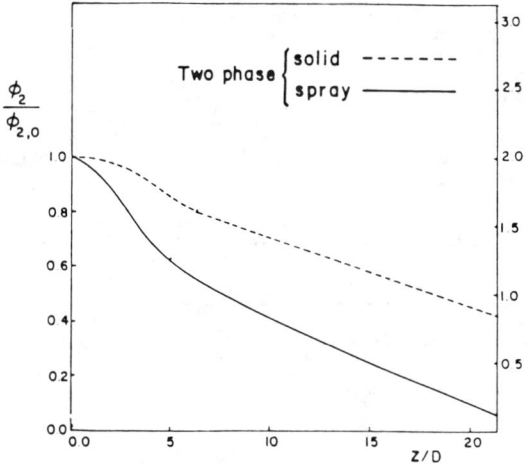

Fig. 9 Volume fraction centerline decay for the solid particles and spray.

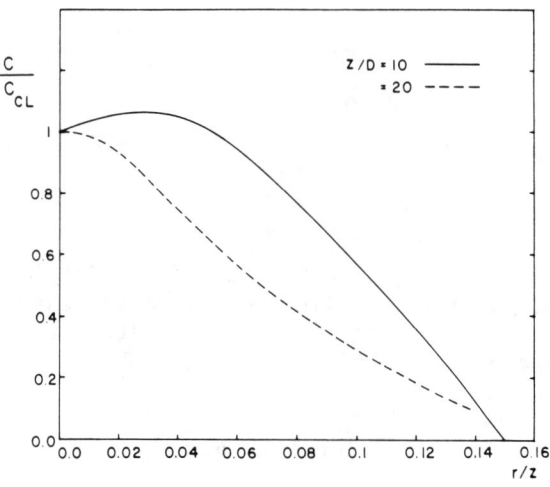

Fig. 10 Vapor concentration profile.

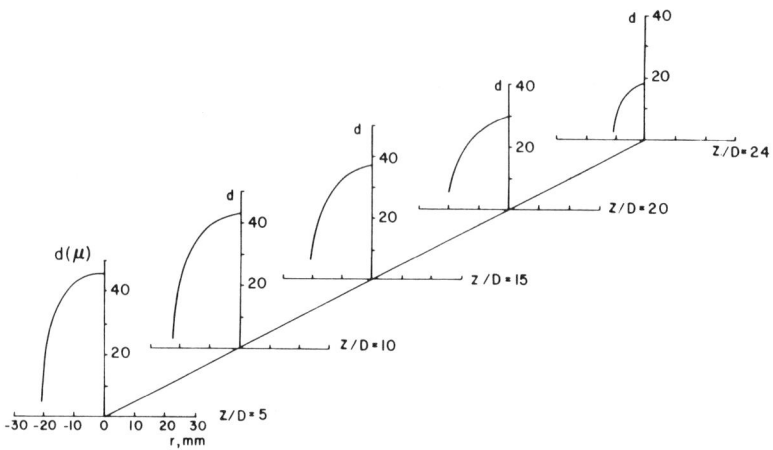

Fig. 11 Radial distributions of the local droplet diameters at different axial stations.

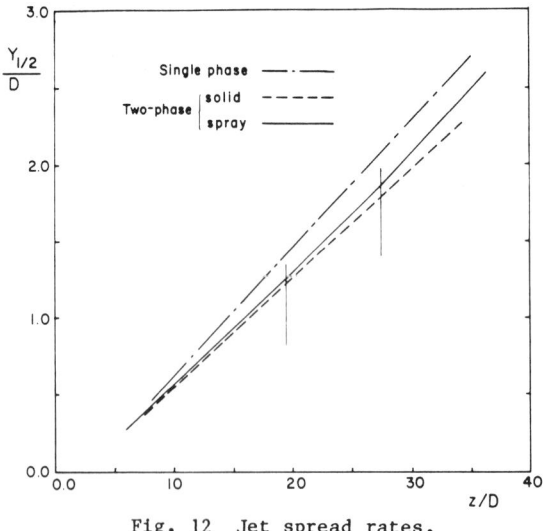

Fig. 12 Jet spread rates.

vaporization proceeds the effects of the droplets on the carrier fluid diminish allowing the fluid behavior to approach that of a single-phase jet. After the complete evaporation of the droplets, the jet behaves exactly as the single-phase one.

Concluding Remarks

A two-equation turbulence model for predicting two-phase flows with phase change is presented. The model is based on the exact transport equations of the turbulence kinetic energy and its dissipation rate derived from the carrier-phase momentum equations. The conservation equations of mass, momentum of each phase, and concentration of vapor are solved to predict the flow of a turbulent jet laden with vaporizing droplets as a test for the mathematical model.

The predictions show significant reductions in the turbulent shear stress and the kinetic energy of turbulence of the carrier phase due to the presence of droplets in the jet. In the developing region the values of those two quantities are nearly the same for both the droplet-laden and solid-laden jets. Due to the continuous vanishing of the dispersed phase by evaporation, the momentum exchange forces between the two-phases become less than that in the case of the solid-laden jet. Thus, the difference between the centerline mean axial velocity of the carrier phase and that of the single phase is less pronounced in the spray-laden jet. The radial distribution of the droplet diameter shows a continuous reduction accompanied by narrowing the radial extent of the dispersed phase. The spreading rate is significantly affected by the evaporation rate. Closer to the pipe exit the spreading rate is almost the same as that of the jet laden with solid particles. Farther downstream, it asymptotically approaches that of a single-phase jet.

More validation testing of the model is needed via well-defined experiments.

Acknowledgments

This work has been supported by NASA Lewis Research Center under Grant No. NAG-3-176. The authors express their thanks to Mrs. Verna Bruce for her expert typing of the manuscript.

References

Abramovich, G. N. (1970) The effect of admixture of solid particles or droplets, on the structure of a turbulent gas jet. Dokl. Akad. Nauk. SSSR (Sov. Phys. Dokl.) 15, 101.

Chao, B. T. (1964) Turbulent transport behavior of small particles in dilute suspension. Osterr. Ing. Arch. 18, 7.

Clift, R., Grace, J. R., and Weber, M. E. (1978) Bubbles, Drops, and Particles. pp. 111-133 Academic Press, New York. pp. 111-133.

Crowe, C. T. (1980) On the dispersed flow equations. Two-Phase Momentum, Heat and Mass Transfer, (edited by T. Durst and Afgan), Vol. 1, 25, London.

Daly, B. J. and Harlow, F. H. (1970) Transport theory of turbulence, University of California, Report AL-DC-11207.

Danon, H., Wolfshtein, M., and Hetsroni, G., (1977) Numerical calculations of two-phase turbulent round jet, Int. J. Multiphase Flow 3, 223.

Elghobashi, S. E. and Abou-Arab, T. W. (1983) A two-equation turbulence model for two-phase flows. Phys. Fluids 25, 931.

Elghobashi, S., Abou-Arab, T., Rizk, M., and Mostafa, A. (1983) A two-equation turbulence model for two-phase jets. Accepted for publication in the 4th Symposium on Turbulent Shear Flows.

Elghobashi, S. E. and Megahed, I. E. A. (1981) Mass and momentum transport in a laminar isothermal two-phase round jet. Numerical Heat Transfer 4, 317.

Genchev, ZH. D. and Karpuzou, D. S. (1980) Effects of the motion of dust particles on turbulence transport. J. Fluid Mech. 101, 833.

Hanjalic, K. and Launder, B. E. (1972) A Reynolds stress model of turbulence and its application to thin shear flows. J. Fluid Mech. 52, 60.

Ishii, M. (1975) Thermo-Fluid Dynamic Theory in Two-Phase Flow. Eyrolles, Paris.

Labowski, M. and Rosner, D. E. (1976) Conditions for "Group" combustion of droplets in fuel clouds: 1. quasi-steady predictions. Symposium on Evaporation - Combustion of Fluid Droplets Division of Petroleum Chemistry, Am. Ch. Soc.

Launder, B. E. (1976) Turbulence, (edited by Bradshaw), Chap. 6, Springer-Verlag, Berlin.

Launder, B. E., Reece, G. J. and Rodi, W. (1975) Progress in the development of a Reynolds-stress turbulence closure. J. Fluid Mech. 68, 537.

Lumley, J. L. (1978) Turbulent transport of passive contaminants and particles: Fundamentals and advanced methods of numerical modeling. Lecture Series at the Von Karman Institute for Fluid Dynamics, Rhode-St-Genese, Belgium.

Modarress, D., Tan, H. and Elghobashi, S. (1983) Two-component LDA measurement in a two-phase turbulent jet. AIAA Paper 83-0052,

Mostafa, A. A. (1984) A Two-Equation Turbulence Model for Dilute Vaporizing Sprays Ph.D. Thesis, University of California, Irvine.

Owen, P. R. (1969) Pneumatic transport. J. Fluid Mech. 39, 407.

Ranz, W. E. and Marshall, W. R. (1952) Evaporation from drops.
   Chem. Eng. Prog. 48(3), 141, 173.

Saffman, P. G. (1969) On the stability of laminar flow of a dusty
   gas. J. Fluid Mech. 39, 407.

Sirignano, W. A. (1983) Fuel droplet vaporization and spray
   combustion theory. (to be submitted to Energy Combust. Sci.)

Soo, S. L. (1967) Fluid Dynamics of Multiphase Systems.
   Blaisdell, Walthan, Mass.

Spalding, D. B. (1979) Numerical computation of multi-phase
   flows, Lecture Notes, Thermal Sciences and Propulsion Centre,
   Purdue University, West Lafayette, In.

Tennekes, H. and Lumley, J. L. (1972) A First Course in
   Turbulence. MIT, Cambridge, Mass.

Yule, A. J., Seng, C. Ah., Felton, P. G., and Chigier, N. A.
   (1982) A study of vaporizing fuel sprays by laser
   techniques. Combustion and Flame 44, 71.

Simulations of Two-Dimensional Fuel Droplet Flows

M. J. Fritts,* D. E. Fyfe,* and E. S. Oran*
Naval Research Laboratory, Washington, D.C.

Abstract

In this paper, two-dimensional Cartesian calculations of internal and external flows for oscillating and deforming kerosene droplets in an air jet are presented. The simulations are performed by a Lagrangian technique which uses a general restructuring mesh. This technique is accurate at material interfaces even though the interfaces undergo convolutions and may evolve into multiconnected surfaces. The numerical method is based on a triangular grid that automatically reconnects triangle vertices to eliminate grid distortions. In the calculations presented, the surrounding air moves relative to an initially cylindrical kerosene droplet. Surface tension is included as a jump in pressure across an interface by casting the surface tension force in the form of a gradient of a potential. The surface tension algorithm is tested by studying the oscillatory behavior of an n = 2 normal mode. The viscosity algorithm for a general mesh is presented and tested through the calculation of the spreading of a viscous shear layer. Results of the air-kerosene calculation are illustrated by sequences of frames from a computer-generated movie of fluid particle positions for both the internal and external flows. Extension of these results to a compressible, fully reactive calculation is discussed, together with the algorithms necessary to extend the method to three dimensions.

Presented at the 9th ICODERS, Poiters, France, July 3-8, 1983. This paper is declared a work of the U.S. Government and therefore is in the public domain.

*Research Scientist, Laboratory for Computational Physics.

Introduction

Particularly severe physical and mathematical approximations are normally made to describe the detailed interactions between droplets and the external flowfield in spray combustion models (Williams 1973; Faeth 1977, 1983). For example, equivalent spheres are used to approximate droplet deformations, and empirical expressions are used to account for drag and convection. The effects of droplet breakup are included by using estimated breakup times and drop sizes after breakup. When quasisteady flow approximations are used, the changes in the flowfield due to droplet deformations, wake effects, and interactions of the flowfield back onto the droplet are neglected. Finally, in most models, the droplet concentration is assumed to be dilute, since little is known about droplet-droplet or droplet-wake interactions.

This paper describes applications of a numerical technique to the study of the individual and collective behavior of burning two-dimensional droplets. The calculations are performed using the fully conservative, two-dimensional Lagrangian finite-difference method developed by Fritts and Boris (1979), which is based on a dynamically restructuring Cartesian triangular grid. A triangular grid avoids the problems of mesh tangling encountered in Lagrangian methods using a quadrilateral mesh. Individual mesh points are continually reconnected to account for the migration of fluid elements in the flowfield. Since the number of grid lines meeting at a vertex is variable, the resolution can be altered nondiffusively where needed (e.g., around a region of droplet distortion) without affecting the resolution in other areas of the computation. This is a major step forward in the computation of droplet flows, because the Lagrangian technique allows for the evolution of the grid to multiply connected regions. Thus, according to the flow conditions, droplets can break up and shatter.

In the past, the Lagrangian restructuring triangular grid technique has been applied to a number of incompressible fluid flow problems, including calculations of flows over solid obstacles, Kelvin-Helmholtz and Rayleigh-Taylor instabilities, Couette flows and Taylor vortex flows. To extend the method to calculate the properties of burning droplets, new algorithms representing additional physical processes must be incorporated. These include, for example, surface tension, viscosity, and evaporation . In addition, the hydrodynamics algorithm should be applicable to compressible flows in order to describe the effects of heat release due to chemical interactions (Oran and Boris 1981). This paper deals primarily with the development of these new algorithms and shows applications using the new surface tension and viscosity terms.

Lagrangian Hydrodynamics on a Triangular Grid

In principle, a Lagrangian formulation of the hydrodynamic equations is particularly attractive for droplet combustion calculations. Each fluid element is tracked as it evolves through the interaction with its changing environment and with external forces. Heat release, contaminant reactions, and soot formation can all be represented locally, without resort to global models and without nonphysical diffusion. Conservation laws are simple to express, since there are no advective fluxes out of the fluid element boundaries, and the paths of the fluid elements themselves provide flow visualization. However, in all but the simplest flows, the individual fluid element must deform, and these deformations are a severe hindrance to actually using a Lagrangian formulation.

In numerical calculations, fluid element distortion appears as stretching, shearing, and eventual tangling of the computational grid. Although the use of a general-connectivity triangular mesh eliminates tangling, the accuracy of a calculation may still deteriorate when there are abrupt local changes in resolution and when the high order effects of deformations are not represented. Therefore, it is very important to pay close attention to how well conservation laws are satisfied. For example, the accuracy of the finite-difference algorithms for a general mesh may not be sufficient to conserve quantities advected with the fluid elements if some flux is allowed to flow out of elements to maintain straight lines in the computational grid. In the following section it is shown how exact conservation may be maintained.

The divergence and curl of the velocity field prescribe the kinetics of the field by specifying the local rate of expansion of the fluid d and local vorticity $\bar{\xi}$ by

$$\nabla \cdot \bar{v} = d$$

$$\nabla \times \bar{v} = \bar{\xi} \tag{1}$$

For incompressible flow, d = 0, and for irrotational flow, $\bar{\xi} = 0$. For incompressible and irrotational flow in two dimensions, the velocity field is specified by a velocity potential ϕ and stream function ψ:

$$v_x = \frac{\partial \phi}{\partial x} = \frac{\partial \psi}{\partial y}$$
$$v_y = \frac{\partial \phi}{\partial y} = -\frac{\partial \psi}{\partial x} \tag{2}$$

These equations automatically satisfy the conservation of vorticity and mass, since

$$\nabla \cdot \bar{v} = \nabla \cdot (\nabla \times k \psi) \equiv 0$$

$$\nabla \times \bar{v} = \nabla \times \nabla \phi \equiv 0 \tag{3}$$

where k is the unit vector in the neglected z direction. Finite-difference operators are defined for divergence, curl, and gradient that have these identical properties, and this requirement restricts the placement of variables. In particular, if ψ and ϕ are to be assigned to the Lagrangian verticies, the velocities $\bar{v}$ must be specified at the centroids of triangles or the midpoints of line segments. Therefore, the Lagrangian vertex velocities must be obtained by local averages.

With this placement of variables, the dynamics of the flow, as well as the kinematics, behave properly. That is,

$$\nabla \cdot \bar{v} = \nabla \cdot \nabla \phi = \nabla^2 \phi \tag{4}$$

is a general triangular grid Poisson equation that may be used to solve for the local pressure. At the same time, the pressures generated by forcing all local divergences to zero cannot by themselves alter the local vorticities, due to Eq. (3).

The accuracy of the numerical algorithms is determined by both the local resolution and connectivity of the grid. For the solution used here, the local connectivity of the grid and the resolution are both determined in part by the requirement that the matrix generated from the Poisson equation, Eq. (4), remains diagonally dominant. With this restriction, convergence of an iterative solver is assured.

In summary, the basic two-dimensional hydrodynamics computer code is constructed in such a way that the finite-difference operators for divergence, curl, and gradient exactly reflect the properties of the continuum operators. This construction assures conservation of vorticity and mass and provides a basis for determining the local grid connectivity. The extensions to this code described below are all made in exactly the same spirit. The finite-difference approximations to both physical and dynamical processes conform to the continuum limit and conserve properly.

Surface tension forces can be included as external forces on interface vertices to accelerate those vertices. With such an algorithm there is no guarantee that the pressure force, as formulated, will provide the correct pressure jump across the interface or that the surface tension forces will not generate unphysical changes in vorticity. However, if the surface tension force is formulated as a gradient of a potential present only at the surfaces, then both the "surface tension potential" and the pressure are dynamically similar, and the physical pressure drop across the interface must exactly cancel the surface tension forces.

The finite-difference algorithms for surface tension are therefore quite simple in form. The surface tension forces are included through Laplace's formula for the jump in pressure across an interface (Landau and Lifshitz 1975):

$$P_i - P_o = \sigma/R \qquad (5)$$

where P_i is the pressure just inside the droplet at the interface; P_o is the pressure just outside the droplet at the interface; σ is the surface tension coefficient associated with the two media that define the interface; and R is the radius of curvature of the cylindrical droplet calculated from a parametric cubic spline fit to the interface vertices. These pressure jumps are included in the Poisson equation for the pressure. The average pressure, $(P_i + P_o)/2$, is computed at an interface vertex. From the average pressure and the pressure jump, a triangle-centered pressure gradient is computed for inclusion in the momentum equation.

The parametric spline fit is also used for regridding. When the regridding algorithm calls for the bisection of a triangle side that borders the two media, a new vertex is added on the spline interpolant between the indicated vertices rather than bisecting the straight line segment. A straight line bisection introduces spurious interface oscillations (Foote 1973), whereas bisecting the spline maintains the general overall shape of the interface.

The effects of viscosity enter into the two-dimensional equations of motion through the additional terms

$$\frac{dV_x}{dt} = \ldots + \nabla \cdot \nu \nabla V_x$$

$$\frac{dV_y}{dt} = \ldots + \nabla \cdot \nu \nabla V_y \qquad (6)$$

Since velocities are triangle-centered quantities, the gradients used in evaluating this expression are vertex centered instead of the usual triangle-centered gradients used elsewhere in this formulation. Therefore, the viscosities at interfaces must be averages in order to evaluate the triangle-centered divergence needed to advance the velocities.

The implementation of Eq. (6) as averages ensures that the gross vorticity generation term will be correctly calculated for shear flows. As shown below, if sufficient resolution is provided by the grid, the algorithm will produce proper velocity profiles at the correct times in the evolution of the shear layer. If there is not enough resolution, the details of the profile will be lost, and the width of the layer will be overestimated. However, the average vorticity generated at any point will still be correct, because the total gradients of the velocity components are accounted for in the source terms in Eq. (6). For calculations of droplet boundary layers having insufficient grid points to resolve the boundary layer, the average vorticity shed to the wake is correct for monotonic velocity profiles, but the wake is too diffuse. Similarly, the internal recirculation within the droplet is correct on average. Any study of the details of the velocity profiles near the droplet interface would require sufficient resolution to resolve those profiles. In the calculation discussed below, for example, the internal recirculation patterns within the drop are clearly evident even though relatively few vertices are used to grid the interior of the droplet.

Numerical Results

Preliminary Tests of Algorithms

Because the differencing techniques used to calculate the viscous and surface tension terms were previously untested, it was necessary to determine the accuracy of the algorithms by comparing numerical results to analytically soluble problems. To test the viscosity algorithm, the spreading of a viscous shear layer of initially zero thickness was examined. The evolution of the velocity distribution across the layer and the growth of the width of the layer are known quantities to be compared to the results of calculations. The calculated layer width was found to agree exactly with the theory, and the shear layer velocities were correct over the entire mesh. The components of the velocity perpendicular to the shear layer remained zero, indicating that the algorithm works well even as the grid distorts.

To test the algorithm for surface tension, calculations of droplets that oscillate under the effect of surface tension were performed. A linear theory for small-amplitude oscillations on cylindrical jets was first given by Rayleigh (1879). When a perturbation is totally in the plane perpendicular to the axis of the cylinder, Rayleigh found the frequency ω for the oscillations is given by

$$\omega^2 = (n^3 - n)\sigma/(\rho a^3) \qquad (7)$$

where the surface of the jet is given in polar coordinates by

$$r = a + \varepsilon \cos(n\theta) \qquad (8)$$

From Eq. (7), it is clear that the lowest oscillating mode is given by n = 2.

In the numerical calculation presented, an n = 2 oscillation was studied with the parameters a = 0.0125 cm and σ = 30 dynes/cm (values that are typical for droplet combustion problems) and with a droplet density of 2 g/cm^3 and an external fluid density of 1 g/cm^3. The results of a calculation with ε = 0.2a = 0.0025 cm are shown in Fig. 1. The numerical oscillation period is approximately 1.25×10^{-3} s. To compare this result with Rayleigh's theory, his result must be first corrected for the effect of the presence of the external fluid. Equation (7) then becomes

$$\omega^2 = (n^3 - n)\sigma/((\rho_d + \rho_e)a^3) \qquad (9)$$

where ρ_d is the droplet density, and ρ_e is the density of the external fluid. With the period defined as $2\pi/\omega$, Eq. (9) gives a period of 1.13×10^{-3} s. The dis-

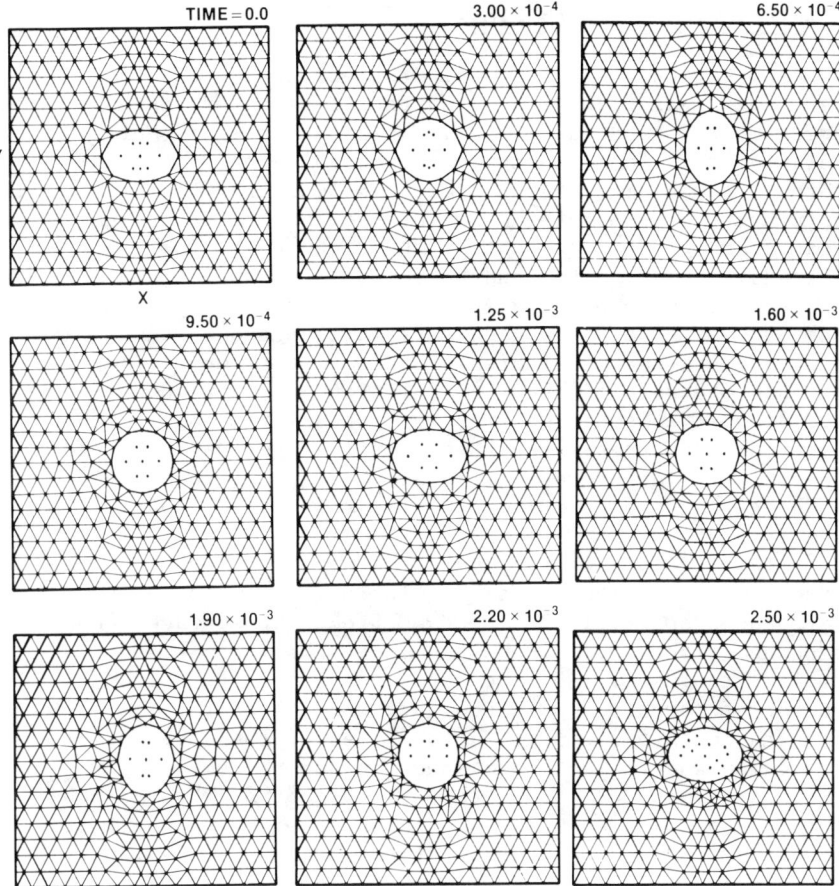

Fig. 1 Computer-generated frames from a Lagrangian simulation of an oscillating droplet for two periods of an n = 2 normal mode.

crepancy between the numerical and theoretical results can be explained by the finite grid spacing. For large-amplitude oscillations Rayleigh found that the experimental frequency diverges from that predicted by the linear theory, and it is reasonable to expect the computational period to differ somewhat from that given by the linear theory. Currently, further calculations are being performed with smaller amplitudes, ε, to see if any of the difference is attributable to nonlinear perturbation effects or if the linear theory is directly applicable in this regime.

Simulations of Droplet Flows

In previous calculations, droplets were accelerated by an external flow with no surface tension or viscosity present (Fritts et al. 1982). In such a case a recirculation zone forms early in the calculation in the external flow behind the droplet. The internal flow is driven by the compression set up between the front and rear stagnation points and by the high shear flow that extends around the top and bottom of the droplet and recirculation zone. The interaction of the droplet back onto the external flow occurs primarily through the enlarged cross-sectional area of the droplet, which increases the size of the recirculation zone. Eventually the droplet is squeezed into a thinned layer coating the recirculation zone. The thinned film then shatters into several smaller pieces, first at the rear of the droplet and later in the more laminar flow toward the front of the droplet.

Figure 2 follows the evolution of the internal and external flowfields for a calculation of airflow past a kerosene droplet, including the effects of surface tension and viscosity. The air velocity is 120 m/s and the droplet radius is 125 μ, corresponding to a Reynolds number of roughly 2000. Boundary conditions for the computation are periodic on the sides of the computational region and reflective at the top and bottom. The first clear indication of the development of the recirculation region is seen in the fourth insert. In the fifth insert, a pair of counter-rotating vortices are evident. The recirculation zone continues to develop throughout the calculation, although at times the vortex pair is not as evident because of the deletion and addition of vertices that interrupt the continuity of the pathlines. By the last insert, it appears that another pair of vortices is forming near the droplet, indicating that the original pair may be shed. There is now a large distortion of the leading face of the droplet, and the droplet is about to enter the wake of the preceding droplet in the periodic array. The internal velocities are small compared to the external flow rates and therefore cannot be distinguished as pathlines. However, an indication of the internal recirculation may be obtained by comparing the internal mesh point positions at various timesteps. It is also evident that vertex addition has occurred primarily where needed, in the developing wake of the droplet and all along the droplet interface.

The final frame shows the droplet as a whole traveling at roughly 30 cm/s, while the flow at the top and

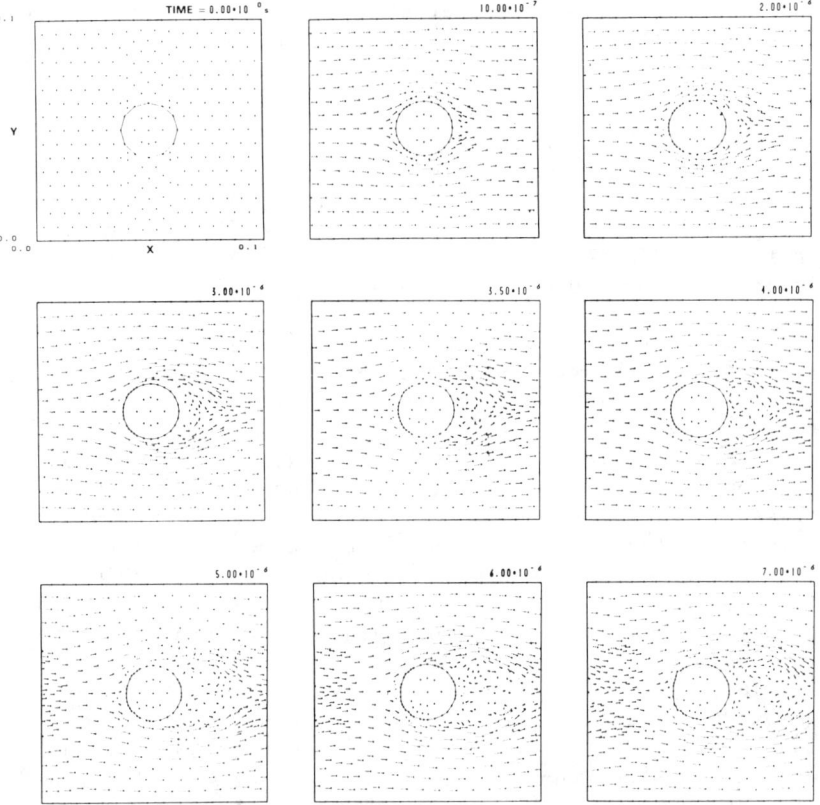

Fig. 2 Pathlines of the fluid flow from a computer-generated movie of incompressible airflow past a deforming kerosene droplet. Heads of the pathlines are the current vertex positions, and the tails are made up of the previous five positions.

bottom of the computational region is 300 times faster. Because the droplet itself has not yet reached terminal velocity in the flow, a calculation of the Weber number based on the droplet velocity results in We = 0.01, an underestimate of the effect of the external flow velocity. However, the external flow velocity cannot be used instead, since the calculation is initialized to potential flow rather than an air blast. In the last frame, the droplet is entering the wake of the preceding droplet, and the deformations at the front surface of the droplet are caused by local shear stress induced by the approaching wake. The major deformations seen at this point in the calculation are therefore not representative of the global flow characteristics, but are caused by the local flowfield in front of the droplet.

Future Developments

The next step in the construction of the droplet combustion model is to include an algorithm for compressibility. The addition of compressibility will occur in two ways depending on the characteristic flow velocities in the calculations. When flow speeds are slow with respect to the sound speed, we do not want the timestep to be limited by the Courant condition. In such cases, the sound waves can be filtered out of the solution by altering the pressure field to account for local divergences on the time scale of the fluid flow (Jones and Boris 1979). These divergences, which arise, for example, from heat release, are introduced into the pressure Poisson equation in a manner similar to that for incompressible flow. However, there is a restriction that the relaxation times occur at the proper time scale. For the triangular mesh, such additions should be easy to implement, since a divergence correction term is already used to account for the effects of maintaining straight triangle sides.

In the case for which sound waves must be included, an energy evolution equation and an equation of state must be included in the finite difference-algorithms as well. The algorithm that will be used for the equation of state expresses the density as a function of the pressure and energy. Given a new internal energy derived from the energy evolution equation and an approximation to the pressure, density is calculated from the equation of state. This equation-of-state density is compared to the density derived from the fluid dynamics, and the difference is iterated to zero. This solution is the inverse of the usual algorithms for the equation of state, which express pressure as a function of density and energy. The method has been tested extensively for a one-dimensional restructuring mesh in the code ADINC (Boris 1979; Fritts et al. 1981). The technique works well in one dimension and exhibits diminished finite-difference error propagation due to the fact that numerical errors in pressure result in small density fluctuations. In the usual technique, small density errors can give rise to large pressure fluctuations, and, hence, a larger error propagation.

An energy evolution equation,

$$\frac{dE}{dt} = - E \, \nabla \cdot \bar{v} - \nabla \cdot (P\bar{v}) + \nabla \cdot \lambda \nabla T \qquad (10)$$

will also be needed to account for the effects of thermal
conductivity, represented by the last term in Eq. (10).
If both energy and temperature are carried as vertex-
centered quantities, then there are no new techniques
required for the first and third terms, since the finite-
difference approximations for similar terms are well
tested. The center term must be carried as an average,
since pressures are vertex centered while velocities are
triangle centered. The incorporation of reactions is
rather straightforward and will follow previously tested
techniques given by Oran and Boris (1981).

The three-dimensional analog of a triangular grid is
a tetrahedral grid in which surfaces are tessellated by
triangles. Although the addition of one more dimension
introduces new complications in the reconnection algorithms, much of what was learned from the two-dimensional
case carries over intact into three dimensions. Vertices
can still be deleted by successive reconnections to isolate a vertex within a single tetrahedron. At that
point, the vertex and its four lines can be eliminated
from the grid. Vertices can be added within tetrahedra,
in the plane of a triangle and on lines without major
modification of the techniques used in two dimensions.
The conservation criteria used for reconnecting, adding,
and deleting cells in three dimensions usually involve
either extending integrals to one higher dimension, or
measuring an angle between planes rather than lines.
Similarly, the hydrodynamics finite-difference algorithms
are logical extensions of the two-dimensional algorithms.
The use of primitive variables allows a simple extension
for the vorticity integrals, and the solution of
Poisson's equation still requires just one matrix
inversion.

Although the hydrodynamics and physical conservation
algorithms can be efficiently extended to three dimensions, additional bookkeeping arrays are needed to define
the connectivity of the tetrahedra. This increases both
the computer storage requirement and the execution time.
The two-dimensional code now runs on a vector processor,
the Texas Instruments ASC, with execution speeds of about
0.01 s per timestep per grid point, including program
diagnostics and output in the form of two three-color
movie films, individual frames on fiche, and fiche list-
ings. Up to 2000 grid points can be accommodated with no
paging and with no virtual memory. A three-dimensional
code configured in the same manner and with the same
total number of grid points as the two-dimensional code
would use at least 50% more memory and take twice as long

to execute, with most of the increase arising from diagnostic output. New solution algorithms are being investigated with smaller overhead for the three-dimensional version.

Conclusions

This paper reports calculations of the flows in and about dense fuel droplets. These preliminary calculations include the effects of convection, surface tension, and viscosity. The new algorithms for surface tension and viscosity that have been implemented and tested have the right dynamic properties and appear to work well in the Lagrangian triangular grid method. The reconnecting mesh has been shown to be capable of accurately tracking interfaces despite transition to multiply connected flows and is appropriate to the tracking of flame fronts as well. Future work will concentrate on extending these results to fully reactive, compressible flows in two and three dimensions.

Acknowledgments

This work has been sponsored by the National Aeronautics and Space Administration, the Naval Material Command, the Naval Research Laboratory, and the Office of Naval Research.

References

Boris, J. P. (1979) ADINC: An implicit Lagrangian hydrodynamics code. NRL Memorandum Report 4022, Naval Research Laboratory, Washington, D. C.

Faeth, G. M. (1977) Current status of droplet and liquid combustion. Prog. Energy Combust. Sci. 3, 191-243.

Faeth, G. M. (1983) Evaporation and combustion of sprays. Prog. Energy Combust. Sci. 9, 1-76.

Foote, G. B. (1973) A numerical method for studying liquid drop behavior: Simple oscillations. J. Comput. Phys. 11, 507-530.

Fritts, M. J. and Boris, J. P. (1979) The Lagrangian solution of transient problems in hydrodynamics using a triangular grid. J. Comput. Phys. 31, 173- 215.

Fritts, M. J., Oran, E. S., and Boris, J. P. (1981) Lagrangian fluid dynamics for combustion modeling. NRL Memorandum Report 4570, Naval Research Laboratory, Washington, D.C.

Fritts, M. J., Fyfe, D. E., and Oran, E. S. (1982) Numerical simulations of droplet flows with surface tension. ASME paper 82-WA/HT-17.

Jones, W. W. and Boris, J. P. (1979) Flame, a slow-flow combustion model. NRL Memorandum Report 3970, Naval Research Laboratory, Washington, D.C.

Landau, L. D. and Lifshitz, L. D. (1975) Fluid Mechanics, pp. 230-234. Pergamon Press, New York.

Oran, E. S. and Boris, J. P. (1981) Detailed modeling of combustion systems. Prog. Energy Combust. Sci. 7, 1-71.

Rayleigh, Lord (1879) On the capillary phenomena of jets. Proc. R. Soc. London 29, 71-97.

Williams, A. (1973) Combustion of droplets of liquid fuels. Combust. Flame 21, 1-31.

Induction Time Measurements for Ignition of Liquid Fuel Jets in Air at High Temperatures and Pressures

V. K. Baev,* A. N. Bazhaikin,† A. A. Buzukov,‡ and B. P. Timoshenko†
University of Novosibirsk, Academy of Sciences, Novosibirsk, USSR

Abstract

The paper reports on experiments conducted to measure the induction time of gasoline and diesel oil sprays injected into a reservoir containing air at high temperature and pressure. The results are presented in terms of Arrhenius relations, covering a range of temperatures from 750 to 1000 K and pressures from 16 up to 75 atm. In particular, they reveal the existence of a critical temperature at which the ignition mechanism undergoes a drastic change. The values of the effective activation energy we obtained suggest that the induction time at such conditions is not so much governed by the chemical reaction of the fuel-air mixture as by the fluid mechanical properties of the jet: the dynamic factors influencing the development of the concentration field and the dispersion fuel, as well as the initial temperature distribution in the mixing zone. As a consequence, the activation energy determined from the measurement of the induction time cannot be considered as an expression for the physicochemical barrier in its usual sense.

Introduction

Sprays of liquid fuels injected into preheated air, as used, for example, in diesel engines, represent a

---

Presented at the 9th ICODERS, Poitiers, France, July 3-8, 1983. Copyright © 1984 American Institute of Aeronautics and Astronautics, Inc. All rights reserved.
*Deputy Director, Professor, Department of Gas Dynamics of Flame and Thermal Processes.
†Research Engineer, Laboratory of Gas Dynamics of Flame.
‡Senior Scientist, Laboratory of Gas Dynamics of Flame.

well-established means in combustion technology. In spite of that, there is no theoretical model available yet to take into account the complex physicochemical processes occurring in the course of ignition. Experimental studies (Mullins 1953; De Zubay 1978) have been concerned primarily with the measurement of what is usually referred to as the ignition delay but should be more properly called the induction time. The results are usually expressed in terms of a great variety of empirical formulas, of which the most popular (Todes, 1933) is the classical Arrhenius relation:

$$\tau \approx P^{1-\nu} \exp(E/RT) \qquad (1)$$

where τ is the induction time; P and T the pressure and temperature, respectively; ν the overall reaction order; R the universal gas constant; and E the effective activation energy.

It should be expected that values of E for a hydrocarbon fuel should not differ appreciably from the level of the chemical kinetic barrier of the reaction occurring in the fuel-air mixture during ignition, which is normally governed by some specific limiting elementary step. However, the activation energy evaluated from experimental measurements of the induction time for ignition of various commonly used fuels ranges from 2 to 50 kcal/mole (Semenov 1970; Mullaney 1959). Considerable deviations in the values of E have been also obtained for these fuels when the experiments were carried out in different installations. Thus, the widespread notion that the effective activation energy expresses the level of a chemical barrier is not applicable in this case.

This work reports experimental measurements of the induction time for ignition of gasoline and diesel oil sprays injected into a reservoir where conditions typical of the combustion chamber in a diesel engine are maintained. The study produced further insights concerning the dependence of the induction time on the temperature and pressure of the air, as well as revealing some interesting peculiarities of the process.

Experiments

The apparatus used for the experiments has been described by Baev et al. (1979). Ignition takes place in an airtight cylindrical chamber that is 150 mm in internal diameter, and whose endplates, fitted with 45-mm-thick quartz windows, are also 150 mm apart. Before the fuel is injected, the chamber was filled with compressed air (up to 80 atm) and heated up to 1000 K. Injection was accomplished with a closed-type spray nozzle that injected approximately 27 mg of gasoline or 46 mg of diesel oil in about 3 to 4 ms. The reservoir and the fuel system were provided with pressure and temperature transducers. A high-speed camera was also used for visual diagnostics of the process.

Experimental results demonstrated that for gasoline there exists a critical temperature, depending on initial pressure in the reservoir, where a distinct change in the self-ignition mechanism takes place. At lower temperatures no effects indicative of the onset of an exothermic process are observable (line 1 in Fig. 1). Close to the critical temperature, the process is influenced by many factors, such as the rate at which the reservoir is pressurized, the presence of negligible amounts of reaction products, and injection instabilities.

When the temperature is increased above the critical level by 10 to 30°C, the exothermic process acquires a two-stage character, as depicted in Fig. 2. Figure 2 represents a series of oscilloscope records: The upper trace displaying the pressure in the reservoir, while the lower displays the pressure in the fuel line. The two-stage behavior is observed for gasoline only within a narrow range of temperatures, while the pressure does not

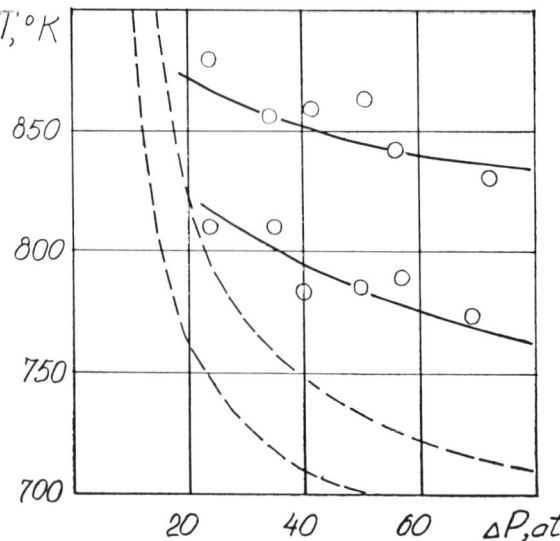

Fig.1 Dependence of critical temperatures on pressure.

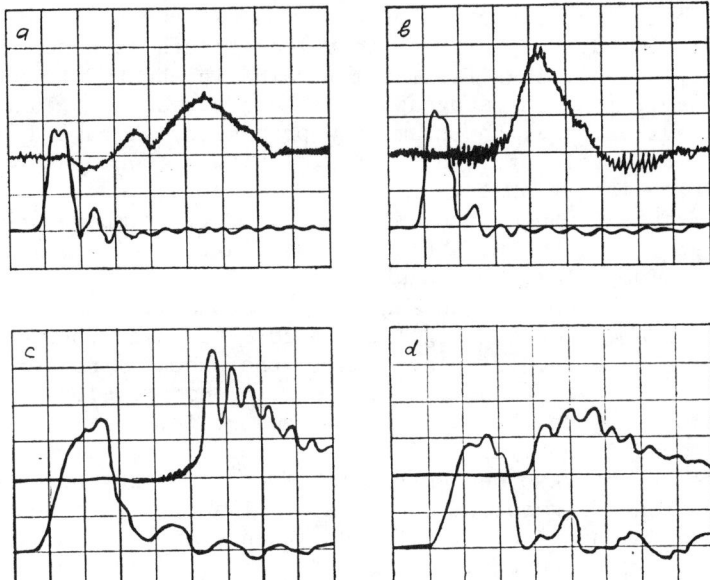

Fig.2 Experimental oscilloscope records of transient pressures in the reservoir (upper traces) and in the fuel line (lower traces):

Fig. No.	ΔP, atm	T, K	Sweep rate, ms/division	Pressure in reservoir, atm/division	Pressure in fuel line, atm/division
2^a	16	840	4	140	0.2
2^b	16	920	4	140	0.4
2^c	32	860	1.6	140	0.4
2^d	80	850	1.6	140	0.6

exceed 40 atm (line 2 in Fig. 1). From high-speed cinematographic records, it has been found that the first pressure pulse is formed as a result of the onset of one or several ignition centers, while the flame has not been established. The second pulse is due to combustion of soot particles formed, as a rule, in the rear portion of the flame following evaporation and thermal fuel decomposition accompanied by the appearance of new ignition centers.

Within the above regime of initial conditions, induction times are quite long, extending to 20 ms. However, at the critical temperature, the character and duration of the induction process are drastically changed. Although, as before, a relatively small pressure pulse due to the initial formation of ignition centers is observed first, immediately after that the pressure increases sharply, achieving its maximum value in about 2 to 3 ms (Fig. 2b). The cinematographic records reveal that, unlike the previous case, the flame now propagates throughout the whole reservoir filled with fuel vapor, whereas soot formation is not observed.

With a further increase in the initial temperature of the air, another significant jump in the process takes place. The flame propagates along the jet at a high velocity, and combustion acquires a detonative character. Oscillogram records, Fig. 2c and 2d, taken in the course of this process, display clearly the shock character of pressure oscillations associated with compression waves bouncing between the walls of the reservoir.

Careful analysis of these records revealed the existence of another critical temperature beyond which combustion of gasoline spray in air leads to detonation. Figures 3 and 4 display the dependence of the induction time on the temperature at different pressures for

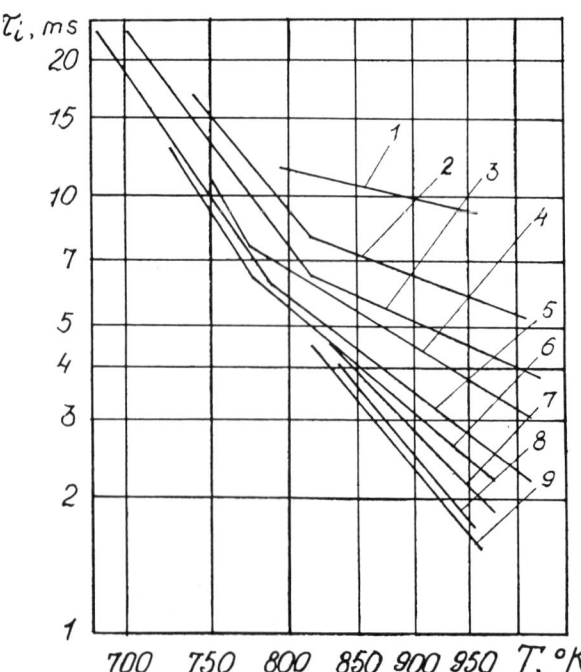

Fig.3 Arrhenius plots of data obtained for gasoline at reservoir pressures of 16, 24, 32, 40, 48, 56, 64, 72, and 80 atm, denoted by numbers 1-9, respectively.

gasoline and diesel oil, respectively. Lines 1-9 in Fig. 3 correspond to gage pressures in the reservoir of 16, 24, 32, 40, 48, 56, 64, 72, and 80 atm, respectively, while lines 1-6 in Fig. 4 correspond to gage pressures of 16, 24, 32, 40, 48, and 56 atm, respectively. The scale of ordinates in these graphs is logarithmic, while that of abscissas is proportional to the inverse of the temperature.

Interpretation

In general, the exact meaning of the induction time, or the ignition delay period, under the specific conditions of a particular experimental study, is not strictly defined. Usually one adopts for this purpose a time interval determined as follows: Immediately after the start of fuel injection into a reservoir, the pressure decreases as a consequence of evaporation. After the onset of the exothermic process, it increases, reaching back to, after a certain amount of time, the initial value before

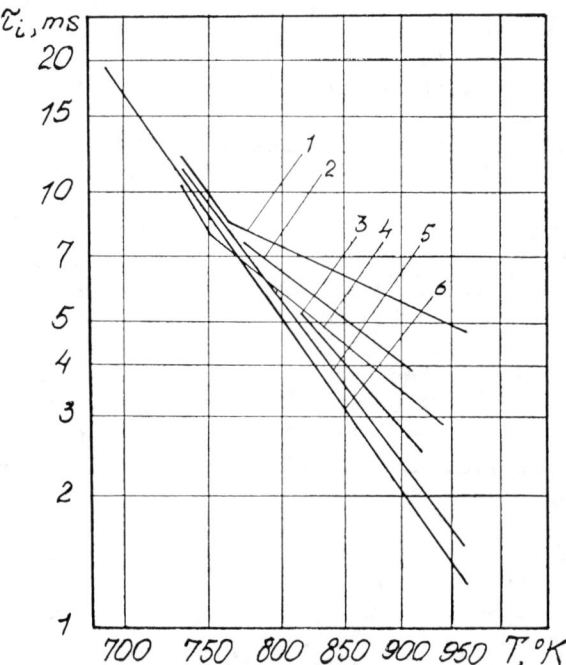

Fig.4 Arrhenius plots of data obtained for diesel oil at reservoir pressures of 16, 32, 40, 48, and 56 atm, denoted by numbers 1-6, respectively.

rising further. The moment of pressure recovery up to the initial value is noted as the end of the induction time (Alekseev 1974). In a number of studies, this period was measured by the use of a photo cell recording the start of glow (Voinov et al. 1964). However, Baturin and co-workers (1975) have shown that measurements of the induction period obtained by different methods are not in mutual agreement--a fact explained by the incompatibility of the pressure signal with that of the onset of glow, which may not be associated with the initiation of combustion in the fuel-air mixture. In this work, the mass of the injected fuel was relatively small, and its evaporation did not lead to a noticeable drop in pressure. Therefore, the end of the induction period was identified with a sharp rise in pressure over its initial value, while the beginning of the period was taken at the start of fuel injection.

The results shown in Figs. 3 and 4 are indicative of the existence of at least two distinct forms of the dependence period of the induction period on pressure and temperature. Up to about 800 K, this dependency is expressed by parallel lines indicating that they can be described by the Arrhenius relation of Eq. (1). However, at higher temperatures, the character of the dependency becomes drastically changed so that the activation energy cannot be expressed in terms of Eq. (1) anymore. The induction time depends, then, not only on the type of fuel, but also on its pressure and temperature. Therefore, one may conclude that under such circumstances the induction period is governed not only by the chemical reaction of the fuel-air mixture and its oxidation kinetics, but to a considerable extent by such characteristic features of the jet as the concentration field and the concomitant dispersion of fuel. Depending on the pressure, the dynamics of the injection process has a profound influence upon the evaporation of fuel and the formation of the temperature field. The start of the chemical reaction and its process depends only in a secondary way on initial conditions in the reservoir. Although some attention has been paid earlier (Sviridov 1972) to the dependence of the induction period on spray characteristics and fuel evaporation, the influence of the fluid mechanic effects established by this study has not been, to the best of our knowledge, stressed before.

The results obtained could, nonetheless, be expressed in the form of modified Arrhenius relations. In this regard, instead of a determination of the activation energy, equivalent parameters were introduced to correlate with experimental data within the range of gage pressures from 16 up to 80 atm and temperatures between 750 and 1000 K. This was accomplished for gasoline and diesel oil, respectively, by the following relations:

$$\tau_g = 22 \exp(36\, P^{1.2}/T - 0.12P) \tag{2}$$

$$\tau_d = 7.6 \exp(113\, P^{1.05}/T - 0.17P) \tag{3}$$

It should be noted that for both fuels the power of P is close to unity, while the temperature T to the first power is in the denominator. Thus, it appears that τ depends mainly on the density of the air--a parameter that is most influential in governing the dynamic properties of the fuel spray and the dispersion of fuel that eventually establishes the temperature field. This conclusion emphasizes the influence of the fluid mechanical effects on the induction processes. Consequently, the use of the effective activation energy to describe the dependence of the induction time on pressure and temperature should be considered basically incorrect.

If this parameter is to be measured, it should be recognized that it essentially expresses the level of the chemical reaction barrier, and that the mixture must be homogeneous and maintained at a uniform pressure and temperature during the induction process. The most appropriate method for this purpose is the use of a shock tube technique (Wierzba et al. 1974). However, the important processes governing the formation of a combustible mixture and its effect on ignition are thereby left out of scope.

References

Alekseev, V.P. (1974) "Equation for Determining the Period of Ignition Delay in a Diesel Engine Combustion Chamber" Izv. Vyssh. Uchebn. Zaved. Mashinostr. (in Russian) 6, 106-109.

Baev, V.K., Bazhaikin, A.N., Boldyrev, I.V., et al. (1981) Gasoline Ignition Delay in a Model Combustion Engine Chamber" Fiz. Gor. Vzr. (in Russian) 1, 28-35.

Baturin, S.A., Dyachenko, N. Ch., and Magidovich, L.E. (1975) "Light Emission Delay and Increased Pressure Delay in Diesel Engines" Transactions of the Altai Polytechnical Institute (in Russian) 47, 17-26.

De Zubay, E.A. (1978) "A Note on the Autoignition of Liquid Fuel, "Combust. Flame 32, 313-315.

Mullaney, G.J. (1959) "Autoigniton of Liquid Fuel Sprays, " Ind. Eng. Chem. 51, (6), 779-782.

Mullins, B.P. (1953) "Studies on the Spontaneous Ignition of Fuels Injected into a Hot Air Stream. 11. Effect of Physical Factors Upon the Ignition Delay of Kerosine-air Mixtures," Fuel 32, 211-217.

Semenov, V.I. (1970) "Investigation of the Induction Period of High RPM, Rich Diesel Ignition Delay in a Porsche Engine" Izv. Vyssh. Uchebn. Zaved. Mashinostr. (in Russian) 1, 81-85.

Sviridov, Yu. B. (1972) "Nature of Diesel Fuel Atomization Combustion" Transactions of NAMI (in Russian) 88, 186-192.

Todes, O. M. (1933) "Positive Mechanism of Three Ignition Limits for Hydrocarbons" Zh. Fiz. Khim. (in Russian 4, 18-32.

Voinov, A.N. Skorodelov, D.I., Sokolov, F.P. (1964) "Temperature and Pressure Dependence of Ignition Delay in Hydrocarbon-air Mixtures During Adiabatic Compression" Kinet. Katal. (in Russian) 5, (3), 388-398.

Wierzba, A.S., Kauffman, C.W., and Nicholls, J.S. (1974) "Ignition of Partially Shattered Liquid Fuel Drops in a Reflected Shock Wave Environment, " Combust. Sci. Technol. 9 (5-6), 233-245.

Spray Characteristics of Simplex Swirl Atomizers

N. K. Rizk* and A. H. Lefebvre†
Purdue University, West Lafayette, Indiana

Abstract

The effects of air and liquid properties, and atomizer dimensions, on the spray characteristics of simplex pressure swirl atomizers are examined. Mean drop size and drop-size distribution are measured using an improved form of light-scattering technique. The test range includes wide variations in air pressure, atomizer flow number, liquid flow rate, liquid viscosity, and injection pressure differential. The effects on mean drop size and drop-size distribution of variation in distance between the atomizer and the plane of measurement are also examined. The results show that increase in distance from the nozzle, beyond the minimum needed to complete the atomization process, produces both an increase in mean drop size and a broader distribution of drop sizes in the spray. In fact, it is generally observed that any change in air properties, liquid properties, and atomizer geometry that lowers the mean drop size, also narrows the drop-size distribution in the spray. It is also shown that conventional power law expressions for describing the dependence of mean drop size on air and liquid properties are basically unsound and have only limited ranges of application.

Nomenclature

A = constant in Eq. (14)
d_o = nozzle discharge orifice diameter, m

Presented at the 9th ICODERS, Poitiers, France, July 3-8, 1983. Copyright © 1984 by N. K. Rizk and A. H. Lefebvre. Published by the American Institute of Aeronautics and Astronautics with permission.
*Visiting Professor, School of Mechanical Engineering.
†Reilly Professor of Combustion Engineering, School of Mechanical Engineering.

FN_1	= flow number, $(kg/hr)/(MPa)^{0.5}$
FN_2	= flow number, $(lb/hr)/(psi)^{0.5}$
$\dot{m}$	= mass flow rate, kg/s
P	= pressure, kPa
ΔP	= pressure differential, Pa (except in definition of FN_1)
q	= drop-size distribution parameter
SMD	= Sauter mean diameter of drops in spray, m
t	= liquid film thickness at discharge from nozzle, m
U	= velocity, m/s
V	= volume (or mass) fraction of spray in drops of diameter less than x
We	= Weber number
x	= drop diameter, m
$\bar{x}$	= drop-size parameter, m
x, y	= constants in Eq. (13)
μ	= liquid dynamic viscosity, kg/ms
ν	= liquid kinematic viscosity, m^2/s
ρ	= density, kg/m^3
σ	= surface tension, kg/s^2

Subscripts

A	= air
L	= liquid

Introduction

In recent years the advantage offered by the airblast type of atomizer in terms of reduced soot formation and smoke has enabled it to supplant the dual-orifice nozzle in most gas turbine applications, especially for advanced engines of high compression pressure ratio. However, considerable interest still remains in pressure atomizers, and especially in the simplex swirl atomizer, due partly to its inherent simplicity and also to the fact that it serves as a pilot fuel injector for both dual-orifice nozzles and hybrid airblast atomizers.

Despite the geometrical simplicity of the simplex swirl atomizer, the hydrodynamic processes leading to the formation of a conical liquid sheet and to the disintegration of this sheet into ligaments and jets, are highly complex. Previous studies have revealed the basic mechanisms involved in the conversion within the swirl chamber of a finite number of discrete liquid jets into a thin conical sheet, and have led to the formulation of quantitative relationships between the main atomizer dimensions and various flow parameters such as discharge coefficient and initial spray cone angle. However, much less is known about mean drop size, drop-size distribution, and radial fuel distribution, all of which can have a strong influence on the burning rate and on the nature of the combustion process. For example, it is well known that mean drop size affects the light-up performance of jet engines and the exhaust emissions of unburned hydrocarbons and soot. Moreover, drop-size distribution has a controlling effect on the rate of spray evaporation, while radial fuel distribution strongly influences soot formation at high pressures and extinction limits at low pressures. Most previous measurements of spray properties have been confined to normal atmospheric pressure and few data are available at other conditions more representative of actual engine operation. Even where tests have been conducted at high air pressures, for example, in the investigation by Jasuja (1978), no information has been gained on the distribution of drop sizes in the spray.

The work reported herein constitutes the first phase of a comprehensive study on the spray characteristics of simplex swirl atomizers. In common with previous studies it will embrace many different liquids and wide ranges of liquid flow rate and fuel injection pressure. However, attention will be focused on the effects of ambient air properties on mean drop size and drop-size distribution. Considerable importance is attached to drop-size distribution due to its marked influence on the evaporation history of a spray. It has been shown elsewhere (Chin et al. 1983) that the initial rate of evaporation is higher for nonuniform (polydisperse) sprays which is very advantageous to the ignition performance of aircraft gas turbines, especially at high altitudes where failure to achieve relight is often due to insufficient fuel vapor in the spark zone. However, an obvious drawback to a polydisperse spray, in comparison with a monodisperse spray, is that a significant proportion of the fuel is contained in a relatively small number of large drops which take a long time to evaporate. If the residence time in the combustion zone is too short to ensure complete evaporation of these large drops, the result will be a lowering of combustion efficiency and an increase in the concentration of unburned hydrocarbons in the exhaust gases. Thus, for the purposes of design and modeling of liquid-fuelled combustion systems, accurate knowledge is needed on the spray characteristics of simplex swirl atomizers, especially mean drop size and drop-size distribution, and the manner and extent to which these characteristics vary with changes in combustion pressure.

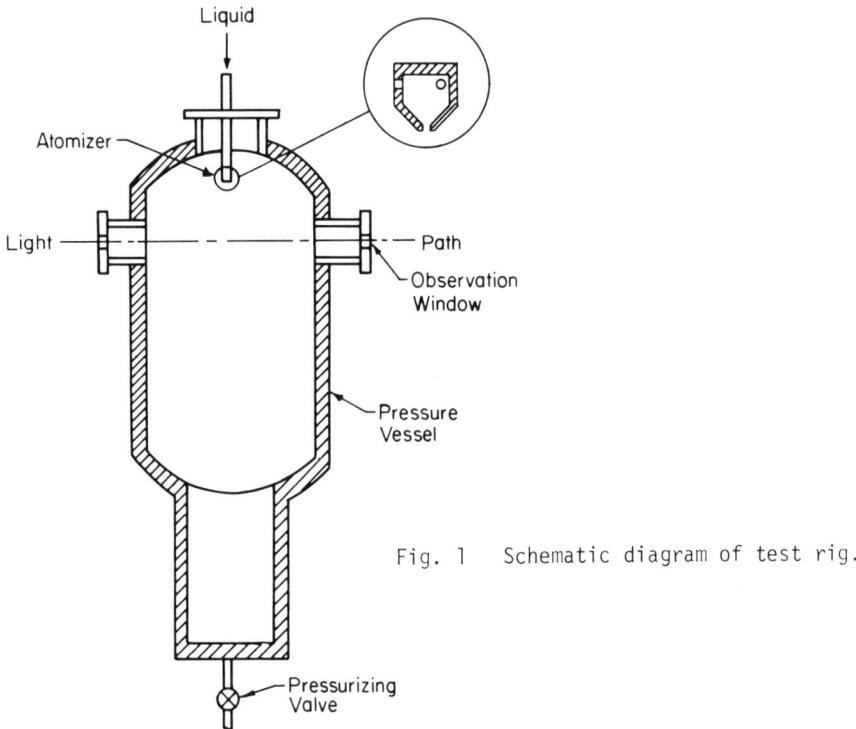

Fig. 1 Schematic diagram of test rig.

Experimental

Two fuels were chosen to cover the range of interest for the aircraft gas turbine, namely, a light aviation kerosine and a light diesel oil. Their physical properties relevant to atomization are:

Aviation kerosine; $\mu = 0.00129$ kg/ms, $\sigma = 0.0275$ kg/s^2, $\rho = 780$ kg/m^3

Light diesel oil; $\mu = 0.003$, kg/ms, $\sigma = 0.0281$ kg/s^2, $\rho = 840$ kg/m^3

A cross-sectional schematic drawing of the type of simplex swirl atomizer employed in this study is shown in Fig. 1. Three geometrically similar atomizers were used, having flow numbers (FN) of 13, 19.5, and 26 where flow number is defined as

$$FN_1 = \text{fuel flow rate, (kg/h)/(fuel pressure drop, MPa)}^{0.5} \quad (1)$$

The above expression was chosen to achieve some measure of consistency with SI units and, at the same time, yield convenient and manageable values of flow number. In the past the most widely used definition of flow number has been

$$FN_2 = \text{fuel flow rate (lb/h)}/(\text{fuel pressure drop, psi})^{0.5} \quad (2)$$

It can readily be shown that

$$FN_1 = 5.4626 \times FN_2 \quad (3)$$

In the following text flow numbers will be quoted in SI units with FN_2 shown alongside in brackets.

The cylindrical pressure vessel for studying spray characteristics is also illustrated in Fig. 1. It is 120 cm long, 75 cm in diameter, and is mounted on a stand with its axis in the vertical position. The atomizer under test is located centrally at the top of the cylinder and sprays downward into the vessel which is pressurized to the desired level with gaseous nitrogen. The only reason for choosing nitrogen instead of air is to avoid the risk of explosion at high pressures. As the physical properties of nitrogen are very similar to those of air, this substitution is considered to have no discernible effect on spray characteristics. The droplets produced by atomization gravitate into a collection tank at the bottom of the chamber, from whence the fuel is returned to the storage tank. The objective is to conserve fuel and to avoid potential pollution problems arising from the escape of fuel droplets into the atmosphere. Two extra nitrogen lines are connected to the tank. One line is used to protect the windows from any contamination by fuel drops or mist, while the other line is connected to a manifold located at the top of the tank which provides a gentle downdraft of nitrogen through a large number of holes. By this means the problem of droplet recirculation is kept to a minimum.

Spray characteristics are measured using an extended form of light-scattering technique whereby the light intensity profile generated by a monochromatic beam of light is analyzed after passage through the spray to determine both mean drop size (SMD), and drop-size distribution [Rizk and Lefebvre (1983)].

It is important not to attempt measurements of mean drop size and drop-size distribution too close to the nozzle. This is because although all the drops leave the nozzle with approximately the same velocity, due to air resistance the smaller drops tend to lose momentum faster than the larger drops. This may lead to overrepresentation of the fine drops in the sampling volume. Further away from the nozzle, where all the drops are moving at roughly the same velocity as the downdraft of nitrogen, the measurements indicate larger values of SMD which are more representative of the actual spray.

Effect of Variables on Mean Drop Size

The main factors governing the atomization quality of pressure atomizers are liquid properties, the physical properties of the air or gas into which the liquid is injected, the liquid injection pressure, and the size of the atomizer as indicated by its flow number.

Liquid Properties

The liquid properties of importance are surface tension, density, and viscosity. In practice the significance of surface tension is diminished by the fact that most commercial fuels exhibit only slight differences in this property. The influence of density is also quite small, since liquid hydrocarbon fuels tend to have similar values of density (typically between 750-900 kg/m^3). Moreover, all available experimental data indicate only a slight effect of liquid density on mean drop size. However, as viscosity varies appreciably from one fuel to another its effect on atomization quality can be quite large.

Empirical drop-size relationships have been derived by many workers, [see Jasuja (1978); Radcliffe (1960); Lefebvre (1983)] and are usually of the form

$$SMD \propto \sigma^a \nu_L^b \dot{m}_L^c \Delta P_L^d \tag{4}$$

One of the earliest and most widely quoted expressions is that of Radcliffe (1960):

$$SMD = 7.3 \, \sigma^{0.6} \, \nu_L^{0.2} \, \dot{m}_L^{0.25} \, \Delta P_L^{-0.4} \tag{5}$$

while a later study by Jasuja (1978) yielded the following expression

$$SMD = 4.4 \, \sigma^{0.6} \, \nu_L^{0.16} \, \dot{m}_L^{0.22} \, \Delta P_L^{-0.43} \tag{6}$$

Although Radcliffe and Jasuja agree on a value of 0.6 for the exponent of surface tension, it is noteworthy that the total range of variation of surface tension in their experiments was only about 20%, and this was accompanied by very large variations in viscosity. Measurements of SMD obtained recently by Simmons and Harding (1980) for water and kerosine, which have roughly the same values of density and viscosity but differ in surface tension by a factor of around 3, revealed a much lower influence of surface tension on mean drop size (SMD α $\sigma^{0.19}$).

There appears to be more consistency in regard to the effect of viscosity on SMD. The published experimental data are usually expressed in the form

$$\text{SMD} \alpha \mu_L^b \qquad (7)$$

Values of b of 0.16, 0.20, and 0.215 have been reported by Jasuja (1978), Radcliffe (1960), and Knight (1955), respectively. An average of these three values is about 0.2 which gives

$$\text{SMD} \alpha \mu_L^{0.2} \qquad (8)$$

Liquid Flow Rate

Few data have been reported on the effect of liquid flow rate on mean drop size. Some of the results obtained in the present study are shown plotted in Fig. 2. By using three geometrically similar atomizers of different flow number it was possible to vary the fuel flow rate while maintaining both the fuel injection pressure differential and the ambient air pressure constant. The measurements show that mean drop size increases slightly with increase in liquid flow rate due, presumably, to the concomitant increase in t, the liquid film thickness at the discharge orifice.

Flow Number

The influence of nozzle flow number on mean drop size is illustrated in Fig. 3. The measurements of SMD shown in this figure were obtained using Jet A fuel at a distance of 125 mm from the atomizer face. The results are shown plotted as SMD vs fuel injection pressure. The two sets of data are drawn for two different levels of ambient air pressure, (roughly one and seven atmospheres) and show the beneficial effect of increase in air pressure on atomization quality. They also indicate that mean drop size can be reduced by reduction in flow number. The effect is not large and stems directly from the influence of FN on t.

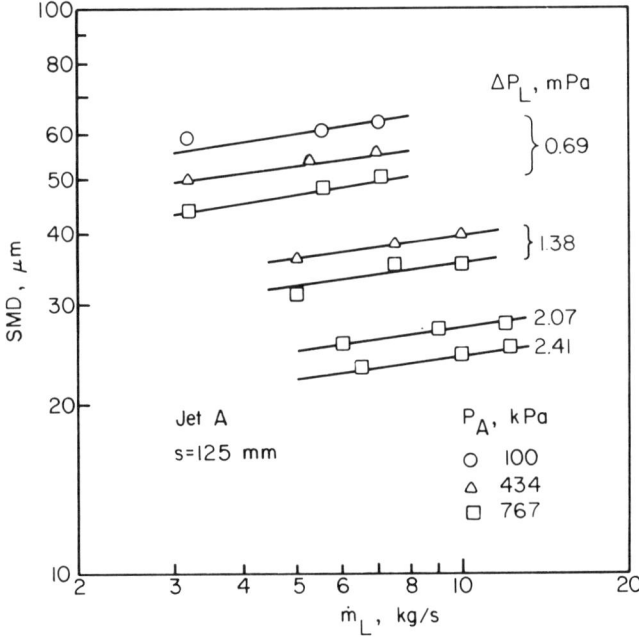

Fig. 2 Variation of mean drop size with liquid flow rate.

Fuel Injection Pressure

Increase in fuel injection pressure differential causes the fuel to be discharged from the atomizer at a higher level of velocity. This raises the Weber number, thereby promoting finer atomization. The effect may be expressed quantitatively as

$$\text{SMD} \propto \Delta P_L^{-c} \tag{9}$$

Reported values of c include 0.275, 0.35, 0.4, and 0.43, due to Simmons (1979), Abou Ellail et al. (1978), Radcliffe (1960), and Jasuja (1978), respectively. Further evidence on the influence of ΔP_L on SMD is contained in Figs. 3 and 4.

Air Properties

The two properties of main interest are pressure and temperature. Experimental studies on the effect of ambient pressure on mean drop size have produced conflicting results. De Corso (1960) found that SMD increased with P_A, and this was confirmed by Neya and Sato (1968) whose experimental data conform to the relationship SMD $\propto P_A^{0.27}$. However, Retel (1938) observed an opposite effect, and Ballal and Lefebvre (1978) also noted a decline in SMD with increase in P_A, according to the relationship SMD $\propto P_A^{-0.1}$. Both Giffen and Lamb (1953) and Miesse (1955) found SMD $\propto P_A^{-0.2}$, while the more recent experimental study by Abou Ellail et al. (1978) gave the result SMD $\propto P_A^{-0.25}$. In this test program it was found that the dependence of SMD on P_A increased with increase in P_A. Figure 5 is typical of the results obtained. It shows that at pressures below around 350 kPa the dependence of SMD on P_A can be expressed as SMD α $P_A^{-0.1}$. At higher pressures the results are better described by SMD α $P_A^{-0.28}$. Clearly these effects warrant further investigation, and more tests are planned for the near future.

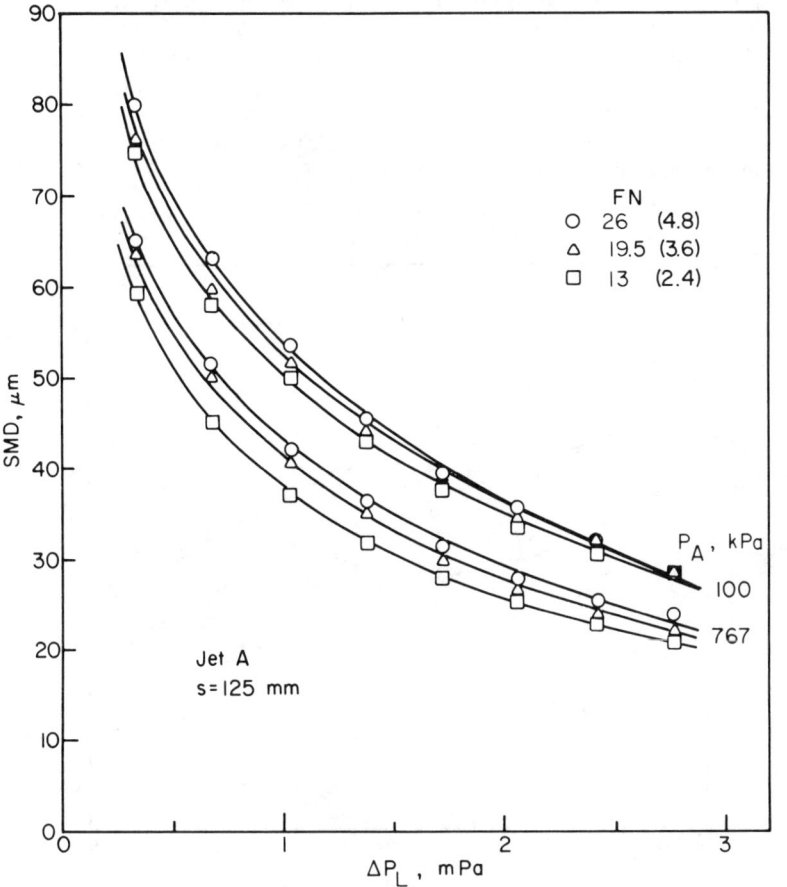

Fig. 3 Variation of mean drop size with liquid injection pressure and nozzle flow number.

Abou Ellail et al. (1978) also studied the effect of ambient air temperature on mean drop size. Their results are interesting because they are contrary to what might be expected from purely theoretical considerations. Since an increase in air temperature lowers the Weber number it would be expected to impede disintegration and give rise to larger drops. However, Abou Ellail et al. found that drop sizes diminished with increase in air temperature according to the relationship

$$SMD \propto T_A^{-0.56} \qquad (10)$$

This somewhat surprising result is attributed to the fact that "as the fuel is injected in the hot air it is heated by convection and radiation which lowers its surface tension and hence increases its Weber number. Therefore, to keep the Weber number below a certain critical value, the large droplets break up into smaller droplets to balance the effect of the decreasing surface tension as T_A increases." Although not mentioned by these workers, these same mechanisms of heat transfer to the fuel should also assist atomization by lowering the fuel viscosity.

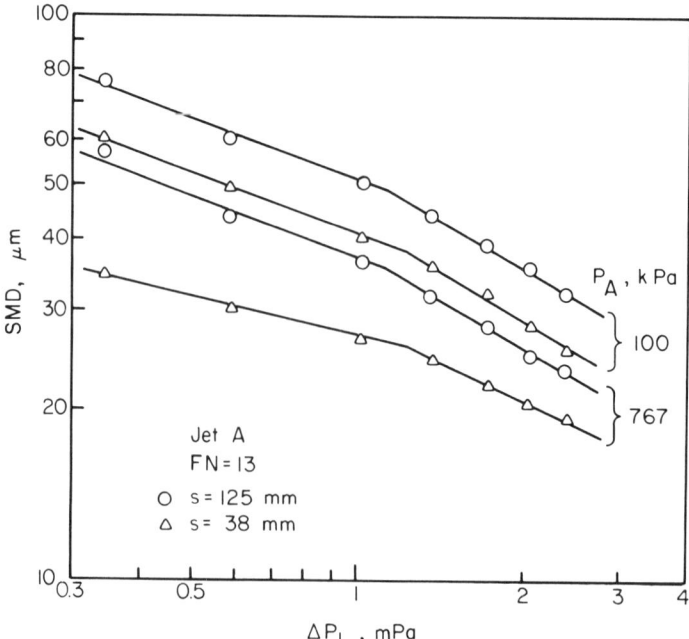

Fig. 4 Influence of air pressure and liquid injection pressure on mean drop size.

Drop-Size Analysis

The atomization process in simplex fuel nozzles is governed basically by two opposing forces, 1) the aerodynamic forces which tend to break up the liquid sheet formed at the nozzle exit into ligaments and drops, and 2) the surface tension forces that oppose sheet disintegration. The relative magnitude of these opposing forces may be expressed quantitatively as the Weber number, in which the liquid sheet thickness at the final discharge orifice t should be designated as the characteristic dimension.

Thus, we have

$$\frac{SMD}{t} = \alpha \, We^{-x} \, \alpha \left[\frac{U_L^2 \, \rho_A \, t}{\sigma}\right]^{-x} \tag{11}$$

This equation suggests that SMD is independent of liquid viscosity. However, Rizk and Lefebvre (1980) have shown that t varies with viscosity. For airblast atomizers they found that

$$t/d_0 \, \alpha \left[U_L \, \rho_L \, d_0/\mu_L\right]^{-y} \tag{12}$$

Substituting for t from Eq. (12) into Eq. (11) gives

$$SMD \, \alpha \left[\frac{\sigma}{U_L^2 \, \rho_A}\right]^x \left[\frac{\mu_L}{U_L \, \rho_L}\right]^{y(1-x)} d_0^{(1-x)(1-y)} \tag{13}$$

Substitution of $U_L^2 = 2\Delta P_L/\rho_L$ and $d_0 \, \alpha \, \dot{m}_L^{0.5}/(\Delta P_L \, \rho_L)^{0.25}$ along with "best" values for x and y (as suggested by analysis of the available experimental data) of 0.25 and 0.33, respectively, leads to the following dimensionally correct expression for mean drop size.

$$SMD = A \, \sigma^{0.25} \, \mu_L^{0.25} \, \dot{m}_L^{0.25} \, \Delta P_L^{-0.5} \, \rho_A^{-0.25} \tag{14}$$

The value of 0.25 for the surface tension exponent is reasonably close to Simmons' estimates of 0.16 to 0.19 (1980). The exponent of 0.25 for liquid viscosity agrees fairly well with all reported values. Equation (14) suggests that SMD is independent of liquid density. Although few data exist against which to check this result it is certainly true for airblast atomizers, at least for liquids of low viscosity (El Shanawany and Lefebvre 1980; Rizk and Lefebvre 1983).

The exponents for injection pressure differential and air density of -0.5 and -0.25, respectively, are identical to the values reported by Abou Ellail et al. (1978). [Note that Abou Ellail et al.'s value of -0.35 becomes -0.5 when transposed to the form of Eq. (14).] If Jasuja's (1978) data is used to determine the value of A, Eq. (14) becomes

$$SMD = 2.25 \, \sigma^{0.25} \, \mu_L^{0.25} \, \dot{m}_L^{0.25} \, \Delta P_L^{-0.5} \, \rho_A^{-0.25} \qquad (15)$$

Although drop-size relationships of this kind are extremely useful from a practical viewpoint, they should be treated with great caution. For example, Simmons and Harding (1980) have demonstrated that in the relationship $SMD \propto \Delta P_L^{-d}$ the value of d is not constant but is higher at lower levels of fuel injection pressure. The data obtained in the present study also dictate a change in d above a certain value of ΔP_L, as illustrated in Fig. 4. However, whereas Simmons and Harding found that the dependence of SMD on ΔP_L diminished with increase in ΔP_L, the opposite effect was observed here. Clearly, this is another area that warrants

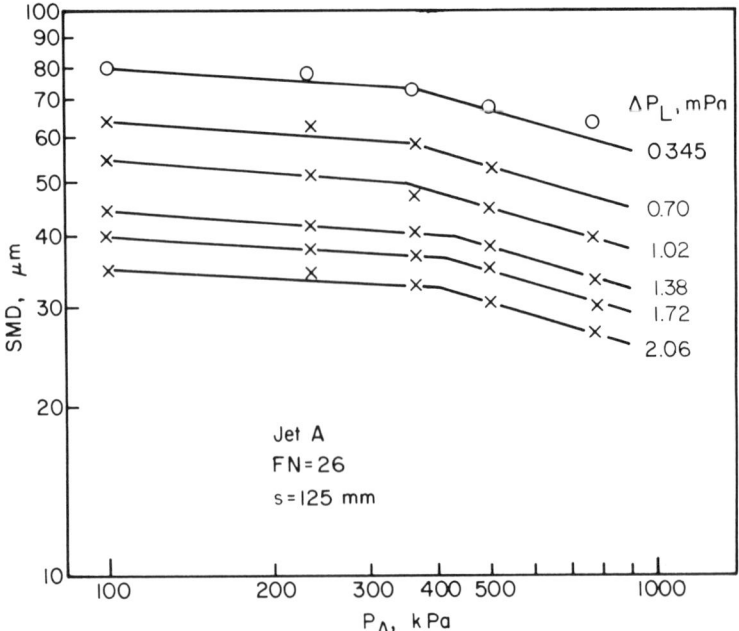

Fig. 5 Influence of ambient air pressure on mean drop size.

further investigation. In the meantime it should be recognized that although relationships such as Eq. (4) provide a simple and convenient means for describing quantitatively any given set of experimental data, they have no general validity and cannot be applied to fluid properties and test conditions that lie outside the ranges covered in the investigation.

Another cause for concern is that almost all the published data on mean drop size for pressure swirl atomizers were obtained by carrying out measurements at a fixed distance downstream of the atomizer. Now for airblast atomizers it has been shown that measured values of SMD are initially very high in the region close to the atomizer, and then diminish with increase in downstream distance, reaching a minimum value when the atomization process is complete. Further downstream the mean drop size gradually increases due to evaporation and, possibly also, droplet coalescence.

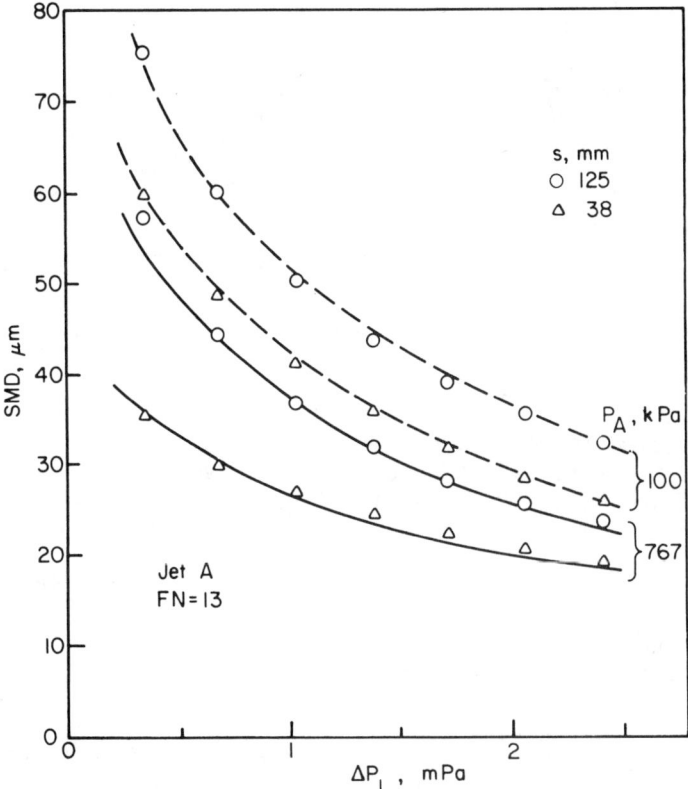

Fig. 6 Graphs illustrating variation of measured values of SMD with distance from nozzle.

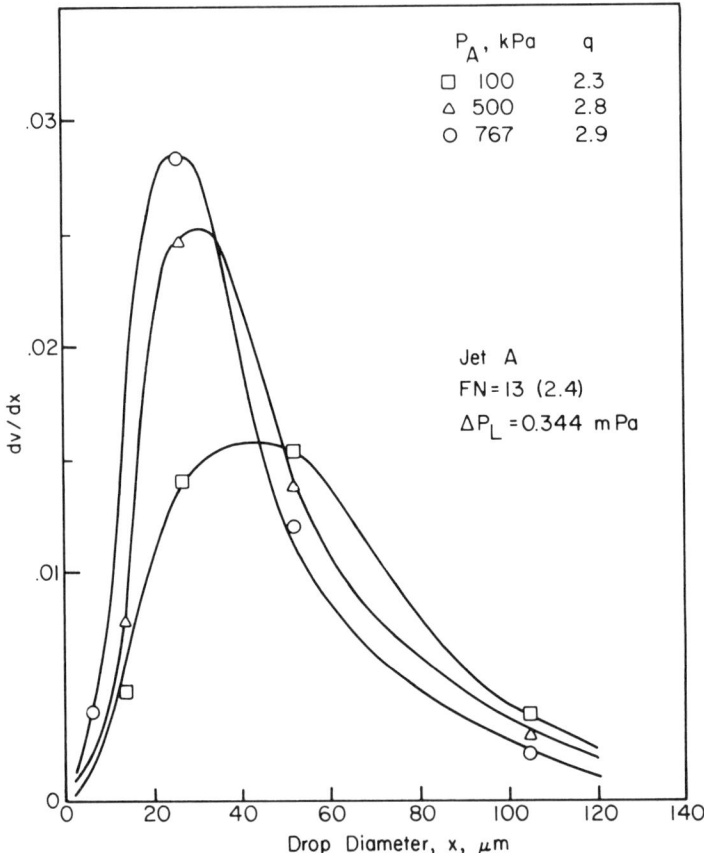

Fig. 7 Influence of ambient air pressure on drop-size distribution.

Although the effect of downstream distance on SMD has not been studied for pressure swirl atomizers a similar situation must exist, as evidenced by the curves drawn in Fig. 6. This figure demonstrates a significant rise in SMD with increase in downstream distance from 38 mm to 125 mm.

Effect of Variables on Drop-Size Distribution

Owing to the heterogeneous nature of the atomization process, the threads and ligaments formed by the various mechanisms of liquid sheet disintegration vary widely in diameter, and the resulting main drops and satellite drops vary in size correspondingly. The drop-size distribution of a spray is of considerable practical importance because it determines its evaporation history. However, it is extremely difficult to predict theoretically and determine

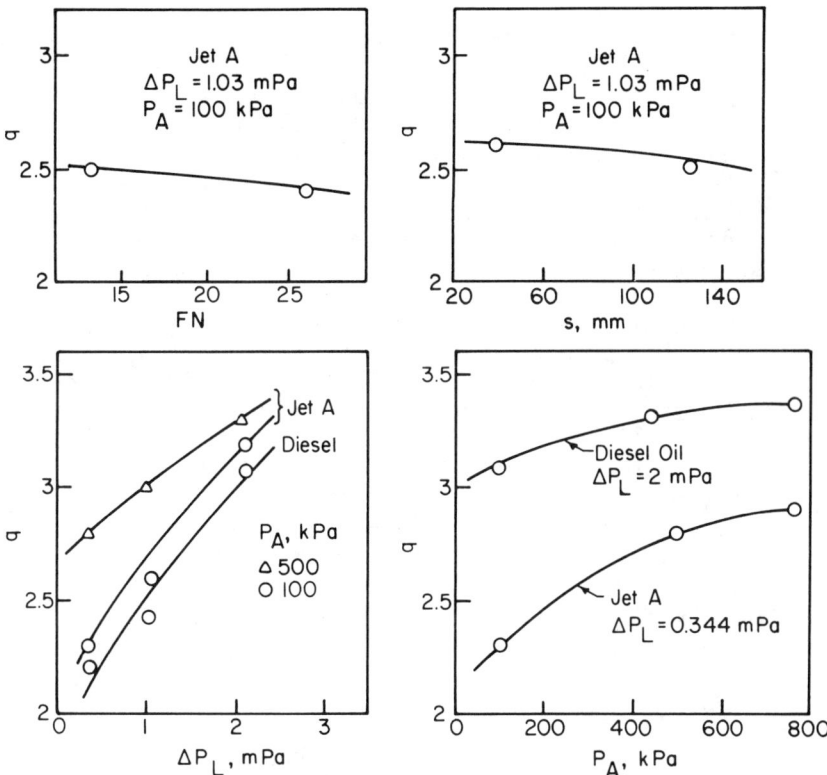

Fig. 8 Influence of nozzle flow number, ambient air pressure, liquid injection pressure, and distance from nozzle on drop size distribution parameter, q.

experimentally. In consequence very little information on drop-size distribution is available in the literature.

Several mathematical expressions have been derived to represent drop-size distribution. Usually they contain two independent parameters, one of which is a mean diameter of some kind and the other is a measure of the dispersion of the spray, or the deviation from the mean. The Rosin-Rammler expression (1933) is perhaps the most widely used at the present time. Although developed originally for powders it has been applied with success to liquid drops. It can be expressed in the form

$$1 - v = \exp - (x/\bar{x})^q \qquad (16)$$

where v is the volume fraction of the spray occurring in drops of diameter less than x, $\bar{x}$ is a size parameter, and q is a distribution parameter. The higher the value of q, the more uniform is the spray. Values of q were determined over the entire range of test conditions.

The beneficial effect on atomization quality of increase in ambient air pressure from 100 to 500 kPa, and then to 767 kPa, while maintaining the liquid flow rate constant, is illustrated for Jet A fuel in Fig. 7. The number of large drops in the spray decreases rapidly with increase in air pressure. At the same time the proportion of fuel contained in the smaller drops rises rapidly. The overall effect is to reduce SMD and raise q.

The results of many similar studies on the influence of air and liquid properties on the value of q are summarized in Fig. 8. They show that any change in liquid properties, air properties, and atomizer size (flow number) that tends to produce a finer spray will also yield a higher value of q, i.e., a more uniform spray. This point is clearly illustrated in Fig. 8 which shows that increases in air pressure and fuel injection pressure both produce a more uniform drop-size distribution. However, increase in atomizer size and downstream distance (beyond the distance required to complete the atomization process) both produce a coarser spray.

Conclusions

1) From analysis of the basic processes governing the mechanisms of sheet formation and disintegration in swirl pressure atomizers, it is found that the effects of the main air and liquid properties on mean drop size can be expressed in a dimensionally correct form as

$$\text{SMD} = A \, \sigma^{0.25} \, \mu_L^{0.25} \, \dot{m}_L^{0.25} \, \Delta P_L^{-0.5} \, \rho_A^{-0.25} \quad (14)$$

The various exponents given in this equation exhibit good agreement with published data and with the experimental results obtained in the present study. However, it must be recognized that the various exponents in this equation represent average values only. In practice these exponents may vary appreciably with changes in operating conditions. This is especially true for the exponents of air pressure and liquid injection pressure. Thus Eq. (14) and all similar expressions, are valid only over a limited range of conditions.

2) Beyond a certain minimum distance needed to complete the atomization process, the mean drop size of the spray gradually increases with increase in downstream distance from the atomizer.

3) Any change in air properties, liquid properties, and atomizer size or geometry that tends to produce a finer spray will also reduce the range of drop sizes in the spray.

References

Abou-Ellail, M. M. M., Elkotb, M. M., and Rafat, N. M. (1978) Effect of fuel pressure, air pressure and air temperature on droplet size distribution in hollow-cone kerosene sprays. 1st International Conference on Liquid Atomization and Spray Systems, pp. 85-92, Tokyo.

Ballal, D. R. and Lefebvre, A. H. (1978) Ignition of liquid fuel sprays at subatmospheric pressure. Combustion and Flame 31 (2), 115-126.

Chin, J. S., Durrett, R., and Lefebvre, A. H. The interdependence of spray characteristics and evaporation history of fuel sprays in stagnant air. ASME Paper 83-GT-7, 28th International Gas Turbine Conference, Phoenix, Az., March 1983.

Decorso, S. M. (1960) Effect of ambient and fuel pressure on spray drop size. J. Engrg. Pwr. 82, 10-18.

El-Shanawany, M. S. M. R. and Lefebvre, A. H. (1980) Airblast atomization: The effect of linear scale on mean drop size. J. Energy 4 (4), 184-189.

Giffen, E. and Lamb, T. A. J. (1953) The effect of air density on spray atomization. Motor Industry Research Association, Rep. No. 1953/5.

Jasuja, A. K. Atomization of crude and residual fuel oils. ASME Paper 78/GT/83, Gas Turbine Conference, London, England, April 1978.

Knight, B. E. (1955) Communication on the performance of a type of swirl atomizer, by A. Radcliffe. Proceedings Institution of Mechanical Engineers 169, 104.

Lefebvre, A. H. (1983) Gas Turbine Combustion, McGraw-Hill, New York.

Miesse, C. C. (1955) The effect of ambient pressure oscillations on the disintegration and dispersion of a liquid jet. Jet Propulsion 25, 525.

Neya, K. and Sato, S. (1968) Effect of ambient air pressure on the spray characteristics of swirl atomizers. Paper No. 27, Papers on the Ship Research Institute, Tokyo, Japan.

Radcliffe, A. (1960) Fuel injection. High Speed Aerodynamics and Jet Propulsion. Section D, Vol. XI, Princeton University Press, Princeton, New Jersey.

Retel, R. (1938) Publications scientifiques et techniques du ministere de l' air, B. S. T., No. 81.

Rizk, N. K. And Lefebvre, A. H. (1980) Influence of liquid film thickness on airblast atomization. Trans. ASME, Journ. Eng. Power 102, 706-710.

Rizk, N. K. and Lefebvre, A. H. Spray characteristics of plainjet airblast atomizers. ASME Paper 83-GT-138, 28th International Gas Turbine Conference, Phoenix, Az., March 1983.

Rosin, P. and Rammler, E. (1933) The laws governing the fineness of powdered coal. Journal of the Institute of Fuel 7 (31), 29-36.

Simmons, H. C. The prediction of Sauter mean diameter for gas turbine fuel nozzles of different types. ASME Paper No. 79-WA/GT-5. International Gas Turbine Conference, San Diego, March 1979.

Simmons, H. C. and Harding, C. F. Some effects of using water as a test fluid in fuel nozzle spray analysis. ASME Paper No. 80-GT-90, Gas Turbine Conference and Products Show, New Orleans, La., March 1980.

Chapter VI. Nonequilibrium Flows

Flows in Laval Nozzles
with a High-Temperature Diatomic Gas

Nimai K. Mitra* and Martin Fiebig†
Ruhr-Universität, Bochum, Federal Republic of Germany

Abstract

Supersonic nozzle flows with high-temperature oxygen have been calculated for vibrational dissociational relaxation. Inviscid quasi-one-dimensional flow equations are used. For relaxation models, vibrational equilibrium, uncoupled and coupled (preferential and nonpreferential) vibrational dissociational relaxations have been used. Results show that, for moderate values of stagnation chamber pressure, temperature, and throat height (corresponding to small Damkoehler numbers), the assumption of equilibrium up to the throat is quite unsatisfactory and may produce wrong results. For small Damkoehler numbers, the frozen value of the degree of dissociation at the exit is nearly independent of the vibrational relaxation model. The vibrational temperature is strongly dependent on the model chosen. For the coupled vibrational dissociational relaxation, the vibrational energy on the axis goes down before and in the throat region, and then goes up in the vibrationally-frozen recombination-dominated zone.

Introduction

When a partially dissociated diatomic gas such as oxygen expands in a Laval nozzle, recombination of atoms and vibrational relaxation of molecules become important physical phenomena. There exists a coupling between dissociation recombination and vibrational relaxation which was first discussed by Hammerling et al. (1959) and further improved by Treanor and Marrone (1962), Marrone and Treanor (1963), Treanor (1965), and Tirumalesa (1967). The coupling of dis-

---

Presented at the 9th ICODERS, Poitiers, France, July 3-8, 1983.
Copyright 1984 by N. K. Mitra. Published by the American Institute of Aeronautics and Astronautics, Inc. with permission.
*Akademischer Rat, Institut für Thermo- und Fluiddynamik.
†Professor, Institut für Thermo- und Fluiddynamik.

sociational and vibrational nonequilibrium is obtained by modification of: 1) the dissociation-rate constant by a coupling factor V, which takes into account the vibrational nonequilibrium, and 2) the vibrational-rate equation to take into account the fact that only vibrationally excited molecules dissociate and that energy released by recombination is pumped into a vibrational mode. The coupling can be non-preferential, meaning that the probability of dissociation from all the vibrational levels is equal and preferential, which assumes higher probability for dissociation from higher vibrational levels.

Tirumalesa (1967) computed partially-dissociated oxygen flows in Laval nozzles from quasi-one-dimensional inviscid-flow equations for vibrational equilibrium and nonequilibrium (with and without coupling). In so doing, he assumed that the flow stayed in equilibrium up to the throat. Thus, the problem of finding the critical mass flow rate was avoided. For vibrational relaxation and atomic recombination, characteristic Damkoehler numbers Q_V and Q_R, respectively, can be defined from the following formulas

$$Q_V = r_*/\sqrt{2h_0} \; \tau_0 \qquad (1)$$

and

$$Q_R = r_* \, k_{ro} \, \rho_0^2 \, \alpha_0/\sqrt{2h_0} \; m_A^2 \qquad (2)$$

where r_*, h, τ, ρ, α, m_A are throat height, enthalpy, vibrational relaxation time, density, degree of dissociation, and atomic mass, respectively, and o stands for the stagnation chamber conditions (Mitra and Fiebig 1974, 1979). These Damkoehler numbers become larger as the stagnation chamber temperature T_0, or pressure p_0, or r_* increases. With $Q_V = 0$ and $Q_R = 0$ a fully frozen flow will result, and with $Q_V = \infty$ and $Q_R = \infty$ the flow will stay always in equilibrium. So only for large Q_R and Q_V (i.e., for large p_0, T_0, and r_*) the assumption of equilibrium up to the throat is acceptable. For a typical hypersonic arc tunnel nozzle flow with $r_* = 2$ mm, $p_0 = 2$ bar, $T_0 = 4000$ K, and oxygen as flow medium, one obtains $Q_R = 0.008$, $Q_V = 0.856$. The equivalent sudden freezing point for recombination should appear before the throat for this case. The departure from equilibrium before the throat should be considered. Mitra and Fiebig (1979) used slender channel equations for fully-viscous flow in conjunction with the relaxation model of Tirumalesa (1967) to calculate the above arc tunnel nozzle flow to investigate the effects of viscosity and wall cooling. The preliminary results show that the nonequilibrium departure appears before the throat. It also shows an unexpected behavior of the vibrational dissociational relaxation. The vibrational energy initially goes down (as expected), but starts going up downstream of the geometrical throat as recombination pumps energy into the vibrational mode. This is probably peculiar to low Damkoehler number flows and has not been reported previously.

The purpose of the present work is to investigate numerically in detail nonequilibrium nozzle flows of low Damkoehler numbers with partially-dissociated oxygen as the flow medium.

Since viscous-flow computations are highly time consuming, the quasi-one-dimensional inviscid-flow model will be used. This will substantially reduce the computational effort but will keep the nonequilibrium chemistry intact. A time-dependent solution of the basic unsteady equations will be obtained so that the nonequilibrium departure just downstream of the inlet of the nozzle can be handled. The relaxation model of Tirumalesa (1967) will be used. This model assumes oxygen molecules to be anharmonic oscillators with cutoff vibrational levels where a Boltzmann distribution exists. Treanor (1965) questioned the use of Boltzmann distribution for vibration in coupled relaxation. Unfortunately, to avoid the use of Boltzmann distribution, one has to solve a large number of equations (master equations) with many unknown rate constants. This will increase the complexity of the problem without necessarily increasing the credibility of computational results and will not be followed here.

Basic Equations

The basic equations for quasi-one-dimensional inviscid unsteady flows are

continuity $\quad A \frac{\partial \rho}{\partial t} + \frac{\partial}{\partial z} (\rho u A) = 0 \quad$ (3)

momentum $\quad \rho \frac{\partial u}{\partial t} + \rho u \frac{\partial u}{\partial z} + \frac{\partial p}{\partial z} = 0 \quad$ (4)

energy $\quad \rho \frac{\partial e}{\partial t} + \rho u \frac{\partial e}{\partial z} + p \frac{\partial u}{\partial z} + pu \frac{d \ln A}{dz} = 0 \quad$ (5)

vibrational rate $\quad \frac{\partial e_v}{\partial t} + u \frac{\partial e_v}{\partial z} = \frac{e_{v*} - e_v}{\tau} + (\bar{e}_r \dot{\alpha}_r - \bar{e}_d \dot{\alpha}_d) \quad$ (6)

dissociational rate $\quad \frac{\partial \alpha}{\partial t} + u \frac{\partial \alpha}{\partial z} = \dot{\alpha}_d - \dot{\alpha}_r \quad$ (7)

The equations of state are

$$p = R(1 + \alpha)\rho T \qquad (8)$$

and

$$e = \frac{RT}{2}(5 + \alpha) + \alpha e_d + (1 - \alpha)e_v \qquad (9)$$

Here e is the internal energy; e_v is the vibrational energy; e_{v^*} is the equilibrium vibrational energy; e_d is the dissociation energy; R is the specific gas constant; t is the time coordinate; z is the nozzle axial coordinate; A is the nozzle cross-sectional area; and u, p, and T are velocity, pressure, and static (translational rotational) temperature, respectively (see Fig. 1). Furthermore $\bar{e}_r$ and $\bar{e}_d$ are vibrational energies associated with recombination and dissociation, respectively; $\dot{\alpha}_r$ and $\dot{\alpha}_d$ are production rate of molecules and atoms, respectively. In the computations of $\bar{e}_r$ and $\bar{e}_d$ and $\dot{\alpha}_d$ and $\dot{\alpha}_r$, the effects of coupling are taken into account. Following Marrone and Treanor's (1963) and Tirumalesa's (1967) model one can write

$$\dot{\alpha}_d = (k_r \rho / 2\, m_a)(1 - \alpha + 2\lambda\alpha)(1 - \alpha) K_c V \qquad (10)$$

$$\dot{\alpha}_r = k_r \rho^2 \alpha^2 (1 - \alpha + 2\lambda\alpha)/m_a^2 \qquad (11)$$

$$\bar{e}_d = \frac{E - e_v}{1 - \alpha} \; ; \; \bar{e}_r = \frac{G - e_v}{1 - \alpha} \qquad (12)$$

The recombination rate k_r is obtained from experiments. λ is the relative efficiency of atoms to molecules as collision partner for dissociation ($\lambda = 35\, kT/e_d$ with k = Boltzmann's constant); K_c is the equilibrium constant. The dissociation rate at vibrational equilibrium k_{de} is given by $k_{de} = K_c k_r$ and for the vibrational nonequilibrium the dissociation rate

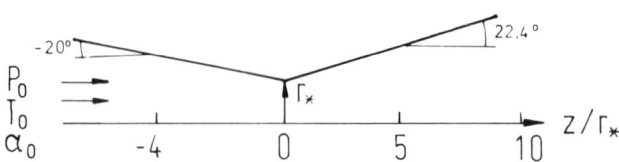

Fig. 1 Nozzle geometry and coordinate system.

is $k_d = k_{de} V = K_c k_r V$. The coupling factor V is calculated
as

$$V = \{Q(T) Q(T_f)\}/\{Q(T_v) Q(-U)\} \quad (13)$$

where T_v is the vibrational temperature, and U is the coupling parameter. An equal probability of dissociation (nonpreferential coupling) from all vibrational levels will mean $U = \infty$. A preferential coupling will require $U = e_d/kn$. Tirumalesa (1967) used n = 6. Furthermore we have

$$T_f = (\frac{1}{T} - \frac{1}{T_v} - \frac{1}{U})^{-1} \quad (14)$$

Q = vibrational partition function; $E = e_v (T_f)$, which is the vibrational energy at temperature T_f; and $G = e_v (-U)$.

Method of Solution

A time dependent technique due to McCormack (1969) and modified by Anderson (1970) and Griffin and Anderson (1977) has been used here to solve the nondimensional form of Eqs. (3-7).

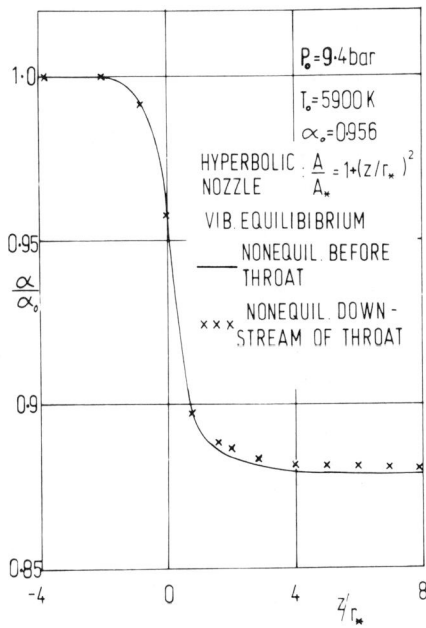

Fig. 2 Normalized atom mass fraction (degree of dissociation) α/α_0 distribution along the hyperbolic nozzle for vibrational equilibrium flow.

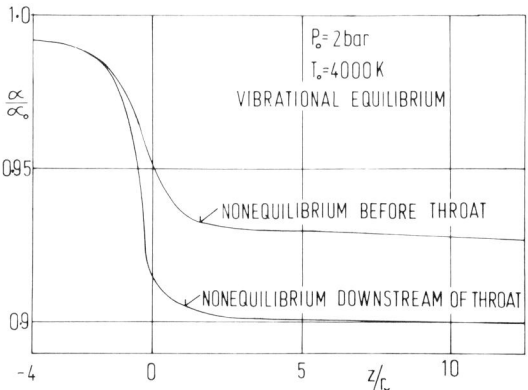

Fig. 3   Normalized atom mass fraction α/α_0 distribution along the nozzle of Fig. 1 for vibrational equilibrium flow.

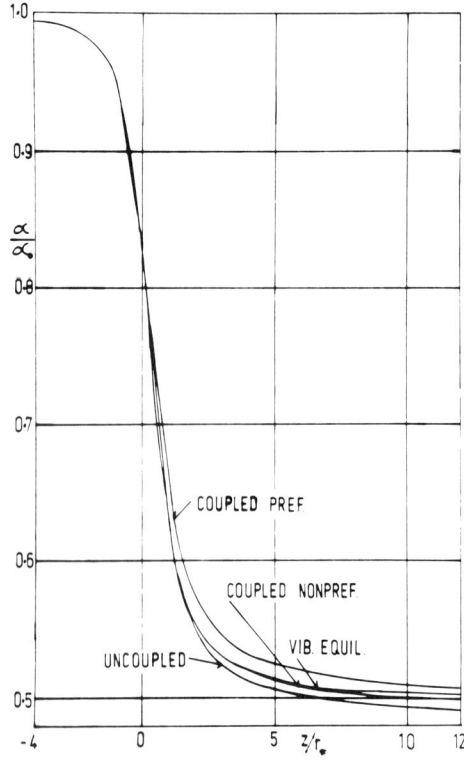

Fig. 4   Normalized atom mass fraction α/α_0 distribution along the nozzle of Fig. 1 for all vibrational relaxation models.

During the time iterations, the dependent variables at the exit are extrapolated. At the inlet, all the variables except the velocity are kept fixed. The velocity is upgraded by extrapolation from two interior points at each iteration. In this way the unknown critical mass flow rate is calculated as a part of the solution.

Results and Discussion

Since the main objective is to calculate nozzle flows at moderate temperature and pressure, the assumption of equilibrium up to the throat was tested. Figure 2 shows the atom mass fraction distribution along a hyperbolic nozzle for p_o = 9.4 bar and T_o = 5900 K. The vibrational equilibrium has been assumed. The computational results of α are nearly the same whether the nonequilibrium computation is started downstream of the throat or at the inlet of the nozzle. Figure 3 shows similar computations in the nozzle of Fig. 1 but with T_o = 4000 K and p_o = 2 bar. The effect of the starting point used in nonequilibrium computations is quite important. The results of α can differ by almost 5%.

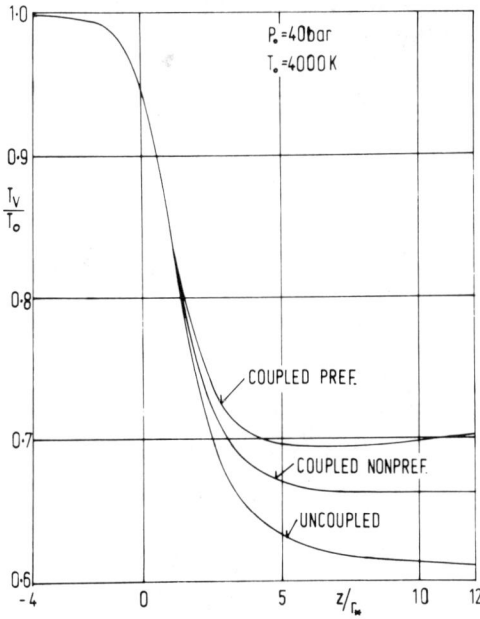

Fig. 5   Nondimensionalized vibrational temperature T_v/T_o along the nozzle of Fig. 1 for uncoupled, nonpreferentially-coupled and preferentially-coupled vibrational relaxation.

Figure 4 shows distributions of α for all four vibrational relaxation models for p_0 = 40 bar and T_0 = 4000 K in the nozzle of Fig. 1. Results up to the throat are the same for every case. Only a slight difference appears at the exit. Figure 5 shows vibrational temperatures for the same case of high-pressure flow. Upstream of the throat the results are the same, but downstream substantial differences appear. The uncoupled case has naturally the lowest temperature. For the coupled cases, the vibrational temperatures show a slight increase after a 6 or 7 throat height downstream.

Figure 6 shows the α distribution for the low-pressure (p_0 = 2 bar) flows in the nozzle of Fig. 1. Differences appear even before the throat, although at the exit the differences are of the same order as those in the high-pressure flows. Figure 7 shows the vibrational temperature distributions for the low-pressure flow. The vibrational temperatures for the coupled cases start going up after one throat height downstream of the throat. The reason for this behavior lies in the last term of the vibrational-rate equation which acts as a source term. Downstream of the throat the dissociation rate ($\dot{\alpha}_d$) becomes negligible, but the recombination rate ($\dot{\alpha}_r$), though small, still persists and pumps recombination energy into the vibrational mode. Since the vibrational relaxation time τ is quite large here, this excess vibrational energy is not distributed among active modes. This is the recombination dominated, vibrationally-frozen zone which is a characteristic of low Damkoehler number flows. Tirumalesa (1967) had not found this behavior of vibrational temperature in the recombination-dominated zone

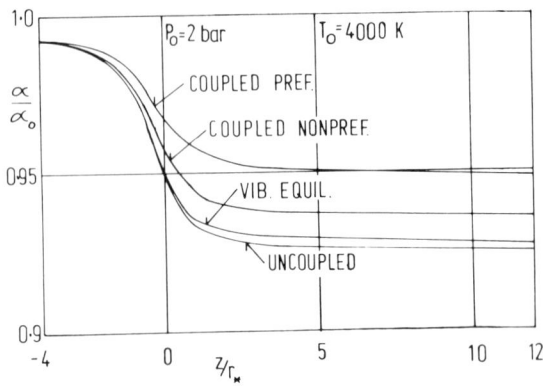

Fig. 6   Normalized atom mass fraction α/α_0 distribution along the nozzle of Fig. 1 for all vibrational relaxation models.

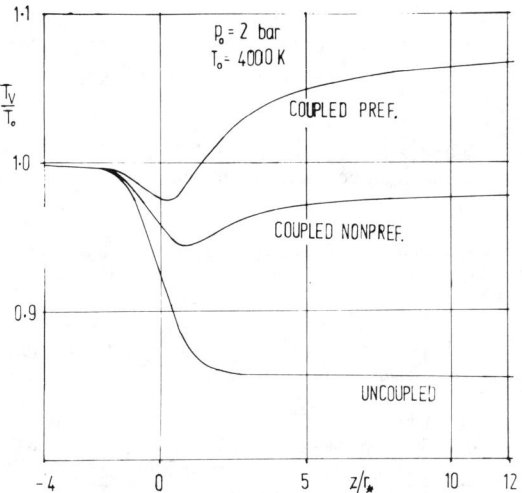

Fig. 7 Nondimensionalized vibrational temperature T_v/T_o along the nozzle of Fig. 1 for uncoupled, nonpreferentially-coupled and preferentially-coupled vibrational relaxation.

because he computed only flows with high Damkoehler numbers. Even if one avoids the Boltzmann distribution in vibration and solves the master equations instead of the simple rate equation, one should obtain an increase in the total vibrational energy of the molecules at the nozzle exit because of the source terms in master equations.

Conclusions

For vibrational dissociational nonequilibrium flows of moderate p_o, T_o and r_* (corresponding to low Damkoehler numbers) nonequilibrium departure appears before the throat. The vibrational relaxation model has only minor influences on the frozen value of degree of dissociation at the nozzle exit. But the vibrational temperatures are strongly dependent on the relaxation model. For coupled vibrational dissociational relaxation, the vibrational temperature may, at the supersonic part of the nozzle, increase for flows with small Damkoehler numbers.

References

Anderson, J. D., Jr. (1970) A time-dependent analyses for vibrational and chemical nonequilibrium nozzle flows. <u>AIAA J.</u> 8 (3), 545-550.

Griffin, M. D. and Anderson, J. D., Jr. (1977) On the application of boundary conditions to time-dependent computations for quasi-one-dimensional fluid flows. Computers & Fluids 5, 127-137.

Hammerling, P., Teare, J. D., and Kievel, B. (1959) Theory of radiation from luminous shock waves in nitrogen. Phys. of Fluids 2, 422-427.

Marrone, P. V. and Treanor, C. E. (1963) Chemical relaxation with preferential dissociation from excited vibrational levels. Phys. of Fluids 6, 1215-1221.

McCormack, R. W. (1969) The effect of viscosity in hypervelocity impact cratering. AIAA Paper 69-354.

Mitra, N. K. and Fiebig, M. (1974) Low Reynolds number vibrational nonequlibrium flow in laval nozzle. Proceedings of the 9th International Symposium on Rarefied Gas Dynamics, B 21, pp. 1-12, DFVLR Press, Göttingen, W. Germany.

Mitra, N. K. and Fiebig, M. (1979) Low Reynolds number nozzle flows with vibrational dissocianional nonequilibrium. Proceedings of the 11th International Symposium on Rarefied Gas Dynamics, pp. 857-868, CEA-Paris.

Tirumalesa, D. (1967) Nozzle flows with coupled vibrational dissociational nonequilibrium, AIAA J. 5, 254-260.

Treanor, C. E. and Marrone, P. V. (1962) Effect of dissociation on the rate of vibrational relaxation. Phys. of Fluids 5, pp. 1022-1026.

Treanor, C. E. (1965) Coupling of vibration and dissociation in gas dynamic flows. AIAA Paper 65-29.

Unsteady Aerodynamics of Chemically Reacting Flows Past Oscillating Thin Bodies

Liviu Librescu*
Tel-Aviv University, Tel-Aviv, Israel

Abstract

This paper deals with the determination of the unsteady pressure field on elastic thin bodies harmonically oscillating in a reactive flow environment. In this context, the cases of planar and circular cylindrical panels, flown by a supersonic reactive gas are separately treated.

For flat thin bodies, two-closed form solution of the unsteady pressure are derived, viz.: 1)quasistatic and 2) piston-theory aerodynamics, both of them being of a wide use in flutter analyses. The features of the solutions, involving qualitative and quantitative differences with respect to their frozen and equilibrium counterparts are discussed, and some implications, for associated flutter problems are underlined.

For the case of circular cylindrical shells of finite length, the pressure expression in Laplace space is derived. This solution proves useful in obtaining <u>piston-theory applicability conditions</u>. These provide the criteria under which, piston theory aerodynamics derived for the case of flat panels, may be applied to circular cylindrical panels.

This paper constitutes both a contribution to chemically reacting unsteady aerodynamics and a basis for consistent evaluation of aeroelastic phenomena confronting spacecraft and rocket nozzles in a reactive flowfield.

Presented at the 9th ICODERS, Poitiers, France, July 3-8, 1983. Copyright © 1984 by L. Librescu. Published by the American Institute of Aeronautics and Astronautics with permission.

*Professor, Faculty of Engineering, Department of Solid Mechanics, Materials, and Structures.

Introduction

During the last two decades there has been substantial research activity devoted to the theory of compressible flows involving nonequilibrium processes. However, a survey of the literature[†] in the field reveals that, little attention was given to some important problems involving mutual interaction among <u>reactive flows</u> fields and oscillating elastic bodies.[§] This paper investigates some elements related to this topic and is principally concerned with the determination of unsteady pressure loads on oscillating thin bodies in a supersonic chemically-reacting flow environment. This analysis makes it possible to obtain more accurate information about aeroelastic behavior of advanced spacecrafts and rocket engines oscillating in reactive flowfields.

General Equations

The State Equation

As distinguished from classical gasdynamics, in chemically-reacting flows the thermodynamic state of each fluid particle is described by at least three parameters, two of them <u>external</u> - state variables (e.g., the pressure and the density) and the third an <u>internal</u> - state variable (also termed a <u>chemical</u> - state variable).

If the state of the system is in chemical equilibrium, the chemical variables are a function of the external variables. For the systems which are not in an equilibrium state, the chemical variables may be considered to be independent. This is due to the fact that during sudden changes of external variables, a delay occurs in the adjustment of the chemical variables to the new external conditions. The internal rate process affecting this adjustment is a <u>relaxation process</u>. For a single relaxation

---

[†] See exhaustive reviews by Becker (1967), (1972); Broer (1969); Clarke (1969), (1978). See also monographs by Vincenti and Kruger (1965); Clarke and McChesney (1976); Pai (1981).

[§] The paper by Sislyan (1966) appears to be the only paper available in the literature dealing with this topic.

process [π], and consequently a single chemical variable ξ, the following well-known rate equation pertains:

$$\xi - \xi_o = (\xi_{t=o} - \xi_o) \exp(-t/\tau) \qquad (1)$$

which states the path by which the system tends to equilibrium. Appropriate selection of the internal variable (see Groot and Mazur 1962), makes it possible to apply Eq. (1) to a large variety of nonequilibrium processes (involving, among others vibrational excitation, chemical reaction, dissociation, and ionization). Here t denotes the time variable while $\tau (0 \leq \tau \leq \infty)$, denotes the chemical relaxation time considered as a constant quantity throughout the reacting flowfield. (For a discussion of this statement see the paper by Wu and Turner, 1974.) In addition ξ_o and $\xi_{t=o}$ correspond to the equilibrium and initial values of ξ, respectively. Neglecting transport phenomena, eliminating from the equation of state of the reacting system by means of the rate equation (see Becker 1967; Rudenko and Soluian 1975), and using the small-disturbance concept,[∫] the relationship between the pressure and density in the relaxing medium is governed by:

$$\frac{d\hat{p}}{dt} + \frac{\hat{p}}{\tau} = c_f^2 \frac{d\hat{\rho}}{dt} + c_e^2 \frac{\hat{\rho}}{\tau} \qquad (2)$$

where $\hat{p}$ and $\hat{\rho}$ denote the disturbances of the pressure p_o and density ρ_o of the undisturbed medium, and c_f and c_e denote the frozen and equilibrium speeds of sound, respectively. For their definition see Becker 1972, Clarke 1969, and Vincenti and Kruger 1965. It should be noted that the frozen speed of sound is always greater than the equilibrium speed c_e. In its turn c_e is greater than the speed of sound for conventional nonrelaxing fields, $c_ = (\kappa p_o/\rho_o)^{1/2}$, where κ is the appropriate polytropic gas coefficient.

Equation (2), the <u>dynamic</u> equation of state, takes into account the finite reaction rate and is valid for non quasistatic and noninstantaneous reaction processes.

[π]More complex models involving multiple relaxation processes are considered e.g. by Napolitano (1964) and Becker and Bohme (1969).

[∫]In this framework each local flow variable denoted generically by f should be expressed as: $f(x_i,t) = f_o + \hat{f}$, (i=1,2,3), f_o being a reference quantity held to be constant and corresponding to the undisturbed equilibrium conditions while $\hat{f} \equiv \hat{f}(x_i,t)$, the small disturbance field fulfilling the requirement $\hat{f}/f_o \ll 1$.

Integration of Eq. (2) yields another form of the dynamic state equation

$$\hat{p} = c_e^2 \hat{\rho} + \frac{c_f^2 - c_e^2}{\tau} \int_{-\infty}^{t} \frac{d\hat{\rho}(t')}{dt'} e^{-(t-t')/\tau} dt' \quad (3)$$

which in its turn may be modified as under

$$\hat{p} = c_f^2 \hat{\rho} - \frac{c_f^2 - c_e^2}{\tau} \int_{-\infty}^{t} \hat{\rho}(t') e^{-(t-t')/\tau} dt' \quad (4a)$$

A more compact form of Eq. (4a) is:

$$\hat{p} = \int_{-\infty}^{t} k(t-t') \hat{\rho}(t') dt' \quad (4b)$$

where the kernel k is expressed by

$$k = c_f^2 \delta(t - t') - \frac{c_f^2 - c_e^2}{\tau} e^{-(t-t')/\tau}$$

while $\delta(t)$ stands for Dirac's distribution.

Equations (3) and (4) are identical to those obtained by Becker 1967 and Rudenko and Soluian 1975.

For simple harmonic motion, implying the following generic representation of field variables:

$$\hat{f}(x_i, t) = \bar{f}(x_i) \exp(j\omega t) \qquad j = (-1)^{1/2} \quad (5)$$

Equation (4) becomes

$$\hat{p} = c^2 \hat{\rho} \quad (6)$$

where $c^2 = c^2(\omega)$ is the dispersed speed of sound in the chemically reacting gas. Its expression

$$c^2(\omega) = c_f^2 - \frac{c_f^2 - c_e^2}{1 + j\omega\tau} \quad (7)$$

is identical to that alternatively derived by Barrère 1964.

The Governing Equation

The state equation [Eqs. (3) or (4)] is to be supplemented by the following linearized equations

$$\frac{D\hat{v}_i}{Dt} = -\frac{1}{\rho_o}[c_f^2 \frac{\partial \hat{\rho}}{\partial x_i} - \frac{c_f^2 - c_e^2}{\tau}\frac{\partial}{\partial x_i}\int_{-\infty}^{t}\hat{\rho}(t')e^{-(t-t')/\tau}dt'], \quad (8)$$
$$(i=1,2,3)$$

$$\frac{D\hat{\rho}}{Dt} + \rho_o \operatorname{div} \vec{\hat{v}} = 0 \quad (9)$$

the former being the Eulerian equation of motion (in which the dynamic equation of state was used) and the latter one being the continuity equation. Here $D/Dt \equiv \partial/\partial t + U\partial/\partial x_1$ denotes the substantial time derivative; $\vec{U} \equiv U\vec{I}_1$ is the undisturbed velocity vector; and $\vec{I}_i$ stands for the unit-vectors along x_i (i=1,2,3) where x_1, x_2 are the streamwise and spanwise coordinates while x_3 is the upward coordinate normal to the mid-surface of the panel (or lifting surface, L.S.). In addition, $\vec{\hat{v}} \equiv \hat{v}_i \vec{I}_i$ denotes the disturbance flow velocity vector.

At this stage, two steps will be performed to determine the equation governing the small disturbances in a compressible reacting gas flow. The first step concerns the acoustical case, $U \equiv 0$. Elimination of $\hat{v}_i$ between Eqs. (8) and (9) (both specialized for U=0) yields

$$\frac{\partial^2 \hat{\rho}}{\partial t^2} - c_e^2 \Delta\hat{\rho} - (c_f^2 - c_e^2)\Delta \int_{-\infty}^{t} \frac{\partial \hat{\rho}}{\partial t'} e^{-(t-t')/\tau}dt' = 0 \quad (10)$$

where Δ denotes the three-dimensional Laplace operator. Multiplication of Eq. (10) by $\tau\partial/\partial t$, followed by the substraction of Eq. (10), yields the equation

$$\tau\frac{\partial}{\partial t}(\frac{\partial^2 \hat{\rho}}{\partial t^2} - c_f^2 \Delta\hat{\rho}) + \frac{\partial^2 \hat{\rho}}{\partial t^2} - c_e^2 \Delta\hat{\rho} = 0 \quad (11)$$

The next step involves the transformation of Eq. (11) to a set of coordinates moving with constant velocity U in the** negative x_1 direction (i.e., a Galilean transformation of Eq. (11)):

**For its definition see Pai (1981), p. 513.

$$\tau \frac{D}{Dt}\left(\frac{D^2\hat{\rho}}{Dt^2} - c_f^2\Delta\hat{\rho}\right) + \frac{D^2\hat{\rho}}{Dt^2} - c_e^2\Delta\hat{\rho} = 0 \qquad (12)$$

Equation (12) governs the disturbances field produced by the unsteady motion of a slender body in a chemically-reacting gas. The perfect similarity between Eq. (12) expressed in terms of $\hat{\rho} \equiv \hat{\rho}(x_i,t)$ (playing the role of a potential) and the one expressed in terms of the small-disturbance velocity potential (derived independently by Vincenti 1959, and Moore and Gibson 1960) is worthy of mention. However, the former representation appears more appropriate for approaching aerolastic problems†† and in addition its introduction does not imply an explicit statement of irrotationality of the flowfield.

The solution to Eq. (12) is subject to the constraints imposed by the impenetrability condition expressed by

$$\hat{v}_3\big|_{x_3=0} = -\frac{DZ}{Dt} \qquad (13)$$

and the finiteness condition at infinity. The quantity $Z \equiv Z(x_1,x_2,t)$ denotes the transverse deflection of the panel (or of L.S.) immersed in the flowing gas, while $\hat{v}_3$ denotes the downwash velocity. The nondimensional variables are introduced into the governing equations and constraints:

$$\overset{v}{x}_i \equiv x_i/\ell$$
$$(i=1,3)$$
$$\overset{v}{x}_2 \equiv x_2/b \qquad (14)$$
$$\overset{v}{\omega} \equiv \omega\ell/U$$
$$\overset{v}{t} \equiv tU/\ell$$

where ℓ and b denote, respectively, a reference chord and the span of the panel (or of L.S.).

If the density $\hat{\rho}$ is expressed as:

$$\hat{\rho} = \bar{\rho}(\overset{v}{x}_1, \overset{v}{x}_2, \overset{v}{x}_3)\exp(j\overset{v}{\omega}\overset{v}{t}) \qquad (15)$$

---

††In this sense, see papers by Librescu (1977a,b) where, in the context of ionized, nonreactive flows, the representation in terms of $\hat{\rho}$ was preferred to the latter one.

Equation (12) in conjunction with Eq. (14) may be transcribed under the form

$$[(\hat{\tau} c_f^2/c_e^2)(\frac{\partial S_f}{\partial \overset{v}{x}_1} + j\overset{v}{\omega}S_f) + S_e]\hat{\rho} = 0 \qquad (16)$$

where $\hat{\tau} \equiv \tau U/\ell$ is a nondimensional relaxation time, S_f and S_e are differential operators defined as

$$S_A \equiv B_A^2 \frac{\partial^2}{\partial \overset{v}{x}_1^2} + 2j\overset{v}{\omega}M_A^2 \frac{\partial}{\partial \overset{v}{x}_1} - M_A^2\overset{v}{\omega}^2 - \phi^2 \frac{\partial^2}{\partial x_2^2} - \frac{\partial^2}{\partial \overset{v}{x}_3^2} \qquad (17)$$

In addition,

$$\phi \equiv \ell/b$$
$$B_A^2 \equiv M_A^2 - 1$$
$$M_A^2 \equiv U^2/c_A^2 \qquad (18)$$
$$(A \equiv f,e)$$

while M_f and M_e are, respectively, the frozen and equilibrium Mach numbers of the freestream.

Equation (16) constitutes an alternative form to Eq. (12). It describes three-dimensional unsteady flow regimes. Those regimes are: 1) supersonic for $B_e^2 \geq B_f^2 \geq 0$, 2) subsonic for $0 \geq B_e^2 \geq B_f^2$, and 3) transonic for $B_e^2 \geq 0 \geq B_f^2$.

The Two-Dimensional Flow

The problem is now restricted to the case of a panel (or L.S.) which extends to infinity in the $\overset{v}{x}_2$ direction and is flown supersonically in the $\overset{v}{x}_1$ direction by a chemically-reacting gas. Since in this case the disturbances do not propagate in $\overset{v}{x}_1 < 0$ direction, Eq. (16) (specialized for the two-dimensional case) will be subjected to a Laplace Transformation (L.T.) with respect to $\overset{v}{x}_1$. Since $\hat{\rho} = \partial\hat{\rho}/\partial\overset{v}{x}_1 \equiv 0$ for $\overset{v}{x}_1 \leq 0$, the 2-D governing equation in the Laplace space is:

$$\frac{d^2\hat{\rho}^L}{d\overset{v}{x}_3^2} - \mu^2\hat{\rho}^L = 0 \qquad (19)$$

where $\hat{\rho}^L(s, \overset{v}{x}_3, \overset{v}{t}) \equiv L\{\hat{\rho}(\overset{v}{x}_1, \overset{v}{x}_3, \overset{v}{t})\}$ and

$$\mu^2 = \frac{B_e^2 s^2 + 2j\overset{v}{\omega} M_e^2 s - M_e^2 \overset{2v^2}{\omega} + T(j\overset{v}{\omega} + s)(B_f^2 s^2 + 2j\overset{v}{\omega} M_f^2 s - M_f^2 \overset{2v^2}{\omega})}{1 + T(s + j\overset{v}{\omega})} \quad (20)$$

where s is the L.T. variable while $T \equiv \hat{\tau} c_f^2 / c_e^2$ defines a modified relaxation time.

The general solution to Eq. (19), which fulfills the impenetrability condition Eq. (13) (converted in terms of $\hat{\rho}$ with the aid of Eq.(8) specialized for i=3) and the finiteness condition at infinity, is

$$\hat{\rho}^L \Big|_{\overset{v}{x}_3 = 0_\pm} = - \frac{\rho_o U^2}{\ell} \frac{(s + j\overset{v}{\omega})^2}{\mu c^2(\overset{v}{\omega})} Z^L \text{sign} \overset{v}{x}_3 . \quad (21)$$

Here, $Z_v^L \equiv Z^L(s, \overset{v}{t}) = L[Z(\overset{v}{x}_1, \overset{v}{t})]$ while $\text{sign} \overset{v}{x}_3 (=1$ for $\overset{v}{x}_3 > 0$; and -1 for $\overset{v}{x}_3 < 0)$ denotes the signum distribution.

Equation (21) shows that $\hat{\rho}^L$ is antisymmetrical with respect to $\overset{v}{x}_3 = 0$. This property proves very useful for the evaluation of the pressure jump on lifting surfaces. On the basis of Eqs. (21), (20), and (6), $\hat{p}^L\Big|_{\overset{v}{x}_3=0_+}$ is given by:

$$\hat{p}^L \Big|_{\overset{v}{x}_3 = 0_+} = - \frac{U^2 \rho_o}{\ell M_e} (s + j\overset{v}{\omega})^2$$

$$\times \frac{(1 + T(s + j\overset{v}{\omega}))^{1/2}}{[B_e^2/M_e^2 \overset{v}{\omega}^2 s^2 + 2j\overset{v}{\omega} s - \overset{v^2}{\omega} + (s + j\overset{v}{\omega})(B_f^2/M_f^2 \overset{v}{\omega}^2 s^2 + 2j\overset{v}{\omega} s - \overset{v^2}{\omega})]^{1/2}} Z^L \quad (22)$$

The inverse Laplace transform of Eq. (22) yields the two-dimensional unsteady pressure distribution of the chemically-reacting gas flow. However, due to the complexity of Eq. (22), the analysis is restricted to cases which play a special role in flutter analyses and for which closed-form solutions may be obtained.

The Quasistatic Supersonic Approximation

For this case it is assumed that the panel (or L.S.) performs harmonic oscillations with the very small frequency $\overset{v}{\omega}$. In the limit $\overset{v}{\omega} = 0$ Eq. (22) becomes

$$\hat{p}^L\Big|_{\overset{v}{x}_3=0_+} = -\frac{\rho_o U^2}{\ell B_f} \frac{(T^{-1}+s)^{1/2}}{s(T^{-1}B_e^2/B_f^2+s)^{1/2}} s^2 z^L \quad (23)$$

If the convolution theorem and the inversion relationship (see Erdelyi 1954, p. 254)

$$L^{-1}\left[\frac{1}{s}\left(\frac{b+s}{c+s}\right)^{1/2}\right] = \exp\left(-\frac{c+b}{2}\overset{v}{x}_1\right) I_o\left(\frac{c-b}{2}\overset{v}{x}_1\right) \quad (24)$$

$$+ b\int_o^{\overset{v}{x}_1} \exp\left(-\frac{c+b}{2}\overset{v}{\xi}_1\right) I_o\left(\frac{c-b}{2}\right)\overset{v}{\xi}_1)d\overset{v}{\xi}_1$$

are invoked the following pressure distribution is derived from Eq. (23)

$$\hat{p}\Big|_{\overset{v}{x}_3=0_+} = -\frac{\rho_o U^2}{\ell B_f}\left[\frac{\partial z}{\partial \overset{v}{x}_1}\Big|_{\overset{v}{x}_1=0} \tilde{K}(\overset{v}{x}_1) + \int_o^{\overset{v}{x}_1}\frac{\partial^2 z}{\partial \overset{v}{\xi}_1^2}\tilde{K}(\overset{v}{x}_1-\overset{v}{\xi}_1)d\overset{v}{\xi}_1\right] \quad (25)$$

where the aerodynamic Kernel $\tilde{K}$ is expressed by

$$\tilde{K}(\overset{v}{x}_1, T) = \exp\left(-\frac{1+B_e^2/B_f^2}{2T}\overset{v}{x}_1\right) I_o\left(\frac{B_e^2/B_f^2-1}{2T}\overset{v}{x}_1\right)$$

$$+ \frac{1}{T}\int_o^{\overset{v}{x}_1} \exp\left(-\frac{1+B_e^2/B_f^2}{2T}\overset{v}{\chi}_1\right) I_o\left(\frac{B_e^2/B_f^2-1}{2T}\overset{v}{\chi}_1\right)d\overset{v}{\chi}_1 \quad (26)$$

In Eq. (26) I_o denotes the zero-order modified Bessel function of the first kind, while $\overset{v}{\xi}_1$ and $\overset{v}{\chi}_1$ are dummy variables related to $\overset{v}{x}_1$. The function $\tilde{K}$ as defined by Eq. (26) also occurs in the analysis of nonequilibrium aerodynamics, but in a different context (see papers by Clarke 1960, and Der 1961). In the limits $T\to\infty$ and $T\to\infty$ Eq. (26) reduces to

$$\lim_{T\to\infty}\tilde{K} = 1 \text{ and } \lim_{T\to\infty}\tilde{K} = B_f/B_e$$

The pressure for these two limiting cases may be expressed under the following unitary form

$$\hat{p}\bigg|_{\overset{v}{x}_3=0_+} = -\frac{\rho_o U^2}{\ell B_A} \frac{\partial \bar{Z}}{\partial \overset{v}{x}_1} e^{j\omega \overset{v}{t}}$$

(27)

$$(Z \equiv \bar{Z}(\overset{v}{x}_1) e^{j\omega \overset{v}{t}})$$

where $(A \equiv e,f)$, the subscripts e or f denoting equilibrium or frozen flow conditions obtained for $T \to 0$ and $T \to \infty$, respectively.

For finite reaction-rates $(T \neq 0, \infty)$, it may be inferred from Eq. (25) that the pressure at a certain point $\overset{v}{x}_1$ depends upon the values of Z at all other points $(0 \leq \overset{v}{\xi}_1 \leq \overset{v}{x}_1)$. In contrast to this dependence, in the limiting cases $T \to 0$ and $T \to \infty$ (as for classical nonreacting flows), the pressure is a local function of $\overset{v}{x}_1$ in the sense that $\hat{p}_{x_3=0}$ at point $\overset{v}{x}_1$ depends upon Z at the same point.

As in the case of classical nonreacting flows (see Dowell and Voss 1965) the quasistatic approximation of reacting flowfields may constitute an efficient tool for panel flutter analyses at low supersonic velocities.

The Case of High Supersonic Mach Numbers

Another case deserving attention is that of high supersonic flight velocities. The velocity is assumed to be sufficiently high so that $M_A^2/B_A^2 \to 1$, $(A \equiv e,f)$, or alternatively, that $M_A^2 \gg 1$. Hence, Eq. (20), which defines μ^2, becomes

$$\mu^2 = \frac{M_e^2(j\overset{v}{\omega} + s)^2[1 + \hat{\tau}(j\overset{v}{\omega} + s)]}{1 + T(s + j\overset{v}{\omega})}$$

(28)

from which by virtue of Eqs. (21) and (6), $\hat{p}^L\bigg|_{\overset{v}{x}_3=0_+}$ becomes

$$\hat{p}^L\bigg|_{\overset{v}{x}_3=0_+} = -\frac{\rho_o U c_f}{\ell} s(s + j\overset{v}{\omega}) \frac{1}{s} \left(\frac{s + j\overset{v}{\omega} + T^{-1}}{s + j\overset{v}{\omega} + \hat{\tau}^{-1}}\right)^{1/2} Z^L$$

(29)

Employment of Eq.(24) and of the convolution theorem leads to the following pressure expression

$$\hat{p}\Big|_{\overset{v}{x}_3=0_+} = -\frac{\rho_o U}{\ell} c_{\hat{\tau}} \left[\frac{\partial Z}{\partial \overset{v}{x}_1}\Big|_{\overset{v}{x}_1=0_+} \tilde{\tilde{K}}(\overset{v}{x}_1) + \int_o^{\overset{v}{x}_1} \left(\frac{\partial^2 Z}{\partial \xi_1^{v2}}\right) \right.$$

$$\left. + j\overset{v}{\omega} \frac{\partial Z}{\partial \xi_1^v} \right) \tilde{\tilde{K}}(\overset{v}{x}_1 - \overset{v}{\xi}_1) d\overset{v}{\xi}_1] \tag{30}$$

The aerodynamic Kernel is

$$\tilde{\tilde{K}}(\overset{v}{x}_1; \hat{\tau}, \overset{v}{\omega}) = \exp[-(j\overset{v}{\omega} + \frac{1+c_e^2/c_f^2}{2\hat{\tau}}) \overset{v}{x}_1] I_o(\frac{1-c_e^2/c_f^2}{2\hat{\tau}} \overset{v}{x}_1)$$

$$+ (j\overset{v}{\omega} + \frac{1}{\hat{\tau}} \frac{c_e^2}{c_f^2}) \int_o^{\overset{v}{x}_1} \exp[-(j\overset{v}{\omega} + \frac{1+c_e^2/c_f^2}{2\hat{\tau}}) \overset{v}{X}_1]$$

$$\times I_o(\frac{1-c_e^2/c_f^2}{2\hat{\tau}} \overset{v}{X}_1) d\overset{v}{X}_1 \tag{31}$$

Its limits are $\tilde{\tilde{K}} = 1$ (for $\hat{\tau} \to \infty$) and $\tilde{\tilde{K}} = c_e/c_f$ (for $\hat{\tau} \to \infty$), while the corresponding limiting pressures may be expressed under the unified form as

$$\hat{p}\Big|_{\overset{v}{x}_3=0_+} = -\frac{\rho_o U c_A}{\ell} (\frac{\partial \bar{Z}}{\partial \overset{v}{x}_1} + j\overset{v}{\omega} \bar{Z}) e^{j\overset{v}{\omega}t} \tag{32}$$

($A \equiv e, f$, for $\hat{\tau} \to 0$ and $\hat{\tau} \to \infty$, respectively).

It is easily seen that Eqs. (32) and (30) represent the frozen (for $A \equiv f$) and equilibrium (for $A \equiv e$), nonequilibrium counterparts of the classical non-reacting linear piston theory aerodynamics as derived by Ashley and Zartarian 1956. It may be remarked that in the case of finite-rate reactions, the Kernel is a complex-valued function dependent on ω, c_e^2/c_f^2, $\overset{v}{x}_1$. The integral term in Eq. (30) describes the spatial memory effect, in the sense that the pressure at a certain point $\overset{v}{x}_1$ depends not only on Z at $\overset{v}{x}_1$ but on all the values of Z in the space interval $[0, \overset{v}{x}_1)$.

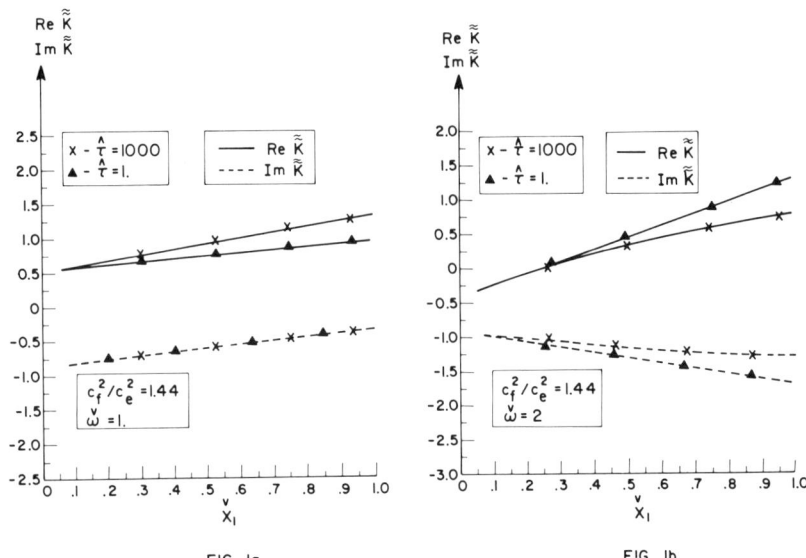

FIG. 1a   FIG. 1b

Fig. 1 Depictions of $\tilde{\tilde{K}}(\overset{v}{x}_1; \overset{v}{\omega}, \hat{\tau})(\equiv \mathrm{Re}K + j\mathrm{Im}K)$, vs $\overset{v}{x}_1$ for two values of $\hat{\tau}$ and $\overset{v}{\omega}$. For $\hat{\tau}(0, \infty)$, K turns out to be a real valued quantity, i.e., $\lim_{\hat{\tau} \to \infty} K \to 1$; $\lim_{\hat{\tau} \to 0} \tilde{\tilde{K}} \to c_e/c_f (= 0.833)$.

In contrast to this behavior, in the equilibrium and frozen flows (and also within the classical framework) K turns out to be a real-valued quantity independent of $\overset{v}{x}_1$ and $\overset{v}{\omega}$. In these cases $\hat{p}$ has a <u>local</u> character (which is equivalent to asserting that the spatial memory effect is zero).

The depiction of $\tilde{K}$ for several parameters is given in Figs. 1a and 1b.

The large variability in $\tilde{K}$, as shown in these figures, may induce large differences in the obtained flutter speeds. To predict the relationship between the flutter velocities arriving in the two extreme nonequilibrium flight conditions, the case of bending-torsion flutter of an airfoil flown by a high supersonic chemically-reacting gas at the velocity U will be considered.

Extending Ashley and Zartarian 1956 treatment for reacting flows, Eq. (32) is used separately for the two extreme non-equilibrium conditions. The resulted ratio of the flutter speeds, corresponding to frozen ($\tilde{U}$) and equilibrium ($\tilde{U}$) flight conditions is $(U/\tilde{U})_F \approx c_e/c_f$. This estimate is obtained for zero frequency ratio and large mass ratio parameters. Since c_f and c_e can differ by a substantial amount (see Anderson 1982 pp. 411–412), the flutter velocity, which is determined on the basis of frozen flows, may be 20-25% lower (i.e., more critical) than that predicted on the basis of equilibrium (or of ordinary chemically nonreacting) flow conditions.

Linear piston theory aerodynamics deduced in this broader context could play an important role in the determination of the flutter instability boundary of panel and lifting surfaces flown by a high supersonic reactive gas. A full representation of reacting piston-theory aerodynamics should also include appropriate nonlinear terms. These terms (together with the ones appropriate to the structural non-linearities) would make it possible to determine the conditions which define the <u>character</u> of the flutter boundary and to determine implicitly the influence of aerodynamic (and structural) non-linearities. For pertinent analyses of this topic (developed in the framework of the nonreacting, nonlinear piston theory aerodynamics), see works by Bolotin 1963; Librescu 1965, 1975; and McIntosh Jr. 1973.

Circular Cylindrical Shells Flown (Internally or Externally) by a Supersonic Reactive Gas

The case of a circular cylindrical shell of finite length flown externally (or internally) by a supersonic reactive gas shall now be considered. The undisturbed velocity vector $U(\equiv U\,I_1)$ of the gas is parallel to the cylinder axis. The governing equation is developed along the lines discussed in previous sections. In this sense Eq. (12) will be converted into a cylindrical coordinate system $(\overset{v}{x}_1, \theta, r)$ and the nondimensional variables defined by $\overset{v}{x}_1 = x_1/\ell$, $\overset{v}{x}_3 = r/R$; $\overset{v}{t} = tU/\ell$; $\overset{v}{\omega} = \omega\ell/U$ will be introduced, where R denotes the radius of curvature of the cylinder midsurface. Consistent with transverse deflection Z expressed as $Z(\overset{v}{x}_1, \theta, \overset{v}{t}) = \bar{Z}_n(\overset{v}{x}_1)\exp(j\overset{v}{\omega}\overset{v}{t})\cos n\theta$, the remaining unknown functions will be represented as

$$\begin{array}{l} \hat{f}(\overset{v}{x}_1, \theta, \overset{v}{x}_3, \overset{v}{t}) \\ \hat{g}(\overset{v}{x}_1, \theta, \overset{v}{x}_3, \overset{v}{t}) \end{array} = \begin{array}{l} \bar{f}_n(\overset{v}{x}_1, \overset{v}{x}_3)\cos n\theta \\ \bar{g}_n(\overset{v}{x}_1, \overset{v}{x}_3)\sin n\theta \end{array} e^{j\overset{v}{\omega}\overset{v}{t}} \qquad (33)$$

where $\hat{f}$ and $\hat{g}$ denote generically one of the disturbance functions $\hat{v}_1$, $\hat{v}_3$, $\hat{\rho}$ and $\hat{v}_2$, respectively, while n denotes the circumferential wave number.

If the previous developments are followed and a L.T. with respect to $\overset{v}{x}_1$ is performed the following governing equation in the image space results

$$\bar{\rho}^{-L}_{\overset{o}{x}_3\overset{o}{x}_3} + \frac{1}{\overset{o}{x}_3}\bar{\rho}^{-L}_{\overset{o}{x}_3} - (1 + \frac{n^2}{\overset{o2}{x}_3})\bar{\rho}^{-L}=0 \quad (\bar{\rho}^{-L}_{\overset{o}{x}_3} \equiv \partial\bar{\rho}^{-L}/\partial\overset{o}{x}_3) \quad (34)$$

where $\overset{o}{x}_3 = \nu\overset{v}{x}_3$, while

$$\nu^2 = (R^2/L^2)[M_e^2(\frac{B_e^2}{M_e^2}\overset{v}{s}^2 + 2j\overset{v}{\omega}\overset{v}{s} - \overset{v}{\omega}^2) + T(j\overset{v}{\omega}+s)M_f^2(\frac{B_f^2}{M_f^2}\overset{v}{s}^2 + 2j\overset{v}{\omega}\overset{v}{s} - \overset{v}{\omega}^2)]$$

$$\div 1 + T(j\overset{v}{\omega} + s) \qquad (35)$$

For the sake of brevity, in Eq. (34) the subscript n affecting $\bar{\rho}^{-L}$ was dropped.

The solution to Eq. (34) (considered as a modified Bessel equation), satisfying to the impenetrability condition and the finiteness conditions at $\overset{v}{x}_3 \to \infty$ (for external flow conditions) or at $\overset{v}{x}_3 = 0$ (for internal flow), provides with aid of Eq. (6) the disturbance pressure (in Laplace space)

$$\hat{p}_n^L(s,\theta,\overset{v}{\ell})\Big|_{\overset{v}{x}_3=1} = \frac{\rho_o U^2(s+j\overset{v}{\omega})^2}{\ell}\left(\frac{\Gamma_n(\nu\overset{v}{x}_3)}{\nu\Gamma_n'(\nu\overset{v}{x}_3)}\right)\Big|_{\overset{v}{x}_3=1} Z_n^L(s,\theta,\overset{v}{\ell})$$

(36)

where

$$\Gamma_n(x) = \begin{matrix} K_n(x) \text{ (for external flow)} \\ I_n(x) \text{ (for internal flow)} \end{matrix}$$

and K_n and I_n denote the modified Bessel functions of second and first kind, respectively.

Due to the intricacy of $\psi_n(\nu\overset{v}{x}_3) \equiv \Gamma_n(\nu\overset{v}{x}_3)/(\nu\Gamma_n'(\nu\overset{v}{x}_3))$ is would be not possible to obtain a closed-form expression for the inverse transform of Eq. (36).

For that is the reason the pressure can be evaluated only in an approximate sense. For large values of ν (implying small ℓ/R) and arbitrary n, the asymptotic evaluation of $\psi_n(\nu\overset{v}{x}_3)$ (see Miles 1957 and Dzygadlo 1969) is

$$\psi_n(\nu\overset{v}{x}_3)\Big|_{\overset{v}{x}_3=1} = \begin{matrix} \dfrac{1}{(n^2+\nu^2)^{1/2}} \text{ for external flow} \\ \\ -\dfrac{1}{(n^2+\nu^2)^{1/2}} \text{ for internal flow} \end{matrix} \qquad (37)$$

It is easily seen that with the introduction of Eq.(37) the difficulties in obtaining the original of $\hat{p}^L$ are still maintained. However, a relationship expressing the applicability conditions of linear piston theory aerodynamics (derived for planar surfaces) to cylindrical shells may be deduced. To this end, Eq. (29) shall be compared to (36) [considered in conjunction with Eq. (37)]. As it may easily be seen, the two expressions coincide whenever conditions a) and b) are fulfilled, namely

$$\text{a) } M_A^2/B_A^2 \to 1 \text{ and b) } \left(\frac{n\ell}{\pi R}\right)^2 \ll \frac{M_e^2}{\pi^2} \frac{1 + \hat{\tau}(j\overset{V}{\omega} + s)}{1 + \hat{\tau}(c_f^2/c_e^2)(j\overset{V}{\omega} + s)}$$

where large s implies small $\overset{V}{x}_1$ and vice-versa. For the limiting cases ($_2\hat{\tau}\to 0$ and $\hat{\tau}\to\infty$), condition b) becomes: $(n\ell/\pi R)^2 \ll M_A^2/\pi^2$, $(A \equiv e, f)$.

It may be concluded that linear piston theory aerodynamics (generalized to include the nonequilibrium phenomena) deduced for planar surfaces may be applied at high supersonic flight Mach numbers to short cylndrical shells with an arbitrary circumferential wave number n.

These conclusions obtained here on the basis of a qualitative analysis (and subsisting in the nonreacting case also), are similar to those derived by numerical means within the classical framework (see Parthan and Johns 1972). In the case of finite-rate-reactions, the applicability of planar piston theory aerodynamics is enhanced for the front region of the cylindrical panel.

Finally, it should be mentioned that the results obtained constitute a basis for a consistent evaluation of aeroelastic phenomena confronting spacecraft and rocket nozzles in reactive flowfield. Nevertheless, Eqs. (16), (22), and (34) may furnish additional results, which could enlarge the field applicability both as concerns the velocity range and the shape of considered panels (or L.S.).

Acknowledgement

The author expresses his thanks to the reviewers for their helpful comments in improving the first version of the manuscript.

References

Anderson, J. D. Jr. (1982) Modern Compressible Flow; With Historical Perspective, McGraw-Hill Book Co., New York.

Ashley, H. Zartarian, G., (1956) Piston theory-a new aerodynamic tool for the aeroelastician. J. Aeronaut. Sci. 23 (12), 1109-1118.

Barrere, M. (1964) Influence des reactions chimiques sur l'ecoulement, Supersonic Flow, Chemical Processes and Radiative Transfer (edited by H. Olfe and U. Zakkay), The McMillan Co., New York, pp.31-77.

Becker, E. (1967) Neuere Probleme der Dynamik realer Gase, ZAMM 47, T3-15.

Becker, E. (1972) Chemically reacting flows, Ann. Rev. Fluid Mech. 4, 155-194.

Becker, E. and Bohme, G. (1969) Steady one-dimensional flow; structure of compression waves. Gasdynamics, Vol. 1; Nonequilibrium Flows, Part I (edited by P. P. Wegener), pp. 73-117. Marcel Dekker, New York.

Bolotin, V. V. (1963) Nonconservative problems of the theory of elastic stability, pp. 265-280 Pergamon Press, Oxford, England.

Broer, L. J. F. (1969) Some basic properties of relaxation gasdynamics. Gasdynamics; Vol. 1, Nonequilibrium Flows, Part II (edited by P. P. Wegener), pp. 1-31, Marcel Dekker, New York.

Clarke, J. F. (1960) The linearized flow of a dissociating gas. J. Fluid Mech. Part 4, 7(4), 577-595.

Clarke, J. F. (1969) Small-disturbance theories. Gasdynamics, Vol. 1; Nonequilibrium Flows, Part I.(edited by P. P. Wegener), pp. 1-70. Marcel Dekker, New York.

Clarke, J. F. (1978) Gas dynamics with relaxation effects. Rep. Prog. Phys. 41, 807-869.

Clarke, J. F. and McChesney, M. (1976) Dynamics of Relaxing Gases Butterworth, Washington, D.C.

Der, J. J. (1961) Linearized supersonic nonequilibrium flow past an arbitrary boundary. NACA TN-R-119.

Dowell, E. H. and Voss H. M. (1965) Theoretical and experimental panel flutter studies in the Mach number range 1.0 to 1.5. AIAA Journal 3 (12), 2292-2304.

Dzygadlo, Z. (1969) Asymptotic theory of the pressure on a cylindrical shell performing unsteady oscillation in external or internal supersonic flow. Proc. Vibrat. Prob. 1 (10), 41-54.

Tables of Integral Transforms, (edited by A. Erdelyi, 1954) Vol. 1, p. 254. McGraw-Hill Book Co., New York.

de Groot, S. R. and Mazur, P. (1969) Non-Equilibrium Thermodynamics North-Hollan Publ. Co., Amsterdam, The Netherlands.

Librescu, L (1965) Aeroelastic stability of orthotropic heterogeneous thin panels in the vicinity of the flutter critical boundary. J. de Mecanique, 4, 1, 51-76.

Librescu, L. (1975) Aeroelastic stability of anisotropic multilayered thin panels. Elastostatics and Kinetics of Anisotropic and Heterogeneous Shell-Type Structures, Ch. 1, pp. 79-82; 106-178 Noordhoff Internat. Publ., Leyden, The Netherlands.

Librescu, L. (1977a) Recent contributions concerning the flutter problem of elastic thin bodies in an electrically conducting gas flow, a magnetic field being present. S. M. Archives 2, 11-108.

Librescu, L. (1977b) Unsteady magnetoaerodynamic forces on an oscillating circular cylindrical shell of finite length, II transient solution, Rendiconti dei Lincei. Sci. Fis. Mat. e Nat. LXIII, 6(12), 538-543.

McIntosh, S. C. Jr. (1973) Effect of hypersonic nonlinear aerodynamic loading on panel flutter. AIAA Journal 11 (1), 29-32.

Miles, J. W. (1957) Supersonic flutter of a cylindrical shell. J. Aeronaut. Sci. 24 (2), 107-118.

Napolitano, L. G. (1964) Generalized velocity potential equations for pluri-reacting mixtures Archiwum Mechaniki Stosowanej 2 (16), 373-390.

Moore F. K. and Gibson W. E. (1960) Propagation of weak disturbances in a gas subject to relaxation effects. J. Aero/Space Sci. 27 (2), 117-127.

Pai, S. I. (1981) Aerothermochemistry: flow with chemical reactions. Modern Fluid Mechanics, Ch. 5, Science Press, (distributed by Van Nostrand, Reinhold Co., New York.)

Parthan, S. and Johns, D. J. (1972) Aerodynamic generalized forces for supersonic shell flutter. AIAA Journal, 10 (10), 1369-1371.

Rudenko, O. V., Soluian, S. I. (1975) Theoretical Basis of Non-Linear Acoustics, pp. 82-88 (in Russian), Nauka, Moscow, U.S.S.R.

Sislyan, D. S. (1966) Linearized unsteady nonequilibrium flows of a compressible gas. Mechanika Zhidkosti i Gaza, 1(2), 23-28.

Vincenti, W. G. (1959) Nonequilibrium flow over a wavy wall. J. Fluid Mech. Pt. 4, 6, 481-496.

Vincenti, W. G. and Kruger, C. H. (1965) Introduction to Physical Gas Dynamics, pp. 254-310, John Wiley and Sons, Inc., New York.

Wu, J. C. and Turner, L. III (1974) Linear theory for chemically reacting flows. AIAA Journal 12 (4), 468-474.

Uniform Solutions for Characteristics and Weak Shock Waves in a Reactive Medium

Bishun D. Pandey* and David C. Chou†
University of New Mexico, Albuquerque, New Mexico

Abstract

Problems concerning the propagation of weak axisymmetric shock waves in a fluid that may undergo internal nonequilibrium transformation are theoretically investigated. The interesting, singular behavior of a general nonequilibrium system, i.e., the transition from near-field frozen characteristics to far-field equilibrium characteristics, is fully described by the composite solution of the problem through a combination of the characteristic perturbation and the matched asymptotic expansion methods. Nonlinear effects and the nonequilibrium singular phenomena associated with the weak wave propagation are presented. The results of the investigation are compared with some existing experimental data for a weak conical wave propagating in a typical reactive mixture.

Introduction

The unsteady or supersonic steady inviscid nonequilibrium flows are usually described by a set of hyperbolic partial differential equations. The characteristics or the corresponding Mach lines of the system are obtained in terms of local velocities and the frozen speed of sound; i.e., they are the so-called "frozen" characteristics (Chou 1982; Chou and Maa, 1975, 1979; Chou and Chu 1971). However, for either large-time behavior or the far-field prop-

---
Presented at the 9th ICODERS, Poitiers, France, July 3-8, 1983. Copyright American Institute of Aeronautics and Astronautics, Inc., 1984. All rights reserved.
*Research Scholar, Mechanical Engineering.
†Professor and Consultant to the Air Force Weapons Laboratory, Kirtland Air Force Base, N.M.

agation, it is known that the flow would eventually approach either near-equilibrium or equilibrium when the disturbances are attenuated by the nonequilibrium effects. Therefore, the correct characteristics in these regions should be the "equilibrium" ones, even though the governing equations produce only the frozen characteristics. This contradictory phenomenon is due to the singular nature of the nonequilibrium system. In the method of characteristics (MOC) numerical calculation of reacting flows, the frozen characteristics are the ones used.

Sedney et al. (1962) pointed out the need for direct evidence of the reliability of a frozen Mach line scheme in calculating a flowfield going from frozen to near-equilibrium conditions. This need arises not only because the Courant-Freidrichs-Lewy (CFL) principle [see Hahn (1958)] would not assure the global convergence and stability, but also because of a question about the loss of accuracy due to the deviation of characteristics from frozen to equilibrium ones as the flow approaches equilibrium (Sedney 1970). However, a distinctly different difficulty may appear when the MOC is applied to internal flows, where the supersonic region often begins close to the assumed location of equilibrium (Bray 1970).

Chu's (1970) earlier work of propagating discrete acoustic waves in a general nonequilibrium flow provides some physical insights into the singular behavior of the reacting flow system. He indicates that the wave front travels at the frozen speed of sound, irrespective of the value of relaxation time τ. However, the wave front decays exponentially with t/τ_0, and the bulk of the disturbance is traveling eventually more or less at the equilibrium speed of sound, further indicating the seminal importance of resolving this difficult problem.

The present investigation is aimed at removing the error introduced by the singular nature of the nonequilibrium system. Chou and Chu's earlier work (1971) on the decay of weak shock waves in an axisymmetric nonequilibrium flow is extended to obtain the composite solutions for characteristics and shock wave locations. The system of nonequilibrium equations is first transformed into frozen characteristics coordinates to form the "inner" (or the near-field) region description of the problem. The far-field equilibrium equations are also transformed into their corresponding equilibrium coordinates system. Then, the method of matched asymptotic expansions (MAE) is used to construct uniformly valid solutions for characteristics and shock waves up to the second order of the initial disturbance. In the analysis, the characteristics perturbation technique is used to retain the important nonlinear properties of the waves, and the method of matched asymptotic expansion is employed to remove any error introduced by the singular nature of the nonequilibrium system. The expressions for uniformly valid characteristics can be used in the MOC nonequilibrium numerical schemes.

Description of the Problem

Because of the availability of both theoretical and experimental near-field results, the propagation of an axisymmetric weak shock wave in a nonequilibrium fluid was chosen to be investigated. However, the expressions of uniformly valid characteristics and shock waves obtained in the study can be easily modified to suit the corresponding planar and spherical cases.

The basic equations governing the motion of a steady axisymmetric nonequilibrium supersonic flow over a projectile are well known. Neglecting the various transport effects and assuming only one nonequilibrium mode, they may be written in the form (Chou and Chu 1971)

$$\rho\left(u\frac{\partial u}{\partial x} + v\frac{\partial u}{\partial r}\right) + \frac{\partial p}{\partial x} = 0 \qquad \rho\left(u\frac{\partial v}{\partial x} + v\frac{\partial v}{\partial r}\right) + \frac{\partial p}{\partial r} = 0$$

$$\frac{\partial \rho u}{\partial x} + \frac{\partial \rho v}{\partial r} + \frac{\rho v}{r} = 0 \qquad u\frac{\partial q}{\partial x} + v\frac{\partial q}{\partial r} = \dot{q}(p,\rho,q) \qquad (1)$$

$$\frac{1}{a_f^2}\left(u\frac{\partial p}{\partial x} + v\frac{\partial p}{\partial r}\right) - \left(u\frac{\partial \rho}{\partial x} + v\frac{\partial \rho}{\partial r}\right) = \frac{h_q}{h_\rho}\dot{q} \qquad h = h(p,\rho,q)$$

Here, x is the distance along the axis of the symmetry measured from the tip of the projectile in the direction of the oncoming flow, and r is the radial distance from the axis; u and v denote respectively the velocity components along the x and r axis; p, ρ, h, q, and $\dot{q}$ are respectively the pressure, density, specific enthalpy, a variable characterizing the progress of the nonequilibrium process, and the reaction rate (describing the rate of change of the progress variable q); $\dot{q} \equiv \dot{q}(p,\rho,q)$ is a known function of p, ρ, and q. Finally,

$$h_q = \frac{\partial[h(p,\rho,q)]}{\partial q} \qquad h_\rho = \frac{\partial h}{\partial \rho} \qquad h_p = \frac{\partial h}{\partial p}$$

and the frozen sound speed a_f is given by

$$a_f^2 = \left(\frac{\partial p}{\partial \rho}\right)_{s,q} = \frac{-h_\rho}{h_p - 1/\rho} \qquad (2)$$

(Vincenti and Kruger 1965). When the frozen Mach number $M_f = [(u^2 + v^2)/a_f^2]^{1/2}$ is greater than 1, system (1) possesses three families of real characteristics: the outgoing and incoming Mach waves and the streamlines. In terms of u, v, and a_f, an outgoing Mach wave has a slope determined by

$$\frac{dx}{dr} = \frac{-uv + a_f(u^2 + v^2 - a_f^2)^{1/2}}{a_f^2 - v^2} = \lambda \qquad (3)$$

where $\lambda = \cot(\theta + \mu_f)$; $\theta = \tan^{-1}(v/u)$ is flow angle; and μ_f is the frozen Mach angle.

In studying the shape of the nose shock, it is essential [see, e.g., Chu (1970)] to introduce a new coordinate system (α, β) defined as follows: α is constant along an outgoing Mach line so that if this line intersects the surface of the projectile at a point $x = x^*$, the line will be labeled $\alpha = x^*$; β is simply r. It follows that in the region of the flowfield affected by the projectile motion, $x = x(\alpha, \beta)$ and $r = \beta$. The transformation relationships between (x, r) on the one hand and (α, β) on the other can be deduced immediately from $dx = x_\alpha d\alpha + x_\beta d\beta$, where subscripts α and β signify partial differentiation with respect to α and β, respectively; for example, $\partial\alpha/\partial x = 1/x_\alpha$, $\partial\alpha/\partial r = -x_\beta/x_\alpha$, etc. In terms of α and β, Eqs. (1) and (3) become

$$\rho[(u - \lambda v) u_\alpha + vu_\beta x_\alpha] = -p_\alpha$$

$$\rho[(u - \lambda v) v_\alpha + vv_\beta x_\alpha] = \lambda p_\alpha - p_\beta x_\alpha$$

$$(\rho u)_\alpha - \lambda(\rho v)_\alpha + x_\alpha(\rho v)_\beta = -\rho v x_\alpha/\beta \qquad (4)$$

$$(u - \lambda v) q_\alpha + v q_\beta x_\alpha = \dot{q} x_\alpha$$

$$(1/a_f^2)[(u - \lambda v) p_\alpha + v p_\beta x_\alpha] - [(u - \lambda v) \rho_\alpha + v \rho_\beta x_\alpha]$$

$$= (h_q/h_\rho) \dot{q} x_\alpha, \quad x_\beta = \lambda$$

Let the projectile be described by the equation $r = \varepsilon R(x)$, where ε may be taken as the slope dr/dx of the projectile at $x = 0$. We shall be concerned here with slender and smooth projectiles; that is, ε is assumed small and $R(x)$ is a differentiable function for $x > 0$. Generalization to cases where dr/dx may have jump discontinuities along the projectile presents no basic difficulties.

The requirement that the flow should be tangential to the projectile implies that $v/u \to \varepsilon R'(x)$ as $r \to \varepsilon R(x)$. In the α, β plane, we have

$$v/u \to \varepsilon R'(\alpha) \quad \text{as} \quad \beta \to \varepsilon R(\alpha)$$
$$x = \alpha \quad \text{at} \quad \beta = \varepsilon R(\alpha) \tag{5}$$

The last condition follows directly from the particular way in which the characteristics are labeled. In addition to these boundary conditions, the usual jump conditions must be satisfied at each point on the nose shock. Thus, if δ is the shock angle and subscript 0 denotes the freestream condition, which is assumed to be in thermodynamic equilibrium, we have

$$\rho(u - v \cot \delta) = \rho_0 u_0, \quad h + \tfrac{1}{2} u^2 = h_0 + \tfrac{1}{2} u_0^2 \quad q = q_0$$
$$p - p_0 = \rho_0 u_0 (u_0 - u), \quad v = (u_0 - u) \cot \delta \tag{6}$$

representing respectively the continuity equation, the energy equation, the continuity of q, and the momentum balance in the normal and tangential directions. The position of the nose shock is, of course, not known <u>a priori</u>. It must assume a form so that $dx/dr = \cot \delta$ at every point on the nose shock. In the α, β plane, the system of boundary conditions (6) must be applied along the shock locus $\alpha = \alpha(\beta, \varepsilon)$ determined by integrating.

$$x_\alpha \frac{d\alpha}{d\beta} + x_\beta = \cot \delta \quad \text{or} \quad \frac{d\alpha}{d\beta} = \frac{\cot \delta - x_\beta}{x_\alpha} \tag{7}$$

subject to the initial condition $\alpha = 0$ at $\beta = 0$.

Obviously the problem is a very complicated one. Nevertheless, as will be seen, a systematic calculation of the nose shock can be carried out to any degree of accuracy for the case of a slender body.

Inner Solutions

"Inner" Perturbation Equations

The near-field solutions will be labeled as "inner" ones in the investigation. Following the notations used by Van Dyke (1964), $x^i(\alpha,\beta)$ can be any of the dependent variables in system (4), and it can be expressed as

$$x^i(\alpha,\beta) = x_0(\alpha,\beta) + \varepsilon x_1^i(\alpha,\beta) + \varepsilon^2 x_2^i(\alpha,\beta) + \cdots \qquad (8)$$

where x_0 is the value of x in the undisturbed flow stream; therefore, the superscript "i" is unnecessary for the zeroth-order terms. In particular, p_0, ρ_0, q_0, and u_0 are respectively the freestream pressure, density, progress variable, and velocity; moreover, $v_0 = 0$. Writing the dependent variables of the system (4) in the form (8) and collecting terms of like order in ε, results of a set of equations governing terms of zeroth-, first-, second-,... order variables. System (4) gives

$$\rho_0 u_0 u_{1\alpha}^i = -p_{1\alpha}^i, \quad \rho_0 u_0 v_{1\alpha}^i = \lambda_0 p_{1\alpha}^i - p_{1\beta}^i x_{0\alpha}$$

$$\rho_0 u_{1\alpha}^i + \rho_{1\alpha}^i u_0 - \lambda_0 \rho_0 v_{1\alpha}^i + x_{0\alpha} \rho_0 v_{1\beta}^i = \frac{-\rho_0 v_1^i x_{0\alpha}}{\beta} \qquad (9)$$

$$u_0 q_{1\alpha}^i = \dot{q}_1^i x_{0\alpha} \quad \frac{1}{a_{f0}^2}\left(p_{1\alpha}^i - \rho_{1\alpha}^i\right) = \left(\frac{h_q}{h_\rho u}\right)_0 x_{0\alpha} \dot{q}_1^i$$

etc., where

$$\dot{q}_1^i = \dot{q}_{q0}(q_1^i - q_1^*), \quad q_1^* = \left(\frac{\partial q^*}{\partial p}\right)_0 p_1^i + \left(\frac{\partial q^*}{\partial p}\right)_0 \rho_1^i$$

and $q^* = q^*(p,\rho)$ is the equilibrium value of the progress variable q at a pressure p and density ρ.
Also

$$x_{0\beta} = \lambda_0, \quad x_{1\beta}^i = \lambda_1^i = -M_{f0}^2 \frac{v_1^i}{u_0} + \frac{M_{f0}^2}{\lambda_0}\left(\frac{u_1^i}{u_0} - \frac{a_1}{a_{f0}}\right) \cdots \qquad (10)$$

where

$$\lambda_0 = (M_{f0}^2 - 1)^{1/2} = \cot \mu_{f0}, \quad a_f = a_{f0} + \varepsilon a_1 + \ldots$$

$$a_1 = \left(\frac{\partial a_f}{\partial p}\right)_0 p_1^i + \left(\frac{\partial a_f}{\partial \rho}\right)_0 \rho_1^i + \left(\frac{\partial a_f}{\partial q}\right)_0 q_1^i$$

In a similar manner, the boundary condition (5) can be decomposed into a set of conditions, according to different powers of ε, with the help of Taylor's expansion. For example, Eq. (5) gives

$$x_0(\alpha,0) = \alpha, \quad x_1^i(\alpha,0) + x_{0\beta}(\alpha,0) R(\alpha) = 0 \tag{11}$$

$$x_2^i(\alpha,0) + x_{1\beta}^i(\alpha,0) R(\alpha) + \frac{1}{2} x_{0\beta\beta}(\alpha,0) R^2(\alpha) = 0 \ldots$$

and

$$\lim_{\beta \to 0} \beta v_1^i = 0$$

$$\lim_{\beta \to 0} \left(\beta v_2^i + R(\alpha) \frac{d(\beta v_1^i)}{d\beta}\right) = u_0 R(\alpha) R'(\alpha) \ldots \tag{12}$$

The perturbed jumping conditions at the frontal shock are

$$\varepsilon^1: \quad \rho_0(u_1^i - v_1^i \cot \delta_0) + \rho_1^i u_0 = 0$$

$$h_{p0} p_1^i + h_{\rho 0} \rho_1^i + h_{q0} q_1^i + u_0 u_1^i = 0 \tag{13}$$

$$q_1^i = 0, \quad p_1^i + \rho_0 u_0 u_1^i = 0, \quad v_1^i + u_1^i \cot \delta_0 = 0$$

Unlike Eqs. (6), this is a homogeneous system in p_1^i, ρ_1^i, u_1^i, v_1^i, and q_1^i. If these are not simultaneously zero at the shock, the determinant of the system must vanish. It follows that

$$\cot \delta_0 = (M_{f0}^2 - 1)^{1/2} = \lambda_0 \quad \text{or} \quad \delta_0 = \mu_{f0}$$

the freestream frozen Mach angle. Moreover,

$$p_1^i = a_{f0}^2 p_1^i = -\rho_0 u_0 u_1^i = (\rho_0 u_0/\lambda_0) v_1^i, \quad q_1^i = 0 \qquad (14)$$

at the shock. Likewise, collecting terms of ε^2 gives

$$\rho_0(u_2^i - v_2^i \cot \delta_0) + \rho_2^i u_0 = -\rho_0 M_{f0}^2 \delta_1 v_1^i + \frac{\rho_0 u_0 (M_{f0}^2/\lambda_0^2) v_1^{i2}}{a_{f0}^2}$$

$$h_{p0} p_2^i + h_{\rho 0} \rho_2^i + h_{q0} q_2^i + u_0 u_2^i = -\frac{1}{2}(M_{f0}^2/\lambda_0^2) v_1^i$$

$$\times \{1 + \rho_0^2 a_{f0}^2 [h_{pp} + (2/a_f^2) h_{p\rho} + (1/a_f^4) h_{\rho\rho}]_0\}$$

$$q_2^i = 0, \quad p_2^i + \rho_0 u_0 u_2^i = 0, \quad v_2^i + u_2^i \cot \delta_0 = \frac{-M_{f0}^2 \delta_1 v_1^i}{\lambda_0}$$

(15)

where subscript 0 appended to a bracket signifies that all quantities in the bracket are evaluated at the undisturbed state of the oncoming freestream. By eliminating the second-order quantities from the system (15),

$$\delta_1 = \frac{1}{4}(A + 1)\left(\frac{M_{f0}^2}{\lambda_0^2}\right) v_1^i / u_0$$

where

$$A = 1 + \frac{a_{f0}^2}{\rho_0 h_{\rho 0}} \left[1 + \rho^2 a_f^2 \left(h_{pp} + \frac{2}{a_f^2} h_{p\rho} + \frac{1}{a_f^4} h_{\rho\rho}\right)\right]_0 \qquad (16)$$

For an ideal gas, $h = [\gamma/(\gamma - 1)]p/\rho$ so that A is just γ. Since

$$\cot \delta^i = \cot(\delta_0 + \varepsilon \delta_1^i + \cdots) \quad \text{and} \quad \cot \delta_0 = \cot \mu_{f0} = \lambda_0$$

$$\cot \delta^i = \lambda_0 - \varepsilon M_{f0}^2 \delta_1^i + O(\varepsilon^2)$$

(17)

$$= \lambda_0 - \frac{1}{4}(A+1)\,\varepsilon\left(\frac{M_{f0}^4}{\lambda_0^2}\right) v_1^i/u_0 + O(\varepsilon^2)$$

Zeroth- and First-Order "Inner" Solutions

The zeroth- and first-order solutions can be written down by inspection, as shown in Chou and Chu (1971). From Eq. (10) and the first equation of Eq. (11),

$$x_0 = \lambda_0 \beta + \alpha \qquad (18)$$

However, the system of Eqs. (9), (11), (12), and (14) do not have any inhomogeneous terms. Therefore,

$$p_1^i = 0,\ \rho_1^i = 0,\ q_1^i = 0,\ u_1^i = 0,\ v_1^i = 0 \qquad (19)$$

The only "inner" first-order quantity that is nonzero is $x_1^i(\alpha,\beta)$ which, according to the second parts of Eqs. (10) and (11), is given by

$$x_1^i = -\lambda_0 R(\alpha) \qquad (20)$$

The second-order "inner" solution assumes the same form as the second-order solution in Chou and Chu's (1971) work, which was constructed by the method of Laplace transform. It has demonstrated that for characteristics located near the shock waves, the following facts do exist, i.e.,

$$\alpha/u_0\tau_0 \quad \text{and} \quad \alpha/\lambda_{f0}\beta \ll 1 \qquad (21)$$

The asymptotic expression for the second-order quantities are

$$v_2^i = u_0(\lambda_0/2\beta)^{1/2} \exp(-\kappa\beta)\, F(\alpha)$$

$$u_2^i = -(v_2^i)/\lambda_0,\quad p_2 = -\rho_0 u_0 u_2^i = a_{f0}^2 \rho_2^i$$

$$(22)$$

$$q_2^i = -\frac{\rho_0 h_{\rho 0}}{h_{q0}} \frac{\sigma_0^M M_{f0}^2}{\lambda_0} \frac{1}{u_0 \tau_0} \left(\frac{\lambda_0}{2\beta}\right)^{1/2} \exp(-\kappa\beta) \frac{1}{2\pi} \int_0^\alpha \frac{S'(\xi) \, d\xi}{(\alpha - \xi)^{1/2}}$$

$$x_2^i = -\left(\frac{\pi}{2}\right)^{1/2} \left(\frac{A+1}{2}\right) \frac{M_{f0}^4}{(\lambda_0^3 \kappa)^{1/2}} \operatorname{erf}[(\kappa\beta)^{1/2}] \, F(\alpha)$$

where

$$F(\alpha) = \frac{1}{2\pi} \int_0^\alpha \frac{S''(\xi) \, d\xi}{(\alpha - \xi)^{1/2}} \quad \text{and} \quad \kappa = \frac{\sigma^M M_{f0}^2}{2\lambda_0 u_0 \tau_0} \tag{23}$$

Inner Characteristics and Shock Locus

Finally, by combining the zeroth-, first-, and second-order solutions listed above, the "inner" characteristics can be given as

$$x^i = \alpha + \lambda_0 \beta - \epsilon\lambda_0 R(\alpha) - \epsilon^2 F(\alpha) G(\beta) \tag{24}$$

where

$$G(\beta) = \left(\frac{\pi}{2}\right)^{1/2} \left(\frac{A+1}{2}\right) \frac{M_{f0}^4}{(\lambda_0^3 \kappa)^{1/2}} \operatorname{erf}(\kappa\beta)^{1/2} \tag{25}$$

since

$$\frac{d\alpha}{d\beta} = \frac{\epsilon^2}{2} \frac{F(\alpha) G'(\beta)}{1 - \epsilon\lambda_0 R'(\alpha) - \epsilon^2 G(\beta) F'(\alpha)} \tag{26}$$

which may be readily integrated to given the equation of the front shock wave in the α, β plane as

$$\frac{1}{2} \epsilon^2 F^2(\alpha) G(\beta) = \int_0^\alpha F(\alpha)[1 - \epsilon\lambda_0 R'(\alpha)] \, d\alpha \tag{27}$$

The equation of the shock in the physical plane may be obtained by eliminating α from Eqs. (27) and (26) and replacing β in these equations by r; i.e.,

$$x^i = \lambda_0 r - \frac{3}{4} \epsilon^4 G^2(r) \tag{28}$$

where

$$G(r) = \frac{1}{2} \pi^{1/2} Kk^{-1/2} \text{erf}(\kappa r)^{1/2}, \quad K = \frac{(A+1)M_{f0}^4}{(2\lambda_0^3)^{1/2}}$$

"Outer" or Far-Field Solutions

"Outer" Variable and "Outer" Equations

The uniform solution of the problem can be obtained by constructing the composite solutions from the "inner" region and the "outer" region that is away from the source of the disturbance. First, an "outer" variable η is defined by

$$\eta = \varepsilon\sigma\beta \tag{29}$$

where

$$\sigma = (a_f^2/a_e^2)_0 = 1 + \varepsilon\sigma_1 + \varepsilon^2\sigma_2 + \cdots \tag{30}$$

The governing equations of the flow in the outer region in the transformed $\alpha\eta$ plane are

$$\begin{gathered}
\rho u u_\alpha + \rho v v_\alpha + p_\alpha = 0, \quad \rho u v_\alpha + \lambda\rho u u_\alpha + \varepsilon\sigma x_\alpha p_\eta = 0 \\
(\rho u)_\alpha + \varepsilon\sigma(\rho v)_\eta x_\alpha - \lambda(\rho v)_\alpha + \varepsilon\sigma\rho v x_\alpha/\eta = 0 \\
u(p_\alpha - a_f^2 \rho_\alpha) = x_\alpha a_f^2 (h_q/h_\rho)\dot{q} \\
u q_\alpha = \dot{q} x_\alpha, \quad \varepsilon\sigma x_\eta = \lambda
\end{gathered} \tag{31}$$

whereas the equilibrium Mach wave slope is given by

$$\lambda_e = \frac{-uv + a_e^2(M_e^2 - 1)^{1/2}}{a_e^2 - v^2} \tag{32}$$

"Outer" Perturbation Equations

Again, ϕ can be any one of the dependent variables that appear in the system of Eq. (31) and it is assumed that it can be expressed as

$$\phi(\alpha,\eta) = \phi_0^0(\alpha,\eta) + \epsilon\phi_1^0(\alpha,\eta) + \epsilon^2\phi_2^0(\alpha,\eta) + \cdots \quad (33)$$

where the superscript "0" signifies outer variables. Writing the dependent variables of Eq. (31) in the form of Eq. (33) and collecting terms of like order in ϵ, a set of outer equations is obtained that governs terms of various order variables. The zeroth- and first-order equations are

$$u_0 q_{1\alpha}^0 = x_{0\alpha}^0 \dot{q}_1^0, \quad x_{0\eta}^0 = \lambda_{e1} \quad (34)$$

$$u_0(p_{1\alpha}^0 - a_{e0}^2 \rho_{1\alpha}^0) = a_{e0}^2 x_{0\alpha}^0 (h_q/h_\rho)_0 q_1^0$$

$$\rho_0 u_0 u_{1\alpha}^0 + p_{1\alpha}^0 = 0, \quad \rho_0 u_0 v_{1\alpha}^0 + \lambda_{e0}\rho_0 u_0 u_{1\alpha}^0 = 0 \quad (35)$$

$$\rho_0 u_{1\alpha}^0 + u_0 \rho_{1\alpha}^0 - \lambda_{e0}\rho_0 v_{1\alpha}^0 = 0, \quad x_{1\eta}^0 = \lambda_{e2} - \lambda_{e0}\sigma_2$$

where

$$\lambda_{e0} = (M_e^2 - 1)_0^{1/2}, \quad \sigma_1 = (2a_{f0}a_{f1}/a_{e0}^2)$$

$$a_{e1} = \left(\frac{\partial a_e}{\partial p}\right)_0 p_1^0 + \left(\frac{\partial a_e}{\partial \rho}\right)_0 \rho_1^0 \quad (36)$$

The second-order equations are

$$u_0 q_{2\alpha}^0 = x_{0\alpha}^0 \dot{q}_2^0, \quad u_0 p_{2\alpha}^0 - a_{e0}^2 \rho_{2\alpha}^0 = a_{e0}^2 x_{0\alpha}^0 \left(\frac{h_q}{h_\rho}\right)_0 \dot{q}_2^0$$

$$\rho_0 u_0 u_{2\alpha}^0 + p_{2\alpha}^0 = 0, \quad \rho_0 u_0 v_{2\alpha}^0 + u_0 \lambda_{e0}\rho_0 u_{2\alpha}^0 = 0 \quad (37)$$

$$-\lambda_{e0}\rho_0 v_{2\alpha}^0 + u_0 \rho_{2\alpha}^0 + \rho_0 u_{2\alpha}^0 = 0, \quad x_{2\eta}^0 = \sigma_2 \lambda_{e1}$$

where

$$\sigma_2 = (2a_{f0}a_{f2})/a_{e0}^2 \quad (38)$$

Zeroth- and First-Order "Outer" Solutions

One can integrate Eq. (34) to obtain $x_0^0 = f_1(\alpha)$, since

$$q_1^0 = u_1^0 = v_1^0 = p_1^0 = \rho_1^0 = \lambda_{e1} = 0$$

However, by matching "inner" first-order solution, i.e., Eq. (18), it is obvious that the function $f_1(\alpha)$ should be α. Hence,

$$x_0^0 = \alpha \tag{39}$$

Accordingly,

$$x_1^0 = \lambda_{e2}\eta - \lambda_{e0}\sigma_2\eta - \lambda_{e0}R(\alpha) \tag{40}$$

Second-Order "Outer" Solutions

The system of Eqs. (37) represents the governing equations for an equilibrium flow system. In other words, the transformation of $\eta = \varepsilon\sigma\beta$ reduced the nonequilibrium system into an equilibrium approximated system of the second order of ε^2 by direct integrating and matching with the corresponding inner solutions. The second-order "outer" solutions are found to be

$$v_2^0 = u_0 F(\alpha) \, (\varepsilon\lambda_{e0}/2\eta)^{1/2}$$

$$u_2^0 = -v_2^0/\lambda_{e0}$$

$$p_2^0 = -\rho_0 u_0 u_2^0, \quad x_2^0 = 0 \tag{41}$$

where

$$A = 1 + 2\rho_0 a_{f0} \left(\frac{\partial a_f}{\partial p} + \frac{1}{a_f^2} \frac{\partial a_f}{\partial \rho} \right)_0$$

$$A' = 1 + 2\rho_0 a_{e0} \left(\frac{\partial a_e}{\partial p} + \frac{1}{a_e^2} \frac{\partial a_e}{\partial \rho} \right)_0 \tag{42}$$

$$\lambda_{e2} = \frac{M_e^4(A' + 1)}{2u_0 \lambda_{e0}} u_2^0 + (\lambda_\sigma)_0 \sigma$$

Now, the outer characteristics can be expressed as

$$x^0 = \alpha + \lambda_{e0}\beta - \epsilon\lambda_{e0}R(\alpha) - \epsilon^2[G_1(\beta) + G_2(\beta)] F(\alpha) \quad (43)$$

where

$$G_1(\beta) = \frac{M_{e0}^4(A - 1) \beta^{1/2}}{\left(2\lambda_{f0}^3\right)^{1/2}}, \quad G_2(\beta) = \frac{M_{e0}^4(A' + 1) \beta^{1/2}}{\left(2\lambda_{e0}^3\right)^{1/2}} \quad (44)$$

"Composite" or Uniformly Valid Solutions

Composite Solutions for Characteristics

In constructing the uniformly valid or the so-called "composite" solution, Van Dyke's (1964) notation and formulation can be used; i.e.,

$$x^c = x^0 + x^i - [x^i]^0 \quad (45)$$

where the superscript c indicates the composite natural and $[x^i]^0$ denotes the outer expansion of the inner expression.

By combining Eqs. (24), (43) and (45), one has

$$x^c = \alpha + \lambda_0\beta - \epsilon\lambda_0 R(\alpha) - \epsilon^2 G(\beta)F(\alpha) - \epsilon^2 G_1(\beta)F(\alpha) \quad (46)$$

Composite Solutions for the Nose Shock Waves

The equation of the shock angle is governed by

$$\frac{dx^c}{dr} = \cot \delta^c \quad (47)$$

Substituting Eqs. (27) and (46) into Eq. (47) and replacing β by r, the equation of the shock angle in the physical plane is obtained:

$$\delta^c = \mu_{e0} + \left(\epsilon^4/M_{e0}^2\right)\{[1.5G(r) + H(r)]G'(r) + H'(r)G(r)\} \quad (48)$$

where

$$H(r) = \left(1 + \frac{M_f^4 \lambda_e}{2M_e^4}\right)_0 G_1(r)$$

Verification of the Theory by the Limiting Process

A quick verification of the theory developed in this investigation can be obtained if one takes "frozen" and "equilibrium" limits on characteristics. It is easy to show that by applying the limits to Eq. (46) at $\beta \to 0$, i.e., very close to the body, δ^c is approaching the frozen Mach angle, i.e.,

$$\frac{dx^c}{d\beta} \to \frac{dx^i}{d\beta} \to \mu_{f0} \tag{49}$$

and at $\beta \to \infty$, i.e., very far from the body, δ_c is approaching the equilibrium Mach angle, i.e.,

$$\frac{dx^c}{d\beta} \to \frac{dx^0}{d\beta} \to \mu_{e0} \tag{50}$$

Comparison With Experimental Results of Wegener-Klikoff

Some interesting experiments on the propagation of weak conical waves in a reactive mixture were reported in Wegener et al. (1965). A supersonic projectile was fired between two metal strips on which a line of small holes was drilled parallel to the flight path. As the projectile traverses the firing range at supersonic speeds, a strong bow shock is generated that sweeps across the line of holes at supersonic speeds, producing a traveling pressure disturbance over the metal strips. The increased pressure behind the shock produces a succession of gas puffs through the holes, which, in turn, generates weak conical shock waves on the far side of the metal strips. The decay of these weak conical shock waves was measured. The variation of the shock strength with radial distance is expressed in curves giving the shock angle as a function of r. The model gas used in the experiments was a well-understood reacting mixture:

$$N_2 + N_2O_4 \rightleftarrows N_2 + 2NO_2 \tag{51}$$

with reactant mole fractions varying from 0 (i.e., pure nitrogen) to 0.15. Experiments conducted in pure nitrogen yield results that are indistinguishable from those obtained in air. The test conditions of this specific experiment are illustrated in Table 1. However, it should be pointed out that these experiments are measured for a relatively short distance from the origin of the disturbance.

One of the typical comparisons between the theoretical and empirical data is presented in Fig. 1. The curve on the top of the figure represents the variation of shock angles, which are calculated by the frozen characteristics, with respect to the radial distances r. The bottom curve in the figure is calculated by the uniformly valid characteristics. It is noted that the shock angles decays into the frozen Mach angle for the top curve and into the equilibrium one in the lower case. However, upperbound, the

Table 1 Test condition of experiments from Wegener et al. (1965)

Experiment No.	n_r, mole fraction	U_0, m/sec	M_{f0}	M_{e0}	τ_0, μsec	$U_0\tau_0$, cm	T_0, K	P_0, atm	μ_{f0}, deg	σ
146	0.054	1230	3.64	3.86	3.82	0.470	0.296	1.01	15.95	0.1245
143	0.081	1230	3.70	3.96	2.58	0.317	0.297	1.00	15.68	0.1455

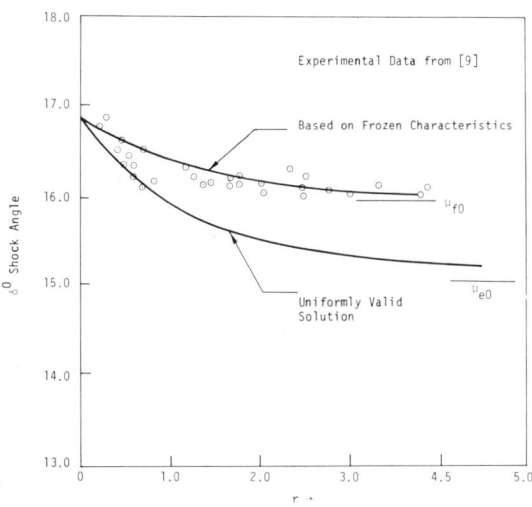

Fig. 1 Comparison of theoretical and experimental shock wave attenuation.

experimental data spread is obtained in the near field, which agrees well with the frozen shock. The correct uniformly valid shock angle distribution agrees well with the lower parts of the experimental data. The interesting aspect of this comparison is that a large discrepancy

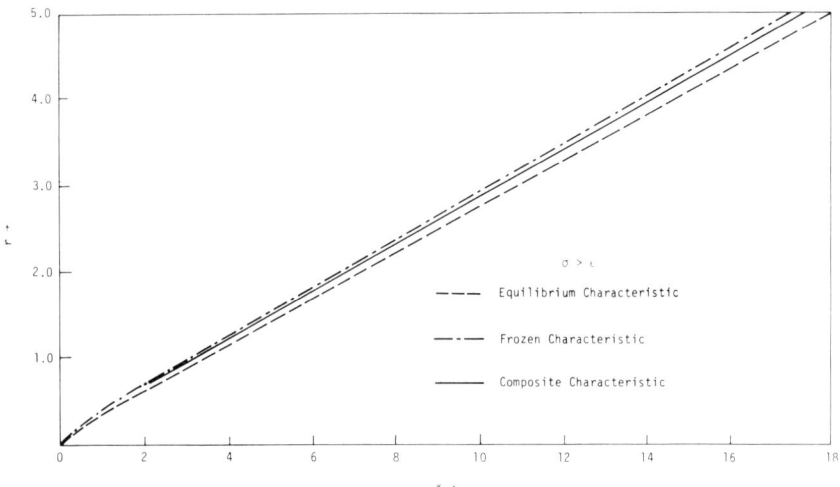

Fig. 2 Comparisons of various characteristics: Case $\sigma > \varepsilon$.

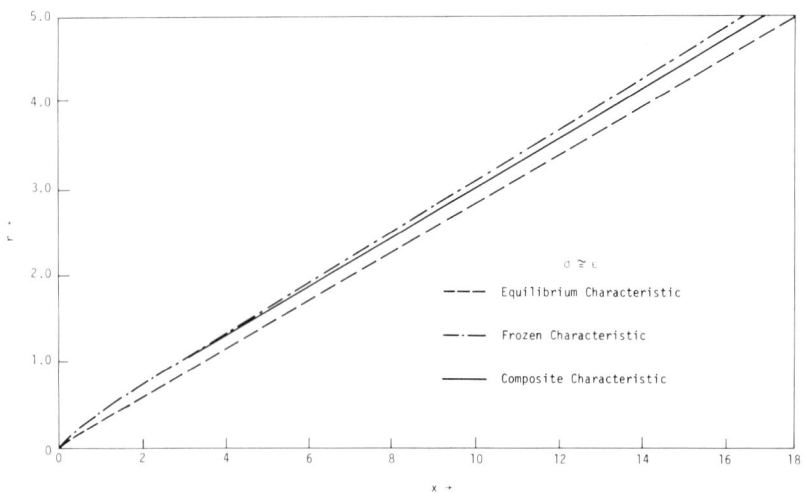

Fig. 3 Comparisons of various characteristics: Case $\sigma = \varepsilon$.

UNIFORM SOLUTIONS FOR WEAK SHOCK WAVES

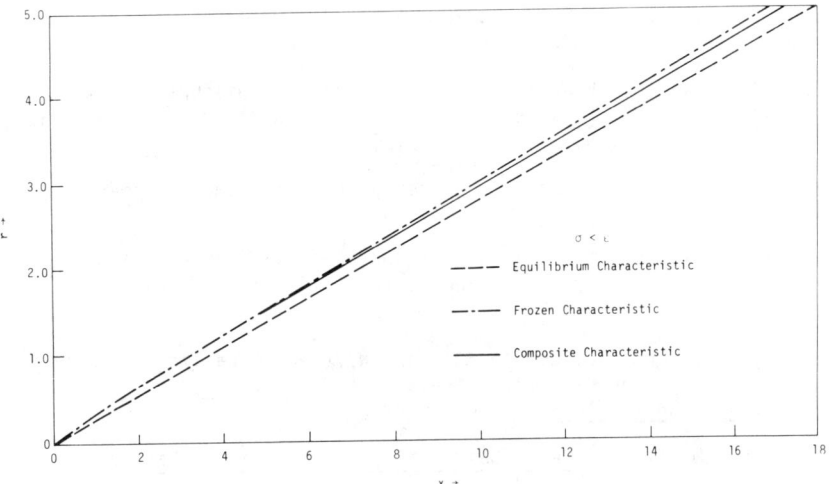

Fig. 4 Comparisons of various characteristics: Case $\sigma < \varepsilon$.

exists between these two solutions, and care must be taken in calculating the nonequilibrium flow when the equilibrium condition is approached. Next, the conditions for experiment 146 of Wegener et al. (1965) were used to demonstrate the deviation of composite characteristics from the frozen characteristics for the cases of $\sigma < \varepsilon$, $\sigma = \varepsilon$, and $\sigma > \varepsilon$. Results of typical nose characteristics for these cases are presented in Figs. 2-4. It is noted that the uniformly valid solution, i.e., the composite characteristics, emerge from the nose of the projectile with a similar slope to the frozen characteristic and end at the far field with a slope similar to the equilibrium one.

References

Bray, K. N. C. (1970) The method of characteristics. <u>Nonequilibrium Flow</u> (edited by P. P. Wegener), Vol. II. Marcel Dekker, New York, pp. 60-154.

Chou, D. C. (1982) Weak shock wave propagation in a relaxing gas <u>Astronaut. Acta</u> Vol. 17, No. 2, pp. 625-632.

Chou, D. C. and Chu, B. T. (1971) On the decay of weak shock waves in axisymmetric non-equilibrium flow. <u>J. Fluid Mech</u>. Vol. 50, No. 2, pp. 355-367.

Chou, D. C. and Maa, S. Y. (1975) Propagation of weak shock waves in a vibrational nonequilibrium, nonuniform gas. <u>Trans. ASME J. Appl. Mech</u>. Vol. 42, No. 3, pp. 564-568.

Chou, D. C. and Maa, S. Y. (1979) Self-similar shock waves in a relaxing gas. Acta Astronaut. Vol. 6, Nos. 5-6, pp. 669-683.

Chu, B. T. (1970) Weak nonlinear waves in nonequilibrium flow. Nonequilibrium Flow (edited by P. P. Wegener), Part II, Marcel Dekker, New York, pp. 33-59.

Hahn, S. G. (1958) Stability criteria for difference schemes. Comm. Pure Appl. Math. Vol. 11, No. 2, pp. 243-255.

Sedney, R. (1970) The method of characteristics. Nonequilibrium Flow (edited by P. P. Wegener), Vol. II. Marcel Dekker, New York, pp. 159-225.

Sedney, R., South, J. C., and Gerber, N. (1962) Ballistic Research Laboratory Report 1173; see also The High Temperature Aspects of Hypersonic Flows. Pergamon Press, Oxford.

Van Dyke, M. (1964) Perturbation Methods in Fluid Mechanics. Academic Press, New York.

Vincenti, W. G. and Kruger, C. H (1965) Introduction to Physical Gas Dynamics. John Wiley, New York.

Wegener, P. P., Chu, B. T., and Klikoff, W. A. (1965) Weak waves in relaxing flows. J. Fluid Mech. Vol. 23, part 4, pp. 787-800.

# Chapter VII. Combustion Diagnostics

CARS Instrument for Practical Combustion Measurements

Gregory M. Dobbs,* John H. Stufflebeam,† and Alan C. Eckbreth‡
United Technologies Research Center, East Hartford, Connecticut

and

P. A. Tellex§
Pratt & Whitney Aircraft Group, West Palm Beach, Florida

Abstract

This paper describes a coherent anti-Stokes Raman spectroscopy (CARS) instrument, capable of completely remote operation in hostile environments. Instantaneous measurements of properties at 20 Hz may be made in temporally fluctuating media. Rapid data reduction methods have been developed to assemble the single-shot measurements into the probability distribution function (pdf) and time history for the property. The average value and the magnitude of turbulent fluctuation may thus be obtained. The instrument's first use has been in a gas turbine engine exhaust. Also, to test the data processing procedures, nitrogen spectra were recorded using an isothermal source of known temperature. The resultant temperature pdf has finite width. The utility of the CARS technique is determined in part by how this width compares with the width of the pdf from the fluctuating system of interest. The spatial resolution, range, sensitivity, and required data processing techniques are summarized.

Introduction

Coherent anti-Stokes Raman spectroscopy (CARS) is of particular importance in laser diagnosis of hostile, yet sensitive, combustion environments. Spatially precise laser diagnostics have the potential for remote, nonperturbing, temporally resolved, in situ measurement of tempera-

---

Presented at the 9th ICODERS, Poitiers, France, July 3-8, 1983. Copyright © American Institute of Aeronautics and Astronautics, Inc., 1984. All rights reserved.
*Senior Research Scientist, Chemical Physics.
†Research Scientist, Chemical Physics.
‡Manager, Chemical Physics.
§Project Engineer--Optics, Instrumentation.

ture and species concentrations in combustion gases. CARS is a coherent, wave-mixing process in which the signal emerges in a laser-like beam. Although its single-shot measurement capability was demonstrated in prior work, routine, sustained recording of repetitive, single-pulse signals from a practical device and the reduction of the data to probability distribution functions (pdf) is new. The design of a remotely operated instrument, whose first use was for measurements in jet engine exhausts, is described. The data analysis techniques are also summarized.

The details of the theory and application of CARS are reviewed elsewhere (Eckbreth and Schrieber 1981; Taran 1982). As illustrated in Fig. 1, incident laser beams at frequencies ω_1 and ω_2 (termed the pump and Stokes beams, respectively) interact to generate coherent radiation at frequency $\omega_3 = 2\omega_1 - \omega_2$. When the frequency difference ($\omega_1 - \omega_2$) is close to the frequency of a Raman-active molecular transition, resonant enhancement results in a signature characteristic of that species. The signal shapes are both density and temperature dependent, providing the basis for diagnostics. In air-fed combustion N_2 is usually used for thermometry. Concentration measurements are made using either the integrated area if the background spectra is small or the shape of the combination of resonant spectrum and background.

Reported here are some initial demonstration measurements in which spectra from individual laser pulses are averaged before recording. The instrument was then upgraded to be able to record the entire CARS spectrum from each laser pulse on magnetic tape. Histogram representions of the pdf's for N_2 thermometry are presented.

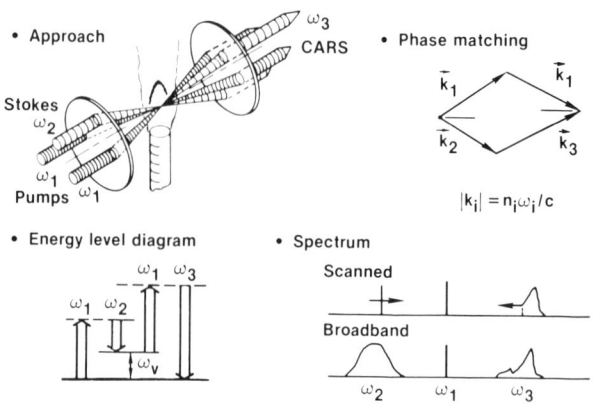

Fig. 1 Coherent anti-Stokes Raman spectroscopy (CARS).

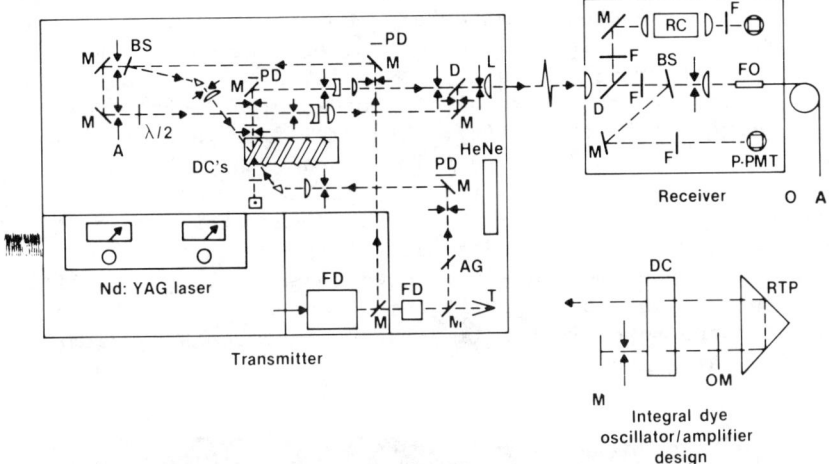

Fig. 2 Schematic layout of CARS instrument. Code: FD, frequency doubler; M, mirror; AG, absorbing glass; PD, photodiode; HeNe, helium-neon laser; DC, dye cell; A, aperture; BS, beamsplitter; λ/2, half-wave plate; D, dichroic; L, lens; F, filter; RC, reference cell; FO, fiberoptic; PMT, photomultiplier tube; RTP, rooftop prism; T, trap; and OM, output mirror.

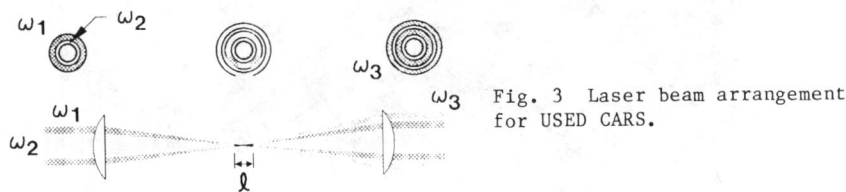

Fig. 3 Laser beam arrangement for USED CARS.

CARS Instrument

The instrument consists of a transmitter, receiver, and X-Y traversing system located about the device under test, as well as remotely located spectrographic detection and computer equipment. To our knowledge, this is the first time an entire CARS instrument has been operated remotely.

Figure 2 is a schmatic diagram of the device. A frequency-doubled Nd:YAG laser generates a primary ω_1 pump beam at 532 nm. This beam is separated from residual 1.06-μm radiation by a dichroic mirror and is routed to a field lens that focuses the beam to the measurement point. The residual 1.06-μm light is frequency doubled to produce a second beam that pumps the dye laser oscillator to form the ω_2 Stokes beam. An enlargement of the dye oscillator/amplifier is in the lower right of Fig. 2. A planar, Fabry-Perot broadband cavity is employed. A carriage with dye cells (one for each molecule of interest) can move any one of them into the cavity. The amplifier is optically pumped by about 30% of the primary beam, which has been coupled off by a beam splitter. Galilean telescopes and half-wave plates are inserted in the beams for control of the focal waist sizes and polarizations. Remotely controlled stepper motors are mounted on several of the critical optical mounts to permit remote adjustment. Detectors monitor the power in the laser and signal beams to permit optimum alignment.

Fig. 4 CARS instrument and traversing system in jet engine test stand.

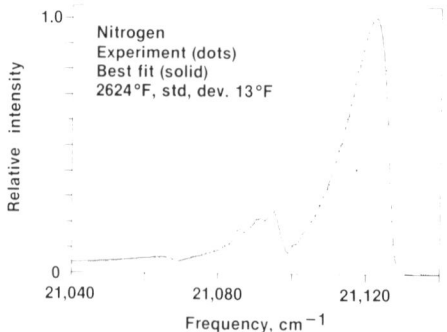

Fig. 5 Time-averaged N_2 CARS spectrum in jet engine exhaust at full augmentation.

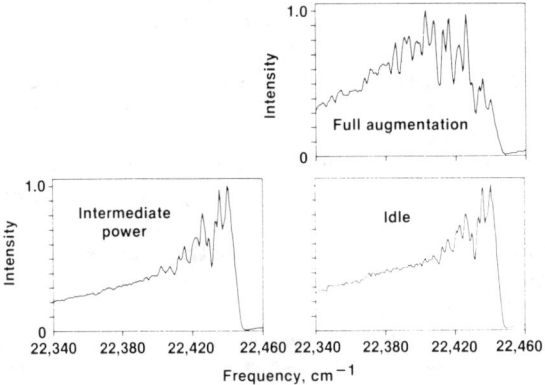

Fig. 6 Time-averaged H_2O CARS spectra in jet engine exhaust at various power levels.

The lasers are focussed into the jet exhaust generally using a two-beam, three-dimensional phase-matching configuration (Marko and Rimai 1979). In this approach, the ω_2 Stokes beam is coaxial with and inside the annular ω_1 pump beam as shown in Fig. 3. The annular beam results from the unstable resonator employed on the neodymium laser; thus, we use the acronym USED CARS for unstable-resonator spatially enhanced detection. The pump and Stokes beams do not intersect until the focal region. The signal is generated in a cylindrical interaction volume approximately 0.1-mm diam by 50-mm long, with the axis oriented traverse to the 1-m-diam exhaust flowfield. The pump beams are split off from the signal beam by a dichroic mirror and the CARS beam is focussed into a fiber-optic cable. The fiber transports the signal approximately 30 m to the remote spectrograph. Just prior to focussing into the fiber, a small part of the CARS beam is split off into a photomultiplier tube (PMT), which monitors the spectrally integrated signal. This is used for absolute intensity measurement and transmitter alignment checks. The ω_1 and ω_2 beams, are refocussed in a cell to produce a reference CARS signal which is monitored by a second PMT. The cell contains a calibrating test gas mixture at a known pressure and temperature that is used as a normalizing signal.

Figure 4 shows the CARS instrument mounted about the engine. The mounting framework is translatable, permitting two-dimensional mapping of the exhaust. The transmitter and receiver are separated by 2.2 m and are housed in acoustic isolation boxes. The CARS signal emerges from the fiber directly into a spectrograph fitted with an optical multichannel analyzer (OMA). Initially, a 0.6-m, f/6.8

commercial spectrograph was employed. Later a homemade 0.75-m, f/3.5 spectrograph was used that had a 20-cm-diam concave holographic grating. Variable, neutral-density filter wheels or wedges are used to keep the signal level in the linear response region of the detector. For fluctuating environments, an optical splitter is inserted within the spectrograph to increase dynamic range by creating multiple attenuated images of the spectrum.

An EG&G/PARC model 1254/1216 vidicon detector system was used for initial time-averaged measurements due to its availability. For later single-shot work, a model 1420/1218 diode array system was used. Data are sent from either OMA to a DEC PDP-11/34a computer by custom interfaces. The vidicon is scanned at about 33 Hz and the diode array at 88 Hz. For the former, the averaged spectrum is stored on the computer disk. For the latter, single-shot spectra are stored on magnetic tape. After the CARS data have been acquired at a spatial point of interest, a mechanical shutter blocks the Stokes beam, which precludes further CARS generation. As the translation system is being scanned to a new point, data is recorded in the same manner to acquire "dark" counts, which are subtracted from the signal counts later. While the translation system is moving, the computer samples analog-to-digital converters that monitor low-frequency outputs from boxcar averagers, PMT's, and other instruments for storage with the CARS data.

Time-averaged Measurements

In the initial demonstrations fine time-averaged spectra were obtained from N_2 for thermometry and from O_2, H_2O, and CO for concentration. Figure 5 shows a N_2 spectrum from a fully augmented jet engine exhaust averaged for 15 s at 10 Hz. The computed least-squares fit gives a temperature of 1713 K with a standard error of 13 K. Figure 6 shows spectra of H_2O vapor on the centerline at several engine powers. With the full-augmentation

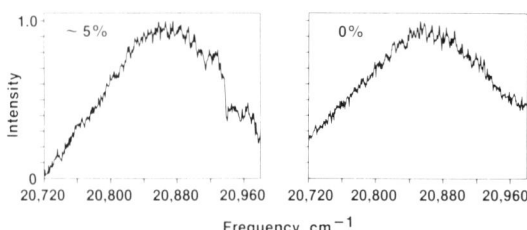

Fig. 7 Time-averaged CARS measurements of CO at various locations in jet exhaust at full augmentation.

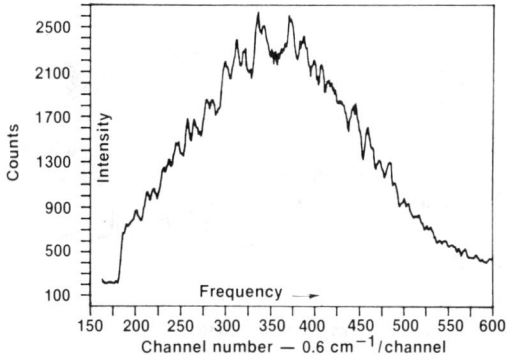

Fig. 8 Single-pulse dye laser spectral profile illustrating amplitude random noise.

temperature at this point taken from N_2 data, the H_2O vapor concentration was computed to be about 15%. A CO spectrum is shown in Fig. 7. The modulation of the nonresonant background is easily distinguished, and the spectrum indicates a 5% concentration at 1400 K. In the other location shown, no CO was detected. O_2 concentration measurements were also performed. As expected, these were difficult at full augmentation due to the consumption of the residual, i.e., bypassed O_2, in the afterburner.

Repetitive, Single-Pulse Measurements

For the single-shot work, a 20-Hz Quanta Ray DCR-2A laser and a fast, one-dimensional, linear diode array detector were used. An optical splitter is used in the spectrograph for dynamic range control to create multiple, spectrally dispersed, nonoverlapping images differing in intensity by a factor of about 30. The average data rate to magnetic tape is 56 K-byte-s^{-1}. One reel can hold data for 400-shot (20-s) samples taken at 30 locations.

Regression analysis could be applied, but would only be tractable for data reduction on the fastest of computers. Instead, a faster method was developed using "quick fitters." They also have an additional benefit. One noise source in the experiment is the pulse-to-pulse fine-scale irregularity in the dye laser profile. A typical dye profile is shown in Fig. 8, obtained as a CARS spectrum generated in a nonresonant reference gas. The fine structure leads to irregularities in the single-pulse CARS spectrum and the distortion leads to temperature measurement errors. Fortunately, the general shape, is reproducible from shot to shot and the average shape can be used in the analysis.

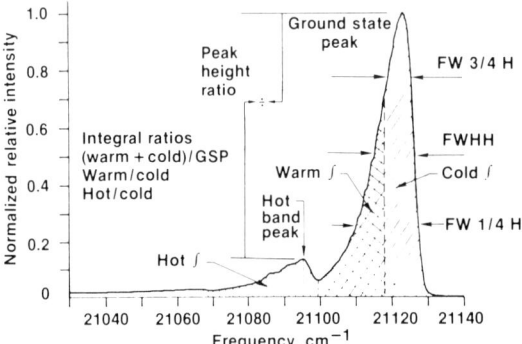

Fig. 9 Various quick-fitting N_2 CARS temperature measurement methods.

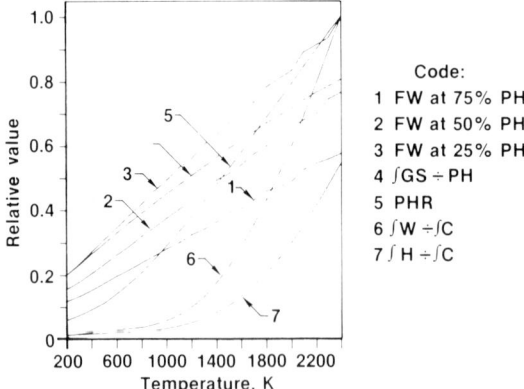

Fig. 10 Temperature dependence of N_2 CARS temperature quick-fitters. Code: 1, FW 3/4H; 2, FWHH; 3, FW 1/4H; 4, (warm integral + cold integral)/ground state peak; 5, peak height ratio; 6, warm integral/cold integral; 7, hot integral/cold integral.

The "quick fit" routines that also address the dye laser noise problem rely on examining only parts of the spectrum for determination of the temperature. Theoretical CARS spectra for the instrument's parameters are calculated at temperature intervals and the temperature dependence of some parameter is determined. Nine different parameters were chosen and their their temperature curves shown as Figs. 9 and 10. Each experimental spectrum has the same parameters determined for it and the temperatures are then obtained by interpolation. Methods 1-3 use the width of the fundamental band (v = 0 to 1 transitions) at three different heights. Method 4 uses the spectrally integrated

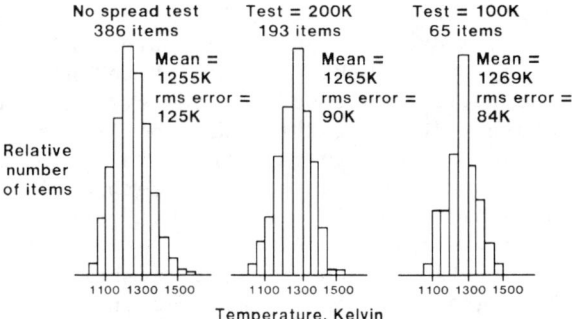

Fig. 11 Effect of temperature spread test on tube furnace histograms.

area of the entire fundamental band divided by its height. Method 5 uses the ratio of the height of the hot band to the height of the cold band. Methods 6 and 7 are integral methods. In method 6, the fundamental band is divided into two regions and their areas ratioed. In method 7, the area of the hot band is divided by the area of the high-frequency part of the fundamental band. Method 8 is an average of methods 1-3. Method 9 is an average of methods 6 and 7. Additional methods similar to the above may be easily devised. A DEC VAX 11/750 computer is able to reduce a single spectrum for all nine methods in less than 2 s. Hence, a 30-point contour map requires about 7 h of computer time for 400-shot pdfs.

For a low-noise shot (or averaged spectrum) the temperatures from all of the quick-fit methods should be in close agreement. Noise sources cause a spread in the answers. Conditional sampling using a "spread test" may be used to reject those shots in which there is poor agreement. It thus allows one to infer the quality of the broadband dye laser for each pulse, causing the rejection of a shot if the disagreement is greater than some threshold. Figure 11 shows that application of a spread test narrows the histogram considerably, with little shift in the center temperature; there is, however, a reduction in the number of points included as the allowable spread is decreased. There is obviously a point of diminishing returns. The expected pdf width for the device under test determines how narrow the isothermal pdf should be made.

Spectra were recorded at several temperatures in a tube furnace to assess the sensitivity of the quick-fit methods to the random dye laser noise. Radiation-corrected thermocouples were used to obtain the approximate tube furnace temperature. While this provided a comparison with the center temperature of the pdf, it was more useful in determining how the dye noise created a broadening of the pdf. The utility of the CARS measurement technique is determined by the ratio of the pdf for the system of interest to the isothermal width. If the device pdf is narrow enough, one may be able to average the data instead. Because the noise is random, the integral methods (4, 6, 7) were found to be best. Table 1 shows that method 9, the average of these, is most useful.

As an illustration of the present capabilities, a CARS temperature histogram is shown in Fig. 12 in an engine exhaust at a low level of augmentation. Note that the standard deviation in temperature, with a 200-K spread test, is larger by a factor of 2 than previously observed in the tube furnace. In Fig. 13, the temperature is displayed as a function of the sequential laser shot number. No low-frequency variation is apparent and the engine operation

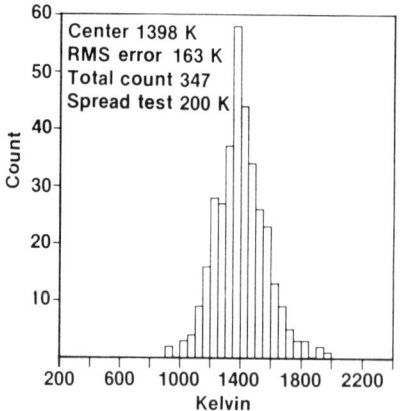

Fig. 12 CARS temperature histogram in augmented jet engine exhaust using average value of integral quick-fitters and 200-K spread test.

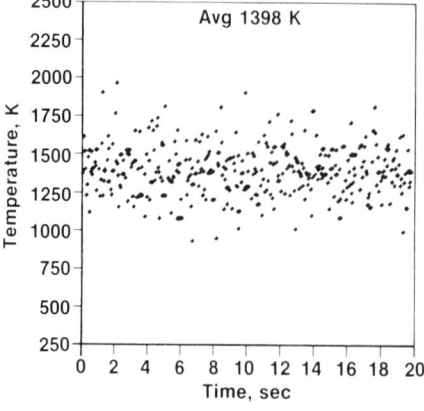

Fig. 13 Time dependence of temperature measurements in augmented jet engine exhaust used to assemble the Fig. 17 histogram.

Table I Comparison of several quick fit methods without a spread test for a 1253-K tube furnace measurement. Results from averaging the spectra are compared to single-shot pdf analysis (temperatures are Kelvin)

Integral method	Averaged spectrum	pdf average	pdf width
Fundamental	1238	1156	167
Warm/cold	1251	1255	125
Hot/cold	1325	1324	135
Average	1271	1243	110

was stable, i.e., constant throttle setting, during the 20-s duration of the data taking.

Discussion

In the short period of time that CARS has been under active development, it has progressed rapidly from the laboratory environment to practical application in internal combustion engines, gas turbine combustion zone simulations, and, as reported here, jet engine exhausts. More work is required to automate the technique, e.g., autoalignment systems, software development to expedite data reduction for other molecular species, etc. More research is required in the area of broadband dye laser technology to compensate for or reduce the random dye laser noise. Clearly, modest improvements in the random noise will result in reduced isothermal histogram widths. Coupled with "smart" data reduction approaches, the outlook is optimistic for accurate single-shot measurements without the requirement and complications of referencing. Data reduction approaches can also improve measurement capabilities considerably.

References

Eckbreth, A. C. and Schreiber, P. (1981) "Coherent anti-Stokes Raman spectroscopy (CARS): application to combustion and gas-phase diagnostics," Chemical Applications of Nonlinear Raman Spectroscopy (edited by A. B. Harvey), pp. 27-87 and references therein. Academic Press, New York.

Marko, K. A. and Rimai, L. (1979) Space- and time-resolved Coherent anti-Stokes Raman spectroscopy for combustion diagnostics. Opt. Lett. 4, 211-213.

Taran, J. P. (1982) Coherent anti-Stokes Raman scattering, Non-Linear Raman spectroscopy and Its Chemical Applications (edited by W. Kiefer and D. A. Long), pp. 281-323 and references therein. D. Reidel Publishing Co., Dordrecht, the Netherlands.

Study of OH-Saturated Laser-Induced Fluorescence in Low-Pressure Flame

D. Stepowski* and M. J. Cottereau†
Faculté des Sciences de Rouen, Mont Saint Aignan, France

Abstract

The purpose of this work is to study the behavior of fluorescence from the OH radical in a low-pressure flame as the excitation power is increased and the molecule is driven towards saturation. Analysis indicated that, although the coupling between neighboring rotational levels complicates the dynamics of excitation, a steady near saturation state can be reached between the laser coupled levels. The fluorescence signal was observed through a narrow-band spectrometer, which was tuned to the appropriate line when the laser excited the $Q_1 7$ line of the $2\ \Sigma^+\ (v'=0) \leftarrow 2\pi(v''=0)$ transition. A boxcar was used for signal-averaging with its gate set to capture the peak of the fluorescence pulse. The observed fluorescence signal did not completely saturate with respect to laser power. Analysis of the data indicated that lack of saturation was due to the nonuniform intensity distribution across the laser beam as previously calculated by Daily (1976). Satisfactory comparison between our results and Daily's analysis leads to the conclusion that the benefits of the saturation method are significantly reduced by this geometrical effect.

Introduction

Laser-induced fluorescence (LIF) provides a new tool for local and instantaneous nonintrusive measurement of radical species found at low concentrations in flames. In LIF the emission of fluorescent light due to spontaneous

---

Presented at the 9th ICODERS, Poitiers, France, July 3-8, 1983. Copyright by the American Institute of Aeronautics and Astronautics, Inc., 1984. All rights reserved.
*Chargé de Recherche au C.N.R.S. LA 230.
†Professeur. LA 230.

relaxation of the species excited by a tuned laser pulse is measured. To avoid the dependence of the fluorescence signal on the quenching due to collisional de-excitation Stepwoski and Cottereau (1979) have used a technique based on fast linear excitation. Fast excitation is difficult to achieve at atmospheric pressure due to high quenching rates. To avoid these problems and to make the fluorescence signal insensitive to the dye laser-power fluctuations, both Piepmeier (1972) and Daily (1976) have proposed to saturate the transition with a strong laser pulse.

The saturation method can be described with a simple level model. The exciting laser pulse is characterized by its spectral energy density U_ν. If N_1 is the population of the lower fundamental level of energy E_1, under the action of the laser pulse tuned so that $h\nu = E_2-E_1$, the population N_2 of the upper level of energy E_2 increases according to

$$\frac{dN_2}{dt} = N_1 B_{12} U_\nu - N_2 B_{21} U_\nu - N_2 (A+Q) \qquad (1)$$

The first term of this rate equation describes the absorption of photons with the probability coefficient B_{12}, the second term describes the stimulated emission of photons with the probability coefficient B_{21}, and the last term describes the spontaneous radiative and collisional (quenching) relaxations with the rates A and Q, respectively. For the experimental situation discussed herein, the process can be described by the previous classical Einstein balance equation.

Since the upper level is nearly empty at flame temperature, the sum $N_1+N_2 = N^\circ_1$ (the initial population of the lower state). The population N_2 increases during the laser excitation according to

$$N_2(t) = \frac{N_1^\circ B_{12} U_\nu}{(B_{12}+B_{21})U_\nu + A+Q} \left[1 - e^{-\left|(B_{12}+B_{21})U_\nu + A+Q\right| t} \right] \qquad (2)$$

The fluorescence flux collected in a solid angle Ω from a volume portion V of the exciting laser beam is given by

$$d(t) = N_2(t) (A V \Omega/4\pi) \qquad (3)$$

Usually this florescence signal depends on the quenching rate Q but if the steady state is reached, $\text{Exp} -(2BU + A+Q) t \ll 1$. If the laser is so strong that

$(B_{12} + B_{21}) U_\nu \gg A+Q$, then $N_2 \simeq N°_1 |B_{12}/(B_{12} + B_{21})|$ no longer depends on Q or on U_ν, and the transition is saturated.

For most of the investigated species, the term $(B_{12} + B_{21}) U_\nu$ is usually not sufficiently larger than $A + Q$ at atmospheric pressure thus it is necessary (as first suggested by Baronavski and McDonald, 1977a) to write the exact relation in the form

$$\frac{1}{N_2} = \frac{1}{N_1°} \quad \frac{B_{12} + B_{21}}{B_{12}} + \frac{A + Q}{B_{12} U_\nu} \quad (4)$$

The inverse of the fluorescence signal is always a linear function of $1/U_\nu$. Extrapolation to the origin $(1/U_\nu) \to 0$ yields the exact value of $N°_1$.

Interpretation of the results given by such a two-level model has been made for molecular species by Baronavski and McDonald (1977a) for C_2, by Bonczyk and Shirley (1979) for CH an CN, by Pasternack et al. (1978) for M_gO, and by Lucht et al. (1979) for OH in atmospheric pressure flames. These studies have lead to somewhat inconsistent conclusions. Mailander (1978) and particularly Lucht (1981) have made similar studies with low-pressure flames. Under low-pressure situation, the consequent lowering of the quenching rate permits a time-resolved analysis of the fluorescence signal (Stepowski and Cottereau, 1979) and reduces in the same proportion the laser energies needed to reach the saturation. To test the advantages of the saturation method for concentration measurement, a systematic study of the processes in a low pressure flame has been made with a dye laser pumped by YAG laser.

Saturation of a Multilevel System

The previous two-level model analysis is not sufficient to study molecular species in a case where many rotational sublevels are involved. A more rigorous analysis is made possible, when the following experimental conditions are met.

Only two rotational levels are directly affected by the laser pumping, one belonging to the fundamental state, the other to the first excited electronic state. Because of the many transitions in the OH spectrum, it is necessary to find an isolated line and to fit the laser spectral width to the absorption line ($\Delta\lambda \simeq 3 \cdot 10^{-3}$ nm).

A transition of the (0,0) vibration band is used so that vibrational transfers to other vibrational levels can be neglected at the flame temperature. Depopulation of the excited species by chemical reaction is negligible for OH.

Under these conditions (Fig. 1), depopulation of the upper rotational level excited by the laser is made by radiative and collisional relaxation to the ground state and, also, by rotational energy transfer (RET) to the neighboring rotational levels of the upper state. Similarly, the pumped rotational level of the ground state is partially repopulated by the neighboring rotational levels of the ground state. The following equations are applicable

$$\frac{dNe'}{dt} = NeBee'U\nu - Ne'Be'eU\nu - Ne'(\Sigma_i Ae'i + \Sigma_i Qe'i + \Sigma_{i'} re'i') + \Sigma_{i'} Ni'ri'e' \quad (5)$$

$$\frac{dNi'}{dt} = Ne're'i' - Ni'(\Sigma Ai'i + \Sigma Qi'i + \Sigma ri'j') + \Sigma Nj'rj'i' \quad (6)$$

$$\frac{dNe}{dt} = -NeBee'U_\nu + Ne'Be'eU_\nu + \Sigma_{i'} Ni'(Ai'e + Qi'e) - Ne\Sigma_i rei + \Sigma_i Nirie \quad (7)$$

$$\frac{dNi}{dt} = -Ni\Sigma_j rij + \Sigma_j Njrji + \Sigma_{i'} Ni'(Ai'i + Qi'i) \quad (8)$$

Where e' is the upper rotational level directly excited ; i' or j' is any other rotational level of this upper state ; e is the lower rotational level directly pumped ; and i or j is any other rotational level of the ground state. In these equations the terms A and B are the Einstein probability coefficients of the radiative transitions specified

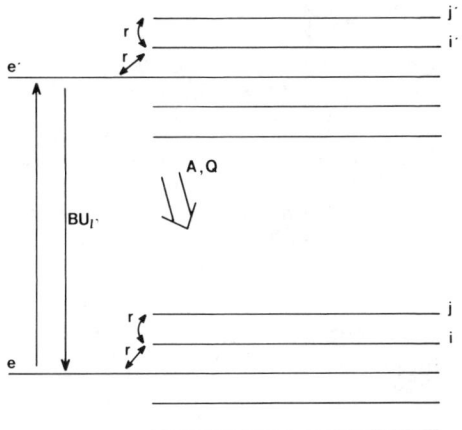

Fig. 1   Energy level diagram.

by the subscripts ; the Q terms are the de-excitation rates
by collision to the ground state (such transfers from the
lower state to the upper state are negligible) ; and the r
terms are the rates of rotational energy transfer by collision within a given electronic state.

Such a system is very difficult to solve because most
of the collisional rates are not well known, especially the
RET rates. A four-level system is one of the simplifications
which can be used to minimize the number of collisional-rate
terms to be determined. In such a model the two laser
coupled levels are each counted as one, and all other rotational sublevels in each electronic state are lumped together to form two more pseudostates. This model is well
suited for situations in which the steady state is reached
during the laser excitation, as first pointed out by Berg
and Shackleford (1979). However, it requires some particular
excitation conditions to describe the pumping kinetics
(Stepowski and Cottereau 1981).

An important simplification occurs when the laser power
is so high that BU_ν terms are dominant in the above equation
system. A steady, near-saturation state is then quickly
reached between the levels e and e' directly coupled by the
laser pumping. It the rotational transfer out of the e'
upper level is balanced by the rotational transfer into the
e lower level, the population $N_e + N_{e'}$ will be nearly
constant. This is the basic assumption of the balance cross-
rate model first developed by Lucht (1981). Lucht has shown
that the assumption applies to the OH radical in flames. It
is possible to study the saturation of level e', as in the
simplified two level model, but with the spontaneous relaxation rate A&Q term replaced by $A+Q+ \sum_{i'=e'} r_{e'i'}$.

Experimental Apparatus

The tunable pulsed dye laser is pumped by the second
harmonic of a YAG laser. This Q-Switched Nd : YAG laser
with an oscillator followed by two amplifier stages is mono-
mode TEM00 and produces a 350-mJ pulse at 1.064 m every
0.1 s. Frequency-narrowing and tuning of the dye laser is
accomplished by means of a Littrow grating at one end of the
cavity. The red-dye laser beam is amplified and frequency
doubled through a KDP crystal to generate $\sim$ 1-mJ pulses of
8-ns duration over a spectral width of 6 pm around 310 nm.
This laser beam is focused in the burnt gas zone of a low
pressure flat flame on a spot approximately 0.5 mm in
diameter.

The flat flame is set up in a low-pressure enclosure
(18 < p < 150 Torr) by a cylindrical porous burner
(ϕ = 2 cm) supplied with a near-stochiometric premixed
oxygen-propane flow. In the investigated zone just
downstream of the luminous front, the temperature was
measured by thermocouple to be about 1900 K.

Fluorescent light emanating in a direction 90 deg. from
the exciting radiation is collected with F/10 optics, focused on the entrance slit of a spectrometer (F=0.6 m) and
detected by a photomultiplier (RTC XP 2020) connected to a
boxcar averaging over 100 shots with a time resolution
of 1-ns.

Study of OH Laser Saturation

The objective is to study the intensity S_F of the fluorescence emitted by a given volume of excited gases as a
function of the laser spectral energy density U_ν. To attenuate the laser energy without modification of either the
beam geometry or its spectral or polarization features, a
number of calibrated glass plates have been interposed. At
310 nm each plate has a transmission of 45%. The calibration
curve in Fig. 2 shows how the transmitted intensity increases linearly with the expected theoretical values. This
agreement proves the validity of the procedure.

In the theoretical discussion the laser spectral energy
density U_ν as a simple datum. Some correction must be made
to take the actual features of the laser pulse into account.

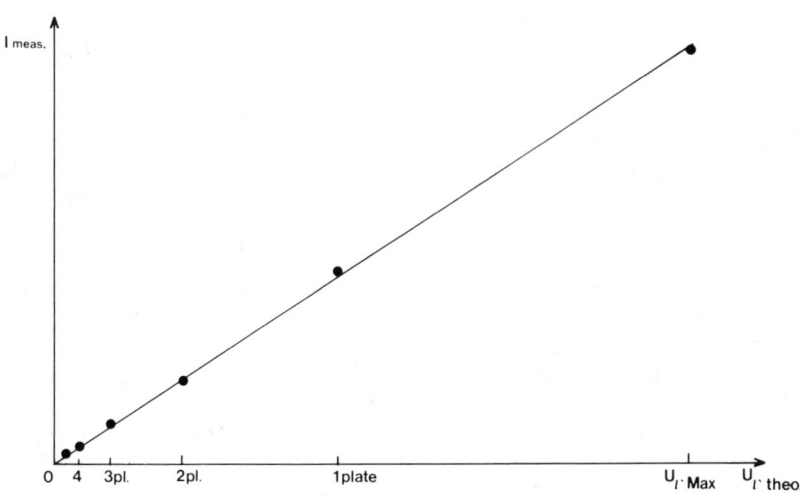

Fig. 2   Calibration curve.

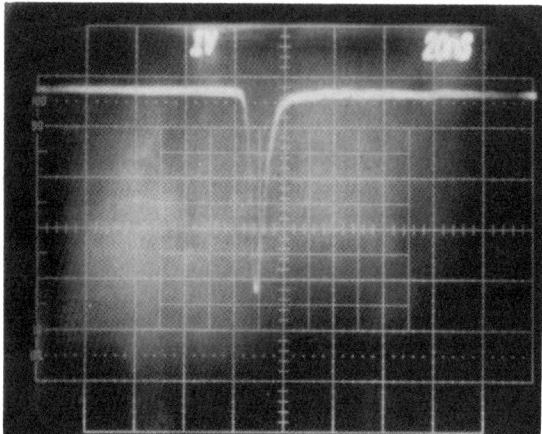

Fig. 3   Oscillogram of the laser pulse.

First, as shown in Fig. 3, the laser intensity is not constant over the pulse duration. The intensity is measured S_F at the peak of the laser pulse with the 1-ns sampling probe of the boxcar. It should be pointed out that a time integrated measurement would lead to a wrong result, a fact noticed by Van Calcar et al. (1979).

Moreover, the energy density is not constant across the laser-beam section. The two common geometrical configurations used to determine the excited analyzed zone are shown in Fig. 4. In the first configuration the entrance slit of the spectrometer is parallel to the laser axis and determines a slab along the beam. In the second configuration the entrance slit of the spectrometer is perpendicular to the laser axis so that a cylindrical-shaped portion of the beam is analyzed. In both cases a signal contribution from the peripheral zones of the beam leads to an alteration of the saturation curve $S_F = f(U_\nu)$. (The contribution and thus the alteration are stronger in the second configuration.) For a given laser intensity the central zone may be saturated, whereas the peripheral zone may not ; an increase of the laser intensity does not change the central saturated zone contribution to the fluorescence, but increases the peripheral zone contribution. As indicated by Daily (1978), for these two geometries and a Gaussian laser profile, the saturation curve does not reach a plateau but continues to grow. This effect has been observed by Sharp and Goldwasser (1976) for atomic species, and particularly by Mailander (1978) for C_2 in a flame, but it has not seemed worth extensive study. Nevertheless, the present work indicates the importance of this geometrical effect.

For a multilevel system the saturation occurs only between the two rotational levels directly coupled by the laser (the other upper sublevels become slowly populated without reaching a steady state). The following procedure has been used.

A given upper rotational level is excited by one of the P, Q, or R associated lines and the fluorescence is collected only from one of these lines. To avoid interferences due to Mie scattering, tow different lines are used for excitation and collection ; many levels of the $2\Sigma^+(v'=0) \leftarrow 2\pi(v''=0)$ transition of OH allow this procedure. In these experiments the laser is tuned on the $Q_1 7$ line and the fluorescence is measured from the $P_1 8$ line isolated by the spectrometer.

Figures 5 and 6 show the fluorescence response curves obtained in the burnt gas zone of flames at different pressures for the two geometries. Each point is obtained from the average of a hundred laser shots with the boxcar. For each geometry, after an initial linear growth the response curve bends over but the slope does not tend to zero.

For each geometry the data obtained at different pressures can be compared if adequate common normalized coordinates are used. To take into account the N°_e dependence on the pressure, the normalized ordinate is

$$S'_F = S_F \left[N_e^\circ (20 \text{ Torr})/N_e^\circ (P \text{ Torr}) \right] \qquad (9)$$

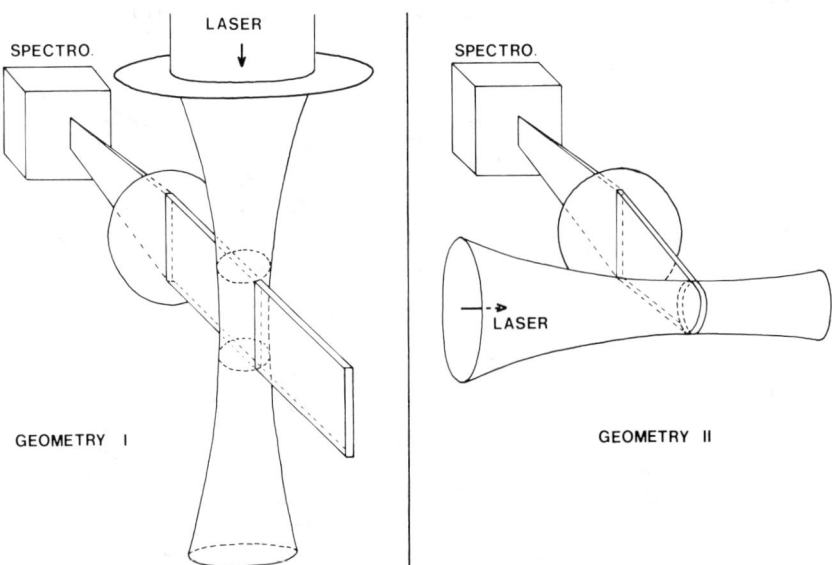

Fig. 4 The two common geometrical configurations.

The ratios N°e (20)/N°e (P Torr) have been easily determined from measurement of the fluorescence ratios under linear low excitation conditions. The ratios are 1.4 and 1.5 for flames burning at 40 and 80 Torr, respectively. Since the collisional relaxation rates Q and $\Sigma\, r_{e_i i_i}$ are proportional to the pressure (for the flame composition and temperature constant), the saturation energy $U\nu_o = A+Q+\Sigma\, r_{e_i i_i}/2B$ can be thought to be also proportional to the pressure (the spon-

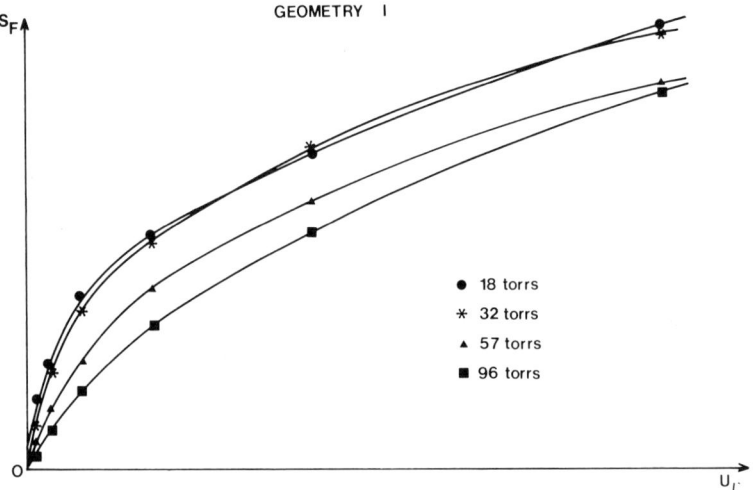

Fig. 5   Fluorescence response curve for geometry I.

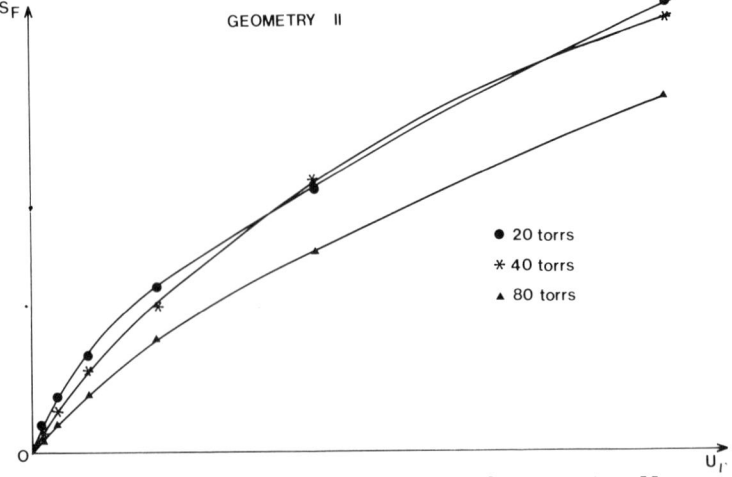

Fig. 6   Fluorescence response curve for geometry II.

taneous radiative rate A is negligible). Thus, the normalized abcissa

$$U'_\nu = U_\nu [20/P \text{ (Torr)}] \qquad (10)$$

has been chosen.

Figures 7 and 8 show all the experimental points provided by geometries I and II, respectively, when plotted in this normalized coordinate system. For each geometry the experimental points are gathered along a single curve, which corroborates our assumptions.

Under the ideal situation of spatially-homogeneous laser excitation $U\nu$, the fluorescence signal would have been

$$S_F = \frac{A \, \Omega \, V \, N e^\circ}{8 \pi} \frac{1}{1 + (U_{\nu o}/U_\nu)} \qquad (11)$$

The curve obtained $1/S_F = f(1/U_\nu)$ (Fig. 9) is not the straight line predicted by the simple theory. Under the actual situation for which the laser energy is not constant across the beam section, the signal is in the form

$$S_F = \frac{A \, \Omega \, V \, N e^\circ}{8 \pi} \, f \, \frac{U_\nu}{U_{\nu o}} \qquad (12)$$

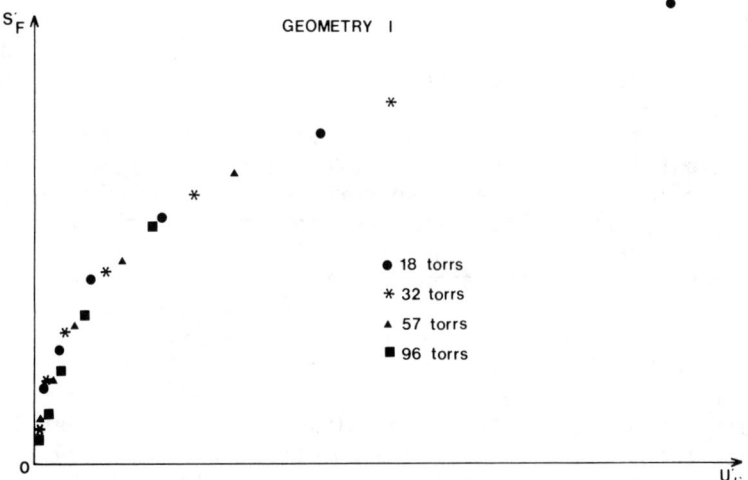

Fig. 7 Fluorescence response curve for geometry I in normalized coordinate system.

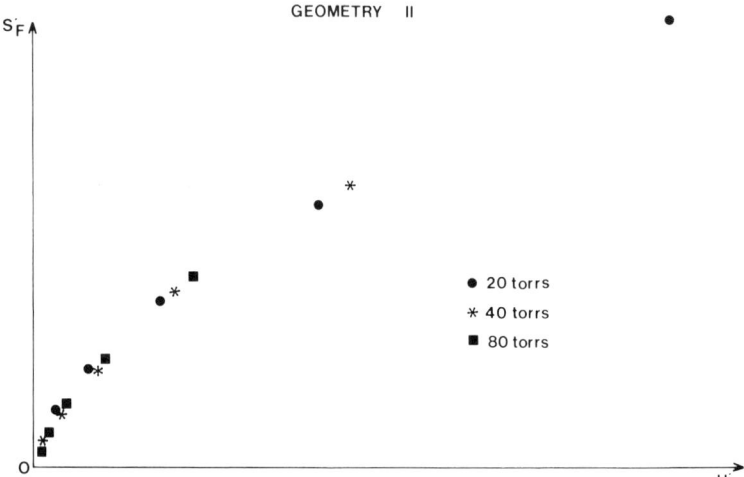

Fig. 8 Fluorescence response curve for geometry II in normalized coordinate system.

where U_ν is the mean value of the laser beam energy. The function f depends on the energy distribution across the beam and also on the collection geometry used.

The data have been compared with those of Daily (1976). The curve

$$1/S'_F = f(1/U'_\nu) \qquad (13)$$

plotted in Fig. 9 for a 20-Torr flame and for the first geometry is linear for the low values of U' as predicted by the simple theory. The discrepancy that can be noted for the high laser energies is due to the contribution of the beam wings. Extrapolation of the S'_F values, which would give the straight line for the high laser energies, produces the ideal fluorescence response curve without any wing effect (Fig. 10). From this curve one can determine the ideal saturated signal and also the saturation energy $U\nu_0$, the laser energy for which the signal is half the full saturated signal. The saturation energy in this instance is about one-twentieth the maximum energy of the laser.

An absolute value of OH concentration may be determined from the ideal response curve defined above. However, difficulty in defining the probe volume leads to a rather large uncertainty (100%). For P = 20 Torr, OH $\simeq 10^{22}$ m^{-3} in reasonable agreement with the mean value determined by absorption.

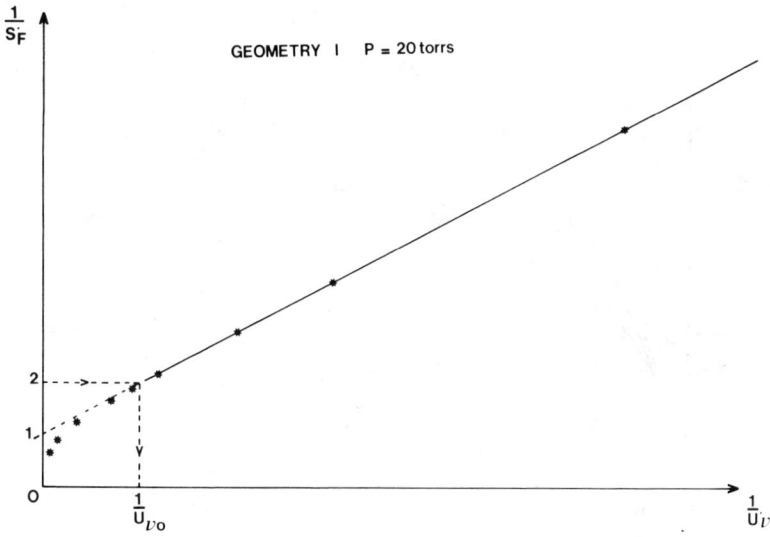

Fig. 9 Reciprocal of fluorescence signal vs reciprocal of laser power.

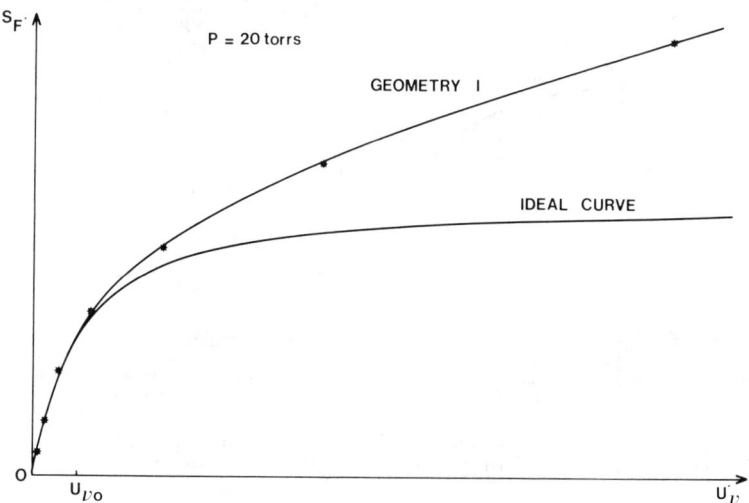

Fig. 10 Fluorescence response curves, the ideal curve is provided by linear extrapolation of Fig. 9.

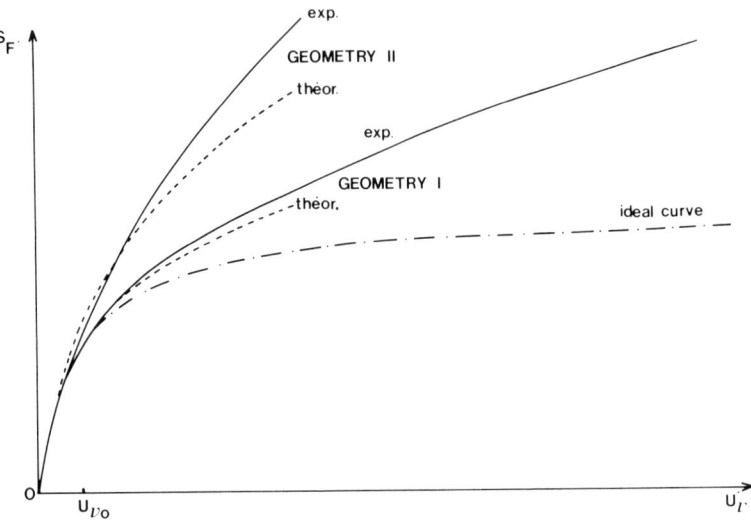

Fig. 11　Comparison between experimental and theoretical response curves.

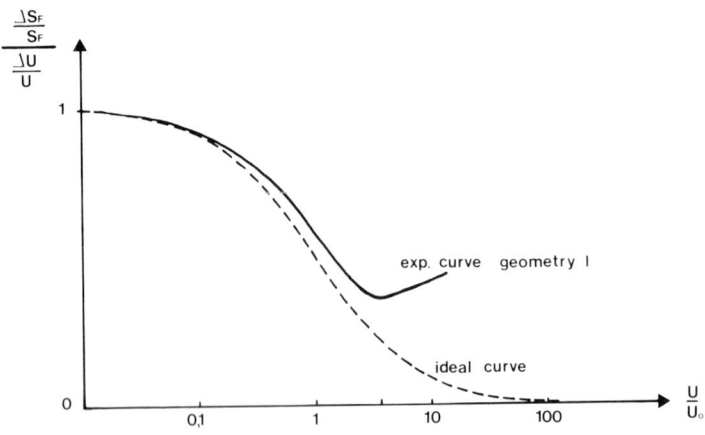

Fig. 12　Relative sensitivity of the fluorescence signal to the laser power fluctuation as a function of laser power.

The two experimental normalized saturation curves can then be compared to the theoretical curves given by Daily for a Guassian laser energy distribution (Fig. 11). Good agreement is found for the low laser energies but the experimental curves grow faster than the theoretical ones possibly because the laser energy profile is broader than a Gaussian profile.

The actual spectral distribution of the laser, which is not a square slit function, could also influence the results. However, as the laser spectral width is about twice the (Doppler) absorption line width, all the velocity classes are excited alike. If this effect were important, the influence of the geometry would not be predominant, as it is in the present results.

Conclusion

The major advantage of the saturation method for concentration measurement is not as obvious as expected. Under actual conditions the fluorescence signal continues to increase as the laser energy increases beyond the saturation energy. Nevertheless, this residual increase of the signal is only 7% and 20% of the initial linear growth under our experimental situations. On Fig. 12 we have plotted the ratio

$$D = \frac{\Delta S_F/S_F}{\Delta U_\nu/U_\nu} \qquad (14)$$

as a function of the relative excitation energy $U_\nu/U_{\nu 0}$.

Contrary to the ideal situation of a homogeneous laser beam where the ratio tends to zero as U_ν increases, the actual curve passes through a minimum, the sharpness of which depends on the geometrical configuration used. This minimum is reached when the laser energy is only five times the saturation energy. Since the fluorescent signal does not become independent of laser energy, for a precise measurement the de-excitation rates and the laser energy must be known, as in the linear fluorescence method. In addition, the measured fluorescence signal is not very intense since the light must be detected only from a single rotational line, in contrast to the linear fluorescence method where the fluorescence must be collected from all the rotational levels of the excited state.

Acknowledgment

This work was supported by DRET under Contract 81-607.

References

Baronavski, A. P. and McDonald, J. R. (1977a) Measurement of C_2 concentrations in an oxygen-acetylene flame: An application of saturation spectroscopy. J. Chem. Phys. 66 (7), 3300-3301.

Baronavski, A. P. and McDonald, J.R. (1977b) Application of saturation spectroscopy to the measurement of C_2 concentration in oxy-acetylene flames. Appl. Opt. 16 (7), 1897-1901.

Berg, J. O. and Shackleford, W. L. (1979) Rotational redistribution effect on saturated laser-induced fluorescence. Appl. Opt. 18 (13), 2093-2094.

Bonczyk, P. A. and Shirley, J. A. (1979) Measurement of CH and CN concentration in flames by laser-induced saturated fluorescence. Combust. Flame 34 (3), 253-264.

Daily, J. W. (1976) Pulsed resonance spectroscopy applied to turbulent combustion flow. Appl. Opt. 15 (4), 955-960.

Daily, J. W. (1978) Saturation of fluorescence in flames with a Gaussian laser beam. Appl. Opt. 17 (2), 225-229.

Lucht, R. P. (1981) Experimental and theoretical investigation of laser saturated OH fluorescence in flames. Ph. D. Thesis, Purdue University, West Lafayette, Ind.

Lucht, R. P. and Laurendau, N. M. (1979) Two-level model for near-saturated fluorescence in diatiomic molecules. Appl. Opt. 18 (16), 856-861.

Mailander, M. (1978) Determination of absolute transition probability and particle densities by saturated fluorescence excitation. J. Appl. Phys. 49 (3), 1256-1259.

Pasternack, L., Baronavski, A. P., and McDonald, J. R. (1978) Application of saturation spectroscopy for measurement of atomic Na and NgO in acetylene flames. J. Chem. Phys. 69 (11), 4830-4837.

Piepmeier, H. (1972) Theory of laser saturated atomic resonance fluorescence. Spectrochimica Acta 27-B, 431-452.

Sharp, B. L. and Goldwasser, A. (1976) Some studies of the laser-excited atomic fluorescence of sodium. Spectrochimica Acta 31 (8-9), 431-457.

Stepowski, D. and Cottereau, M. J. (1979) Direct measurement of OH local concentration in a flame from the fluorescence induced by a single laser pulse. Appl. Opt. 18 (3), 354-356.

Stepowski, D. and Cottereau, M. J. (1981) Study of the collisional lifetime of hydroxyl radicals in flames by time-resolved laser-induced fluorescence. Combut. Flame 40 (1), 65-70.

Stepowski, D. and Cottereau, M. J. (1981) Time resolved study of rotational energy transfer in A 2 Σ state of OH in flame by laser induced fluorescence. J. Chem. Phys. 74 (12), 6674-6679.

Van Calcar, R. A., Van de Ven, M. J. M., Van Vitert, B. K., Biewenya, K. J., Hollander, J. J., and Alkemade, C. T. M. J. (1979) Saturation of sodium fluorescence in a flame irradiated with a pulsed tunable dye laser. J. Quant. Spectrosc. Radiat. Transfer 21 (11), 11-18.

Flame Concentrations and Temperatures by Spontaneous Raman Spectroscopy

R. Michael-Saade,* J. P. Sawerysyn,* L-R. Sochet,*
G. Buntinx,† M. Crunelle-Cras,† F. Grase,† and M. Bridoux†
Université des Sciences et Techniques de Lille, France

Abstract

The spatial distribution of gas flows issuing from a Parker-Wolfhard burner has been determined by pulsed multichannel Raman spectrometry and compared to those determined with conventional probes (sampling quartz microprobe connected to a gas chromatograph, thermocouple). Two different multichannel detector devices, both coupled to an EMI image intensifier tube have been used: a SIT (Silicon Intensified Target) camera tube allowing a three-dimensional investigation or a photodiode array leading to a two-dimensional investigation. This last configuration is more suitable for quantitative measurements. The systems $N_2|CO|O_2|N_2$ and $N_2|CO+N_2|O_2+N_2|N_2$ have been analyzed at 900 mbars either as gas jets or as combustion systems. A direct comparison of the results achieved in the case of a diluted system studied as gas jets shows a fair agreement between the results obtained by conventional and Raman probes.

Introduction

The analysis of reactive and turbulent systems requires more and more sophisticated investigation methods. Among the recently developed methods, spontaneous Raman scattering spectrometry—in spite of its relatively low threshold of detection—is of great interest when it is coupled both to a pulsed laser excitation and to a multichannel spectrometer. The aim of this paper is to compare results obtained by conventional methods of temperature and concentration measurements with those provided by spontaneous Raman spectrometry concerned with both nonreactive and reactive gaseous jets.

Presented at the 9th ICODERS, Poitiers, France, July 3-8, 1983. This paper is declared a work of the U.S. Government and therefore is in the public domain.
*Laboratoire de Cinétique et Chimie de la Combustion. ERA CNRS 1025.
†Laboratoire de Spectrochimie Infrarouge et Raman. LP CNRS 2641.

Experimental Arrangements

The Burner

A Parker-Wolfhard burner made of stainless steel has been used. It consists of four 5-cm-long channels each having a 30x5.2-mm rectangular section. Each channel is separated from others by 0.6-mm-thick wall. The gas flows issuing from the different channels were homogenized in the burner by circulating through glass bead packings located between stainless steel wire meshes. Fuel and oxidant streams —eventually diluted by nitrogen—flow through the central channels, as shown in Fig. 1. The flame stability is achieved by controlling the lateral expansion of gaseous jets by admission of nitrogen through the lateral channels. For each channel, the flow rates have been fixed at 100 l/h, which corresponds to a 20 cm/s velocity at the outlet of burner. In this work, the fuel is carbon monoxide (>99%) taken from Air Liquide cylinders without further purification. A very low quantity of methane (1%) is added to the fuel to improve the flame stability. Oxygen and nitrogen were used without particular purification. The $N_2|CO|O_2|N_2$ and $N_2|CO+N_2|O_2+N_2|N_2$ streams have been more particularly studied either as gas jets or as combustion systems. Such a gaseous system presents a spatial distribution of different products to which a temperature field is superimposed when the gas jets are ignited.

The burner is located in a water-cooled chamber fitted with four ports that permit the insertion of either conventional probes (sampling quartz microprobe and thermocouple) or of optical windows for the laser excitation of medium and the analysis of the scattering Raman light. The pressure in the chamber is measured by means of a classical manometer. Experiments are conducted under 900 mbars.

Conventional Probes

The gaseous samples are locally withdrawn under low pressure by means of a sampling quartz microprobe movable perpendicularly to the gas flows. The withdrawn gases are compressed in a piston, then injected into a gas chromatograph to be analyzed as previously described by Pauwels et al. (1981, 1982). For any position in the flame, the balance in mole fractions of different species ΣX_i is equal to 1±0.03. The spatial resolution of the sampling probe is of the order of 100 μm. The temperature measurements are made with 100-μm-diam. platinum, 30% rhodium-platinum, 6% rhodium wire thermocouple. The thermojunction wires are rectilinear and located normally to the gas jets to reduce conduction cooling of the thermojunction. The wires are insulated with a very thin

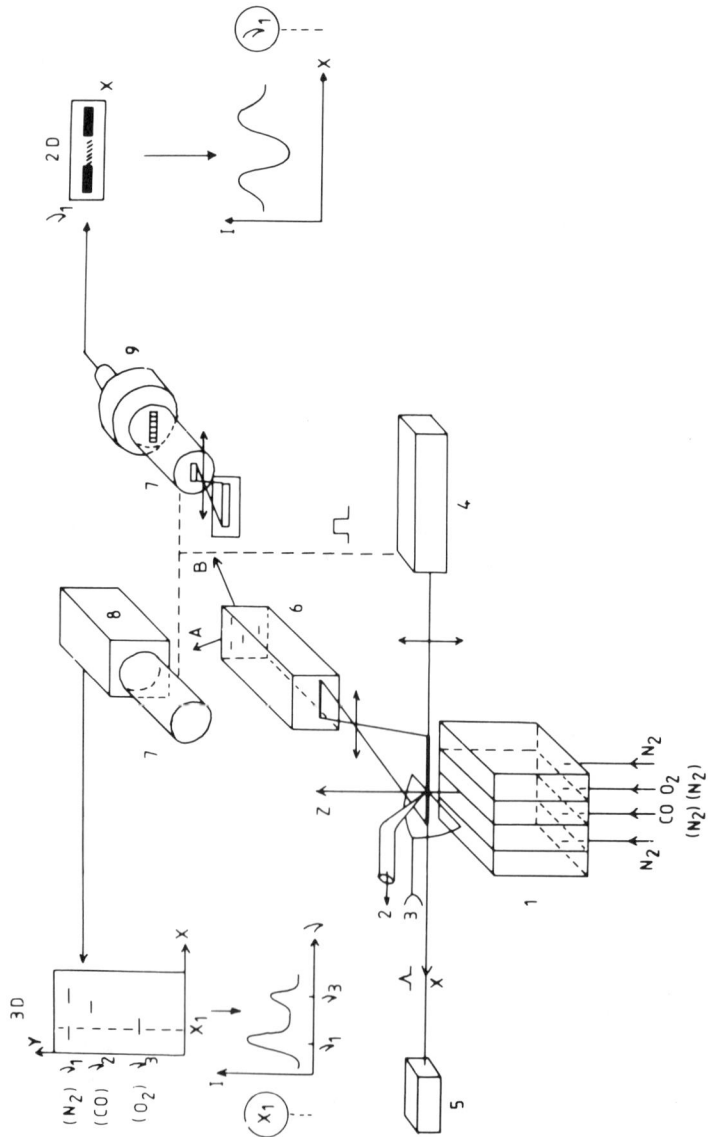

Fig. 1 Block diagram of experimental setup. 1) burner, 2) quartz-probe, 3) thermocouple, 4) Yag laser, 5) power meter, 6) polychromator, 7) image intensifier tube, 8) SIT camera tube, 9) photodiode array detector.

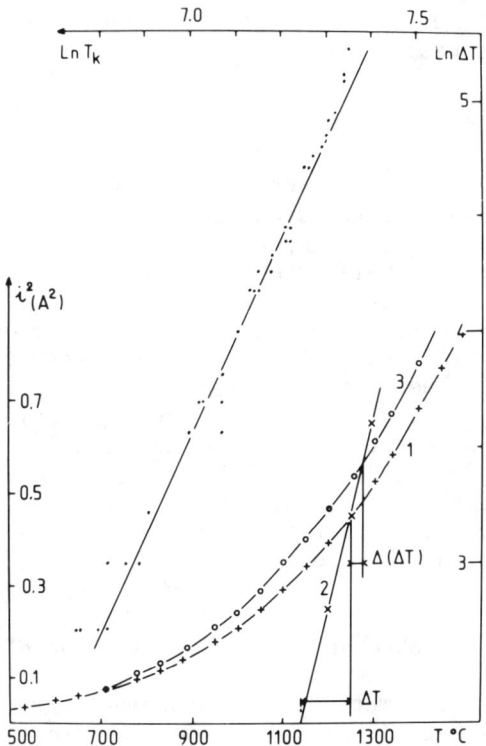

Fig. 2 Temperature correction of radiation loss for the thermocouple. Curves $i^2 = f(T)$ obtained 1) under vacuum + 2) within the flame x 3) under vacuum after the temperature measurements in the flame o. The evolution of $Ln\ (\Delta T) = f(T)$.

beryllium oxide-yttrium oxide to avoid catalytic effects (Kent 1970). The correction of radiation loss is achieved by means of an electrical heating compensation method (Wagner et al. 1960). Between 700 and 1500°C, the actual temperature of gases is determined by the intersection of the curves $i^2 = f(T)$ (Fig. 2, curves 1 and 2) obtained, respectively, under vacuum and within the flame. At higher temperature, because of the fragility of the thermocouple, the corrected temperature is obtained by extrapolation (Fig. 2) of the straight line $Ln\ (\Delta T) = f(T)$ obtained at lower temperatures from 57 experimental points, with a correlation coefficient of 99.3%. The correction can be as much as 400°C, with a precision $\Delta(\Delta T)$ estimated at about 30 K from a second calibration made under vacuum (Fig. 2, curve 3) after the temperature measurements in the flame. The temperature and concentration measurements were not determined for flames of undiluted $CO-O_2$

because of a much higher rise of temperature that causes the fusion of probes.

Optical Probes

In contrast with conventional probes previously described, optical probes are considered as nonintrusive probes. Among different optical methods presently in full development, spontaneous Raman spectrometry can provide instantaneous and local measurements both of major species concentration and temperature (Pealat et al. 1977; Sochet et al. 1979; Lapp and So 1980; Barj et al. 1981; Eckbreth 1981; Sawerysyn et al. 1981). The schematic diagram is exhibited in Fig. 1. The analyzed medium is excited by a light pulse of short duration (20 ns) emitted by a frequency-doubled Q-switched Yag laser (λ = 532 nm; 900 mJ). The scattering Raman light is collected at the right angle of the exciting laser beam. A scattering volume of variable lenght and 0.1-mm diam. is optically focused onto the entrance slit of a stigmatic polychromator. The Raman spectral images, corresponding to different molecular species located in the scattering volume of laser beam, are obtained in the focal plane of the polychromator.

Two different methods of measurements have been used (Fig. 1). 1) Method A: A wide spectral range of the Raman image is focused onto the photocathode of a four-stage image intensifier tube (EMI). The intensified spectral image is stored in the television target camera and transferred into a magnetic multichannel memory. "Playing back" the magnetic memory allows the image to be analyzed parallel to the wavelength coordinate axis $(0,\bar{\nu})$ successively at different locations along the spatial axis (OX). The intensity $I(x) = f(\bar{\nu})$ ... $I(X_n) = f(\bar{\nu})$ can then be extracted from the spectral image (Bridoux and Delhaye 1976; Sochet et al. 1979; Sawerysyn et al. 1981). and 2) Method B: In this case, the multichannel detector is a photodiode array consisting of 1024 25-μm-2.5-mm elements optically coupled to an EMI image intensifier tube (Surbeck et al. 1981). On the photocathode of this tube, only one spectral line of the Raman spectral image included within the focal plane of the polychromator is optically focused by means of an intermediate slit. To increase the signal-to-noise ratio, the photodiode array is cooled and the spectral line is focused onto only 50 photodiodes. The spatial resolution, hence, is about 1 mm. The resulting signal is stored point by point into a data collecting and processing system. The distribution of the signal intensity is thereby obtained along the scattering path of the laser beam for the different molecular species.

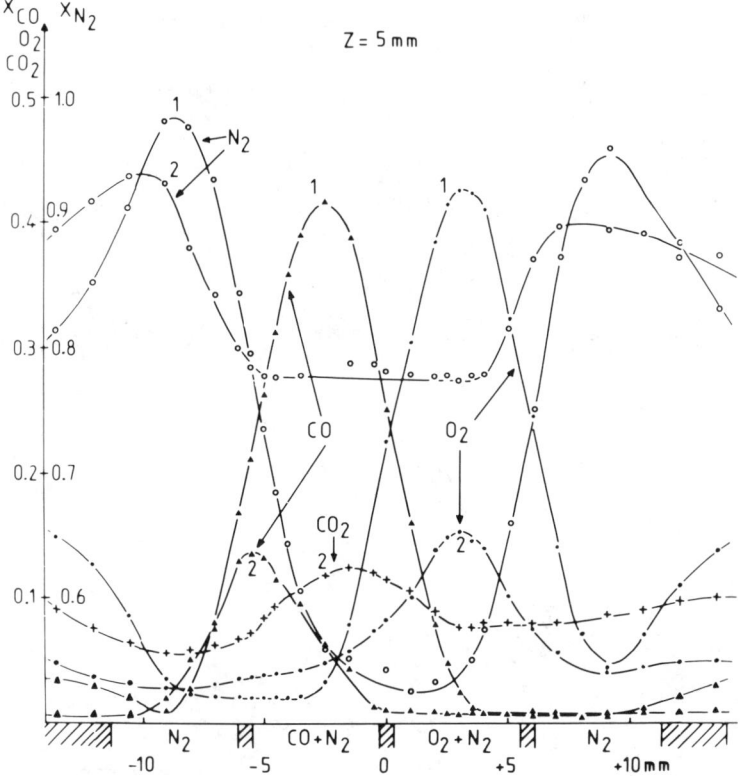

Fig. 3 Spatial distribution of N_2, CO, O_2, and CO_2 for Z = 5 mm in the $N_2|CO+N_2|O_2+N_2|N_2$ streams (mol ratios 1|0.5+0.5|0.5+0.5|1) when the gas jets are ignited (2) or not (1). Jet flow rates: 100 l/h in each channel.

Results and Discussion

Figures 3 and 4 show the results obtained by making use of conventional probes in the case of system $N_2|CO+N_2|O_2+N_2|N_2$ for which concentration and temperature profiles have been determined. The spatial distribution of different molecular species is exhibited for Z = 5 mm above the burner in the case of gas jets and flame. A near-parabolic distribution of species observed in the case of gas jets is substantially modified by chemical reactions of combustion. This effect-much more obvious for the CO distribution than for that of O_2—is due to the initial composition difference from the stoichiometric proportions. The profiles of corrected temperature T_c are given in Fig. 4a, from which isothermal lines of Fig. 4b can be extracted. The zone of maximal temperature approximately corresponds to the luminous front of

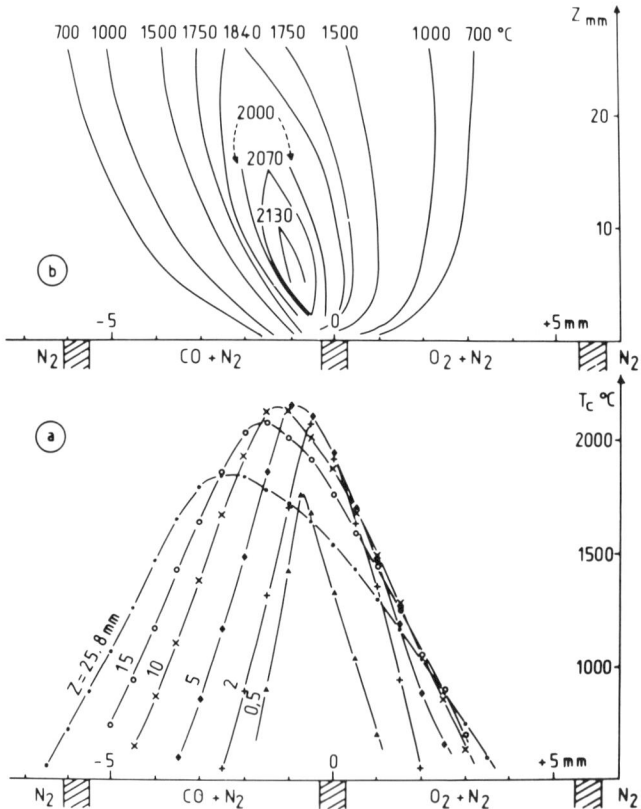

Fig. 4 Temperature profiles (a) at different altitudes above the burner and (b) the corresponding isothermal lines for the $N_2|CO+N_2|O_2+N_2|N_2$ flame (conditions as described in Fig. 3).

flame. These measurements are used for comparison with those obtained by Raman spectroscopy. However, aerodynamic, thermal, and catalytic perturbations cannot be completely neglected. Figure 5 illustrates the spectral Raman image obtained by a single laser pulse of 20 ns in the case of system $N_2|CO|O_2|N_2$ investigated in gas jets at $Z = 2.7$ mm above the burner. The spatial distribution of three species is clearly defined, and the interpenetration of jets is obvious. From this spectral image and for about ten spatial elements, the intensity evolution of different spectral lines, which is in direct relation with concentration evolution of different species, can be obtained. The analyzed range is about 15 mm and, hence, the spatial resolution is about 2 mm.

When the gas flow is ignited (Fig. 6), the spectral images reveal the CO_2 formation characterized by two vibration bands and a progressive disappearance of CO. O_2, initially

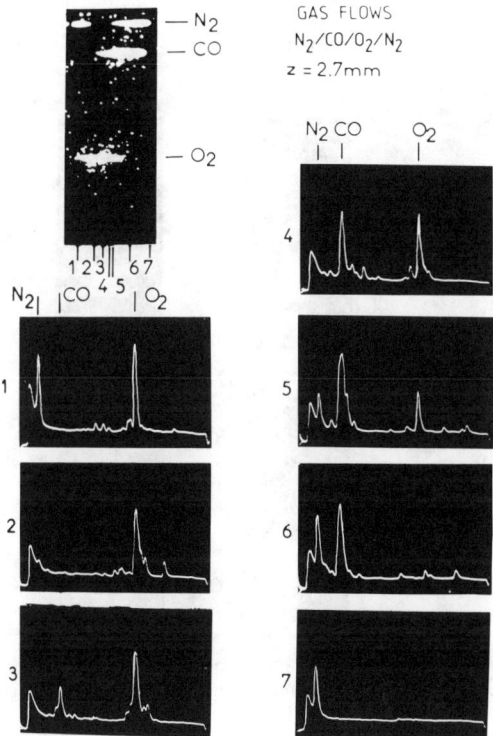

Fig. 5 Analysis of the Raman spectral image obtained at Z = 2.7 mm for the gas jets $N_2|CO|O_2|N_2$ (mol ratios 1|1|1|1). Jet flow rates: 100 l/h in each channel. (900 mJ, spectral resolution 19 Å/mm, spatial investigation ≃ 15 mm).

in excess, remains in the medium for all heights above the burner. In these figures, the homogeneization of system appearing when Z increases, shows the important role played by diffusion processes. The temperature effect cannot be easily observed in this case because the spectrometer dispersion (19 Å/mm) is too low to separate the hot bands of N_2.

With the same spectral dispersion, the analysis of a larger object field (about 30 mm) of gas jets containing CO and O_2, both 50% diluted by N_2, in combustion or not, shows a very important spatial heterogeneity of different spectral lines (Fig. 7). In this configuration, the spectral lines of CO, O_2 and CO_2 appear as dashed lines because of a lack of sensitivity to detect the species in the regions where the mole fractions are much too low (<0.05). Under these experimental conditions, N_2 is anywhere in high concentration, and, consequently, its spectral line appears strong except in the local zone where the temperature is high enough to decrease

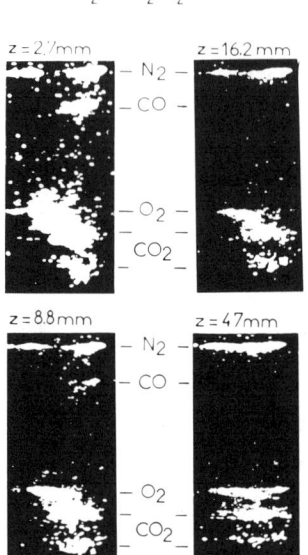

Fig. 6 Raman spectral images obtained at different altitudes for the $N_2|CO|O_2|N_2$ flame (conditions as described in Fig. 5).

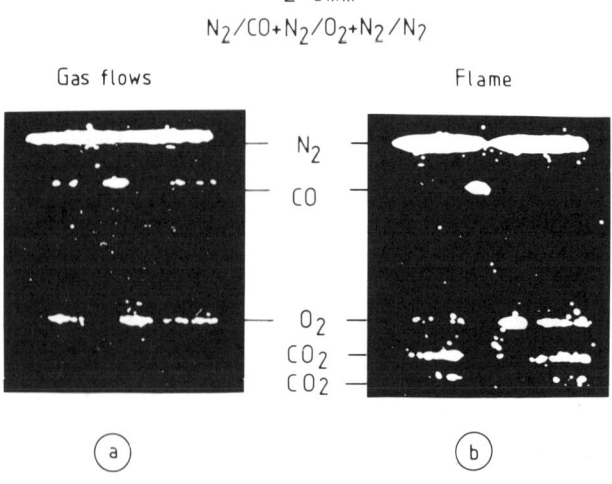

Fig. 7 Raman spectral images obtained at Z = 5 mm in the $N_2|CO+N_2|O_2+N_2|N_2$ streams (mol ratios 1|0.5+0.5|0.5+0.5|1) when the gas jets are ignited (b) or not (a). Jet flow rates: 100 l/h in each channel. (900 mJ, spectral resolution 19 Å/mm, spatial investigation ≈ 30 mm).

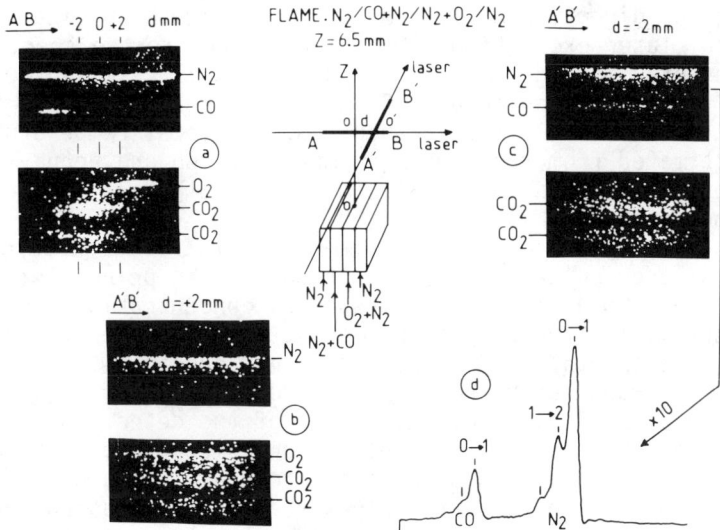

Fig. 8 Raman spectral images obtained at Z = 6.5 mm in the case of a perpendicular (a) or parallel (b), (c) investigation to the flame front (conditions as described in Fig. 7 but with a spectral resolution of 8.6 Å/mm and a spatial investigation of about 15 mm).

the N_2 concentration, which can be divided by a factor of 10. The sensitivity threshold can be lowered by increasing the photon gain of the EMI intensifier image tube. However, in this case, the relatively low dynamic range of the multichannel optical detection device causes a saturation of spectral lines of all species in high concentration. A more detailed investigation of gas jets in combustion has been achieved in the zone of the flame front with an object field of about 15 mm by means of a spectrometer with a better dispersion (8.6 Å/mm). Results are presented in Fig. 8. Two investigations were made: first, with a laser excitation perpendicular to the flame front (analyzed object field AB), then with a laser excitation parallel with the flame front (object field A'B'), consequently, within a near-isothermal plane. In this last case, the spectrometer dispersion is sufficient to resolve different hot bands. As shown in Fig. 8a, the nonhomogeneity of the temperature clearly appears on the spectral line of N_2. The dissymmetry of the spectral line with respect to the axis d=0 is caused by the localization of the flame front in the part of CO jet. However, it is practically impossible by means of only one laser pulse to determine simultaneously in several points the temperature in the region where a large gradient exists ($-2 < d_{mm} < +2$). For measurements of the near-isothermal planes (Fig. 8b and 8c), the intensity of different spectral lines can be integrated

along the direction A'B'. Averaging the signal obtained for
several laser excitations (ten shots) leads to the spectrum
represented in Fig. 8d. This relatively well resolved spectrum shows both for N_2 and CO the fundamental bands and the
first hot bands. From this spectrum, the local temperature
is estimated from the band area method (Lapp and Penney 1977)
to be 1700 K; this is in fair agreement with that obtained
by thermocouple (Fig. 4b).

The analysis of a single Raman spectral image obtained
from a laser pulse of 20 ns duration can, in principle, lead
to the determination of major species concentration and temperature simultaneously at different points. Under the present experimental conditions, it is not possible to obtain
accurate quantitative results. The relatively low detection
threshold and the low dynamic range of the detection chain,
essentially limited by the television camera tube, are important limiting factors. A better accuracy could be obtained in the case of a stationary system with averaging and
treatment of several spectral images. The large dynamic range of photodiodes contributes to the improvement of the con
centration measurements. The use of method B allows to obtain
successively—with a rotation of the polychromator grating—
the intensity of each Raman signal corresponding to the species existing in the laser beam scattering volume, which is
conjugated with the entrance slit. As the signal contains
random statistical errors, the averaging of several data can
increase the signal-to-noise ratio. In the case of oxygen,
under a pressure of 100 mbars, this ratio reaches a value of
about 6 for a sample excited by ten laser shots of 950 mJ. A
direct comparison of the results obtained by conventional
quartz microprobe and Raman probe using photoarray detectors
is shown in Fig. 9 for the system $N_2|CO+N_2|O_2+N_2|N_2$ studied
as gas jets at 900 mbars and for Z = 5 mm. The data for the
Raman probe are provided by acquisition and treatment of the
intensity of N_2, O_2, and CO vibration Raman bands (an average of 20 laser shots of 950 mJ) and by calibrations previously achieved in an homogeneous medium. While spatial distributions of major species obtained by the two methods are in
fair agreement, quantitative differences are apparent for
all species and especially for the species in low concentration (CO and O_2). The mole fraction balance ΣX_i of the species shows a better accuracy for the chromatographic measurements (1 ± 0.03) than that obtained for Raman measurements
(1 ± 0.08). With Raman probe, and method B, the temperature
effects due to the combustion reaction cannot be observed.
However, the concentrations N could be evaluated to a good
approximation from the integrated intensity $I = \alpha NF(\Delta\lambda,T)$
in a spectral band $\Delta\lambda$ sufficiently wide (defined by the
width of the intermediate slit) to make $F(\Delta\lambda,T)$ nearly inde-

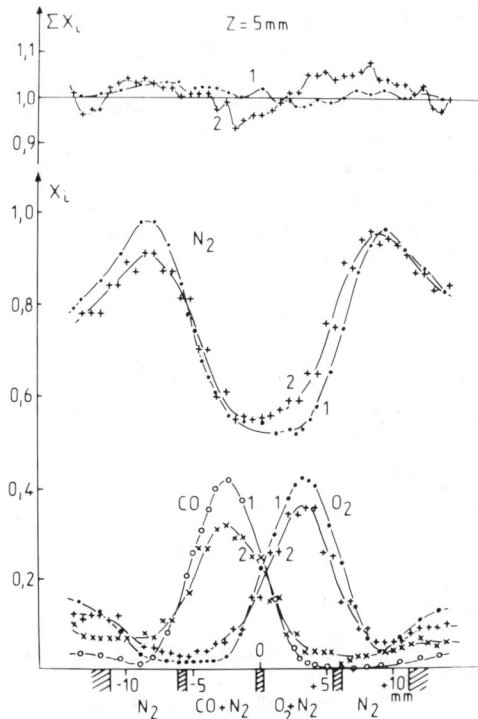

Fig. 9 Comparison of the spatial distribution of N_2, O_2, and CO obtained by gas chromatography (1) and Raman Spectrometry (2) with photodiode array detector in the case of gas jets and with the conditions described in Fig. 3.

pendent from the medium temperature (Setchell 1976). The temperature could be estimated from the ratio of the intensity integrated over all the Stokes Q-branches to that integrated on all the anti-Stokes Q-branches. This could be achieved with a second photodiode array.

Conclusion

The application of multichannel pulsed Raman spectrometry for nonintrusive determination of space-resolved concentration and temperature profiles in reactive systems has been demonstrated for gas jets and diffusion flames. Spatial distribution of major molecular species for N_2-CO-O_2 systems determined with Raman spectrometry has been compared to those determined by conventional probes. Under the present conditions, accurate quantitative measurements cannot be simultaneously obtained, with a single laser excitation, at different points with a television camera tube as a detector device because of its relatively low detection threshold and its low dynamic range. If several Raman spectral images are averaged, a relatively good agreement was observed with the value provided by thermocouple. Because of a wider dynamic range, a photodiode array yields more accurate quantitative measurements of the spatial distribution of species in gas jets. When several measurements are averaged, the accuracy obtained for the mole fraction balance of the species is lower than that determined by chromatographic method.

References

Barj, M., Bridoux, M., Crunelle-Cras, M., Grase, F., Sawerysyn, J.P., and Sochet, L.R. (1981) Space and time-resolved investigations of combustion phenomena by emission and Raman spectroscopy. Combustion in Reactive Systems : AIAA Progress in Astronautics and Aeronautics, vol.76, edited by J.R. Bowen, N. Manson, A.K. Oppenheim, and R.I. Soloukhin, pp.635-645. AIAA, New York.

Bridoux, M., and Delhaye, M. (1976) Advances in Infrared and Raman Spectroscopy (edited by R.J.H. Clark and R.E. Hester), vol.2, pp.140-152. Heyden and Son Ltd., London.

Eckbreth, A.C. (1981) Recent advances in laser diagnostics for temperature and species concentration in combustion. 18th Symposium (International) on Combustion, pp.1471-1488. The Combustion Institute. Pittsburgh, Pa.

Kent, J.H. (1970) A noncatalytic coating for platinum-rhodium thermocouples. Combust. Flame 14, 279-282.

Lapp, M. and Penney, C.M. (1977) Advances in Infrared and Raman Spectroscopy (edited by R.J.H. Clark and R.E. Hester), vol.3, pp.204. Heyden and Son Ltd., London.

Lapp, M. and So, R.M.C. (1980) The study of turbulent diffusion flames : modeling needs and experimental light scattering capabilities. Agard Conference Proceedings on Testing and Measurement Techniques in Heat Transfer and Combustion, pp.19-1, 19-17.

Pauwels, J.F., Carlier, M., and Sochet, L.R. (1981) Couplage d'une technique d'échantillonnage sous basse pression à un chromatographe en phase gazeuse. J. Chromatogr. 211(2), 247-251.

Pauwels, J.F., Carlier, M., and Sochet, L.R. (1982) Analysis by gas-phase electron spin resonance of H, O, OH and halogen atoms in flames. J. Phys. Chem. 86(22), 4330-4335.

Pealat, M., Bailly, R., and Taran, J.P. (1977) Real time study of turbulence in flames by Raman scattering. Opt. Commun. 22(1), 91-94.

Sawerysyn, J.P., Barj, M., Bridoux, M., Chapput, A., and Crunelle-Cras, M. (1981) Time and space-resolved analysis of perturbations caused by a flame propagating through a gas mixture by

C-W laser multichannel Raman spectroscopy. 18th Symposium (International) on Combustion, pp.1703-1707. The Combustion Institute. Pittsburgh Pa.

Setchell, R.E. (1976) Time-averaged measurements in turbulent flames using Raman spectroscopy. AIAA Paper 76-28. 14th Aerospace Science Meeting.

Sochet, L.R., Lucquin, M., Bridoux, M., Crunelle-Cras, M., Grase, F., and Delhaye, M. (1979) Use of multichannel pulsed Raman spectroscopy as a diagnostic technique in flames. Combust. Flame 36(2), 109-116.

Surbeck, H., Hug, W., Gremaud, M., Bridoux, M., Deffontaine, A., and Da Silva, E. (1981) The direct recording of Raman spectra with solid state detectors. Opt. Commun. 38(1), 57-60.

Wagner, H.W., Bonne, U., and Grewer, T. (1960) Messungen in der reaktionszone von wasserstaff-sauerstoff und methan-sauerstoff flamen. Z. Phys. Chem. Neue Folge, Bd 26, 93-110.

The Application of Rotational Raman Spectroscopy to Dynamic Measurements in Gas Flowfields

Perry P. Yaney,* Roger J. Becker,† Paul T. Danset,‡
Michael R. Gallis§ and Juan I. Perez**
University of Dayton, Dayton, Ohio

Abstract

The design and performance of a photon-counting, two-channel spectroscopic system using the 6 W, 488 nm line of a continuous-wave argon ion laser is described. The system is capable of dynamic, simultaneous measurements of either the concentrations of two gas species in cold flow or the temperature and concentration of a single gas species (such as N_2) in a flame. By means of a laser beam retroreflecting multipass cell, the effective laser power incident on the observed volume was magnified by a factor of 13 over a single pass. Other features plus the near order-of-magnitude higher signals [compared to the vibrational (Q-branch) bands] obtained from pure, spontaneous rotational Raman lines of N_2, O_2, and CO_2 provided signals about 1000 times stronger than typically observed. Measurements were carried out on a methane diffusion flame (Re 1600) and a room temperature CO_2 jet (Re 10,000 and 30,000). Up to 4096 samples (reads)^2 from

---
Presented at the 9th ICODERS, Poitiers, France, July 3-8, 1983. Copyright © 1984 by the American Institute of Aeronautics and Astronautics, Inc. All rights reserved.
  *Professor of Physics, Department of Physics. Member AIAA.
  †Research Physicist, Fluid Mechanics Group, Research Institute.
  ‡Presently Research Associate, National Center for Rehabilitation Engineering, Wright State University, Dayton, Ohio.
  §Presently Graduate Assistant, Department of Physics, Pennsylvania State University, University Park, Pennsylvania.
  **Presently Graduate Assistant, Department of Physics, Wright State University, Dayton, Ohio.

each of the two channels were recorded by a dedicated minicomputer. Sampling rates of 20-100 Hz were used in the flame studies, while rates up to 2.0 kHz were used in the CO_2 jet measurements. Analyses were carried out using probability density functions (pdf), power spectral density functions, auto- and cross-correlation functions, and calculations of skewness and kurtosis. The temperature pdf's in the nonreacting regions of the flame showed a strong positive skewness that is in agreement with the prediction of a Gaussian density distribution and the gas law at constant pressure.

Introduction

A wide variety of laser probe techniques have been investigated and employed for the purpose of measuring temperatures and/or gas concentrations in both reacting (Crosley 1980) and nonreacting (Birch et al. 1975, 1978; Black and Chang 1978; Leipertz 1981) gas flowfields. Many of these techniques use high-peak-power lasers with pulse repetition frequencies (prf) of 1-10 Hz (Eckbreth 1981). These lasers produce narrow pulses that are usually short enough to "freeze" the gas movement during the measurement. As long as the integral time scale of the fluctuations is somewhat shorter than the time interval between samples (i.e., the sampling period), low-prf sampling can be used to obtain accurate probability density functions (pdf) of the fluctuations (Bendat and Piersol 1971; George et al. 1979). However, this approach does not permit a measurement of the temporal power spectral density function (psdf) or the temporal autocorrelation function (acf). These functions can be obtained from a real-time record of the fluctuations if the sampling rate is sufficiently high. The importance of these functions to gas flow analysis is that they can provide an insight into the dynamics of the processes being studied, particularly when two or more variables are being measured simultaneously. Another advantage of using a high sampling rate is that a large amount of data can be recorded in a short time, which can significantly reduce the run time.

Real-time records of gasdynamic variables can be obtained most conveniently with an optical technique using a continuous wave (cw) laser. However, the maximum power available from cw lasers having suitable characteristics is generally limited to the 1-10 W range. Since high sampling rates require high signal levels, other means must be employed to increase the signal level if a cw laser is to be used. This paper presents recent results obtained with a measurement system utilizing Raman scattering (RS) wherein the the observed signals were more than 1000 times stronger than typically observed from the N_2 Q-branch using a 1 W cw laser.

Experimental Approach

The approach here was to use the ~6 W power available from the 488 nm line of a Spectra Physics model 171-18 argon ion laser with a laser beam retroreflecting multipass cell (Hill et al. 1977) to achieve a cw excitation level in the observed volume of ~70 W (Yaney et al. 1982). To further enhance the available signals, pure Stokes rotational RS was used because it is generally the strongest Raman signal (about 10 times the vibrational signal) available from the gases of interest (Penney et al. 1974). A similar approach involving a cw laser and a multipass cell was used by Setchell (1976) to make time-averaged measurements on a hydrogen flame in air using the Q-branches of the vibrational spectra. He carried out a thorough analysis of the errors in such measurements introduced by the fluctuations in temperature and concentration. Fortunately, these errors are greatly diminished with pure rotational spectra compared to that obtained with the Q-branches. This is due to the factor of 10-20 smaller energy positions of the rotational levels compared to the vibrational levels and the nearly identical functional forms of the temperature dependences of the intensities of two rotational lines. In reacting flows, rotational RS has the additional advantage of providing temperatures and species concentrations obtained from signals, which are derived from molecular motions, that are in good thermal contact with the translational motions of the molecules. Spontaneous scattering was chosen because it is proportional to the molecular concentration and laser power and, therefore, can be easily used with a cw laser.

This approach has two significant limitations: 1) Stokes spontaneous RS is susceptible to interferences from non-Raman background signals such as laser-induced fluorescence and incandescence; and 2) the choice to use pure rotational RS raises the problem of interferences with the Raman signals from other species due to the high density of lines in this region. This difficulty is easily seen in the rotational RS spectrum of room air given in Fig. 1. Significant control over both of these difficulties can be effected by control of the spectral resolution of the spectrometer used to select the desired lines. Since background signals are usually proportional to the spectral bandwidth, narrow spectral bandwidths can provide higher signal-to-noise ratios when the narrow rotational Raman lines are used. By using this approach we were able to use theoretically derived expressions to calibrate the system. On the other hand, when the background signals are relatively small, the rotational Raman signals can be increased by using wider bandwidths so as to include two or more rotational lines (Lapp 1973). Since this approach is more susceptible to interferences from unwanted Raman signals, the calibration of the system is best done empirically (Williams et al. 1977).

The variables of interest in these studies were temperature and concentration. The temperature T of a diatomic gas is related to the ratio R of the intensities of two pure rotational Raman lines specified by quantum numbers J_A and J_B by the relation

$$T = K_1/\ln(K_2 R) \qquad (1)$$

where K_1 and K_2 are constants dependent on the molecular species and the values of J_A and J_B (Yaney et al. 1982). The values for $J_A = 11$ and $J_B = 8$ of N_2 used in this work are $K_1 = -171.75$ K and $K_2 = 1.5182$. The resulting calibration curves for this line pair and three other pairs are given in Fig. 2. Similarly, in constant-pressure

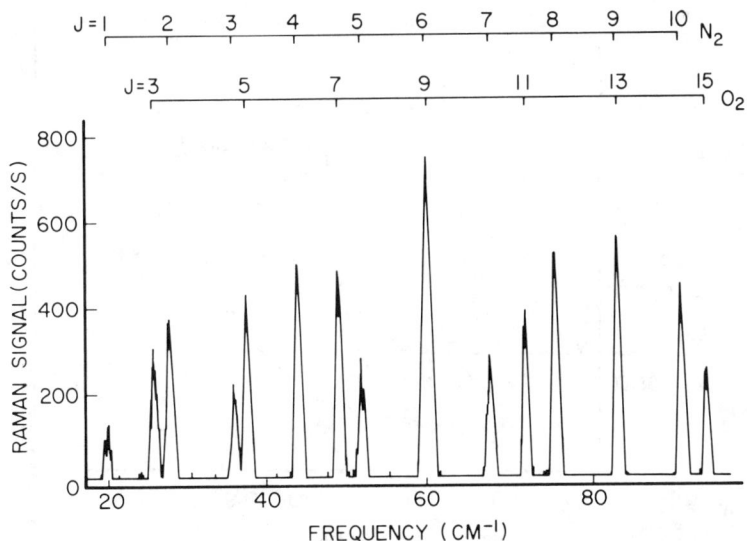

Fig. 1   Typical Stokes rotational Raman spectrum of room air for a single pass of a 2 W, 488 nm laser beam (bandwidth equals 0.8 cm^{-1}).

concentration measurements (Yaney et al. 1983), it can be shown that the mole fraction c of a gas is related to the Raman count rate $\dot{n}_J$ of the rotational line having quantum number J by the expression

$$c = \dot{n}_J \left\{ \frac{\lambda_J^3 \lambda_L T^2}{\varepsilon P_e \ell p} \exp\left(\frac{hc}{k} \cdot \frac{E_J}{T}\right) \right\} \qquad (2)$$

where λ_J is the wavelength of the RS light; λ_L the wavelength of the laser light; ε the calibration constant depending on the collection geometry, the molecular species, and the system detection efficiency; and P_e the excitation power in the observed volume. The length of the observed volume is given by ℓ, which corresponds here to the length of the entrance slit; p is the atmospheric pressure; hc/k are the usual fundamental constants that give 1.4388 K/cm^{-1}; and E_J is the energy separation of the J rotational state from the ground state (J = 0) closely given by $E_J = B_0 J(J + 1)$, where B_0 = 1.9895 cm^{-1} for N_2. The primary empirical factor in the calibration constant ε is the overall quantum detection efficiency (viz., the number of observed counts per scattered photon into the collecting solid angle). For temperature measurements,

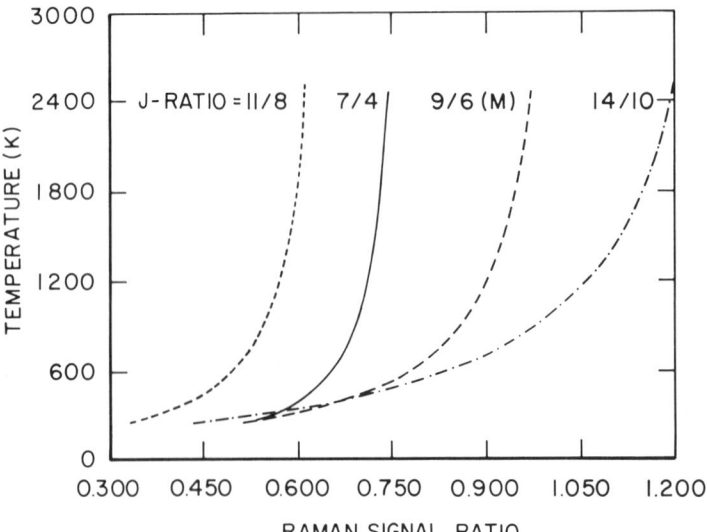

Fig. 2 Theoretical temperature calibration curves for four pairs of Stokes rotational Raman lines of N_2 [the 9/6(M) pair includes the J = 13 and 9 lines of O_2.]

only the ratio of the factors for the two channels is needed. For concentration measurements, an absolute calibration is effectively required. Equation (2) assumes that the ideal-gas law with the pressure constant is applicable to the flow. Also, it is assumed that the measurement includes all pure rotational transitions from state J independent of the vibrational state of the molecule. The former constraint requires low Mach number flows, while the latter sets a limit on the minimum bandwidth (i.e., slit width) that can be used without applying corrections (Drake et al. 1980).

In constant-temperature and constant-pressure flows, that is, in "cold" flows, Eq. (2) simplifies to

$$c = \dot{n}_J/\dot{n}_{Jr} \qquad (3)$$

where $\dot{n}_{Jr}$ is the reference or calibration count rate corresponding to an 100% (c = 1.0) concentration of the particular gas species in the observed volume. To use Eq. (3), all the quantities inside the braces in Eq. (2) must be known and held constant throughout all of the measurements.

In order to apply Eq. (1) successfully, the background signals under the two lines must be measured. If the background spectrum is reasonably steady in magnitude and shape, then its contribution to the two Raman channels A and B can be measured in a separate experiment. This allows temperature measurements using only two simultaneous measurements of the Raman spectrum. Furthermore, having measured T, it is then possible to obtain a simultaneous value of c of the "probe gas" using Eq. (2) and the signal from either channel or a weighted average of the signals from both channels. On the other hand, in cold flow, the

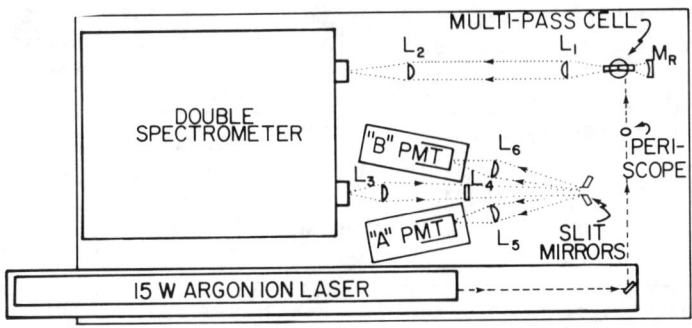

Fig. 3   Layout of the two-channel cw TiLaRS system (from Yaney et al. 1982).

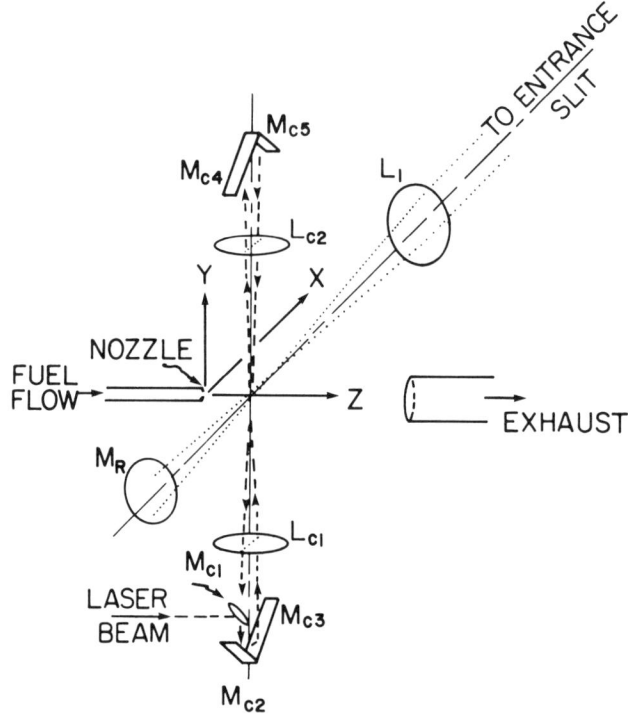

Fig. 4 Schematic diagram of the laser excitation and the Raman collection geometries relative to the gas nozzle (from Yaney et al. 1982).

use of two parallel channels allows simultaneous measurements of the concentrations of two gases in the flow.

The layout of the two-channel Raman system is illustrated in Fig. 3. The configuration of the excitation and collection optics relative to the flow nozzle are shown in Fig. 4. Complete descriptions of this cw time-resolved laser Raman spectroscopy (TiLaRS) system have been previously given by Yaney et al. (1982, 1983).†† An example of the system performance is given by the intensity of the 60 cm^{-1} "air" line in Fig. 1 at room temperature. For 2 mm high by 0.15 mm wide slits (giving a 0.038 nm bandwidth) and 2.0 W laser power at 488 nm, we usually

---

††The signal levels reported by Yaney et al. (1982) were about five times weaker than reported here due to the reduced quantum efficiency caused by an exposure of the photomultipliers to excessive light.

obtained 90 kcounts/s. This is illustrated in the trace of channel A in Fig. 5. The nearly symmetric, sharp line shapes of the second-order spectra shown in Fig. 5 are evidence for the good performance of the dual-channel, exit-slit optical system. Since the scans in Fig. 5 were all carefully started at the same wavelength, this figure also shows the good match of the channel settings with the peaks of the two lines.

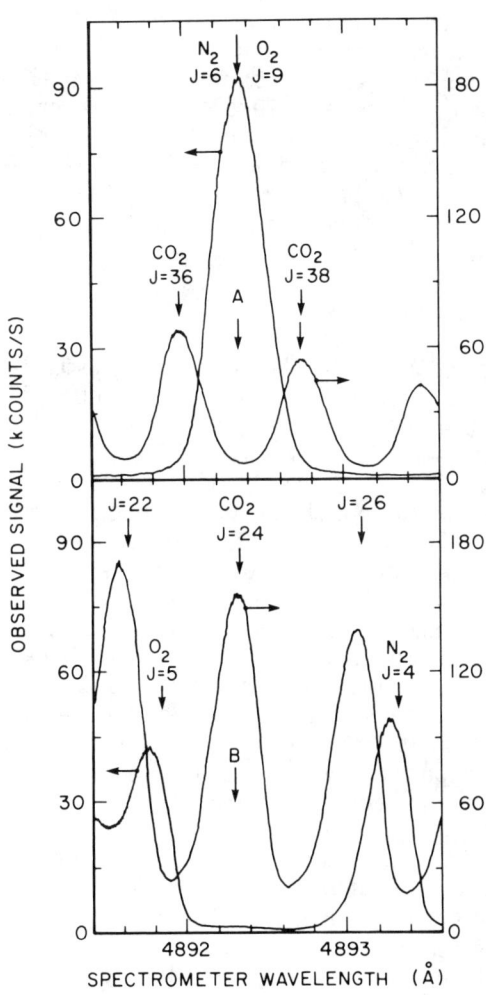

Fig. 5   Superimposed scans of 100% air and 100% CO_2 at room temperature and pressure: laser power 2.0 W, exit slit widths 0.15 mm, and slit heights 2 mm for air and 1 mm for CO_2. The true air and CO_2 wavelengths are shifted up 2.0 Å and down 2.8 Å from the spectrometer setting, respectively.

Measurements and Results

CH$_4$ Flame Studies

The diffusion flame studies used a Delavan 45 deg spray nozzle with a 1.4 mm (0.055 in.) diam orifice mounted on a straight 75 cm long, 1 cm i.d. stainless steel tube. The X-Y-Z position and alignment (see Fig. 5) of the tube and nozzle were accomplished by five translation stages calibrated to 0.05 mm. Methane was chosen because it does not have a pure rotational Raman spectrum (at ordinary pressures) since it is a spherical top molecule (Schrotter and Klockner 1979). The choice of flow conditions was largely determined by the maximum flow rate that could be achieved with a stable flame. A volume flow rate of 1.8 liter/min, corresponding to a mass flow of 0.07 kg/h and an exit velocity of nearly 20 m/s, was used. The cold flow Reynolds number (Re) was about 1600. The horizontal flame extended several centimeters into the exhaust pipe where it was a bright orange. The upstream portion of the flame was blue except for a ridge of orange along the top edge. The exhaust and a small, stabilizing airflow was provided by a small centrifugal blower connected to the exhaust pipe shown in Fig. 4.

The measurements were made along the radius (i.e., along X) at 2.5 nozzle diameters downstream (i.e., Z = 2.5d). This axial position was chosen primarily because there was little evidence of incandescent soot at this location. Thus, the background signals were due to flame luminosity, leakage of the laser light through the spectrometer, and possibly some fluorescence. The background was measured by shifting the spectrometer wavelength up 0.08 nm, corresponding to 3.4 cm^{-1} in Fig. 1. The outer dimensions of the "bow-tie-shaped" observed volume were about 0.5 x 2 x 0.05 mm^3 in the X-Y-Z directions, respectively.

Since the spectrometer constrained $\Delta J = J_A - J_B$ to three, the precision of the measurements rapidly deteriorated as the temperature increased above 1200 K. Therefore, the measurements were limited to the cooler parts of the flame. This is clearly evident from the high slope of the calibration curve above 1200 K given in Fig. 2 for J_A/J_B = 11/8. [This difficulty can be reduced with the use of a spectrometer or gratings that permit J_A/J_B = 14/10 (ΔJ = 4) to be used.] The apparent high-temperature excursions above the maximum adiabatic equilibrium temperature of 2250 K were handled by clipping these values at 2250 K or deleting them. The derived temperature variances σ'^2 were determined by subtracting the Poisson temperature variance s_T^2, computed from the various signal levels, from the observed variance σ^2 in the temperature data. The concentration data were analyzed in an analogous manner. In effect, the measuring process of photon counting introduces intrinsic random fluctuations characterized by Poisson statistics wherein the variance of these intrinsic fluctuations equals the mean of the counts accumulated over the measurement intervals. These "detection" fluctuations are superimposed on the intensity fluctuations of the measured light signal (Bertolotti 1974; Birch et al. 1975, 1978).

In the case of negligible background signal levels, the Poisson temperature standard deviation becomes

$$s_T \approx \overline{T}^2 \sqrt{f_r[(1/\bar{n}_A) + (1/\bar{n}_B)]} / |K_1| \qquad (4)$$

where f_r is the measurement sampling or read rate (Yaney et al. 1982). That is, the time interval of each read is f_r^{-1}. Equation (4) clearly shows the desirability of having high Raman count rates when a high read rate is used. Although s_T increases with increasing background, the change is not excessive. For example, a Raman-to-background ratio of unity increases s_T over Eq. (4) by a factor of only two (Yaney et al. 1983).

The Poisson concentration standard deviation s_c for negligible background signals is given for channel B values by

$$s_c \approx \overline{c} \sqrt{(f_r/\bar{n}_B) + [(K_B/\overline{T}) - 2]^2 (s_T/\overline{T})^2} \qquad (5)$$

where $\overline{c}$ is the mean mole fraction, $K_B = 1.4388 \, B_0 J_B (J_B + 1)$, and s_T is given by Eq. (4) (Yaney et al. 1983). Equation (5) shows the contribution of the temperature measurement uncertainty to s_c. It also suggests that this contribution can be reduced by selection of J_B such that $K_B = 2\overline{T}$. However, this must be balanced against the signal levels available from the flame and at room temperature for a given J value to achieve a functional choice.

The results of the pdf analyses of the data for the X = 0.0, 0.4 (5.7r), and 0.5 cm (7.1r) positions are presented in the pdf's given in Fig. 6. Both the temperature pdf and the N_2 concentration pdf given in terms of mole fraction are presented for each of the three

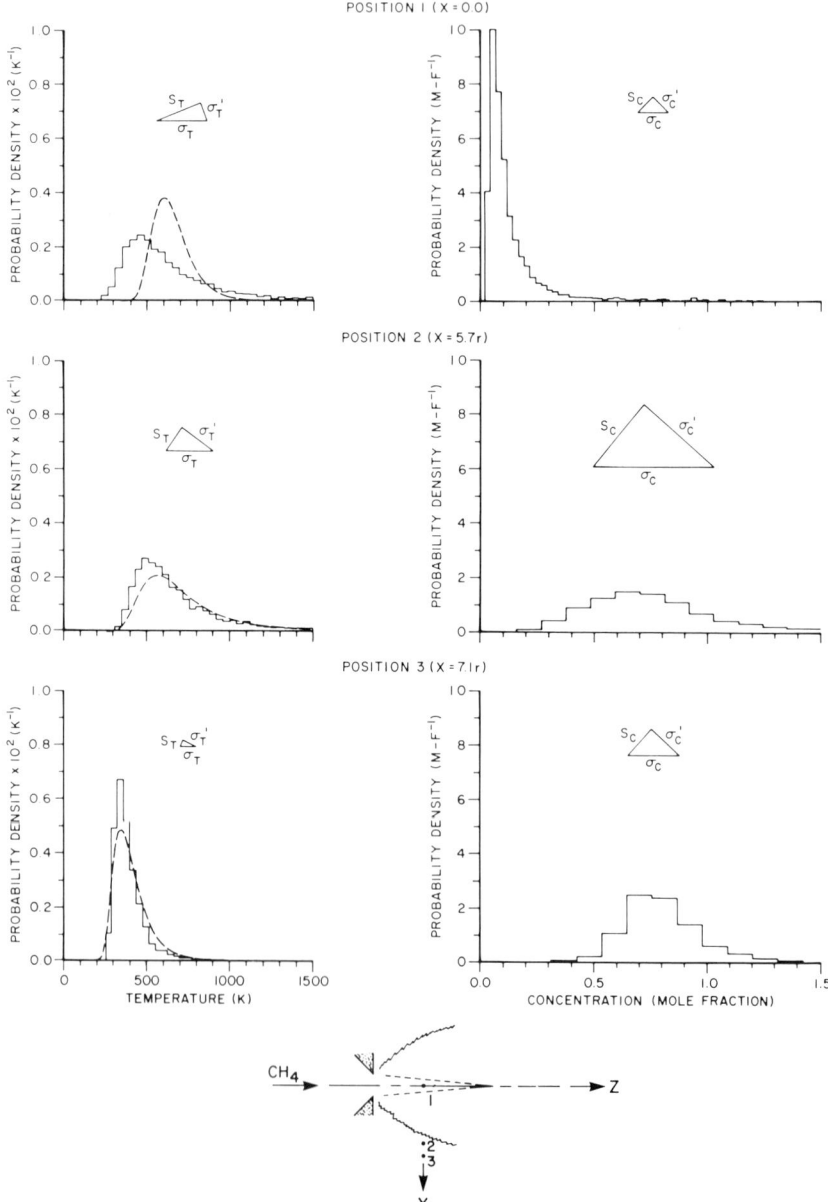

Fig. 6 Simultaneously measured pdf's (solid line) of temperature (left) and N_2 concentration (right) at Z = 2.5d for the potential core (position 1) and near the flame front (positions 2 and 3) of the diffusion methane flame. (See text for explanation.)

positions. The on-axis data were recorded at a 20 Hz read rate for 2048 reads, while the 5.7r and 7.1r data were taken at 50 and 100 Hz, respectively, for 4096 reads. The number of reads exceeding 2250 K for the three positions were 2.1, 1.9, and 0% for X = 0.0, 4.7r, and 7.1r, respectively. The pdf analyses were carried out after these reads had been deleted. The influence of these deletions was to slightly lower the mean temperatures and reduce the measured standard deviations.

Associated with each plot in Fig. 6 is a triangle graphically depicting the relationship between the measured, Poisson, and derived standard deviations. All of the quantities presented in these plots are summarized in Table 1. Although the measured standard deviations (σ_T and σ_c) are quite large, particularly for the concentration data, the data still permitted determination of the derived values (σ_T' and σ_c'). The extension of the broad concentration pdf's beyond unity mole fraction is not supported on physical grounds; however, these extensions are wholly due to the large Poisson uncertainties (s_c) and the fact that a constant background was subtracted from each data set. Since the background levels actually fluctuate from read to read, there is an additional smearing of the limit of unity mole fraction over and above the tail due to the Poisson uncertainty. These broadening contributions, plus the slight positive skewness of the concentration pdf's seen in Fig. 6, may be responsible for the slightly high mean concentration values at the X = 5.7r and 7.1r positions given in Table 1.

The striking feature in the Fig. 6 plots is the strong positive skewed character of the three temperature pdf's. This is in contrast to the concentration pdf's at the edge of flame, which are reasonably symmetrical. The on-axis concentration pdf is skewed due to the proximity of the zero mole fraction limit. The positive skewness in the temperature pdf's can be shown to be the consequence of a

Table 1 Temperature and N_2 concentration data from CH_4 flame studies

x/r	x, cm	f_r, Hz	$\bar{T}$, K	σ_T, K	s_T, K	σ_T', K	$\bar{c}$,	σ_c,	s_c,	σ_c',
							\multicolumn{4}{c}{mole fraction}			
0.0	0.0	20	641	308	288	109	0.13	0.14	0.10	0.10
5.7	0.4	50	676	279	162	227	0.85	0.53	0.35	0.40
7.1	0.5	100	392	103	50	90	0.82	0.23	0.16	0.17

Gaussian total number density distribution function and the ideal-gas law. Since the axial position was mostly in the potential core and the two off-axis positions were outside the flame front, the measurements were made largely on hot, nonreacting gases. Thus, if a Gaussian pdf is assumed for the total number density η, the temperature pdf can be computed by a straightforward coordinate transformation (Bendat and Piersol 1971) using the gas law $\eta = p/kT$ with the pressure p constant (Yaney et al. 1982). The distribution function resulting from this calculation is given by

$$P(T) = (\sqrt{2\pi}\ T^2 \Sigma)^{-1} \exp[-(T^{-1} - \overline{T}^{-1})^2 / 2\Sigma^2] \qquad (6)$$

where $\Sigma \equiv \sigma_\eta / \overline{\eta}\overline{T}$. Making the additional assumption that $\sigma_\eta/\overline{\eta} = \sigma_T'/\overline{T}$ permits us to generate temperature pdf's. These functions are shown in Fig. 6 as dotted plots. The agreement between the measured and calculated plots is quite good for $X = 5.7r$ and $7.1r$. The measured plot for 5.7r shows a broadened low-temperature edge, which is consistent with the contribution of the Poisson uncertainty.

The discrepancy in the on-axis case is due to the very large Poisson uncertainty dominating the measured pdf. Thus, in this case, the calculated pdf is a more accurate representation of the distribution. The shape of the measured pdf for this case arises from the fact that the Poisson pdf is nearly Gaussian, which, when transformed, will also produce a positively skewed distribution function.

Useful representations of psdf's or acf's are derivable from real-time data using the TiLaRS system when the read rate is higher than twice the maximum frequency of interest and when the number of measurements is very large. The latter condition was reasonably fulfilled by the capability of recording up to 4096 pairs of samples, while the former condition could be fulfilled in the flame studies only where excessive clipping was not needed. The runs at $X = 0.0$, 5.7r, and 7.1r had small enough numbers of clipped points to permit psdf's and acf's to be determined. The psdf's were obtained by squaring the magnitude of the Fourier transform coefficients generated by a fast Fourier transform (FFT) routine. Since the read rates for the 5.7r and 7.1r data were 50 and 100 Hz, respectively, then the spectra were calculated out to the Nyquist frequency, namely 25 and 50 Hz. Inasmuch as these spectra are essentially flat beyond 5 Hz, only the low-frequency

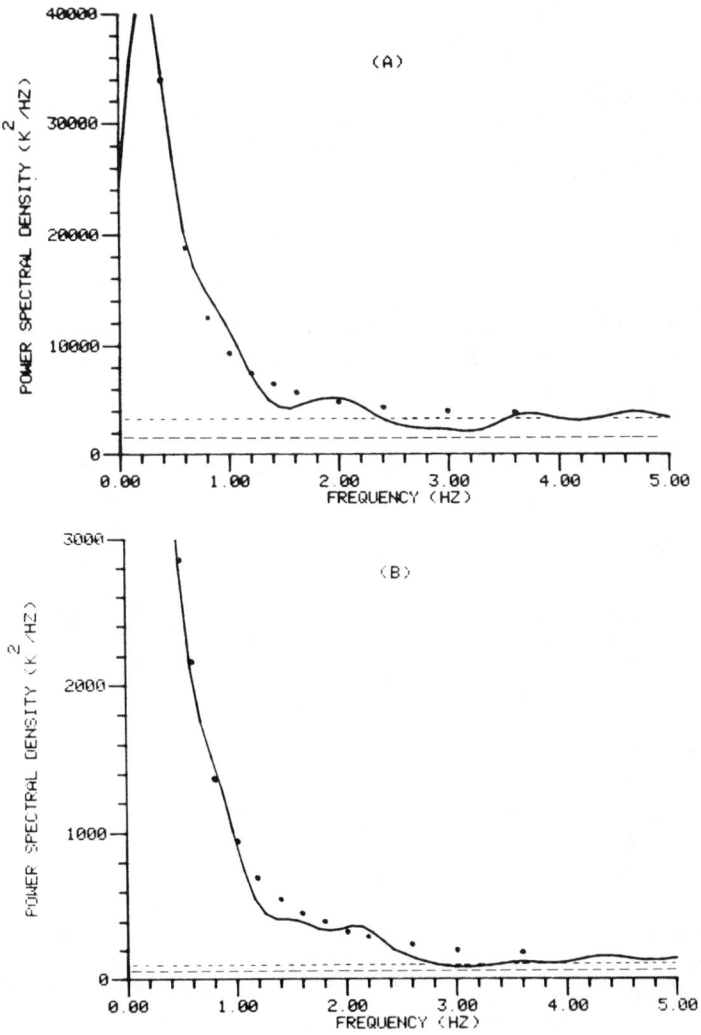

Fig. 7 Low-frequency regions of the methane flame psdf's showing fits to the "$1/f^2$" spectra: a) X = 0.4 cm (5.7r) and b) X = 0.5 cm (7.1r) (dots are the calculated fits, the short dashed lines the estimated total "white noise" levels, and the long dashed lines the "Poisson noise").

portions are shown in Fig. 7. The spectra were smoothed with a cosine filter of 0.7 Hz bandwidth.

The long dashed lines in Fig. 7 identify the power spectral densities due to the "Poisson noise" arising from the photon-counting process. These spectra give the total power summed over both positive and negative frequencies.

Hence, the Poisson level is computed by taking the Poisson variance and dividing by the bandwidth of the spectrum. As can be seen in Fig. 7, the spectra show an additional "white noise" contribution, shown by the short dashed lines, plus a low-frequency feature. This feature has been consistently observed in the temperature psdf's of earlier flame measurements (Yaney et al. 1982). So far, it has been observed only near the flame front and it gets stronger as the front is approached. The psdf's of the concentration data of these runs, as well as the temperature psdf at X = 0.0, do not show any low-frequency feature of this sort. The spectra given in Fig. 7 are quite similar to the psdf's reported by Boyer et al. (1981) for the fluctuations of the flame front position and the longitudinal velocity obtained from studies of a premixed turbulent flame front. In particular, they obtain the integral time scale for the flame front spectrum to be 0.33 s, which is in the same range as the values given in Table 2.

These spectra in Fig. 7 were fitted to the power spectral density function that can be described as white noise passing through a low-pass filter (Bendat and Piersol 1971). This function can be written as

$$G(f) = K_G / [f^2 + (1/2\pi t_c)^2] \qquad (7)$$

where K_G and t_c are the constants fitted to the data. The plots in Fig. 7 show the resulting fits as solid points. These fits permitted the total variance due to the "$1/f^2$" spectrum to be computed. Subtracting this and the Poisson variance $s_p^2 = G_p f_r / 2$ from the observed variance, the non-Poisson variance or "thermal" white noise level G_T could be found. These results are summarized in Table 2.

A further check on the $1/f^2$ analysis can be carried out using the acf. The acf can be obtained using the Fourier transform of Eq. (7). This gives

Table 2 Results of $1/f^2$ analysis

X, cm	K_G, $K^2 \cdot Hz$	t_c, s	G_p, K^2/Hz	G_T, K^2/Hz	C(0)	$C_{exp}(0+)$
0.4	6280	0.78	1640	1390	0.24	0.28
0.5	914	0.58	60	32	0.36	0.30

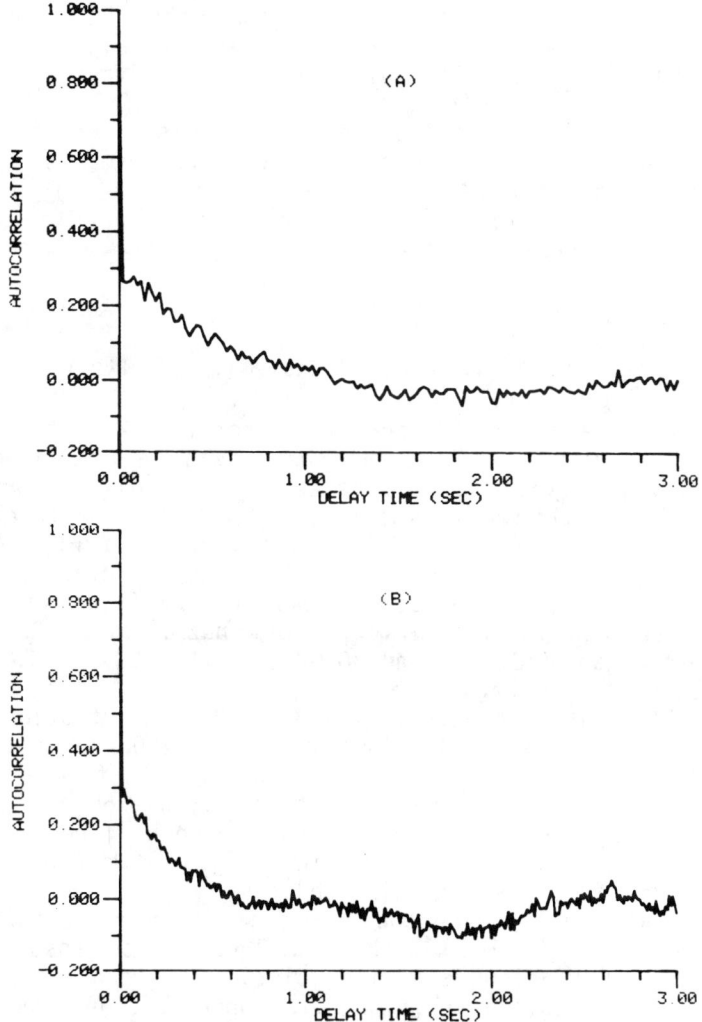

Fig. 8 Temperature acf coefficients for the methane diffusion flame at $Z = 2.5d$: a) for $X = 0.4$ cm (5.7r); b) for $X = 0.5$ cm (7.1r) (averaged over eight nonoverlapping segments of 512 points each).

$$R(\tau) = \pi^2 K_G t_c \exp(-\tau/t_c) \qquad (8)$$

The "acf coefficients" $C(\tau)$ for the 5.7r and 7.1r data were obtained by normalizing the acf's against the observed total variances s_{tot}^2. These curves given in Fig. 8 show an exponential-like decay with time constants consistent with the values of the integral time scale t_c given in Table 2. The values at one read interval greater than $\tau = 0$ given by $C_{exp}(0+)$ in Table 2 also compare favorably with the acf coefficients computed from $C(\tau) = R(\tau)/s_{tot}^2$. As seen in Fig. 8, the Poisson and "white noise" contributions are uncorrelated and therefore appear only as a "spike" at $\tau = 0$. Also, the clipped points contributed only to the white noise level, with negligible effect on the other features of these functions.

CO_2 Jet Studies

Three sets of runs were carried out. Two were axial (Z direction) profiles at 2.0 and 6.0 kg/h CO_2 flow rates and one was a radial (X direction) profile taken at Z = 2.0 cm (4.2d) with 6 kg/h. The nozzle was an ASME flat-top velocity profile, stainless steel nozzle with an orifice diameter of 4.77 mm and was supplied by the same tube used in the flame studies. (The exhaust pipe in Fig. 4 was removed.) The nozzle was mounted on a 9 cm diam aluminum disk so that the face of the nozzle was flush with the outer surface of the disk. This nozzle gave Reynolds numbers of about 10,000 and 30,000 for the 2.0 and 6.0 kg/h flows, respectively.

The calibration count rate at 100% concentration with an 1 mm high observed volume was about 350 kcounts/s for the $J = 24$ line of CO_2. Measurements of the "air" concentration were accomplished by using the 60 cm^{-1} line, which is due to an accidental coincidence of lines from N_2 and O_2. These lines are shown in Fig. 5. We neglect the trace constituents in air and assume N_2 and O_2 are indistinguishable. The total mole fraction represented by N_2 and O_2 is 0.99 in pure air. Thus, these assumptions provide a convenient starting point. The read rates were 2.0 kHz in regions of high CO_2 concentrations and were dropped down to as low as 0.2 kHz at the larger radial displacements.

The mean and rms profiles for the axial and radial runs are shown in Figs. 9 and 10. The mean data showed a systematic error in the total mole fraction due to spectrometer drift of about 1-2% except for the Z = 10.0 and 14.0 cm data, which gave sums of ~ 95%. (These discrepancies can be easily reduced in future measurements.) The rms values were obtained by subtracting the calculated Poisson variance from the measured variance at each position (Arecchi 1969).

Similar studies using a single channel, photon-correlation measurement on a nonreacting, free

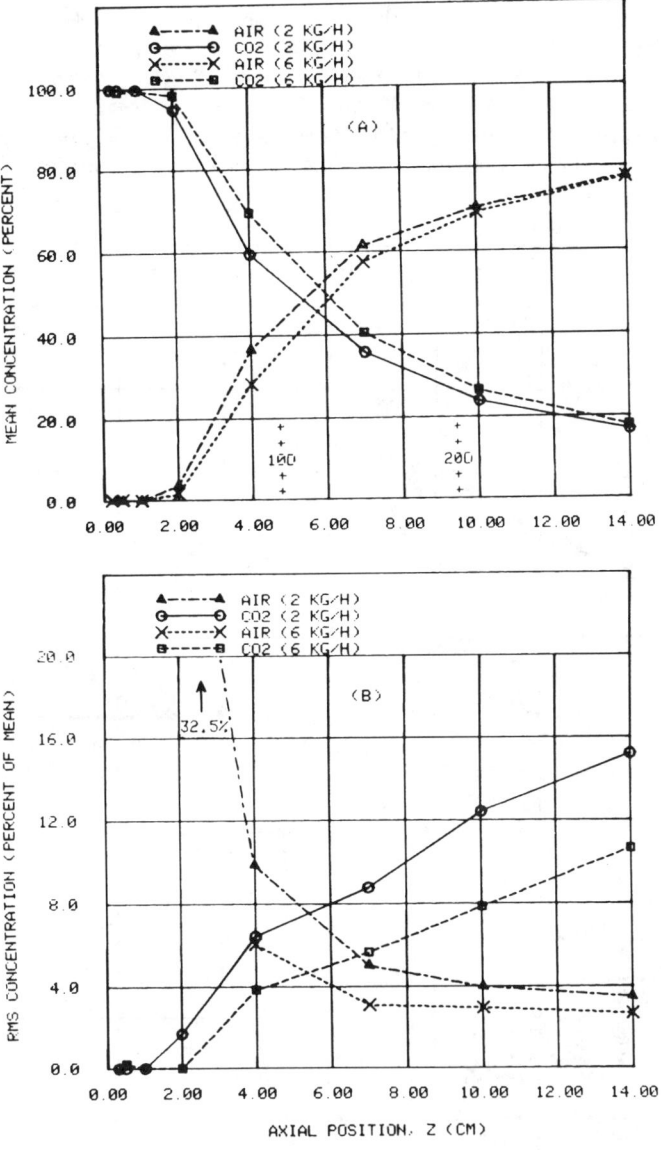

Fig. 9 Axial profiles of air and CO_2 concentrations measured simultaneously for 2.0 and 6.0 kg/h flow rates and 0.5 kHz read rate: a) mean concentrations; b) relative rms concentrations.

isothermal natural gas jet were carried out by Birch et al. (1978). They made extensive use of the fact that the factorial moments of the measured count distribution can be used to compute the moments of the pdf of the fluctuations of the light incident on the detector and thereby eliminate the contributions of the "Poisson noise" (Mandel 1959; Arecchi 1969; Bertolotti 1974). We also carried out similar calculations by which the skewness and kurtosis were obtained for each of the CO_2 data records. Generally, our values for the 2 kg/h flow were in good agreement with their results; namely, skewness values between -0.3 and

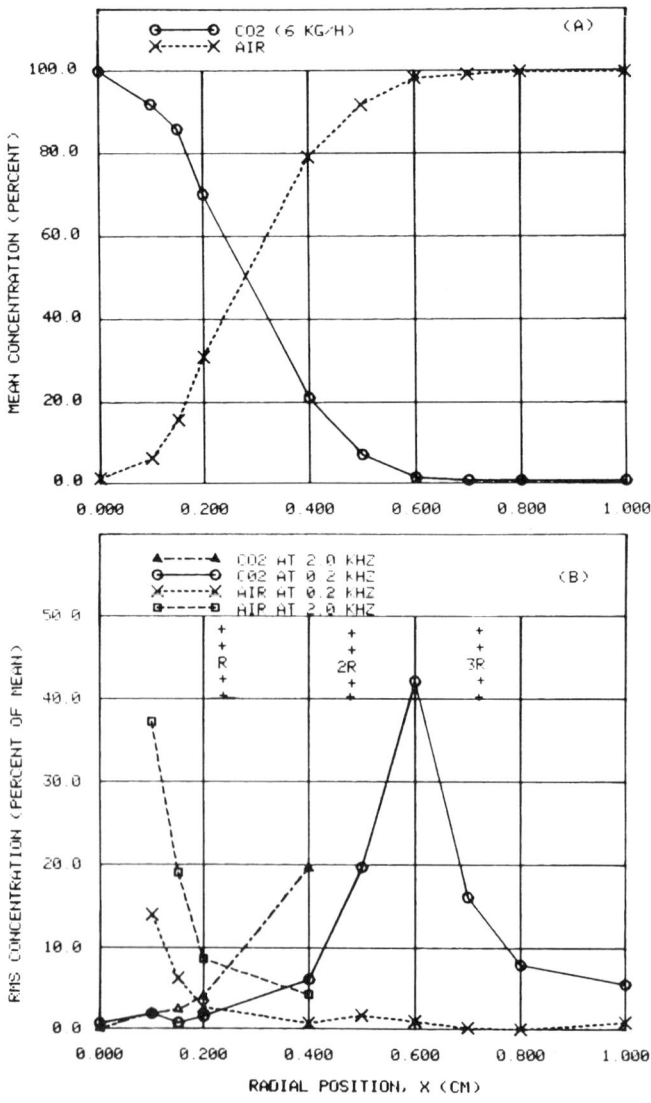

Fig. 10   Radial profiles of air and CO_2 concentrations measured simultaneously for 6.0 kg/h flow rate at Z = 2.0 cm (4.2d): a) mean concentrations; b) relative rms concentrations.

-0.5 and kurtosis values slighly above 3.0 were found beyond 10d. Curiously, the 6 kg/h flow rate gave about twice these magnitudes for these parameters near 20d. The 2 kg/h values near the nozzle (about 4d) were found to be sharply dependent on the read rate, with the 2 kHz data more than tripling the above downstream magnitudes. This latter result is reasonable due to the higher frequency fluctuations expected near the nozzle and the impact of the limitation on the total mole fraction of unity.

The axial decay of the mean CO_2 concentration was computed in terms of the mass fraction following the prescription of Birch et al. (1978) and fitted to the hyperbolic function given by

$$\bar{\theta}/\bar{\theta}_0 = k_1 \sqrt{\rho_0/\rho_a} / [(z/d) + (a/d)] \qquad (9)$$

where $\bar{\theta}$ is the mean mass fraction concentration, $\bar{\theta}_0$ the orifice value, d the orifice diameter, and ρ_c and ρ_a^0 the specific gravities of CO_2 and air, respectively. This is really an approximation to a more general exponential decay function, which holds for fractional concentrations somewhat less than unity (Hinze 1975).

Least-square fits carried out for the data given in Fig. 9 are plotted as reciprocal mass fraction concentration in Fig. 11. Although the quantity of data is small and there are values outside the approximation range, the fits are very good. The values for the decay constant k_1 at 2.0 and 6.0 kg/h are 6.3 and 6.2, respectively, which compares favorably with the 6.1 obtained in velocity studies using the same flat-top velocity profile nozzle (Lightman et al. 1983). Birch et al. (1978) studied fully developed pipe flow from a 12.65 mm diam tube and obtained k_1 = 4.7 in the near-field region, which also compared well with the velocity decay.

The rms profiles of air and CO_2 should be equal (i.e., cross) at the same position that the mean profiles cross. A close examination of Figs. 9 and 10 with this in mind suggests that, while a read rate of 2.0 kHz was adequate to give self-consistent rms values for the 2.0 kg/h flow, it was not high enough for the 6.0 kg/h flow. In particular, the air rms values appear to be too low in the latter case.

A more precise analysis of the fluctuation measurements can be carried out using the covariance between the air and CO_2 fluctuations. The assumed constancy of temperature and pressure dictate that the total of the mole fractions is unity throughout the series of measurements. Since the air and CO_2 mole fractions constitute at least 99% of the gas in the observed volume, we can write

$$c_a + c_c = 1 \tag{10}$$

where c_a and c_c are the air and CO_2 mole fractions, respectively. Then the variances in these quantities given by s_a^2 and s_c^2 are related as

$$s_a^2 + s_c^2 + 2s_{ac}^2 = 0 \tag{11}$$

or if we define a normalized covariance T as

$$T \equiv 2s_{ac}^2/(s_a^2 + s_c^2) \tag{12}$$

then, if the three variance values are correct, $T = -1$, where s_{ac}^2 is the covariance between c_a and c_c and is estimated using

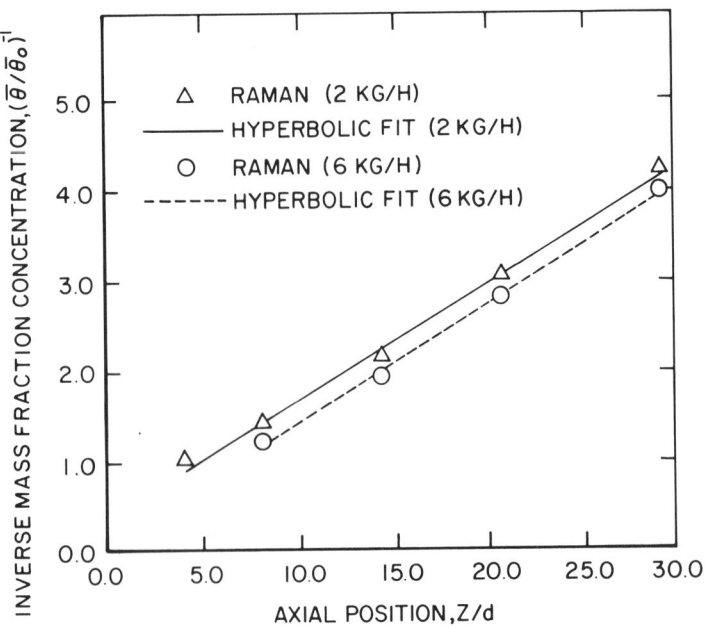

Fig. 11 Mean CO_2 concentration profile along the jet centerline given in Fig. 9 in terms of mass fraction.

$$s_{ac}^2 = \sum_{i=1}^{N} [(c_{ai} - \bar{c}_a)(c_{ci} - \bar{c}_c)]/N \qquad (13)$$

where N is the number of data pairs or reads (N = 4096 here).

The measured values of covariance are plotted in Fig. 12. The axial data in Fig. 12a show, in particular, a lower covariance at 6.0 kg/h than at 2.0 kg/h, which ties in with the lower than expected air rms values at the higher flow rate suggested above. Similarly, the radial data at the 0.2 kHz read rate in Fig. 12b show low covariance values consistent with the low air rms data in Fig. 10b. These characteristics can be made quantitative by calculating T given in Eq. (12). This quantity can be used as a test for the self-consistency of the data. For both air and CO_2, the rms values (i.e., s_a and s_c) were obtained from the data by a subtraction of constant background levels as well as the Poisson variances. In addition, the "true" air values had to be obtained by a subtraction of a fraction of the CO_2 concentrations from the apparent air values. This "interference" of CO_2 with the air channel can be seen in Fig. 5. These manipulations, plus the fact that the background levels were really not constant and that the instantaneous CO_2 fraction in the air channel was probably not constant, all indicate the possibility of $T \neq -1$. It should be noted here that the covariance between the Poisson noise signals in the two variables are completely uncorrelated and, therefore, do not contribute to s_{ac}. Thus, the closeness of T to -1 is an indication of the correctness of both the measurement of the fluctuations and the calculations of the variances and the covariance.

The meaning of $T = -1$ is that the two variables are perfectly anticorrelated. In this instance, we have that $s_a = s_c$. This result allows us to define a "quotient factor" q as

$$q_{ac} = 2s_a s_c / (s_a^2 + s_c^2) \qquad (14)$$

which equals unity only when $s_a = s_c$ and is less than unity whenever $s_a \neq s_c$, regardless of which quantity is small. This factor allows us to test the rms values of concentration independent of the covariance. The values of T and q are given in Table 3 for positions where the air or

CO_2 signal levels are not excessively low. We note that the ratio T/q is the linear correlation coefficient that cannot exceed the magnitude of unity. The values of T and q, as well as their ratio, shown in Table 3 are quite reasonable for the 2 kg/h flow rate, while the values for the 6 kg/h flow show numerous deviations from the appropriate values. The low values of q for the high flow

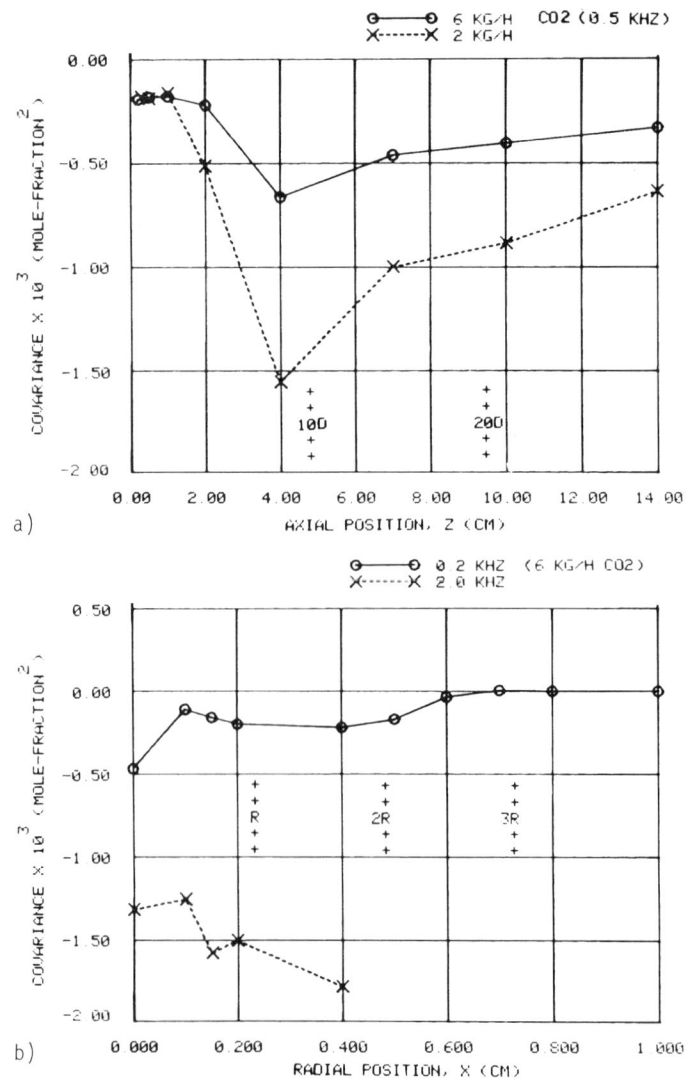

Fig. 12 Air-CO_2 concentration covariance profiles: a) axial; b) radial at Z = 2.0 cm (4.2d).

rate are due to low air rms values. The suggestion is that the air fluctuations are not being adequately resolved. This is supported in the radial data by the closer approach of q to unity at the higher read rate. It is to be noted that Rayleigh scattering studies of two component flows (Leipertz 1982; Pitts and Kashiwagi 1983) do not permit this type of analysis since a single measurement is used to compute the concentration of each component.

Spectral and correlation analyses carried out on the CO_2 and air data showed that the fluctuations consisted of a $1/f^2$ contribution with an integral time scale very close

Table 3 Normalized covariance T and quotient factors q

CO_2 flow, kg/h	Read rate, kHz	Profile	Position, cm	T	q
2.0	0.5	Axial (Z)	4.0	-1.11	0.998
			7.0	-1.01	1.000
			10.0	-1.05	0.998
			14.0	-0.91	0.998
				$\overline{T}$ = -1.02 ±0.066	$\overline{q}$ = 0.999 ±0.001
6.0	0.5	Axial (Z)	4.0	-1.31	0.907
			7.0	-1.09	0.970
			10.0	-0.94	1.000
			14.0	-0.84	0.997
				$\overline{T}$ = -1.05 ±0.16	$\overline{q}$ = 0.969 ±0.043
6.0	0.2	Radial (X) at Z = 2.0 cm	0.1	-0.53	0.79
			0.15	-2.07	0.94
			0.20	-2.02	0.97
			0.40	-2.13	0.74
			0.50	-0.76	0.99
			0.60	-0.53	0.89
				$\overline{T}$ = -1.19 ±0.99	$\overline{q}$ = 0.89 ±0.10
6.0	2.0	Radial (X) at Z = 2.0 cm	0.1	-2.90	0.978
			0.15	-2.38	0.936
			0.20	-1.99	0.999
			0.40	-1.25	0.980
				$\overline{T}$ = -2.13 ±0.70	$\overline{q}$ = 0.973 ±0.027

to the read-time interval, an apparently flat or white noise contribution, and the usual Poisson contribution. For example, in the CO_2 data at 2.0 kg/h and Z = 4.0 cm (8.4d), the apparent white noise was about 20% of the total measured variance, while the $1/f^2$ contribution was about 33%. The remainder was due to the Poisson contribution. As observed in the flame studies, the auto- and cross-correlation functions (acf's and ccf's) had a "spike" at zero delay corresponding to the Poisson and white noise contributions, with the weak $1/f^2$ contribution showing up as a fast-decay curve. Also, as expected, the ccf was strongly negative near zero delay.

Conclusions

It is evident from the two simple studies presented here that high-data rate, two-or-more channel, laser probe measurements of reacting and nonreacting flows may provide an opportunity for obtaining new information and insight not possible with single-channel dynamic measurements and not easily acquired with any kind of low-prf measurement scheme. We have shown good agreement between the temperature pdf's computed from a simple model and the observed pdf's of hot, nonreacting gases. These results show that these pdf's have positive skewness due to a significant high-temperature tail.

To the best of our knowledge, the CO_2 jet studies represent the first time the concentration fluctuations of two gases have been measured directly and simultaneously at kilohertz data rates.

A revealing result from these studies is the power of the correlation analysis to discriminate between purely random fluctuations such as those due to "Poisson noise" and the correlated fluctuations of the measured variable, be it concentration in cold flow (Birch et al. 1975 and 1978) or the temperature in a flame. This feature is being examined in greater detail and will be the subject of future papers.

The multichannel continuous-wave TiLaRS technique as developed here was by no means optimized in all aspects. A new system can be assembled using commercially available apparatus that would have four or five channels, measure temperatures and concentrations in the flame to at least 150 Hz, and cold-flow concentrations in excess of 6 kHz. Furthermore, a channel can be added to simultaneously measure the fuel concentration in a CH_4 flame using the vibrational Q-branch that gives an RS signal similar in strength to the rotational lines used in this work. The integration of such a system with laser Doppler anemometry and other high-data-rate optical techniques such as Rayleigh scattering (Pitz et al. 1976; Dibble and Hollenbach 1981; Talbot and Roblen 1982; Driscoll et al. 1983; Dibble et al. 1984) or flame emission (Magill et al. 1982) should provide an unique opportunity for gaining additional insight into the processes of gas mixing, intermittancy, and energy transfer in both reacting and nonreacting flowfields.

Acknowledgments

The authors wish to thank Dr. Mel Roquemore for his support and encouragement, Peter D. Magill for the initial computer interfacing and software development, and Prof. James Schneider, Chairman of the Department of Physics. This work was part of a larger program managed by Dr. Eugene H. Gerber of the University of Dayton Research Institute and sponsored by the Aero Propulsion Laboratory, Wright-Patterson Air Force Base, Ohio, through Contract F33615-78-C-2005.

References

Arecchi, F. T. (1969) Photocount distributions and field statistics. <u>Proceedings of the International School of Physics Course XLII Quantum Optics</u>, edited by R. J. Glauber, pp. 57-110. Academic Press, New York.

Bendat, J. S. and Piersol, A. G. (1971) <u>Random Data: Analysis and Measurement Procedures</u>. Wiley-Interscience, New York.

Bertolotti, M. (1974) Photon statistics. <u>Photon Correlation and Light Beating Spectroscopy</u>, edited by H. Z. Cummins and E. R. Pike, pp. 41-74. Plenum Press, New York.

Birch, A. D., Brown, D. R., Dodson, M. G., and Thomas, J. R. (1975) The determination of gaseous turbulent concentration fluctuations using Raman photon correlation spectroscopy. <u>J. Phys. D: Appl. Phys.</u> 8, L167-L170.

Birch, A. D., Brown, D. R., Dodson, M. G., and Thomas, J. R. (1978) The turbulent concentration field of a methane jet. <u>J. Fluid Mech.</u> 88, 431-449.

Black, P. C. and Chang, R. K. (1978) Laser-Raman optical multichannel analyzer for transient gas concentration profile and temperature determination. <u>AIAA J.</u> 16, 295.

Boyer, L., Clavin, P., and Sabathier, F. (1981) Dynamic behavior of a premixed turbulent flame front. <u>18th Symposium (International) on Combustion</u>, pp. 1041-1049. The Combustion Institute, Pittsburgh, Pa.

Crosley, D. R. (1980) Laser Probes for Combustion Chemistry. American Chemical Society, Washington, D.C.

Dibble, R. W. and Hollenbach, R. E. (1981) Laser Rayleigh thermometry in turbulent flames. 18th Symposium (International) on Combustion, pp. 1489-1499. The Combustion Institute, Pittsburgh, Pa.

Dibble, R. W., Kollmann, W., and Schefer, R. W. (1984) Conserved scalar fluxes measured in a turbulent nonpremixed flame by combined laser Doppler velocimetry and laser Raman scattering. Combust. Flame 55, 307-321.

Drake, M. C., Asawaroengchai, C., and Rosenblatt, G. M. (1980) Temperature from rotational and vibrational Raman scattering: effects of vibrational-rotational interactions and other corrections. Laser Probes for Combustion Chemistry, edited by D. R. Crosley, pp. 231-237. American Chemical Society, Washington, D.C.

Driscoll, J. F., Schefer, R. W., and Dibble, R. W. (1983) Mass fluxes $\overline{\rho'u'}$ and $\overline{\rho'v'}$ measured in a turbulent nonpremixed flame. 19th Symposium (International) on Combustion, pp. 477-485. The Combustion Institute, Pittsburgh, Pa.

Eckbreth, A. C. (1981) Recent advances in laser diagnostics for temperature and species concentration in combustion. 18th Symposium (International) on Combustion, pp. 1471-1488. The Combustion Institute, Pittsburgh, Pa.

George, Jr., W. K., Beuther, P. D., and Lumley, J. L. (1979) Processing of random signals. Proceedings of the Dynamic Flow Conference, edited by B. W. Hansen, pp. 757-793. Sijthoff and Noordhoff, Alphen aan den Rijn, The Netherlands.

Hill, R. A., Mulac, A. J., and Hackett, C. E. (1977) Retroreflecting multipass cell for Raman scattering. Appl. Opt. 16, 2004-2006.

Hinze, J. A. (1975) Turbulence. 2nd Ed. McGraw-Hill Book Co., New York.

Lapp, M. (1973) Flame temperatures from vibrational Raman scattering. Laser Raman Gas Diagnostics, edited by Marshall Lapp and C. M. Penney, pp. 107-145. Plenum Press, New York.

Leipertz, A. (1981) Nondestructive probing of free jets using cw laser Raman spectroscopy. Opt. Laser Techno. 13 (2), 21-25.

Leipertz, A. (1982) Rayleigh measurements in a CO_2 jet. Appl. Opt. 21, 2872-2874.

Lightman, A., Magill, P. D., and Andrews, R. J. (1983) Laser diagnostic development and measurement and modeling of turbulent flowfields of jets and wakes. Final Technical Rept., Pt. I, AFWAL-TR-83-2044 on Contract F33615-78C-2005, Air Force Wright-Aeronautical Laboratories, Aero Propulsion Laboratory, Wright-Patterson AFB, Ohio.

Magill, P. D., Lightman, A. J., Orr, C. E., Bradley, R. P., and Roquemore, W. M. (1982) Simultaneous velocity and emission measurements in a bluff-body combustor. AIAA Paper 82-0883.

Mandel, L. (1959) Fluctuations of photon beams: the distribution of photo-electrons. Proc. Phys. Soc. London 74, 233-243.

Penney, C. M., St. Peters, R. L., and Lapp M. (1974) Absolute rotational Raman cross sections for N_2, O_2 and CO_2. J. Opt. Soc. Am. 64, 712-716.

Pitts, W. M. and Kashiwagi, T. (1983) The application of laser-induced Rayleigh light scattering to the study of turbulent mixing. National Bureau of Standards Rept. NBSIR 83-2641. National Technical Information Service, Springfield, Va.

Pitz, R. W., Cattolica, R., Robben, F., and Talbot, L. (1976) Temperature and density in a hydrogen-air flame from Rayleigh scattering. Combust. Flame 27, 313-320.

Schrotter, H. W. and Klockner, K. W. (1979) Raman scattering cross sections in gases and liquids. Raman Spectroscopy of Gases and Liquids, edited by A. Weber, pp. 123-202. Springer-Verlag, New York.

Setchell, R. E. (1976) Time-averaged measurements in turbulent flames using Raman spectroscopy. AIAA Paper 76-28.

Talbot L. and Roblen, F. (1982) Applications of Rayleigh scattering to turbulent flows with heat transfer and combustion. Final Tech. Rept. AFOSR-TR-82-0667, Dept. of Mechanical Engineering, Univ. of California, Berkeley (Gov't Ascension ADA 118576).

Williams, W. D., Powell, H. M., McGuire, R. L., Price, L. L., Jones, J. H., Weaver, D. P., and Lewis, J. W. L. (1977) Diagnostics of rocket plum-airstream turbulent mixing using laser-Raman scattering. Turbulent Combustion, AIAA Progress in Astronautics and Aeronautics, Vol. 58, edited by L. A. Kennedy, pp. 273-289. AIAA, New York.

Yaney, P. P., Becker, R. J., Magill, P. D., and Danset, P. (1982) Dynamic temperature measurements of flames using spontaneous Raman scattering. Temperature, Its Measurement and Control in Science and Industry, edited by J. F. Schooley, pp. 639-645. American Institute of Physics, New York.

Yaney, P. P., Becker, R. J., Danset, P. T., and Gallis, M. R. (1983) Laser diagnostic development and measurement and modeling of turbulent flowfields of jets and wakes. Final Tech. Rept., Pt. III, AFWAL-TR-83-2044 on Contract F33615-78C-2005, Air Force Wright-Aeronautical Laboratories, Aero Propulsion Laboratory, Wright-Patterson AFB, Ohio.

Flash X-Ray Tomographic System for Diagnostics of Microsecond Phenomena

Csaba K. Zoltani* and Kevin J. White*
Ballistic Research Laboratory
Aberdeen Proving Ground, Maryland

Abstract

This paper describes the design of a flash x-ray experimental facility for the study of cross sections of objects with the capability of detecting density differences much smaller than conventional radiography and with a spatial resolution better than 4 mm. Such a system, under development at BRL and based on the tomographic principle exploited with success for medical diagnosis (CAT Scans), enables the study of cross section of objects tens of centimeters on a side and a few millimeters in depth. Its chief advantage is that it is noninvasive and can capture transient ballistic events in the microsecond range. The first part of the paper describes the preliminary studies and static experiments which established the least number of x-ray sources and detectors needed to generate the required information for the image reconstruction algorithm. Next, the system components are discussed and the chosen reconstruction algorithm outlined. The following section discusses the physical layout of the BRL flash tomographic system. The paper concludes by giving the plans and results of on-going experiments obtained on static mock-ups of ballistic events.

Introduction

The determination of density differences in a ballistic environment, such as, for example, the two-phase flow in a gun tube, poses a serious challenge to the

---

Presented at the 9th ICODERS, Poitiers, France, July 3-8, 1983. Copyright © 1984 by the American Institute of Aeronautics and Astronautics, Inc. All rights reserved.
 *Research Physicist, U.S. Army Armament Research and Development Command.

experimentalist. Conventional techniques have reached their limit of refinement and little, if any, progress has been recorded in the last decade in this area. This paper describes a new technique with the potential of adding considerable new information on ballistic processes which, heretofore, were either unavailable or very difficult to obtain. The primary advantage of the method is that it allows one to look at transient events, even in the microsecond regime, with excellent spatial resolution.

With the invention of computed tomography (CAT Scans), the last ten years have witnessed a phenomenal advance in medical diagnostic capability in the determination of density differences in human tissue, as low as a fraction of a percent with a spatial resolution of less than a millimeter. The method is noninvasive and relatively fast. In contrast to a conventional x-ray, which gives a two-dimensional picture, a shadowgraph of the average absorption of x-rays through a given object, computed tomography gives the density distribution of cross sections within the object, a few millimeters in thickness, taken at any chosen angle. A further advantage of this technique is that it is far superior in density resolution to conventional x-ray shadowgraphs, indeed, details become visible which, heretofore, were inaccessible to measurement. In addition, possibilities exist of extending the technique to obtaining truly three-dimensional representations of the density profiles within an object for the first time.

Thus, the idea suggests itself for using computed tomography for the investigation of ballistic events whenever it is desirable to detect density differences within an object, or within a region of interest, such as fragmentation and spall patterns, measuring sabot separation from a projectile, interior ballistic processes in solid-fuel ramjets, rocket motors and guns, dynamic deformation of solids, mixing in binary filled projectiles, detonation processes, and blast effects.

In contrast to static objects, a resting patient, for example, ballistic events are highly transitory and induce a hostile environment (Table 1). Consequently, rather severe design constraints are imposed on a tomographic system for ballistic applications and, indeed, such a system requires the incorporation of new technologies in novel designs.

The design philosophy of the BRL multiflash x-ray tomography system is detailed here. Zoltani et al. (1983) describe how the system parameters were determined and point out some of the potential areas of application.

What is Tomography?

A typical CT layout, see Fig. 1, consists of an x-ray source, an object to be studied, and an x-ray detector, usually a scintillation counter. Each detector element produces a signal proportional to the absorbed x-ray energy indicating the attenuation along the ray path between the detector and the x-ray source. Any detector system, including photographic film, can, in principle, be used to record the data for the projections. These data are then digitized and subsequently used in the reconstruction of the cross-sectional image. The x-ray beam is collimated into a thin (1.5-10 mm) fan that lies at an angle to the major axis of the object being studied. The experiment proceeds by making measurements of the transmitted x-ray signal as the source is moved in a semicircle around the object. Typically, exposures are taken at 1-deg intervals. This will yield 180 different projections. The source is far enough from the detector array so that the object is completely enveloped by the fan of the x-ray beam.

One of the chief advantages of tomography over conventional radiography is the greatly increased density resolution that can be achieved. Since, basically, tomography uses a series of radiographs taken at different angles, how is it possible to have higher resolution than on the original radiographs? The following explanation will

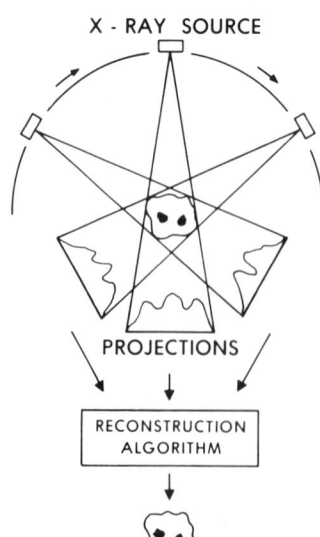

Fig. 1 Typical tomographic configuration.

Table 1 Medical vs ballistic problem

	MEDICAL	BALLISTIC
Time	5-20 s	1-100 µs
Spatial resolution	< 1 mm	Several mm
Temperature	Ambient	3000 K
Pressure	Ambient	1-300 MPa (Blast)
Number of views	180	~20
Contrast	Bone-tissue	Propellant-air
Scattering	1 view	20 views
Sequencing	Mechanical	Electronic

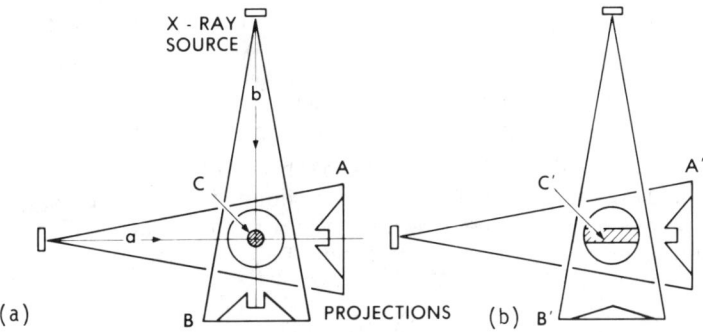

Fig. 2 Sample projections for two different objects.

not prove this point but, hopefully, will give an appreciation of the basis of this fact. Consider the object shown in Fig. 2a which consists of a solid material with a lower density inclusion C. Contours of radiographs taken at two different angles are shown at A and B. One of the techniques of computed tomography is an iterative calculation of the absorption characteristics of the object that will make all of the radiographs taken at different angles self-consistent. For example, the signal on A produced by the x-ray transmission along path "a" must be consistent with the signal on B produced by the transmission along path "b." Since in an ordinary tomograph there are not just two, but perhaps 180 views, it can be seen that the self-consistency requirement will produce an increase in resolution of the object compared with that of a single radiograph. Consider now Fig. 2b which has a region C' that has an absorption coefficient slightly less than the rest of the object and less than inclusion C. The radiograph A' will have the same essential features as A. However, radiograph B' will be substantially different from B. The iterative reconstruction will then generate the object in Fig. 2b which must have characteristics that will produce radiographs A' and B'. Thus, although A and A' are identical, the difference in B and B' contribute information in determining the properties of the object. In practical tomography, the 180 views require extensive iterations to yield self-consistent projections.

For its realization, tomography relies on two important ideas. An arbitrary function, Radon (1917), which is bounded, continuous, and has continuous first partial derivatives, can be uniquely reconstructed from an infinite set of parallel line integrals. Second, it must be remembered that x-ray radiation is attenuated when it traverses an object. The change in radiation intensity is given by

$$dI = - I \mu ds \qquad (1)$$

where I is the intensity, ds the path length, and μ a proportionality coefficient. μ can depend on the energy of the x-ray as well as the composition of the material in each path length of the object. Solving Eq. (1) we get,

$$- \ln (I/I_0) = \int_0^X \mu \, ds \qquad (2)$$

where the integral is taken from the source to the detector. The x-ray shadowgraph is a measure of I/I_0. The problem of tomography then is to invert this equation and solve for μ of the object.

Tomography in a Ballistic Environment

Since the tens of microseconds available to record a ballistic event is too short to mechanically move an x-ray source around an object and to record the transmitted radiation, consideration has to be given to using results from several x-ray heads and flashing them simultaneously, or in a time-frame short compared with the ballistic event. To determine the x-ray photon energy needed and to see if scattered radiation, which can significantly degrade signal-to-noise ratios, poses a problem, a series of radiation transport calculations were carried out on a typical gun tube propellant combination. The calculations showed that at 1 MeV about one fourth of the radiation is transmitted through the object and arrives at the detector. Moreover, the ratio of scattered to transmitted radiation was more favorable at the higher x-ray energies.

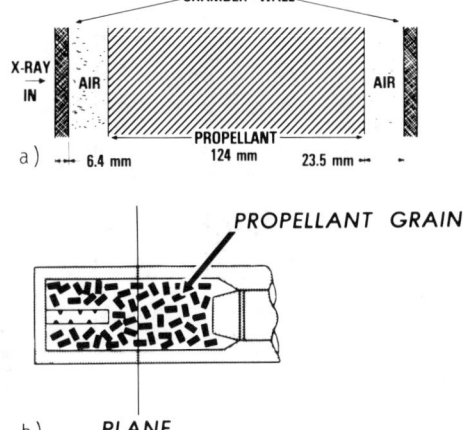

Fig. 3a) Chamber/propellant configuration. b) Interior ballistic cycle indicating tomographic plane.

Many processes of interest to the ballistician take place inside steel chambers which have a large x-ray absorption coefficient. Indeed, calculations show that tomography is not feasible in thick-walled steel tubes.

A useful assessment of materials as chambers for ballistic investigation can be carried out by doing radiation transport calculations on a propellant-chamber mock-up. A one-dimensional Monte Carlo code, TIGER, developed by the Sandia Laboratories, Halbleib and Vandevender (1974), was used to carry out the transport calculations. From 1000 to 50,000 photon histories were followed with the code. Monoenergetic x-rays were used as input. The chamber/propellant configuration is shown in Fig. 3a. A monolithic propellant grain is used to simulate the granular propellant bed, with a thickness (124 mm) equivalent to that found in a relatively high loading density configuration. Using the known geometry, density, and chemical composition of the object, the code calculates the percentage of energy absorbed, forward-scattered, back-scattered and attenuated within the object material. The output also includes the percentage of photons absorbed, forward-scattered, back-scattered and unattenuated, and the energy distribution of the scattered photons. An example of this output is given in Fig. 4, in which the chamber was made of 6-mm steel. The transmitted radiation is defined as the energy of the photons emerging from the object having the same energy as the input photons. The bar chart shown in Fig. 4 reflects the fraction of photons over an energy range for the forward-scattered radiation. This is a one-dimensional code and does not give an angular distribution

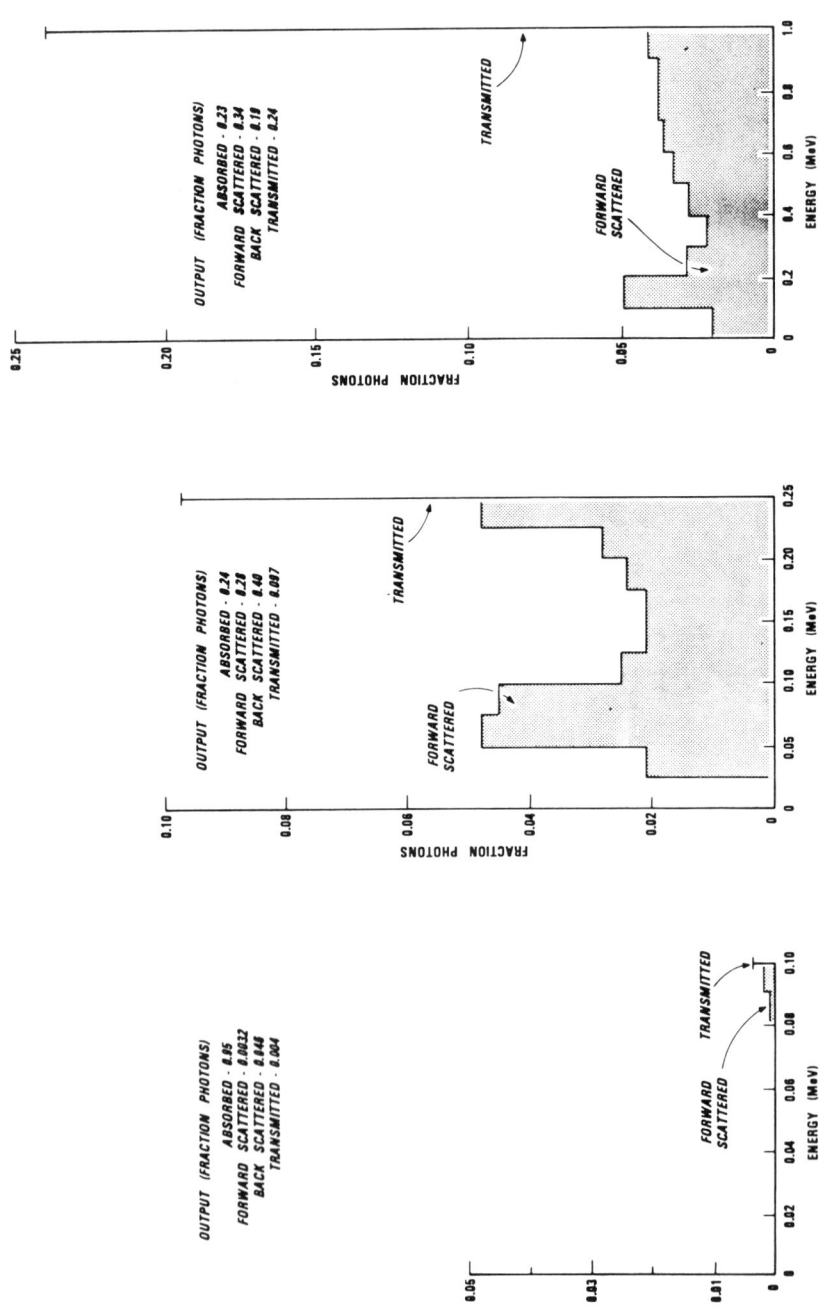

Fig. 4 Radiation transport calculations for a steel chamber.

of the scattered radiation but only an energy
distribution. The presence of forward-scattered radiation
in a radiograph is undesirable as it contributes fog or
background noise to the signal. In addition, the percentage
of energy transmitted is also important since this
determines the strength of the signal at the detector.
Hence, in the absence of the knowledge of the angular
distribution of the scattered radiation, a good figure of
merit is the ratio of unattenuated to forward-scattered
radiation. These calculations also demonstrated that the
high-energy x-rays are superior, whether the chamber is made
of steel or fiberglass. Since the x-ray sources under
consideration are polychromatic, further tests must be
carried out to determine the significance of radiation below
0.1 MeV on the quality of these radiographs.

Early in the study, it was realized that a limited
number of static experiments on a mock-up of Fig. 3b would
be useful in helping to establish the feasibility of
building a tomographic system for ballistic applications.
The experiments were designed to answer the question on the
least number of sources that could be used for an acceptable
reconstruction and whether sufficient contrast could be
obtained from x-ray sources in the sub-MeV regime. As it
turns out, both of these questions could be answered in an
acceptable manner. A series of experiments simulating the
interior ballistics of a gun were carried out at the Los
Alamos National Laboratory (LANL) which under idealized
conditions showed that, indeed, S/N ratios are high enough
to assure good reconstructed pictures. The experiments
evinced that eighteen (18) projections or radiographs were
adequate for a reconstructed image with a spatial resolution
of the order of millimeters. The experimental setup (Fig.
5) consisted of a source, a detector, a rotating table, and
ancillary electronics required for the subsequent data
reduction.

The source was Iridium with most gamma radiation at 316
and 468 keV. The source to detector distance was 0.6 m. A
detector consisting of a NaI(Tl) scintillator fronting a
single photomultiplier tube was used to record the
transmitted radiation. It was placed behind a detector
collimator having movable plates which permits the
collimator aperture to be varied in size from 0.5 mm up to
several millimeters on a side. A 12-bit analog-to-digital
converter with a 40-ms digitizing time was used for the
recording of the x-ray transmission through the object. The
study module was placed on a table allowing 3 deg of freedom
of motion. The axes were driven by direct current stepping
motors controlled by an LSI-11 computer. The projection
data was recorded on floppy disks.

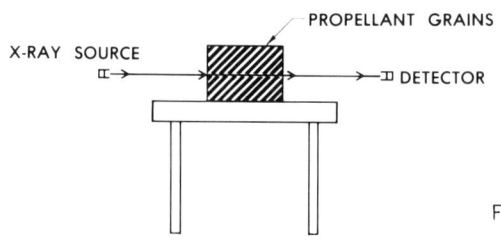
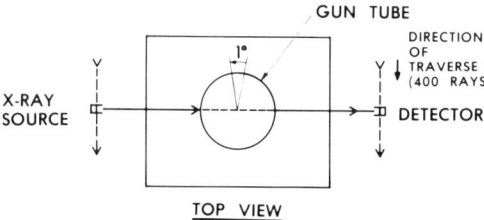

Fig. 5 LANL experimental setup.

Two different test phantoms were used. In the first, a 20.0-cm-diam fiberglass cylindrical tube, with a wall thickness of 3 mm was filled with inert, 7-perforation (diam 1 mm) propellant grains, of 1.0 cm in diam and 2.0 cm in length. This material contained 31% Pb_3O_4. For the second test, the test pellets, the same size as before but now consisting of lucite, were embedded in a styrofoam matrix and oriented at random angles (Fig. 6c). This approximates the fluidized regime within a gun tube. In both cases, x-ray absorption measurements were made with 800 data points per view angle and the experiment was repeated 180 times, stepped at 1-deg intervals. The scan aperture was 0.5 x 1 mm and the step over between samples was 0.25 mm. It took approximately 6 h to acquire the data for one reconstruction.

The result of the study is illustrated in Figs. 6a-i. In Fig. 6a, the first object, a propellant filled chamber was reconstructed with an algorithm (filtered back projection, FBP) using 180 views and 400 detectors per view angle. Thirty-six views were used for the reconstruction shown in Fig 6b. Note that the artifacts become discernible and detail is lost. A photograph of the second object is shown in Fig. 6c with a 180 view FBP reconstruction shown in Fig. 6d and an 18 view reconstruction in Fig. 6e. When the reconstruction is performed using MENT, Minerbo (1979), a maximum entropy algorithm, for the same number of views and 200 detectors, the image seen in Fig. 6f exhibits fewer artifacts. Finally, in Fig. 6g, nine views were used and

the results show the clustering of the propellant grains and
the general topology of the flow, but quantitative
evaluation is no longer feasible. Further analysis of the
experimental data was carried out to determine the minimum
number of detectors per view angle that must be used for a
satisfactory picture. Reconstructions were done when only
100 (see Fig. 6h), 50 (see Fig. 6i), and 25 sets of
detectors per view were used. Below 50 detectors, the
fuzziness of the obtained picture precludes any quantitative
use of the results.

It is clear from this study that a minimum of 15-20
views will be required in order to reconstruct a reasonable
image, even for a low-loading density configuration shown in
Fig. 6c. These experiments were carried out under idealized
conditions so that scattering did not degrade the data
obtained. All of these projections must be acquired in a
time frame that is short compared with any geometrical
changes in the subject under investigation. For some
ballistic systems, a time frame of 10-100 μs will be
adequate. A chamber which simulates a large caliber
configuration should have a diameter of between 150 and
200 mm.

The BRL Tomographic Facility

The basic design of the BRL Tomographic Facility is as
follows: all detectors and sources are located in a single
plane and each source has its own detector screen. In order
to accommodate 17 sets of sources and detectors, the angle
subtended by the detector from the object must be 10 deg or
less. This necessitates rather large source-to-object and
object-to-detector distances. Representative numbers are
given in Fig. 7. This could give rise to problems with
respect to x-ray intensity, which decreases as $1/r^2$, where r
is the source-detector distance. Only testing can actually
determine if the projections will be adequate for
reconstruction. An insufficient number of x-ray photons
leads to statistical fluctuations in the recorded data which
is known to degrade the quality of the reconstructed
image. The source-to-object distance could be decreased
but, in order to maintain a subtending angle of 10 deg, the
object-to-detector distance would have to be increased
leading to a poor quality shadowgraph because of the finite
effective size of the x-ray source. Calculations show that
the minimum distance from source-to-detector that can be
used is 4.6 m, with the object being midway between the
two. A large object-to-detector distance does have one
advantage; some of the forward-scattered radiation will miss

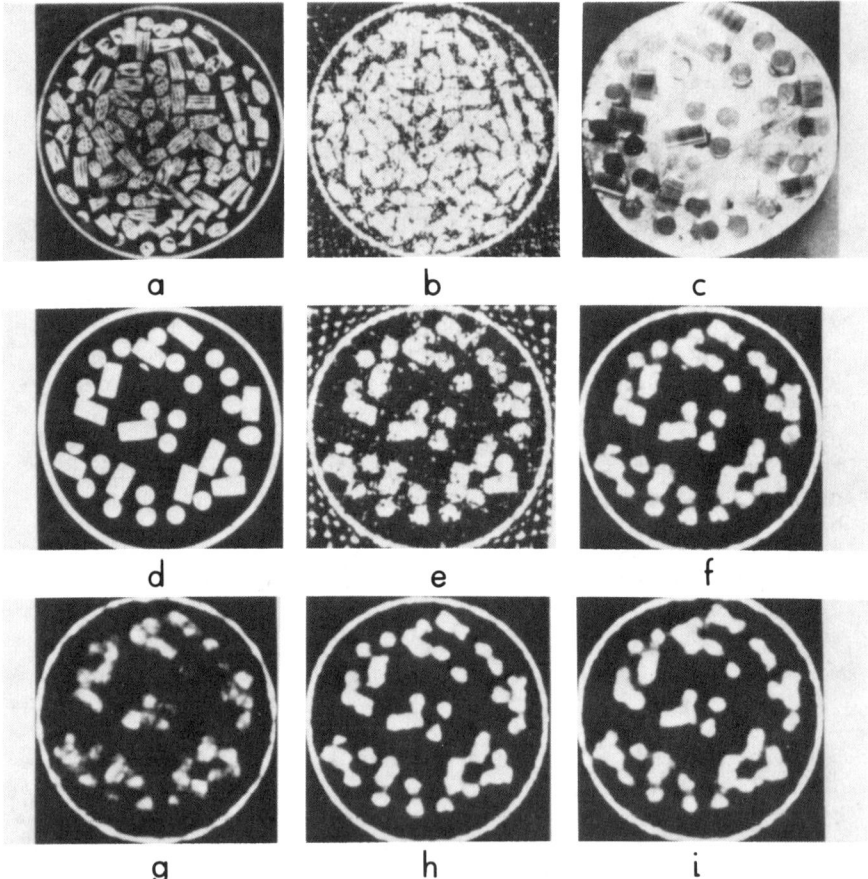

Fig. 6 The BRL mock-up and reconstructed images.

the detector and reduce the unwanted background fogging. Scattering can be further reduced by use of focused grids since each detector array or screen is associated with only one source as in Fig. 7. Scattering from the object from the firings of all 17 sources may reach each detector. This could be a source of difficulty. However, preliminary experiments, using an HP 1-MeV source at BRL indicate that scattering measured through 360 deg around the test object, a propellant-filled tube, is negligible under conditions simulating 17 radiation sources. Forward-scattering, confined to small forward angle, is present but only adds noise to the signal level.

The sources are 1-MeV flash x-ray units. Initially, conventional medical x-ray film with intensifying screens are used as detectors. Use of film cassettes in this

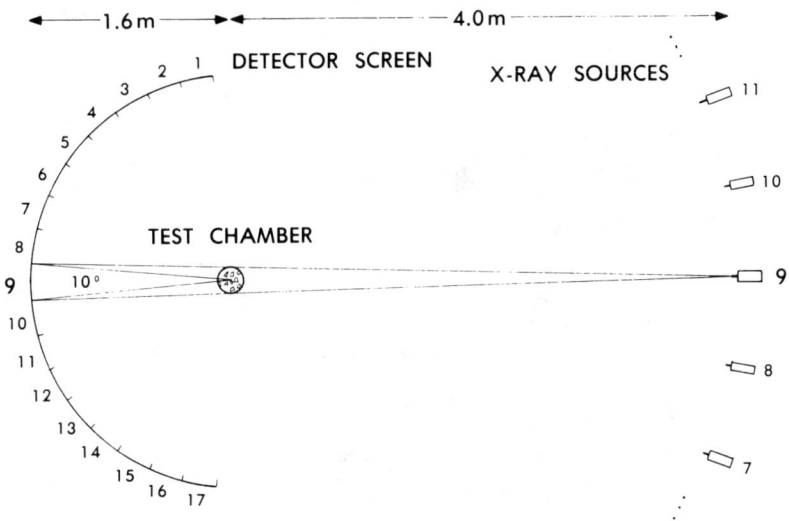

Fig. 7 Schematic of the proposed BRL tomographic facility.

instance is dictated by the potentially hostile ballistic environment which precludes the use of solid state electronic detectors. Film cassettes have been used for many years under these conditions. Moreover the two-dimensional radiographic data that film-cassette detectors supplies can be used for three-dimensional tomographic reconstruction. The radiographic data is subsequently digitized, reconstructed, and displayed on a COMTAL Vision One/20 imaging device. In addition, several types of liquid and solid state detectors are being tested to see if they offer any advantages over film. For some applications CCD's may have to be considered in the future. Using the arrangement shown in Fig. 7, the data is recorded simultaneously with the flashing of the x-ray sources. Each of the filmstrips (for instance at position 9, Fig. 7) fronting one source is subdivided into 100 segments, each approximating a detector. The recorded trace, an example of which is shown in Fig. 8, is reduced by assigning for each position the corresponding signal amplitude. This is repeated for all views. Here another advantage of the film format becomes apparent. The film can be divided into any number of segments after data acquisition depending on the system being studied and the desired spatial resolution. By interpreting over larger segments on the film signal-to-noise ratios can be improved if detected signals are weak. Now the field of interest, i.e., the tube cross section in the example, is subdivided into equilateral pixels (picture

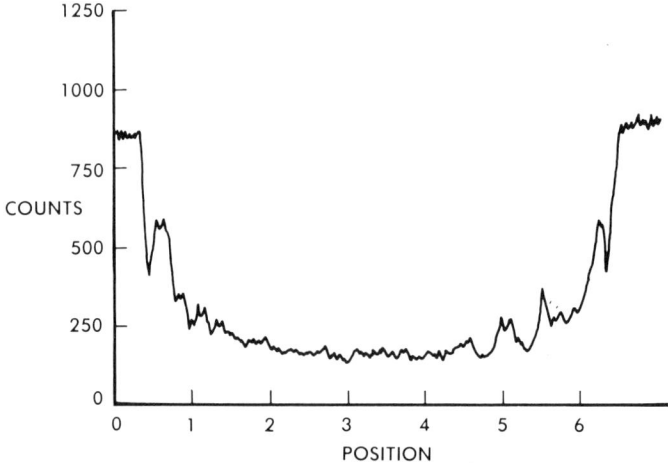

Fig. 8 Typical projection data.

elements) in each of which the absorption coefficient is to be determined. The only known data in the field at this juncture are the source strength and the signal strength at the detector.

The basic idea behind the reconstruction algorithm is to solve for the unknown absorption coefficients in each of the pixels. It is assumed that locally, i.e., within the individual pixels, the absorption is uniform. An algorithm then consists in solving a number of linear equations relating the known transmitted radiation levels to the unknown absorption in each of the pixels. Constraints can be applied, such as maximum entropy, when the system is undetermined. Typically, the grid is 256 x 256 and the system of equations is underdetermined. The solution then proceeds, from an initial guess, iteratively with an appropriate convergence criteria. Due to the amount of the data, running times an the Cyber 76 are of the order of 30 min.

The quality of the radiographic data and subsequently the reconstruction is influenced by among other things the presence of scattered radiation. An experimental and computational effort is underway to determine the significance of this effect and to see whether collimation at the detectors is needed. Work is also progessing on the developement of an improved reconstructive algorithm which is more tolerant to missing information, i.e., gives less artifact prone reconstructions when the number of sources flashed drops below 18.

Conclusions

From this study it can be concluded that a tomography system for transient events in a ballistic environment would require 17, 1-MeV flash x-ray sources. Such units are commercially available and are commonly used to record shadowgraphs of ballistic events. Using available detector technology, spatial resolution of 4 mm is quite feasible, 2-mm resolution possible but quite expensive. Such a system, when operational, will find application in a large number of fields ranging from quality assurance, product design to ballistic modeling.

Acknowledgments

It is a pleasure to thank Dr. R. P. Kruger for his excellent cooperation in carrying out the scanning experiment reconstructions at LANL. We would also like to express our thanks to Dr. Norman Banks for his continuing help and support in this project.

References

Halbleib, J. A. and Vandevender, W. H. (1974) TIGER: A one-dimensional multilayer electron/photon Monte Carlo transport code. SLA-73-1026, Sandia Laboratories, Albuquerque, NM.

Radon, J. (1917) Ueber die Bestimmung von Funktionen durch ihre Integralwerte laengs gewisser Mannigfaltigkeiten. Berichte Saechsische Akad. Wiss. 69, 262-277.

Zoltani, C. K., White, K. J., and Kruger, R. P. (1983) Results of feasibility study on computer assisted tomography for ballistic applications. BRL Report, ARBRL-TR-02513, USA ARRADCOM/Ballistic Research Laboratory, Aberdeen Proving Ground, MD.

Two-Dimensional Imaging of Flame Temperature Using Laser-Induced Fluorescence

R. J. Cattolica* and D. A. Stephenson†
Sandia National Laboratories, Livermore, California

Abstract

The temperature field in the postflame gas of a premixed methane-air flat flame was determined using two-dimensional imaging of the laser-induced fluorescence from OH molecules. A frequency-doubled Nd:YAG pumped-dye laser produced a 48 x 1.0-mm sheet of uv light for excitation of OH molecules in the $A^2\Sigma - X^2\Pi$ electronic transition. The broadband fluorescence from the OH molecules was detected with a vidicon camera with two stages of image intensification. By exciting the $Q_1(5)$ line in the (1,0) and (1,1) vibrational bands, two-dimensional (100 x 100 pixel) images were obtained of the relative concentration of the v"=0 and v"=1 vibrational levels. The ratio of these two data sets gave a two-dimensional image of the vibrational population distribution and the OH temperature in the combustion gas. With the current experimental configuration, OH temperatures above 1800 K could be observed with 10% precision and with a spatial resolution of 0.6 x 0.6 mm. Below this temperature the rapid decrease in OH concentration with temperature and limited dynamic range of the present detection system precluded temperature measurement.

Introduction

Laser diagnostics have become powerful tools for studying combustion phenomena. These laser diagnostic techniques

---

Presented at the 9th ICODERS, Poitiers, France, July 3-8, 1983. This paper is declared a work of the U.S. Government and therefore is in the public domain.
*Member of Technical Staff, Combustion Physics Division.
+Member of Technical Staff, Advanced Systems Division.

have been largely confined to single point measurements, due to limited signal levels and insensitive detection capability. With the combination of strong laser scattering processes and low-light-level detector technology, linear and planar imaging of combustion processes are now possible.

Using Rayleigh scattering, Smith (1978) was able to measure the flame temperature in a diffusion flame with an image intensified one-dimensional array. Utilizing a multi-pass technique and an image intensified vidicon, Escoda and Long (1983) were able to obtain two-dimensional density measurements. Alden et al. (1982), using laser-induced fluorescence (LIF) of OH, have mapped the OH fluorescence signal in a flame with a one-dimensional intensified array. Two-dimensional detection of LIF from OH in flames has been demonstrated independently by Dyer and Crosley (1982) with an image intensified vidicon and by Kychakoff et al. (1982) with an image intensified diode array.

In order to extract quantitative estimates of OH concentration from these images, the temperature at each point must also be determined. The purpose of this paper is to demonstrate the feasibility of using the ratio of two distinct fluorescence images to determine temperature distributions in flames.

Experimental Technique

The measurement of flame temperature using laser-induced fluorescence is based on a two-line excitation technique which yields vibrational level population ratios in the OH molecule. An analysis of the fundamental aspects

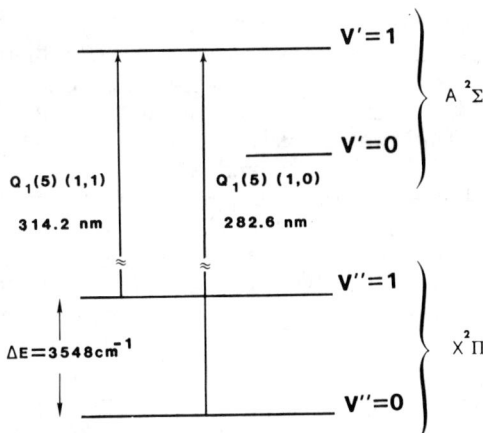

Fig. 1 Two-line laser-induced fluorescence excitation scheme for the OH molecule.

of this method has been described in detail by Cattolica (1981) for rotational temperature. For vibrational temperature the analysis is similar. The principal advantages of this two-line excitation method are that it eliminates problems encountered with quenching, fluorescence trapping, nonequilibrium excited state populations, and sensitivity to the spectral bandwidth of the detection system.

Two-line LIF excitation of the OH molecule is illustrated schematically in Fig. 1. A specific rotational state in the $v'=1$ vibrational level in the $A\,^2\Sigma$ state is excited from the two lowest vibrational levels $v''=0$ and $v''=1$ in the $X^2\Pi$ ground state. The ratio R of the fluorescence signals S_{fL} from alternately exciting the two different vibrational levels yields the relative vibrational population ratio and the vibrational temperature T_v through the Boltzmann factor,

$$R = \frac{S_{fL}(v''=1)}{S_{fL}(v''=0)} = \frac{(qB_{12}n_{v''})_{v''=1}}{(qB_{12}n_{v''})_{v''=0}}$$

$$= \frac{(qB_{12}g_1)_{v''=1}}{(qB_{12}g_1)_{v''=0}}\, e^{-[(E_{v''=1}-E_{v''=0})/(KT_v)]} \quad (1)$$

where $(E_{v''=1} - E_{v''=0})$ is the vibrational energy level difference, q is a laser line shape factor, B_{12} is the absorption coefficient, and g_1 is the degeneracy. Although B_{12} and g_1 are known, q must be calculated from the convolution of the laser line shape and absorption line shape. In lieu of such a computation the pre-exponential term in Eq. (1) can be determined from calibration at a known temperature. The temperature dependence of the line shape factors should be very similar and should be eliminated in their ratio. The two transitions used in the excitation scheme [$Q_1(5)$ in the (1,0) band at 282.6 nm and $Q_1(5)$ in the (1,1) band at 314.2 nm] were chosen because their separation from adjacent transitions is greater than the bandwidth of the laser source ($\Delta\nu = 0.6$ cm^{-1}).

The sensitivity of the temperature with variation in fluorescence ratio can be calculated from the relation

$$\frac{\Delta T_v}{T_v} = \frac{k}{hc}\, \frac{T_v}{\Delta E}\, \frac{\Delta R}{R} \quad (2)$$

For the energy level separation between the ground states of these two transitions, 3548 cm^{-1}, a 10% variation in fluorescence intensity ratio represents a 4% variation in temperature at 2000 K. At this temperature the v"= 1 level contains 8% of the OH population.

The experimental configuration for the two-dimensional fluorescence measurement is illustrated in Fig. 2. uv laser light from a Nd:YAG pumped-dye laser with frequency doubling (Quanta Ray: DCR1, PDL1, WEX) was formed into a planar sheet 48 mm in height with a measured width of 1.0 mm. This laser sheet was passed through the postflame gas of a premixed methane-air flat flame stabilized on a capillary tube burner with a 25.4-mm diam. Two (1024-element) diode arrays (Reticon model 1024S) of 25-mm length were used to monitor the laser intensity distribution and attenuation of the laser light sheet passing through the postflame gas. For each laser pulse the output of the arrays was digitized and stored in a computer for normalization of the fluorescence signal. LIF signals from OH in the flame gas were measured with a computer controlled low-light-level vidicon with a uv scintillator and two stages of image intensification (ISIT, PAR model 1257). In order to minimize the effect of uv emission from the flame, the image intensifier was gated

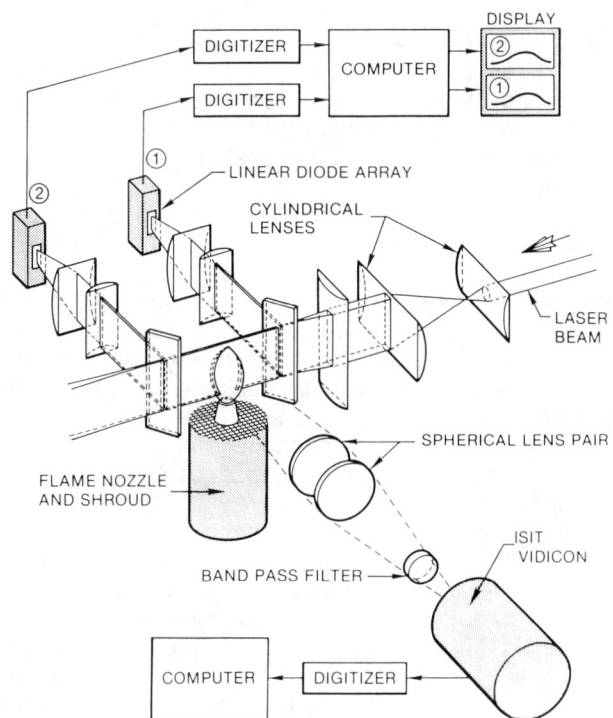

Fig. 2 Two-dimensional laser fluorescence imaging experiment.

(400 ns) to monitor only the OH fluorescence excited with the 5-ns laser pulse. The light collection system consisted of a pair of 50-mm-diam quartz lenses; a 500-mm focal length lens closest to the flame and a 100-mm focal length lens imaging onto the vidicon through a uv filter to block any ambient light. With this optical system the 100 x 100 format of the vidicon provided a pixel spatial resolution of 0.6 x 0.6 mm.

For each laser pulse the digitized signal from the vidicon was stored in the computer. Subsequent frames were accumulated in the computer memory. To correct the fluorescence data for any flame emission effects and thermal noise on the silicon target in the vidicon, a background measurement without the laser operating was subtracted from the raw data. Typical vidicon background signal levels of 250 counts/pixel/pulse were encountered. This level could change by a factor of 2 during operation. To eliminate this drift the vidicon was thermally stabilized with a water jacket (15 C).

Results and Discussion

The stoichiometric methane-air flat flame used in this experiment has been previously characterized for OH concentration (Cattolica 1982) by laser absorption and temperature (Cattolica 1981) by thermocouple and laser fluorescence of OH. These point measurements were obtained along the flame centerline near the burner surface and provide convenient reference measurements. In Plate 1 a photograph of the visible luminosity from the postflame gas illustrates the structure of this methane-air flat flame.

For the measurement of the OH fluorescence resulting from excitation of the $Q_1(5)$ (1,1) transition, the laser system was operated using DCM dye with a pulse energy of 3.75 mj at 314.2 nm. The maximum background corrected signal was 200 counts/pixel/pulse. In Plate 2 the two-dimensional image of the fluorescence signal from 10 laser pulses is displayed. The fluorescence signal has been corrected by the laser sheet intensity distribution. In Plate 2 the color scale is linear and normalized to the maximum value in the frame. The two-dimensional fluorescence image has not been corrected for vidicon sensitivity variation. This sensitivity variation is self cancelling in taking the ratio of the fluorescence signals to obtain vibrational population ratios.

In Plate 3 the OH fluorescence image from 5 laser pulses with excitation of the $Q_1(5)$ (1,0) transition is shown. The dye laser was operated with R590 dye with a

IMAGING OF FLAME TEMPERATURE

Plate 1 Visible luminosity from postflame gas of methane-air flame.

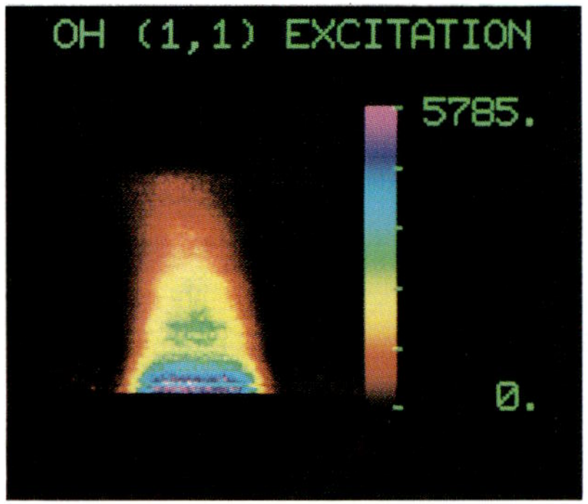

Plate 2 Image of OH fluorescence with excitation of Q1(5) in the (1,1) band.

Plate 3 Image of OH fluorescence with excitation of Q1(5) in the (1,0) band.

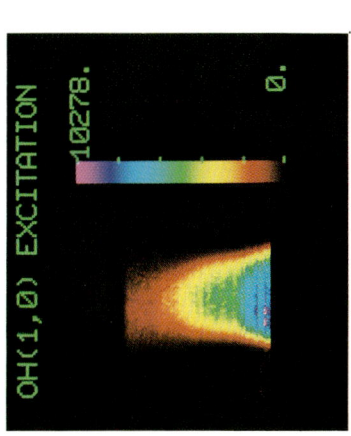

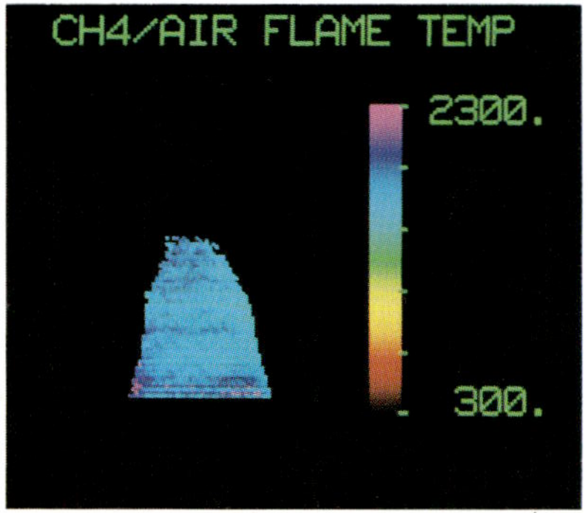

Plate 4 Image of temperature from OH vibrational population ratio.

pulse energy of 1.2 mj at 282.6 nm. It was necessary to operate the laser at this lower energy level to restrict the peak signal level to 1200 counts/pixel/pulse to maintain signal linearity. On a single-pulse basis the dynamic range of the vidicon was limited to (10^3). For the image in Plate 3 the color scale is again normalized to the peak signal. In this flame the maximum OH concentration was 2.3 x 10^{16} cm^{-3}. Although the image in Plate 3 is not strictly proportional to OH concentration (because temperature and quenching corrections have not been included), it does represent a good qualitative view of the variation of the total OH concentration.

To obtain an image of the temperature field in the postflame gas of the methane-air flame, the image in Plate 3 was divided by the image in Plate 4 on an individual pixel basis, yielding an image of the relative vibrational population ratio $n(j",v"=1)/n(j",v"=0)$. This vibrational population ratio was put on an absolute basis from point measurements of flame temperature obtained previously (see Cattolica 1981). OH vibrational temperature was then calculated from the vibrational population ratio using the Boltzmann relation expressed in Eq. (1).

The two-dimensional temperature field obtained from the vibrational population ratio is shown in Plate 4. The central core of the postflame gas is observed to be at 1900 ± 150 K. Error limits on the temperature observed in the central core region exceed that expected from SNR by a factor of 2. This increased uncertainty is the result of systematic effects introduced by changing the laser system configuration to pump the two dyes used in the experiment. The edge of the temperature image was truncated where the fluorescence signal with the (1,1) excitation has a SNR less than 3. At the edge of the high-temperature gas core region, the OH concentration drops approximately a factor of 10 for every 100 K near 1800 K. For an OH concentration of 2.3 x 10^{16} cm^{-3} a maximum signal level of 200 counts/pixel was obtained for a single laser pulse exciting the $Q_1(5)$ (1,1) transition. Baseline fluctuations due to dark current on the vidicon target were approximately 4% of this value. At 1800 K the OH equilibrium concentration was 1.0 x 10^{15} cm^{-3}, the corresponding SNR under these conditions was near unity. For ten laser pulses this SNR improved to 3. The truncation of the temperature field image in Plate 4 for data below this SNR has precluded determination of OH temperature below 1800 K. The uniform high-temperature core region above 1800 K is consistent with previous point measurements of the temperature (see Cattolica 1981).

Although the two-dimensional OH temperature measurement as implemented in this work was limited to temperatures above 1800 K, the OH concentration detectability was much better. Using the $Q_1(5)$ (1,0) excitation a minimum detectable OH concentation (i.e., SNR greater than 1) of 3×10^{14} cm^{-3} could be observed on a single-pulse basis with the current system. This minimum concentration limit could easily be lowered a factor of 50 by increasing the laser energy a factor of 10 and using a stronger transition in the (0,0) band. However, the range of observable OH concentration would be limited by the dynamic range (10^3) of the vidicon system.

Conclusions

The capability to obtain two-dimensional measurements of flame temperature in a planar slice through a flame has been demonstrated using two-line laser-induced fluorescence applied to the OH molecule. The OH vibrational temperature measurement is applicable to temperatures above 1800 K. Below this temperature the decrease in OH concentration, particularly in the v"=1 vibrational level, results in insufficient signal levels to obtain an adequate signal-to-noise ratio. For temperatures below 1800 K it is possible to increase signal strength by using an OH rotational temperature measurement or using a fluorescence species such as NO, which is stable over a wide temperature range.

Acknowledgment

Research sponsored by the Department of Energy, Office of Basic Energy Sciences.

References

Alden, M., Edner, H., Holmsted, G., Svanberg, S., and Hoegberg, T. (1982), Single pulse laser-induced OH fluorescence in an atmospheric flame with a diode array detector. Appl. Opt. 21, 1236-1240.

Cattolica, R. J. (1982). OH rotational temperature from two-line laser-excited fluorescence. Appl. Opt. 7, 1156-1166.

Cattolica, R. J. (1981). OH radical nonequilibrium in methane-air flat flames. Comb. and Flame 44, 43-59.

Dyer M. J. and Crosley, D. R. (1982). Two-dimensional imaging of OH fluorescence in a flame. Opt. Lett. 7, 382-384.

Escoda, M. C. and Long, M. B. (1983). Rayleigh scattering measurements of the gas concentration field in turbulent jets. AIAA Journal 21, 81-84.

Kychakoff, G., Howe, R. D., Hanson, R. K., and McDaniel, J. C. (1982). Quantitative visualization of combustion species in a plane. App. Opt. 21, 3225-3227.

Smith, J. R. (1978). Rayleigh temperature profiles in a hydrogen diffusion flame. Laser Spectroscopy, SPIE 158, 84-90.

LDV Measurements of Gas Flow Behind Reflected Shocks

M. Frenklach* and C. K. Li Kwok Cheong†
Louisiana State University, Baton Rouge, Louisiana
and
E. S. Oran‡
Naval Research Laboratory, Washington, D.C.

Abstract

In this paper, the results of velocity measurements behind reflected shocks using laser Doppler velocimetry (LDV) are reported. Horizontal and vertical velocities have been measured for three cases. The first set of measurements was in the inert gas, argon, which was used as a baseline for the succeeding experiments. In the second set of measurements, in nitrogen, effects were observed that may be attributed to boundary layers and sound waves. The experiments with toluene provided a natural seeding through the early formation of soot particles and were used to test the conventional seeding with aluminum oxide particles. The final set of experiments was in undiluted stoichiometric hydrogen-oxygen mixtures. The interesting observation in this case is the organized gas motion along the centerline of the shock tube near the end wall. This motion starts substantially before the rapid rise in pressure observed at the end of the induction period. For the particular mixture chosen, these results were highly repeatable, and the induction times as determined by the pressure measurements agreed well with those obtained by other investigators. Since no equivalent motion was observed in the argon, nitrogen, or toluene-argon mixtures, the observed effect is due to expansion caused by the small

---

Presented at the 9th ICODERS, Poitiers, France, July 3-8, 1983. Copyright © by the American Institute of Aeronautics and Astronautics, Inc., 1984. All rights reserved.
*Associate Professor, Department of Chemical Engineering.
†Research Assistant. Permanent address: Alberta Sulphur Research Ltd., Canada.
‡Senior Research Scientist, Laboratory for Computational Physics.

but finite energy release that occurs during the preignition period.

Introduction

Research in the last 25 years has shown that shock tubes provide one of the best experimentally available environments for studying high-temperature chemical kinetics (Lifshitz 1981). They have also provided an excellent environment for studying reaction initiation leading to the development of flames and detonations (Voevodsky and Soloukhin 1965; Strehlow 1968; Meyer and Oppenheim 1971; Surrete and Gerstein 1975; Bradley et al. 1980; Oran et al. 1982; Oran and Boris 1982; Kailasanath and Oran 1983). The reflected shock wave technique is particularly attractive because the material behind the reflected shock is at rest. This, of course, is an idealization that assumes ideal gases, no sound waves, and no boundary layers. In fact, in the present work, these nonidealities are present in profusion, and unless precautions are taken, interpretation of the data is questionable (Gardiner and Wakefield 1970; Bowman and Hanson 1979; Khandewal and Skinner 1981; Gardiner et al. 1981). Thus, it would be extremely useful to be able to evaluate the deviations from the ideal environment and perhaps use these deviations themselves as a probe of the reacting system.

In this paper, the results of velocity measurements behind reflected shocks made by means of laser Doppler velocimetry are presented. This technique was used in conjunction with pressure measurements to determine horizontal and vertical velocities for three distinct cases. Measurements in the inert gas (argon) were used as a baseline for the succeeding experiments. Measurements in nitrogen revealed effects that may be attributed to boundary layers and sound waves. The experiments with toluene provided a natural seeding through the early formation of soot particles and were used to test the conventional seeding with aluminum oxide particles.

The final set of experiments was in an undiluted stoichiometric hydrogen-oxygen mixture. The interesting observation in this case is the organized gas motion along the centerline of the shock tube near the end wall. This motion starts substantially before the rapid rise in pressure observed at the end of the induction period. For the particular mixture chosen, these results were highly repeatable, and the induction times as determined by the pressure measurements agreed well with those obtained by Meyer and Oppenheim (1971).

In the following sections, the experimental arrangement and the seeding procedure are discussed in some detail. Then the experimental results are presented for the three cases outlined above. Finally, the results are summarized, and the significance of the observations is discussed.

Experimental Arrangement and Procedure

The experiments were conducted in a conventional stainless steel shock tube: 7.62 cm i.d., 3-m driver section, and 7.3-m driven section. The double diaphragm burst technique was employed, using Mylar and aluminum foils as diaphragm materials. Both mechanical (Edwards ED-500) and diffusion (Edwards Speedivac E-04) pumps were used in the shock tube gas-handling and vacuum systems. The systems could be evacuated to less than 10^{-5} Torr. The driver gas was helium. The test gas mixtures, 66.7% hydrogen-33.3%oxygen and 0.311%toluene-argon, were prepared manometrically in a stainless steel tank and allowed to mix for at least 24 hours prior to experimental runs. A Wallace and Tiernan model FA-145 manometer was used for precise pressure readings. The stated purities of the gases were argon-99.995%, helium-99.99%, nitrogen-99.995%, hydrogen-99.94%, and oxygen-99.5%. The toluene was of Matheson Coleman & Bell Reagent grade and was purified by repeated freezing and evacuation. The shock tube was cleaned after every run.

The state of the gas behind reflected shock waves was calculated in a standard manner assuming full relaxation and no chemical reactions and using the measured incident shock velocity extrapolated to the end wall of the shock tube (Gardiner et al. 1981). The observed shock wave attenuation was approximately 2%/m. Shock velocities were measured using four piezoelectric pressure transducers to trigger the start and stop channels of an interval timer. All pressure transducers and optical windows were mounted flush with the surface of the shock tube to minimize distortion of the flow.

Velocities of the suspended particles were determined by laser Doppler velocimetry (LDV). A Thermo-Systems Inc., 9100-3 general purpose LDV system was employed in this study. The laser velocimeter was operated in a dual-beam on-axis forward scattering mode with a 15-mW He-Ne laser source (632.8 nm). The laser beam was divided into two parallel beams of equal intensity, 50 mm apart, by a beam splitter. One of the two beams was fed directly to the transmitting lens (120 mm focal length), while the other passed first through a Bragg cell (TSI 9180) where the frequency of the beam was shifted by 40 MHz. Acoustic frequency shifting was employed to increase the accuracy in measurements of small velocities and to determine the direction of the gas flow. The two laser beams leaving the transmitting lens were focused to cross each other, usually at the midpoint of the shock tube. High-quality fused quartz (Suprasil-1) was used for the shock tube windows. The scattered light was collected by a receiving lens and focused on a photomultiplier tube (RCA 4526). The high-frequency signals from the photomultiplier tube were downmixed to lower frequencies and then processed by a counter-type signal processor (TSI 1980), where they were filtered and analyzed for amplitude and frequency. The high-amplitude scattering signals were automatically rejected, thus reducing the probability of making velocity measurements of large particles that may not follow the flow faithfully. Only signals generated from relatively symmetrical Doppler burst envelopes were considered. The processed signals were displayed on a digital oscilloscope (Nicolet 2090-3). The specified calibration of the velocimeter was tested in the backward scattering mode by using a series of rotating disks and in the forward scattering mode by using a seeded laminar steady gas flow controlled by a flow meter.

Homogeneous dispersion of seed particles in the test section of the shock tube has always been a problem for researchers (Lowenstein and von Rosenberg 1978). Several

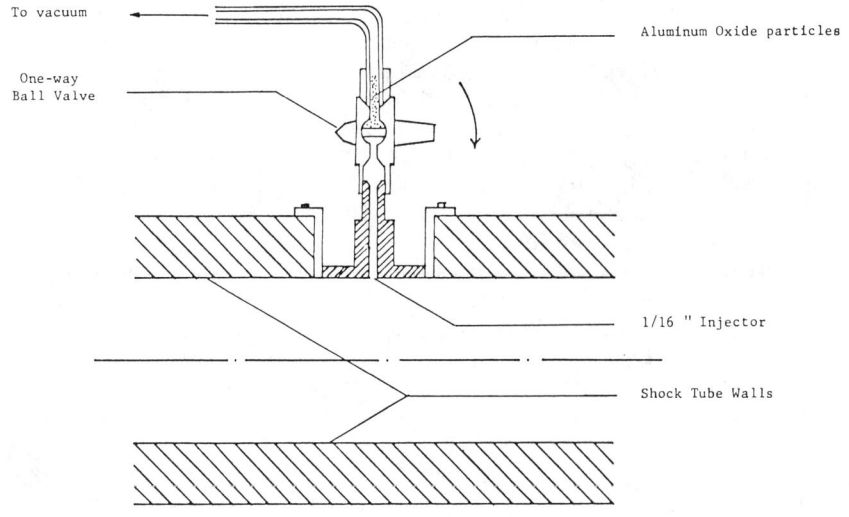

Fig. 1. A schematic diagram of the seeder device.

techniques were tried in this study, and by a trial-and-error approach a satisfactory system was developed. Figure 1 shows a schematic diagram of the seeder device. The main feature of this device is the on/off mechanism of the one-way ball valve. The valve was mounted on a detachable plug (for easy cleaning) that was screwed into the shock tube wall at some distance from the LDV observation station. A 1.6-mm bore was drilled inside the plug and acted as an injector to spray the particles. The size of the hole was chosen so that it did not seriously affect the shock wave structure.

Prior to starting an experiment, 5 to 6 mg of Al_2O_3 particles (Microgrit GB-2000, nominal size 1 µm) were placed into the upper part of the ball valve (see Fig. 1), and the vacuum was slowly drawn so that the particles were not removed from the valve. After a test gas mixture was introduced into the driven section of the shock tube, the same gas mixture was also introduced into the upper part of the ball valve but at a higher pressure than that in the shock tube. Owing to the pressure difference, the particles were dispersed within the test section of the shock tube upon opening the ball valve. The valve was opened only for a very short period of time (~1 s) to minimize introduction of surplus gas into the tube (the additional amount of gas introduced during seeding caused a maximum pressure rise of approximately 0.3 Torr inside the shock tube, where P_1 was around 40 Torr). The best LDV signals were obtained when the experiments were conducted within about 30 s after particle seeding.

The seeding procedure was subjected to a number of tests. First, the frequent appearance of particles in the LDV measuring volume indicated relatively homogeneous dispersion of the particles. Of course, no LDV signals were observed in blank tests without seeding. Second, the size of aluminum oxide particles was determined under postshock conditions by the visibility method (Farmer 1972) and it was found to be within the range of from 0.7 to 1.6 µm (Frenklach et al. 1983). The relaxation time (particle lag) for a micron-sized particle is of the order of a microsecond (Vom Stein and Pfeifer 1973; Smeets and George 1980). The measured particle velocities were assumed therefore to be equal to local velocities of gas. Third, the Doppler shift generated by the dispersed particles prior to the arrival of the incident shock wave was observed to be identical to the "effective" frequency shift induced by the Bragg cell, thus indicating that the injected particles had enough time to relax. Finally, the velocities of seeded flows were observed to be practically the same as those determined by the nascent-soot-particle technique that was developed in this study. This technique relies on the fact that small amounts of aromatic hydrocarbons (0.311% of toluene in this work), when heated, may generate sufficiently large soot particles within a short period of time (Frenklach et al. 1983). By definition, nascent soot particles provide an ideal seeding: homogeneous dispersion without any particle lag. Hence, the positive comparison between the two techniques validates the seeding procedure.

The gas velocities were determined at a number of locations within the shock tube and at a variety of conditions. The pressure behind reflected shock waves was measured by a calibrated piezoelectric pressure transducer located at the LDV observation stations. The instant of ignition in hydrogen-oxygen mixtures was defined by the point of maximum curvature in the pressure trace. Further experimental details can be found elsewhere (Li Kwok Cheong 1981).

Results and Discussion

Velocity measurements in pure argon and nitrogen were performed first in order to establish a baseline for

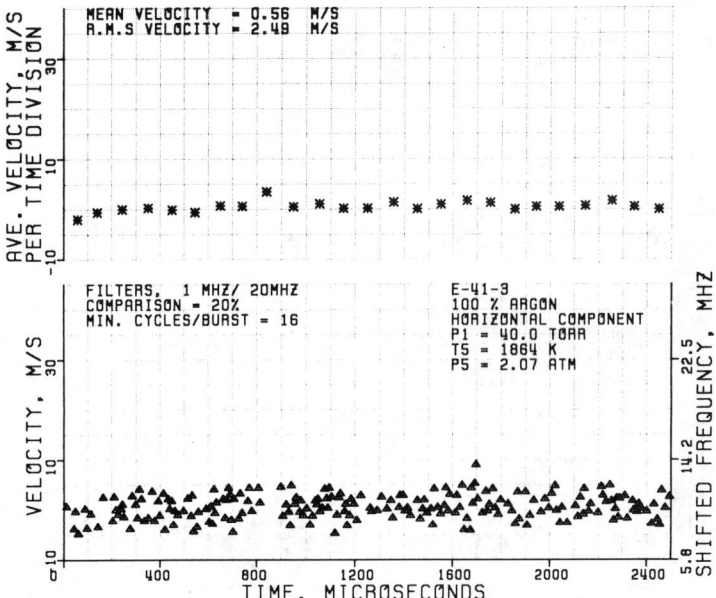

Fig. 2 The horizontal velocities of Al_2O_3 particles determined in argon at T_5 = 1864 K and P_5 = 2.07 atm. The LDV measuring volume was located on the centerline of the shock tube, 10 mm from the end wall.

hydrogen-oxygen mixtures. Most observations were made at the centerline of the shock tube, 10 mm away from the end wall. Although both horizontal and vertical velocities were obtained, they were not measured simultaneously but in separate experiments. Figures 2 and 3 show typical horizontal and vertical velocities determined at similar conditions in experiments with argon. In each figure, the lower graph displays the actual data points. The positive direction is defined by the motion of the reflected shock wave for the horizontal velocities and by the gravitational force for the vertical velocities. Statistical treatment of these results is not a trivial task because the sampling times in the LDV technique are Poisson distributed (Gaster and Roberts 1975; Edwards 1981). The present case is particularly complex: A sufficient number of data points cannot be collected because it is practically impossible to reproduce exactly a set of shock tube conditions, and, more importantly, conditions behind reflected shock waves can hardly be regarded as stationary, which constitutes a necessary condition for a spectral analysis. As a consequence, a spectral analysis, which may be misleading under the circumstances was not made, and the experimental

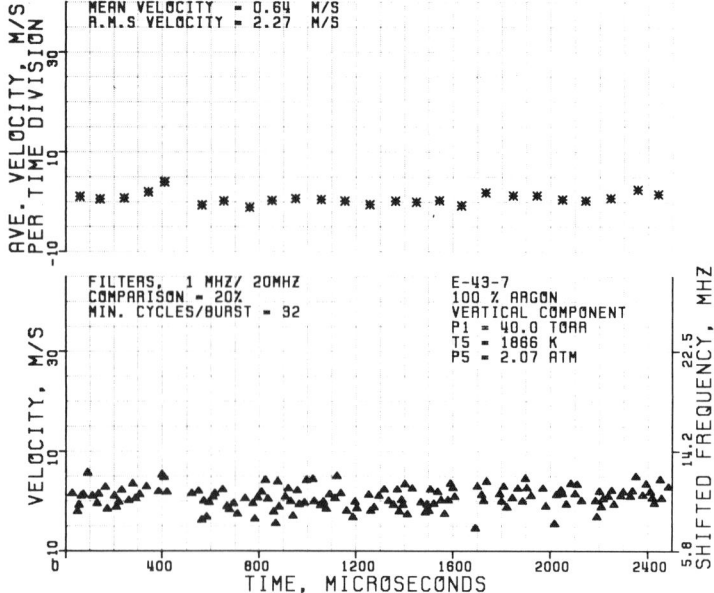

Fig. 3 The vertical velocities of Al_2O_3 particles determined in argon at T_5 = 1866 K and P_5 = 2.07 atm. The LDV measuring volume was at the same location as in Fig. 2.

data were smoothed by averaging the velocities over 100-μs periods. These averaged velocities are displayed on the upper graph in each figure.

Analysis of Figs. 2 and 3 indicates a significant level of scatter in data points around zero velocity. It is hard to decide whether this scatter is due to high-frequency turbulence or to experimental and instrumental errors. It is definitely not a consequence of the seeding procedure, because similar scatter was also obtained with the nascent-soot-particle technique, as indicated in Fig. 4. The slowly rising average velocities observed in Fig. 4 may be attributed to the gas motion resulting from exothermicity during soot formation.

The results obtained in argon were practically unaffected by the state of the shocked gas for temperatures from 1154 to 1866 K and pressures from 1.76 to 2.53 atm. Significantly different results, however, were obtained when the LDV measuring volume was set at different locations. For example, Fig. 5 shows that relatively large velocities were measured near the side wall of the shock tube, indicating positive direction of flow in a

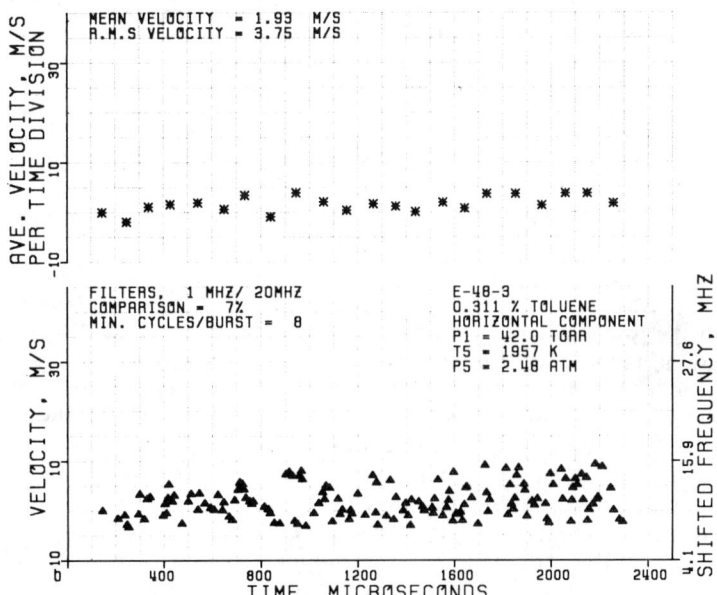

Fig. 4 The horizontal velocities of the nascent soot particles determined in 0.311% toluene-argon mixture at T_5 = 1957 K and P_5 = 2.48 atm. The LDV measuring volume was at the same location as in Fig. 2.

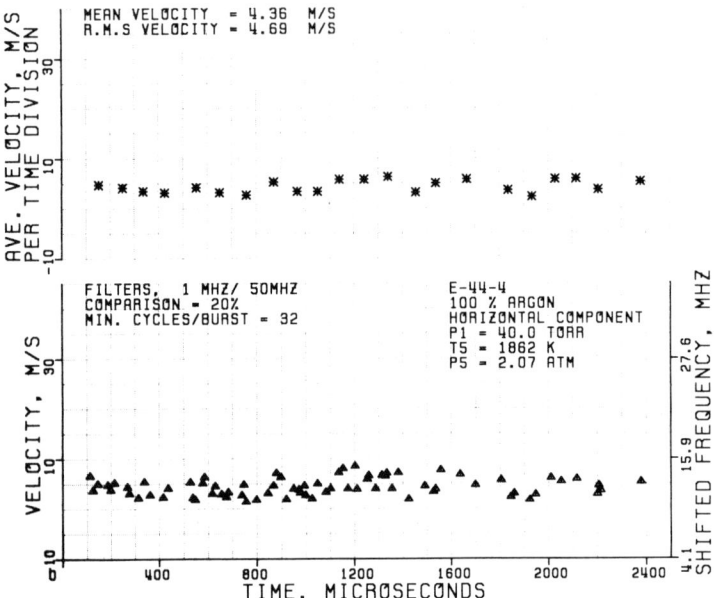

Fig. 5 The horizontal velocities of Al_2O_3 particles determined in argon at T_5 = 1862 K and P_5 = 2.07 atm. The LDV measuring volume was located 10 mm from the end wall of the shock tube, 2 mm from the window.

boundary layer at this location. This result is explained by the well-known phenomenon of bifurcation (Bradley 1962). Also, in agreement with previous works [see for example, Dewey and Whitten (1975); Khandewal and Skinner (1981)], relatively large negative velocities were observed 85 mm away from the end wall of the shock tube (Fig. 6). At this location, the low-frequency component was clearly pronounced. These results are indicative of sound waves traveling in the gas.

The results obtained with nitrogen as a test gas were similar to those observed with argon, except that the average velocity was slowly increasing (Fig. 7). This rise may be attributed to the fact that the boundary layer in the case of nitrogen is thicker than in the case of argon (Bradley 1962). Thus, there is a larger distortion of the ideal flow for the larger boundary layer (Khandewal and Skinner 1981). The experiments with nitrogen covered a temperature range of from 1063 to 1399 K and a pressure range of from 1.24 to 2.53 atm.

The results have shown the presence of relatively low-frequency components in the horizontal velocity

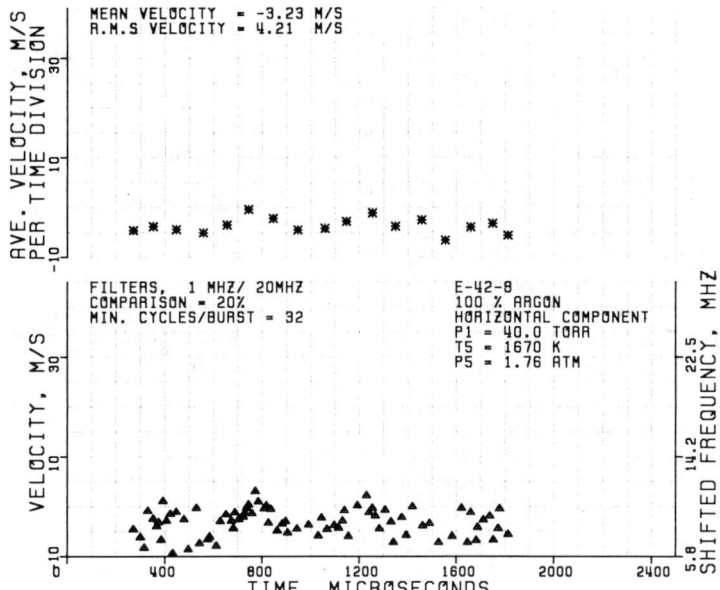

Fig. 6 The horizontal velocities of Al_2O_3 particles determined in argon at T_5 = 1670 K and P_5 = 1.76 atm. The LDV measuring volume was located on the centerline of the shock tube, 85 mm from the end wall.

profiles (e.g., Figs. 5 and 6). These are likely to be the results of sound waves bouncing back and forth between the end wall and the reflected shock. An effect such as this was demonstrated computationally by Oran et al. (1982) in Fig. 14 of their paper. The sound wave pattern is complicated by the fact that the shocked region is constantly increasing in volume.

The velocity profiles determined in undiluted stoichiometric hydrogen-oxygen mixtures were remarkably different from those obtained in argon and nitrogen. The experiments were conducted in the weak ignition regime. Only very narrow temperature and pressure ranges were covered (948-966 K and 1.18-1.25 atm, respectively) to test the repeatability of the results. The ignition delay times, determined by the pressure traces, were found to be in a good agreement with those reported by Meyer and Oppenheim (1971) at similar conditions.

Figure 8 demonstrates a typically observed phenomenon of accelerated gas movement prior to the instant of ignition. Since no equivalent motion was observed with argon, nitrogen, or toluene-argon mixtures, it is unlikely that this effect is due to boundary layers. The effect is then due to energy release during the induction period. One should also notice that the gas movement did not occur immediately after the shock reflection, but initiated at about the midpoint of the induction period, which was typical for all the runs. This behavior is explained as follows: The gas motion is induced by the energy released from the ongoing exothermic reaction. It is well known, however, that the initial phase of the oxidation process is slightly endothermic. Thus, the onset of the notable gas movement occurs when the reaction reaches the state of sufficient exothermicity. The expansion increases following the accelerated energy release of the chain reaction and becomes very large when the pressure jumps indicating the instant of ignition. The abrupt cessation in the velocity profile (as, e.g., in Fig. 8) coincides with the instant of ignition at the observation station determined by the pressure transducer. The absence of further data points indicates that the velocities developed after ignition were too high to be processed by the present experimental arrangement.

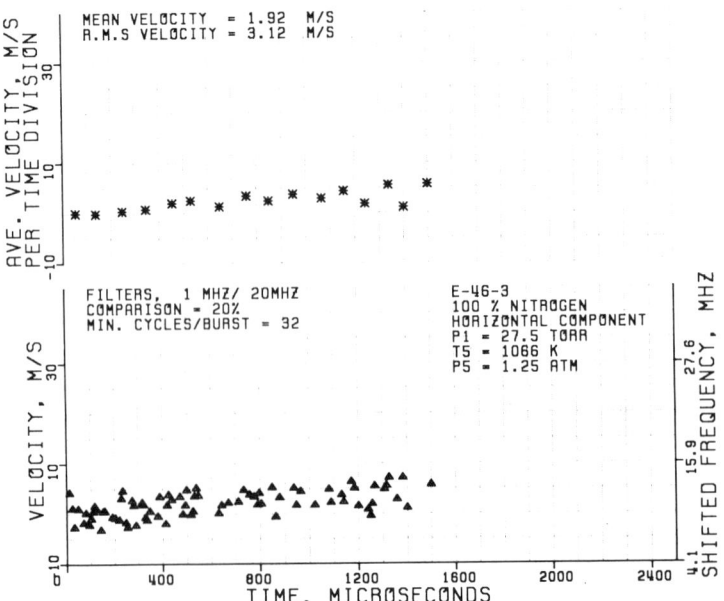

Fig. 7 The horizontal velocities of Al_2O_3 particles determined in nitrogen at T_5 = 1066 K and P_5 = 1.25 atm. The LDV measuring volume was at the same location as in Fig. 2.

GAS FLOW BEHIND SHOCK WAVES

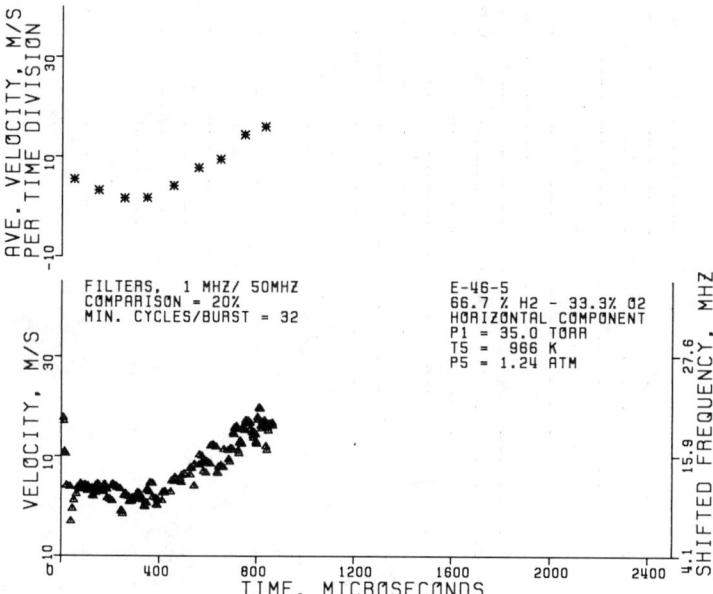

Fig. 8 The horizontal velocities of Al_2O_3 particles determined in undiluted stoichiometric hydrogen-oxygen mixture at T_5 = 966 K and P_5 = 1.24 atm. The LDV measuring volume was at the same location as in Fig. 2. The ignition delay, determined by the pressure transducer located on the side wall of the shock tube 10 mm from the end wall, was 892 μs.

Conclusions

The work presented in this paper has shown that LDV measurements can provide new and valuable information about the flows behind shock waves. This technique can be used to provide information about nonidealities in the medium for chemical kinetic studies, and it can also be used to probe chemical kinetic-fluid dynamic interactions in exothermic systems. The experimental technique developed appears to be well suited for flow measurements behind reflected shock waves where the velocities are presumably small.

The experimental results have shown some unexpected and perhaps important phenomena. The measurements reported above were of horizontal and vertical velocity components at various locations. The velocity measurements off the centerline have indicated the presence of boundary layer. Thus LDV may be used to examine the structure of boundary layers behind shock waves.

The most unexpected observation is of the organized velocities determined during the induction period in the experiments with undiluted stoichiometric hydrogen-oxygen mixtures.

References

Bowman, C. T. and Hanson, R. K. (1979) Shock tube measurements of rate coefficients of elementary gas reactions. J. Phys. Chem. 83(6), 757-763.

Bradley, J. N. (1962) Shock Waves in Chemistry and Physics. Methuen, London.

Bradley, J. N., Capey, W. D., and Farajii, F. (1980) The effect of reaction exothermicity on shock propagation. Proceedings of the 12th International Symposium on Shock Tubes and Waves, pp. 524-532. Magness, Jerusalem.

Dewey, J. M. and Whitten, B. T. (1975) Calibration of a shock tube flow by analysis of the particle trajectories. Phys. Fluids 18(4), 437-445.

Edwards, R. A. (1981) A new look at particle statistics in laser-anemometer measurements. J. Fluid Mech. 105, 317-325.

Farmer, W. M. (1972) Measurement of particle size, number density, and velocity using a laser interferometer. Appl. Opt. 11(11), 2603-2612.

Frenklach, M., Taki, S., Li Kwok Cheong, C. K., and Matula, R. A. (1983) Soot particle size and soot yield in shock tube studies. Combust. Flame 51(1), 37-43.

Gardiner, W. C. Jr. and Wakefield, C. B. (1970) Influence of gas dynamic processes on the chemical kinetics of the hydrogen-oxygen explosion at temperatures near 1000°K and pressures of several atmospheres. Astronaut. Acta 15(5-6), 399-409.

Gardiner, W. C. Jr., Walker, B. F., and Wakefield, C. B. (1981) Mathematical methods for modeling chemical reactions in shock waves. Shock Waves in Chemistry (edited by A. Lifshitz), Chap. 7. Marcel Dekker, New York.

Gaster, M. and Roberts, J. B. (1975) Spectral analysis of randomly sampled signals. J. Inst. Math. Its Appli. 15(2), 195-216.

Kailasanath, K. and Oran, E. S. (1983) Ignition of flamelets behind incident shock waves and the transition to detonation. Combust. Sci. Technol. 34(1-6), 345-362.

Khandewal, S. C. and Skinner, G. B. (1981) Shock tube studies of hydrogen oxidation. Shock Waves in Chemistry (edited by A. Lifshitz), Chap. 7. Marcel Dekker, New York.

Lifshitz, A. (1981) Shock Waves in Chemistry. Marcel Dekker, New York.

Li Kwok Cheong, C. K. (1981) Application of laser Doppler velocimetry to shock tube studies. M.S. Thesis, Louisiana State University, Baton Rouge, La.

Lowenstein, A. I. and von Rosenberg, C. W. Jr. (1978) Shock tube studies of coal devolatilization. Proceedings of the 11th International Symposium on Shock Tubes and Waves, pp. 366-374. University of Washington, Seattle, Wash.

Meyer, J. W. and Oppenheim, A. K. (1971) On the shock-induced ignition of explosive gases. 13th Symposium (International) on Combustion, pp. 1153-1164. The Combustion Institute, Pittsburgh, Pa.

Oran, E. S. and Boris, J. P. (1982) Weak and strong ignition: II. Sensitivity of the hydrogen-oxygen system. Combust. Flame 48(2), 149-161.

Oran, E. S., Young, T. R., Boris, J. P., and Cohen, A. (1982) Weak and strong ignition: I. Numerical simulations of shock tube experiments. Combust. Flame 48(2), 135-148.

Smeets, G. and George, A. (1980) Novel laser Doppler velocimeter enabling fast instantaneous recordings. Proceedings of the 12th International Symposium on Shock Tubes and Waves, pp. 579-588. Magnes, Israel.

Strehlow, R. A. (1968) Fundamentals of Combustion. Krieger, New York.

Surrete, R. G. and Gerstein, M. (1975) Unsteady shock waves initiation of a detonation wave in a compressible gas mixture. AFOSR Report 70-1918.

Voevodsky, V. V. and Soloukhin, S. I. (1965) On the mechanism and explosion limits of hydrogen-oxygen chain self-ignition in shock waves. Tenth Symposium (International) on Combustion, pp. 279-283. The Combustion Institute, Pittsburgh, Pa.

Vom Stein, H. D. and Pfeifer, H. J. (1973) Acceleration of micron-sized particles in shock tubes. Proceedings of the Ninth International Shock Tube Symposium, pp. 804-814. Stanford University, Stanford, Calif.

Droplet Size Distributions from Diffracted Light Intensities

Alain Tardieu,* Sébastien M. Candel,† and Emile Esposito‡
l'Ecole Centrale des Arts et Manufactures
Châtenay-Malabry, France

Abstract

Precise and fast measurement of droplet size distributions in sprays is of considerable technological importance. Among the many optical methods proposed in recent years, those based on the measurement of the light intensity diffracted by the droplet cloud are in advanced state of development. The main difficulty encountered arises in the inversion operation, which provides the droplet size distribution from the measured diffraction profile. The problem is ill-conditioned, and a small measurement error may lead to entirely different size distributions. In previous work, the difficulty has been overcome by performing the inversion on a low-resolution spatial basis or by imposing the shape of the size distribution at the outset. Our objective has been to redesign a system based on light diffraction that gives more accurate results. This system uses a CCD charge coupled device array made up of 1024 photodiodes to provide a high-resolution sampling of the diffracted light intensity profile. A numerical simulation based on synthetic diffracted light profiles with added random noise indicates that direct inversion of the measurements becomes possible if the dynamic range of the system exceeds 1000 (60 dB). This extended range is obtained by scanning the photodiode array at various frequencies and thus modifying

---

Presented at the 9th ICODERS, Poitiers, France, July 3-8, 1983. Copyright American Institute of Aeronautics and Astronautics, Inc., 1984. All rights reserved.
*Graduate Student, now Engineer, Electricité de France, Paris.
†Professor, Research Scientist at Office National d'Etudes et de Recherches Aérospatiales, Châtillon.
‡Professor, Ecole Centrale, Châtenay-Malabry.

the exposure time of the diodes. The diffraction profile is then obtained by assembling successive diffraction patterns with partial overlap. This new technique of measurement, the signal processing method, and the first results are described in this paper.

Introduction

Precise and fast measurements of droplet sizes insprays of current technological interest. Information on the distribution of droplet sizes is particulary useful in the analysis and design of combustors for aircraft powerplants and rocket engines. Numerous methods have been proposed to obtain such data. The application of optical diagnostics based on light diffraction for this purpose has been the subject of many recent investigations (see Dobbins et al. 1963 ; Swithenbank et al. 1977, 1978 ; Cornillaut 1972 ; Felton 1979 ; Mc Sweeney and Rivers 1972 ; Azzopardi 1979 ; Wang et al. 1981). An instrument based on the studies of Swithenbank and coworkers has been commercialized by Malvern and its characteristics are described by Negus and Azzopardi (1978). One important difficulty encountered in the practical application of light diffraction for size distribution measurements appears in the process of extracting the size distribution $\varphi_N(a)$ or mass distribution $\varphi_M(a)$ from the scattered light intensity profile $I(s)$. This is an ill-conditioned inverse problem, and a small error in the measurement may induce a large error in the size distribution. This difficulty has been overcome by perfor-

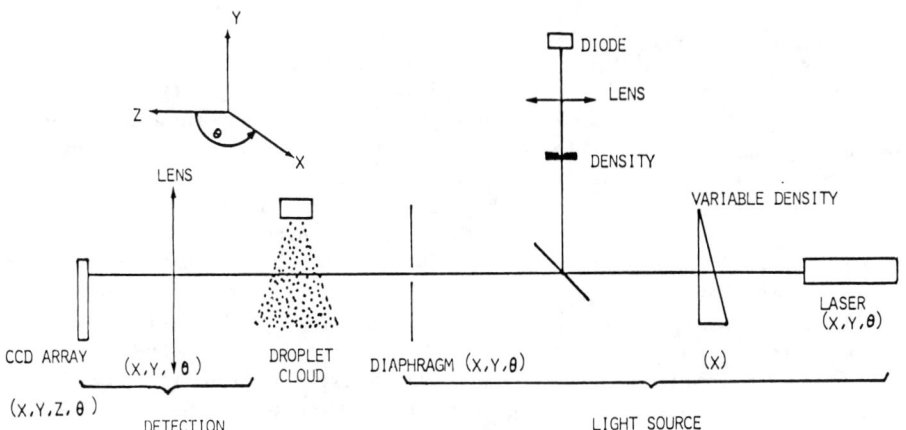

Fig. 1   Schematic diagram of the light diffraction system.

ming the inversion on a low-resolution spatial basis or by imposing the type of size distribution at the outset. The purpose of this work has been to develop a new system based on laser light diffraction that does not require such restrictions.

The detection element in the system is a CCD array of 1024 photodiodes. This array provides a tightly sampled scattered light profile and an extendable dynamic range. This paper presents the system developed around this new technology and describes the characteristics and problems of the scattered light measurement. The results of a systematic simulation performed to define the best inversion method are outlined and some preliminary results are discussed. Additional details on the system may be found in the work of Tardieu (1983).

The Measurement System

The optical setup is shown on Fig. 1. A helium-neon 5-mW laser operating in the TEM_{oo} mode at a wavelenght λ = 0.6328 µm constitutes the light source. The laser beam has an initial diameter of 0.8 mm. To extend the measurement dynamic range, the incident light intensity is adjusted by passing the laser beam through a neutral density filter. The intensity level is measured with a SGD-100 A photodiode on light diverted from the main optical path by a beam splitter. The laser beam then crosses a rectangular test section (200 x 200 x 300 mm). This section may be displaced in two directions at right angles to the laser beam. The tested injectors are double-swirl aerodynamic atomizers developed by the French motorist SNECMA, operating with air at 1.15 bars and water. An f = 300-mm focal length Clairaut-doublet focuses the laser light on a photodiode array RETICON CCPD 1024. This element has 1024 square diodes, each having a side Δs = 16 µm. A very tightly sampled scattered light intensity profile can be obtained if the radius a of the diffracting particles does not exceed $\lambda f/4\pi\Delta s$, which for this system reduces to a $\leq$ 1 mm.

A detailed study of the technical data sheets as well as experiments indicate that the main noise sources associated with the photodiode array are of three types : 1) a constant level additive noise due to charges created by clock signals, 2) thermal noise depending on the array temperature ; and, 3) random noise associated with the scanning electronics. The first two noise components may be corrected ; the third is generally less than 1% of the measurement. To these noise sources, one must add optical noise corresponding to light diffusion in the window protecting the diode array (this window is not treated against reflections).

Under these circumstances the dynamic range for a single sweeping frequency of the diode array cannot exceed 100. This range is too small to allow a proper inversion of the measurements, and measurements at different sweeping frequencies must be performed to determine a single diffracted light intensity profile. In this method, weak light intensities are determined with long exposure times (corresponding to a sweep frequency of the order of 1kHz), while strong light intensities are obtained with short exposures of the diode array (corresponding to a sweep frequency of the order of 25 kHz). With this procedure, the dynamic range can be enhanced and, in principle, can reach a value of 8000 (78 dB), with an additive random noise that does not exceed 10% of each measured value. The numerical simulation described in the next section is based on these two values.

Theoretical Aspects and Simulation

Under certain restrictions on the droplet size and scattering angle, the light intensity in the focal plane of the collecting lens, at a distance s from the object focal point, can be derived from Goodman (1968)

$$I_t(s) = A \int_0^\infty a^4 \left(\frac{J_1(2\pi as/\lambda f)}{2\pi as/\lambda f} \right)^2 \varphi_N(a) \, da$$

where λ is the wavelenght ; f the focal distance ; a the particle radius ; $\varphi_N(a)$ the particle size distribution ; and A a constant accounting for the intensity of the incident beam. The total scattered light appears as the sum over all radii of single particle diffracted light intensity profiles weighted by the particle size distribution. In practice, the measured light intensity I(s) is proportional to the theoretical intensity $I_t(s)$ with added noise

$$I(s) = \alpha I_t(s) + \epsilon(s)$$

The mathematical problem consists of obtaining an estimate $\phi_N(a)$ of the real distribution $\varphi_N(a)$ from a given I(s).

The inversion method must 1) impose minimal constraints and, in particular, not use an a priori particle distribution model ; 2) be sufficiently precise ; 3) be numerically stable and robust ; and 4) be insensitive to mea-

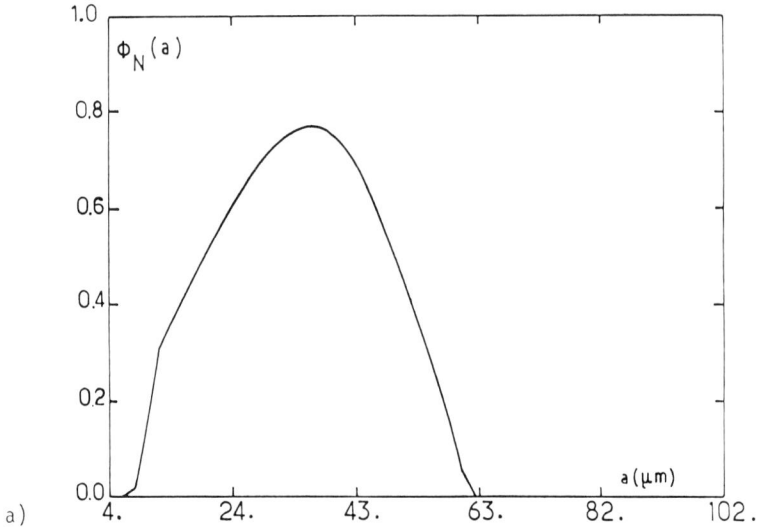

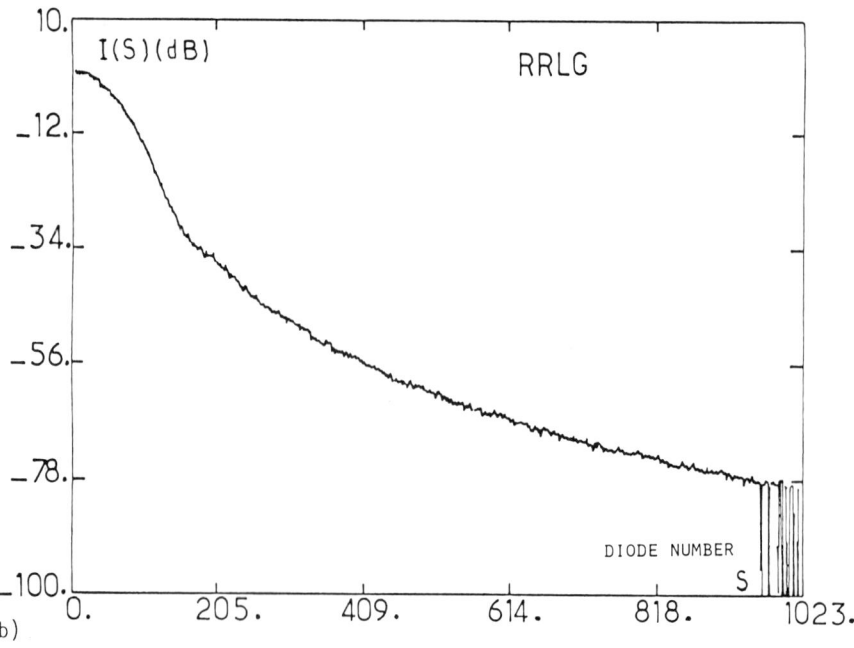

Fig. 2a) A droplet size distribution $\phi_N(a)$ used in he numerical simulation. b) Diffracted light intensity profile corresponding to the size distribution of Fig. 2a. A random error with a maximum amplitude equal to 10% of the local value is added to the theoretical profile. The dynamic range is limited to 8000 (78 db).

surement errors. The choice of a method that satisfies these criteria is based on a numerical simulation of the problem (see Tardieu 1983 for details). The simulation uses five types of particle-size distributions and the corresponding light intensity profiles. Noise is added to the simulated profiles, and the dynamic range is assumed to be finite. Fig. 2a displays one typical distribution, and Fig. 2b shows the corresponding simulated diffracted intensity profile. The dynamic range is set equal to 8000--an upper limit that may be reached with the RETICON CCPD-1024 array.

One of the criteria used to quantify the accuracy of the different inversion methods is the relative norm of the error on the mass distribution

$$E_M = \left(\sum_i \left[(n_{ci} - n_{ri}) a_i^3 \right]^2 \Big/ \sum_i (n_{ri} a_i^3)^2 \right)^{1/2}$$

where n_{ci} and n_{ri}, respectively, designate the calculated and actual numbers of droplets of radius a_i. The value of E_M is given below for the different methods under consideration.

The simulation leads to the following conclusions : 1) Inversion in Fourier space leads to a simple numerical problem involving a single upper triangular matrix inversion, but the computation diverges due to round-off errors ($E_M \sim 100\%$). 2) Good results ($E_M \sim 4\%$) are obtained with a method of inversion under constraints due to Twomey (1977). The constraints prescribe that the solution must be a weakly oscillating function. Unfortunately, the result depends on the straining parameter, which must be chosen at the outset, and the solution degrades with a poor choice. 3) A least-mean-square (LMS) inversion on a fine spatial resolution (typically $\Delta a = 2$ μm) yields oscillatory solutions ($E_M \sim 100\%$). The result is improved if it is obtained with a non-negative least-square (NNLS) algorithm ($E_M \sim 50\%$). 4) A non-negative least-square inversion on a coarser spatial resolution (typically from 5-100 μm with steps $\Delta a = 5$ μm) yields good results ($E_M \sim 4\%$). A calculated distribution obtained from the simulated diffraction pattern of Fig. 2b is shown on Fig. 3. 5) The analytic inversion integral used by Chin et al. (1955) works well if the data are noise-free $E_M = 3\%$. However, this method is very sensitive to measurement errors and dynamic range limitations ($E_M \sim 25\%$). Thus, it appears that the non-negative least-square method is the most appropriate. Simulations have also been performed with other inversion bases made up of a fixed number of classes for a given

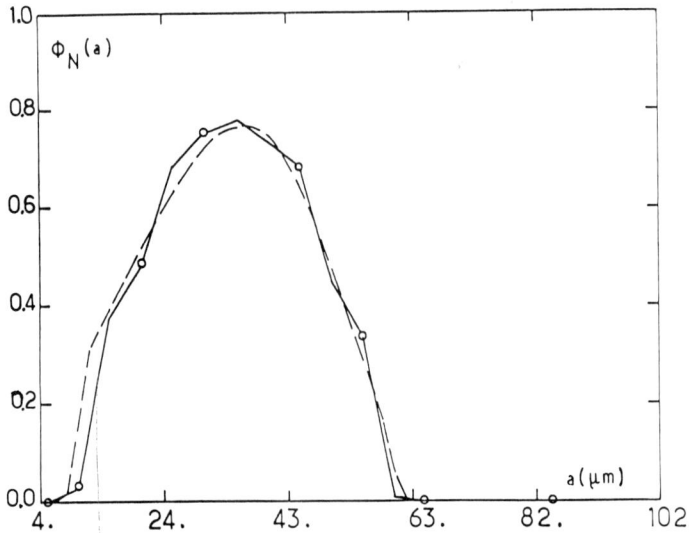

Fig. 3   Result obtained by inverting the simulated profile of Fig. 2b with a non-negative least-mean-square algorithm on a 5-μm basis of radii.

interval of radii. The first basis interval, where $\Delta(a_i^2)$ = constant, corresponds to a uniform distribution of diffracted intensities ; the second, where $\Delta(1/a_i)$ constant, is associated with a uniform distribution locations of the sidelobes of the diffraction diagrams. The solutions obtained with these two bases are less precise than those obtained on a uniform basis $\Delta(a_i)$ = constant, which appears to be the best choice and is used in the remainder of this work.

Further improvements of the inversion calculation may be obtained by intensity matrix preconditioning involving a condensation of adjacent lines. When the dynamic range diminishes below 4000, the quality of the result is maintained by increasing the spatial resolution step size. Another limitation imposed by the detection element is that the first photodiodes of the CCD array must be shielded to prevent saturation by the incident focused laser beam. The simulation indicates that inversion may be performed if the number of shielded diodes is less than 50, which corresponds to diffraction angles smaller than about 3×10^{-3} radian (see Fig. 4 for a result obtained when 40 diodes are assumed to be masked). The solution deteriorates when the number of shielded diodes becomes greater than 50. However, experimentation shows that it is possible to operate with about 30 shielded diodes (1.6×10^{-3} radian).

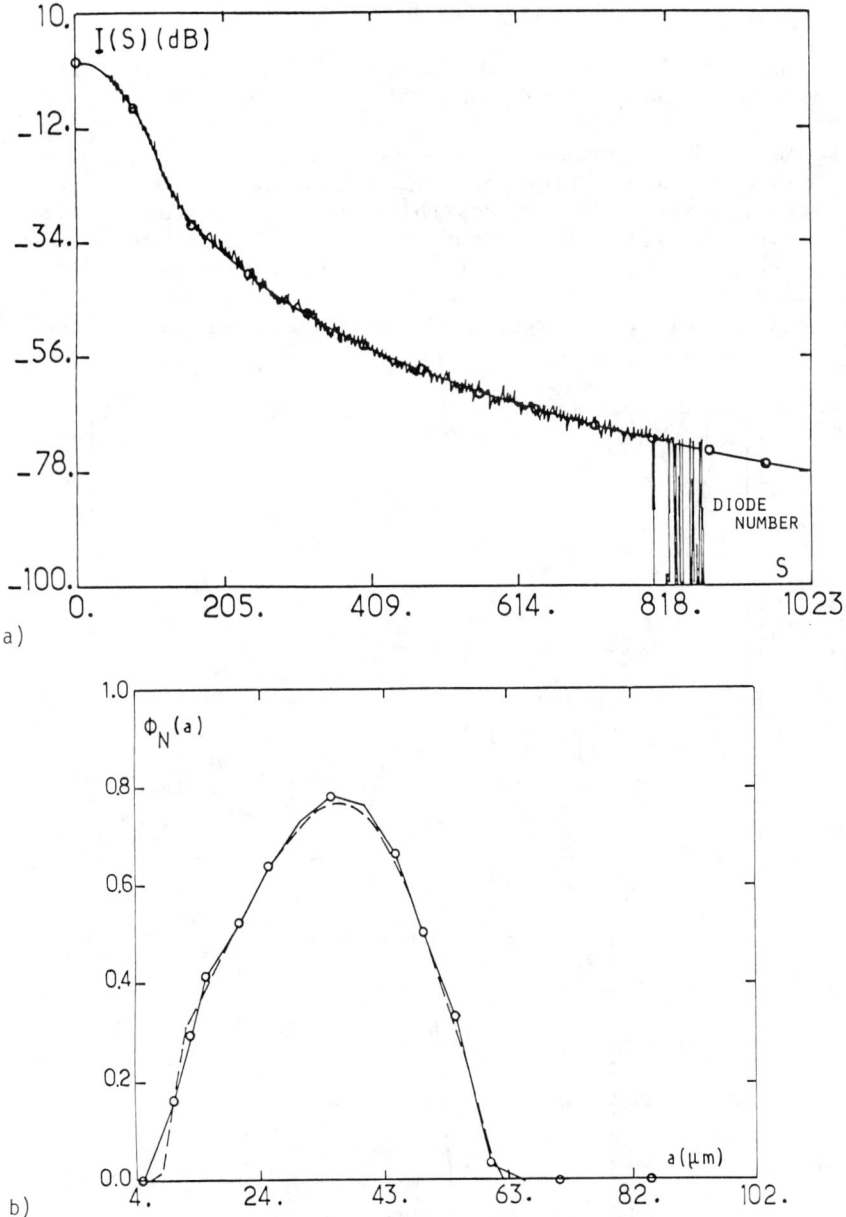

Fig. 4a) Diffracted light intensity profile corresponding to the simulated size distribution of Fig. 2a. The random error has a maximum value of 20% of the local intensity. The dynamic range is limited to 4000 and the first 40 diodes are assumed hidden. b) Result obtained by inverting the simulated profile of Fig. 4a with a non-negative least-square algorithm on a 5-μm basis of radii.

Experimental Results

The numerical simulation of the previous section provides the necessary conditions for a proper inversion of the light diffraction profile. Some experimental results obtained with the system of Fig.1 are now described.

A first validation of the measurement system and signal processing is now possible. As a demonstation our method is used to retrieve the radius of a calibrated pinhole. Fig. 5a shows the diffracted light intensity profile measured when a pinhole of radius a = 50 µm is inserted in the test section. The result obtained on inversion of the

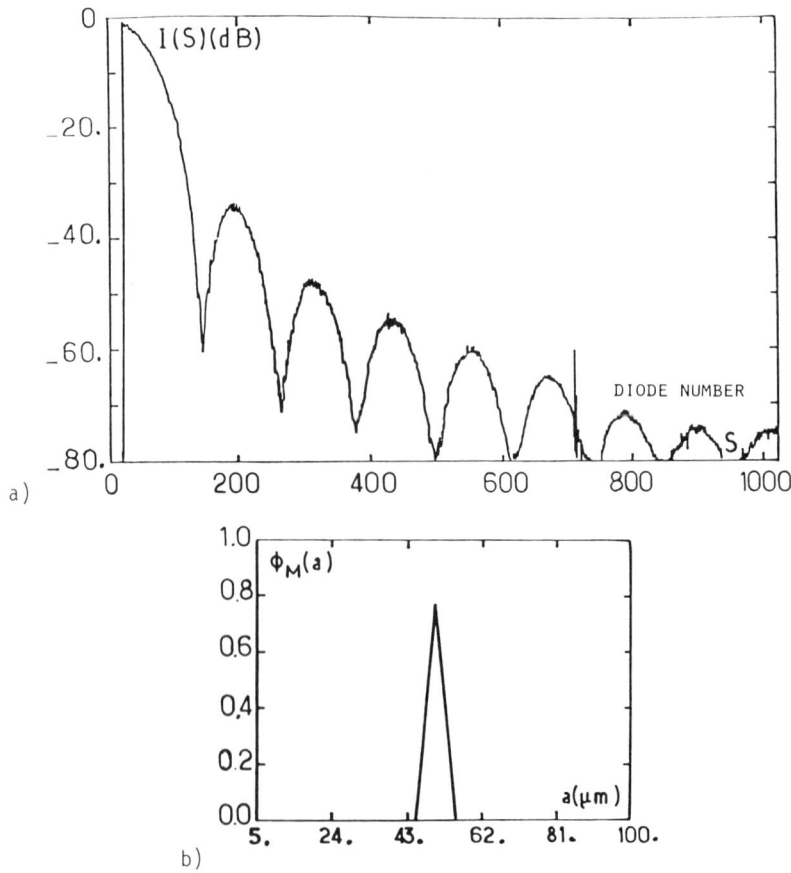

Fig. 5a) Diffracted light intensity profile for a 50-µm pinhole. b) Result obtained by inverting the profile of Fig. 5a on a 5-µm basis of radii.

DROPLET SIZE DISTRIBUTIONS   745

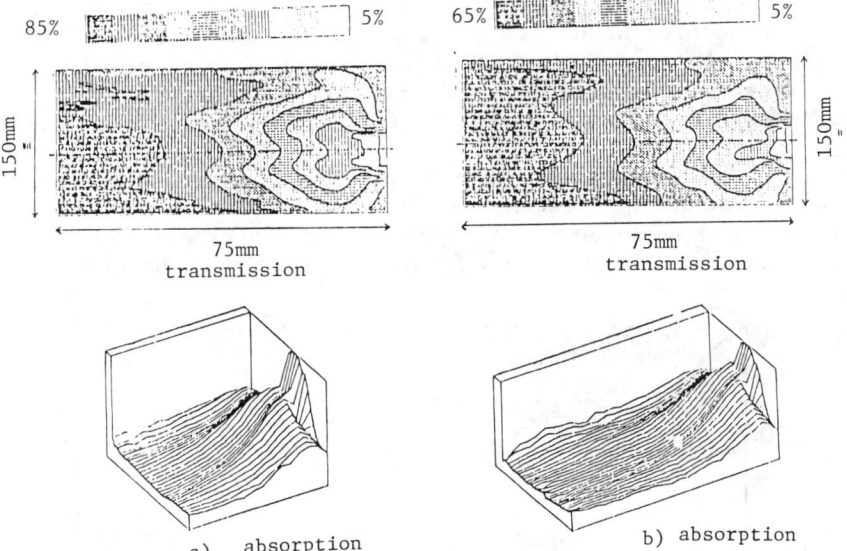

Fig. 6 Transmission and absorption factors of the droplet spray for a mass flow rate of air of 25 g/s. a) $\dot{m}_{WATER}$ = 20 kg/h ; (b) $\dot{m}_{WATER}$ = 40 kg/h.

intensity pattern (Fig. 5a) yields the correct value of the radius (see Fig. 5b).

A first step in the characterization of a droplet spray is the determination of the transmission factor. A transmission map is obtained by a computer controlled displacement of the test ensemble and measurement of the transmitted light by the SGD-100 A diode under uniform illumination of the spray. Two examples of transmission and absorption factors for different liquid spray mass flow rates are given in Fig. 6. The absorption factor is seen to be important in a cone located near the atomizer where the liquid sheet breaks down into droplets. The cone angle is quite open, and the transmission factor becomes uniform at a short distance from the injection plane. The rotational symmetry of the jet is also evident. For the smallest mass flow rate of water, the transmission factor becomes sufficient at about 2 cm from the injector. For the highest mass flow rate of water, the light diffraction method may be used beyond 4 cm.

Correct measurements require a number of additional precautions and operations : 1) A good alignment of the optical setup must be obtained and, in particular, the incidence of the laser beam on the CCD array should be precisely adjusted. 2) The photodiode array must be calibrated to correct heterogeneous responses of individual diodes. This calibration is performed under uniform illumination and for all sweeping frequencies of the detection array. 3) Optical noise associated with parasitic reflections must be measured and stored in computer memory. 4) The first diodes of the CCD array must be hidden to prevent saturation. 5) The range of scanning frequencies and the integration time must be selected to ensure a good statistical averaging of the light diffraction profiles. 6) The raw data must be correc-

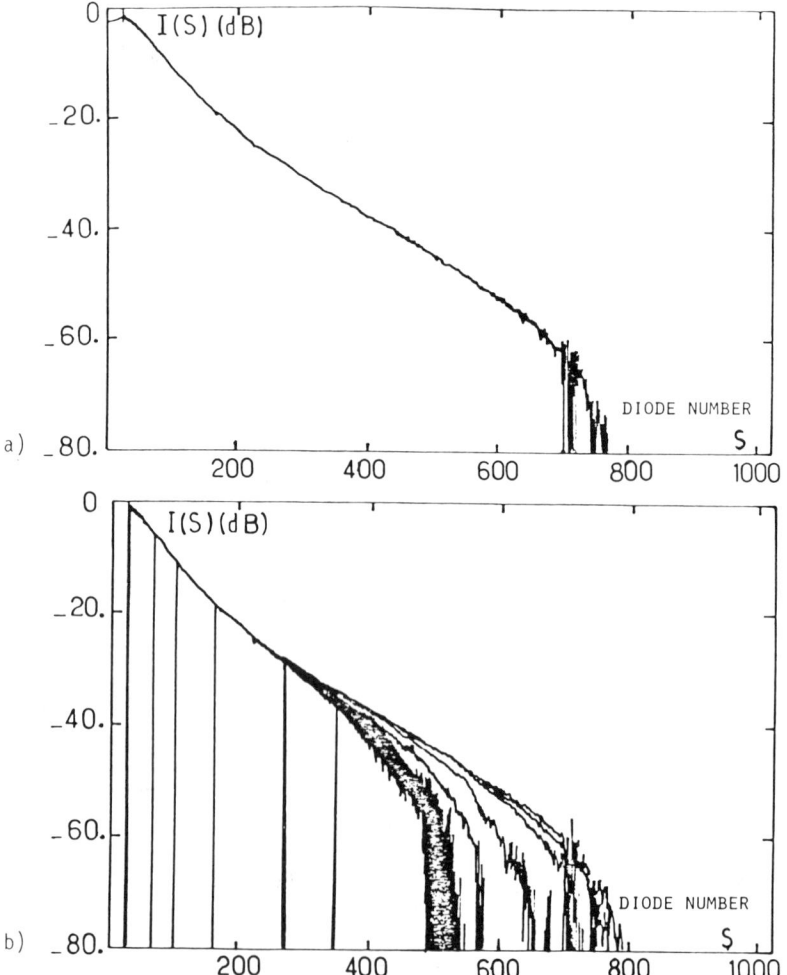

Fig. 7 Diffracted light intensity profile measured on the injector axis at 2.5 cm of the exhaust plane $\dot{m}_{AIR}$ = 25 g/s, $\dot{m}_{WATER}$ = 20 kg/h. a) Final profile ; b) Measurements used to construct the final light intensity profile.

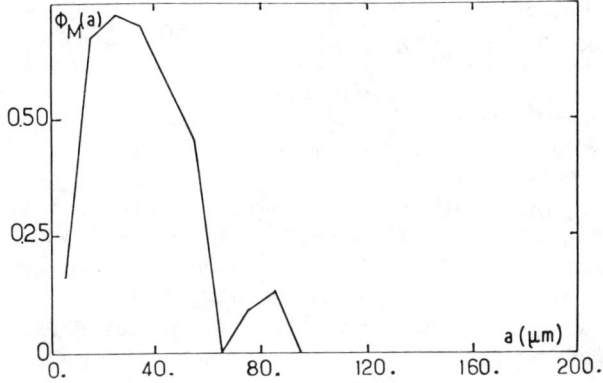

Fig. 8 Result obtained by inverting the measurement of Fig. 7a on a 10-μm basis of radii after preconditioning of the inversion matrix.

ted for optical noise and interaction between adjacent photodiodes.

A typical light intensity profile (Fig.7) is obtained by scanning the diode array with frequencies of 25 to 0.5 kHz. Fig. 7b shows the collection of intensity profiles that compose the final pattern of Fig. 7a. The light intensity diminishes progressively down to -60 dB, below which a rapid change and strong oscillations in the light intensity are observed. In this region, the thermal noise fluctuations and the low-intensity levels measured by the photodiodes become comparable§.

To augment the precision and dynamic range beyond - 60 dB, thermal noise level is reduced by cooling the photodiode array with Peltier elements. Without cooling, the dynamic range remains inferior to our initial expectations. Inversion remains possible for a basis of 20 droplet radii extending from 5 to 200 μm with a step Δa = 10 μm. The mass distribution function deduced from the light intensity profile of Fig. 7 is displayed on Fig. 8. The mean droplet radius and the Sauter mean radius are, respectively, a = 7.5 μm and SMR = 24.7 μm. These results were obtained at 2.5 cm

---

§ It has been suggested that a photodiode array consisting of about 100 equally spaced rings would permit measurements of the small diffracted intensities with larger detection areas and thereby might be a solution to the dynamic range problem. While this is an attractive idea, unfortunately such arrays are not currently available. Furthermore their thermal noise would be higher as it increases with the detection area.

from the injection plane on the axis and for mass flow rates $\dot{m}_{AIR}$ = 25 g/s, $\dot{m}_{WATER}$ = 20 kg/h. The distribution is typical of the aerodynamic injector operating at low mass flow rates of water.

Conclusion

The use of light diffraction for particle sizing poses difficult problems which may now be solved with modern mathematical methods and the recently developed CCD photodiode arrays. These allow a direct inversion of the measured data with minimal constraints on the expected size distribution.

Acknowledgment

Financial support for this research provided by the Société Nationale d'Etudes et de Construction de Moteurs d'Aviation (SNECMA) is gratefully acknowledged.

References

Azzopardi, B. J. (1979) Measurement of drop sizes. Int. J. Heat and Mass Trans. 22, No. 9, 1245.

Chin, J. H., Slieocevich, C. M., and Tribus, M. (1955) Determination of particle size distributions in polydisperse systems by means of measurements of angular variation of intensity of forward scattered light at very small angles, J. Phys. Chem. 5, 841.

Cornillaut, J. (1972) Particle size analyzer. Appl. Opt. 11(2),265.

Dobbins, R. A., Crocco, L., and Glassman, I. (1963) Measurment of mean particle sizes of sprays from diffractively scattered light. AIAA J. (8),1882.

Felton, P. G. (1979) Measurement of particle/droplet size distributions by a laser diffraction technique. Paper presented at 2nd European Symposium on Particle Characterization, PARTEC, Nurenberg.

Goodman, J. W. (1968) Introduction to Fourier Optics. McGraw-Hill, New York.

Negus, C. and Azzopardi, B. J. (1978) The Malvern particle size distribution analyzer: its accuracy and limitations report. Report AERE-R-9075, U.K. Atomic Energy Authority, Harwell. Also Malvern Ltd. document.

McSweeny, A. and Rivers, N. (1972) Optical fiber array for measuring radial distributions of light intensity for particle size analysis. Appl. Opt. 11(9), 1210.

Swithenbank, J., Beer, J. M., Tayor, D. S., Abbot, D., and McCreath, C. G. (1977) A laser diagostic technique for the measurement of droplet and particle size distribution. Experimental Diagnostics in Gas Phase Combustion Systems: AIAA Progress in Astronatuics and Aeronautics (edited by B. T. Zinn), Vol. 53. AIAA, New York.

Swithenbank, J. M., Prior, D. S., and Felton, P. G. (1978) Stirred reactor modeling of a low pollution liquid fueled combustor. Turbulent Combustion: AIAA Progress in Astronatuics and Aeronautics (edited by M. Summerfield), Vol. 58. AIAA, New York.

Tardieu, A. (1983) Dévelopment d'une méthode optique de diffraction pour la détermination des tailles de gouttelettes dans un jet pulvérisé. Thèse se Docteur-Ingénieur, Ecole Centrale des Arts et Manufactures, Châtenay-Malabry, France.

Twomey, S. (1977) Introduction to the Mathematics of Inversion on Remote Sensing and Indirect Measurements. Elsevier, New York.

Wang, T. I., Lerfald, G. M., and Derr, V. E. (1981) Simple inversion technique to obtain cloud droplet size parameters using solar aureole data. Appl. Opt. 20(9), 1511.

Measurement of NO in Methane-Air Flames by Tunable Atomic Line Molecular Spectroscopy

E. Cuellar* and N. J. Brown†
University of California, Berkeley, California

Abstract

An <u>in situ</u> technique called tunable atomic line molecular spectroscopy (TALMS) has been applied to the measurement of NO in the postflame region of atmospheric methane-air flames. The TALMS technique utilizes the Zeeman effect and is based on the splitting and polarization of atomic emission lines induced by an external magnetic field. Concentrations of NO were determined by doping the fuel and oxidizer with known amounts of NO. The highly selective and sensitive TALMS technique permitted measurement of NO in concentrations ranging from a few ppm to those typical of high nitrogen content fuels.

Introduction

Characterization of nitrogenous emissions from combustion sources and determination of the atmospheric fate of nitrogen compounds are crucial for an assessment of air quality. Generic sources of nitrogen oxides are classified according to how emissions are generated and released to the environment. On a global basis 90% of the emissions of oxides of nitrogen are attributable to natural sources, and these are emissions from the nitrogen cycle and lightning. The most preponderate compound

---

Presented at the 9th ICODERS, Poitiers, France, July 3-8, 1983. Copyright 1984 by the American Institute of Aeronautics and Astronautics, Inc. All rights reserved.

* Senior Staff Member, Present address: Raychem Corporation, Research and Development Laboratory, Menlo Park, California.

† Senior Scientist, Applied Science Division, Lawrence Berkeley Laboratory.

associated with the natural source emissions is N_2O. The second source type of oxides of nitrogen is anthropogenic, and the major contribution is due to combustion with mobile and stationary sources contributing equally. In urban environments the distribution between natural and anthropogenic sources is reversed with anthropogenic (and mainly combustion) sources responsible for 90% of the emissions of oxides of nitrogen, mainly as nitric oxide.

Two mechanistic paths are responsible for nitric oxide production during combustion, one involves the oxidation of atmospheric nitrogen (thermal and prompt NO) and the other results from NO production from the oxidation of nitrogen chemically bound to fuel molecules (fuel NO). In order to understand mechanistic details of NO formation and destruction mechanisms, it is necessary to have reliable methods for NO quantification in combustion mixtures.

The most frequently utilized technique for NO quantification in combustion environments is chemiluminescent analysis following sample extraction with a probe. There are intrinsic errors associated with such a technique. Since the method is intrusive, it is important that the disturbance to the flowfield and temperature field in the vicinity of the microprobe be small to prevent sampling biases. Additional sources of error are homogeneous and heterogeneous reactions in the sampling system which alter the concentration of NO and NO_2. Furthermore, there are some potential errors associated with the analyzer, namely those attributable to quenching effects which have been described by Matthews et al.(1977) and those associated with changes in sample viscosity with sample composition as described by Zabielski et al. (1984).

Nonintrusive optical techniques for NO quantification are highly desirable since they eliminate the potential errors associated with chemiluminescent analysis. Optical methods are not without problems of their own; for instance, problems associated with high-temperature calibration require a great deal of ingenuity to solve. A uniform temperature field is required along a line-of-site absorption measurement which is difficult to achieve for most burner systems. For some optical techniques, (e.g., laser fluorescence) proper accounting for deactivation of the excited state by collisions with chaperone gases is necessary yet difficult to quantify experimentally.

In this paper we report on the application of the nonintrusive technique of tunable atomic line molecular spectroscopy TALMS to the measurement of NO in the postcombustion environment of atmospheric pressure, premixed methane-air stoichiometric flames. Although the TALMS technique is relatively new for combustion applications, utilization of the Zeeman effect to achieve differential absorption measurements has been available for some time. The magnetic scanning of a single Zeeman component of an atomic emission line was first used by Bitter and co-workers (1953; 1954) for investigating the hyperfine structure and isotope shift of the resonance radiation of mercury. This technique was utilized by Hadeishi and McLaughlin (1971) to develop a new type of atomic absorption spectrometer to detect trace mercury, and utilizing the Zeeman effect for background correction. The technique has been extended to the detection of small molecules which exhibit sharp rotational structure (Koizumi et al. 1979a), and has been used to detect NO, NO_2, SO_2, and HCHO (Koizumi et al. 1979b). Cuellar and Brown (1981; 1982) have recently used the technique to detect S_2 in the presence of SO_2 and to resolve several closely spaced lines belonging to different branches which result from the triplet splitting in the (7,2) band of the $B^3\Sigma_u^- - X^3\Sigma_g^-$ system of S_2.

Experimental

Flat Flamer Burner

A schematic diagram of the experimental system is shown in Fig. 1. Methane-air flames were stabilized at atmospheric pressure on a water cooled porous plug flat flame burner. The burner was constructed from a 6.0-cm-diam sintered bronze disk with a porosity of 100 um. Water cooling was provided by a copper coil imbedded in the sintered disk with the plane of the coil parallel to the burner surface. The burner was mounted below the optical path of the tunable atomic line molecular spectrometer (see Fig. 1) through an opening in the optical bench. A micrometer driven x-y-z translation stage permitted motion of the burner relative to the fixed focused optical beam. Stable CH_4/air flat flames were obtained over a wide range of equivalence ratios and total gas flows. We elected to operate the burner at a total flow Q_T = 30,000 standard cubic centimeters/minute (sccm), corresponding to a linear velocity of 17.7 cm/s. At this flow, stable flames could be obtained at equivalence ratios ϕ ranging from 0.75 to 1.5.

Methane (Matheson C.P. grade, 99.0% minimum purity) and air were supplied in high-pressure gas cylinders.

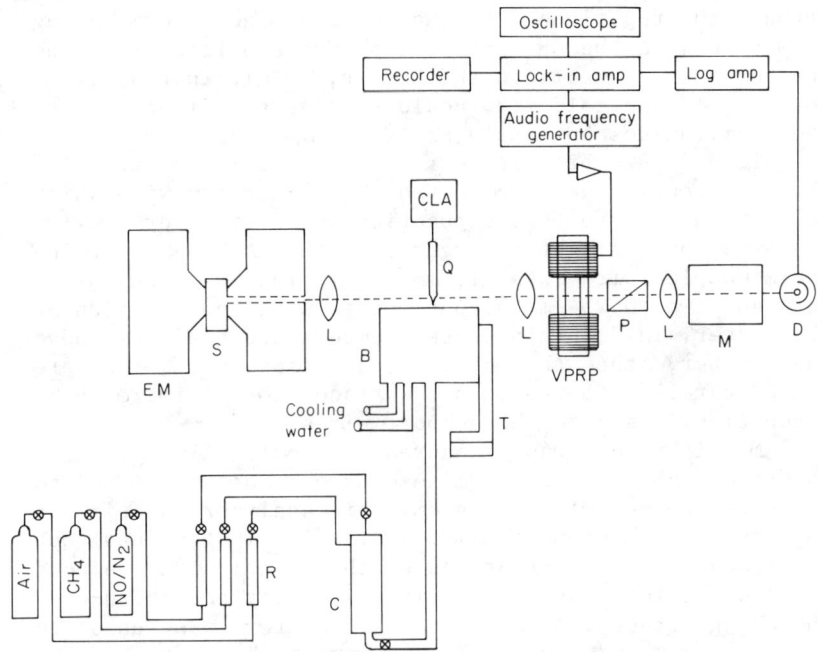

Fig. 1 Schematic diagram of the experimental apparatus: EM electromagnet; S atomic light source; L quartz lens; B porous plug burner; T x-y-z translation stage; Q uncooled quartz probe; CLA chemiluminescent gas analyzer; VPRP variable phase retardation plate; M monochromator; D photomultiplier tube; C mixing chamber; R rotameters.

Calibration gases consisting of various concentrations of NO in N_2 were obtained from AIRCO (97 ppm, 511 ppm, and 4.41% NO in N_2). The calibration gases were checked against NBS standards by the Air and Industrial Hygiene Laboratory of the State of California Department of Health. Gas flows were metered separately using rotameters (Matheson, 600 Series) which were calibrated using a wet test meter. These calibrations were checked against calibrations obtained using a rotating vane dry test meter (Singer American Meter Division, Model DTM-325) and Brooks Vol-U-Meter gas calibrators.

A cylindrical stainless steel mixing chamber 30-cm long and 4.8-cm i.d. was packed with glass beads to provide adequate mixing of the gases prior to flowing into the flat flame burner. Methane and air were introduced near the top of the chamber through two 3/8-in. ports placed opposite to each other and perpendicular to the axis of the mixing chamber. The NO/N_2 dopant was introduced via a 1/8-in. stainless steel tubing penetrating through the top flange of the chamber and terminating midway between the CH_4/air ports. Gases exiting from the chamber then flowed into the burner, or alternatively, a portion of the cold flow could be diverted directly into the chemiluminescent analyzer.

Temperature profiles were obtained using a Pt/Pt-13% Rh thermocouple constructed from 0.076-mm-diam wire which was butt welded to form a junction whose diameter (0.102 mm) was only slightly larger than that of the wire. The thermocouple wire was 12-mm long and was supported between two 0.254-mm support wires with the junction at the center. In determining the temperature profiles above the burner, the thermocouple junction lead wires were placed parallel to the burner surface to minimize heat conduction losses to the support wires.

Nitric oxide was measured in cold flow with a laboratory built chemiluminescent gas analyzer (CLA) to evaluate the gas mixing system. This analyzer is similar in design to commercially available instruments, but incorporated several features which significantly reduced the corrections associated with viscosity and third-body quenching effects. A calibrated rotameter was used to maintain a constant and known flow rate of oxygen plus ozone from the ozone generator. The sample flow rate was measured using a mass flow meter (Tylan Corporation), and the ratio of ($O_2 + O_3$) to the sample was maintained at 0.94. The CLA was operated at a reaction chamber pressure of 3 Torr. By adjusting the flow of the sample stream such that 6% of the gas in the reaction chamber was due to probed sample while 94% was due to the ($O_2 + O_3$) stream, the effects of third-body quenching can be significantly reduced. A commercially built stainless steel NO_x to NO converter (Thermo Electron Corporation) was interfaced with the instrument. Light emitted by the chemiluminescent reaction of NO with O_3 was detected by a photomultiplier tube (EMI 9558A) filtered through a CS2-63 Corning filter to prevent the transmission of light 6000 A. The output of the PMT was measured on a Keithley electrometer and displayed on a strip chart recorder.

The chemiluminescent gas analyzer (CLA) and the gas mixing system were tested by introducing known amounts of the 4.41% NO in N_2 into a N_2 stream (metered through calibrated rotameters). Part of this flow was sampled by the CLA and quantified using the 97- and 511-ppm NO/N_2 calibration gases. The response curve of the CLA shown in Fig. 2 is linear in NO concentration to at least 2400-ppm NO. No significant differences were found between the NO determined with the rotameters and the mixing system and with the CLA since the two measurements agreed to within 5%.

TALM Spectrometer

The measurement of NO in the postflame region of a stoichiometric CH_4/air flat flame was accomplished using tunable atomic line molecular spectroscopy (TALMS). This technique has been pioneered by Koizumi and his co-workers (1979a; 1979b), and has been utilized to determine low concentrations of NO in a room-temperature absorption cell with high selectivity and sensitivity (Koizumi et al. 1980). These authors used the accidental near coincidence between the Cd II emission between the $^2P_{3/2}$ and $^2S_{1/2}$ states yielding a line in emission at 214.438 nm and discrete rotational-vibrational lines in the γ bands of NO. The lower state designation in NO is the ground electronic state $X^2\Pi_{1/2}$, $v''=0$, and $J''=12\ 1/2$, and the upper electronic state is designated by $A^2\Sigma^+$ with $v'=1$ and $J'=13\ 1/2$. The Cd II emission line at 214.438 nm is coincident with an R branch transition in the γ bands of NO. By using a single isotope ^{114}Cd Hadeishi and co-workers reported a detection limit of NO in N_2 of 180 ppb in a 20-cm cell.

The basis for detection of molecules by tunable atomic line molecular spectroscopy is the splitting and polarization of atomic emission lines by an external mag-

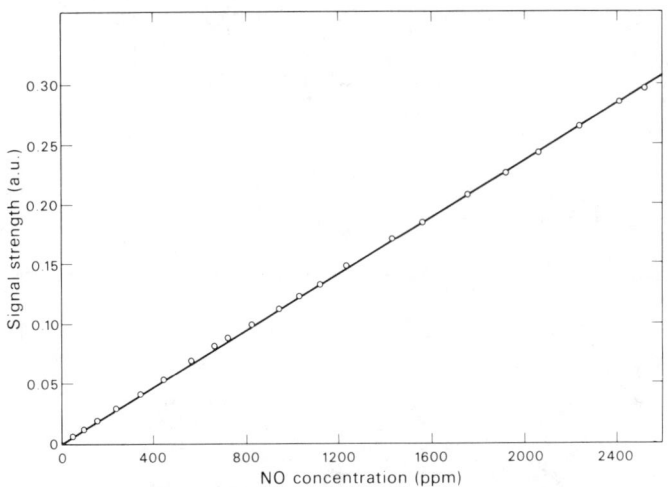

Fig. 2 Response curve for the chemiluminescent gas analyzer in cold flow where mixtures of methane, air, and NO/N_2 are blended in the mixing apparatus.

netic field. In a direction parallel to the magnetic field, the Cd line at 214.4 nm is split into σ+ and σ- circularly polarized components: the high-frequency σ+ component is circularly polarized in a counterclockwise direction, while the low-frequency σ- component is circularly polarized in the opposite direction. By varying the strength of the magnetic field, one of the Zeeman components of the Cd emission line is tuned into exact coincidence with a discrete rotational-vibrational line in the γ band of NO. The matching Zeeman component indicates the extent of absorption by NO, while the unmatched component indicates background absorption only. A differential measurement of the matched and unmatched Zeeman components of the Cd emission line provides a quantitative measurement of the NO in the optical path.

A schematic diagram of the TALM spectrometer is shown in Fig. 1. Except for the light source, the instrument is identical to that described earlier (Cuellar and Brown, 1982). The lamp consists of a sealed-off quartz U-tube containing a small amount of cadmium metal (natural abundance Cd) and an inert buffer gas. Nichrome wire heaters wrapped around the U-tube are used to heat the lamp to a few hundred degrees Celsius and provide a small amount of vaporized Cd ($\approx$ 1 Torr) which is excited by an electric discharge through the gas. The lamp is housed between the poles of a Varian electromagnet, and the light emitted parallel to the magnetic field is focused with a quartz lens to a 2-mm-diam spot above the burner surface. The variable phase retardation plate in combination with the linear polarizer allows for the

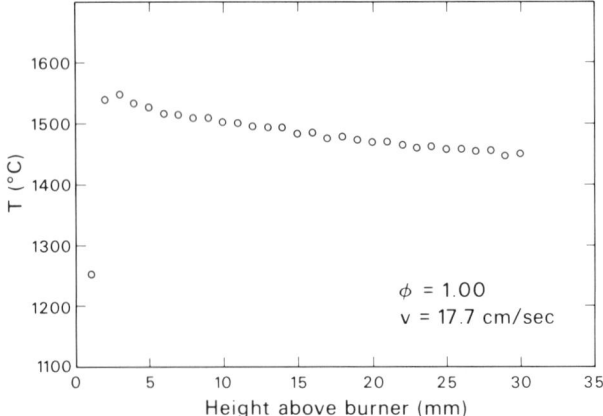

Fig. 3 Vertical temperature profile of the CH_4/air flat flame measured at the center of the burner.

alternate transmission of the matched and unmatched Zeeman components through the monochromater to the detector. The output of the photomultiplier tube is processed electronically and displayed on a strip chart recorder.

Results

We elected to measure NO in the postflame region of a stoichiometric methane-air flame with the NO concentration determined by a constant amount of NO produced by the thermal mechanism and an amount determined by doping the flame with the 4.41% NO in N_2 mixture. The mixing system was proven to be adequate by checking its reliability in cold flow with the CLA. It was also important to verify that we were measuring in the region of the flame where thermal gradients are small and where the thermal NO is nearly constant. Vertical and radial temperature profiles are shown in Figs. 3 and 4. The postflame region begins at approximately 3.5 mm above the burner and is most certainly characterized by small gradients as seen from examination of Fig. 3. The temperature variation along a horizontal profile is approximately 150 deg. For certain types of transitions this could be a significant problem; however, the Boltzmann distribution for the J states of interest does not change appreciably over this temperature interval. Temperatures reported here are not corrected for radiative losses which we estimate to be on the order of 50 K in the postflame region. Radial profiles of NO concentration in the undoped flame are shown in Fig. 5. These profiles

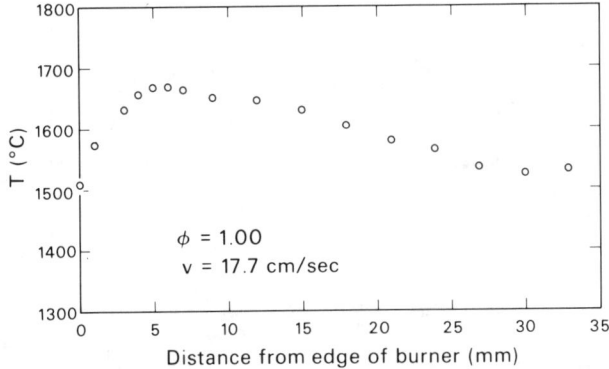

Fig. 4 Horizontal temperature profile of the CH_4/air flat flame measured at a height of 5.0 mm above the burner surface.

were determined by sample extraction with an uncooled quartz sampling probe and analysis by the CLA. No quenching corrections were made which we estimate to increase the concentrations by approximately 4%. In addition, the concentrations were increased by approximately 6 ppm when measurements were made in the NOx mode of the analyzer. As seen from the figure, the horizonal profiles are nearly flat and at distances above the burner of 5 and 15 mm differ only by a few ppm.

To determine the conditions of maximum sensitivity, the differential absorption between $\sigma+$ and $\sigma-$ components was measured as a function of magnetic field strength and is shown in Fig. 6. Maximum sensitivity was obtained at 11 kG, and all measurements reported here were taken at this field strength. The dependence of the differential absorption signal due to NO on magnetic field strength shown here is considerably broader than that reported in Koizumi et al. (1980). This is because we used natural abundance Cd rather than the single isotope ^{114}Cd, resulting in a broadening of the emission line by the hyperfine structure of the odd isotopes of naturally occurring cadmium.

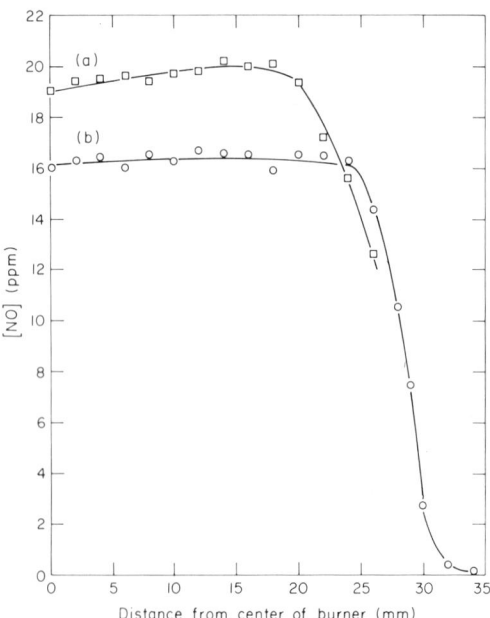

Fig. 5 Radial profile of NO concentrations measured with an uncooled quartz sampling probe. a) 15.0 mm above the burner surface; b) 5.0 mm above the burner surface.

The response of the TALM spectrometer to increasing NO concentration is shown in Fig. 7. In these experiments the CH_4 and air were mixed in the gas mixing chamber and varying amounts of 4.41% NO/N_2 gas were added. A portion of the gas stream was passed through a 7.5-cm-long flow cell placed in the optical path of the spectrometer, while the remainder of the flow was exhausted to a hood. The observed signal was linear in NO concentration to about 1200-ppm NO. At 2000-ppm NO, the percent transmission through the absorption cell was about 28% and the observed signal was 10% low with respect to the extrapolated linear response. This deviation from linearity at high concentrations will not be a problem in combustion applications where a density decrease of approximately a factor of 5 will obtain. In other words, measurement of the number of molecules/cm^3 associated with 1200-ppm NO in cold flow will correspond to the measurement of 6000 ppm of NO under combustion conditions along the same optical path. Concentrations of NO in fuels of high nitrogen content are well within the linear range.

Optical measurements of NO in the postflame region of the stoichiometric methane-air flame using tunable atomic line molecular spectroscopy were obtained by doping the fuel and oxidizer mixture with known amounts of NO. The cadmium atomic emission line was focused to a 2-mm-diam spot 5.0 mm above the burner surface with the optical axis along the centerline of the burner. A typical recorder trace of the differential absorption signal

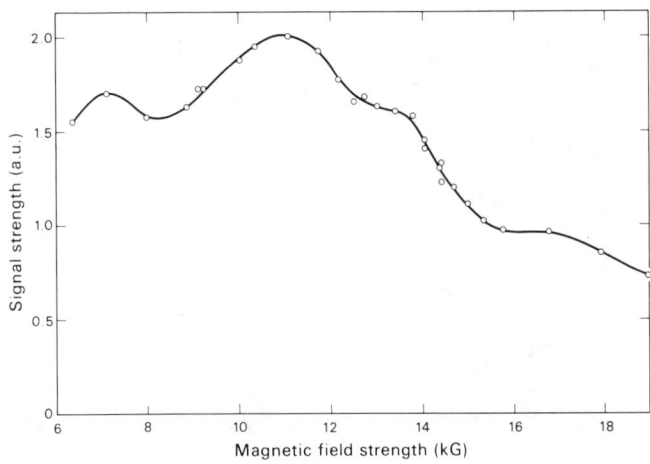

Fig. 6 Differential absorption signal due to NO as a function of magnetic field strength.

observed by thermal NO and added NO is shown in Fig. 8. The chart drive of the recorder was stopped at the point when NO was added, and restarted once the signal stabilized (approximately 1 min later). The maximum differential absorption signal corresponds to 256 ppm added NO plus thermal NO. The decay of this signal corresponds to halting the added NO, and turning the burner off corresponds to the loss of the thermal NO signal. The sensitivity of this technique is such that concentrations of NO corresponding to thermal NO levels can be detected.

A range of NO concentrations from 50 to 1000 ppm was added to the burner and a plot of the observed signal strength as a function of added NO is shown in Fig. 9. A least-square fit of the data points results in the equation:

$$S = 3.07 \times 10^{-4} [NO]_{added} + 1.87 \times 10^{-2} \quad (1)$$

The intercept of the graph represents the amount of thermal NO present.

A series of calibration experiments was attempted using heated quartz cells. These cells were 14.3-cm long and were filled with 152 Torr of 97-ppm NO in N_2 calibration gas. Upon heating in a Lindbergh furnace to

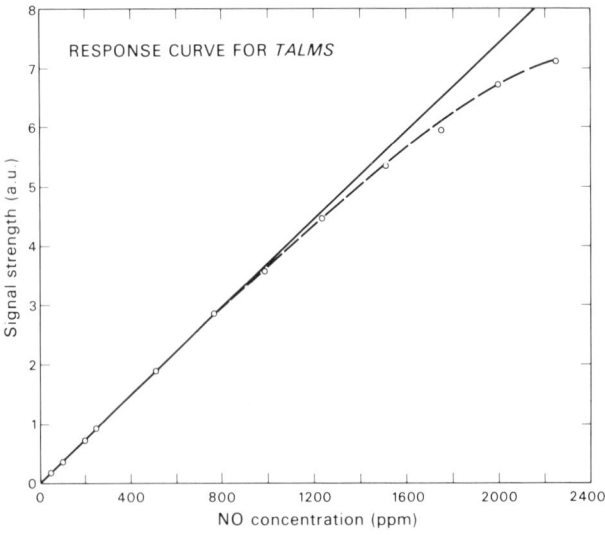

Fig. 7 TALMS signal as a function of increasing concentration of NO. The solid line represents the linear response.

1000°C, the TALMS signal due to NO disappeared and was not recovered by cooling the cell to room temperature.

Discussion

It is important to review briefly other research concerned with the quantification of NO in combustion environments in order to assess the advantages and disadvantages of the TALMS approach. Often NO is measured

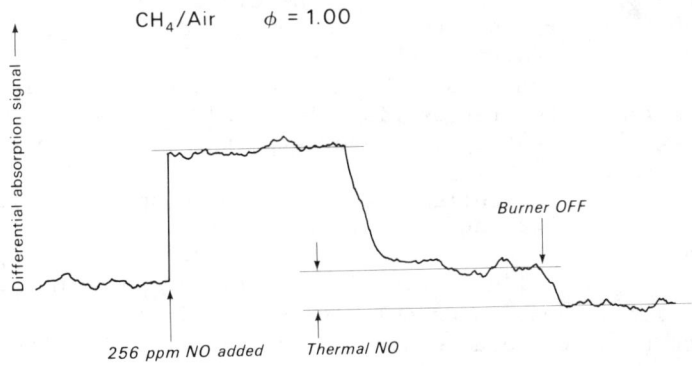

Fig. 8 Strip chart recorder trace showing the differential absorption signal obtained by TALMS.

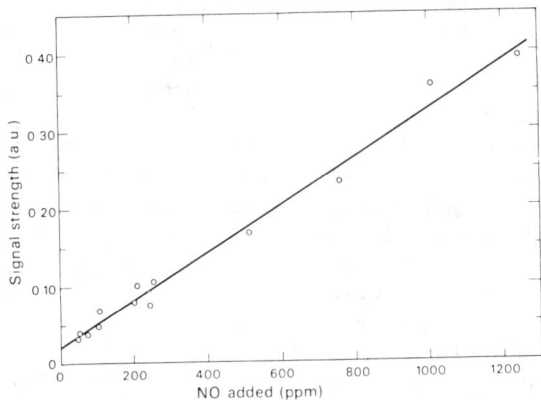

Fig. 9 Signal strength measured by TALMS vs NO added for a stoichiometric CH_4/air flame ($\phi = 1.00$).

with a CLA following sample extraction with quartz microprobes. Bowman (1977) has reviewed probe measurements in flames for a variety of probe types and combustion conditions. He indicates that the major fluid mechanical disturbance to the subsonic flowfield by the probe is streamline distortion, which in turn, results in a perturbation of concentration gradients in the vicinity of the probe. A potentially more serious problem discussed in his review regards composition changes due to inadequate quenching of homogeneous and/or heterogeneous reactions in the probe. Cernansky (1977) has reviewed sampling and measurement of NO and NO_2 in combustion systems, and, in particular, has discussed the effects of probe materials, the chemiluminescent analyzer, and combustion conditions which enhance the concentration of NO_2. The role of uncooled quartz probes in altering measured NO/NO_2 ratios is discussed, as well as the behavior of catalytic converters in effecting NO_2 to NO conversions.

Meinel and colleagues[8] report on comparative studies on NO measurement in combustion systems in a series of papers. A new nondispersive uv analyzer was described which had a limit of detectability of approximately 0.5 ppm for an absorption cell of 39 cm. This is not an in situ technique and requires probe sampling. A study comparing the measurement of NO in propane-air flames and hydrogen-air flames using an in situ differential absorption technique and a nondispersive uv technique of a sample extracted with a cooled quartz microprobe was described in the last paper (Meinel and Krauss 1978) of the series. Good agreement was obtained for the two types of NO measurements in lean flames, but for stoichiometric and rich flames the in situ measurements were between 20 and 30% lower than the probe measurements. The authors speculate that NO may have been produced within the probe.

Falcone (1981) measured NO in lean and rich atmospheric methane-air flames. The concentration of NO was adjusted through doping. Three types of NO measurements were compared. Two involved sample extraction with uncooled quartz microprobes and one was an in situ method. In a series of experiments, samples were extracted from both lean and rich flames and then passed through an analytical train into a CLA analyzer or into an absorption cell where NO was detected by laser absorption spectroscopy utilizing a tunable diode laser as a

---

[8] See Meinel 1975; Meinel and Just 1976; Meinel and Just 1977; Meinel and Krauss 1978.

light source. After quenching corrections for the CLA were made, agreement between the measurements using the two detectors was very good. In another series of experiments, NO concentrations measured in situ with laser absorption and by CLA following sample extraction were compared for both lean and rich flames. Although discrepancies among various data reduction procedures for in situ measurements prevail throughout the experiments, no discrepancy greater than 20% was claimed for the measurements in lean flames. No destruction of NO was detected in the lean flames, within experimental error. Although significant scatter was observed in the in situ measurements, it appeared that a portion of the NO was destroyed in the rich flames. Concentrations of NO measured in the rich flames by laser absorption and probe sampling/CLA agreed to within 20%. It is important to note that the lower level of detection of NO for the in situ technique was 200 ppm.

One of the most extensive studies of optical and probe CLA measurements of NO in combustion systems has been described in a series of papers of Zabielski et al. (1981; 1984) and reports by Dodge et al. (1979), Colket et al. (1979), and Zabielski et al. (1980). Many of the measurements were performed on NO seeded, laminar atmospheric CH_4-O_2-N_2 flames. Probes were water cooled quartz or stainless steel, and the spectroscopic absorption measurements were accomplished using a hollow cathode resonant lamp or a high-pressure Xe lamp as a light source. The optical and probe measurements agreed to within 25%.

Although there is disagreement regarding whether optical or probe measurements of NO in doped CH_4/air laminar flames are higher, there is general agreement that discrepancies between them are less than 30%. Most of these measurements required relatively high concentrations of added NO due to the lack of sensitivity of the various optical techniques used.

The TALMS technique employed here offers greater sensitivity than previous in situ optical measurements of NO in combustion environments. We can easily detect NO concentrations on the order of a few ppm along the optical path of a 6.0-cm-diam burner operating at atmospheric pressure. This is a most attractive technique to explore discrepancies between optical and probe measurements. The relationship between signal intensity obtained with the TALM spectrometer and NO concentration in the flame has not been established. We used the method of differences to record signal changes resulting from the addition of fixed concentrations of NO to the flame. The signal response was linear with NO concentration indicating either complete survival through the flame or a constant loss mechanism which is first-order in NO concentration.

The TALMS technique offers another real advantage over some of the newer laser techniques. The spectrometer is relatively inexpensive and is extremely versatile. By changing the lamp, one can detect a variety of other molecules important in combustion environments. Research is currently in progress in our laboratory for the detection of SO_2 and NO_2. Moreover, the extensive quenching corrections required for some of the more popular laser fluorescence techniques are not required for TALMS.

We plan to perform a series of calibration experiments in the future to reliably relate signal strength to NO concentration in the flame. We shall calibrate in very lean flames after determining flame conditions where there is no prevalent loss mechanism for NO.

Conclusions

A study of nitric oxide quantification in the postcombustion environment of an atmospheric premixed CH_4/air stoichiometric one-dimensional flame has been conducted. The in situ optical technique employed is called tunable atomic line molecular spectroscopy TALMS and is based on the splitting and polarization of atomic emission lines induced by an external magnetic field. The method of differences was used whereby NO concentrations were fixed by doping the fuel and oxidizer mixture prior to combustion with known concentrations of NO. Absolute calibration was not accomplished in the study and remains to be accomplished. The signal response was linear with NO concentration indicating either complete survival through the flame or a constant loss mechanism which is first-order in NO concentration. The highly selective and sensitive TALMS technique permitted measurement of NO in concentrations ranging from a few ppm to those typical of high nitrogen content fuels.

Acknowledgments

We thank Drs. Tetsuo Hadeishi and Donald Lucas for many helpful discussions during the course of this work. We also wish to thank Dr. Robert Schefer for his participation in the early stages of this work. The light source was designed and built by Dr. Tetsuo Hadeishi of

Lawrence Berkeley Laboratory. We are especially grateful to him for supplying us with this lamp. This research was supported by the Director's Office of Energy Research, Office of Health and Environmental Research, Physical and Technological Research of the U.S. Department of Energy under Contract No. DE-AC03-76SF00098.

References

Bitter, F., Plotkin, H., Richter, B., Teviotdale, A., and Young, J.E.R. (1953) A "magnetic scanning"method for investigating hyperfine structure and isotope shift. Phys. Rev. 91, 421; Bitter, F., Davis, S.P., Richter, B., and Young, J.E.R.(1954) Optical sutdies of radioactive mercury. Phys. Rev. 96, pp. 1531-1539.

Bowman, C.T. (1977) Probe measurements in flames. Experimental Diagnostics in Gas Phase Combustion Systems:AIAA Progress in Astronautics and Aeronautics (edited by B.T. Zinn) Vol. 53, pp. 3-24. AIAA, New York.

Cernansky, N.P. (1977) Sampling and measuring for NO and NO_2 in combustion systems. Experimental Diagnostics in Gas Phase Combustion: AIAA Progress in Astronautics and Aeronautics (edited by B. T. Zinn) Vol. 53, pp. 83-102. AIAA, New York.

Colket, M.B. III, Zabielski, M.F., Chiappetta, L.J., Dodge, L.G., Guile, R.N., and Serry, D.J. (1979) Nitric oxide measurement study: probe methods Task II report. United Technologies Research Center Report R79-994150-2.

Cuellar, E. and Brown, N.J. (1981) Combustion diagnostics by tunable atomic line molecular spectroscopy. First Specialists Meeting (International) of the Combustion Institute, 2, pp. 545-550.

Cuellar, E. and Brown, N.J. (1982) Detection of S_2 by magnetic tuning of a chromium (I) atomic emission line. J. Phys. Chem. 86, pp. 1966-1969.

Dodge, L.G., Colket, M.B. III, Zabielski, M.F., Dusek, J., and Serry, D.J. (1979) Nitric oxide measurement study: optical calibration task I report. United Technologies Research Center Report R79-994150-1.

Falcone, P.K. Absorption Spectroscopy of Combustion Gases Using a a Tunable Diode Laser, PH.D Thesis, HTGL Report No. 121, Stanford University, 1981.

Hadeishi, T. and McLaughlin, R.D. (1971) Hyperfine Zeeman effect atomic absorption spectrometer for mercury. Science 174, pp. 404-407.

Koizumi, H., Hadeishi, T., and McLaughlin, R.D. (1979a) Detection of small molecules by magnetically tuned frequency-modulated atomic line source. Appl. Phys. Lett. 34, pp. 382-384.

Koizumi, H., Hadeishi, T., and McLaughlin, R.D. (1979b) A new technique for the determination of isotopic species using Zeeman scanning of an atomic line. Appl. Phys. Lett. 34, pp. 277-279.

Koizumi, H., Hadeishi, T., and McLaughlin, R.D. (1980) Nitric oxide determination by a Zeeman-tuned frequency-modulated atomic line source. Anal. Chem. 52, pp. 500-504.

Matthews, R.D., Sawyer, R.F., Schefer, R.W. (1977) Interferences in chemiluminescent measurement of NO and NO_2 emissions from combustion systems. Env. Science and Tech. 11, pp. 1092-1096.

Meinel, H. (1975) Detection of nitric oxide by the resonance absorption technique. Z. Naturforsch 30A, pp. 323-328.

Meinel, H. and Just, Th. (1976) AIAA 14th Aerospace Sciences Meeting, Measuremtn of NO_x exhaust emissions by a new NDUV analyzer. AIAA paper No. 76-137.

Meinel, H. and Just, Th. (1977) Measurement of NO_x exhaust emissions by a new NDUV analyzer. Experimental Diagnostics in Gas Phase Combustion Systems: AIAA Progress in Astronautics and Aeronautics (edited by B.T. Zinn) Vol. 53, pp. 177-186. AIAA, New York.

Meinel, H. and Krauss, L. (1978) Monitoring of nitric oxide in flames by UV differential resonance absorption. Comb. and Flame 33, pp. 69-77.

Zabielski, M.F., Dodge, L.G., Colket, M.B. III, and Seery, D.J. (1981) The optical and probe measurement of NO: a comparative study. Eighteenth Symposium (International) on Combustion pp. 1591-1598. The Combustion Institute, Pittsburgh, PA.

Zabielski, M.F., Dodge, L.G. Colket, M.B. III, and Seery, D.J. (1980). Nitric oxide measurement study: comparison of optical and probe methods task III report. United Technologies Research Center Report R80-994154-3.

Zabielski, M.F., Seery, D.J., and Dodge, L.G. (1984) The influence of mass transport and quenching on nitric oxide chemiluminescent analysis. Env. Science and Tech. 18, pp. 88-92.

Author Index for Volume 95

Abdalla, A. Y. 356
Andersson, L. L. 164
Baev, V. K. 554
Bahadori, M. Y. 261
Bazhaikin, A. N. 554
Becker, R. J. 672
Ben Aïm, R. I. 367
Bilger, R. W. 293
Boyer, L. 103, 129
Bracco, F. V. 484
Bradley, D. 356
Bray, K. N. C. 305
Bridoux, M. 658
Brown, N. J. 750
Buntinx, G. 658
Buzukov, A. A. 554
Candel, S. M. 736
Cant, R. S. 142
Cattolica, R. J. 714
Chin, S. B. 356
Chou, D. C. 610
Choudhury, P. R. . . . 455
Christenson, B. 164
Clarke, J. F. 142
Clavin, P. 1, 103
Cottereau, M. J. 642
Crunelle-Cras, M. . . . 658
Cuellar, E. 750
Danset, P. T. 672
Dobbs, G. M. 631
Dumas, G. M. L. 367
Dwyer, H. A. 464
Eckbreth, A. C. 631
Elghobashi, S. E. 513
Esposito, E. 736
Fiebig, M. 583
Fisson, F. 433

Frenklach, M. 722
Fritts, M. J. 540
Fyfe, D. E. 540
Gallis, M. R. 672
Gardiner, W. C. 198
Gerstein, M. 455
Girard, A. 433, 443
Givi, P. 384
Gökalp, I. 367
Grase, F. 658
Gülder, O. L. 181
Higuera, F. J. 248
Höglund, A. 164
Hustad, J. 320
Hwang, S. -M. 198
Kennedy, E. M. 261
Kidin, N. 343
Lam, C. 356
Lefebvre, A. H. 563
Leyer, J. C. 433, 443
Librescu, L. 593
Librovich, V. 343
Liew, S. K. 305
Li Kwok Cheong, C. K. . 722
Liñán, A. 248
Ludford, G.S.S. . . . 75, 92
Martinelli, L. 484
Masri, A. R. 293
Michael-Saade, R. . . . 658
Mitra, N. K. 583
Monreal, J. 103
Moss, J. B. 305
Mostafa, A. A. 513
O'Connor, S. J. 421
Ohta, Y. 236
Olsson, J. O. 164
Oran, E. S. 540, 722

Pandey, B. D. 610
Penner, S. S. 261
Perez, J. I. 672
Peters, N. 37, 61, 75
Pitz, W. J. 211
Quinard, J. 129
Ramos, J. I. 384
Reitz, R. D. 484
Rizk, N. K. 563
Roberts, J. 343
Rosengren, L. G. . . . 164
Sabathier, F. 103
Sanders, B. R. 464
Sawerysyn, J. P. . . . 658
Searby, G. 103, 129
Sen, A. K. 92
Sirignano, W. A . . . 384
Sochet, L. -R. 658
Sønju, O. K. 320
Stårner, S. H. 293
Stephenson, D. A. . . 714
Stepowski, D. 642
Stufflebeam, J. H. . . 631
Takahashi, H. 236
Tardieu, A. 736
Tellex, P. A. 631
Timoshenko, B. P. . . 554
Tromans, P. S. 421
Vuillermoz, M. 343
Warnatz, J. . . . 61, 198
Westbrook, C. K. . . 211
White, K. J. 700
Williams, F. A. 37
Yaney, P. P. 672
Ybarra, P. G. 115
Zarrad, N. 367
Zoltani, C. K. 700

PROGRESS IN ASTRONAUTICS AND AERONAUTICS SERIES VOLUMES

VOLUME TITLE/EDITORS

*1. **Solid Propellant Rocket Research** (1960)
Martin Summerfield
Princeton University

*2. **Liquid Rockets and Propellants** (1960)
Loren E. Bollinger
The Ohio State University
Martin Goldsmith
The Rand Corporation
Alexis W. Lemmon Jr.
Battelle Memorial Institute

*3. **Energy Conversion for Space Power** (1961)
Nathan W. Snyder
Institute for Defense Analyses

*4. **Space Power Systems** (1961)
Nathan W. Snyder
Institute for Defense Analyses

*5. **Electrostatic Propulsion** (1961)
David B. Langmuir
Space Technology Laboratories, Inc.
Ernst Stuhlinger
NASA George C. Marshall Space Flight Center
J.M. Sellen Jr.
Space Technology Laboratories, Inc.

*6. **Detonation and Two-Phase Flow** (1962)
S.S. Penner
California Institute of Technology
F.A. Williams
Harvard University

*7. **Hypersonic Flow Research** (1962)
Frederick R. Riddell
AVCO Corporation

*8. **Guidance and Control** (1962)
Robert E. Roberson
Consultant
James S. Farrior
Lockheed Missiles and Space Company

*9. **Electric Propulsion Development** (1963)
Ernst Stuhlinger
NASA George C. Marshall Space Flight Center

*10. **Technology of Lunar Exploration** (1963)
Clifford I. Cummings and Harold R. Lawrence
Jet Propulsion Laboratory

*11. **Power Systems for Space Flight** (1963)
Morris A. Zipkin and Russell N. Edwards
General Electric Company

*12. **Ionization in High-Temperature Gases** (1963)
Kurt E. Shuler, Editor
National Bureau of Standards
John B. Fenn, Associate Editor
Princeton University

*13. **Guidance and Control—II** (1964)
Robert C. Langford
General Precision Inc.
Charles J. Mundo
Institute of Naval Studies

*14. **Celestial Mechanics and Astrodynamics** (1964)
Victor G. Szebehely
Yale University Observatory

*15. **Heterogeneous Combustion** (1964)
Hans G. Wolfhard
Institute for Defense Analyses
Irvin Glassman
Princeton University
Leon Green Jr.
Air Force Systems Command

*16. **Space Power Systems Engineering** (1966)
George C. Szego
Institute for Defense Analyses
J. Edward Taylor
TRW Inc.

*17. **Methods in Astrodynamics and Celestial Mechanics** (1966)
Raynor L. Duncombe
U.S. Naval Observatory
Victor G. Szebehely
Yale University Observatory

*18. **Thermophysics and Temperature Control of Spacecraft and Entry Vehicles** (1966)
Gerhard B. Heller
NASA George C. Marshall Space Flight Center

*19. **Communication Satellite Systems Technology** (1966)
Richard B. Marsten
Radio Corporation of America

---

*Out of print.

*20. **Thermophysics of Spacecraft and Planetary Bodies: Radiation Properties of Solids and the Electromagnetic Radiation Environment in Space** (1967)
Gerhard B. Heller
NASA George C. Marshall Space Flight Center

*21. **Thermal Design Principles of Spacecraft and Entry Bodies** (1969)
Jerry T. Bevans
TRW Systems

*22. **Stratospheric Circulation** (1969)
Willis L. Webb
Atmospheric Sciences Laboratory, White Sands, and University of Texas at El Paso

*23. **Thermophysics: Applications to Thermal Design of Spacecraft** (1970)
Jerry T. Bevans
TRW Systems

24. **Heat Transfer and Spacecraft Thermal Control** (1971)
John W. Lucas
Jet Propulsion Laboratory

25. **Communication Satellites for the 70's: Technology** (1971)
Nathaniel E. Feldman
The Rand Corporation
Charles M. Kelly
The Aerospace Corporation

26. **Communication Satellites for the 70's: Systems** (1971)
Nathaniel E. Feldman
The Rand Corporation
Charles M. Kelly
The Aerospace Corporation

27. **Thermospheric Circulation** (1972)
Willis L. Webb
Atmospheric Sciences Laboratory, White Sands, and University of Texas at El Paso

28. **Thermal Characteristics of the Moon** (1972)
John W. Lucas
Jet Propulsion Laboratory

29. **Fundamentals of Spacecraft Thermal Design** (1972)
John W. Lucas
Jet Propulsion Laboratory

30. **Solar Activity Observations and Predictions** (1972)
Patrick S. McIntosh and Murray Dryer
Environmental Research Laboratories, National Oceanic and Atmospheric Administration

31. **Thermal Control and Radiation** (1973)
Chang-Lin Tien
University of California at Berkeley

32. **Communications Satellite Systems** (1974)
P.L. Bargellini
COMSAT Laboratories

33. **Communications Satellite Technology** (1974)
P.L. Bargellini
COMSAT Laboratories

34. **Instrumentation for Airbreathing Propulsion** (1974)
Allen E. Fuhs
Naval Postgraduate School
Marshall Kingery
Arnold Engineering Development Center

35. **Thermophysics and Spacecraft Thermal Control** (1974)
Robert G. Hering
University of Iowa

36. **Thermal Pollution Analysis** (1975)
Joseph A. Schetz
Virginia Polytechnic Institute

37. **Aeroacoustics: Jet and Combustion Noise; Duct Acoustics** (1975)
Henry T. Nagamatsu, Editor
General Electric Research and Development Center
Jack V. O'Keefe, Associate Editor
The Boeing Company
Ira R. Schwartz, Associate Editor
NASA Ames Research Center

38. **Aeroacoustics: Fan, STOL, and Boundary Layer Noise; Sonic Boom; Aeroacoustic Instrumentation** (1975)
Henry T. Nagamatsu, Editor
General Electric Research and Development Center
Jack V. O'Keefe, Associate Editor
The Boeing Company
Ira R. Schwartz, Associate Editor
NASA Ames Research Center

39. **Heat Transfer with Thermal Control Applications** (1975)
M. Michael Yovanovich
University of Waterloo

SERIES LISTING

40. **Aerodynamics of Base Combustion** (1976)
S.N.B. Murthy, Editor
Purdue University
J.R. Osborn, Associate Editor
Purdue University
A.W. Barrows and J.R. Ward, Associate Editors
Ballistics Research Laboratories

41. **Communications Satellite Developments: Systems** (1976)
Gilbert E. LaVean
Defense Communications Agency
William G. Schmidt
CML Satellite Corporation

42. **Communications Satellite Developments: Technology** (1976)
William G. Schmidt
CML Satellite Corporation
Gilbert E. LaVean
Defense Communications Agency

43. **Aeroacoustics: Jet Noise, Combustion and Core Engine Noise** (1976)
Ira R. Schwartz, Editor
NASA Ames Research Center
Henry T. Nagamatsu, Associate Editor
General Electric Research and Development Center
Warren C. Strahle, Associate Editor
Georgia Institute of Technology

44. **Aeroacoustics: Fan Noise and Control; Duct Acoustics; Rotor Noise** (1976)
Ira R. Schwartz, Editor
NASA Ames Research Center
Henry T. Nagamatsu, Associate Editor
General Electric Research and Development Center
Warren C. Strahle, Associate Editor
Georgia Institute of Technology

45. **Aeroacoustics: STOL Noise; Airframe and Airfoil Noise** (1976)
Ira R. Schwartz, Editor
NASA Ames Research Center
Henry T. Nagamatsu, Associate Editor
General Electric Research and Development Center
Warren C. Strahle, Associate Editor
Georgia Institute of Technology

46. **Aeroacoustics: Acoustic Wave Propagation; Aircraft Noise Prediction; Aeroacoustic Instrumentation** (1976)
Ira R. Schwartz, Editor
NASA Ames Research Center
Henry T. Nagamatsu, Associate Editor
General Electric Research and Development Center
Warren C. Strahle, Associate Editor
Georgia Institute of Technology

47. **Spacecraft Charging by Magnetospheric Plasmas** (1976)
Alan Rosen
TRW Inc.

48. **Scientific Investigations on the Skylab Satellite** (1976)
Marion I. Kent and Ernst Stuhlinger
NASA George C. Marshall Space Flight Center
Shi-Tsan Wu
The University of Alabama

49. **Radiative Transfer and Thermal Control** (1976)
Allie M. Smith
ARO Inc.

50. **Exploration of the Outer Solar System** (1976)
Eugene W. Greenstadt
TRW Inc.
Murray Dryer
National Oceanic and Atmospheric Administration
Devrie S. Intriligator
University of Southern California

51. **Rarefied Gas Dynamics, Parts I and II** (two volumes) (1977)
J. Leith Potter
ARO Inc.

52. **Materials Sciences in Space with Application to Space Processing** (1977)
Leo Steg
General Electric Company

53. **Experimental Diagnostics in Gas Phase Combustion Systems** (1977)
Ben T. Zinn, Editor
Georgia Institute of Technology
Craig T. Bowman, Associate Editor
Stanford University
Daniel L. Hartley, Associate Editor
Sandia Laboratories
Edward W. Price, Associate Editor
Georgia Institute of Technology
James G. Skifstad, Associate Editor
Purdue University

54. **Satellite Communications: Future Systems** (1977)
David Jarett
TRW Inc.

55. **Satellite Communications: Advanced Technologies** (1977)
David Jarett
TRW Inc.

56. **Thermophysics of Spacecraft and Outer Planet Entry Probes** (1977)
Allie M. Smith
ARO Inc.

57. **Space-Based Manufacturing from Nonterrestrial Materials** (1977)
Gerard K. O'Neill, Editor
Princeton University
Brian O'Leary, Assistant Editor
Princeton University

58. **Turbulent Combustion** (1978)
Lawrence A. Kennedy
State University of New York at Buffalo

59. **Aerodynamic Heating and Thermal Protection Systems** (1978)
Leroy S. Fletcher
University of Virginia

60. **Heat Transfer and Thermal Control Systems** (1978)
Leroy S. Fletcher
University of Virginia

61. **Radiation Energy Conversion in Space** (1978)
Kenneth W. Billman
NASA Ames Research Center

62. **Alternative Hydrocarbon Fuels: Combustion and Chemical Kinetics** (1978)
Craig T. Bowman
Stanford University
Jorgen Birkeland
Department of Energy

63. **Experimental Diagnostics in Combustion of Solids** (1978)
Thomas L. Boggs
Naval Weapons Center
Ben T. Zinn
Georgia Institute of Technology

64. **Outer Planet Entry Heating and Thermal Protection** (1979)
Raymond Viskanta
Purdue University

65. **Thermophysics and Thermal Control** (1979)
Raymond Viskanta
Purdue University

66. **Interior Ballistics of Guns** (1979)
Herman Krier
University of Illinois at Urbana-Champaign
Martin Summerfield
New York University

67. **Remote Sensing of Earth from Space: Role of "Smart Sensors"** (1979)
Roger A. Breckenridge
NASA Langley Research Center

68. **Injection and Mixing in Turbulent Flow** (1980)
Joseph A. Schetz
Virginia Polytechnic Institute and State University

69. **Entry Heating and Thermal Protection** (1980)
Walter B. Olstad
NASA Headquarters

70. **Heat Transfer, Thermal Control, and Heat Pipes** (1980)
Walter B. Olstad
NASA Headquarters

71. **Space Systems and Their Interactions with Earth's Space Environment** (1980)
Henry B. Garrett and Charles P. Pike
Hanscom Air Force Base

72. **Viscous Flow Drag Reduction** (1980)
Gary R. Hough
Vought Advanced Technology Center

73. **Combustion Experiments in a Zero-Gravity Laboratory** (1981)
Thomas H. Cochran
NASA Lewis Research Center

74. **Rarefied Gas Dynamics, Parts I and II** (two volumes) (1981)
Sam S. Fisher
University of Virginia at Charlottesville

75. **Gasdynamics of Detonations and Explosions** (1981)
J.R. Bowen
University of Wisconsin at Madison
N. Manson
Université de Poitiers
A.K. Oppenheim
University of California at Berkeley
R.I. Soloukhin
Institute of Heat and Mass Transfer, BSSR Academy of Sciences

76. **Combustion in Reactive Systems** (1981)
J.R. Bowen
University of Wisconsin at Madison
N. Manson
Université de Poitiers
A.K. Oppenheim
University of California at Berkeley
R.I. Soloukhin
Institute of Heat and Mass Transfer, BSSR Academy of Sciences

77. **Aerothermodynamics and Planetary Entry** (1981)
A.L. Crosbie
University of Missouri-Rolla

78. **Heat Transfer and Thermal Control** (1981)
A.L. Crosbie
University of Missouri-Rolla

SERIES LISTING

79. Electric Propulsion and Its Applications to Space Missions (1981)
Robert C. Finke
NASA Lewis Research Center

80. Aero-Optical Phenomena (1982)
Keith G. Gilbert and Leonard J. Otten
Air Force Weapons Laboratory

81. Transonic Aerodynamics (1982)
David Nixon
Nielsen Engineering & Research, Inc.

82. Thermophysics of Atmospheric Entry (1982)
T.E. Horton
The University of Mississippi

83. Spacecraft Radiative Transfer and Temperature Control (1982)
T.E. Horton
The University of Mississippi

84. Liquid-Metal Flows and Magnetohydrodynamics (1983)
H. Branover
Ben-Gurion University of the Negev
P.S. Lykoudis
Purdue University
A. Yakhot
Ben-Gurion University of the Negev

85. Entry Vehicle Heating and Thermal Protection Systems: Space Shuttle, Solar Starprobe, Jupiter Galileo Probe (1983)
Paul E. Bauer
McDonnell Douglas Astronautics Company
Howard E. Collicott
The Boeing Company

86. Spacecraft Thermal Control, Design, and Operation (1983)
Howard E. Collicott
The Boeing Company
Paul E. Bauer
McDonnell Douglas Astronautics Company

87. Shock Waves, Explosions, and Detonations (1983)
J.R. Bowen
University of Washington
N. Manson
Université de Poitiers
A.K. Oppenheim
University of California at Berkeley
R.I. Soloukhin
Institute of Heat and Mass Transfer, BSSR Academy of Sciences

88. Flames, Lasers, and Reactive Systems (1983)
J.R. Bowen
University of Washington
N. Manson
Université de Poitiers
A.K. Oppenheim
University of California at Berkeley
R.I. Soloukhin
Institute of Heat and Mass Transfer, BSSR Academy of Sciences

89. Orbit-Raising and Maneuvering Propulsion: Research Status and Needs (1984)
Leonard H. Caveny
Air Force Office of Scientific Research

90. Fundamentals of Solid-Propellant Combustion (1984)
Kenneth K. Kuo
The Pennsylvania State University
Martin Summerfield
Princeton Combustion Research Laboratories, Inc.

91. Spacecraft Contamination: Sources and Prevention (1984)
J.A. Roux
The University of Mississippi
T.D. McCay
NASA Marshall Space Flight Center

92. Combustion Diagnostics by Nonintrusive Methods (1984)
T.D. McCay
NASA Marshall Space Flight Center
J.A. Roux
The University of Mississippi

93. The INTELSAT Global Satellite System (1984)
Joel Alper
COMSAT Corporation
Joseph Pelton
INTELSAT

94. Dynamics of Shock Waves, Explosions, and Detonations (1984)
J.R. Bowen
University of Washington
N. Manson
Universite de Poitiers
A.K. Oppenheim
University of California
R.I. Soloukhin
Institute of Heat and Mass Transfer, BSSR Academy of Sciences

95. Dynamics of Flames and Reactive Systems (1984)
J.R. Bowen
University of Washington
N. Manson
Universite de Poitiers
A.K. Oppenheim
University of California
R.I. Soloukhin
Institute of Heat and Mass Transfer, BSSR Academy of Sciences

(Other Volumes are planned.)